Student's Solutions Manual

to accompany

College Algebra
Graphs and Models

Third Edition

Raymond A. Barnett
Merritt College

Michael R. Ziegler
Marquette University

Karl E. Byleen
Marquette University

and

Dave Sobecki
Miami University Hamilton

Prepared by
Dave Sobecki

McGraw-Hill
Higher Education

Boston   Burr Ridge, IL   Dubuque, IA   New York   San Francisco   St. Louis
Bangkok   Bogotá   Caracas   Kuala Lumpur   Lisbon   London   Madrid   Mexico City
Milan   Montreal   New Delhi   Santiago   Seoul   Singapore   Sydney   Taipei   Toronto

The McGraw·Hill Companies

McGraw-Hill
Higher Education

Student's Solutions Manual to accompany
COLLEGE ALGEBRA: GRAPHS AND MODELS, THIRD EDITION
RAYMOND A. BARNETT, MICHAEL R. ZIEGLER, KARL E. BYLEEN, AND DAVE SOBECKI

Published by McGraw-Hill Higher Education, an imprint of The McGraw-Hill Companies, Inc., 1221 Avenue of the Americas, New York, NY 10020. Copyright © 2009, 2005, and 2000 by The McGraw-Hill Companies, Inc. All rights reserved.

1 2 3 4 5 6 7 8 9 0 QPD/QPD 0 9 8

ISBN: 978-0-07-334185-9
MHID: 0-07-334185-1

www.mhhe.com

CHAPTER 1   Functions, Graphs, and Models

Section 1-1

1. Since the x coordinate, 0, is between Xmin and -7 and Xmax = 9, and the y coordinate, 0, is between Ymin = -4 and Ymax = 11, (0, 0) lies in the viewing window.

3. Since the x coordinate, 10, is not between Xmin = -7 and Xmax = 9, (10, 0) does not lie in the viewing window.

5. Since the x coordinate, -5, is between Xmin = -7 and Xmax = 9, and the y coordinate, -3, is between Ymin = -4 and Ymax = 11, (-5, -3) lies in the viewing window.

7. (A) Since the x coordinates vary between a minimum of -7 and a maximum of 6, Xmin = -7 and Xmax = 6. Since the y coordinates vary between a minimum of -9 and a maximum of 14, Ymin = -9 and Ymax = 14.

   (B)

   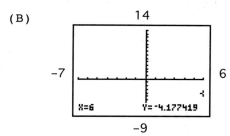

   The viewing window is shown. The cursor can, in general, show the coordinates of the points only approximately. For example, the point (6, -4) is approximated by the cursor location (6, -4.177419) in the screen shown. Pixels are geometrical rectangles whose coordinates are intervals, specified by the coordinates of the point at their center. Since a limited number of pixels are available on the calculator, it is not possible to show points exactly, since a point is a geometric object with no dimensions, and there are an infinite number of points.

9. Answers will vary.

11.   13.   15.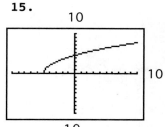

17. Answers can vary. One possible answer is Xmin = -20, Xmax = 10, Ymin = -2, Ymax = 10.

19. Answers can vary. One possible answer is Xmin = -20, Xmax = 10, Ymin = 0, Ymax = 20.

21.

x	$y = 4 + 4x - x^2$
-2	-8
0	4
2	8
4	4
6	-8

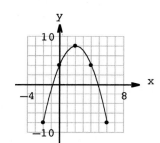

23.

x	$y = 2\sqrt{2x + 10}$
−5	0
−3	4
−1	5.7
1	6.9
3	8
5	8.9

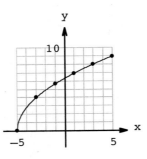

25.

x	$y = 0.5x(4 - x)(x + 2)$
−3	10.5
−1	−2.5
1	4.5
3	7.5
5	−17.5

27. The graph of $y = 4 - 3\sqrt[3]{x + 4}$ is shown:

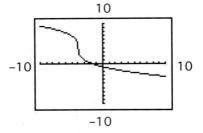

(A) After applying ZOOM and TRACE, the following is obtained. $x = -6.37$.

(B) After applying ZOOM and TRACE, the following is obtained. $x = 0.63$.

29. The graph of $y = 3 + x + 0.1x^3$ is shown.

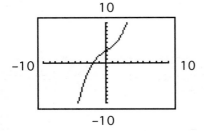

(D) The company should raise the price $5 to increase the revenue.

Section 1-2

1. False. Only relations where each element in the domain corresponds to exactly one element in the range are functions.

3. Answers will vary.

5. This is a function: each element in the domain corresponds to exactly one element in the range. $\{(-1, 3), (0, 2), (1, 1)\}$

7. This is not a function since the domain elements 3 and 5 both correspond to more than one range element. $\{(1, 3), (3, 5), (3, 7), (5, 7), (5, 9)\}$

9. This is a function: each element in the domain corresponds to exactly one element in the range. $\{(-1, 3), (0, 3), (1, 3), (2, 3)\}$

11. Domain: $\{2, 3, 4, 5\}$; Range: $\{4, 6, 8, 10\}$; function

13. Domain: $\{0, 5, 10\}$; Range: $\{-10, -5, 0, 5, 10\}$; not a function

15. Domain: $\{0, 1, 2, 3, 4, 5\}$; Range: $\{1, 2\}$; function

17. A function

19. Not a function (fails vertical line test since the y-axis hits the graph 3 times.)

21. Not a function (fails vertical line test since the graph is vertical itself!)

23. (A) This is a function. Each congressional district has only one representative, so all residents correspond to exactly one representative.

    (B) This is not a function. Every state has two senators that represent the entire state, so every resident corresponds to two senators. This violates the definition of function.

25. This equation does define y as a function of x. If you plug in any number for x, there is only one choice of y that will then make the equation true.

27. This equation does not define y as a function of x. For many choices of x, there are two values of y that will make the equation true. For example, if $x = 0$ the equation is $y^2 = 4$ and y can be either 2 or -2. Since we found a value for x that corresponds to two different values of y, the equation is not a function.

29. $f(-1) = 3(-1) - 5 = -8$

31. $G(-2) = (-2) - (-2)^2 = -6$ 33. $F(-1) + f(3) = 3(-1)^2 + 2(-1) - 4 + 3(3) - 5$
$= 3 - 2 - 4 + 9 - 5 = 1$

35. $2F(-2) - G(-1) = 2[3(-2)^2 + 2(-2) - 4] - [(-1) - (-1)^2]$
$= 2[12 - 4 - 4] - [-1 - 1] = 10$

37. $\dfrac{f(0) \cdot g(-2)}{F(-3)} = \dfrac{[3(0) - 5] \cdot [4 - (-2)]}{3(-3)^2 + 2(-3) - 4} = \dfrac{(-5)(6)}{27 - 6 - 4} = -\dfrac{30}{17}$

39. $\dfrac{f(4) - f(2)}{2} = \dfrac{[3(4) - 5] - [3(2) - 5]}{2} = \dfrac{7 - 1}{2} = \dfrac{6}{2} = 3$

41. Since $f(x)$ is a polynomial, the domain is all real numbers; $(-\infty, \infty)$.

43. If the denominator of the fraction is zero, the function will be undefined, since division by zero is undefined. For any other values of x, $h(x)$ represents a real number. Solve the equation $4 - x = 0$; $x = 4$. The domain is all real numbers except 4 or $(-\infty, 4) \cup (4, \infty)$.

45. The square root of a number is defined only if the number is nonnegative. The expression $\sqrt{t - 4}$ is defined if $t - 4 \geq 0$ and undefined if $t - 4 < 0$. Note that $t - 4 \geq 0$ when $t \geq 4$, so the domain is all real numbers greater than or equal to 4 or $[4, \infty)$.

47. The formula $\sqrt{7 + 3w}$ is defined only if $7 + 3w \geq 0$, since the square root is only defined if the number inside is nonnegative.
We solve the inequality: $7 + 3w \geq 0$
$$3w \geq -7$$
$$w \geq -\frac{7}{3}$$

The domain is all real numbers greater than or equal to $-\dfrac{7}{3}$ or $\left[\dfrac{7}{3}, \infty\right)$.

49. The fraction $\dfrac{u}{u^2 + 4}$ is defined for any value of u that does <u>not</u> make the denominator zero, so we solve the equation $u^2 + 4 = 0$ to find values that make the function undefined. $u^2 + 4 = 0$
$$u^2 = -4$$
This equation has no solution since the square of a number can't be negative. So there are no values of u that make the fraction undefined and the domain is all real numbers or $(-\infty, \infty)$.

51. The fraction $\dfrac{v + 2}{v^2 - 16}$ is defined for any value of v that does <u>not</u> make the denominator zero, so we solve the equation $v^2 - 16 = 0$ to find values that make the function undefined.
$$v^2 - 16 = 0$$
$$v^2 = 16$$
$$v = 4, -4$$
The domain is all real numbers except 4 and -4, or $(-\infty, -4) \cup (-4, 4) \cup (4, \infty)$.

53. There are two issues to consider: we need to make certain that $x + 4 \geq 0$ so that the number inside the square root is nonnegative, and we need to avoid x-values that make the denominator zero.

First, $x + 4 \geq 0$ whenever $x \geq -4$. So x must be greater than or equal to -4 to avoid a negative under the root. Also, $x - 1 = 0$ when $x = 1$, so x cannot be 1. The domain is all real numbers greater than or equal to -4 except 1, or $[-4, 1) \cup (1, \infty)$.

55. There are two issues to consider: the number inside the square root has to be nonnegative, and the denominator of the fraction has to be nonzero. Since we just have t inside the square root, t has to be greater than or equal to zero. Next, we solve $3 - \sqrt{t} = 0$ to find any t-values that make the denominator zero.
$$3 - \sqrt{t} = 0$$
$$3 = \sqrt{t}$$
$$9 = t$$
So t cannot be 9. The domain is all real numbers greater than or equal to zero except 9, or $[0, 9) \cup (9, \infty)$.

57. $f(x) = 2x - 3$

59. $f(x) = 4x^2 - 2x + 9$

61.
$$F(s) = 3s + 15$$
$$F(2 + h) = 3(2 + h) + 15$$
$$F(2) = 3(2) + 15$$
$$\frac{F(2 + h) - F(2)}{h} = \frac{[3(2 + h) + 15] - [3(2) + 15]}{h} = \frac{[6 + 3h + 15] - [21]}{h}$$
$$= \frac{3h + 21 - 21}{h} = \frac{3h}{h} = 3$$

63.
$$g(x) = 2 - x^2$$
$$g(3 + h) = 2 - (3 + h)^2$$
$$g(3) = 2 - (3)^2$$
$$\frac{g(3 + h) - g(3)}{h} = \frac{[2 - (3 + h)]^2 - [2 - (3)^2]}{h} = \frac{[2 - 9 - 6h - h^2] - [-7]}{h}$$
$$= \frac{-6h - h^2}{h} = \frac{h(-6 - h)}{h} = -6 - h$$

65.
$$L(w) = -2w^2 + 3w - 1$$
$$L(-2 + h) = -2(-2 + h)^2 + 3(-2 + h) - 1$$
$$L(-2) = -2(-2)^2 + 3(-2) - 1$$
$$\frac{L(-2 + h) - L(-2)}{h} = \frac{[-2(-2 + h)^2 + 3(-2 + h) - 1] - [-2(-2)^2 + 3(-2) - 1]}{h}$$
$$= \frac{[-2(4 - 4h + h^2) - 6 + 3h - 1] - [-15]}{h}$$
$$= \frac{-8 + 8h - 2h^2 - 6 + 3h - 1 + 15}{h}$$
$$= \frac{11h - 2h^2}{h} = \frac{h(11 - 2h)}{h} = 11 - 2h$$

67. $g(x) = 3x + 1$

69. $F(x) = \dfrac{x}{8 + \sqrt{x}}$

71. Function f multiplies the domain element by 2 and then subtracts the product of 3 and the square of the element.

73. Function F takes the square root of the sum of the fourth power of the domain element and 9.

75. In this case, the domain element is the expression $(x + h)$, so we identify the operations that are performed on $(x + h)$. The function multiplies the square of $(x + h)$ by 2, then subtracts 4 times $(x + h)$, and finally adds 6. The function equation is $f(x) = 2x^2 - 4x + 6$.

77. In this case, the domain element is the expression $(x + h)$, so we identify the operations that are performed on $(x + h)$. The function multiplies 4 by $(x + h)$, then subtracts 3 times the square root of $(x + h)$, and finally adds 9. The function equation is $f(x) = 4x - 3\sqrt{x} + 9$.

79. (A)

$$f(x) = 3x - 4$$
$$f(x + h) = 3(x + h) - 4$$
$$\frac{f(x + h) - f(x)}{h} = \frac{[3(x + h) - 4] - [3x - 4]}{h}$$
$$= \frac{3x + 3h - 4 - 3x + 4}{h}$$
$$= \frac{3h}{h}$$
$$= 3$$

(B)

$$f(x) = 3x - 4$$
$$f(a) = 3a - 4$$
$$\frac{f(x) - f(a)}{x - a} = \frac{(3x - 4) - (3a - 4)}{x - a}$$
$$= \frac{3x - 4 - 3a + 4}{x - a}$$
$$= \frac{3x - 3a}{x - a}$$
$$= \frac{3(x - a)}{x - a}$$
$$= 3$$

81. (A)

$$f(x) = x^2 - 1$$
$$f(x + h) = (x + h)^2 - 1$$
$$\frac{f(x + h) - f(x)}{h} = \frac{[(x + h)^2 - 1] - [x^2 - 1]}{h}$$
$$= \frac{x^2 + 2xh + h^2 - 1 - x^2 + 1}{h}$$
$$= \frac{2xh + h^2}{h}$$
$$= \frac{h(2x + h)}{h}$$
$$= 2x + h$$

(B)

$$f(x) = x^2 - 1$$
$$f(a) = a^2 - 1$$
$$\frac{f(x) - f(a)}{x - a} = \frac{(x^2 - 1) - (a^2 - 1)}{x - a}$$
$$= \frac{x^2 - 1 - a^2 + 1}{x - a}$$
$$= \frac{x^2 - a^2}{x - a}$$
$$= \frac{(x - a)(x + a)}{x - a}$$
$$= x + a$$

83. (A)

$$f(x) = -3x^2 + 9x - 12$$
$$f(x + h) = -3(x + h)^2 + 9(x + h) - 12$$
$$\frac{f(x + h) - f(x)}{h} = \frac{[-3(x + h)^2 + 9(x + h) - 12] - [-3x^2 + 9x - 12]}{h}$$
$$= \frac{-3(x^2 + 2xh + h^2) + 9x + 9h - 12 + 3x^2 - 9x + 12}{h}$$
$$= \frac{-3x^2 - 6xh - 3h^2 + 9x + 9h - 12 + 3x^2 - 9x + 12}{h}$$
$$= \frac{-6xh - 3h^2 + 9h}{h}$$
$$= \frac{h(-6x - 3h + 9)}{h}$$
$$= -6x - 3h + 9$$

(B)

$$f(x) = -3x^2 + 9x - 12$$
$$f(a) = -3a^2 + 9a - 12$$
$$\frac{f(x) - f(a)}{x - a} = \frac{(-3x^2 + 9x - 12) - (-3a^2 + 9a - 12)}{x - a}$$
$$= \frac{-3x^2 + 9x - 12 + 3a^2 - 9a + 12}{x - a}$$
$$= \frac{-3x^2 + 3a^2 + 9x - 9a}{x - a}$$
$$= \frac{(x - a)(-3x - 3a) + 9(x - a)}{x - a}$$
$$= \frac{(x - a)(-3x - 3a + 9)}{x - a}$$
$$= -3x - 3a + 9$$

85. (A)

$$f(x) = x^3$$
$$f(x + h) = (x + h)^3$$
$$\frac{f(x + h) - f(x)}{h} = \frac{(x + h)^3 - x^3}{h}$$
$$= \frac{x^3 + 3x^2h + 3xh^2 + h^3 - x^3}{h}$$
$$= \frac{3x^2h + 3xh^2 + h^3}{h}$$
$$= \frac{h(3x^2 + 3xh + h^2)}{h}$$
$$= 3x^2 + 3xh + h^2$$

(B)

$$f(x) = x^3$$
$$f(a) = a^3$$
$$\frac{f(x) - f(a)}{x - a} = \frac{x^3 - a^3}{(x - a)}$$
$$= \frac{(x - a)(x^2 + xa + a^2)}{x - a}$$
$$= x^2 + ax + a^2$$

87. First, let's look at the graph of $f(x)$ in a window that shows only x-values close to 1.

Notice that the heights of points on the graph with x-coordinates very close to 1 are very close to 2.

Next, we'll use the TABLE command to evaluate $f(x)$ for some numbers close to $x = 1$. We start at $x = 0.7$ and increment by 0.1 as shown on the TABLE SETUP screen:

Notice that as x gets closer to 1, the result gets closer to 2. Finally, we'll simplify the fraction by factoring the numerator.

$$\frac{x^2 - 1}{x - 1} = \frac{(x + 1)(x - 1)}{x - 1} = x + 1 \text{ as long as } x \neq 1.$$

It's easy to see that if you plug x-values close to 1 into the expression $x + 1$, the results will be close to 2.

89. (A)

x	0	5,000	10,000	15,000	20,000	25,000	30,000
$B(x)$	212	203	194	185	176	167	158

(B) The boiling point drops 9°F for each 5,000 foot increase in altitude.

91. (A) $S(0) = 16(0)^2 = 0$; $S(1) = 16(1)^2 = 16$; $S(2) = 16(2)^2 = 16(4) = 64$;
$S(3) = 16(3)^2 = 16(9) = 144$ (Note: Remember that the order of operations requires that we apply the exponent first then multiply by 16.)

(B)
$$\frac{S(2 + h) - S(2)}{h} = \frac{16(2 + h)^2 - 16(2)^2}{h} = \frac{16(4 + 4h + h^2) - 16(4)}{h}$$
$$= \frac{64 + 64h + 16h^2 - 64}{h} = \frac{64h + 16h^2}{h} = \frac{h(64 + 16h)}{h} = 64 + 16h$$

(Note: Be careful when evaluating $S(2 + h)$! You need to replace the variable t in the function with $(2 + h)$. $S(2 + h)$ is <u>not</u> the same thing as $S(2) + h$!)

(C) $h = 1$: $64 + 16(1) = 80$; $h = -1$: $64 + 16(-1) = 64 - 16 = 48$
$h = 0.1$: $64 + 16(0.1) = 64 + 1.6 = 65.6$;
$h = -0.1$: $64 + 16(-0.1) = 64 - 1.6 = 62.4$
$h = 0.01$: $64 + 16(0.01) = 64 + 0.16 = 64.16$:
$h = -0.01$: $64 + 16(-0.01) = 64 - 0.16 = 63.84$
$h = 0.001$: $64 + 16(0.001) = 64 + 0.016 = 64.016$
$h = -0.001$: $64 + 16(-0.001) = 64 - 0.016 = 63.984$

(D)   The smaller h gets the closer the result is to 64. The numerator of the fraction, $S(2 + h) - S(2)$, is the difference between how far an object has fallen after $2 + h$ seconds and how far it's fallen after 2 seconds. This difference is how far the object falls in the small period of time from 2 to $2 + h$ seconds. When you divide that distance by the time (h), you get the average velocity of the object between 2 and $2 + h$ seconds. Part (C) shows that this average velocity approaches 64 feet per second as h gets smaller.

93. The rental charges are $20 per day plus $0.25 per mile driven.

95. For detailed help on how to use your graphing calculator on this problem, see Example 10 in Section 1-2.

(A) We begin by entering our lists under the STAT EDIT screen. In this model, the independent variable is years after 1997 and the dependent variable is the average admission price. So our first list is 0, 2, 4, 6, and 8, representing the years from 1997 to 2005. The second list is the corresponding prices.

The second screen shows the appropriate plot settings; the third shows a good choice of viewing window; the last is the scatter plot of the data from the table.

To compare the algebraic function, we enter $y_1 = 0.23x + 4.6$ in the equation editor, choose the "ask" option for independent variable in the TABLE SETUP window, then enter x values 0, 2, 4, 6, 8.

Comparing the two tables, we see that some of the average prices are very close and none differ by more than 14 cents. Finally, we look at the graphs of both the function and the scatter plot:

As expected, the graph of the function matches the scatter plot points closely.

(B) To estimate the average price in 2006, we need to evaluate the given function for $x = 9$, since 2006 is 9 years after 1997.
$A(9) = 0.23(9) + 4.6 = 6.67$
The average price in 2006 is $6.67. Similarly, $A(10) = 0.23(10) + 4.6 = 6.90$, so the average price in 2007 is $6.90.

97. For detailed help on how to use your graphing calculator on this problem, see Example 10 in Section 1-2.

(A) In this example, the independent variable is years after 1997, so the first list under the STAT EDIT screen is 0, 2, 4, 6, and 8. The dependent variable is sales, so the second list is the sales row: 14.0, 17.3, 21.2, 22.5, and 22.0.

The second screen shows the appropriate plot settings; the third shows a good choice of viewing window; the last is the scatter plot of the data from the table.

To compare to the algebraic function, we enter $y_1 = -0.18x^2 + 2.5x + 14$ in the equation editor, choose the "ask" option for independent variable in the TABLE SETUP window, than enter x values 0, 2, 4, 6, 8.

Comparing the two tables, we see that all of the sales figures are similar — the largest difference is less than 1.

Finally, we look at the graphs of both the function and the scatter plot:

As expected, the graph of the function matches the scatter plot points closely.

(B) To estimate sales in 2006, we need to evaluate the given function for $x = 9$, since 2006 is 9 years after 1997.

$$S(9) = -0.18(9)^2 + 2.5(9) + 14 = 21.92.$$

The estimated sales figure for 2006 is $22 billion (rounded to 2 significant digits). Similarly, $S(11) = -0.18(11)^2 + 2.5(11) + 14 = 19.72$, so the estimated sales figure for 2008 is $20 billion (rounded to 2 significant digits).

(C) For the period 1997-2001, Merck's sales increased steadily by an average of approximately $6 billion per year.

99. For detailed help on how to use your graphing calculator on this problem, see Example 10 in Section 1-2.

(A) In this example, the independent variable (r) is R & D expenses, so the first list under the STAT EDIT screen comes from the R & D row of the table: 1.7, 2.1, 2.5, 3.2, 3.8. The dependent variable(s) is sales, so the second list is the sales row: 14.0, 17.3, 21.2, 22.5, and 22.0.

The second screen shows the appropriate plot settings; the third shows a good choice of viewing window; the last is the scatter plot of the data from the table.

To compare to the algebraic function, we enter $y_1 = -3.54x^2 + 23.3x - 15.5$ in the equation editor, choose the "ask" option for independent variable in the TABLE SETUP window, then enter x values 1.7, 2.1, 2.5, 3.2, 3.8.

Comparing the two tables, we see that the corresponding sales figures are very close.

Finally, we look at the graphs of both the function and the scatter plot:

As expected, the graph of the function matches the points in the scatter plot very well.

(B) To estimate the sales with $1.2 billion in research and development spending, we evaluate the function for $r = 1.2$:

$$S(1.2) = -3.54(1.2)^2 + 23.3(1.2) - 15.5 = 7.3624$$

To 2 significant digits, the estimated sales figure is $7.4 billion with $1.2 billion spent on research and development.

Similarly, $S(4.2) = -3.54(4.2)^2 + 23.3(4.2) - 15.5 = 19.9144$, so the estimated sales figure is $20 billion with $4.2 billion spent on research and development.

Section 1-3

1. Answers will vary.　　　**3.** Answers will vary.　　　**5.** Answers will vary.

7. (A) [−4, 4)　　(B) [−3, 3)　　(C) 0　　(D) 0　　(E) [−4, 4)　　(F) None
(G) None　　　(H) None

9. (A) (−∞, ∞)　　(B) [−4, ∞)　　(C) −3, 1　　(D) −3　　(E) [−1, ∞)　　(F) (−∞, −1]
(G) None　　　(H) None

11. (A) $(-\infty, 2) \cup (2, \infty)$ (The function is not defined at $x = 2$.)
    (B) $(-\infty, -1) \cup [1, \infty)$ (C) None   (D) 1   (E) None   (F) $(-\infty, -2] \cup (2, \infty)$
    (G) $[-2, 2)$ (H) $x = 2$

13. $f(-4) = -3$ since the point $(-4, -3)$ is on the graph; $f(0) = 0$ since the point $(0, 0)$ is on the graph; $f(4)$ is undefined since there is no point on the graph at $x = 4$.

15. $h(-3) = 0$ since the point $(-3, 0)$ is on the graph; $h(0) = -3$ since the point $(0, -3)$ is on the graph; $h(2) = 5$ since the point $(2, 5)$ is on the graph.

17. $p(-2) = 1$ since the point $(-2, 1)$ is on the graph; $p(2)$ is undefined since there is no point on the graph at $x = 2$; $p(5) = -4$ since the point $(5, -4)$ is on the graph.

19. The graph of $f(x)$ is shown below in a standard window.

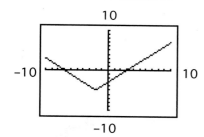

The function is increasing on $[-2, \infty)$ and decreasing on $(-\infty, -2]$. It is not constant on any interval shown.

21. The graph of $j(x)$ is shown below in a standard window.

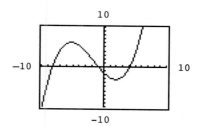

The function is increasing on $(-\infty, -5]$ and $[2, \infty)$ and decreasing on $[-5, 2]$. It is not constant on any interval shown.

23. The graph of $m(x)$ is shown below in a standard window.

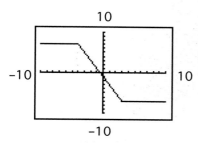

The function is decreasing on $[-4, 3]$ and constant on $[-\infty, -4]$ and $[3, \infty]$. It is not increasing on any interval shown.

25. The graph of $r(x)$ is shown below in a standard window.

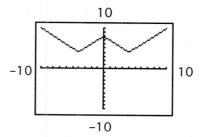

The function is increasing on $[-4, 0]$ and $[4, \infty]$ and decreasing on $[-\infty, -4]$ and $[0, 4]$. It is not constant on any interval shown.

27. To find the x-intercepts we'll use a graphing calculator to draw the graph then use the ZERO command from the calculate menu. (For detailed instructions on using the ZERO command, see Example 1 in Section 1-3.)

The two x-intercepts are $x = -1.405$ and $x = 6.405$. The y-intercept is easier to find if we just plug zero in for x in the given function: $f(0) = 0^2 - 5(0) - 9 = -9$, so the y-intercept is -9. It appears that the graph of f has a local minimum near $x = 2$ and no local maximum. (To be on the safe side, you may want to look at some bigger viewing windows to make sure there are no other extrema.)

We'll use the feature from the calculate menu. (For detailed instructions on using the MINIMUM command, see Example 4 in Section 1-3.)

We see from the graph that has a local minimum value of -15.25 at $x = 2.5$.

29. To find the x-intercepts we'll use a graphing calculator to draw the graph then use the ZERO command from the calculate menu. (For detailed instructions on using the ZERO command, see Example 1 in Section 1-3.)

Notice that a fairly large viewing window is necessary to see all of the key features of the graph. We see that the only x-intercept is $x = 3.377$. The y-intercept is easier to find if we just plug zero in for x in the given function: $h(0) = -(0)^3 + 4(0) + 25 = 25$, so the y-intercept is 25. We can see that the graph has one local minimum and one local maximum, which we'll find using the MINIMUM and MAXIMUM commands from the calculate menu. (For detailed instructions on using the MINIMUM and MAXIMUM commands see Example 4 in Section 1-3.)

The local minimum of 21.921 occurs at $x = -1.155$; the local maximum of 28.079 occurs at $x = 1.155$.

31. To find the x-intercepts we'll use a graphing calculator to draw the graph then use the ZERO command from the calculate menu. (For detailed instructions on using the ZERO command, see Example 1 in Section 1-3.)

This is a tricky one because the graph is a bit deceiving onscreen. It looks like the graph does not hit the x-axis. But if you look at the formula for the function, you can see that the height of the graph will be zero when $x^2 - 12 = 0$; this happens when $x = \sqrt{12}$ or $-\sqrt{12}$, so in fact there are 2 intercepts which the ZERO command will find.

The two x-intercepts are $x = -3.464$ and $x = 3.464$. (These are $-\sqrt{12}$ and $\sqrt{12}$ as we expected.)

The y-intercept is <u>easier</u> to find if we just plug zero in for x in the given function: $m(0) = \sqrt{|0^2 - 12|} = \sqrt{12} \approx 3.464$ so the y-intercept is 3.464.

From the graph we see that there is one local minimum and two local maxima. The good news is that we've found all three already! The local maximum is the y-intercept so it's 3.464 and occurs at $x = 0$. The two local minima are the x-intercepts: both are zero and occur at $x = -3.464$ and $x = 3.464$.

33. One possible answer:

35. One possible answer:

37. One possible answer:

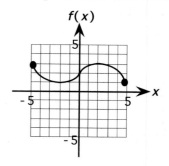

39. Answers will vary.

41. With the aid of a graphing calculator, the graph is seen to be as follows:

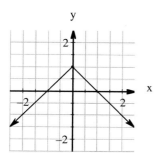

Notice that the top formula, $x + 1$, corresponds to x-values less than or equal to zero, so we need to plug $x = -2$ and $x = -1$ into the top formula.
$f(-2) = -2 + 1 = -1$;
$f(-1) = -1 + 1 = 0$
The bottom formula, $-x + 1$, corresponds to x-values greater than zero, so we need to plug $x = 1$ and $x = 2$ into the bottom formula.
$f(1) = -1 + 1 = 0$; $f(2) = -1$

43. With the aid of a graphing calculator the graph is seen to be as follows:

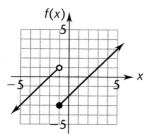

Notice that only x-values less than 1 are supposed to be plugged into the top formula, $x + 2$. So we plug $x = -2$ into the top formula $f(-2) = -2 + 2 = 0$. The remaining three x-values, -1, 1, and 2, are all greater than or equal to -1, so we plug into the bottom formula, $x - 2$.
$f(-1) = -1 - 2 = -3$;
$f(1) = 1 - 2 = -1$; $f(2) = 2 - 2 = 0$

45. With the aid of a graphing calculator the graph is seen to be as follows:

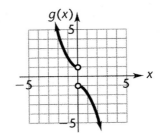

Notice that all x-values less than zero are supposed to be plugged into the top formula, $x^2 + 1$. So we plug $x = -2$ and $x = -1$ into the top formula:

$f(-2) = (-2)^2 + 1 = 4 + 1 = 5$

$f(-1) = (-1)^2 + 1 = 1 + 1 = 2$

The remaining two x-values, $x = 1$ and $x = 2$, are greater than zero so we plug them into the bottom formula, $-x^2 - 1$.

$f(1) = -(1)^2 - 1 = -1 - 1 = -2;\; f(2) = -(2)^2 - 1 = -4 - 1 = -5$

Common Error: When evaluating -1^2, order of operations tells us to square 1 first **then** multiply by -1 so the result is -1 not +1.

47. First we look at the graph in an appropriate viewing window:

Because such a large viewing window is necessary to see all the features, ZoomFit is very helpful here.

Domain: All real numbers or $(-\infty, \infty)$. (The function is a polynomial.)

Range: All real numbers or $(-\infty, \infty)$. (Every height gets hit eventually.)

y-intercept: Plug in zero for x. $m(0) = 0^3 + 45(0)^2 - 30 = -30$

x-intercepts: There are three of them. It's not clear from the graph, but since $(0, -30)$ is on the graph, it must cross the x-axis twice close to $x = 0$.

The x-intercepts are -44.99, -0.82, and 0.81.

49. First we look at the graph in an appropriate viewing window:

Because such a large viewing window is necessary to see all the features, ZoomFit is very helpful here.

Domain: All real numbers (The function is a polynomial.)

Range: $(-\infty, 10,200]$ The highest points on the graph are at height 10,200 and there is no lowest point.

y-intercept: Plug in zero for x. $n(0) = 200 + 200(0)^2 - (0)^4 = 200$
x-intercept: The graph crosses the x-axis twice. (It looks like it might hit
the x-axis at zero, but we just found that it's at height 200 there.)

The x-intercepts are −14.18 and 14.18.

51. First we look at the graph in an appropriate viewing window:

Domain: $[0, \infty)$. This makes sense because any negative numbers will make the
root undefined.
Range: $(-\infty, 16]$. The highest point on the graph is at height 16 and there is no
lowest point.
y-intercept: Since $(0, 0)$ is on the graph, the y-intercept is 0.
x-intercepts: The graph hits the x-axis twice:

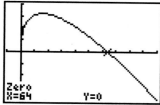

The x-intercepts are 0 and 64.

53. First we look at the graph in an appropriate viewing window:

Domain: $[-5, \infty)$. There's nothing to the left of −5.
Range: $[-134.02, \infty)$. The lowest point on the graph is at height −134.02 and
there is no highest point.
y-intercept: Using the "value" feature on a graphing calculator we see that
$k(0) = -111.80$, so the y-intercept is −111.80.
x-intercepts: The graph crosses the x-axis twice.

The x-intercepts are −4.79 and 14.94.

55.

The graph of f is rising and f is increasing on $(-\infty, 0.13]$ and $[7.87, \infty)$. The graph of f is falling and f is decreasing on $[0.13, 7.87]$. Note that these are x-values! $f(0.13) = -9.81$ is a local maximum and $f(7.87) = -242.19$ is a local minimum.

57.

The graph of m is rising and m is increasing on $(-\infty, -12]$ and $[0, 12]$. The graph of m is falling and m is decreasing on $[-12, 0]$ and $[12, \infty)$. Note that these are x-values! $m(-12) = 144$ and $m(12) = 144$ are local maxima and $m(0) = 0$ is a local minimum.

59.

The graph of g is rising and g is increasing on $[-15, 2.5]$ and $[20, \infty)$. The graph of g is falling and g is decreasing on $(-\infty, -15]$ and $[2.5, 20]$. Note that these are x-values! $g(2.5) = 306.25$ is a local maximum and $g(-15) = 0$ and $g(20) = 0$ are local minima.

61. The graph of $f(x)$ is shown as well as the result of applying the MINIMUM command to find the local minimum.

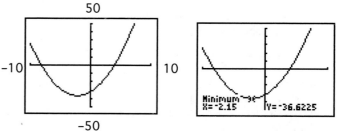

The graph of f decreases on $[-10, -2.15]$ to a local minimum value, $f(-2.15) = -36.62$, and then increases on $[-2.15, 10]$.

63. The results of applying the MAXIMUM and MINIMUM commands to find the local maximum and the local minimum are shown.

The graph of *h* increases on [-10, -4.64] to a local maximum value, *f*(-4.64) ≈ 281.93, decreases on [-4.64, 5.31] to a local minimum value, *f*(5.31) ≈ -211.41, and then increases on [5.31, 10].

65. The results of applying the MAXIMUM and MINIMUM commands to find the local minima and the local maximum are shown.

The graph of *p* decreases on [-10, -3.77] to a local minimum value, *f*(-3.77) ≈ 0, increases on [-3.77, 0.50] to a local maximum value, *f*(0.50) ≈ 18.25, decreases on [0.50, 4.77] to a local minimum value, *f*(4.77) ≈ 0, and then increases on [4.77, 10].

67. One possible answer:

69. One possible answer:

71. One possible answer:

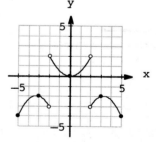

73. The graph of *f* is shown below.

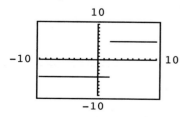

Domain: all real number except *x* = 2.
Range: The only possible values of *y* are -5 and 5. The range consists of these two values in set notation: {-5, 5}. (This does not mean the interval (-5, 5).)
Discontinuous: at *x* = 2.

75. The graph of f is shown below.

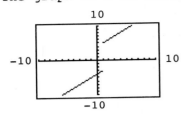

Domain: All real numbers
except $x = 1$;
Range: $(-\infty, -3) \cup (5, \infty)$;
Discontinuous at $x = 1$

77. The graph of f is shown below.

Domain: All real numbers
except $x = 3$;
Range: $(0, \infty)$;
Discontinuous at $x = 3$

79. We note: $f(0) = [\![0/2]\!] = [\![0]\!] = 0$

$f(1) = [\![1/2]\!] = 0$

$f(2) = [\![2/2]\!] = [\![1]\!] = 1$

$f(3) = [\![3/2]\!] = 1$ $f(x) = [\![x/2]\!]$ appears to jump at intervals of 2 units.

Generalizing, we can write: $f(x) = \begin{cases} \vdots & & \vdots \\ -2 & \text{if} & -4 \le x < -2 \\ -1 & \text{if} & -2 \le x < 0 \\ 0 & \text{if} & 0 \le x < 2 \\ 1 & \text{if} & 2 \le x < 4 \\ 2 & \text{if} & 4 \le x < 6 \\ \vdots & \text{if} & \vdots \end{cases}$

Domain: All real numbers
Range: All integers
Discontinuous at the even integers

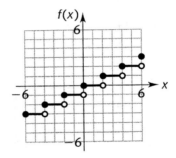

81. Having noted in problem 79 that $[\![\frac{x}{2}]\!]$ jumps at intervals of 2 units, we can

investigate $[\![3x]\!] = [\![x \div \frac{1}{3}]\!]$ to see if it jumps at intervals of $\frac{1}{3}$ unit.

$f(0) = [\![3 \cdot 0]\!] = [\![0]\!] = 0$

$f(\tfrac{1}{6}) = [\![3 \cdot \tfrac{1}{6}]\!] \quad = [\![\tfrac{1}{2}]\!] \quad = 0$

$f(\tfrac{1}{3}) = [\![3 \cdot \tfrac{1}{3}]\!] \quad = [\![1]\!] \quad = 1$

$f(\tfrac{1}{2}) = [\![3 \cdot \tfrac{1}{2}]\!] \quad = [\![\tfrac{3}{2}]\!] \quad = 1$

$f(\tfrac{2}{3}) = [\![3 \cdot \tfrac{2}{3}]\!] \quad = [\![2]\!] \quad = 2$

$f(\tfrac{5}{6}) = [\![3 \cdot \tfrac{5}{6}]\!] \quad = [\![\tfrac{5}{2}]\!] \quad = 2$

Generalizing, we can write: $f(x) = \begin{cases} \vdots & & \vdots \\ -2 & \text{if} & -\frac{2}{3} \le x < -\frac{1}{3} \\ -1 & \text{if} & -\frac{1}{3} \le x < 0 \\ 0 & \text{if} & 0 \le x < \frac{1}{3} \\ 1 & \text{if} & \frac{1}{3} \le x < \frac{2}{3} \\ 2 & \text{if} & \frac{2}{3} \le x < 1 \\ \vdots & \text{if} & \vdots \end{cases}$

Domain: All real numbers
Range: All integers
Discontinuous at rational numbers of the
form $\dfrac{k}{3}$ where k is an integer

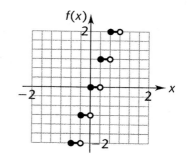

83. This function is discontinuous at every integer value of x. If x is an integer,
then $x - [\![x]\!] = x - x = 0$. If x is between n and $n + 1$ for some integer n, then
$[\![x]\!] = n$, and $x - [\![x]\!] = x - n$. So the graph between $x = n$ and $x = n + 1$ is a
line segment of the form $y = x - n$, connecting $(n, 0)$ and $(n + 1, 1)$. But
notice that $n + 1$ is always an integer, so $f(n + 1) = 0$, not 1. We can write:

$f(x) = \begin{cases} \vdots & & \vdots \\ x + 2 & \text{if} & -2 \le x < -1 \\ x + 1 & \text{if} & -1 \le x < 0 \\ x & \text{if} & 0 \le x < 1 \\ x - 1 & \text{if} & 1 \le x < 2 \\ x - 2 & \text{if} & 2 \le x < 3 \\ \vdots & \text{if} & \vdots \end{cases}$

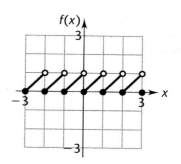

Domain: all real numbers
Range: [0, 1)

85. (A) One possible answer:

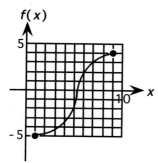

(B) This graph crosses the x axis once.
To meet the conditions specified a graph
must cross the x axis exactly once. If
it crossed more times the function would
have to be decreasing somewhere; if it
did not cross at all the function would
have to be discontinuous somewhere.

87. (A) One possible answer:

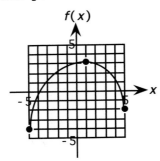

(B) This graph crosses the *x* axis twice. To meet the conditions specified a graph must cross the *x* axis at least twice. If it crossed fewer times the function would have to be discontinuous somewhere. However, the graph could cross more times; in fact there is no upper limit on the number of times it can cross the *x* axis.

89. (A) One possible answer:

(B) This graph crosses the *x* axis once. If $f(0)$ were negative, a graph could meet the given conditions and cross the *x* axis 0 times. If $f(0)$ remained positive, while $f(10)$ were positive, a graph could meet the given conditions and cross the *x* axis twice. It could not cross more than twice without having another local extremum on the interval.

91. Look at the graph of the revenue function on the given interval and use the MAXIMUM command.

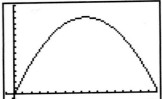

The variable *x* represents the number of car seats; the *x*-coordinate of the maximum is 857.14273, so 857 car seats must be sold. (It doesn't make sense to sell .14273 car seats.) The second coordinate represents revenue so the maximum revenue is $25,714.

93.

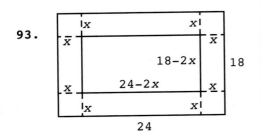

We need to write a function that describes the volume of the box then find the maximum of that function to find the volume of a rectangular box. We multiply the three dimensions.
According to the diagram, the dimensions of the base are 24 − 2*x* and 18 − 2*x* while the height is *x*. (The height comes from turning up the flaps.)
So the volume is $V(x) = x(24 - 2x)(18 - 2x)$.
Look at the graph of this function and use the MAXIMUM command.

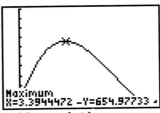

The maximum occurs at the point (3.39, 654.98) and the *y*-coordinate represents the volume. So the maximum volume is 654.98 cubic inches.

95. Look at the graph of the function $C(x)$ on the given interval and use the MINIMUM command.

The minimum cost occurs when $x = 6.41$. But notice that the first part of the question asks for the land portion of the pipe, which according to the diagram is $20 - x$. So the land portion should be $20 - 6.41 = 13.59$ miles. The minimum value of $C(x)$ rounds to 300. Since the units for C is thousands of dollars the minimum cost is $300,000.

97.

$$
\begin{array}{rcll}
f(4) = & 10\|0.5 + 0.4\| = & 10(0) = & 0 \\
f(-4) = & 10\|0.5 - 0.4\| = & 10(0) = & 0 \\
f(6) = & 10\|0.5 + 0.6\| = & 10(1) = & 10 \\
f(-6) = & 10\|0.5 - 0.6\| = & 10(-1) = & -10 \\
f(24) = & 10\|0.5 + 2.4\| = & 10(2) = & 20 \\
f(25) = & 10\|0.5 + 2.5\| = & 10(3) = & 30 \\
f(247) = & 10\|0.5 + 24.7\| = & 10(25) = & 250 \\
f(-243) = & 10\|0.5 - 24.3\| = & 10(-24) = & -240 \\
f(-245) = & 10\|0.5 - 24.5\| = & 10(-24) = & -240 \\
f(-246) = & 10\|0.5 - 24.6\| = & 10(-25) = & -250
\end{array}\Bigg\}
$$
rounds numbers to the tens place

99. Since $f(x) = [\![10x + 0.5]\!]/10$ rounds numbers to the nearest tenth, (see text Example 7) we try $[\![100x + 0.5]\!]/100 = f(x)$ to round to the nearest hundredth.

$$f(3.274) = [\![327.9]\!]/100 \quad = 3.27$$
$$f(7.846) = [\![785.1]\!]/100 \quad = 7.85$$
$$f(-2.8783) = [\![-287.33]\!]/100 = -2.88$$

A few examples suffice to convince us that this is probably correct. (A proof would be out of place in this book.)
$$f(x) = [\![100x + 0.5]\!]/100.$$

101. Since 100 miles are included, only the daily charge of $32 applies for mileage between 0 and 100. So if $R(x)$ is the daily cost of rental where x is miles driven, $R(x) = 32$ if $0 \le x \le 100$. After 100 miles, the charge is an extra $0.16 for each mile: the mileage charge will be 0.16 times the number of miles over 100 which is $x - 100$. So the mileage charge is $0.16(x - 100)$ or $0.16x - 16$. The $32 charge still applies so when $x \ge 100$ the rental charge is $32 + 0.16x - 16$, or $16 + 0.16x$.

$$R(x) = \begin{cases} 32 & 0 \le x \le 100 \\ 16 + 0.16x & x > 100 \end{cases}$$

103. (A) $C(x) = \begin{cases} 15 & 0 < x \le 1 \\ 18 & 1 < x \le 2 \\ 21 & 2 < x \le 3 \\ 24 & 3 < x \le 4 \\ 27 & 4 < x \le 5 \\ 30 & 5 < x \le 6 \end{cases}$

(B) The two functions appear to coincide, for example
$$C(3.5) = 24 \quad f(3.5) = 15 + 3[\![3.5]\!] = 15 + 3 \cdot 3 = 24$$
However,
$$C(1) = 15 \quad f(1) = 15 + 3[\![1]\!] = 15 \cdot 3 \cdot 1 = 18$$
The functions are not the same, therefore.
In fact, $f(x) \neq C(x)$ at $x = 1, 2, 3, 4, 5, 6$.

105. If $0 \le x \le 3,000$, $E(x) = 200$

	Base + Salary	Commission on Sales over $3,000
If $3,000 < x < 8,000$, $E(x)$ = 200	+	$0.04(x - 3,000)$
= 200	+	$0.04x - 120$
= 80	+	$0.04x$

Common Error:
Commission is not $0.04x$
(4% of sales) nor is it
$0.04x + 200$
(base salary plus 4% of sales).

There is a point of discontinuity at $x = 8,000$.

	Salary	+	Bonus
If $x \ge 8,000$, $E(x)$ = $80 + 0.04x$	+	100
=	180	+	$0.04x$

Summarizing,
$$E(x) = \begin{cases} 200 & \text{if } 0 \le x \le 3,000 \\ 80 + 0.04x & \text{if } 3,000 < x < 8,000 \\ 180 + 0.04x & \text{if } 8,000 \le x \end{cases}$$

$$E(5,750) = 80 + 0.04(5,750) = \$310$$
$$E(9,200) = 180 + 0.04(9,200) = \$548$$

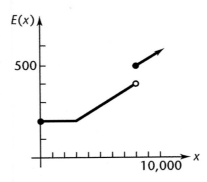

x	$y = 200$	x	$y = 80 + 0.04x$	x	$y = 180 + 0.04x$
0	200	3,000	200	8,000	500
2,000	200	7,000	360	10,000	580

107. A tax of 5.35% on any amount up to $19,890 is calculated by multiplying .0535 by the income, which is represented by x. So the tax due, $t(x)$, is $0.0535x$ if x is between 0 and 19,890. For x-values between 19,890 and 65,330, the percentage is computed only on the portion over 19,890. To find that portion we subtract 19,890 from the income $(x - 19,890)$; then multiply by 0.0705 to get the percentage portion of the tax. The total tax is $1,064 plus the percentage portion, so $t(x) = 1,064 + 0.0705(x - 19,890)$ if x is between 19,890 and 65,330. The tax for incomes over $65,330 is computed in a similar manner: $4,268 plus 7.85% of the portion over 65,330, which is $x - 65,330$. We get $t(x) = 4,268 + 0.0785(x - 65,330)$ if x is over 65,330. Combined we get

$$t(x) = \begin{cases} 0.0535x & 0 \le x \le 19,890 \\ 1,064 + 0.0705(x - 19,890) & 19,890 < x \le 65,330 \\ 4,268 + 0.0785(x - 65,330) & x > 65,330 \end{cases}$$

or if we multiply out parentheses to simplify,

$$t(x) = \begin{cases} 0.0535x & 0 \le x \le 19,890 \\ 0.0705x - 338.25 & 19,890 < x \le 65,330 \\ 0.0785x - 860.41 & x > 65,330 \end{cases}$$

$t(10,000) = 0.0535(10,000) = 535$; the tax is $535
$t(30,000) = 1,064 + 0.0705(30,000 - 19,890) = 1,776.75$; the tax is $1,776.75
$t(100,000) = 4,268 + 0.0785(100,000 - 65,330) = 6,989.60$; the tax is $6,989.60

109. (A) For detailed instructions on inputting the data see Example 10 in Section 1-2. The data input, viewing window, and scatter plot are as follows:

Next, we input the given function and use the table feature to construct a table with the same x values:

Notice that the y-values are very similar to those in the data table. Finally we compare the scatter plot and the graph of $f(x)$:

As expected, the graph of $f(x)$ fits the points in the scatter plot well.

(B) $f(31) = -0.518(31)^2 + 33.3(31) - 481 = 53.50$, so the mileage for a tire pressure of 31 lb/m^2 will be 53,500. $f(35) = 49.95$ so the mileage for a tire pressure of 35 lb/in^2 will be 49,950.

(C) As the tire pressure increases from 28 to 32 lb/in^2 the mileage increases from 45,000 mi to 55,000 mi. A local maximum occurs when the tire pressure is 32 lb/in^2 and the mileage is 55,000 mi. As the tire pressure continues to increase from 32 lb/in^2 to 36 lb/in^2 the mileage decreases from 55,000 mi to 47,000 mi.

Section 1-4

1. Answers will vary

3. Answers will vary.

5. The graph of $f(x)$ is shifted up 2 units

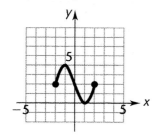

7. The graph of $g(x)$ is shifted up 2 units

9. The graph of $f(x)$ is shifted right 2 units.

11. The graph of $g(x)$ is shifted left 2 units.

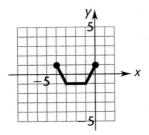

13. The graph of $f(x)$ is reflected in the x axis.

15. The graph of $g(x)$ is vertically stretched by multiplying each y coordinate by 2.

> **Common Errors:**
> Confusing the -2 in $f(x - 2)$ with a shift left.
> Confusing the positive 2 in $g(x + 2)$ with a shift right.

17. $g(2x)$ is a horizontal shrinking of the graph of $g(x)$ by a factor of $\frac{1}{2}$.

19. The graph of $f(-x)$ is a reflection in the y-axis of the graph of $f(x)$.

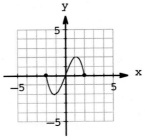

21. $g(-x) = (-x)^3 + (-x) = -x^3 - x = -(x^3 + x) = -g(x)$. Odd

23. $m(-x) = (-x)^4 + 3(-x)^2 = x^4 = 3x^2 = m(x)$. Even

25. $F(-x) = (-x)^5 + 1 = -x^5 + 1$
$-F(x) = -(x^5 + 1) = -x^5 - 1$
Therefore $F(-x) \neq F(x)$. $F(-x) \neq -F(x)$. $F(x)$ is neither even nor odd.

27. $G(-x) = (-x)^4 + 2 = x^4 + 2 = G(x)$. Even

29. $q(-x) = (-x)^2 + (-x) - 3 = x^2 - x - 3$.
$-q(x) = -(x^2 + x - 3) = -x^2 - x + 3$
Therefore $q(-x) \neq q(x)$. $q(-x) \neq -q(x)$. $q(x)$ is neither even nor odd.

31. The graph of $y = x^2$ is shifted 2 units to the right: $y = (x - 2)^2$.
Check: $y = (x - 2)^2$ is graphed on a graphing calculator.

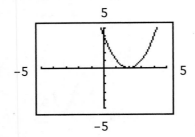

33. The graph of $y = x^3$ is shifted down 2 units: $y = x^3 - 2$.
Check: $y = x^3 - 2$ is graphed on a graphing calculator.

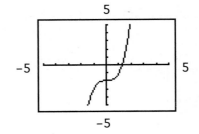

35. The graph of $y = |x|$ is shrunk vertically by a factor of 0.25: $y = 0.25|x|$.
Check: $y = 0.25|x|$ is graphed on a graphing calculator.

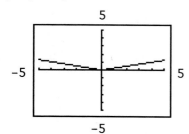

37. The graph of $y = x^3$ is reflected in the x axis (or the y axis): $y = -x^3$.
Check: $y = -x^3$ is graphed on a graphing calculator.

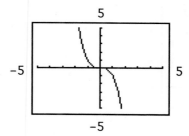

39. $g(x) = \sqrt[3]{x + 4} - 5$.
The graphs of $f(x) = \sqrt[3]{x}$ (thin curve) and $g(x)$ (thick curve) are shown.

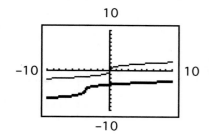

41. $g(x) = -0.5(6 + \sqrt{x})$
The graph of $f(x) = \sqrt{x}$ (thin curve) and $g(x)$ (thick curve) are shown.

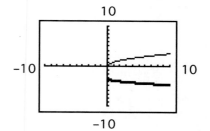

43. $g(x) = -2(x + 4)^2 - 2$
The graph of $f(x) = x^2$ (thin curve) and $g(x)$ (thick curve) are shown.

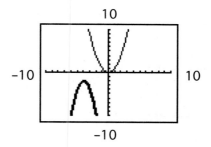

45. $g(x) = \sqrt{-\dfrac{1}{2}(x + 2)}$. The graph of $f(x) = \sqrt{x}$ and $g(x)$ are shown.

47. The graph of $y = x^2$ is shifted 7 units left and 9 units up.

49. The graph of $y = |x|$ is shifted 8 units right and reflected in the x axis.

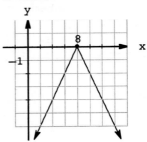

51. The graph of $y = \sqrt{x}$ is reflected in the x axis and shifted 3 units up.

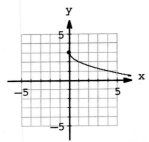

53. The graph of $y = x^2$ is stretched vertically by a factor of 4 and reflected in the x axis.

55. The graph of $y = |x|$ is shifted 2 units left and 2 units up:
$y = |x + 2| + 2$.
Check: $y = |x + 2| + 2$ is graphed on a graphing calculator.

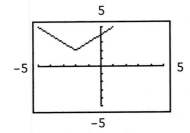

57. The graph of $y = \sqrt{x}$ is reflected in the x axis and shifted 4 units up:
$y = 4 - \sqrt{x}$
Check: $y = 4 - \sqrt{x}$ is graphed in a graphing calculator.

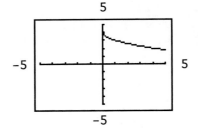

59. The graph of $y = x^2$ is shifted 1 unit right, reflected in the x axis, and shifted 4 units up:
$y = 4 - (x - 1)^2$
Check: $y = 4 - (x - 1)^2$ is graphed in a graphing calculator.

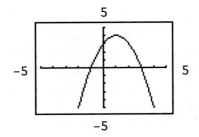

61. The graph of $y = x^3$ is shifted 3 units right, vertically shrunk by a factor of 0.5, and shifted 1 unit up:
$y = 0.5(x - 3)^3 + 1$
Check: $y = 0.5(x - 3)^3 + 1$ is graphed in a graphing calculator.

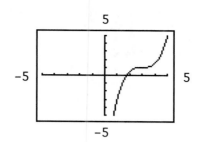

63. (A) The function f is a horizontal shrink of $y = \sqrt[3]{x}$ by a factor of 1/8, while g is a vertical stretch of $y = \sqrt[3]{x}$ by a factor of 2.

(B) The graphs are shown below in a standard window: they are identical.

    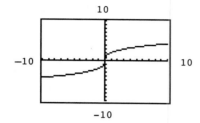

(C) $\sqrt[3]{8x} = \sqrt[3]{8} \cdot \sqrt[3]{x} = 2\sqrt[3]{x}$

65. (A) The graphs are shown below.

    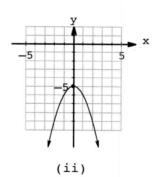

(i)   (ii)

The graphs are different, so order is significant when performing multiple transformations.

(B) (i): $y = -\left(x^2 - 5\right)$; (ii): $y = -x^2 - 5$

These functions are different. In the second one, order of operations tells us to first multiply by -1, then subtract 5. In the first, the parentheses indicate that this order should be reversed.

67. The graph of $y = f(x - h)$ represents a horizontal shift from the graph of $y = f(x)$. The graph of $y = f(x) + k$ represents a vertical shift from the graph of $y = f(x)$. The graph of $y = f(x - h) + k$ represents both a horizontal and a vertical shift but the order does not matter:
Vertical first then horizontal: $y = f(x) \rightarrow y = f(x) + k \rightarrow y = f(x - h) + k$
Horizontal first then vertical: $y = f(x) \rightarrow y = f(x - h) \rightarrow y = f(x - h) + k$
The same result is achieved; reversing the order does not change the result.

69. Consider the graph of $y = x^2$.
If a vertical shift is performed the equation becomes $y = x^2 + k$.
If a reflection is now performed the equation becomes
$y = -(x^2 + k)$ or $y = -x^2 - k$.
If the reflection is performed first the equation becomes $y = -x^2$.
If the vertical shift is now performed the equation becomes $y = -x^2 + k$.
Since $y = -x^2 - k$ and $y = -x^2 + k$ differ (unless $k = 0$), reversing the order
changes the result.

71. The graph of $y = f(x - h)$ represents a horizontal shift from the graph of
$y = f(x)$. The graph of $y = -f(x)$ represents a reflection of the graph of
$y = f(x)$ in the x axis. The graph of $y = -f(x - h)$ represents both a horizontal
shift and a reflection but the order does not matter:
Shift first then reflection: $y = f(x) \rightarrow y = f(x - h) \rightarrow y = -f(x - h)$
Reflection first then shift: $y = f(x) \rightarrow y = -f(x) \rightarrow y = -f(x - h)$
The same result is achieved; reversing the order does not change the result.

73.

75.

77. (A) $E(x) = \dfrac{1}{2}[f(x) + f(-x)]$

$E(-x) = \dfrac{1}{2}[f(-x) + f\{-(-x)\}] = \dfrac{1}{2}[f(-x) + f(x)] = \dfrac{1}{2}[f(x) + f(-x)] = E(x)$.

$E(x)$ is even.

(B) $O(x) = \dfrac{1}{2}[f(x) - f(-x)]$

$O(-x) = \dfrac{1}{2}[f(-x) - f\{-(-x)\}] = \dfrac{1}{2}[f(-x) - f(x)] = -\dfrac{1}{2}[f(x) - f(-x)] = -O(x)$.

$O(x)$ is odd.

(C) $E(x) + O(x) = \dfrac{1}{2}[f(x) + f(-x)] + \dfrac{1}{2}[f(x) - f(-x)]$

$= \dfrac{1}{2}f(x) + \dfrac{1}{2}f(-x) + \dfrac{1}{2}f(x) - \dfrac{1}{2}f(-x)$

$= f(x)$

Conclusion: Any function can be written as the sum of two other functions, one
even and the other odd.

79. Graph of $f(x)$

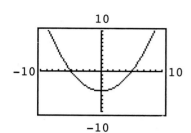

Graph of
$$|f(x)| = |0.2x^2 - 5|$$

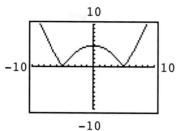

Graph of
$$-|f(x)| = -|0.2x^2 - 5|$$

81. Graph of $f(x)$

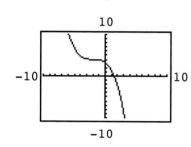

Graph of
$$|f(x)| = |4 - 0.1(x + 2)^3|$$

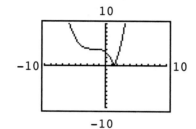

Graph of
$$-|f(x)| = -|4 - 0.1(x + 2)^3|$$

83. The graph of $y = |f(x)|$ is the same as the graph of $y = f(x)$ whenever $f(x) \geq 0$ and is the reflection of the graph of $y = f(x)$ with respect to the x axis whenever $f(x) < 0$.

85. The graph of the function $C(x) = 30,000 + f(x)$ is the same as the given graph of the function $f(x)$ shifted up 30,000 units ($).

87. $y = 10 + 0.004(x - 10)^3$,
$y = 15 + 0.004(x - 10)^3$,
$y = 20 + 0.004(x - 10)^3$.
Each graph is a vertical translation of the graph of $y = 0.004(x - 10)^3$.

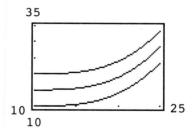

89. $y = \dfrac{1}{2}x^2 - 2$, $y = \dfrac{1}{3}x^2 - 3$, $y = \dfrac{1}{4}x^2 - 4$, $y = \dfrac{1}{5}x^2 - 5$.

Each graph is a vertical shrink followed by a vertical translation of the graph of $y = x^2$.

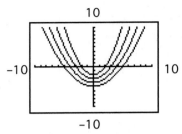

91. $V(t) = \dfrac{64}{c^2}(C - t)^2 \qquad 0 \le t \le C$

t	$C = 1$ $V = 64(1 - t)^2$	$C = 2$ $V = 16(2 - t)^2$	$C = 4$ $V = 4(4 - t)^2$	$C = 8$ $V = (8 - t)^2$
0	64	64	64	64
1	0	16	36	49
2	not defined for $t>1$	0	16	36
4		not defined for $t>0$	0	16
8			not defined for $t > 4$	0

Each graph is a portion of the graph of a horizontal translation followed by a vertical shrink (except for $C = 8$) of the graph of $y = t^2$.

The height of the graph represents the volume of water left in the tank, so we see that for larger values of C, the water stays in the tank longer. We can conclude that larger values of C correspond to a smaller opening.

Section 1-5

1. Answers will vary.

3. No. Elements that make the output of g equal to zero are excluded from the domain of f/g.

5. No. Explanations will vary.

7. Construct a table of values for $f(x)$ and $g(x)$ from the graph, then add to obtain $(f + g)(x)$.

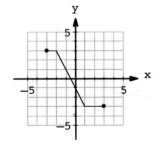

x	-3	-2	-1	0	1	2	3
$f(x)$	1	0	-1	-2	-3	-2	-1
$g(x)$	2	3	2	1	0	-1	-2
$(f + g)(x)$	3	3	1	-1	-3	-3	-3

9. Construct a table of values for $f(x)$ and $g(x)$ from the graph, then multiply to obtain $(fg)(x)$.

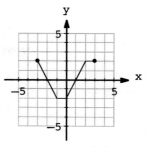

x	-3	-2	-1	0	1	2	3
$f(x)$	1	0	-1	-2	-3	-2	-1
$g(x)$	2	3	2	1	0	-1	-2
$(f + g)(x)$	2	0	-2	-2	0	2	2

11. $(f \circ g)(-1) = f[g(-1)]$. From the graph of g, $g(-1) = 2$. From the graph of f, $f[g(-1)] = f(2) = -2$.

13. $(g \circ f)(-2) = g[f(-2)]$. From the graph of f, $f(-2) = 0$. From the graph of g, $g[f(-2)] = g(0) = 1$.

15. From the graph of g, $g(1) = 0$. From the graph of f, $f[g(1)] = f(0) = -2$.

17. From the graph of f, $f(2) = -2$. From the graph of g, $g[f(2)] = g(-2) = 3$.

19. Using values from the table, $g(-7) = 4$, so $(f \circ g)(-7) = f(g(-7)) = f(4) = 3$. Similarly, $(f \circ g)(0) = f(g(0)) = f(-2) = 9$, and $(f \circ g)(4) = f(g(4)) = f(6) = -10$.

21.
$(f + g)(x) = f(x) + g(x) = 4x + x + 1 = 5x + 1$ Domain: $(-\infty, \infty)$
$(f - g)(x) = f(x) - g(x) = 4x - (x + 1) = 3x - 1$ Domain: $(-\infty, \infty)$
$(fg)(x) = f(x)g(x) = 4x(x + 1) = 4x^2 + 4x$ Domain: $(-\infty, \infty)$
$\left(\dfrac{f}{g}\right)(x) = \dfrac{f(x)}{g(x)} = \dfrac{4x}{x + 1}$ Domain: $\{x \mid x \neq -1\}$, or

$(-\infty, -1) \cup (-1, \infty)$

> **Common Error:**
> $f(x) - g(x) \neq 4x - x + 1$. The parentheses are necessary.

23.
$(f + g)(x) = f(x) + g(x) = 2x^2 + x^2 + 1 = 3x^2 + 1$ Domain: $(-\infty, \infty)$
$(f - g)(x) = f(x) - g(x) = 2x^2 - (x^2 + 1) = x^2 - 1$ Domain: $(-\infty, \infty)$
$(fg)(x) = f(x)g(x) = 2x^2(x^2 + 1) = 2x^4 + 2x^2$ Domain: $(-\infty, \infty)$
$\left(\dfrac{f}{g}\right)(x) = \dfrac{f(x)}{g(x)} = \dfrac{2x^2}{x^2 + 1}$ Domain: $(-\infty, \infty)$

(since $g(x)$ is never 0.)

25.
$(f + g)(x) = f(x) + g(x) = 3x + 5 + x^2 - 1$
$\qquad = x^2 + 3x + 4$ Domain: $(-\infty, \infty)$
$(f - g)(x) = f(x) - g(x) = 3x + 5 - (x^2 - 1)$
$\qquad = 3x + 5 - x^2 + 1 = -x^2 + 3x + 6$ Domain: $(-\infty, \infty)$
$(fg)(x) = f(x)g(x) = (3x + 5)(x^2 - 1)$
$\qquad = 3x^3 - 3x + 5x^2 - 5 = 3x^3 + 5x^2 - 3x - 5$ Domain: $(-\infty, \infty)$
$\left(\dfrac{f}{g}\right)(x) = \dfrac{f(x)}{g(x)} = \dfrac{3x + 5}{x^2 - 1}$ Domain: $\{x \mid x \neq \pm 1\}$, or

$(-\infty, -1) \cup (-1, 1) \cup (1, \infty)$

27. $(f \circ g)(x) = f[g(x)] = f(x^2 - x + 1) = (x^2 - x + 1)^3$ Domain: $(-\infty, \infty)$

$(g \circ f)(x) = g[f(x)] = g(x^3) = (x^3)^2 - x^3 + 1 = x^6 - x^3 + 1$ Domain: $(-\infty, \infty)$

29. $(f \circ g)(x) = f[g(x)] = f(2x + 3) = |2x + 3 + 1| = |2x + 4|$ Domain: $(-\infty, \infty)$

$(g \circ f)(x) = g[f(x)] = g(|x + 1|) = 2|x + 1| + 3$ Domain: $(-\infty, \infty)$

31. $(f \circ g)(x) = f[g(x)] = f(2x^3 + 4) = (2x^3 + 4)^{1/3}$ Domain: $(-\infty, \infty)$

$(g \circ f)(x) = g[f(x)] = g(x^{1/3}) = 2(x^{1/3})^3 + 4 = 2x + 4$ Domain: $(-\infty, \infty)$

33. $(f \circ g)(x) = f[g(x)] = f(2x - 2) = \dfrac{1}{2}(2x - 2) + 1 = x - 1 + 1 = x$

$(g \circ f)(x) = g[f(x)] = g\left(\dfrac{1}{2}x + 1\right) = 2\left(\dfrac{1}{2}x + 1\right) - 2 = x + 2 - 2 = x$

Graphing f, g, $f \circ g$, and $g \circ f$,
we obtain the graph at the right.

The graphs of f and g are
reflections of each other in the
line $y = x$, which is the graph of
$f \circ g$ and $g \circ f$.

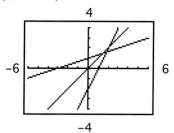

35. $(f \circ g)(x) = f[g(x)] = f\left(-\dfrac{3}{2}x - \dfrac{5}{2}\right) = -\dfrac{2}{3}\left(-\dfrac{3}{2}x - \dfrac{5}{2}\right) - \dfrac{5}{3} = x + \dfrac{5}{3} - \dfrac{5}{3} = x$

$(g \circ f)(x) = g[f(x)] = g\left(-\dfrac{2}{3}x - \dfrac{5}{3}\right) = -\dfrac{3}{2}\left(-\dfrac{2}{3}x - \dfrac{5}{3}\right) - \dfrac{5}{2} = x + \dfrac{5}{2} - \dfrac{5}{2} = x$

Graphing f, g, $f \circ g$, and $g \circ f$, we
obtain the graph at the right.

The graphs of f and g are
reflections of each other in the
line $y = x$, which is the graph of
$f \circ g$ and $g \circ f$.

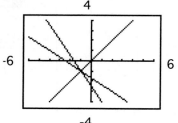

37. $(f + g)(x) = f(x) + g(x) = \sqrt{2 - x} + \sqrt{x + 3}$

$(f - g)(x) = f(x) - g(x) = \sqrt{2 - x} - \sqrt{x + 3}$

$(fg)(x) = f(x)g(x) = \sqrt{2 - x}\sqrt{x + 3} = \sqrt{(2 - x)(3 + x)} = \sqrt{6 - x - x^2}$

$\left(\dfrac{f}{g}\right)(x) = \dfrac{f(x)}{g(x)} = \dfrac{\sqrt{2 - x}}{\sqrt{x + 3}} = \sqrt{\dfrac{2 - x}{x + 3}}$

The domains of f and g are:
Domain of $f = \{x \mid 2 - x \geq 0\} = (-\infty, 2]$

Domain of $g = \{x \mid x + 3 \geq 0\} = [-3, \infty)$
The intersection of these domains is $[-3, 2]$.
This is the domain of the functions of $f + g$, $f - g$, and fg.

Since $g(-3) = 0$, $x = -3$ must be excluded from the domain of $\dfrac{f}{g}$, so its domain
is $(-3, 2]$.

39. $(f + g)(x) = f(x) + g(x) = \sqrt{x} + 2 + \sqrt{x} - 4 = 2\sqrt{x} - 2$

$(f - g)(x) = f(x) - g(x) = \sqrt{x} + 2 - (\sqrt{x} - 4) = \sqrt{x} + 2 - \sqrt{x} + 4 = 6$

$(fg)(x) = f(x)g(x) = (\sqrt{x} + 2)(\sqrt{x} - 4) = x - 2\sqrt{x} - 8$

$\left(\dfrac{f}{g}\right)(x) = \dfrac{f(x)}{g(x)} = \dfrac{\sqrt{x} + 2}{\sqrt{x} - 4}$

The domains of f and g are both $\{x \mid x \geq 0\} = [0, \infty)$

This is the domain of $f + g$, $f - g$, and fg. We note that in the domain of $\dfrac{f}{g}$,

$g(x)$ cannot be 0, so $\sqrt{x} - 4$ cannot be 0. To find which values to exclude, we solve

$\sqrt{x} - 4 = 0$

$\sqrt{x} = 4$ $\boxed{\text{Common Error: } x \neq \sqrt{4}}$

$x = 16$

So 16 must be excluded from $\{x \mid x \geq 0\}$ to find the domain of $\dfrac{f}{g}$.

Domain of $\dfrac{f}{g} = \{x \mid x \geq 0, x \neq 16\} = [0, 16) \cup (16, \infty)$.

41. $(f + g)(x) = f(x) + g(x) = \sqrt{x^2 + x - 6} + \sqrt{7 + 6x - x^2}$

$(f - g)(x) = f(x) - g(x) = \sqrt{x^2 + x - 6} - \sqrt{7 + 6x - x^2}$

$(fg)(x) = f(x)g(x) = \sqrt{x^2 + x - 6}\sqrt{7 + 6x - x^2} = \sqrt{-x^4 + 5x^3 + 19x^2 - 29x - 42}$

$\left(\dfrac{f}{g}\right)(x) = \dfrac{f(x)}{g(x)} = \dfrac{\sqrt{x^2 + x - 6}}{\sqrt{7 + 6x - x^2}} = \sqrt{\dfrac{x^2 + x - 6}{7 + 6x - x^2}}$

The domains of f and g are:

Domain of $f = \{x \mid x^2 + x - 6 \geq 0\} = \{x \mid (x + 3)(x - 2) \geq 0\} = (-\infty, -3] \cup [2, \infty)$

Domain of $g = \{x \mid 7 + 6x - x^2 \geq 0\} = \{x \mid (7 - x)(1 + x) \geq 0\} = [-1, 7]$

The intersection of these domains is $[2, 7]$.

This is the domain of the functions $f + g$, $f - g$, and fg.

Since $g(x) = 7 + 6x - x^2 = (7 - x)(1 + x)$, $g(7) = 0$ and $g(-1) = 0$, so 7 must be excluded from the domain of $\dfrac{f}{g}$. Its domain is $[2, 7)$.

43. $(f \circ g)(x) = f[g(x)] = f(x - 4) = \sqrt{x - 4}$ Domain: $\{x \mid x \geq 4\}$ or $[4, \infty)$

$(g \circ f)(x) = g[f(x)] = g(\sqrt{x}) = \sqrt{x} - 4$ Domain: $\{x \mid x \geq 0\}$ or $[0, \infty)$

45. $(f \circ g)(x) = f[g(x)] = f\left(\dfrac{1}{x}\right) = \dfrac{1}{x} + 2$ Domain: $\{x \mid x \neq 0\}$ or $(-\infty, 0) \cup (0, \infty)$

$(g \circ f)(x) = g[f(x)] = g(x + 2) = \dfrac{1}{x + 2}$ Domain: $\{x \mid x \neq -2\}$ or $(-\infty, -2)$ $\cup (-2, \infty)$

47. $(f \circ g)(x) = f[g(x)] = f\left(\dfrac{1}{x - 1}\right) = \left|\dfrac{1}{x - 1}\right| = \dfrac{1}{|x - 1|}$ Domain: $\{x \mid x \neq 1\}$ or $(-\infty, 1) \cup (1, \infty)$

$(g \circ f)(x) = g[f(x)] = g(|x|) = \dfrac{1}{|x| - 1}$

To find the domain we must exclude any x values for which $|x| - 1 = 0$ from the domain of f.

$|x| - 1 = 0$

$|x| = 1$

$x = -1$ or 1

Domain of $g \circ f$: $\{x \mid x \neq -1 \text{ or } 1\} = (-\infty, -1) \cup (-1, 1) \cup (1, \infty)$

In Problems #49 through 51, f and g are linear functions. f has slope -2 and y intercept 2, so f(x) = -2x + 2. g has slope 1 and y intercept -2, so g(x) = x - 2.

49. $(f + g)(x) = f(x) + g(x) = (-2x + 2) + (x - 2) = -x.$
The graph of $f + g$ is a straight line with slope -1 passing through the origin. This corresponds to graph (d).

51. $(g - f)(x) = g(x) - f(x) = (x - 2) - (-2x + 2) = x - 2 + 2x - 2 = 3x - 4.$
The graph of $g - f$ is a straight line with slope 3 and y intercept -4. This corresponds to graph (a).

In Problems #53 through 59, there may be other possible solutions.

53. If we let $g(x) = 2x - 7$, then
$h(x) = [g(x)]^4$
Now if we let $f(x) = x^4$, we have
$h(x) = [g(x)]^4 = f[g(x)] = (f \circ g)(x)$

55. If we let $g(x) = 4 + 2x$, then
$h(x) = \sqrt{g(x)}$
Now if we let $f(x) = x^{1/2}$, we have
$h(x) = \sqrt{g(x)} = [g(x)]^{1/2}$
$\qquad\qquad = f[g(x)] = (f \circ g)(x).$

57. If we let $f(x) = x^7$, then
$h(x) = 3f(x) - 5$
Now if we let $g(x) = 3x - 5$, we have
$h(x) = 3f(x) - 5 = g[f(x)] = (g \circ f)(x)$

59. If we let $f(x) = x^{-1/2}$, then
$h(x) = 4f(x) + 3$
Now if we let $g(x) = 4x + 3$, we have
$h(x) = 4f(x) + 3 = g[f(x)] = (g \circ f)(x)$

61. Yes, the function $g(x) = x$ satisfies these conditions.
$(f \circ g)(x) = f(g(x)) = f(x)$, so $f \circ g = f$
$(g \circ f)(x) = g(f(x)) = f(x)$, so $g \circ f = f$

63. $(f + g)(x) = f(x) + g(x) = x + \dfrac{1}{x} + x - \dfrac{1}{x} = 2x$

$(f - g)(x) = f(x) - g(x) = x + \dfrac{1}{x} - \left(x - \dfrac{1}{x}\right) = \dfrac{2}{x}$

$(fg)(x) = f(x)g(x) = \left(x + \dfrac{1}{x}\right)\left(x - \dfrac{1}{x}\right) = x^2 - \dfrac{1}{x^2}$

$\left(\dfrac{f}{g}\right)(x) = \dfrac{f(x)}{g(x)} = \dfrac{x + \frac{1}{x}}{x - \frac{1}{x}} = \dfrac{x^2 + 1}{x^2 - 1}$

> **Common Error:**
> Domain is not $(-\infty, \infty)$. See below.

The domains of f and g are both $\{x \mid x \neq 0\} = (-\infty, 0) \cup (0, \infty)$
This is therefore the domain of $f + g$, $f - g$, and fg. To find the domain of $\dfrac{f}{g}$, we must exclude from this domain the set of values of x for which $g(x) = 0$.

$x - \dfrac{1}{x} = 0$
$x^2 - 1 = 0$
$x^2 = 1$
$x = -1, 1$

The domain of $\dfrac{f}{g}$ is $\{x \mid x \neq 0, -1, \text{ or } 1\}$ or

$(-\infty, -1) \cup (-1, 0) \cup (0, 1) \cup (1, \infty).$

65. $(f + g)(x) = f(x) + g(x) = 1 - \dfrac{x}{|x|} + 1 + \dfrac{x}{|x|} = 2$

$(f - g)(x) = f(x) - g(x) = 1 - \dfrac{x}{|x|} - \left(1 + \dfrac{x}{|x|}\right) = 1 - \dfrac{x}{|x|} - 1 - \dfrac{x}{|x|} = \dfrac{-2x}{|x|}$

$(fg)(x) = f(x)g(x) = \left(1 - \dfrac{x}{|x|}\right)\left(1 + \dfrac{x}{|x|}\right) = (1)^2 - \left(\dfrac{x}{|x|}\right)^2 = 1 - \dfrac{x^2}{|x|^2}$

$$= 1 - \dfrac{x^2}{x^2} = 1 - 1 = 0$$

$\left(\dfrac{f}{g}\right)(x) = \dfrac{f(x)}{g(x)} = \dfrac{1 - \frac{x}{|x|}}{1 + \frac{x}{|x|}} = \dfrac{|x| - x}{|x| + x}$. This can be further simplified

however, when we examine the domain of $\dfrac{f}{g}$ below.

The domains of f and g are both
$\{x \mid x \neq 0\} = (-\infty, 0) \cup (0, \infty)$

This is therefore the domain of $f + g$, $f - g$, and fg. To find the domain of $\dfrac{f}{g}$, we must exclude from this domain the set of values of x for which $g(x) = 0$.

$1 + \dfrac{x}{|x|} = 0$

$|x| + x = 0$

$\quad |x| = -x$

This is true when x is negative.

The domain of $\dfrac{f}{g}$ is the positive numbers, $(0, \infty)$. On this domain,

$|x| = x$, so $\left(\dfrac{f}{g}\right)(x) = \dfrac{|x| - x}{|x| + x} = \dfrac{x - x}{x + x} = \dfrac{0}{2x} = 0$

67. $(f \circ g)(x) = f[g(x)] = f(x^2) = \sqrt{4 - x^2}$

The domain of f is $(-\infty, 4]$. The domain of g is all real numbers. The domain of $f \circ g$ is therefore the set of those real numbers x for which $g(x)$ is in

$(-\infty, 4]$, that is, for which $x^2 \leq 4$, or $-2 \leq x \leq 2$.
Domain of $f \circ g = \{x \mid -2 \leq x \leq 2\} = [-2, 2]$

$(g \circ f)(x) = g[f(x)] = g(\sqrt{4 - x}) = (\sqrt{4 - x})^2 = 4 - x$

The domain of $g \circ f$ is the set of those numbers x in $(-\infty, 4]$ for which $f(x)$ is in $(-\infty, \infty)$, that is, $(-\infty, 4]$.

69. $(f \circ g)(x) = f[g(x)] = f\left(\dfrac{x}{x - 2}\right) = \dfrac{\frac{x}{x-2} + 5}{\frac{x}{x-2}} = \dfrac{x + 5(x - 2)}{x} = \dfrac{x + 5x - 10}{x} = \dfrac{6x - 10}{x}$

The domain of f is $\{x \mid x \neq 0\}$. The domain of g is $\{x \mid x \neq 2\}$. The domain of $f \circ g$ is therefore the set of those numbers in $\{x \mid x \neq 2\}$ for which $g(x)$ is in $\{x \mid x \neq 0\}$. We must exclude from $\{x \mid x \neq 2\}$ those numbers x for which $\dfrac{x}{x - 2} = 0$, or $x = 0$. The domain of $f \circ g$ is $\{x \mid x \neq 0, \; x \neq 2\}$, or $(-\infty, 0) \cup (0, 2) \cup (2, \infty)$.

$(g \circ f)(x) = g[f(x)] = g\left(\dfrac{x + 5}{x}\right) = \dfrac{\frac{x+5}{x}}{\frac{x+5}{x} - 2} = \dfrac{x + 5}{x + 5 - 2x} = \dfrac{x + 5}{5 - x}$

The domain of $g \circ f$ is the set of those numbers in $\{x \mid x \neq 0\}$ for which $f(x)$ is in $\{x \mid x \neq 2\}$. We must exclude from $\{x \mid x \neq 0\}$ those numbers x for which $\dfrac{x + 5}{x} = 2$, or $x + 5 = 2x$, or $x = 5$. The domain of $g \circ f$ is $\{x \mid x \neq 0, \; x \neq 5\}$ or $(-\infty, 0) \cup (0, 5) \cup (5, \infty)$.

71. $(f \circ g)(x) = f[g(x)] = f(\sqrt{9 + x^2}) = \sqrt{25 - (\sqrt{9 + x^2})^2} = \sqrt{25 - (9 + x^2)} = \sqrt{16 - x^2}$
The domain of f is $[-5, 5]$. The domain of g is $(-\infty, \infty)$. The domain of $f \circ g$ is therefore the set of those real numbers x for which $g(x)$ is in $[-5, 5]$, that is, $\sqrt{9 + x^2} \leq 5$, or $9 + x^2 \leq 25$, or $x^2 \leq 16$, or $-4 \leq x \leq 4$. The domain of $f \circ g$ is $\{x \mid -4 \leq x \leq 4\}$ or $[-4, 4]$.

$(g \circ f)(x) = g[f(x)] = g(\sqrt{25 - x^2}) = \sqrt{9 + (\sqrt{25 - x^2})^2} = \sqrt{9 + 25 - x^2} = \sqrt{34 - x^2}$
The domain of $g \circ f$ is the set of those numbers x in $[-5, 5]$ for which $g(x)$ is real. Since $g(x)$ is real for all x, the domain of $g \circ f$ is $[-5, 5]$.

> **Common Error:** The domain of $g \circ f$ is not evident from the final form $\sqrt{34 - x^2}$. It is not $[-\sqrt{34}, \sqrt{34}]$.

73. $f(x) = \sqrt{5 - x^2}$ Domain of f: $[-\sqrt{5}, \sqrt{5}]$
$g(x) = \sqrt{3 - x}$ Domain of g: $(-\infty, 3]$
$(f \circ g)(x) = f(\sqrt{3 - x}) = \sqrt{5 - (\sqrt{3 - x})^2}$
Graph of $f \circ g$, not simplified:

Simplifying $(f \circ g)(x)$, we obtain $\sqrt{5 - (\sqrt{3 - x})^2}$
$= \sqrt{5 - (3 - x)} = \sqrt{5 - 3 + x} = \sqrt{2 + x}$. Apparent graph of $f \circ g$, simplified:

The domain of $f \circ g$ is not apparent from the final form $\sqrt{2 + x}$. It is not $[-2, \infty)$. The domain of $f \circ g$ is the set of those numbers in $(-\infty, 3]$ for which $g(x)$ is in $[-\sqrt{5}, \sqrt{5}]$, that is, $-\sqrt{5} \leq \sqrt{3 - x} \leq \sqrt{5}$, of $3 - x \leq 5$, or $x \geq -2$. The domain of $f \circ g$ is $[-2, 3]$. This is clearly shown in the first graph so the first graph is correct.

75. $f(x) = \sqrt{x^2 + 5}$ Domain of f: $(-\infty, \infty)$
$g(x) = \sqrt{x^2 - 4}$ Domain of g: $(-\infty, -2] \cup [2, \infty)$
$(f \circ g)(x) = f(\sqrt{x^2 - 4}) = \sqrt{(\sqrt{x^2 - 4})^2 + 5}$
Graph of $f \circ g$, not simplified:

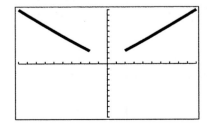

Simplifying $(f \circ g)(x)$, we obtain

$$\sqrt{(\sqrt{x^2 - 4})^2 + 5} = \sqrt{x^2 - 4 + 5} = \sqrt{x^2 + 1}.$$

Apparent graph of $f \circ g$, simplified:

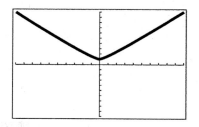

The domain of $f \circ g$ is not apparent from the final form $\sqrt{x^2 + 1}$. It is not $(-\infty, \infty)$. The domain of $f \circ g$ is the set of those numbers in $(-\infty, -2] \cup [2, \infty)$ for which $g(x)$ is real, that is, all of $(-\infty, -2] \cup [2, \infty)$. This is clearly shown in the first graph, so the first graph is correct.

77. $f(x) = \sqrt{x^2 + 7}$ Domain of f: $(-\infty, \infty)$

   $g(x) = \sqrt{9 - x^2}$ Domain of g: $[-3, 3]$

   $(f \circ g)(x) = f(\sqrt{9 - x^2}) = \sqrt{(\sqrt{9 - x^2})^2 + 7}$

   Graph of $f \circ g$, not simplified:

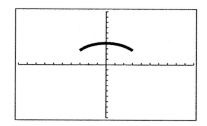

Simplifying $(f \circ g)(x)$, we obtain $\sqrt{(\sqrt{9 - x^2})^2 + 7}$ $= \sqrt{9 - x^2 + 7} = \sqrt{16 - x^2}$. Apparent graph of $f \circ g$, simplified:

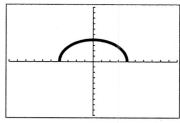

The domain of $f \circ g$ is not apparent from the final form $\sqrt{16 - x^2}$. It is not $[-4, 4]$. The domain of $f \circ g$ is the set of those numbers in $[-3, 3]$ for which $g(x)$ is real, that is, all of $[-3, 3]$. This is clearly shown in the first graph, so the first graph is correct.

79. Profit is the difference of the amount of money taken in (Revenue) and the amount of money spent (Cost), so $P(x) = R(x) - C(x)$

$$= \left(20x - \frac{1}{200}x^2\right) - (2x + 8,000)$$

$$= 20x - \frac{1}{200}x^2 - 2x - 8,000 \quad \text{(Distribute!)}$$

$$= 18x - \frac{1}{200}x^2 - 8,000$$

We have a profit function, but it's a function of the demand (x), not the price (p). We were given $x = 4,000 - 20p$, so we can substitute $4,000 - 200p$ in for x to get the desired function:

$$P(x) = P(4,000 - 200p) = 18(4,000 - 200p) - \frac{1}{200}(4,000 - 200p)^2 - 8,000$$

$$= 72,000 - 3,600p - \frac{1}{200}(16,000,000 - 1,600,000p + 40,000p^2) - 8,000$$

$$= 72,000 - 3,600p - 80,000 + 8,000p - 200p^2 - 8,000$$

$$= -16,000 + 4,400p - 200p^2$$

Now we'll graph the function and use the MAXIMUM command to find the largest profit.

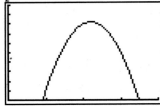

```
WINDOW
 Xmin=0
 Xmax=20
 Xscl=5
 Ymin=0
 Ymax=10000
 Yscl=1000
 Xres=1
```

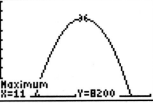

The maximum is (11, 9,200), so the largest profit occurs when the price is \$11.

81. We are given $V(r) = 0.1A(r) = 0.1\pi r^2$ and $r(t) = 0.4t^{1/3}$.
We can use composition to express V as a function of the time.

$$(V \circ r)(t) = V[r(t)]$$
$$= 0.1\pi[r(t)]^2$$
$$= 0.1\pi[0.4t^{1/3}]^2$$
$$= 0.1\pi[0.16t^{2/3}]$$
$$= 0.016\pi t^{2/3}$$

So $V(t) = 0.016\pi t^{2/3}$

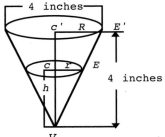

83.

(A) Note: In the figure, triangles VCE and $VC'E'$ are similar. Also

R = radius of cup = $\frac{1}{2}$ diameter of cup = $\frac{1}{2}(4)$ = 2 inches.

So $\frac{r}{2} = \frac{h}{4}$ or $r = \frac{1}{2}h$. We can write $r(h) = \frac{1}{2}h$.

(B) Since $V = \frac{1}{3}\pi r^2 h$ and $r = \frac{1}{2}h$, $V = \frac{1}{3}\pi\left(\frac{1}{2}h\right)^2 h = \frac{1}{3}\pi\frac{1}{4}h^2 h = \frac{1}{12}\pi h^3$.

We can write $V(h) = \frac{1}{12}\pi h^3$.

(C) Since $V(h) = \frac{1}{12}\pi h^3$ and $h(t) = 0.5\sqrt{t}$, we can use composition to express V as a function of t.

$$(V \circ h)(t) = V[h(t)]$$
$$= \frac{1}{12}\pi \ [h(t)]^3$$
$$= \frac{1}{12}\pi \ [0.5\sqrt{t}]^3$$
$$= \frac{1}{12}\pi(0.125)(t^{1/2})^3$$
$$= \frac{0.125}{12}\pi t^{3/2}$$

Then $V(t) = \frac{0.125}{12}\pi t^{3/2}$

Section 1-6

1. This is a one-to-one function. All of the first coordinates are distinct, and each first coordinate is paired with a different second coordinate. If all of the ordered pairs are reversed, the situation is the same.

3. This is a function but it is not one-to-one. First coordinates 5 and 2 are both paired with 4, and first coordinates 4 and 3 are both paired with 3. If the ordered pairs are reversed, the result is not a function since first coordinates 3 and 4 will each be paired with two different second coordinates.

5. This is not a function: 1 is a first coordinate that is paired with 2 different second coordinates (as is -3). If the ordered pairs are reversed, the result is also not a function since 4 will be a first coordinate paired with 2 different second coordinates (as will 2).

7. One-to one

9. The range element 7 corresponds to more than one domain element. Not one-to-one.

11. One-to-one

13. Some range elements (0, for example) correspond to more than one domain element. Not one-to-one.

15. One-to-one   **17.** One-to-one

19. This is a one-to-one function, so the inverse function exists.
$f^{-1} = \{(3, 2), (4, 3), (5, 4), (6, 5)\}$

21. This is not a one-to-one function (the range element -3 has two domain elements corresponding to it), so the inverse function does not exist.

23. This is a one-to-one function, so the inverse function exists.
$F^{-1} = \{(7, a), (11, c), (-9, e), (-13, g)\}$

25. Answers will vary.   **27.** Answers will vary.

29. Answers will vary.

31. Assume $F(a) = F(b)$
$$\frac{1}{2}a + 1 = \frac{1}{2}b + 1$$
Then $\frac{1}{2}a = \frac{1}{2}b$
$$a = b$$
Therefore F is one-to-one.

33. $H(x) = 4 - x^2$
Since $H(1) = 4 - 1^2 = 3$ and $H(-1) = 4 - (-1)^2 = 3$, both $(1, 3)$ and $(-1, 3)$ belong to H. H is not one-to-one.

35. Assume $M(a) = M(b)$
$$\sqrt{a + 1} = \sqrt{b + 1}$$
Then $a + 1 = b + 1$
$$a = b$$
M is one-to-one.

37. $f(x) = \dfrac{7}{x}$

This graph passes the horizontal line test. f is one-to-one.

39. $f(x) = 0.3x^3 - 3.8x^2 + 12x + 11$

In a standard window (first graph), the graph appears to pass the horizontal line test, but with an expanded view (second graph) we see that it actually does not. f is not one-to-one.

41. $f(x) = \dfrac{x^2 + |x|}{x}$

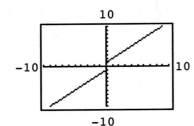

This graph passes the horizontal line test. f is one-to-one.

43. $f(x) = \dfrac{x^2 - 4}{|x - 2|}$

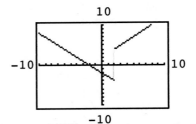

This graph does not pass the horizontal line test. f is not one-to-one.

45. $f(g(x)) = f\left(\dfrac{1}{3}x - \dfrac{5}{3}\right) = 3\left(\dfrac{1}{3}x - \dfrac{5}{3}\right) + 5 = x - 5 + 5 = x$

$g(f(x)) = g(3x + 5) = \dfrac{1}{3}(3x + 5) - \dfrac{5}{3} = x + \dfrac{5}{3} - \dfrac{5}{3} = x$

f and g are inverses

47. $f(g(x)) = f(\sqrt[3]{3 - x} - 1) = 2 - ((\sqrt[3]{3 - x} - 1) + 1)^3 = 2 - (\sqrt[3]{3 - x})^3$
$$= 2 - (3 - x) = -1 + x$$
f and g are not inverses since $(f \circ g)(x)$ is not x.

49. $f(g(x)) = f\left(\dfrac{3 + 4x}{2 - x}\right) = \dfrac{2\left(\frac{3+4x}{2-x}\right) - 3}{\frac{3+4x}{2-x} + 4} = \dfrac{\frac{6+8x}{2-x} - 3}{\frac{3+4x}{2-x} + 4} = \dfrac{\frac{6+8x}{2-x} - \frac{3(2-x)}{2-x}}{\frac{3+4x}{2-x} + \frac{4(2-x)}{2-x}}$

$$= \dfrac{\frac{6+8x-6+3x}{2-x}}{\frac{3+4x+8-4x}{2-x}} = \dfrac{\frac{11x}{2-x}}{\frac{11}{2-x}} = \dfrac{11x}{2-x} \cdot \dfrac{2-x}{11} = x$$

$g(f(x)) = g\left(\dfrac{2x - 3}{x + 4}\right) = \dfrac{3 + 4\left(\frac{2x-3}{x+4}\right)}{2 - \left(\frac{2x-3}{x+4}\right)} = \dfrac{3 + \frac{8x-12}{x+4}}{2 - \frac{2x-3}{x+4}} = \dfrac{\frac{3(x+4)}{x+4} + \frac{8x-12}{x+4}}{\frac{2(x+4)}{x+4} - \frac{2x-3}{x+4}}$

$$= \dfrac{\frac{3x+12+8x-12}{x+4}}{\frac{2x+8-2x+3}{x+4}} = \dfrac{\frac{11x}{x+4}}{\frac{11}{x+4}} = \dfrac{11x}{x+4} \cdot \dfrac{x+4}{11} = x$$

f and g are inverses.

51. The function h multiplies the input by 3 and then subtracts 7. The opposite of this is adding 7 to the input and then dividing by 3. The equation that corresponds to these steps is $h^{-1}(x) = \dfrac{x + 7}{3}$. (Remember that the order of the steps must be reversed.)

53. The function m adds 11 to the input then takes the cube root. The opposite of this is cubing the input, then subtracting 11. The equation that corresponds to these steps is $m^{-1}(x) = x^3 - 11$. (Remember that the order of the steps must be reversed.)

55. The function s multiplies the input by 3, adds 17, than raises the resulting expression to the fifth power. The opposite of this is taking the fifth root of the input, subtracting 17, then dividing by 3. The equation that corresponds to these steps is $s^{-1}(x) = \dfrac{\sqrt[5]{x} - 17}{3}$. (Remember that the order of the steps must be reversed.)

57. From the given graph of f, we see
Domain of $f = [-4, 4]$ Range of $f = [1, 5]$
Then
Range of $f^{-1} = [-4, 4]$ Domain of $f^{-1} = [1, 5]$
The graph of f^{-1} is drawn by reflecting the given graph of f in the line $y = x$.

59. From the given graph of f, we see
Domain of $f = [-5, 3]$ Range of $f = [-3, 5]$
Then
Range of $f^{-1} = [-5, 3]$ Domain of $f^{-1} = [-3, 5]$
The graph of f^{-1} is drawn by reflecting the given graph of f in the line $y = x$.

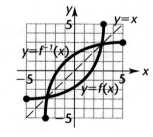

61. $f(x) = 3x + 6$ $g(x) = \dfrac{1}{3}x - 2$

$$g[f(x)] = g(3x + 6) = \frac{1}{3}(3x + 6) - 2 = x + 2 - 2 = x$$

$$f[g(x)] = f\left(\frac{1}{3}x - 2\right) = 3\left(\frac{1}{3}x - 2\right) + 6 = x - 6 + 6 = x$$

f and g are inverses

63. $f(x) = 4 + x^2 \qquad x \geq 0 \quad g(x) = \sqrt{x - 4}$

$$g[f(x)] = g(4 + x^2) = \sqrt{4 + x^2 - 4} = \sqrt{x^2} = x \text{ since } x \geq 0$$

$$f[g(x)] = f(\sqrt{x - 4}) = 4 + (\sqrt{x - 4})^2 = 4 + x - 4 = x$$

f and g are inverses

65. $f(x) = -\sqrt{x - 2} \quad g(x) = x^2 + 2, \quad x \leq 0$

$$f[g(x)] = f(x^2 + 2) = -\sqrt{x^2 + 2 - 2} = -\sqrt{x^2} = -(-x) = x$$

(Note: $\sqrt{x^2} = -x$ since $x \leq 0$)

$$g[f(x)] = g(-\sqrt{x - 2}) = (-\sqrt{x - 2})^2 + 2 = x - 2 + 2 = x$$

f and g are inverses

Common Error: $(-\sqrt{x - 2})^2 \neq -(x - 2)$ nor $-x - 2$

67. $f(x) = 3x$

Solve $y = f(x)$ for x:

$$y = 3x$$
$$x = \frac{1}{3}y = f^{-1}(y)$$

Interchange x and y:

$$y = f^{-1}(x) = \frac{1}{3}x$$

Domain of f^{-1} = Range of $f = (-\infty, \infty)$

Check: $f^{-1}[f(x)] = f^{-1}(3x) = \frac{1}{3}(3x) = x$

$$f[f^{-1}(x)] = f\left(\frac{1}{3}x\right) = 3\left(\frac{1}{3}x\right) = x$$

$f^{-1}(x) = \frac{1}{3}x$

69. $f(x) = 4x - 3$

Solve $y = f(x)$ for x:

$$y = 4x - 3$$
$$y + 3 = 4x$$
$$x = \frac{y + 3}{4} = f^{-1}(y)$$

Interchange x and y:

$$y = f^{-1}(x) = \frac{x + 3}{4}$$

Domain of f^{-1} = Range of $f = (-\infty, \infty)$

Check: $f^{-1}[f(x)] = f^{-1}(4x - 3)$

$$= \frac{4x - 3 + 3}{4}$$
$$= \frac{4x}{4}$$
$$= x$$

$$f[f^{-1}(x)] = f\left(\frac{x + 3}{4}\right)$$
$$= 4\left(\frac{x + 3}{4}\right) - 3$$
$$= x + 3 - 3$$
$$= x$$

$$f^{-1}(x) = \frac{x + 3}{4}$$

71. $f(x) = \dfrac{1}{10}x + \dfrac{3}{5}$

Solve $y = f(x)$ for x:

$$y = \dfrac{1}{10}x + \dfrac{3}{5}$$

$$10y = x + 6$$

$$x = 10y - 6 = f^{-1}(y)$$

Interchange x and y:

$$y = f^{-1}(x) = 10x - 6$$

Domain of f^{-1} = Range of $f = (-\infty, \infty)$

Check: $f^{-1}[f(x)] = f^{-1}\left(\dfrac{1}{10}x + \dfrac{3}{5}\right)$

$$= 10\left(\dfrac{1}{10}x + \dfrac{3}{5}\right) - 6$$

$$= x + 6 - 6$$

$$= x$$

$$f[f^{-1}(x)] = f(10x - 6)$$

$$= \dfrac{1}{10}(10x - 6) + \dfrac{3}{5}$$

$$= x - \dfrac{6}{10} + \dfrac{3}{5} = x$$

$$f^{-1}(x) = 10x - 6$$

73. $f(x) = \dfrac{2}{x - 1}$

Solve $y = f(x)$ for x:

$$y = \dfrac{2}{x - 1} \quad x \neq 1$$

$$y(x - 1) = 2$$

$$xy - y = 2$$

$$xy = y + 2$$

$$x = \dfrac{y + 2}{y} = f^{-1}(y)$$

Interchange x and y:

$$y = f^{-1}(x) = \dfrac{x + 2}{x}$$

Domain of $f^{-1} = (-\infty, 0) \cup (0, \infty)$

Check: $f^{-1}[f(x)] = f^{-1}\left(\dfrac{2}{x - 1}\right)$

$$= \dfrac{\frac{2}{x-1} + 2}{\frac{2}{x-1}}$$

$$= \dfrac{2 + 2(x - 1)}{2}$$

$$= \dfrac{2 + 2x - 2}{2} = \dfrac{2x}{2} = x$$

$$f^{-1}(x) = \dfrac{x + 2}{x}$$

Graph of $y = f(x)$:

Range of $f = (-\infty, 0) \cup (0, \infty)$

$$f[f^{-1}(x)] = \dfrac{2}{\frac{x+2}{x} - 1}$$

$$= \dfrac{2x}{x + 2 - x} = \dfrac{2x}{2} = x$$

75. $f(x) = \dfrac{x}{x + 2}$

Solve $y = f(x)$ for x:

$$y = \dfrac{x}{x + 2} \quad x \neq -2$$

$$y(x + 2) = x$$

$$xy + 2y = x$$

$$2y = x - xy$$

$$2y = x(1 - y)$$

$$x = \dfrac{2y}{1 - y} = f^{-1}(y)$$

Interchange x and y:

$$y = f^{-1}(x) = \dfrac{2x}{1 - x}$$

Domain of $f^{-1} = (-\infty, 1) \cup (1, \infty)$

Graph of $y = f(x)$:

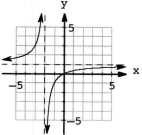

Range of $f = (-\infty, 1) \cup (1, \infty)$

Check: $f^{-1}[f(x)] = \dfrac{2\,\frac{x}{x+2}}{1 - \frac{x}{x+2}}$

$= \dfrac{2x}{x + 2 - x}$

$= \dfrac{2x}{2}$

$= x$

$f[f^{-1}(x)] = \dfrac{\frac{2x}{1-x}}{\frac{2x}{1-x} + 2}$

$= \dfrac{2x}{2x + 2(1 - x)}$

$= \dfrac{2x}{2x + 2 - 2x}$

$= \dfrac{2x}{2}$

$= x$

$f^{-1}(x) = \dfrac{2x}{1 - x}$

77. $f(x) = \dfrac{2x + 5}{3x - 4}$

Solve $y = f(x)$ for x:

$y = \dfrac{2x + 5}{3x - 4}$

$(3x - 4)y = 2x + 5$

$3xy - 4y = 2x + 5$

$3xy - 2x = 4y + 5$

$x(3y - 2) = 4y + 5$

$x = \dfrac{4y + 5}{3y - 2} = f^{-1}(y)$

Interchange x and y:

$y = f^{-1}(x) = \dfrac{4x + 5}{3x - 2}$

Domain of $f^{-1} = \left(-\infty, \dfrac{2}{3}\right) \cup \left(\dfrac{2}{3}, \infty\right)$

Graph of $y = f(x)$:

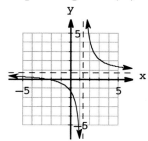

Range of $f = \left(-\infty, \dfrac{2}{3}\right) \cup \left(\dfrac{2}{3}, \infty\right)$

Check: $f^{-1}[f(x)] = \dfrac{4\,\frac{2x+5}{3x-4} + 5}{3\,\frac{2x+5}{3x-4} - 2}$

$= \dfrac{4(2x + 5) + 5(3x - 4)}{3(2x + 5) - 2(3x - 4)}$

$= \dfrac{8x + 20 + 15x - 20}{6x + 15 - 6x + 8}$

$= \dfrac{23x}{23}$

$= x$

$f^{-1}(x) = \dfrac{4x + 5}{3x - 2}$

$f[f^{-1}(x)] = \dfrac{2\,\frac{4x+5}{3x-2} + 5}{3\,\frac{4x+5}{3x-2} - 4}$

$= \dfrac{2(4x + 5) + 5(3x - 2)}{3(4x + 5) - 4(3x - 2)}$

$= \dfrac{8x + 10 + 15x - 10}{12x + 15 - 12x + 8}$

$= \dfrac{23x}{23}$

$= x$

79. $f(x) = x^3 + 1$

Solve $y = f(x)$ for x:

$y = x^3 + 1$

$y - 1 = x^3$

$x = \sqrt[3]{y - 1} = f^{-1}(y)$

Interchange x and y:

$y = f^{-1}(x) = \sqrt[3]{x - 1}$

Domain of f^{-1} = Range of

$f = (-\infty, \infty)$

Check: $f^{-1}[f(x)] = \sqrt[3]{x^3 + 1 - 1} = \sqrt[3]{x^3} = x$

$f[f^{-1}(x)] = (\sqrt[3]{x - 1})^3 + 1 = x - 1 + 1 = x$

$f^{-1}(x) = \sqrt[3]{x - 1}$

81. $f(x) = 4 - \sqrt[5]{x + 2}$

Graph of f:

Range of $f = (-\infty, \infty)$

Solve $y = f(x)$ for x:

$$y = 4 - \sqrt[5]{x + 2}$$

$$y - 4 = -\sqrt[5]{x + 2}$$

$$4 - y = \sqrt[5]{x + 2}$$

$$(4 - y)^5 = x + 2$$

$$x = (4 - y)^5 - 2 = f^{-1}(y)$$

Interchange x and y:

$y = f^{-1}(x) = (4 - x)^5 - 2$ Domain of f^{-1}: $(-\infty, \infty)$

Check:

$$f^{-1}[f(x)] = [4 - (4 - \sqrt[5]{x + 2})]^5 - 2 \qquad f[f^{-1}(x)] = 4 - \sqrt[5]{(4 - x)^5 - 2 + 2}$$

$$= [4 - 4 + \sqrt[5]{x + 2}]^5 - 2 \qquad\qquad\qquad = 4 - \sqrt[5]{(4 - x)^5}$$

$$= [\sqrt[5]{x + 2}]^5 - 2 \qquad\qquad\qquad\qquad = 4 - (4 - x)$$

$$= x + 2 - 2 \qquad\qquad\qquad\qquad\qquad = 4 - 4 + x$$

$$= x \qquad\qquad\qquad\qquad\qquad\qquad = x$$

$$f^{-1}(x) = (4 - x)^5 - 2$$

83. $f(x) = \dfrac{1}{2}\sqrt{16 - x}$

Solve $y = f(x)$ for x:

$$y = \frac{1}{2}\sqrt{16 - x}$$

$$\left.\begin{array}{l} 2y = \sqrt{16 - x} \\ 4y^2 = 16 - x \end{array}\right\} \text{these are equivalent only if } y \geq 0$$

$$x = 16 - 4y^2 \quad y \geq 0 \quad f^{-1}(y) = 16 - 4y^2$$

Interchange x and y:

$y = f^{-1}(x) = 16 - 4x^2,\ x \geq 0$ Domain of f^{-1} = Range of $f = [0, \infty)$

Check: $f^{-1}[f(x)] = 16 - 4\left[\dfrac{1}{2}\sqrt{16 - x}\right]^2 \qquad f[f^{-1}(x)] = \dfrac{1}{2}\sqrt{16 - (16 - 4x^2)}$

$$= 16 - 4\left[\frac{1}{4}\sqrt{16 - x}\right] \qquad\qquad\qquad = \frac{1}{2}\sqrt{16 - 16 + 4x^2}$$

$$= 16 - (16 - x) \qquad\qquad\qquad\qquad = \frac{1}{2}\sqrt{4x^2} \quad \sqrt{4x^2} = 2x \text{ since } x \geq 0$$

$$= 16 - 16 + x \qquad\qquad\qquad\qquad\qquad = \frac{1}{2}(2x)$$

85. $f(x) = 3 - \sqrt{x - 2}$
Solve $y = f(x)$ for x:

$$y = 3 - \sqrt{x - 2}$$
$$y - 3 = -\sqrt{x - 2}$$
$$\left.\begin{array}{l} 3 - y = \sqrt{x - 2} \\ (3 - y)^2 = x - 2 \end{array}\right\} \text{ these are equivalent only if } y \le 3$$
$$x = (3 - y)^2 + 2 \quad y \le 3 \quad f^{-1}(y) = (3 - y)^2 + 2$$

Interchange x and y
$y = f^{-1}(x) = (3 - x)^2 + 2 \quad x \le 3 \quad$ Domain of f^{-1} = Range of $f = (-\infty, 3]$

Check:

$$f^{-1}[f(x)] = [3 - (3 - \sqrt{x - 2})]^2 + 2 \qquad f[f^{-1}(x)] = 3 - \sqrt{(3 - x)^2 + 2 - 2}$$
$$= (\sqrt{x - 2})^2 + 2 \qquad\qquad\qquad = 3 - \sqrt{(3 - x)^2}$$

$$\sqrt{(3 - x)^2} = 3 - x \text{ since } x \le 3$$

$$= x - 2 + 2 \qquad\qquad\qquad = 3 - (3 - x)$$
$$= x \qquad\qquad\qquad\qquad\quad = 3 - 3 + x$$
$$\qquad\qquad\qquad\qquad\qquad\quad = x$$

$f^{-1}(x) = (3 - x)^2 + 2, \ x \le 3$

87. Since in passing from a function to its inverse, x and y are interchanged, the x intercept of f is the y intercept of f^{-1} and the y intercept of f is the x intercept of f^{-1}.

89. They are not inverses. The function $f(x) = x^2$ is not one-to-one, so does not have an inverse. If that function is restricted to the domain $[0, \infty)$, then it is one-to-one, and $g(x) = \sqrt{x}$ (which has range $[0, \infty)$) is its inverse.

91. $f(x) = (x - 1)^2 + 2 \quad x \ge 1$
Solve $y = f(x)$ for x:

$$y = (x - 1)^2 + 2 \quad x \ge 1$$
$$y - 2 = (x - 1)^2 \quad x \ge 1$$
$$\sqrt{y - 2} = \sqrt{(x - 1)^2} \quad \text{Since } x \ge 1, \ \sqrt{(x - 1)^2} = x - 1$$
$$\sqrt{y - 2} = x - 1$$
$$x = 1 + \sqrt{y - 2} = f^{-1}(y)$$

Interchange x and y:
$y = f^{-1}(x) = 1 + \sqrt{x - 2} \quad$ Domain: $x \ge 2$

Check:

$$f^{-1}[f(x)] = 1 + \sqrt{(x - 1)^2 + 2 - 2} \qquad f[f^{-1}(x)] = (1 + \sqrt{x - 2} - 1)^2 + 2$$
$$= 1 + \sqrt{(x - 1)^2} \qquad\qquad\qquad\qquad = (\sqrt{x - 2})^2 + 2$$

$$\sqrt{(x - 1)^2} = x - 1 \text{ since } x \ge 1 \qquad = x - 2 + 2$$
$$= 1 + x - 1 \qquad\qquad\qquad\qquad\qquad = x$$
$$= x$$

$f^{-1}(x) = 1 + \sqrt{x - 2}, \ x \ge 2$

93. $f(x) = x^2 + 2x - 2 \quad x \le -1$

Solve $y = f(x)$ for x:

$$y = x^2 + 2x - 2 \qquad x \le -1$$
$$y + 3 = x^2 + 2x + 1 \qquad x \le -1$$
$$y + 3 = (x + 1)^2 \qquad x \le -1$$
$$\sqrt{y + 3} = \sqrt{(x + 1)^2} \quad \text{Since } x \le -1 \quad \sqrt{(x + 1)^2} = -(x + 1)$$
$$\sqrt{y + 3} = -(x + 1)$$
$$-\sqrt{y + 3} = x + 1$$
$$x = -1 - \sqrt{y + 3} = f^{-1}(y)$$

Interchange x and y:

$$y = f^{-1}(x) = -1 - \sqrt{x + 3} \quad x \ge -3 \quad \text{Domain: } [-3, \infty)$$

Check: $f^{-1}[f(x)] = -1 - \sqrt{x^2 + 2x - 2 + 3} \qquad\qquad x \le -1$

$$= -1 - \sqrt{x^2 + 2x + 1} \qquad\qquad x \le -1$$
$$= -1 - \sqrt{(x + 1)^2} \qquad\qquad x \le -1$$
$$\sqrt{(x + 1)^2} = -(x + 1) \text{ since } x \le -1$$
$$= -1 - [-(x + 1)]$$
$$= -1 + x + 1$$
$$= x$$

$$f[f^{-1}(x)] = (-1 - \sqrt{x + 3})^2 + 2(-1 - \sqrt{x + 3}) - 2$$
$$= 1 + 2\sqrt{x + 3} + x + 3 - 2 - 2\sqrt{x + 3} - 2$$
$$= 4 + x - 4$$
$$= x$$

$$f^{-1}(x) = -1 - \sqrt{x + 3}, \quad x \ge -3$$

95. $f(x) = -\sqrt{9 - x^2} \quad 0 \le x \le 3 \qquad f$ is one-to-one. See graph below.

Solve $y = f(x)$ for x:

$$y = -\sqrt{9 - x^2} \quad 0 \le x \le 3$$
$$y^2 = 9 - x^2 \qquad y \le 0$$
$$y^2 - 9 = -x^2 \qquad y \le 0$$
$$9 - y^2 = x^2 \qquad y \le 0$$
$$x = \underbrace{\sqrt{9 - y^2}} \qquad y \le 0 \quad f^{-1}(y) = \sqrt{9 - y^2}$$

positive square root only because $0 \le x$

Interchange x and y:

$$y = f^{-1}(x) = \sqrt{9 - x^2} \quad \text{Domain: } -3 \le x \le 0$$

Check: $f^{-1}[f(x)] = \sqrt{9 - (-\sqrt{9 - x^2})^2}$

$$= \sqrt{9 - (9 - x^2)}$$
$$= \sqrt{9 - 9 + x^2}$$
$$= \sqrt{x^2}$$
$$\sqrt{x^2} = x \text{ since } x \ge 0 \text{ } \textit{in the domain of } f$$
$$= x$$

$$f[f^{-1}(x)] = -\sqrt{9 - (-\sqrt{9 - x^2})^2}$$

$$= -\sqrt{9 - (9 - x^2)}$$

$$= -\sqrt{9 - 9 + x^2}$$

$$= -\sqrt{x^2}$$

$\sqrt{x^2}$ = $-x$ since $x \le 0$ *in the domain of* f^{-1}

$$= -(-x)$$

$$= x$$

$f^{-1}(x) = \sqrt{9 - x^2}$ Domain of f^{-1} = $[-3, 0]$

Range of f^{-1} = Domain of f = $[0, 3]$

97. $f(x) = \sqrt{9 - x^2}$ $-3 \le x \le 0$ f is one-to-one. See graph below.

Solve $y = f(x)$ for x

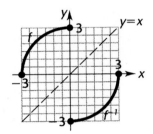

$y = \sqrt{9 - x^2}$ $-3 \le x \le 0$
$y^2 = 9 - x^2$ $y \ge 0$
$y^2 - 9 = -x^2$ $y \ge 0$
$9 - y^2 = x^2$ $y \ge 0$
$x = -\sqrt{9 - y^2}$ $y \ge 0$ $f^{-1}(y) = -\sqrt{9 - y^2}$

negative square root only because $x \le 0$

Interchange x and y:

$y = f^{-1}(x) = -\sqrt{9 - x^2}$ Domain: $0 \le x \le 3$

Check: $f^{-1}[f(x)] = -\sqrt{9 - (\sqrt{9 - x^2})^2}$

$$= -\sqrt{9 - (9 - x^2)}$$

$$= -\sqrt{9 - 9 + x^2}$$

$$= -\sqrt{x^2}$$

$\sqrt{x^2}$ = $-x$ since $x \le 0$ *in the domain of* f

$$= -(-x)$$

$$= x$$

$$f[f^{-1}(x)] = \sqrt{9 - (\sqrt{9 - x^2})^2}$$

$$= \sqrt{9 - (9 - x^2)}$$

$$= \sqrt{9 - 9 + x^2}$$

$$= \sqrt{x^2}$$

$\sqrt{x^2}$ = x since $x \ge 0$ *in the domain of* f^{-1}

$$= x$$

$f^{-1}(x) = -\sqrt{9 - x^2}$ Domain: f^{-1} = $[0, 3]$
Range of f^{-1} = Domain of f = $[-3, 0]$

99. $f(x) = 1 + \sqrt{1 - x^2}$ $0 \le x \le 1$ f is one-to-one. See graph below.

Solve $y = f(x)$ for x:

$$y = 1 + \sqrt{1 - x^2} \quad 0 \le x \le 1$$
$$y - 1 = \sqrt{1 - x^2}$$
$$(y - 1)^2 = 1 - x^2 \qquad y \ge 1$$
$$y^2 - 2y + 1 = 1 - x^2$$
$$y^2 - 2y = -x^2$$
$$2y - y^2 = x^2$$
$$x = \underline{\sqrt{2y - y^2}} \qquad y \ge 1 \quad f^{-1}(y) = \sqrt{2y - y^2}$$

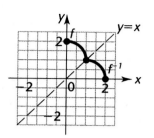

positive square root only because $0 \le x$
Interchange x and y:

$$y = f^{-1}(x) = \sqrt{2x - x^2} \quad \text{Domain: } 1 \le x \le 2$$

Check: $f[f^{-1}(x)] = 1 + \sqrt{1 - (\sqrt{2x - x^2})^2}$

$$= 1 + \sqrt{1 - (2x - x^2)}$$
$$= 1 + \sqrt{1 - 2x + x^2}$$
$$= 1 + \sqrt{(1 - x)^2}$$
$$= 1 + [-(1 - x)] \text{ because } 1 \le x \text{ in the domain of } f^{-1}$$
$$= 1 - 1 + x$$
$$= x$$

$$f^{-1}[f(x)] = \sqrt{2(1 + \sqrt{1 - x^2}) - (1 + \sqrt{1 - x^2})^2}$$

$$= \sqrt{2 + 2\sqrt{1 - x^2} - (1 + 2\sqrt{1 - x^2} + 1 - x^2)}$$

$$= \sqrt{2 + 2\sqrt{1 - x^2} - 1 - 2\sqrt{1 - x^2} - 1 + x^2}$$

$$= \sqrt{x^2}$$

$$= x$$

because $x \ge 0$ in the domain of f

$f^{-1}(x) = \sqrt{2x - x^2}$ Domain of $f^{-1} = [1, 2]$
Range of f^{-1} = Domain of $f = [0, 1]$

101. $f(x) = (2 - x)^2$

(A) $x \le 2$
Solve $y = f(x)$ for x:
$$y = (2 - x)^2 \qquad x \le 2$$
$$\sqrt{y} = 2 - x \quad \text{Positive square root only}$$
$$\qquad \qquad \text{since } 2 - x \ge 0$$
$$\sqrt{y} - 2 = -x$$
$$x = 2 - \sqrt{y} = f^{-1}(y)$$

Interchange x and y:

$$y = f^{-1}(x) = 2 - \sqrt{x} \quad \text{Domain: } x \ge 0$$

Check: $f^{-1}[f(x)] = 2 - \sqrt{(2 - x)^2}$

$\qquad\qquad\quad = 2 - (2 - x)$ since $2 - x \geq 0$ *in the domain of* f

$\qquad\qquad\quad = 2 - 2 + x$

$\qquad\qquad\quad = x$

$\qquad f[f^{-1}(x)] = [2 - (2 - \sqrt{x})]^2$

$\qquad\qquad\quad = [2 - 2 + \sqrt{x}]^2$

$\qquad\qquad\quad = [\sqrt{x}]^2$

$\qquad\qquad\quad = x$

$f^{-1}(x) = 2 - \sqrt{x}$

(B) $x \geq 2$
Solve $y = f(x)$ for x:

$\quad y = (2 - x)^2 \qquad x \geq 2$

$\qquad\quad -\sqrt{y} = 2 - x \qquad$ Negative square root only

$\qquad\qquad\qquad\qquad\qquad$ since $2 - x \leq 0$

$\sqrt{y} = -2 + x$

$\quad x = 2 + \sqrt{y} = f^{-1}(y)$

Interchange x and y:

$y = f^{-1}(x) = 2 + \sqrt{x}$ Domain: $x \geq 0$

Check: $f^{-1}[f(x)] = 2 + \sqrt{(2 - x)^2}$

$\qquad\qquad\quad = 2 + [-(2 - x)]$ since $2 - x \leq 0$ *in the domain of* f

$\qquad\qquad\quad = 2 - 2 + x$

$\qquad\qquad\qquad = x \qquad\qquad\qquad\quad f[f^{-1}(x)] = [2 - (2 + \sqrt{x})]^2$

$\qquad\qquad\qquad\qquad\qquad\qquad\qquad\qquad\qquad = [2 - 2 - \sqrt{x}]^2$

$\qquad\qquad\qquad\qquad\qquad\qquad\qquad\qquad\qquad = [-\sqrt{x}]^2$

$\qquad\qquad\qquad\qquad\qquad\qquad\qquad\qquad\qquad = x$

$\qquad f^{-1}(x) = 2 + \sqrt{x}$

103. $f(x) = \sqrt{4x - x^2}$
(A) $0 \leq x \leq 2$
Solve $y = f(x)$ for x:

$\qquad\quad y = \sqrt{4x - x^2}$

$\qquad\quad y^2 = 4x - x^2 \qquad\quad y \geq 0$

$\qquad\quad -y^2 = x^2 - 4x$

$\quad 4 - y^2 = x^2 - 4x + 4$

$\quad 4 - y^2 = (x - 2)^2$

$\quad -\sqrt{4 - y^2} = x - 2$

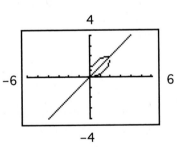

negative square root only because $x \leq 2$

$x = 2 - \sqrt{4 - y^2} \quad y \geq 0 \quad f^{-1}(y) = 2 - \sqrt{4 - y^2}$

Interchange x and y:

$y = f^{-1}(x) = 2 - \sqrt{4 - x^2}$ Domain: $0 \leq x \leq 2$

Check: $f^{-1}[f(x)] = 2 - \sqrt{4 - (\sqrt{4x - x^2})^2}$

$\qquad\qquad\qquad = 2 - \sqrt{4 - (4x - x^2)}$

$\qquad\qquad\qquad = 2 - \sqrt{4 - 4x + x^2}$

$\qquad\qquad\qquad = 2 - \sqrt{(2 - x)^2}$

$\qquad\qquad\qquad = 2 - (2 - x)$ since $2 - x \geq 0$

$\qquad\qquad\qquad = x$

$$f[f^{-1}(x)] = \sqrt{4(2 - \sqrt{4 - x^2}) - (2 - \sqrt{4 - x^2})^2}$$

$$= \sqrt{8 - 4\sqrt{4 - x^2} - (4 - 4\sqrt{4 - x^2} + 4 - x^2)}$$

$$= \sqrt{8 - 4\sqrt{4 - x^2} - 4 + 4\sqrt{4 - x^2} - 4 + x^2}$$

$$= \sqrt{x^2}$$

$$= x \text{ since } 0 \le x$$

$f^{-1}(x) = 2 - \sqrt{4 - x^2}$, $0 \le x \le 2$

(B) $2 \le x \le 4$

Solve $y = f(x)$ for x:

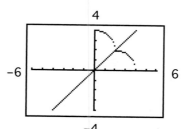

$$y = \sqrt{4x - x^2}$$

$$y^2 = 4x - x^2 \quad y \ge 0$$

$$-y^2 = x^2 - 4x$$

$$4 - y^2 = x^2 - 4x + 4$$

$$4 - y^2 = (x - 2)^2$$

$$\underbrace{\sqrt{4 - y^2}} = x - 2$$

positive square root only because $x \ge 2$

$x = 2 + \sqrt{4 - y^2}$ $y \ge 0$ $f^{-1}(y) = 2 + \sqrt{4 - y^2}$

Interchange x and y:

$y = f^{-1}(x) = 2 + \sqrt{4 - x^2}$ Domain: $0 \le x \le 2$

Check: $f^{-1}[f(x)] = 2 + \sqrt{4 - (\sqrt{4x - x^2})^2}$

$$= 2 + \sqrt{4 - (4x - x^2)}$$

$$= 2 + \sqrt{4 - 4x + x^2}$$

$$= 2 + \sqrt{(2 - x)^2}$$

$$= 2 + [-(2 - x)] \text{ since } 2 \le x \text{ in the domain of } f$$

$$= 2 - 2 + x$$

$$= x$$

$$f[f^{-1}(x)] = \sqrt{4(2 + \sqrt{4 - x^2}) - (2 + \sqrt{4 - x^2})^2}$$

$$= \sqrt{8 + 4\sqrt{4 - x^2} - (4 + 4\sqrt{4 - x^2} + 4 - x^2)}$$

$$= \sqrt{8 + 4\sqrt{4 - x^2} - 4 - 4\sqrt{4 - x^2} - 4 + x^2}$$

$$= \sqrt{x^2}$$

$$= x \text{ since } 0 \le x \text{ in the domain of } f^{-1}$$

$f^{-1}(x) = 2 + \sqrt{4 - x^2}$, $0 \le x \le 2$

105. (A) Graph $q = d(p)$ on the interval $[10, 70]$. Examining the graph, we see that d is a decreasing function, and has its maximum value on $[10, 70]$ at $p = 10$, where

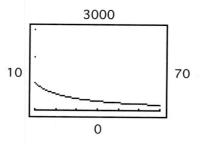

$d(10) = \dfrac{3000}{0.2(10) + 1} = 1{,}000$, and its minimum

value on $(10, 70)$ at $p = 70$ where

$d(70) = \dfrac{3000}{0.2(70) + 1} = 200$. Therefore the range

of d is $[200, 1000]$, or $200 \le q \le 1{,}000$.

(B) Solve for p:

$$q = \frac{3,000}{0.2p + 1}$$

$$q(0.2p + 1) = 3,000$$

$$0.2pq + q = 3,000$$

$$0.2pq = 3,000 - q$$

$$p = \frac{3,000 - q}{0.2q}$$

$$p = \frac{3,000}{0.2q} - \frac{q}{0.2q}$$

$$p = \frac{15,000}{q} - 5$$

$$d^{-1}(q) = \frac{15,000}{q} - 5$$

The domain of d^{-1} is the range of d, or $[200, 1,000]$, or $200 \le q \le 1,000$.
The range of d^{-1} is the domain of d, or $[10, 70]$, or $10 \le p \le 70$.

(C) p and q should not be interchanged. That step is necessary only to maintain the conventional roles of x and y.

107. To find the revenue, we multiply the price (p) and the demand (x). We're asked to find the revenue in terms of x, so we begin by solving the given equation for p:

$$x = 2,000 - 40p$$

$$-40p = x - 2,000$$

$$p = \frac{x - 2,000}{-40} = \frac{-1}{40}x + 50$$

Now we multiply by the demand (x):

$$R(x) = x\left(\frac{-1}{40}x + 50\right) = \frac{-1}{40}x^2 + 50x \quad \text{or} \quad R(x) = -0.025x^2 + 50x$$

CHAPTER 1 REVIEW

1. Since the x coordinates vary between a minimum of -4 and a maximum of 9, Xmin $= -4$ and Xmax $= 9$. Since the y coordinates vary between a minimum of -6 and a maximum of 7, Ymin $= -6$ and Ymax $= 7$. *(1-1)*

2. (A) All of the first coordinates are distinct, so this is a function with domain $\{1, 2, 3\}$. All of the second coordinates are distinct, so the function is one-to-one. The range is $\{1, 4, 9\}$. The inverse function is obtained by reversing the order of the ordered pairs: $\{(1, 1), (4, 2), (9, 3)\}$. It has domain $\{1, 4, 9\}$ and range $\{1, 2, 3\}$.

(B) This is not a function: both 1 and 2 are first coordinates that get matched with two different second coordinates.

(C) All of the first coordinates are distinct, so this is a function with domain $\{-2, -1, 0, 1, 2\}$. The range is $\{2\}$. It's most definitely not one-to-one since every first coordinate gets matched with the same second coordinate.

(D) All of the first coordinates are distinct, so this is a function with domain $\{-2, -1, 0, 1, 2\}$. All of the second coordinates are distinct, so the function is one-to-one. The range is $\{-2, -1, 1, 2, 3\}$.
The inverse function is obtained by reversing the order of the ordered pairs: $\{(2, -2), (3, -1), (-1, 0), (-2, 1), (1, 2)\}$. It has domain $\{-2, -1, 1, 2, 3\}$ and range $\{-2, -1, 0, 1, 2\}$. *(1-2)*

3. (A) Not a function (fails vertical line test)   (B) A function
   (C) A function   (D) Not a function (fails vertical line test)   *(1-2)*

4. (A) $f(1) = (1)^2 - 2(1) = -1$

   (B) $f(-4) = (-4)^2 - 2(-4) = 24$

   (C) $f(2) \cdot f(-1) = [(2)^2 - 2(2)] \cdot [(-1)^2 - 2(-1)] = 0 \cdot 3 = 0$

   (D) $\dfrac{f(0)}{f(3)} = \dfrac{(0)^2 - 2(0)}{(3)^2 - 2(3)} = \dfrac{0}{3} = 0$ *(1-2)*

5. Construct a table of values of $f(x)$ and $g(x)$ from the graph, then subtract to obtain $(f - g)(x)$.

x	-3	-2	-1	0	1	2	3
$f(x)$	3	2	1	0	1	2	3
$g(x)$	-4	-3	-2	-1	0	1	2
$(f - g)(x)$	7	5	3	1	1	1	1

(1-5)

6. Use the top 3 rows of the table above and multiply to get a table of values for $(fg)(x)$. (Note that from the graphs of f and g, we can see that $f\left(\dfrac{1}{2}\right) = \dfrac{1}{2}$ and $g\left(\dfrac{1}{2}\right) = -\dfrac{1}{2}$,

so $(fg)\left(\dfrac{1}{2}\right) = -\dfrac{1}{4}$.)

(1-5)

7. $(f \circ g)(-1) = f[g(-1)]$. From the graph of g, $g(-1) = -2$. From the graph of f, $f[g(-1)] = f(-2) = 2$.   *(1-5)*

8. $(g \circ f)(-2) = g[f(-2)]$. From the graph of f, $f(-2) = 2$. From the graph of g, $g[f(-2)] = g(2) = 1$.   *(1-5)*

9. From the graph of g, $g(1) = 0$. From the graph of f, $f[g(1)] = f(0) = 0$.   *(1-5)*

10. From the graph of f, $f(-3) = 3$. From the graph of g, $g[f(-3)] = g(3) = 2$.   *(1-5)*

11. Some range elements (1 for example) correspond to more than one domain element. Not one-to-one.   *(1-6)*

12. Yes, one-to-one.   *(1-6)*

13. Using values from the table, $g(-11) = -4$, so $(f \circ g)(-11) = f(g(-11)) = f(-4) = 8$. Similarly, $(f \circ g)(1) = f(g(1)) = f(-11) = 12$, and $(f \circ g)(6) = f(g(6)) = f(9) = 1$.   *(1-5)*

14. (A) $f(-x) = (-x)^5 + 6(-x) = -x^5 - 6x = -(x^5 + 6x) = -f(x)$. Odd

   (B) $g(-t) = (-t)^4 + 3(-t)^2 = t^4 + 3t^2 = g(t)$. Even

   (C) $h(-z) = (-z)^5 + 4(-z)^2 = -z^5 + 4z^2$
      $-h(z) = -(z^5 + 4z^2) = -z^5 - 4z^2$
   Therefore, $h(-z) \neq h(z)$. $h(-z) \neq -h(z)$. h is neither even nor odd.   *(1-4)*

15. When $x = -4$, the corresponding y value on the graph is 4. $f(-4) = 4$.

When $x = 0$, the corresponding y value on the graph is -4. $f(0) = -4$.

When $x = 3$, the corresponding y value on the graph is 0. $f(3) = 0$.

When $x = 5$, there is no corresponding y value on the graph.

$f(5)$ is not defined. *(1-2, 1-3)*

16. Two values of x correspond to $f(x) = -2$ on the graph. They are $x = -2$ and $x = 1$. *(1-2, 1-3)*

17. Domain: $[-4, 5)$. Range: $[-4, 4]$ *(1-3)*

18. The graph is increasing on $[0, 5)$ and decreasing on $[-4, 0]$. *(1-3)*

19. The graph is discontinuous at $x = 0$. *(1-3)*

20. The graph of $f(x)$ is shifted up 1 unit.

(1-4)

21. The graph of $f(x)$ is shifted left 1 unit.

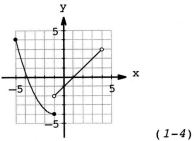

(1-4)

22. The graph of $f(x)$ is reflected in the x axis.

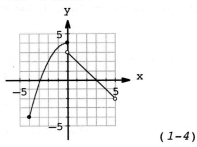

(1-4)

23. The graph of $f(x)$ is contracted by a factor of 0.5.

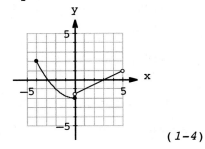

(1-4)

24. The graph is horizontally contracted by a factor of $\frac{1}{2}$.

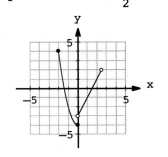

(1-4)

25. The graph is reflected about both axes.

(1-4)

26. (A) The graph that is decreasing, then increasing, and has a minimum at $x = 2$ is g.

(B) The graph that is increasing, then decreasing, and has a maximum at $x = -2$ is m.

(C) The graph that is increasing, then decreasing, and has a maximum at $x = 2$ is n.

(D) The graph that is decreasing, then increasing, and has a minimum at $x = -2$ is f.

(1-4)

27. (A) $\left(\dfrac{f}{g}\right)(x) = \dfrac{f(x)}{g(x)} = \dfrac{x^2 - 4}{x + 3}$ Domain: $\{x \mid x \neq -3\}$, or $(-\infty, -3) \cup (-3, \infty)$

(B) $\left(\dfrac{g}{f}\right)(x) = \dfrac{g(x)}{f(x)} = \dfrac{x + 3}{x^2 - 4}$ Domain: $\{x \mid x^2 - 4 \neq 0\}$, or $\{x \mid x \neq -2, 2\}$ or

$(-\infty, -2) \cup (-2, 2) \cup (2, \infty)$

(C) $(f \circ g)(x) = f[g(x)] = f(x + 3) = (x + 3)^2 - 4 = x^2 + 6x + 5$ Domain: $(-\infty, \infty)$

(D) $(g \circ f)(x) = g[f(x)] = g(x^2 - 4) = x^2 - 4 + 3 = x^2 - 1$ Domain: $(-\infty, \infty)$

(1-5)

28. (A) When $x = 0$, the corresponding y value on the graph is 0, to the nearest integer.

(B) When $x = 1$, the corresponding y value on the graph is 1, to the nearest integer.

(C) When $x = 2$, the corresponding y value on the graph is 2, to the nearest integer.

(D) When $x = -2$, the corresponding y value on the graph is 0, to the nearest integer.

(1-3)

29. (A) Two values of x correspond to $y = 0$ on the graph. To the nearest integer, they are -2 and 0.

(B) Two values of x correspond to $y = 1$ on the graph. To the nearest integer, they are -1 and 1.

(C) No value of x corresponds to $y = -3$ on the graph.

(D) To the nearest integer, $x = 3$ corresponds to $y = 3$ on the graph. Also, every value of x such that $x < -2$ corresponds to $y = 3$.   *(1-3)*

30. Domain: $(-\infty, \infty)$. Range: $(-3, \infty)$.   *(1-3)*

31. $[-2, -1]$, $[1, \infty)$.   *(1-3)*   **32.** $[-1, 1)$ *(1-3)*   **33.** $(-\infty, -2)$ *(1-3)*

34. The graph of g is discontinuous at $x = -2$ and $x = 1$.   *(1-3)*

35. The graph and viewing window are shown below:

Using the INTERSECT command, we find that the graphs cross at (0.64, -12.43) and (23.36, 78.43).   *(1-1)*

36. The graph and viewing window are shown below:

Using the INTERSECT command, we find that the graphs cross at $(-4.26, 2.72)$ and $(88.70, 30.61)$. $(1-1)$

37. (A) We begin by graphing $y_1 = 0.1x^3 - 2x^2 - 6x + 80$. When $b = 0$, the solutions are the x-values where the graph is at height zero, so we use the ZERO command:

The graph is at height zero three times so there are three solutions:
$x = -6.71$, $x = 5.67$, and $x = 21.04$

(B) Graph $y_1 = 0.1x^3 - 2x^2 - 6x + 80$, $y_2 = 100$; then use the INTERSECT command to find where the graphs cross.

The graphs intersect at $(22.99, 100)$ so the only solution is $x = 22.99$.

(C) Graph $y_1 = 0.1x^3 - 2x^2 - 6x + 80$, $y_2 = -50$, then use the INTERSECT command to find where the graphs cross.

The graphs cross three times, at $(-7.97, -50)$, $(8.28, -50)$, and $(19.70, -50)$. There are three solutions: $x = -7.97$, $x = 8.28$, and $x = 19.70$.

(D) Graph $y_1 = 0.1x^3 - 2x^2 - 6x + 80$, $y_2 = -150$; then use the INTERSECT command to find where the graphs cross.

The graphs cross at $(-9.84, -150)$, so the only solution is $x = -9.84$.
                                                                  $(1-1, 1-3)$

38. This equation defines a function. If you solve the equation for y, you get $y = 5 - 0.5x$. This tells us that for any choice of x, we can calculate a unique y that corresponds to it.   (1-2)

39. This equation does not define a function since most choices of x will result in two corresponding values of y. For example, the pairs (2, 2) and (2, -2) both make the equation a true statement.   (1-2)

40. (A) The domain of $f(x) = x^2 - 4x + 5$ is all real numbers.

(B) The fraction $\dfrac{t + 2}{t - 5}$ represents a real number for all replacements of t by real numbers except when $t - 5 = 0$ or $t = 5$, since division by 0 is not defined. The domain of g is the set of all real numbers except 5.

(C) $2 + 3\sqrt{w}$ represents a real number only for replacements of w by nonnegative real numbers, since the square root of a negative number is not defined. The domain of h is the set of all real numbers $w \geq 0$, or $[0, \infty)$.   (1-2)

41. $\dfrac{g(2 + h) - g(2)}{h} = \dfrac{[2(2 + h)^2 - 3(2 + h) + 6] - [2(2)^2 - 3(2) + 6]}{h}$

$= \dfrac{2(4 + 4h + h^2) - 6 - 3h + 6 - [8 - 6 + 6]}{h}$

$= \dfrac{8 + 8h + 2h^2 - 3h - 8}{h}$

$= \dfrac{5h + 2h^2}{h} = \dfrac{h(5 + 2h)}{h}$

$= 5 + 2h$

> Common Errors:
> 1. $g(2 + h)$ is not $g(2) + g(h)$.
> 2. $g(2 + h)$ is not $g(2) + h$.
> 3. Forgetting to enclose $2(2)^2 - 3(2) + 6$ in parentheses or brackets to indicate the unperformed subtraction.

(1-2)

42. $f(x) = 4x^3 - \sqrt{x}$ (1-2)

43. The function f multiplies the square of the domain element by 3, adds 4 times the domain element, and then subtracts 6.   (1-2)

44. The graph is shown below.

Use the ZERO command to find the x-intercepts:

The x-intercepts are $x = 0$ and $x = 3.30$. Since (0, 0) is on the graph, $y = 0$ is the y-intercept.

Te point (0,0) is lower than any nearby points, so is a local minimum. There is a local maximum somewhere near $x = 1$, so we use the MAXIMUM command.

The maximum occurs at $x = 1.31$, $y = 5.15$. The domain is $[0, \infty)$ and the range is $(-\infty, 5.15]$. (The highest point on the graph is at height 5.15 and there is no lowest point.).   *(1-3)*

45. The graph is shown below.

Use the ZERO command to find the x-intercepts and the value feature with $x = 0$ to find the y-intercept.

The x-intercepts are -26.58, -3.58, and 3.15. The y-intercept is -300. There is a local maximum and a local minimum, so we use the MAXIMUM and MINIMUM commands to locate the local extrema.

The local maximum occurs at $x = -18$, $y = 2,616$, and the local minimum occurs at $x = 0$, $y = -300$.
The function is a polynomial, so the domain is $(-\infty, \infty)$. There is no highest or lowest point, so the range is also $(-\infty, \infty)$.   *(1-3)*

46. **(A)**

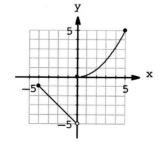

(B) Domain: $[-4, 5]$, range: $(-5, -1] \cup [0, 5]$

(C) The graph is discontinuous at $x = 0$.

(D) Decreasing on $[-4, 0)$, increasing on $[0, 5]$

(1-3)

47. The graph of f is shown as well as the result of applying the MAXIMUM and MINIMUM commands to find the local maximum and the local minimum.

The graph of f increases on $[-\infty, -4.47]$ to a local maximum value, $f(-4.47) \approx 22.89$, decreases on $[-4.47, 4.47]$ to a local minimum value, $f(4.47) \approx -12.89$, and then increases on $[4.47, \infty]$. *(1-3)*

48. (A) The graph of $y = x^2$ is reflected across the x axis.

(B) The graph of $y = x^2$ is shifted down 3 units.

(C) The graph of $y = x^2$ is shifted left 3 units.

(D) The graph of $y = x^2$ is shrunk horizontally by $\frac{1}{2}$. *(1-4)*

49. (A) The graph of $y = x^2$ is shifted 2 units to the right, reflected in the x axis, and shifted up 4 units: $y = -(x - 2)^2 + 4$
Check: $y = -(x - 2)^2 + 4$ is graphed on a graphing calculator.

(B) The graph of $y = \sqrt{x}$ is reflected in the x axis, stretched by a factor of 4, and shifted up 4 units: $y = 4 - 4\sqrt{x}$
Check: $y = 4 - 4\sqrt{x}$ is graphed on a graphing calculator.

 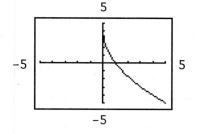

(1-4)

50. $g(x) = 8 - 3|x - 4|$
The -4 shifts 4 units right, the 3 stretches vertically by a factor of 3, and the 8 shifts 8 units up.

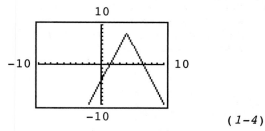

(1-4)

51. The equation is $t(x) = \left(\frac{1}{2}(x + 2)\right)^2 - 4$. The $\frac{1}{2}$ stretches horizontally by 2, the $+2$ shifts 2 units left, and the -4 shifts 4 units down. This function can be simplified: $\left(\frac{1}{2}(x + 2)\right)^2 - 4 = \left(\frac{1}{2}x + 1\right)^2 - 4 = \frac{1}{4}x^2 + x + 1 - 4 = \frac{1}{4}x^2 + x - 3$.

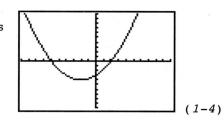

(1-4)

52. Find the composition of the 2 functions in both orders.
$(u \circ v)(x) = u(v(x)) = u(0.25x + 2) = 4(0.25x + 2) - 8 = x + 8 - 8 = x$
$(v \circ u)(x) = v(u(x)) = v(4x - 8) = 0.25(4x - 8) + 2 = x - 2 + 2 = x$
The functions are inverses.   (1-6)

53. The function k cubes the input and then adds 5. The opposite of this is subtracting 5 and taking the cube root. (Remember to reverse the order of the steps!) The inverse function is $k^{-1}(x) = \sqrt[3]{x} - 5$. To verify we find the composition of the two functions in both orders.
$(k \circ k^{-1})(x) = k(k^{-1}(x)) = k(\sqrt[3]{x} - 5) = (\sqrt[3]{x} - 5)^3 + 5 = x - 5 + x = x$

$(k^{-1} \circ k)(x) = k^{-1}(k(x)) = k^{-1}(x^3 + 5) = \sqrt[3]{(x^3 + 5) - 5} = \sqrt[3]{x^3} = x.$ (1-6)

54. $\dfrac{x}{\sqrt{x} - 3}$ represents a real number only for nonnegative values of x, except when $\sqrt{x} - 3 = 0$, that is, when $\sqrt{x} = 3$ or $x = 9$. The domain of f is the set of all nonnegative real numbers except 9: $x \geq 0$, $x \neq 9$, or $[0, 9) \cup (9, \infty)$.   (1-2)

55. $f(x) = \sqrt{x} - 8 \quad g(x) = |x|$
(A) $(f \circ g)(x) = f[g(x)] = f(|x|) = \sqrt{|x|} - 8$
   $(g \circ f)(x) = g[f(x)] = g(\sqrt{x} - 8) = |\sqrt{x} - 8|$

(B) The domain of f is $\{x \mid x \geq 0\}$. The domain of g is all real numbers. So the domain of $f \circ g$ is the set of those real numbers x for which $g(x)$ is non-negative, that is, all real numbers.

The domain of $(g \circ f)$ is the set of all those non-negative numbers x for which $f(x)$ is real, that is all $\{x \mid x \geq 0\}$ or $[0, \infty)$. (The output of g is never negative!)   (1-5)

56. (A) $f(x) = x^3$. The graph passes the horizontal line test, so f is one-to-one.

Alternatively, if we assume $f(a) = f(b)$
$$a^3 = b^3$$
$$a^3 - b^3 = 0$$
$$(a - b)(a^2 + ab + b^2) = 0$$
The only real solutions of this equation are those for which $a - b = 0$, so $a = b$. We conclude that $f(x)$ is one-to-one.

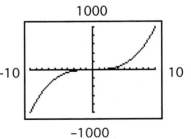

(B) $g(x) = (x - 2)^2$. Since $g(3) = g(1) = 1$, g is not one-to-one.

(C) $h(x) = 2x - 3$
Assume $h(a) = h(b)$
$\quad 2a - 3 = 2b - 3$
Then $\quad 2a = 2b$
$\quad\quad a = b$
We conclude that h is one-to-one.

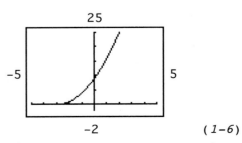

(D) $F(x) = (x + 3)^2 \quad x \geq -3$
The graph passes the horizontal line test, so F is one-to-one.   (1-6)

(E) When viewed in a standard window (first graph), it looks like the graph passes the horizontal line test. But an expanded window shows that it does not (second graph), so f is not one-to-one.

   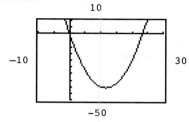

57. $f(x) = 3x - 7$
Assume $f(a) = f(b)$
$$3a - 7 = 3b - 7$$
$$3a = 3b$$
$$a = b$$

f is one-to-one. Solve $y = f(x)$ for x:
$$y = 3x - 7$$
$$y + 7 = 3x$$
$$x = \frac{1}{3}y + \frac{7}{3} = f^{-1}(y)$$

Interchange x and y: $y = f^{-1}(x) = \frac{1}{3}x + \frac{7}{3}$

Domain of f^{-1}: $(-\infty, \infty)$
Range of f^{-1} = Domain of f: $(-\infty, \infty)$

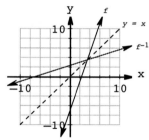

Check: $f^{-1}[f(x)] = \frac{1}{3}(3x - 7) + \frac{7}{3}$
$$= x - \frac{7}{3} + \frac{7}{3} = x$$
$$f[f^{-1}(x)] = 3\left(\frac{1}{3}x + \frac{7}{3}\right) - 7$$
$$= x + 7 - 7 = x$$

$f^{-1}(x) = \frac{1}{3}x + \frac{7}{3} = \frac{x + 7}{3}$ $(1\text{-}6)$

58. $f(x) = \sqrt{x - 1}$.
Assume $f(a) = f(b)$
$$\sqrt{a - 1} = \sqrt{b - 1}$$
$$a - 1 = b - 1$$
$$a = b$$
f is one-to-one.

Solve $y = f(x)$ for x
$$y = \sqrt{x - 1}$$
$$y^2 = x - 1 \qquad y \geq 0$$
$$x = 1 + y^2 \quad y \geq 0 \quad f^{-1}(y) = 1 + y^2$$

Interchange x and y:
$y = f^{-1}(x) = 1 + x^2$
Domain: $x \geq 0$, $[0, \infty)$

Check:
$f^{-1}[f(x)] = 1 + (\sqrt{x-1})^2 = 1 + x - 1 = x$
$f[f^{-1}(x)] = \sqrt{1 + x^2 - 1} = \sqrt{x^2} = x$
since $x \geq 0$ in the domain of f^{-1}

Range of f^{-1} = domain of $f = [1, \infty)$

(1-6)

59. $f(x) = x^2 - 1$ $x \geq 0$
Assume $f(a) = f(b)$
$a^2 - 1 = b^2 - 1$ $a \geq 0$, $b \geq 0$
$a^2 - b^2 = 0$
$(a - b)(a + b) = 0$
$a - b = 0$ or $a + b = 0$
     $a = b$ $a = -b$ impossible, since $a \geq 0$ and $b \geq 0$,
                                     unless $a = b = -b = 0$.

     $a = b$
f is one-to-one.

Solve $y = f(x)$ for x:
$y = x^2 - 1$ $x \geq 0$
$x^2 = y + 1$ $x \geq 0$

$x = \underbrace{\sqrt{y + 1}}\ = f^{-1}(y)$

positive square root since $x \geq 0$

Interchange x and y:
$y = f^{-1}(x) = \sqrt{x + 1}$ Domain: $[-1, \infty)$
Range of f^{-1} = domain of $f = [0, \infty)$

Check:
$f^{-1}[f(x)] = \sqrt{x^2 - 1 + 1}$ $x \geq 0$
$\qquad\qquad = \sqrt{x^2}$ $x \geq 0$
$\qquad\qquad = x$ since $x \geq 0$

$f[f^{-1}(x)] = (\sqrt{x + 1})^2 - 1$
$\qquad\qquad = x + 1 - 1$
$\qquad\qquad = x$

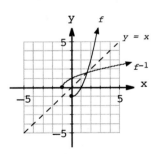

(1-6)

60. (A) One possible answer:

(B) One possible answer:

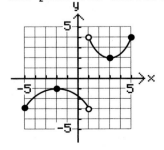

61. The function squares the input, multiplies the result by 2, then subtracts 4 times the input and adds 5. The equation is $g(t) = 2t^2 - 4t + 5$.   *(1-2)*

62.

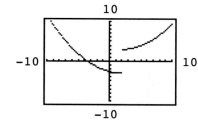

Domain: $x \neq 2$ or $(-\infty, 2) \cup (2, \infty)$.
Range: $y > -3$ or $(-3, \infty)$
Discontinuous at $x = 2$.

(1-3)

63. (A)

(B)

(1-4)

64. (A) $f(x) = 3x^2 - 5x + 7$

$f(x + h) = 3(x + h)^2 - 5(x + h) + 7$

$$\frac{f(x + h) - f(x)}{h} = \frac{[3(x + h)^2 - 5(x + h) + 7] - [3x^2 - 5x + 7]}{h}$$

$$= \frac{3(x^2 + 2xh + h^2) - 5x - 5h + 7 - 3x^2 + 5x - 7}{h}$$

$$= \frac{3x^2 + 6xh + 3h^2 - 5x - 5h + 7 - 3x^2 + 5x - 7}{h}$$

$$= \frac{6xh + 3h^2 - 5h}{h}$$

$$= \frac{h(6x + 3h - 5)}{h}$$

$$= 6x + 3h - 5$$

(B) $f(x) = 3x^2 - 5x + 7$

$f(a) = 3a^2 - 5a + 7$

$$\frac{f(x) - f(a)}{x - a} = \frac{(3x^2 - 5x + 7) - (3a^2 - 5a + 7)}{x - a}$$

$$= \frac{3x^2 - 5x + 7 - 3a^2 + 5a - 7}{x - a}$$

$$= \frac{3x^2 - 3a^2 - 5x + 5a}{x - a}$$

$$= \frac{(x - a)(3x + 3a) - 5(x - a)}{x - a}$$

$$= \frac{(x - a)(3x + 3a - 5)}{x - a}$$

$$= 3x + 3a - 5$$

(1-2)

65. (A) One of many possible graphs is shown. The graph can cross the axis exactly once.

(B) A possible discontinuous graph is shown. The graph can cross the axis once, as in part (A), or not at all, as shown.

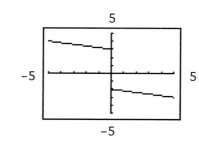

(*1–3*)

66. (A) We'll begin by computing some values for *f*.

$f(-3) = [\![\,|-3|\,]\!] = [\![\,3\,]\!] = 3$ $f(3) = [\![\,|3|\,]\!] = [\![\,3\,]\!] = 3$

$f(-2.5) = [\![\,|-2.5|\,]\!] = [\![\,2.5\,]\!] = 2$ $f(2.5) = [\![\,|2.5|\,]\!] = [\![\,2.5\,]\!] = 2$

$f(-2) = [\![\,|-2|\,]\!] = [\![\,2\,]\!] = 2$ $f(2) = [\![\,|2|\,]\!] = [\![\,2\,]\!] = 2$

$f(-1.5) = [\![\,|-1.5|\,]\!] = [\![\,1.5\,]\!] = 1$ $f(1.5) = [\![\,|1.5|\,]\!] = [\![\,1.5\,]\!] = 1$

$f(-1) = [\![\,|-1|\,]\!] = [\![\,1\,]\!] = 1$ $f(1) = [\![\,|1|\,]\!] = [\![\,1\,]\!] = 1$

$f(-0.5) = [\![\,|-0.5|\,]\!] = [\![\,0.5\,]\!] = 0$ $f(0.5) = [\![\,|0.5|\,]\!] = [\![\,0.5\,]\!] = 0$

$f(0) = [\![\,|0|\,]\!] = [\![\,0\,]\!] = 0$

Generalizing, we can write:

$$f(x) = \begin{cases} 2 & \text{for } -3 < x \le -2 \\ 1 & \text{for } -2 < x \le -1 \\ 0 & \text{for } -1 < x < 1 \\ 1 & \text{for } 1 \le x < 2 \\ 2 & \text{for } 2 \le x < 3 \end{cases}$$

(B)

(C) Range: Nonnegative integers

(D) Discontinuous at all integers except 0.

(E) $f(-x) = [\![\,|-x|\,]\!] = [\![\,x\,]\!] = [\![\,|x|\,]\!] = f(x)$. Even

(*1–3, 1–4*)

67. (A) Examining the graph of *p*, we obtain

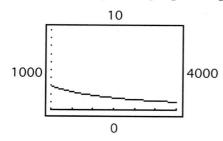

Clearly, *d* is a decreasing function, passing the horizontal line test, so *d* is one-to-one.

When $q = 1000$, $p = d(1000) = \dfrac{9}{1 + 0.002(1000)} = 3$

When $q = 4000$, $p = d(4000) = \dfrac{9}{1 + 0.002(4000)} = 1$

The range of *d* is $1 \le p \le 3$ or $[1, 3]$.

(B) Solve $p = d(q)$ for q.

$$p = \frac{9}{1 + 0.002q}$$

$$p(1 + 0.002q) = 9$$

$$1 + 0.002q = \frac{9}{p}$$

$$0.002q = \frac{9}{p} - 1$$

$$q = \frac{4,500}{p} - 500 = d^{-1}(p)$$

Domain d^{-1} = range $d = [1, 3]$
Range d^{-1} = domain $d = [1000, 4000]$.

(C) Revenue is price times number sold, so we need to multiply the variable p (price) by the function q (number sold) that we obtained in part B:

$$p \cdot q = p \cdot \left(\frac{4,500}{p} - 500\right) = 4,500 - 500p$$

(D) Revenue is price times number sold, so we need to multiply the function p (price) times the variable q (numbers sold):

$$p \cdot q = \frac{9}{1 + 0.002q} \cdot q = \frac{9q}{1 + 0.002q}$$
(1-6)

68. Profit is revenue minus cost so we start by subtracting the revenue and cost functions:

$$P(x) = R(x) - C(x) = (50x - 0.1x^2) - (10x + 1,500)$$
$$= 50x - 0.1x^2 - 10x - 1,500 = -0.1x^2 + 40x - 1,500$$

The variable is x but we're asked to find profit in terms of price (p) so we need to substitute $500 - 10p$ in for x.

$$p(500 - 10p) = -0.1(500 - 10p)^2 + 40(500 - 10p) - 1,500$$
$$= -0.1(250,000 - 10,000p + 100p^2) + 20,000 - 400p - 1,500$$
$$= -25,000 + 1,000p - 10p^2 + 20,000 - 400p - 1,500$$
$$= -10p^2 + 600p - 6,500$$

so $P(p) = -10p^2 + 600p - 6,500$

We graph the function and use the MAXIMUM command to find the largest profit.

The maximum occurs at the point (30, 2,500), so the maximum profit is $2,500 when the price is $30.
(1-5)

69. (A) The graph of $y = 55x - 4.88x^2$ is shown.

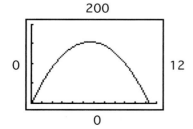

The time the arrow is airborne is the length of the interval between the two x intercepts of the graph, since y which represents height is positive throughout this interval. The first intercept is clearly $x = 0$. Applying the ZERO command to find the second intercept, the result shown is obtained.

The arrow is airborne 11.3 seconds.

(B) The result of applying the MAXIMUM command to find the maximum value of y is shown.

The maximum altitude of the arrow is 155 meters. *(1-3)*

70. (A) We begin by graphing $y_1 = 9.5x^2 - 152x + 708.7$ and $y_2 = 300$ on the interval $0 \le x \le 10$. The first coordinate of the point where the two graphs intersect is the time when the jumper is at height 300 feet. Using the INTERSECT command,

We find that this occurs after 3.4 seconds.

(B) This time we use the MINIMUM command:

The jumper reaches a minimum height of 100.7 feet after 8 seconds. *(1-3)*

71. The dimensions that will make up the length, width, and height of the box are labeled on the diagram below:

The volume is the product of the three dimensions,
so $V(x) = 2x(36 - 2x)(48 - 2x)$.

From the graph, applying the MAXIMUM command we find that the maximum volume is approximately 10,480 in^3 when the flap is 6.8 in wide.

(B) We need to find x when the output of our volume function from part (A) is 9,200, so we add $y_2 = 9,200$ to the graph and use the INTERSECT command.

The volume will be 9,200 cubic inches if the flap is either 4.2″ or 9.7″.

72. (A) This is easy to do using the TBLSET and TABLE commands on a graphing calculator:

The table of values is:

Elevation	Oxygen Level(%)
0	21.6
2,000	20.1
4,000	18.5
6,000	17.0
8,000	15.5

(B) The oxygen level in the air decreases by about 1.5% for every 2,000 foot increase in altitude.

(C) Evaluating f for 29,035, we get
$$f(29,035) = -0.000767(29,035) + 21.5 = -0.7$$
This answer is not reasonable — there can't be less than 0% oxygen in the air. This happened because the domain of the function f is $0 \leq x \leq 8,000$, and the altitude 29,035 is well outside of the domain.   (*1-2*)

73. The 19 in the equation does not depend on the number of minutes x, so that part is the fixed monthly cost for access regardless of minutes used. The $0.012x$ tells us that each minute of calls costs $0.012, or 1.2 cents. So we could say that the provider charges $19 per month for access, plus 1.2 cents per minute of usage.   (1-2)

74. (A) The graphs are shown below. The lowest graph corresponds to $C = 200$, and the highest to $C = 1,500$.

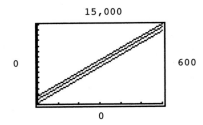

This collection of functions is a series of parallel lines with y intercept equal to the particular value of C.

(B) To find this facility's fee for 300 guests, we can find $R(300)$:

$$R(300) = C + 22.50(300) = C + 6,750$$

To find the flat fee to keep the price at $7,500, we set $R(300)$ equal to 7,500 and solve:

$C + 6,750 = 7,500$ (Subtract 6,750 from both sides)

$C = 750$

The maximum flat fee is $750.   (*1-3*)

75. (A) The graph of $f(x)$ is shown.

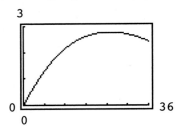

(B) Using the MAXIMUM command to find the local maximum yields the result shown.

The function increases on $[0, 24.8]$ to a local maximum of 2.8 cc/sec, and then decreases on $[24.8, 36]$.   (*1-3*)

76. (A) $f(1) = 1 - (\llbracket \sqrt{1} \rrbracket)^2 = 1 - 1 = 0$
      $f(2) = 2 - (\llbracket \sqrt{2} \rrbracket)^2 = 2 - 1 = 1$
      $f(3) = 3 - (\llbracket \sqrt{3} \rrbracket)^2 = 3 - 1 = 2$
      $f(4) = 4 - (\llbracket \sqrt{4} \rrbracket)^2 = 4 - 4 = 0$
      $f(5) = 5 - (\llbracket \sqrt{5} \rrbracket)^2 = 5 - 4 = 1$

(B) $f(n^2) = n^2 - (\llbracket \sqrt{n^2} \rrbracket)^2$

   $= n^2 - (\llbracket n \rrbracket)^2$ since $\sqrt{n^2} = n$ if n is positive
   $= n^2 - (n)^2$ since $\llbracket n \rrbracket = n$ if n is a (positive) integer.
   $= 0$

   $f(6) = 6 - (\llbracket \sqrt{6} \rrbracket)^2 = 6 - 4 = 2$ $f(12) = 12 - (\llbracket 12 \rrbracket)^2 = 12 - 9 = 3$
   $f(7) = 7 - (\llbracket 7 \rrbracket)^2 = 7 - 4 = 3$ $f(13) = 13 - (\llbracket 13 \rrbracket)^2 = 13 - 9 = 4$
   $f(8) = 8 - (\llbracket 8 \rrbracket)^2 = 8 - 4 = 4$ $f(14) = 14 - (\llbracket 14 \rrbracket)^2 = 14 - 9 = 5$
   $f(9) = 9 - (\llbracket 9 \rrbracket)^2 = 9 - 9 = 0$ $f(15) = 15 - (\llbracket 15 \rrbracket)^2 = 15 - 9 = 6$
   $f(10) = 10 - (\llbracket 10 \rrbracket)^2 = 10 - 9 = 1$ $f(16) = 16 - (\llbracket 16 \rrbracket)^2 = 16 - 16 = 0$
   $f(11) = 11 - (\llbracket 11 \rrbracket)^2 = 11 - 9 = 2$

(C) It determines if an integer is a perfect square integer. If $f(x) = 0$, then x is a perfect square and if $f(x) = 10$, x is not a perfect square integer. (1-3)

77. In this example, the independent variable is years since 1900 so the first list under the STAT EDIT screen comes from the top row of Table 1. (Note that x is years after 1900, so the values we input are 12, 32, 52, 72 and 92. The dependent variable is time, so the second list is 5.41, 4.81, 4.51, 4.00, 3.75.)

The second screen shows the appropriate plot settings; the third shows a good choice of viewing window; the last is the scatter plot of the data from the table.

To compare the algebraic function, we enter $y_1 = -0.021x + 5.57$ in the equation editor, choose the "ask" option for independent variable in the table setup window, then enter x-values 12, 32, 52, 72, and 92.

Comparing the two tables, we see that the corresponding times match very closely. Finally, we look at the graphs of both the function and the scatter plot:

As expected, the graph of the function matches the points from the scatter plot very well.

(B) Since 2010 is 110 years after 1900, we need to plug in 110 for x:
$$f(110) = -0.021(100) + 5.57 = 3.26$$
The estimated time is 3.26 seconds.   (1-2)

78. Amounts less than $3,000: Tax is 2% of income $(x) = 0.02x$

Amounts between $3,000 and $5,000: Tax is $60 + 3% of income over 3,000 $(x - 3,000)$ so tax $= 60 + 0.03(x - 3,000) = 0.03x - 30$

Amounts between $5,000 and $17,000: Tax is $120 + 5% of income over 5,000 $(x - 5,000)$ so tax $= 120 + 0.05(x - 5,000) = 0.05x - 130$

Amounts over $17,000: Tax is $720 + 5.75% of income over 17,000 $(x - 17,000)$ so tax $= 720 + 0.0575(x - 17,000) = 0.0575x - 257.5$

Combined, we get $t(x) = \begin{cases} 0.02x & 0 \leq x \leq 3,000 \\ 0.03x - 30 & 3,000 < x \leq 5,000 \\ 0.05x - 130 & 5,000 < x \leq 17,000 \\ 0.0575x - 257.5 & x > 17,000 \end{cases}$

$t(2,000) = 0.02(2,000) = 40$. The tax on \$2,000 is \$40.

$t(4,000) = 0.03(4,000) - 30 = 120 - 30 = 90$. The tax on \$4,000 is \$90.

$t(10,000) = 0.05(10,000) - 130 - 500 - 130 = 370$. The tax on \$10,000 is \$130.

$t(30,000) = 0.0575(30,000) - 257.5 - 1,725 - 257.5 = 1,467.5$. The tax on \$30,000 is \$1,467.50.   (1-3)

CHAPTER 2   Modeling with Linear and Quadratic Functions

Section 2-1

1. Answers will vary.   3. Answers will vary.

5. Answers will vary.

7. The vertical segment has length 3 so rise = 3. The horizontal segment has length 5 so run = 5. Slope = $\dfrac{rise}{run} = \dfrac{3}{5}$. (2, 2) is on the graph.
   Use the point-slope form.

$$y - 2 = \frac{3}{5}(x - 2)$$ (Multiply both sides by 5)
$$5y - 10 = 3(x - 2)$$ Common error: Multiplying both $\dfrac{3}{5}$ and
$$5y - 10 = 3x - 6$$ $(x - 2)$ by 5 on the right side.
$$-3x + 5y = +4$$
$$3x - 5y = -4$$

9. The vertical segment has length 2 so rise = 2. The horizontal segment has length 8 so run = 8.

$$slope = \frac{rise}{run} = \frac{2}{8} = \frac{1}{4}$$ (-4, 1) is on the graph. Use the point-slope form.

$$y - 1 = \frac{1}{4}(x - (-4))$$ (Multiply both sides by 4)
$$4y - 4 = x + 4$$
$$-x + 4y = 8$$
$$x - 4y = -8$$

11. The vertical segment has length 3 and goes downward so rise = -3.
    The horizontal segment has length 5 so run = 5.

$$slope = \frac{rise}{run} = \frac{-3}{5}$$ (-4, 2) is on the graph. Use the point-slope form.

$$y - 2 = \frac{-3}{5}(x - (-4))$$ (Multiply both sides by 5)
$$5y - 10 = -3(x + 4)$$
$$5y - 10 = -3x - 12$$
$$3x + 5y = -2$$

13. The x intercept is -2. The y intercept is 2. From the point (-2, 0) to the point(0, 2), the value of y increases by 2 units as the value of x increases by 2 units; slope = $\dfrac{change\ in\ y}{change\ in\ x} = \dfrac{2}{2} = 1$. Using $y = mx + b$, we get $y = x + 2$.

15. The x intercept is -2. The y intercept is -4. From the point (-2, 0) to the point(0, -4) the value of y decreases by 4 units as the value of x increases by 2 units; slope = $\dfrac{change\ in\ y}{change\ in\ x} = \dfrac{-4}{2} = -2$. Using $y = mx + b$, we get
    $y = -2x - 4$.

17. The x intercept is 3. The y intercept is -1. From the point (0, -1) to the point (3, 0) the value of y increases by 1 unit as the value of x increases by 3 units; slope = $\dfrac{change\ in\ y}{change\ in\ x} = \dfrac{1}{3}$. Using $y = mx + b$, we get
    $y = \dfrac{1}{3}x - 1$.

19. The equation $y = 2x^2$ is not linear because in linear equations both variables appear only to the first power.

21. The equation $y = \dfrac{x - 5}{3}$ is linear. It can be rewritten as $y = \dfrac{1}{3}x - \dfrac{5}{3}$ which matches the form $y = mx + b$.

23. Simplify the equation:
$$y = \frac{2}{3}(x - 7) - \frac{1}{2}(3 - x) \qquad \text{(Multiply both sides by 6)}$$
$$6y = 4(x - 7) - 3(3 - x)$$
$$6y = 4x - 28 - 9 + 3x$$
$$-7x + 6y = -37$$
The equation can be rewritten in the form $Ax + By = C$, so it is a linear equation.

25. Simplify the equation:
$$y = \frac{1}{4}(2x + 2) + \frac{1}{2}(4 - x) \qquad \text{(Multiply both sides by 4)}$$
$$4y = 2x + 2 + 4 - x$$
$$4y = x + 6$$
$$-x + 4y = 6$$
The equation can be written in the form $Ax + By = C$, so it is a linear equation.

27. This equation is not linear. If we attempt to move x to the numerator by multiplying by $x - 5$, we get
$$y(x - 5) = 3$$
$$xy - 5y = 3$$
In a linear equation, the two variables appear in different terms.

29. $y = -\dfrac{3}{5}x + 4$

    slope: $-\dfrac{3}{5}$

    y intercept: 4

    To find the x intercept, set $y = 0$ and solve for x:
$$0 = -\frac{3}{5}x + 4$$
$$\frac{3}{5}x = 4$$
$$x = \frac{20}{3} \quad (x \text{ intercept})$$
    To graph, use the intercepts or find a few points by substituting values.

x	y
0	4
5	1
−5	7

31. $y = -\dfrac{3}{4}x$ or $y = -\dfrac{3}{4}x + 0$

slope: $-\dfrac{3}{4}$

y intercept: 0

To find the x intercept, set $y = 0$ and solve for x:

$0 = -\dfrac{3}{4}x$

$0 = x$ (x intercept)

To graph, use the intercept and find one or more other points by substituting values.

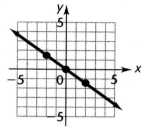

x	y
0	0
4	-3
-4	3

33. $2x - 3y = 15$

$-3y = -2x + 15$

$y = \dfrac{2}{3}x - 5$

slope: $\dfrac{2}{3}$

y intercept: -5

To find the x intercept, set $y = 0$ in $2x - 3y = 15$ and solve for x:

$2x - 3(0) = 15$

$2x = 15$

$x = \dfrac{15}{2}$ (x intercept)

To graph, use the intercepts or find a few points by substituting values.

x	y
0	-5
3	-3
-3	-7

35. $\dfrac{y}{8} - \dfrac{x}{4} = 1$

$\dfrac{y}{8} = \dfrac{x}{4} + 1$

$y = 8\left(\dfrac{x}{4} + 1\right)$

$y = 2x + 8$

slope: 2

y intercept: 8

To find the x intercept, set $y = 0$ in $\dfrac{y}{8} - \dfrac{x}{4} = 1$ and solve for x:

$\dfrac{0}{8} - \dfrac{x}{4} = 1$

$-\dfrac{x}{4} = 1$

$x = -4$ (x intercept)

To graph, use the intercepts or find other points by substituting values.

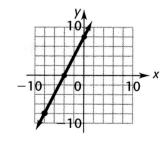

x	y
0	8
-4	0
-8	-8

37. $x = -3$

This is a vertical line.
The slope is undefined.
There is no y intercept.
The x intercept is -3.
Graph a vertical line with all x coordinates equal to -3.

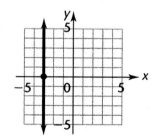

39. $y = 3.5$
This is a horizontal line.
slope = 0
The y intercept is 3.5.
There is no x intercept.
Graph a horizontal line with all
y coordinates equal to 3.5.

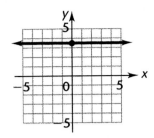

41. The equation of any vertical line looks like $x = k$, where k is the x coordinate of any point on the line. In this case, we get $x = 2$.

43. The equation of any horizontal line looks like $y = k$, where k is the y coordinate of any point on the line. In this case, we get $y = -4$.

45. First, we find the slope:
$$m = \frac{4 - 4}{5 - (-3)} = \frac{0}{8} = 0$$

This tells us that the line is horizontal, and the y coordinate of every point is 4, so the equation is $y = 4$. (You may have noticed that the line is horizontal by observing that both points have the same y coordinate.)

47. First, we find the slope:
$$m = \frac{6 - (-3)}{4 - 4} = \frac{9}{0}, \text{ which is undefined.}$$

This tells us that the line is vertical, and the x coordinate of every point is 4, so the equation is $x = 4$. (You may have noticed that the line is vertical by observing that both points have the same x coordinate.)

49. Slope and y intercept are given; we use slope-intercept form.
$y = 1x + 0$
$y = x$

Subtract x from each side to write in standard form:

$Y = x$

$-x + y = 0$

51. Slope and y intercept are given; we use slope-intercept form.
$y = -\frac{2}{3}x + (-4)$
$y = -\frac{2}{3}x - 4$

Now we write in standard form.

$y = -\frac{2}{3}x - 4$ Multiply both sides by 3

$3y = -2x - 12$ Add $2x$ to both sides

$2x + 3y = -12$

53. Use the point-slope form:
$y - 4 = -3(x - 0)$
$y = -3x + 4$

55. Use the point-slope form:

$$y - 4 = -\frac{2}{5}\big(x - (-5)\big)$$

$$y - 4 = -\frac{2}{5}\big(x + 5\big)$$

$$y - 4 = -\frac{2}{5}x - 2$$

$$y = -\frac{2}{5}x + 2$$

57. First, find the slope:

$$m = \frac{-2 - 6}{5 - 1} = \frac{-8}{4} = -2$$

Now use the point-slope form: (You can use either point; we chose (1,6))

$$y - 6 = -2(x - 1)$$

$$y - 6 = -2x + 2$$

$$y = -2x + 8$$

59. First, find the slope:

$$m = \frac{0 - 8}{2 - (-4)} = \frac{-8}{6} = -\frac{4}{3}$$

Now use the point-slope form: (You can use either point; we chose (2,0))

$$y - 0 = -\frac{4}{3}\big(x - 2\big)$$

$$y = -\frac{4}{3}x + \frac{8}{3}$$

61. The intercepts are the points (-4,0) and (0,3). First, find the slope:

$$m = \frac{3 - 0}{0 - (-4)} = \frac{3}{4}$$

Now use the point-slope form: (You can use either point; we chose (0,3))

$$y - 3 = \frac{3}{4}\big(x - 0\big)$$

$$y = \frac{3}{4}x + 3$$

63. The line $y = 3x - 5$ has slope 3 (coefficient of x), so the line we're looking for does as well (since they are parallel). Use the point-slope form:

$$y - 4 = 3\big(x - (-3)\big)$$

$$y - 4 = 3\big(x + 3\big)$$

$$y - 4 = 3x + 9$$

$$y = 3x + 13$$

65. The line $y = -\frac{1}{3}x$ has slope -1/3 (coefficient of x); the line perpendicular to it will have slope 3 (negative reciprocal of -1/3). Use the point-slope form:

$$y - (-3) = 3\big(x - 2\big)$$

$$y + 3 = 3x - 6$$

$$y = 3x - 9$$

67. To find the slope of the given line, solve for y:

$3x - 2y = 4$

$-2y = -3x + 4$

$y = -\dfrac{1}{2}(-3x + 4)$

$y = \dfrac{3}{2}x - 2$

This line has slope 3/2 (coefficient of x), so the line parallel to it does as well. Use the point-slope form:

$y - 0 = \dfrac{3}{2}(x - 5)$

$y = \dfrac{3}{2}x - \dfrac{15}{2}$

69. To find the slope of the given line, solve for y:

$x + 3y = 9$

$3y = -x + 9$

$y = -\dfrac{1}{3}x + 3$

This line has slope –1/3 (coefficient of x), so the line perpendicular to it will have slope 3 (negative reciprocal). Use the point-slope form:

$y - (-4) = 3(x - 0)$

$y + 4 = 3x$

$y = 3x - 4$

71. The graphs are straight lines; all have the same y intercept, 2, but their slopes vary as m varies.

73. (A) First find the slope:

$m = \dfrac{2 - (-3)}{7 - (-1)} = \dfrac{5}{8}$

Now use the point-slope form

$y - 2 = \dfrac{5}{8}(x - 7)$

$y - 2 = \dfrac{5}{8}x - \dfrac{35}{8}$

$y = \dfrac{5}{8}x - \dfrac{35}{8} + \dfrac{16}{8}$

$y = \dfrac{5}{8}x - \dfrac{19}{8}$

$f(x) = \dfrac{5}{8}x - \dfrac{19}{8}$

(B) $m = \dfrac{7 - (-1)}{2 - (-3)} = \dfrac{8}{5}$

$y - 7 = \dfrac{8}{5}(x - 2)$

$y - 7 = \dfrac{8}{5}x - \dfrac{16}{5}$

$y = \dfrac{8}{5}x - \dfrac{16}{5} + \dfrac{35}{5}$

$y = \dfrac{8}{5}x + \dfrac{19}{5}$

$g(x) = \dfrac{8}{5}x + \dfrac{19}{5}$

(C) The graphs are mirror images about the line $y = x$, so the two functions are inverses.

93. (A) If V is linearly related to t, then we are looking for an equation whose graph passes through $(t_1, V_1) = (0, 8,000)$ and $(t_2, V_2) = (5, 0)$. We find the slope, and then we use the point-slope form to find the equation.

$$m = \frac{V_2 - V_1}{t_2 - t_1} = \frac{0 - 8,000}{5 - 0} = \frac{-8,000}{5} = -1,600$$

$$V - V_1 = m(t - t_1)$$
$$V - 8,000 = -1,600(t - 0)$$
$$V - 8,000 = -1,600t$$
$$V = -1,600t + 8,000 \quad 0 \le t \le 5$$

(B) We are asked for V when $t = 3$
$$V = -1,600(3) + 8,000$$
$$V = -4,800 + 8,000$$
$$V = 3,200$$

(C) The slope is $m = -1,600$; the value decreases by \$1,600 per year.

95. (A) The question asks us to use x to represent the number of golf clubs produced, so we can set up ordered pairs as follows: (golf clubs produced, total daily cost). That gives us $(80, 8,147)$ and $(100, 9,647)$.

$$m = \frac{9,647 - 8,147}{100 - 80} = \frac{1,500}{20} = 75$$
$$y - 9,647 = 75(x - 100)$$
$$y - 9,647 = 75x - 7,500$$
$$y = 75x + 2,147$$
$$C(x) = 75x + 2,147$$

(B) The slope represents the variable cost per golf club manufactured; the cost per club is \$75. The y-intercept indicates that the fixed daily costs are \$2,147.

97. (A) If R is linearly related to C, then we are looking for an equation whose graph passes through $(C_1, R_1) = (210, 0.160)$ and $(C_2, R_2) = (231, 0.192)$. We find the slope, and then we use the point-slope form to find the equation.

$$m = \frac{R_2 - R_1}{C_2 - C_1} = \frac{0.192 - 0.160}{231 - 210} = \frac{0.032}{21} = 0.00152$$

$$R - R_1 = m(C - C_1)$$
$$R - 0.160 = 0.00152(C - 210)$$
$$R - 0.160 = 0.00152C - 0.319$$
$$R = 0.00152C - 0.159, \quad C \ge 210$$

(B) We are asked for R when $C = 260$.
$$R = 0.00152(260) - 0.159$$
$$R = 0.236$$

(C) The slope is $m = 0.00152$; coronary risk increases 0.00152 per unit increase in cholesterol above the 210 cholesterol level.

99. (A) The temperature is 70 degrees at altitude zero, so the point $(0, 70)$ is on our line. At 1,000 feet, the temperature will have dropped by 5 degrees, to 65, so the point $(1, 65)$ is on the line as well. We can find the slope and use the point-slope form to find the equation:

$$m = \frac{65 - 70}{1 - 0} = -5$$

$$T - 70 = -5(A - 0)$$

$$T = -5A + 70$$

(B) An altitude of 10,000 feet corresponds to $A = 10$, so we plug in 10 for A:

$T = -5(10) + 70 = 20$

The temperature at 10,000 feet would be 20 degrees.

(C) The slope is -5. This indicates that the temperature decreases at the rate of 5 degrees for each 1,000 feet of altitude.

Section 2-2

1. Answers will vary.   3. Answers will vary.

5. Answers will vary.   7. Answers will vary.

9. $u(x) = 0$ where the graph of u intersects the x axis, at $x = c$ and $x = f$.

11. $u(x) = v(x)$ for values of x that correspond to the intersection of the graphs of u and v, at $x = b$ and $x = e$.

13. Simplify the left side:
$3(x - 2) - 2(x + 1) = 3x - 6 - 2x - 2 = x - 8$
This is equal to the right side, so the equation is an identity.

15. Simplify the left side:
$2(x - 1) - 3(2 - x) = 2x - 2 - 6 + 3x = 5x - 8$
This is not equal to the right side, so the equation is not an identity.
$5x - 8 = x - 8$
$4x = 0$
$x = 0$
This is a conditional equation with solution $x = 0$.

17. Simplify the left side:
$5(x + 2) - 3(x - 1) = 5x + 10 - 3x + 3 = 2x + 13$
This is not equal to the right side, so the equation is not an identity.
$2x + 13 = 2x + 4$
$0 = -9$
This is never true, so the original equation is a contradiction.

19. $10x - 7 = 4x - 25$
$6x - 7 = -25$
$6x = -18$
$x = -3$

21. $3(x + 2) = 5(x - 6)$
$3x + 6 = 5x - 30$
$-2x = -36$
$x = 18$

23. $5 + 4(t - 2) = 2(t + 7) + 1$
$5 + 4t - 8 = 2t + 14 + 1$
$4t - 3 = 2t + 15$
$2t = 18$
$t = 9$

25. $\dfrac{x}{2} - 2 = \dfrac{x}{5} + \dfrac{2}{5}$ LCD = 10

$10 \cdot \dfrac{x}{2} - 10 \cdot 2 = 10 \cdot \dfrac{x}{5} + 10 \cdot \dfrac{2}{5}$
$5x - 20 = 2x + 4$
$3x = 24$
$x = 8$

27. $3 + 5(x - 7) = 3x - 2(16 - x)$
$3 + 5x - 35 = 3x - 32 + 2x$
$5x - 32 = 5x - 32$
The equation is an identity, so the solution is all real numbers.

29.
$$5 - \frac{2x - 1}{4} = \frac{x + 2}{3} \qquad \text{LCD} = 12$$

$$12 \cdot 5 - 12 \frac{(2x - 1)}{4} = 12 \frac{(x + 2)}{3}$$

$$12 \cdot 5 - 3(2x - 1) = 4(x + 2)$$

$$60 - 6x + 3 = 4x + 8$$

$$-6x + 63 = 4x + 8$$

$$-10x = -55$$

$$x = \frac{-55}{-10}$$

Solution: $\frac{11}{2}$ or 5.5

Check:

10

10 — 10

Intersection
X=5.5   Y=2.5

-10

Common Error:
After line 2,
students often write

$$60 - \overset{3}{\cancel{12}} \frac{2x - 1}{\cancel{4}} = \dots$$

$$60 - 6x - 1 = \dots$$

Forgetting to
distribute the -3.
Put compound
numerators in
parentheses to avoid
this.

31. $\dfrac{7}{t} + 4 = \dfrac{2}{t}$

Excluded value: $t \neq 0 \quad \text{LCD} = t$

$$t \cdot \frac{7}{t} + t \cdot 4 = t \cdot \frac{2}{t}$$

$$7 + 4t = 2$$

$$4t = -5$$

$$t = -\frac{5}{4}$$

33. $\dfrac{1}{m} - \dfrac{1}{9} = \dfrac{4}{9} - \dfrac{2}{3m}$

Excluded value: $m \neq 0 \quad \text{LCD} = 9m$

$$9m \cdot \frac{1}{m} - 9m \cdot \frac{1}{9} = 9m \cdot \frac{4}{9} - 9m \cdot \frac{2}{3m}$$

$$9 - m = 4m - 6$$

$$-5m = -15$$

$$m = 3$$

35. $(x - 2)(x + 3) = (x - 4)(x + 5)$

$$x^2 + x - 6 = x^2 + x - 20$$

$$-6 = -20$$

There is no solution.

37. $(x + 2)(x - 3) = (x - 4)(x + 5)$

$$x^2 - x - 6 = x^2 + x - 20$$

$$-x - 6 = x - 20$$

$$-2x = -14$$

$$x = 7$$

39. $(x - 2)^2 = (x - 1)(x + 2)$

$$x^2 - 4x + 4 = x^2 + x - 2$$

$$-4x + 4 = x - 2$$

$$-5x = -6$$

$$x = \frac{6}{5}$$

41. $\dfrac{2x}{x - 3} = 2 + \dfrac{6}{x - 3}$

Excluded value: $x \neq 3 \quad \text{LCD} = x - 3$

$$(x - 3)\frac{2x}{x - 3} = (x - 3)2 + (x - 3)\frac{6}{x - 3}$$

$$2x = 2x - 6 + 6$$

$$2x = 2x$$

The solution is all real numbers except 3.

43. $\dfrac{2x}{x - 3} = 7 + \dfrac{4}{x - 3}$

Excluded value: $x \neq 3 \quad \text{LCD} = x - 3$

$$(x - 3)\frac{2x}{x - 3} = 7(x - 3) + (x - 3)\frac{4}{x - 3}$$

$$2x = 7x - 21 + 4$$

$$2x = 7x - 17$$

$$-5x = -17$$

$$x = \frac{17}{5} \text{ or } 3.4$$

45. $\dfrac{2x}{x - 3} = 7 + \dfrac{6}{x - 3} \qquad \text{LCD: } x - 3$

$$(x - 3)\frac{2x}{x - 3} = (x - 3)7 + (x - 3)\frac{6}{x - 3}$$

$$2x = 7x - 21 + 6$$

$$-5x = -15$$

$$x = 3$$

3 is excluded since it makes the original equation
undefined, so there is no solution.

47. $P = 2\ell + 2w$

$P - 2\ell = 2w$

$\frac{1}{2}(P - 2\ell) = w$

$w = \frac{P - 2\ell}{2} = \frac{1}{2}P - \ell$

49. $a_n = a_1 + (n - 1)d$

$a_1 + (n - 1)d = a_n$

$(n - 1)d = a_n - a_1$

$d = \frac{a_n - a_1}{n - 1}$

51.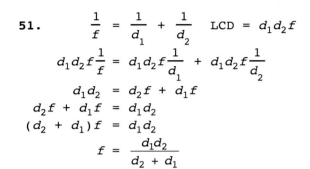

$\frac{1}{f} = \frac{1}{d_1} + \frac{1}{d_2}$ LCD $= d_1 d_2 f$

$d_1 d_2 f \frac{1}{f} = d_1 d_2 f \frac{1}{d_1} + d_1 d_2 f \frac{1}{d_2}$

$d_1 d_2 = d_2 f + d_1 f$

$d_2 f + d_1 f = d_1 d_2$

$(d_2 + d_1)f = d_1 d_2$

$f = \frac{d_1 d_2}{d_2 + d_1}$

53.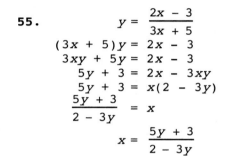

$A = 2ab + 2ac + 2bc$

$2ab + 2ac + 2bc = A$

$2ab + 2ac = A - 2bc$

$a(2b + 2c) = A - 2bc$

$a = \frac{A - 2bc}{2b + 2c}$

55.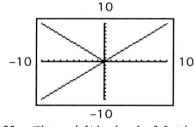

$y = \frac{2x - 3}{3x + 5}$

$(3x + 5)y = 2x - 3$

$3xy + 5y = 2x - 3$

$5y + 3 = 2x - 3xy$

$5y + 3 = x(2 - 3y)$

$\frac{5y + 3}{2 - 3y} = x$

$x = \frac{5y + 3}{2 - 3y}$

57. The graphs of $y_1 = x$ and $y_2 = \sqrt{x^2}$ are shown at the right. The graphs are not identical, so the equation $x = \sqrt{x^2}$ is a conditional equation, and not an identity.

59. The length is twice the width

$\ell = 2 \cdot w$

$\ell = 2w$

61. The width is half the length.

$w = \frac{1}{2} \cdot \ell$

$w = \frac{1}{2}\ell$

63. The length is 3 more than the width

$\ell = 3 + w$

$\ell = 3 + w$

65. The length is 4 less than the width

$\ell = w - 4$

67. (A) For detailed instructions on using a graphing calculator for linear regression, see Example 8 in Section 2-2 of your text.
Enter the x column (1, 5) in list L_1, and the y column (-1, 1) in list L_2.

Use the STAT CALC menu to find the linear regression:

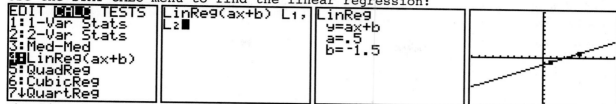

The line is $y = 0.5x - 1.5$

(B) Repeat the steps above using -1, 1 as L_1 and 1, 5 as L_2.

The line is $y = 2x + 3$.

The two lines are symmetrical with respect to the line $y = x$ so the two linear functions are inverses. (This is expected because the difference between tables A and B is that the order of x and y were switched.)

69. $\dfrac{x - \frac{1}{x}}{1 + \frac{1}{x}} = 1$ (Multiply both sides by $1 + \frac{1}{x}$)

$x - \dfrac{1}{x} = 1 + \dfrac{1}{x}$ LCD $= x$

$x\left(x - \dfrac{1}{x}\right) = x\left(1 + \dfrac{1}{x}\right)$

$x^2 - 1 = x + 1$

$x^2 - x - 2 = 0$

$(x - 2)(x + 1) = 0$

$x - 2 = 0,\ x + 1 = 0$

$x = 2,\qquad x = -1$

Check: $x = 2$: $\dfrac{2 - \frac{1}{2}}{1 + \frac{1}{2}} = \dfrac{\frac{3}{2}}{\frac{3}{2}} = 1$

$x = -1$: $\dfrac{-1 - \frac{1}{-1}}{1 + \frac{1}{-1}} = \dfrac{0}{0}$ Not a solution

Solution: $x = 2$

71. $\dfrac{x + 1 - \frac{2}{x}}{1 - \frac{1}{x}} = x + 2$ (Multiply left side by $\frac{x}{x}$)

$\dfrac{x\left(x + 1 - \frac{2}{x}\right)}{x\left(1 - \frac{1}{x}\right)} = x + 2$

$\dfrac{x^2 + x - 2}{x - 1} = x + 2$ (Multiply both sides by $x - 1$)

$x^2 + x - 2 = (x + 2)(x - 1)$

$x^2 + x - 2 = x^2 + x - 2$

This is an identity provided that $x \neq 0$ or 1. (Both 0 and 1 make the left side of the original equation undefined.)

73. Let x = the least of the integers. Then $x + 1$ and $x + 2$ are the others. Their sum looks like $x + (x + 1) + (x + 2)$, so

$$x + (x + 1) + (x + 2) = 84$$
$$3x + 3 = 84$$
$$3x = 81$$
$$x = 27$$

The three consecutive integers are 27, 28, and 29.

75. Let x = the least of the integers. Then the other three are $x + 2$, $x + 4$, and $x + 6$ (they are all separated by two since all are even). Translate the question into an equation:

The sum of the first three is 2 more than twice the fourth
$$x \quad + (x + 2) + (x + 4) = 2 \quad + \quad\quad 2(x + 6)$$
$$3x + 6 = 2 + 2x + 12$$
$$x = 8$$

The four even integers are 8, 10, 12, and 14.

77. Let ℓ = the length, w = the width. Translate the statements into symbols:

The length is twice the width
$$\ell \quad = \quad 2 \quad \cdot \quad w$$

The perimeter is 60 inches
$$2\ell + 2w = 60 \qquad \text{We know } 2w = \ell, \text{ so}$$
$$2\ell + \ell = 60$$
$$3\ell = 60$$
$$\ell = 20$$
$$w = 10$$

The length is twice the width, so $\ell = 20$.
The dimensions are 10 in $\times$ 20 in.

79. Total cost = Fixed cost + variable cost. If x = the number of doughnuts produced, variable cost is $0.12x$ ($0.12 per doughnut times the number of doughnuts produced.) If $C(x)$ represents total cost,
$$C(x) = 124 + 0.12x$$
Plug in 250 for $C(x)$:
$$250 = 124 + 0.12x$$
$$126 = 0.12x$$
$$1,050 = x$$
The shop can produce 1,050 doughnuts for $250.

81. Let x = sales of employee
Then $x - 7,000$ = sales on which 8% commission is paid
$0.08(x - 7,000)$ = (rate of commission) $\times$ (sales) = (amount of commission)
$2,150 + 0.08(x - 7,000)$ = (base salary) + (amount of commission) = earnings
$2,150 + 0.08(x - 7,000) = 3,170$

We're looking for the sales that will earn our employee $3,170, so we set our expression for earnings equal to 3,170 and solve for x.

$$2,150 + 0.08x - 560 = 3,170$$
$$0.08x + 1,590 = 3,170$$
$$0.08x = 1,580$$
$$x = \frac{1,580}{0.08}$$
$$x = \$19,750$$

83. Let x = the speed of the current, and arrange the given information into a table.

	distance	rate	time
upstream	1,000	3 − x	
downstream	1,200	3 + x	

We can use the formula $d = r \cdot t$ to find an expression for each time:

$$1,000 = (3 - x) \cdot \text{(time upstream)}; \text{ time upstream} = \frac{1,000}{3 - x}$$

$$1,200 = (3 + x) \cdot \text{(time downstream)}; \text{ time downstream} = \frac{1,200}{3 + x}$$

The problem tells us that these two times are equal, so we set them equal to each other and solve.

$$\frac{1,000}{3 - x} = \frac{1,200}{3 + x} \qquad \text{LCD} = (3 + x)(3 - x)$$

$$(3 + x)(3 - x)\frac{1,000}{3 - x} = (3 + x)(3 - x)\frac{1,200}{3 + x}$$

$$(3 + x)1,000 = (3 - x)1,200$$

$$3,000 + 1,000x = 3,600 - 1,200x$$

$$2,200x = 600$$

$$x = \frac{600}{2,200} \approx 0.27$$

The speed of the current is about 0.27 meters per second.

85. Let d = distance flown north, and arrange the given information in a table. The two distances are the same; the trip north is against a 30 mph wind, so the overall speed is 120 mph; the trip south is with that same wind, so the speed is 180 mph.

	distance	rate	time
north	d	120	
south	d	180	

We can use the formula $d = r \cdot t$ to find an expression for each time:

$$d = 120 \cdot \text{(time flying north)}; \text{ time flying north} = \frac{d}{120}$$

$$d = 180 \cdot \text{(time flying south)}; \text{ time flying south} = \frac{d}{180}$$

The two times add up to 3 hours, so we set their sum equal to 3 and solve.

$$\frac{d}{120} + \frac{d}{180} = 3 \qquad \text{LCD} = 360$$

$$360 \cdot \frac{d}{120} + 360 \cdot \frac{d}{180} = 360 \cdot 3$$

$$3d + 2d = 1,080$$

$$5d = 1,080$$

$$d = \frac{1,080}{5} = 216$$

He should fly 216 miles north.

(B) We still use the above ideas, except that rate flying north = rate flying south = 150 miles per hour.

$$\frac{d}{150} + \frac{d}{150} = 3$$

$$\frac{2d}{150} = 3$$

$$\frac{d}{75} = 3$$

$$d = 225 \text{ miles}$$

87. Let D = distance from earthquake to station

Then $\dfrac{D}{5}$ = time of primary wave.

$\dfrac{D}{3}$ = time of secondary wave.

Time difference = time of *slower* secondary wave - time of *faster* primary wave

$$12 = \dfrac{D}{3} - \dfrac{D}{5}$$

$$15(12) = 15\left(\dfrac{D}{3}\right) - 15\left(\dfrac{D}{5}\right)$$

$$180 = 5D - 3D$$

$$180 = 2D$$

$$D = 90 \text{ miles}$$

> **Common Error:**
> Time difference is *not* time of fast
> wave (short) - time of slow wave (long)

89.
Let x = amount of distilled water

50 = amount of 30% solution

Then $50 + x$ = amount of 25% solution

acid in 30% solution + acid in distilled water = acid in 25% solution

$$0.3(50) + 0 = 0.25(50 + x)$$
$$0.3(50) = 0.25(50 + x)$$
$$15 = 12.5 + 0.25x$$
$$2.5 = 0.25x$$
$$x = 10 \text{ gallons}$$

91.
Let x = amount of 50% solution

5 = amount of distilled water

Then $x - 5$ = amount of 90% solution

acid in 90% solution + acid in distilled water = acid in 50% solution

$$0.9(x - 5) + 0 = 0.5x$$
$$0.9x - 4.5 = 0.5x$$
$$-4.5 = -0.4x$$
$$x = 11.25 \text{ liters}$$

93. (A) Note: The temperature increased 2.5°C for each additional 100 meters of depth, and one kilometer is ten times as far as 100 meters. So the temperature increased 25 degrees for each additional kilometer of depth.

Let x = the depth (in kilometers); then $x - 3$ = the depth beyond 3 kilometers, and $25(x - 3)$ = the temperature increase for $x - 3$ kilometers of depth.

T = temperature at 3 kilometers + temperature increase beyond 3 km

$$T = 30 + 25(x - 3)$$

(B) We are asked to find T when $x = 15$. Plug $x = 15$ into the formula from part (A):

$$T = 30 + 25(15 - 3)$$
$$= 330°C$$

(C) We're asked to find depth (x) when $T = 280$, so we set our expression for T equal to 280 and solve.

$$280 = 30 + 25(x - 3)$$
$$280 = 30 + 25x - 75$$
$$280 = 25x - 45$$
$$325 = 25x$$
$$13 = x$$

The depth is 13 kilometers.

95. For detailed instructions on using a graphing calculator for linear regression, see Example 8 in Section 2-2 of your text.

Enter the years after 1999 (0,1,2,3,4,5, and 6) as L_1, and the Public column as L_2.

Use the STAT CALC menu to find the linear regression command:

The linear model is $y = 468x + 6,832$.

The year 2008 corresponds to $x = 9$, so plug $x = 9$ into our model:

$y = 468(9) + 6,832 = 11,044$

The estimated average tuition in 2008 is \$11,044. Finally, to find when average tuition will reach \$15,000, set y equal to 15,000 and solve for x:

$15,000 = 468x + 6,832$

$8,168 = 468x$

$$x = \frac{8,168}{468} \approx 18$$

Average tuition is predicted to reach \$15,000 about 18 years after 1999, or in 2017.

97. For detailed instructions on using a graphing calculator for linear regression, see Example 8 in Section 2-2 of your text.

The independent variable is years since 1968, so enter 0, 4, 8, 12, 16, 20, 24, 28, 32, and 36 as L_1 for the years in the table. The first dependent variable is the winning time in the men's 100M freestyle, so enter the times in that column as L_2. Also enter the women's 100M freestyle times as list L_3.

Note that the entire table of values does not fit on one screen, so the first screen shot above is a partial. The first scatter plot is the men's times; the second is the women's.

Use the STAT CALC menu to find the linear regression, first with lists L_1 and L_2, then with lists L_1 and L_3.

The line for men's times is $y = -0.103x + 51.509$. The line for women's times is $y = -0.145x + 58.286$.

To see if the models indicate that the women's times will catch up, we graph both lines and see if they ever meet. We'll need a considerably larger viewing window.

We see that the lines do meet near $x = 161$; 161 years after 1968 is 1968 + 161 = 2129, so in 2129 the times predicted by these models will be equal. We'll leave it up to you to decide whether or not this will actually occur. (Note: You can also find where the lines meet by setting the two equations equal to each other and solving for x.)

99. For detailed instructions on using a graphing calculator for linear regression, see Example 8 in Section 2-2 of your text.
The independent variable is price, so enter the values in the first price column as list L_1. The dependent variable is supply, so enter the supply column as list L_2.

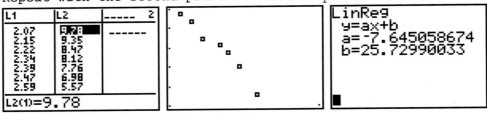

Use the STAT CALC menu to find the linear regression:

LinReg
y=ax+b
a=4.946946947
b=-4.21965251

The price-supply line is $y = 4.95x - 4.22$.

Repeat with the second price column as L_1 and the demand column as L_2.

The demand-supply line is $y = -7.65x + 25.7$.

The equilibrium price is the price at which supply and demand are equal, so we graph the two lines and see where they cross.

The lines meet at (2.37, 7.53), so the equilibrium price is $2.37 per bushel.

Section 2-3

1. Answers will vary.

3. Answers will vary.

5. Answers will vary.

7. Vertex: $(-3, -4)$; axis: $x = -3$

Hand graph:

Graphing calculator graph:

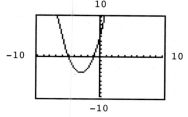

9. Vertex: $\left(\dfrac{3}{2}, -5\right)$; axis: $x = \dfrac{3}{2}$

Hand graph:

Graphing calculator graph:

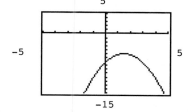

11. Vertex: $(-10, 20)$; axis $x = -10$

Hand graph:

Graphing calculator graph:

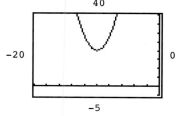

13. The graph of $y = x^2$ is shifted right 2 units and up 1 unit.

15. The graph of $y = x^2$ is shifted left 1 unit and reflected in the x axis.

17. The graph of $y = x^2$ is shifted right 2 units and down 3 units.

19. The graph of $y = x^2$ has been shifted to the right 2 units. This is the graph of $y = (x - 2)^2$, corresponding to the function k.

21. The graph of $y = x^2$ has been shifted to the right 2 units and down 3 units. This is the graph of $y = (x - 2)^2 - 3$, corresponding to the function m.

23. The graph of $y = x^2$ has been reflected in the x axis and shifted to the left 1 unit, corresponding to the function h.

25. Find the number that must be added: $\left(\dfrac{10}{2}\right)^2 = 25$

Complete the square: $x^2 + 10x + 25 = (x + 5)^2$

27. Find the number that must be added: $\left(\dfrac{-7}{2}\right)^2 = \dfrac{49}{4}$

Complete the square: $x^2 - 7x + \dfrac{49}{4} = \left(x - \dfrac{7}{2}\right)^2$

29. Find the number that must be added: $\left(\dfrac{7/5}{2}\right)^2 = \left(\dfrac{7}{10}\right)^2 = \dfrac{49}{100}$

Complete the square: $x^2 + \dfrac{7}{5}x + \dfrac{49}{100} = \left(x + \dfrac{7}{10}\right)^2$

31. Begin by grouping the first two terms with parentheses:

$\begin{aligned} f(x) &= \left(x^2 - 4x\right) + 5 & &\text{Find the number needed to complete the square} \\ &= \left(x^2 - 4x + ?\right) + 5 & &(-4\,/\,2)^2 = 4;\ \text{add and subtract 4} \\ &= \left(x^2 - 4x + 4\right) + 5 - 4 & &\text{Factor parentheses, combine like terms} \\ &= (x - 2)^2 + 1 \end{aligned}$

The vertex form is $f(x) = (x - 2)^2 + 1$. The vertex is $(2,1)$ and the axis is $x = 2$.

33. Begin by grouping the first two terms with parentheses, then factoring -1 out of those two terms so that the coefficient of x^2 is 1:

$\begin{aligned} h(x) &= -1\left(x^2 + 2x\right) - 3 & &\text{Find the number needed to complete the square} \\ &= -1\left(x^2 + 2x + ?\right) - 3 & &(2\,/\,2)^2 = 1;\ \text{add 1 inside the parentheses} \\ &= -1\left(x^2 + 2x + 1\right) - 3 + ? & &\text{We actually added } -1(1),\ \text{so add 1 as well} \\ &= -1\left(x^2 + 2x + 1\right) - 3 + 1 & &\text{Factor parentheses, combine like terms} \\ &= -1(x + 1)^2 - 2 \end{aligned}$

The vertex form is $h(x) = -1(x + 1)^2 - 2$. The vertex is $(-1,-2)$ and the axis is $x = -1$.

35. Begin by grouping the first two terms with parentheses, then factoring 2 out of those two terms so that the coefficient of x^2 is 1:

$\begin{aligned} m(x) &= 2\left(x^2 - 6x\right) + 22 & &\text{Find the number needed to complete the square} \\ &= 2\left(x^2 - 6x + ?\right) + 22 & &(-6\,/\,2)^2 = 9;\ \text{add 9 inside the parentheses} \\ &= 2\left(x^2 - 6x + 9\right) + 22 + ? & &\text{We actually added } 2(9),\ \text{so subtract 18 as well} \\ &= 2\left(x^2 - 6x + 9\right) + 22 - 18 & &\text{Factor parentheses, combine like terms} \\ &= 2(x - 3)^2 + 4 \end{aligned}$

The vertex form is $m(x) = 2(x - 3)^2 + 4$. The vertex is $(3,4)$ and the axis is $x = 3$.

37. Begin by grouping the first two terms with parentheses, then factoring 1/2 out of those two terms so that the coefficient of x^2 is 1:

$$f(x) = \frac{1}{2}\left(x^2 + 6x\right) - \frac{7}{2} \qquad \text{Find the number needed to complete the square}$$

$$= \frac{1}{2}\left(x^2 + 6x + ?\right) - \frac{7}{2} \qquad (6/2)^2 = 9; \text{ add 9 inside the parentheses}$$

$$= \frac{1}{2}\left(x^2 + 6x + 9\right) - \frac{7}{2} + ? \qquad \text{We actually added } \frac{1}{2}(9); \text{ subtract } \frac{9}{2} \text{ as well}$$

$$= \frac{1}{2}\left(x^2 + 6x + 9\right) - \frac{7}{2} - \frac{9}{2} \qquad \text{Factor the parentheses, combine like terms}$$

$$= \frac{1}{2}\left(x + 3\right)^2 - 8$$

The vertex form is $f(x) = \frac{1}{2}(x + 3)^2 - 8$. The vertex is $(-3,-8)$ and the axis is $x = -3$.

39. Begin by grouping the first two terms with parentheses, then factoring 2 out of those two terms so that the coefficient of x^2 is 1:

$$f(x) = 2\left(x^2 - 12x\right) + 90 \qquad \text{Find the number needed to complete the square}$$

$$= 2\left(x^2 - 12x + ?\right) + 90 \qquad (-12/2)^2 = 36; \text{ add 36 inside the parentheses}$$

$$= 2\left(x^2 - 12x + 36\right) + 90 + ? \qquad \text{We actually added 2(36); subtract 72 as well}$$

$$= 2\left(x^2 - 12x + 36\right) + 90 - 72 \qquad \text{Factor the parentheses, combine like terms}$$

$$= 2\left(x - 6\right)^2 + 18$$

The vertex form is $f(x) = 2\left(x - 6\right)^2 + 18$. The vertex is $(6,18)$ and the axis is $x = 6$.

41. $x = -\dfrac{b}{2a} = -\dfrac{8}{2(1)} = -4; \quad f(-4) = (-4)^2 + 8(-4) + 8 = 16 - 32 + 8 = -8$

The vertex is $(-4,-8)$. The coefficient of x^2 is positive, so the parabola opens up. The graph is symmetric about its axis, $x = -4$. It decreases until reaching a minimum at $(-4,-8)$, then increases. The range is $[-8,\infty)$.

43. $x = -\dfrac{b}{2a} = -\dfrac{-7}{2(-1)} = -\dfrac{7}{2}$

$$f\left(-\frac{7}{2}\right) = -\left(-\frac{7}{2}\right)^2 - 7\left(-\frac{7}{2}\right) + 4 = -\frac{49}{4} + \frac{49}{2} + 4 = -\frac{49}{4} + \frac{98}{4} + \frac{16}{4} = \frac{65}{4}$$

The vertex is $\left(-\dfrac{7}{2}, \dfrac{65}{4}\right)$. The coefficient of x^2 is negative, so the parabola opens down. The graph is symmetric about its axis, $x = -7/2$. It increases until reaching a maximum at $(-7/2, 65/4)$, then decreases. The range is $(-\infty, 65/4]$.

45. $x = -\dfrac{b}{2a} = -\dfrac{-18}{2(4)} = \dfrac{18}{8} = \dfrac{9}{4}$

$$f\left(\frac{9}{4}\right) = 4\left(\frac{9}{4}\right)^2 - 18\left(\frac{9}{4}\right) + 25 = 4\left(\frac{81}{16}\right) - \frac{162}{4} + 25 = \frac{81}{4} - \frac{162}{4} + \frac{100}{4} = \frac{19}{4}$$

The vertex is $\left(\dfrac{9}{4}, \dfrac{19}{4}\right)$. The coefficient of x^2 is positive, so the parabola opens up. The graph is symmetric about its axis, $x = 9/4$. It decreases until reaching a minimum at $\left(\dfrac{9}{4}, \dfrac{19}{4}\right)$, then increases. The range is $[19/4,\infty)$.

47. $x = -\dfrac{b}{2a} = -\dfrac{50}{2(-10)} = \dfrac{50}{20} = \dfrac{5}{2}$

$f\left(\dfrac{5}{2}\right) = -10\left(\dfrac{5}{2}\right)^2 + 50\left(\dfrac{5}{2}\right) + 12 = -10\left(\dfrac{25}{4}\right) + 125 + 12 = -\dfrac{125}{2} + \dfrac{250}{2} + \dfrac{24}{2} = \dfrac{149}{2}$

The vertex is $\left(\dfrac{5}{2}, \dfrac{149}{2}\right)$.   The coefficient of x^2 is negative, so the parabola opens down. The graph is symmetric about its axis, $x = 5/2$.   It increases until reaching a maximum at $\left(\dfrac{5}{2}, \dfrac{149}{2}\right)$, then decreases.   The range is $(-\infty, 149/2]$.

49. The vertex of the parabola is at (4, 8). Therefore the equation must have form
   $y = a(x - 4)^2 + 8$
Since the x intercept is 6, (6, 0) must satisfy the equation
   $0 = a(6 - 4)^2 + 8$
   $0 = 4a + 8$
   $a = -2$
 The equation is
   $y = -2(x - 4)^2 + 8$
   $y = -2(x^2 - 8x + 16) + 8$
   $y = -2x^2 + 16x - 32 + 8$
   $y = -2x^2 + 16x - 24$

51. The vertex of the parabola is at (-4, 12). Therefore the equation must have form
   $y = a(x + 4)^2 + 12$
Since the y intercept is 4, (0, 4) must satisfy the equation.
   $4 = a(0 + 4)^2 + 12$
   $4 = 16a + 12$
   $-8 = 16a$
   $a = -0.5$
 The equation is
   $y = -0.5(x + 4)^2 + 12$
   $y = -0.5(x^2 + 8x + 16) + 12$
   $y = -0.5x^2 - 4x - 8 + 12$
   $y = -0.5x^2 - 4x + 4$

53. The vertex of the parabola is at (-5, -25). Therefore the equation must have form
   $y = a(x + 5)^2 - 25$
Since the parabola passes through (-2, 20), these coordinates must satisfy the equation.
   $20 = a(-2 + 5)^2 - 25$
   $20 = 9a - 25$
   $45 = 9a$
   $a = 5$
 The equation is $y = 5(x + 5)^2 - 25$
           $y = 5(x^2 + 10x + 25) - 25$
           $y = 5x^2 + 50x + 125 - 25$
           $y = 5x^2 + 50x + 100$

55. The vertex of the parabola is at (1, -4). Therefore the equation must have form
$$y = a(x - 1)^2 - 4$$
Since the parabola passes through (3, 4), these coordinates must satisfy the equation
$$4 = a(3 - 1)^2 - 4$$
$$8 = 4a$$
$$a = 2.$$
The equation is
$$y = 2(x - 1)^2 - 4$$
$$y = 2(x^2 - 2x + 1) - 4$$
$$y = 2x^2 - 4x + 2 - 4$$
$$y = 2x^2 - 4x - 2$$

57. The vertex of the parabola is at (-1, 4). Therefore the equation must have form
$$y = a(x + 1)^2 + 4$$
Since the parabola passes through (1, 2), these coordinates must satisfy the equation.
$$2 = a(1 + 1)^2 + 4$$
$$2 = 4a + 4$$
$$-2 = 4a$$
$$a = -0.5$$
The equation is
$$y = -0.5(x + 1)^2 + 4$$
$$y = -0.5(x^2 + 2x + 1) + 4$$
$$y = -0.5x^2 - x - 0.5 + 4$$
$$y = -0.5x^2 - x + 3.5$$

59. Notice that the table does not provide the exact coordinates of the vertex, so we can't tell for certain what they are. We know that $f(-1)$ and $f(3)$ are both zero, so the axis of symmetry is halfway between $x = -1$ and $x = 3$. In other words, the x-coordinate of the vertex is 1; the equation looks like
$f(x) = a(x - 1)^2 + k$. Plug in $x = -1$:
$$f(-1) = a(-1 - 1)^2 + k = a(-2)^2 + k = 4a + k = 0 \quad \text{(since } (-1, 0) \text{ is on the graph)}$$
$$4a + k = 0$$
$$k = -4a$$
Substitute $-4a$ in for k: $f(x) = a(x - 1)^2 - 4a$
Plug in $x = 0$:
$$f(0) = a(0 - 1)^2 - 4a = a - 4a = -3a = -3 \text{ (since } (0, -3) \text{ is on the graph)}$$
$$-3a = -3$$
$$a = 1$$
$$f(x) = (x - 1)^2 - 4 \quad \text{or} \quad f(x) = x^2 - 2x - 3$$

61. Notice that the table does not provide the exact coordinates of the vertex, so we can't tell for certain what they are. We know that $f(-1)$ and $f(5)$ are equal, so the axis of symmetry is halfway between $x = -1$ and $x = 5$. In other words, the x-coordinate of the vertex is 2; the equation looks like
$$f(x) = a(x - 2)^2 + k$$
Plug in $x = -1$: $f(-1) = a(-1 - 2)^2 + k = a(-3)^2 + k = 9a + k = 0$
$$\text{(since } (-1, 0) \text{ is on the graph)}$$
$$9a + k = 0$$
$$k = -9a$$
Substitute $-9a$ in for k: $f(x) = a(x - 2)^2 - 9a$
Plug in $x = 0$: $f(0) = a(0 - 2)^2 - 9a = a(-2)^2 - 9a = 4a - 9a = -5a = 2.5$
$$\text{(since } (0, 2.5) \text{ is on the graph)}$$
$$-5a = 2.5$$
$$a = -0.5$$
$$f(x) = 0.5(x - 2)^2 + 4.5 \quad \text{or} \quad f(x) = -0.5x^2 + 2x + 2.5$$

63.
$$a(x - h)^2 + k = a\left(x^2 - 2xh + h^2\right) + k$$
$$= ax^2 - 2axh + ah^2 + k$$
$$= ax^2 - (2ah)x + \left(ah^2 + k\right)$$

65. The graphs shown are $f(x) = x^2 + 6x + 1$, $f(x) = x^2 + 1$, and $f(x) = x^2 - 4x + 1$. These correspond to $f(x) = x^2 + kx + 1$ with $k = 6, 0,$ and -4 respectively.

Note that all have the same shape but a different vertex. In fact, all three are translations of the graph $y = x^2$.

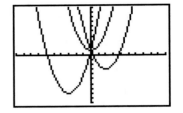

67. The graphs of $f(x) = (x - 1)^2$, $g(x) = (x - 1)^2 + 4$, and $h(x) = (x - 1)^2 - 5$ are shown.

It is clear that $f(x)$ has one x intercept (at $x = 1$), $g(x)$ has no x intercepts and $h(x)$ has two x intercepts.
In general, for $a > 0$, the graph of $f(x) = a(x - h)^2 + k$ can be expected to have no intercepts for $k > 0$, one intercept at $x = h$ for $k = 0$, and two intercepts for $k < 0$.

69. $f(x) = a(x - h)^2 + k$
$f(h + r) = a(h + r - h)^2 + k = ar^2 + k$
$f(h - r) = a(h - r - h)^2 + k = a(-r)^2 + k = ar^2 + k$

For any number r, the inputs $h + r$ and $h - r$ are evenly spaced on either side of h. Since the output is the same for each, we can conclude that the height of the graph is the same for each as well. This tells us that the graph is symmetric about the line $x = h$.

71. The secant line is the line through $(-1, -3)$ and $(3, 5)$. To find its equation we find its slope, then use the point-slope form of the equation of a line.

$$m = \frac{f(x_2) - f(x_1)}{x_2 - x_1} = \frac{5 - (-3)}{3 - (-1)} = \frac{8}{4} = 2$$

$y - f(x_1) = m(x - x_1)$

$y - (-3) = 2[x - (-1)]$

$y + 3 = 2[x + 1]$

$y + 3 = 2x + 2$

$y = 2x - 1$

Graph:

x	$y = 2x - 1$	$y = x^2 - 4$
-2	-5	0
-1	-3	-3
0	-1	-4
1	1	-3
2	3	0
3	5	5

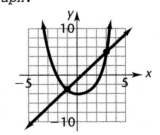

73. (A) The slope of the secant line is given by
$$m = \frac{f(2 + h) - f(2)}{(2 + h) - 2} = \frac{f(2 + h) - f(2)}{h}$$

Since $f(2 + h) = (2 + h)^2 - 3(2 + h) + 5 = 4 + 4h + h^2 - 6 - 3h + 5 = 3 + h + h^2$
and $f(2) = (2)^2 - 3(2) + 5 = 4 - 6 + 5 = 3$
we have

$$m = \frac{f(2 + h) - f(2)}{h} = \frac{(3 + h + h^2) - (3)}{h} = \frac{h + h^2}{h} = \frac{h(1 + h)}{h} = 1 + h$$

(B)

h	slope = 1 + h
1	2
0.1	1.1
0.01	1.01
0.001	1.001

The slope seems to be approaching 1.

75. Let one number = x. Then the other number is $x - 30$. The product is a function of x given by $f(x) = x(x - 30) = x^2 - 30x$. This is a quadratic function with $a > 0$, so it has a minimum value at the vertex of its graph (which is a parabola). Find the vertex:

$$x = -\frac{b}{2a} = -\frac{-30}{2(1)} = 15; \quad f(15) = 15^2 - 30(15) = 225 - 450 = -225$$

The minimum product is -225, when $x = 15$ and $x - 30 = -15$. There is no "highest point" on this parabola, so there's no maximum product.

77. Find the vertex, using $a = -1.2$ and $b = 62.5$:

$$x = -\frac{b}{2a} = -\frac{62.5}{2(-1.2)} \approx 26; \qquad P(26) = -1.2(26)^2 + 62.5(26) - 491 = 322.8$$

The company should hire 26 employees to make a maximum profit of $322,800.

79. (A) Find the first coordinate of the vertex, using $a = -0.19$ and $b = 1.2$:

$$x = -\frac{b}{2a} = -\frac{1.2}{2(-0.19)} = 3.2$$

The maximum box office revenue was three years after 2000, which is 2003.

(B) The function specifies yearly totals for revenue, so the domain should be restricted to whole numbers. The exact vertex occurs at $x = 3.2$, so we needed to round down to 3.

81. (A) Since the four sides needing fencing are x, y, $x + 50$, and y, we have $x + y + (x + 50) + y = 250$. Solving for y, we get $y = 100 - x$.

(Since both x and $100 - x$ must be nonnegative, the domain of $A(x)$ is $0 \le x \le 100$.)

Therefore the area
$A(x) = (x + 50)y = (x + 50)(100 - x) = -x^2 + 50x + 5000$.

(B) This is a quadratic function with $a < 0$, so it has a maximum value at the vertex:

$$x = -\frac{b}{2a} = -\frac{50}{2(-1)} = 25; \quad A(25) = -(25)^2 + 50(25) + 5,000 = -625 + 1,250 + 5,000 = 5,625$$

The maximum area is 5,625 square feet when $x = 25$.

(C) When $x = 25$, $y = 100 - 25 = 75$. The dimensions of the corral are then $x + 50$ by y, or 75 ft by 75 ft.

83. According to Example 10, the function describing the height of the sandbag is $h(t) = 10,000 - 16t^2$ (since the initial height is 10,000 feet). We want to know when it reaches ground level, so plug in zero for $h(t)$, then solve for t.

$$0 = 10,000 - 16t^2$$
$$16t^2 = 10,000$$
$$t^2 = 625$$
$$t = 25$$

It hits the ground 25 seconds after it's dropped.

85. According to Example 10, the function describing the height of the diver is $h(t) = h_0 - 16t^2$, where h_0 is the initial height in feet. (This initial height is the height of the cliff.) We know that $h(2.5) = 0$ since it takes 2.5 seconds to reach the water; we plug in 2.5 for t and 0 for $h(t)$, which allows us to solve for h_0.

$$0 = h_0 - 16(2.5)^2$$
$$0 = h_0 - 100$$
$$100 = h_0$$

The cliff is 100 feet high.

87. (A) Since $d(t)$ is a quadratic function with maximum value 484 when $t = 5.5$, an equation for $d(t)$ must be of the form

$$d(t) = a(t - 5.5)^2 + 484$$

Since $d(0) = 0$,

$$0 = d(0) = a(0 - 5.5)^2 + 484$$
$$0 = 30.25a + 484$$
$$a = -16$$

So $d(t) = -16(t - 5.5)^2 + 484$
$$= -16(t^2 - 11t + 30.25) + 484$$
$$= -16t^2 + 176t - 484 + 484$$
$$= -16t^2 + 176t$$

Since the graph of d must be symmetric with respect to $t = 5.5$, and $d(0) = 0$, $d(11)$ must also equal 0. The distance above the ground will be nonnegative only for values of t between 0 and 11, so the domain of the function is $0 \leq t \leq 11$.

(B) Solve $250 = -16t^2 + 176t$ by graphing $Y_1 = 250$ and $Y_2 = -16x^2 + 176x$ and applying the INTERSECT command.

From the graphs, $t = 1.68$ sec and $t = 9.32$ sec, to two decimal places.

89. (A) If coordinates are chosen with origin at the center of the base, the parabola is the graph of a quadratic function $h(x)$ with maximum value 14 when $x = 0$. The equation must be of form

$$h(x) = ax^2 + 14$$

The parabola is 20 feet wide and symmetrical, so the point where it hits the ground must be 10 feet from the center. This tells us that $h(10) = 0$. We can use this information to find a.

$$0 = h(10) = a(10)^2 + 14$$
$$0 = 100a + 14$$
$$a = -0.14$$
$$h(x) = -0.14x^2 + 14 \quad -10 \leq x \leq 10$$

(B) Suppose the truck were to drive so as to maximize its clearance, that is, in the center of the roadway. Then half its width, or 4 ft, would extend to each side. But if $x = 4$, $h(x) = -0.14(4)^2 + 14 = 11.76$ ft. The arch is only 11.76 feet high, but the truck is 12 feet high. The truck cannot pass through the arch.

(C) From part (B), if $x = 4$, $h(x) = 11.76$ ft is the height of the tallest truck.

(D)   Find x so that $h(x) = 12$.   Solve the equation

$$12 = 0.14x^2 + 14$$

$$-2 = -0.14x^2$$

$$x = \sqrt{\frac{2}{0.14}} = 3.78$$ (The negative solution doesn't make sense.)

The width of the truck is at most $2x = 2(3.78) = 7.56$ feet

91. (A) The independent variable (x) is number of flashlights, so we enter the numbers from the demand column as L_1. Note that demand is in thousands so 45,800 is entered as 45.8. The dependent variable (p) is price, so we enter the prices as L_2.

The linear regression function on a graphing calculator tells us that the linear function $a(x) = -0.129x + 9.27$ describes the price in terms of demand.

(B) To find the revenue, we multiply price times number of flashlights sold: this is $r(x) = p \cdot x = (-0.129x + 9.27)x = -0.129x^2 + 9.27x$. We use the maximum feature to find the maximum:

```
WINDOW
 Xmin=20
 Xmax=50
 Xscl=5
 Ymin=0
 Ymax=200
 Yscl=100
 Xres=1
```

```
Maximum
X=35.914728  Y=166.39294
```

The maximum point is (35.9, 166), so maximum revenue occurs when $x = 35.9$. To find the associated price, we plug back in to our price equation:
    $d(35.9) = -0.129(35.9) + 9.27 = 4.64$
The price that maximizes revenue is $4.64.

Section 2-4

1. Answers will vary.   **3.** Answers will vary.

5. Answers will vary.   **7.** Imaginary, complex

9. Imaginary, pure imaginary, complex   **11.** Real, complex

13. $i^2 = \left(\sqrt{-1}\right)^2 = -1$ **15.** $(4i)^2 = 4^2 \cdot i^2 = 16(-1) = -16$

17. $\left(-i\sqrt{3}\right)^2 = (-1)^2 \cdot i^2 \cdot \sqrt{3}^2 = 1(-1)(3) = -3$ **19.** $(4i)(6i) = 24i^2 = 24(-1) = -24$

21. $(2 + 4i) + (5 + i) = 2 + 4i + 5 + i = 7 + 5i$

23. $(-2 + 6i) + (7 - 3i) = -2 + 6i + 7 - 3i = 5 + 3i$

25. $(6 + 7i) - (4 + 3i) = 6 + 7i - 4 - 3i = 2 + 4i$

27. $(3 + 5i) - (-2 - 4i) = 3 + 5i + 2 + 4i = 5 + 9i$

29. $(4 - 5i) + 2i = 4 - 5i + 2i = 4 - 3i$

31. $-3i(2 - 4i) = -6i + 12i^2 = -6i + 12(-1) = -12 - 6i$

33. $(3 + 3i)(2 - 3i) = 6 - 9i + 6i - 9i^2 = 6 - 3i - 9(-1) = 6 - 3i + 9 = 15 - 3i$

35. $(2 - 3i)(7 - 6i) = 14 - 12i - 21i + 18i^2 = 14 - 33i - 18 = -4 - 33i$

37. $(7 + 4i)(7 - 4i) = 49 - 28i + 28i - 16i^2 = 49 + 16 = 65$

39. $(4 - 3i)^2 = (4 - 3i)(4 - 3i) = 16 - 24i + 9i^2 = 16 - 24i - 9 = 7 - 24i$

41. $\dfrac{1}{2 + i} = \dfrac{1}{(2 + i)}\dfrac{(2 - i)}{(2 - i)} = \dfrac{2 - i}{4 - i^2} = \dfrac{2 - i}{4 + 1} = \dfrac{2 - i}{5} = \dfrac{2}{5} - \dfrac{1}{5}i$

43. $\dfrac{3 + i}{2 - 3i} = \dfrac{(3 + i)}{(2 - 3i)}\dfrac{(2 + 3i)}{(2 + 3i)} = \dfrac{6 + 11i + 3i^2}{4 - 9i^2} = \dfrac{6 + 11i - 3}{4 + 9} = \dfrac{3 + 11i}{13} = \dfrac{3}{13} + \dfrac{11}{13}i$

45. $\dfrac{13 + i}{2 - i} = \dfrac{(13 + i)}{(2 - i)}\dfrac{(2 + i)}{(2 + i)} = \dfrac{26 + 15i + i^2}{4 - i^2} = \dfrac{26 + 15i - 1}{4 + 1} = \dfrac{25 + 15i}{5} = 5 + 3i$

47. $\sqrt{2}\,\sqrt{8} = \sqrt{16} = 4$ **49.** $\sqrt{2}\,\sqrt{-8} = \sqrt{2}\cdot i\sqrt{8} = i\sqrt{16} = 4i$

51. $\sqrt{-2}\,\sqrt{8} = i\sqrt{2}\,\sqrt{8} = i\sqrt{16} = 4i$ **53.** $\sqrt{-2}\,\sqrt{-8} = i\sqrt{2}\cdot i\sqrt{8} = i^2\sqrt{16} = (-1)\,\sqrt{16} = -4$

55. $(2 - \sqrt{-4}) + (5 - \sqrt{-9}) = (2 - i\sqrt{4}) + (5 - i\sqrt{9}) = 2 - 2i + 5 - 3i = 7 - 5i$

57. $(9 - \sqrt{-9}) - (12 - \sqrt{-25}) = (9 - i\sqrt{9}) - (12 - i\sqrt{25})$

$$= (9 - 3i) - (12 - 5i) = 9 - 3i - 12 + 5i = -3 + 2i$$

59. $(3 - \sqrt{-4})(-2 + \sqrt{-49}) = (3 - i\sqrt{4})(-2 + i\sqrt{49})$

$$= (3 - 2i)(-2 + 7i) = -6 + 25i - 14i^2 = -6 + 25i + 14$$

$$= 8 + 25i$$

61. $\dfrac{5 - \sqrt{-4}}{7} = \dfrac{5 - i\sqrt{4}}{7} = \dfrac{5 - 2i}{7} = \dfrac{5}{7} - \dfrac{2}{7}i$

63. $\dfrac{1}{2 - \sqrt{-9}} = \dfrac{1}{2 - i\sqrt{9}} = \dfrac{1}{2 - 3i} = \dfrac{1}{(2 - 3i)}\dfrac{(2 + 3i)}{(2 + 3i)} = \dfrac{2 + 3i}{4 - 9i^2}$

$$= \dfrac{2 + 3i}{4 + 9} = \dfrac{2 + 3i}{13} = \dfrac{2}{13} + \dfrac{3}{13}i$$

65. $\dfrac{2}{5i} = \dfrac{2}{5i}\cdot\dfrac{i}{i} = \dfrac{2i}{5i^2} = \dfrac{2i}{-5} = -\dfrac{2}{5}i$

67. $\dfrac{1 + 3i}{2i} = \dfrac{(1 + 3i)}{(2i)}\cdot\dfrac{i}{i} = \dfrac{i + 3i^2}{2i^2} = \dfrac{i - 3}{-2} = \dfrac{3}{2} - \dfrac{1}{2}i$

69. $(2 - 3i)^2 - 2(2 - 3i) + 9 = 4 - 12i + 9i^2 - 4 + 6i + 9$

$$= 4 - 12i - 9 - 4 + 6i + 9 = -6i$$

71. $f(x) = x^2 - 2x + 2$

(A) $f(1 + i) = (1 + i)^2 - 2(1 + i) + 2$ $f(1 - i) = (1 - i)^2 - 2(1 - i) + 2$

$\qquad\qquad = 1 + 2i + i^2 - 2 - 2i + 2$ $= 1 - 2i + i^2 - 2 + 2i + 2$

$\qquad\qquad = 1 + 2i - 1 - 2 - 2i + 2$ $= 1 - 2i - 1 - 2 + 2i + 2$

$\qquad\qquad = 0$ $= 0$

$1 + i$ and $1 - i$ are both zeros of $f(x)$.

(B) There are no real solutions of $x^2 - 2x + 2 = 0$. The only solutions are the complex numbers $1 + i$ and $1 - i$. Therefore there are no real zeros of f. In fact, since $f(x) = (x - 1)^2 + 1$, the graph is a parabola with vertex at $(1, 1)$, and 1 is a minimum value for f. There can be no x intercepts for the graph of f.

73. $i^{18} = \underbrace{i^{16}}\cdot i^2 = (i^4)^4\cdot i^2 = 1^4(-1) = -1$

largest integer in 18 exactly divisible by 4

$i^{32} = (i^4)^8 = 1^8 = 1$

$i^{67} = \underbrace{i^{64}}\cdot i^3 = (i^4)^{16}\cdot i^2\cdot i = 1^{16}(-1)i = -i$

largest integer in 67 exactly divisible by 4

75. According to the definition of equality for complex numbers

$(2x - 1) + (3y + 2)i = 5 - 4i = 5 + (-4)i$

if and only if

$2x - 1 = 5$ and $3y + 2 = -4$

So $2x = 6$ and $3y = -6$

$x = 3 \quad y = -2$

77.
$$\frac{(1 + x) + (y - 2)i}{1 + i} = 2 - i \qquad \text{Multiply both sides by } 1 + i$$

$(1 + i)\cdot\dfrac{(1 + x) + (y - 2)i}{1 + i} = (1 + i)(2 - i)$ Simplify

$(1 + x) + (y - 2)i = 2 - i + 2i - i^2$

$(1 + x) + (y - 2)i = 2 + i - (-1)$

$(1 + x) + (y - 2)i = 2 + i + 1$

$(1 + x) + (y - 2)i = 3 + i$

According to the definition of equality for complex numbers

$(1 + x) + (y - 2)i = 3 + i$

if and only if

$1 + x = 3$ and $y - 2 = 1$

So $x = 2$ and $y = 3$.

79. $(2 + i)z + i = 4i$

$(2 + i)z = 3i$

$z = \dfrac{3i}{2 + i}$

$z = \dfrac{3i}{(2 + i)}\dfrac{(2 - i)}{(2 - i)} = \dfrac{6i - 3i^2}{4 - i^2} = \dfrac{6i - 3(-1)}{4 + 1} = \dfrac{3 + 6i}{5} = 0.6 + 1.2i$

81. $3iz + (2 - 4i) = (1 + 2i)z - 3i$

$3iz - (1 + 2i)z = -3i - (2 - 4i)$

$3iz - z - 2iz = -3i - 2 + 4i$

$iz - z = -2 + i$

$(-1 + i)z = -2 + i$

$z = \dfrac{-2 + i}{-1 + i}$

$z = \dfrac{(-2 + i)}{(-1 + i)}\dfrac{(-1 - i)}{(-1 - i)} = \dfrac{2 + 2i - i - i^2}{1 - i^2} = \dfrac{2 + i + 1}{1 + 1} = \dfrac{3 + i}{2}$

$= 1.5 + 0.5i$

83. b is a square root of x if $b^2 = x$.

Since $(2 - i)^2 = 2^2 - 2(2)i + i^2 = 4 - 4i - 1 = 3 - 4i$, $2 - i$ is a square root of $3 - 4i$.

Similarly, since $(-2 + i)^2 = (-2)^2 + 2(-2)i + i^2 = 4 - 4i - 1 = 3 - 4i$, $-2 + i$ is also a square root of $3 - 4i$.

85. As noted in the text, some of the properties of radicals that are true for real numbers are not true for complex numbers. In particular, for positive real numbers a and b, $\sqrt{-a}\sqrt{-b} \neq \sqrt{(-a)(-b)}$, and so it is incorrect to write $\sqrt{-1}\sqrt{-1} = \sqrt{(-1)(-1)}$.

87. $(a + bi)(a - bi) = a^2 - abi + abi - b^2 i^2 = a^2 - b^2(-1) = a^2 + b^2$. This is a real number.

89. $(a + bi) + (c + di) = a + bi + c + di = a + c + bi + di = (a + c) + (b + d)i$

91. $(a + bi)(a - bi) = a^2 - b^2 i^2 = a^2 + b^2$ or $(a^2 + b^2) + 0i$

93. $(a + bi)(c + di) = ac + adi + bci + bdi^2 = ac + (ad + bc)i - bd$
$$= (ac - bd) + (ad + bc)i$$

95. $i^{4k} = (i^4)^k = (i^2 \cdot i^2)^k = [(-1)(-1)]^k = 1^k = 1$

97. $S_1 = i$

$S_2 = i + i^2 = -1 + i$

$S_3 = i + i^2 + i^3 = -1 + i + (-i) = -1$

$S_4 = i + i^2 + i^3 + i^4 = -1 + 1 = 0$

Then $S_5 = S_4 + i^5 = S_4 + i = S_4 + S_1 = S_1$

$S_6 = S_5 + i^6 = S_1 + i^2 = S_2$

$S_7 = S_6 + i^7 = S_2 + i^3 = S_3$

$S_8 = S_7 + i^8 = S_3 + i^4 = S_4$

Generalizing, the only possible values for S_n are $S_1 = i$, $S_2 = -1 + i$, $S_3 = -1$, and $S_4 = 0$.

99. 1. Definition of addition

2. Commutative property for addition of real numbers.

3. Definition of addition (read from right to left).

Section 2-5

1. Answers will vary.

3. Answers will vary.

5. Answers will vary.

7. $(x - 8)(2x + 3) = 0$

$x - 8 = 0$ or $2x + 3 = 0$

$x = 8$ or $2x = -3$

$$x = -\frac{3}{2}$$

9. $x^2 - 3x - 10 = 0$

$(x - 5)(x + 2) = 0$

$x - 5 = 0$ or $x + 2 = 0$

$x = 5$ $x = -2$

11.
$$4u^2 = 8u$$
$$4u^2 - 8u = 0$$
$$4u(u - 2) = 0$$
$$4u = 0 \text{ or } u - 2 = 0$$
$$u = 0 \qquad u = 2$$

13.
$$9y^2 = 12y - 4$$
$$9y^2 - 12y + 4 = 0$$
$$(3y - 2)^2 = 0$$
$$3y - 2 = 0$$
$$3y = 2$$
$$y = \frac{2}{3} \text{ (double root)}$$

15.
$$11x = 2x^2 + 12$$
$$0 = 2x^2 - 11x + 12$$
$$2x^2 - 11x + 12 = 0$$
$$(2x - 3)(x - 4) = 0$$
$$2x - 3 = 0 \quad \text{ or } x - 4 = 0$$
$$2x = 3 \quad \text{ or } x = 4$$
$$x = \frac{3}{2}$$

17. $x^2 - 6x - 3 = 0$

$$x^2 - 6x = 3 \qquad\qquad \left(-\frac{6}{2}\right)^2 = 9$$
$$x^2 - 6x + 9 = 3 + 9 \qquad \text{(Add 9 to each side)}$$
$$(x - 3)^2 = 12 \qquad \text{(Factor the left side)}$$
$$x - 3 = \pm\sqrt{12}$$
$$x = 3 \pm 2\sqrt{3} \qquad (\sqrt{12} = \sqrt{4}\sqrt{3} = 2\sqrt{3})$$

19. $t^2 - 4t + 8 = 0$

$$t^2 - 4t = -8 \qquad\qquad \left(-\frac{4}{2}\right)^2 = 4$$
$$t^2 - 4t + 4 = -8 + 4 \qquad \text{(Add 4 to each side)}$$
$$(t - 2)^2 = -4 \qquad \text{(Factor the left side)}$$
$$t - 2 = \pm\sqrt{-4} \qquad (\sqrt{-4} = i\sqrt{4} = 2i)$$
$$t - 2 = \pm 2i$$
$$t = 2 \pm 2i$$

21. $m^2 + 2m + 9 = 0$

$$m^2 + 2m = -9 \qquad\qquad \left(\frac{2}{2}\right)^2 = 1$$
$$m^2 + 2m + 1 = -9 + 1 \qquad \text{(Add 1 to each side)}$$
$$(m + 1)^2 = -8 \qquad \text{(Factor the left side)}$$
$$m + 1 = \pm\sqrt{-8}$$
$$m + 1 = \pm 2\sqrt{2}\,i \qquad \sqrt{-8} = i\sqrt{8} = i\sqrt{4}\sqrt{2} = 2\sqrt{2}\,i$$
$$m = -1 \pm 2\sqrt{2}\,i$$

23. $2d^2 + 5d - 25 = 0$ 　　　　(Divide both sides by 2)

$$d^2 + \frac{5}{2}d - \frac{25}{2} = 0 \qquad \left(\text{Add } \frac{25}{2} \text{ to each side}\right)$$

$$d^2 + \frac{5}{2}d = \frac{25}{2} \qquad \left(\frac{1}{2} \cdot \frac{5}{2}\right)^2 = \left(\frac{5}{4}\right)^2 = \frac{25}{16}$$

$$d^2 + \frac{5}{2}d + \frac{25}{16} = \frac{25}{2} + \frac{25}{16} \qquad \left(\text{Add } \frac{25}{16} \text{ to each side}\right)$$

$$\left(d + \frac{5}{4}\right)^2 = \frac{25}{2} + \frac{25}{16} \qquad \text{(Factor the left side)}$$

$$\left(d + \frac{5}{4}\right)^2 = \frac{200}{16} + \frac{25}{16} = \frac{225}{16}$$

$$d + \frac{5}{4} = \pm\sqrt{\frac{225}{16}}$$

$$d + \frac{5}{4} = \pm\frac{15}{4}$$

$$d = -\frac{5}{4} \pm \frac{15}{4}$$

$$d = \frac{-5 + 15}{4} \quad \text{or} \quad d = \frac{-5 - 15}{4}$$

$$d = \frac{10}{4} = \frac{5}{2} \quad \text{or} \quad d = -\frac{20}{4} = -5$$

25. $2v^2 - 2v + 1 = 0$ (Divide both sides by 2)

$\quad v^2 - v + \frac{1}{2} = 0$ (Subtract ½ from each side)

$$v^2 - v = -\frac{1}{2} \qquad \left(-\frac{1}{2}\right)^2 = \frac{1}{4}$$

$$v^2 - v + \frac{1}{4} = -\frac{1}{2} + \frac{1}{4} \quad \text{(Add } \frac{1}{4} \text{ to each side)}$$

$$\left(v - \frac{1}{2}\right)^2 = -\frac{1}{2} + \frac{1}{4} \quad \text{(Factor the left side)}$$

$$\left(v - \frac{1}{2}\right)^2 = -\frac{2}{4} + \frac{1}{4} = -\frac{1}{4}$$

$$v - \frac{1}{2} = \pm\sqrt{-\frac{1}{4}}$$

$$v - \frac{1}{2} = \pm\frac{1}{2}i \qquad \sqrt{-\frac{1}{4}} = i\sqrt{\frac{1}{4}} = \frac{1}{2}i$$

$$v = \frac{1}{2} \pm \frac{1}{2}i$$

27. $4y^2 + 3y + 9 = 0$ (Divide both sides by 4)

$\quad y^2 + \frac{3}{4}y + \frac{9}{4} = 0$ (Subtract $\frac{9}{4}$ from each side)

$$y^2 + \frac{3}{4}y = -\frac{9}{4} \qquad \left(\frac{1}{2}\cdot\frac{3}{4}\right)^2 = \left(\frac{3}{8}\right)^2 = \frac{9}{64}$$

$$y^2 + \frac{3}{4}y + \frac{9}{64} = -\frac{9}{4} + \frac{9}{64} \quad \text{(Add } \frac{9}{64} \text{ to each side)}$$

$$\left(y + \frac{3}{8}\right)^2 = -\frac{9}{4} + \frac{9}{64} \quad \text{(Factor the left side)}$$

$$\left(y + \frac{3}{8}\right)^2 = -\frac{144}{64} + \frac{9}{64} = -\frac{135}{64}$$

$$y + \frac{3}{8} = \pm\sqrt{-\frac{135}{64}}$$

$$y + \frac{3}{8} = \pm\frac{3i\sqrt{15}}{8} \qquad \sqrt{-\frac{135}{64}} = i\frac{\sqrt{135}}{\sqrt{64}} = \frac{i\sqrt{9}\sqrt{15}}{8} = \frac{3i\sqrt{15}}{8}$$

$$y = -\frac{3}{8} \pm \frac{3\sqrt{15}}{8}i$$

29. $x^2 - 10x - 3 = 0$

$$x = \frac{-b \pm \sqrt{b^2 - 4ac}}{2a} \qquad a = 1, \ b = -10, \ c = -3$$

$$x = \frac{-(-10) \pm \sqrt{(-10)^2 - 4(1)(-3)}}{2(1)}$$

$$x = \frac{10 \pm \sqrt{112}}{2} = \frac{10 \pm 4\sqrt{7}}{2}$$

$$= 5 \pm 2\sqrt{7}$$

> **Common Error:**
> It is incorrect to "cancel" this way:
> $$\frac{\cancel{10} \pm \sqrt{112}}{\cancel{2}} \ne 5 \pm \sqrt{112}$$

31. $\quad x^2 + 8 = 4x$
$x^2 - 4x + 8 = 0$

$$x = \frac{-b \pm \sqrt{b^2 - 4ac}}{2a} \qquad a = 1, \ b = -4, \ c = 8$$

$$x = \frac{-(-4) \pm \sqrt{(-4)^2 - 4(1)(8)}}{2(1)} = \frac{4 \pm \sqrt{-16}}{2}$$

$$x = \frac{4 \pm i\sqrt{16}}{2}$$

$$x = \frac{4 \pm 4i}{2}$$

$$x = 2 \pm 2i$$

33. $\quad 2x^2 + 1 = 4x$
$2x^2 - 4x + 1 = 0$

$$x = \frac{-b \pm \sqrt{b^2 - 4ac}}{2a} \qquad a = 2, \ b = -4, \ c = 1$$

$$= \frac{-(-4) \pm \sqrt{(-4)^2 - 4(2)(1)}}{2(2)}$$

$$x = \frac{4 \pm \sqrt{8}}{4}$$

$$x = \frac{4 \pm 2\sqrt{2}}{4}$$

$$x = \frac{2 \pm \sqrt{2}}{2}$$

> **Common Errors:**
> $$\frac{2 \pm \sqrt{2}}{2} \ne \pm\sqrt{2}$$
> $$\ne 1 \pm \sqrt{2}$$
> These involve incorrect "cancelling".

35. $\quad 5x^2 + 2 = 2x$
$5x^2 - 2x + 2 = 0$

$$x = \frac{-b \pm \sqrt{b^2 - 4ac}}{2a} \qquad a = 5, \ b = -2, \ c = 2$$

$$x = \frac{-(-2) \pm \sqrt{(-2)^2 - 4(5)(2)}}{2(5)}$$

$$x = \frac{2 \pm \sqrt{-36}}{10}$$

$$x = \frac{2 \pm 6i}{10}$$

$$x = \frac{1}{5} \pm \frac{3}{5}i$$

37. $4\left(x^2 + 2x\right) - 4 = 0$

$$4x^2 + 8x - 4 = 0$$

$$x = \frac{-b \pm \sqrt{b^2 - 4ac}}{2a} \qquad a = 4,\ b = 8,\ c = -4$$

$$= \frac{-8 \pm \sqrt{8^2 - 4(4)(-4)}}{2(4)}$$

$$x = \frac{-8 \pm \sqrt{128}}{8}$$

$$= \frac{-8 \pm 8\sqrt{2}}{8}$$

$$= -1 \pm \sqrt{2}$$

(Note: If you begin by dividing both sides by 4, the calculations are simpler)

39. $2.4x^2 + 6.4x - 4.3 = 0 \qquad a = 2.4,\ b = 6.4,\ c = -4.3$
$b^2 - 4ac = 6.4^2 - 4(2.4)(-4.3) = 82.24$
Since the discriminant is positive, there are two real zeros.

41. $6.5x^2 - 7.4x + 3.4 = 0 \qquad a = 6.5,\ b = -7.4,\ c = 3.4$
$b^2 - 4ac = (-7.4)^2 - 4(6.5)(3.4) = -33.64$
Since the discriminant is negative, there are two imaginary zeros.

43. $0.3x^2 + 3.6x + 10.8 = 0 \qquad a = 0.3,\ b = 3.6,\ c = 10.8$
$b^2 - 4ac = 3.6^2 - 4(0.3)(10.8) = 0$
Since the discriminant is zero, there is one real zero.

45. The graph is below:

It looks like there's only one zero, but to be sure we find the minimum value:

The minimum is at height zero so there is one real zero.

47. The graph clearly intersects the x-axis twice, so there are two real zeros.

49. It appears that the graph intersects the *x*-axis once:

To be sure, we find the minimum value:

And it's a good thing we did! The minimum height is positive so the graph does not intersect the *x*-axis and there are two imaginary zeros.

51. $x^2 - 6x - 3 = 0$ The left side doesn't factor, so use the quadratic formula.

$$x = \frac{-b \pm \sqrt{b^2 - 4ac}}{2a} \qquad a = 1, \ b = -6, \ c = -3$$

$$= \frac{6 \pm \sqrt{(-6)^2 - 4(1)(-3)}}{2(1)}$$

$$= \frac{6 \pm \sqrt{48}}{2}$$

$$= \frac{6 \pm 4\sqrt{3}}{2}$$

$$= 3 \pm 2\sqrt{3}$$

Check: $3 + 2\sqrt{3} \approx 6.46$
         $3 - 2\sqrt{3} \approx -0.46$

Graph $y = x^2 - 6x - 3$ and find the intercepts using the ZERO command.

53. $2y^2 - 6y + 3 = 0$ The left side doesn't factor, so use the quadratic formula.

$$y = \frac{-b \pm \sqrt{b^2 - 4ac}}{2a} \qquad a = 2, \ b = -6, \ c = 3$$

$$= \frac{6 \pm \sqrt{(-6)^2 - 4(2)(3)}}{2(2)}$$

$$= \frac{6 \pm \sqrt{12}}{4}$$

$$= \frac{6 \pm 2\sqrt{3}}{4}$$

$$= \frac{3}{2} \pm \frac{1}{2}\sqrt{3}$$

Check: $\dfrac{3}{2} + \dfrac{1}{2}\sqrt{3} \approx 2.366$

$\dfrac{3}{2} - \dfrac{1}{2}\sqrt{3} \approx 0.634$

Graph $y = 2x^2 - 6x + 3$ and find the intercepts using the ZERO command.

55. $3x^2 - 2x - 2 = 0$ The left side doesn't factor, so use the quadratic formula.

$x = \dfrac{-b \pm \sqrt{b^2 - 4ac}}{2a}$ $a = 3,\ b = -2,\ c = -2$

$= \dfrac{2 \pm \sqrt{(-2)^2 - 4(3)(-2)}}{2(3)}$

$= \dfrac{2 \pm \sqrt{28}}{6}$

$= \dfrac{2 \pm 2\sqrt{7}}{6}$

$= \dfrac{1 \pm \sqrt{7}}{3}$

Check: $\dfrac{1 + \sqrt{7}}{3} \approx 1.215$

$\dfrac{1 - \sqrt{7}}{3} \approx -0.549$

Graph $y = 3x^2 - 2x - 2$ and find the intercepts using the ZERO command.

57. $12x^2 + 7x = 10$

$12x^2 + 7x - 10 = 0$

$(4x + 5)(3x - 2) = 0$ Polynomial is
                          factorable.

$4x + 5 = 0$ or $3x - 2 = 0$

$4x = -5$ $3x = 2$

$x = -\dfrac{5}{4}$ $x = \dfrac{2}{3}$

$-\dfrac{5}{4} = -1.25$ $\dfrac{2}{3} \approx 0.67$

Graph $y = 12x^2 + 7x - 10$ and find the intercepts using the ZERO command.

59.
$$x^2 = 3x + 1$$
$$x^2 - 3x - 1 = 0 \quad \text{Polynomial is not factorable, use quadratic formula.}$$

$$x = \frac{-b \pm \sqrt{b^2 - 4ac}}{2a} \qquad a = 1,\ b = -3,\ c = -1$$

$$x = \frac{-(-3) \pm \sqrt{(-3)^2 - 4(1)(-1)}}{2(1)}$$

$$x = \frac{3 \pm \sqrt{13}}{2}$$

Check: $\dfrac{3 + \sqrt{13}}{2} \approx 3.30$

$\dfrac{3 - \sqrt{13}}{2} \approx -0.30$

Graph $y = x^2 - 3x - 1$ and find the intercepts using the ZERO command.

61.
$$s = \frac{1}{2} gt^2$$
$$\frac{1}{2} gt^2 = s$$
$$gt^2 = 2s$$
$$t^2 = \frac{2s}{g}$$
$$t = \sqrt{\frac{2s}{g}}$$

63.
$$P = EI - RI^2$$
$$RI^2 - EI + P = 0$$

$$I = \frac{-b \pm \sqrt{b^2 - 4ac}}{2a} \qquad a = R,\ b = -E,\ c = P$$

$$I = \frac{-(-E) \pm \sqrt{(-E)^2 - 4(R)(P)}}{2(R)}$$

$$I = \frac{E + \sqrt{E^2 - 4RP}}{2R} \quad \text{(positive square root)}$$

65. The quadratic formula is a good choice:

$$x^2 - \sqrt{7}\, x + 2 = 0$$

$$x = \frac{-b \pm \sqrt{b^2 - 4ac}}{2a} \qquad a = 1,\ b = -\sqrt{7},\ c = 2$$

$$x = \frac{-(-\sqrt{7}) \pm \sqrt{(-\sqrt{7})^2 - 4(1)(2)}}{2(1)}$$

$$x = \frac{\sqrt{7} \pm \sqrt{7 - 8}}{2} = \frac{\sqrt{7} \pm \sqrt{-1}}{2} = \frac{\sqrt{7} \pm i}{2}$$

The graph can be used to see that there are two imaginary zeros but not to confirm the exact answer we got.

67. The quadratic formula is a good choice:

$$x^2 - 2\sqrt{3}\,x + 3 = 0$$

$$x = \frac{-b \pm \sqrt{b^2 - 4ac}}{2a} \quad a = 1, \ b = -2\sqrt{3}, \ c = 3$$

$$x = \frac{-(-2\sqrt{3}) \pm \sqrt{(-2\sqrt{3})^2 - 4(1)(3)}}{2(1)}$$

$$x = \frac{2\sqrt{3} \pm \sqrt{12 - 12}}{2} = \frac{2\sqrt{3} \pm 0}{2} = \sqrt{3}$$

We check graphically by graphing the function $y = x^2 - 2\sqrt{3}\,x + 3$ and using the ZERO command:

The only zero is at $x = 1.7320515$ which is easily confirmed to be $\sqrt{3}$.

69. The quadratic formula is a good choice:

$$x^2 + \sqrt{3}\,x - 4 = 0$$

$$x = \frac{-b \pm \sqrt{b^2 - 4ac}}{2a} \quad a = 1, \ b = \sqrt{3}, \ c = -4$$

$$x = \frac{-\sqrt{3} \pm \sqrt{(\sqrt{3})^2 - 4(1)(-4)}}{2(1)}$$

$$x = \frac{-\sqrt{3} \pm \sqrt{3 + 16}}{2}$$

$$x = \frac{-\sqrt{3} \pm \sqrt{19}}{2}$$

We check graphically by graphing the function $y = x^2 + \sqrt{3}\,x - 4$ and using the ZERO command.

The two zeros are at $x = -3.045$ and $x = 1.313$.

$$\frac{-\sqrt{3} + \sqrt{19}}{2} \approx 1.313; \quad \frac{-\sqrt{3} - \sqrt{19}}{2} \approx -3.045$$

71. Begin by multiplying both sides by the least common denominator, x^2.

$$x^2 \cdot 1 + x^2 \cdot \frac{9}{x^2} = x^2 \cdot \frac{5}{x}$$

$$x^2 + 9 = 5x$$

$$x^2 - 5x + 9 = 0$$

$$x = \frac{-b \pm \sqrt{b^2 - 4ac}}{2a} \quad a = 1, \ b = -5, \ c = 9$$

$$x = \frac{-(-5) \pm \sqrt{(-5)^2 - 4(1)(9)}}{2(1)} = \frac{5 \pm \sqrt{25 - 36}}{2} = \frac{5 \pm \sqrt{-11}}{2} = \frac{5 \pm i\sqrt{11}}{2}$$

The graph can be used to see that there are two imaginary zeros but not to confirm the exact answer we got.

73. Begin by multiplying both sides by the least common denominator, x^2.

$$x^2 \cdot 1 + x^2 \cdot \frac{9}{x^2} = x^2 \cdot \frac{6}{x}$$

$$x^2 + 9 = 6x$$

$$x^2 - 6x + 9 = 0 \quad \text{(The left side factors)}$$

$$(x - 3)^2 = 0$$

$$x - 3 = 0$$

$$x = 3$$

We check graphically by graphing the function $y = 1 + \frac{9}{x^2} - \frac{6}{x}$ and using the ZERO command:

The only solution is $x = 3$.

75. Begin by multiplying both sides by the least common denominator, x^2.

$$x^2 \cdot 1 + x^2 \cdot \frac{9}{x^2} = x^2 \cdot \frac{7}{x}$$

$$x^2 + 9 = 7x$$

$$x^2 - 7x + 9 = 0 \quad \text{(Doesn't factor)}$$

$$x = \frac{-b \pm \sqrt{b^2 - 4ac}}{2a} \qquad a = 1, \ b = -7, \ c = 9$$

$$x = \frac{-(-7) \pm \sqrt{(-7)^2 - 4(1)(9)}}{2(1)} = \frac{7 \pm \sqrt{49 - 36}}{2} = \frac{7 \pm \sqrt{13}}{2}$$

We check graphically by graphing the functions $y_1 = 1 + \frac{9}{x^2}$ and $y_2 = \frac{7}{x}$ then using the INTERSECT command.

The two solutions are $x = 1.697$ and $x = 5.303$.

$$\frac{7 + \sqrt{13}}{2} = 5.303, \quad \frac{7 - \sqrt{13}}{2} = 1.697$$

77. We begin by multiplying both sides by the least common denominator which is $(x + 4)(x - 4)$.

$$(x + 4)(x - 4) \cdot 3 + (x + 4)(x - 4)\frac{5}{x - 4} = (x + 4)(x - 4)\frac{7}{x + 4}$$

$$3(x^2 - 16) + 5(x + 4) = 7(x - 4)$$

$$3x^2 - 48 + 5x + 20 = 7x - 28$$

$$3x^2 + 5x - 28 = 7x - 28$$

$$3x^2 - 2x = 0 \qquad \text{(The left side factors.)}$$

$$x(3x - 2) = 0$$

$$x = 0 \quad \text{or} \quad 3x - 2 = 0$$

$$3x = 2$$

$$x = \frac{2}{3}$$

We check graphically by graphing the functions $y_1 = 3 + \dfrac{5}{x - 4}$ and $y_2 = \dfrac{7}{x + 4}$
and using the INTERSECT command:

This confirms our algebraic solution.

79. We begin by multiplying both sides by the least common denominator which is
$(x - 5)(x + 5)$.

$$(x - 5)(x + 5) \cdot \frac{8}{x - 5} = (x - 5)(x + 5)\frac{3}{x + 5} - 2(x - 5)(x + 5)$$

$$8(x + 5) = 3(x - 5) - 2(x^2 - 25)$$
$$8x + 40 = 3x - 15 - 2x^2 + 50$$
$$8x + 40 = -2x^2 + 3x + 35$$
$$2x^2 + 5x + 5 = 0 \qquad \text{(Doesn't factor)}$$

$$x = \frac{-b \pm \sqrt{b^2 - 4ac}}{2a} \qquad a = 2, \ b = 5, \ c = 5$$

$$x = \frac{-5 \pm \sqrt{5^2 - 4(2)(5)}}{4}$$

$$x = \frac{-5 \pm \sqrt{25 - 40}}{4}$$

$$x = \frac{-5 \pm \sqrt{-15}}{4}$$

$$x = \frac{-5 \pm i\sqrt{15}}{4}$$

The graph can be used to see that there are two imaginary zeros but not to
confirm the exact answer we got.

81. In this problem, $a = 1$, $b = 4$, $c = c$. The discriminant is
$b^2 - 4ac = (4)^2 - 4(1)(c) = 16 - 4c$.
If $16 - 4c > 0$, $16 > 4c$ or $c < 4$, then there are two distinct real roots.
If $16 - 4c = 0$, $c = 4$, then there is one real double root,
and if $16 - 4c < 0$, $16 < 4c$ or $c > 4$, then there are two distinct imaginary
roots.

83. $x^2 + 3ix - 2 = 0$

$$x = \frac{-b \pm \sqrt{b^2 - 4ac}}{2a} \qquad a = 1, \ b = 3i, \ c = -2$$

$$x = \frac{-3i \pm \sqrt{(3i)^2 - 4(1)(-2)}}{2(1)}$$

$$x = \frac{-3i \pm \sqrt{-9 + 8}}{2}$$

$$x = \frac{-3i \pm \sqrt{-1}}{2}$$

$$x = \frac{-3i \pm i}{2}$$

$$x = \frac{-3i + i}{2} \quad \text{or} \quad x = \frac{-3i - i}{2}$$

$$x = -i \qquad \quad \text{or} \quad x = -2i$$

85. $x^2 + 2ix = 3$

$x^2 + 2ix - 3 = 0$

$x = \dfrac{-b \pm \sqrt{b^2 - 4ac}}{2a}$ $a = 1,\ b = 2i,\ c = -3$

$x = \dfrac{-2i \pm \sqrt{(2i)^2 - 4(1)(-3)}}{2(1)}$

$x = \dfrac{-2i \pm \sqrt{-4 + 12}}{2}$

$x = \dfrac{-2i \pm \sqrt{8}}{2}$

$x = \dfrac{-2i \pm 2\sqrt{2}}{2}$

$x = \dfrac{2(-i \pm \sqrt{2})}{2}$

$x = -i \pm \sqrt{2}$

$x = \sqrt{2} - i,\ -\sqrt{2} - i$

87. $x^3 - 1 = 0$

$(x - 1)(x^2 + x + 1) = 0$

$x - 1 = 0$ or $x^2 + x + 1 = 0$

$x = 1$ $x = \dfrac{-b \pm \sqrt{b^2 - 4ac}}{2a}$ $a = 1,\ b = 1,\ c = 1$

$x = \dfrac{-1 \pm \sqrt{(1)^2 - 4(1)(1)}}{2(1)}$

$x = \dfrac{-1 \pm \sqrt{1 - 4}}{2}$

$x = \dfrac{-1 \pm \sqrt{-3}}{2}$

$x = \dfrac{-1 \pm i\sqrt{3}}{2}$ or $-\dfrac{1}{2} \pm \dfrac{1}{2}i\sqrt{3}$

89. If a quadratic equation has two roots, they are $\dfrac{-b + \sqrt{b^2 - 4ac}}{2a}$ and

$\dfrac{-b - \sqrt{b^2 - 4ac}}{2a}$. If $a,\ b,\ c$ are rational, then so are $-b$, $2a$, and $b^2 - 4ac$.

There are then three cases: if $\sqrt{b^2 - 4ac}$ is rational, then $\dfrac{-b + \sqrt{b^2 - 4ac}}{2a}$ and

$\dfrac{-b - \sqrt{b^2 - 4ac}}{2a}$ are both rational. If $\sqrt{b^2 - 4ac}$ is irrational, then

$\dfrac{-b + \sqrt{b^2 - 4ac}}{2a}$ and $\dfrac{-b - \sqrt{b^2 - 4ac}}{2a}$ are both irrational. If $\sqrt{b^2 - 4ac}$ is

imaginary, then $\dfrac{-b + \sqrt{b^2 - 4ac}}{2a}$ and $\dfrac{-b + \sqrt{b^2 - 4ac}}{2a}$ are both imaginary. There is
no other possibility. We conclude that one root cannot be rational while the
other is irrational.

91.

$$r_1 = \frac{-b + \sqrt{b^2 - 4ac}}{2a} \qquad r_2 = \frac{-b - \sqrt{b^2 - 4ac}}{2a}$$

$$r_1 r_2 = \frac{(-b + \sqrt{b^2 - 4ac})}{2a} \frac{(-b - \sqrt{b^2 - 4ac})}{2a}$$

$$= \frac{(-b)^2 - (\sqrt{b^2 - 4ac})^2}{4a^2} = \frac{b^2 - (b^2 - 4ac)}{4a^2} = \frac{b^2 - b^2 + 4ac}{4a^2} = \frac{4ac}{4a^2} = \frac{c}{a}$$

93. The ± in front still yields the same two numbers even if a is negative.

95. Let x = one number.
Since their sum is 21, x + the other number = 21, and the other number = $21 - x$.
Then, since their product is 104,

$$x(21 - x) = 104$$
$$21x - x^2 = 104$$
$$0 = x^2 - 21x + 104$$
$$x^2 - 21x + 104 = 0$$
$$(x - 13)(x - 8) = 0$$
$$x - 13 = 0 \quad \text{or} \quad x - 8 = 0$$
$$x = 13 \qquad x = 8$$

The numbers are 8 and 13.

97. Let x = first of the two consecutive even integers.
Then $x + 2$ = second of these integers
Since their product is 168,

$$x(x + 2) = 168$$
$$x^2 + 2x = 168$$
$$x^2 + 2x - 168 = 0$$
$$(x - 12)(x + 14) = 0$$
$$x - 12 = 0 \quad \text{or} \quad x + 14 = 0$$
$$x = 12 \qquad x = -14$$

If $x = 12$, the two consecutive positive even integers must be 12 and 14. We discard the other solution, since the numbers must be positive.

99. Sketch a figure.

Let t = time since 6:00 AM (in hours)
Then (distance of plane 1) = (rate of plane 1) × (time flown by plane 1)
$$= 200\,t$$
(distance of plane 2) = (rate of plane 2) × (time flown by plane 2)
$$= 170(t - 0.5)$$

By the Pythagorean Theorem, the planes will be 500 miles apart, and no longer able to communicate with one another, when

$$(200t)^2 + [170(t - 0.5)]^2 = 500^2$$
$$40{,}000t^2 + 28{,}900(t - 0.5)^2 = 250{,}000$$
$$40{,}000t^2 + 28{,}900(t^2 - t + 0.25) = 250{,}000$$
$$40{,}000t^2 + 28{,}900t^2 - 28{,}900t + 7{,}225 = 250{,}000$$
$$68{,}900t^2 - 28{,}900t - 242{,}775 = 0$$

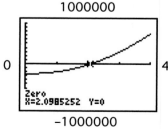

Graphing $y = 68{,}900x^2 - 28{,}900x - 242{,}775$ and applying the ZERO command, we get solution $t = 2.098$ hrs = 126 min, or 2 hrs, 6 min.
The planes will no longer be able to communicate at 8:06 AM.

101. From the diagram, the dimensions of the planting area are $30 - 2x$ and $20 - 2x$.
Then   Area $= 400 = (30 - 2x)(20 - 2x)$

$$400 = 600 - 100x + 4x^2$$
$$0 = 4x^2 - 100x + 200$$
$$0 = x^2 - 25x + 50$$

Graphing $y = x^2 - 25x + 50$ and
applying the ZERO command, we get
$x = 2.19$.

The walkway should be 2.19 ft wide.

(There is another intercept for the graph, but here $x > 20$, so this solution does not make sense.)

103. Begin with a sketch of the garden.

The area is given by $\ell \cdot w$, and
we're told that the area is 1,200,
so we get the equation
$\ell \cdot w = 1,200$.

We need to eliminate one of the variables. The total amount of fence is 150,
so adding up the lengths of the sides, we get
$$w + \ell + w + \ell = 150$$
$$2w + 2\ell = 150$$
Solve for w and substitute the result in for w in the above equation.
$$2w = 150 - 2\ell$$
$$w = 75 - \ell$$
The equation becomes
$$\ell(75 - \ell) = 1,200$$
$$75\ell - \ell^2 = 1,200$$
$$-\ell^2 + 75\ell - 1,200 = 0$$
Graphing $y = -\ell^2 + 75\ell - 1,200$ and applying the ZERO command, we get solutions
$\ell = 23.1$ and 51.9.

If ℓ is 23.1, then $w = 75 - 23.1 = 51.9$, so there is actually only one solution:
23.1 feet by 51.9 feet.

105. (A) Label the text figure as follows:

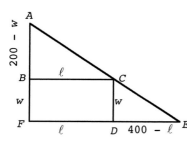

Then, using the hint, in the similar triangles
ABC and *AFE*, we have
$$\frac{\ell}{200 - w} = \frac{400}{200}$$
$$\frac{\ell}{200 - w} = 2$$
$$\ell = 2(200 - w)$$
$$\ell = 400 - 2w$$

Then $A(w) = \ell w = (400 - 2w)w = 400w - 2w^2$.

Since $w \geq 0$ and $\ell = 400 - 2w \geq 0$, the domain of
the function is $0 \leq w \leq 200$.

(B) Solve $150{,}000 \leq 400w - 2w^2$

$$2w^2 - 400w + 15{,}000 \leq 0$$
$$w^2 - 200w + 7{,}500 \leq 0$$
$$(w - 50)(w - 150) \leq 0$$

The solution of this inequality consists of all values of w for which the graph of $y = w^2 - 200w + 7{,}500$ is on or below the w axis. The w intercepts of the graph are 50 and 150.

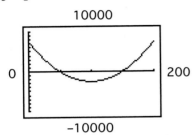

Examining the graph, we see that the solution of the inequality is $50 \leq w \leq 150$.

(C) The cross-sectional area was found in part (A) to be
$$A(w) = 400w - 2w^2$$
Graphing this and applying the MAXIMUM command, we obtain the graph at the right.

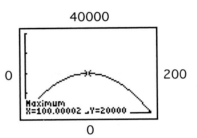

The maximum cross-sectional area of such a building is 20,000 square feet when $w = 100$ ft. A building with area 25,000 square feet is impossible.

107. Let x = distance from the warehouse to Factory A. Since the distance from the warehouse to Factory B via Factory A is known (it is the difference in odometer readings: 52937—52846) to be 91 miles, then

  $91 - x$ = distance from Factory A to Factory B.

The distance from Factory B to the warehouse is known (it is the difference in odometer readings: 53002—52937) to be 65 miles. Applying the Pythagorean theorem, we have

$$x^2 + (91 - x)^2 = 65^2$$
$$x^2 + 8281 - 182x + x^2 = 4225$$
$$2x^2 - 182x + 4056 = 0$$
$$x^2 - 91x + 2028 = 0$$
$$(x - 52)(x - 39) = 0$$
$$x - 52 = 0 \quad \text{or} \quad x - 39 = 0$$
$$x = 52 \text{ mi} \qquad\qquad x = 39 \text{ mi}$$

Since we are told that the distance from the warehouse to Factory A was greater than the distance from Factory A to Factory B, we discard the solution $x = 39$, which would lead to $91 - x = 52$ miles, a contradiction.

  52 miles.

109. (A) The independent variable is years since 1960, so we enter 0, 5, 10, 15, 20, 25, 30, 35, and 40 as L_1. The dependent variable is beer consumption, so we enter the values in the beer column as L_2.

(Note that the whole list is not visible on this figure.)
Next we choose the quadratic regression command from the STAT CALC menu.

The quadratic regression model is $y = -0.000\,587x^2 + 0.0302x + 0.945$.

(B) We need to use our quadratic regression model and find x when $y = 0.99$. We graph $y_1 = -0.000\,587x^2 + 0.0302x + 0.945$ and $y_2 = 0.99$, then use the INTERSECT command:

The intersection occurs at $x = 49.91$: we round to 50 and find that beer consumption will return to 1960 levels 50 years after 1960, or the year 2010.

(C) Use the VALUE command to find that in 2005, when $x = 45$, predicted beer consumption is 1.16 gallons.

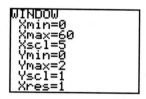

111. (A) The independent variable is years since 1950, so we enter list L_1 with zero for 1950, 5 for 1955, and increment by 5 until reaching 50 for 2000. The dependent variable is production, so we enter the production column as L_2.

(Note that the whole list is not visible on this figure.)

Next we choose the quadratic regression command from the STAT CALC menu

The quadratic regression model is $y = -0.328x^2 + 18.3x + 339$.

(B) We need to use our quadratic regression model and find x when $y = 370$. We graph $y_1 = -0.328x^2 + 18.3x + 339$ and $y_2 = 370$ then use the INTERSECT command:

The intersection occurs at $x = 54.04$; we round to 54 and conclude that production will return to 1950 levels 54 years after 1950, or 2004.

(C) Use the VALUE command to find that in 2005, when $x = 55$, production is predicted to be 353 billion.

113. The independent variable is speed, so we enter the data in the speed column as L_1. The dependent variable is the skid mark length for Auto A, so we enter the length in the second column L_2.

L1	L2	2
20	21	------
30	44	
40	76	
50	114	
60	182	
70	238	
80	305	

L2 ={21,44,76,11…

Next we choose the quadratic regression command from the STAT CALC menu:

The quadratic regression model is $y = 0.0476x^2 + 0.0452x - 0.357$.

(B) We need to use our quadratic regression model and find x when $y = 200$. We graph $y_1 = 0.0476x^2 + 0.0452x - 0.357$ and $y_2 = 200$ then use the INTERSECT command:

The intersection occurs when $x = 64.405$. We round to 64 and conclude that Auto A was traveling 64 mph.

115. (A) The independent variable is speed of Boat A, so we enter the first column as L_1. The dependent variable is the associated mileage, so we enter the second column as L_2.

Next we use the quadratic regression command from the STAT CALC menu:

The quadratic regression model is $y = -0.00217x^2 + 0.140x + 1.90$.

(B) The rental cost is $10 per hour plus $2.30 per gallon for gas. This can be expressed as $C = 10t + 2.3g$ where t is the number of hours and g is the number of gallons. We can write each of t and g in terms of the speed (x) to get a model for cost in terms of speed. The time of the trip will be the distance divided by the speed since distance = speed · time. The trip is 100 miles, so $t = \dfrac{100}{x}$. The gasoline usage can be calculated by dividing the length (100 miles) by the miles per gallon (y), so $g = \dfrac{100}{y}$. But our quadratic model tells us that $y = -0.00217x^2 + 0.140x + 1.90$, so

$$g = \frac{100}{-0.00217x^2 + 0.140x + 1.90}.$$

Our rental cost model is now

$$C = 10\left(\frac{100}{x}\right) + 2.3\left(\frac{100}{-0.00217x^2 + 0.140x + 1.90}\right)$$

$$C = \frac{1000}{x} + \frac{230}{-0.00217x^2 + 0.140x + 1.90}$$

We can use the MINIMUM command to find when this is minimized:

The minimum cost occurs when the speed is 41.5 mph. Plugging $x = 41.5$ back into our quadratic regression model y, we find that the mileage at 41.5 mph is 3.97 mpg. We can use our expressions for t, g, and C to answer the remaining questions.

$$t = \frac{100}{x} = \frac{100}{41.5} = 2.41 \text{ hours}$$

$$g = \frac{100}{y} = \frac{100}{3.97} = 25.2 \text{ gallons}$$

$$C = 10t + 2.3g = 10(2.41) + 2.3(25.2) = 24.1 + 57.96 = \$82.06$$

Exercise 2-6

1. This statement is true since $\sqrt{5}$ and $-\sqrt{5}$ are the only two solutions of the equation $x^2 = 5$.

3. This statement is false. It is not correct to "distribute" exponents. The correct way to simplify the left side is $(2x - 1)^2 = (2x - 1)(2x - 1) = 4x^2 - 4x + 1$.

5. This statement is false.
   $(\sqrt{x - 1} + 1)^2$ is not equal to $x - 1 + 1$ or x, in general.
   In fact
   $(\sqrt{x - 1} + 1)^2 = x - 1 + 2\sqrt{x - 1} + 1 = x + 2\sqrt{x - 1}$.
   This is only equal to x when $x = 1$.

7. This statement is false. If $x^3 = 2$ then $x = \sqrt[3]{2}$ or x is equal to one of two non-real complex numbers whose cube is 2. If $x = 8$, x^3 must equal 512.

9. This is of quadratic type since the power of the first term is twice the power of the second term. With $u = x^{-3}$, the equation becomes
   $$2u^2 - 4u = 0$$

11. This is not of quadratic type: the power of the first term is 3 times the power of the second term.

13. This is of quadratic type since the power of the middle term is half the power of the third term. With the substitution $u = \frac{1}{x^2}$, the equation becomes
   $$\frac{10}{9} + 4u - 7u^2 = 0$$

15. Answers will vary.   17. Answers will vary.

19. $\sqrt{4x - 7} = 5$
    $4x - 7 = 25$
    $4x = 32$
    $x = 8$

    Check:

    $\sqrt{4(8) - 7} = \sqrt{32 - 7} = \sqrt{25} = 5$ √

    Solution: $x = 8$

Graphical check:

Graph $y_1 = \sqrt{4x - 7}$ and $y_2 = 5$ and use the INTERSECT command:

21. $\sqrt{5x + 6} + 6 = 0$

 $\sqrt{5x + 6} = -6$

The left side represents the positive square root of some number and can't be negative, so there is no solution. If we hadn't noticed this, we would have continued:

$5x + 6 = 36$
$5x = 30$
$x = 6$

Check: $\sqrt{5(6) + 6} + 6 = \sqrt{36} + 6 = 12 \neq 0$

There is no solution.

Graphical check:

Graph $y_1 = \sqrt{5x + 6} + 6$:

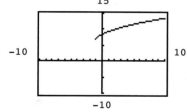

The graph is never at height zero, so there is no solution.

23. $\sqrt[3]{x + 5} = 3$
 $x + 5 = 27$
 $x = 22$
Check:
$\sqrt[3]{22 + 5} = \sqrt[3]{27} = 3$ $\sqrt{}$

Solution: $x = 22$

Graphical check: Graph $y_1 = \sqrt[3]{x + 5}$ and $y_2 = 3$ and use the INTERSECT command to find their intersection.

25. $\sqrt{x + 5} + 7 = 0$

$\sqrt{x + 5} = -7$

The left side represents the positive square root of some number and can't be negative, so there is no solution. If we hadn't noticed this, we would have continued:

$x + 5 = 49$

$x = 44$

Check: $\sqrt{44 + 5} + 7 = \sqrt{49} + 7 = 14 \neq 0$

There is no solution.

Graphical check: Graph $y_1 = \sqrt{x + 5} + 7$:

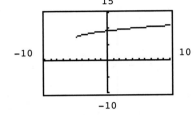

The graph is never at height zero, so there is no solution.

27. $y^4 - 2y^2 - 8 = 0$

Let $u = y^2$, then

$u^2 - 2u - 8 = 0$

$(u - 4)(u + 2) = 0$

$u = 4, -2$

$y^2 = 4 \qquad y^2 = -2$

$y = \pm 2 \qquad y = \pm i\sqrt{2}$

Graphical check: Graph $y = x^4 - 2x^2 - 8$ and use the ZERO command to find the x intercepts.

29. $3x = \sqrt{x^2 - 2}$

$9x^2 = x^2 - 2$

$8x^2 = -2$

$x^2 = -\frac{1}{4}$

$x = \pm\sqrt{-\frac{1}{4}} = \pm\frac{1}{2}i$

Since the solutions are imaginary, we can't check graphically.

31. $\sqrt{x^2 - 5x} = \sqrt{x - 8}$

$x^2 - 5x = x - 8$

$x^2 - 6x + 8 = 0$

$(x - 4)(x - 2) = 0$

$x = 4, 2$

Check:

$\sqrt{4^2 - 5(4)} = \sqrt{-4};$ $\sqrt{4 - 8} = \sqrt{-4}$ √

$\sqrt{2^2 - 5(2)} = \sqrt{-6};$ $\sqrt{2 - 8} = \sqrt{-6}$ √

Solution: $x = 4, 2$

Both solutions result in imaginary values for the equation, so we can't check graphically.

33. $\sqrt{5n + 9} = n - 1$

$5n + 9 = n^2 - 2n + 1$

$0 = n^2 - 7n - 8$

$n^2 - 7n - 8 = 0$

$(n - 8)(n + 1) = 0$

$n = 8, -1$

Check:

$\sqrt{5(8) + 9} = \sqrt{49} = 7;$ $8 - 1 = 7$ √

$\sqrt{5(-1) + 9} = \sqrt{4} = 2;$ $-1 - 1 = -2$ Not a solution

Solution: $n = 8$

Graphical check:
Graph $y_1 = \sqrt{5x + 9}$ and
$y_2 = x - 1$ and use the
INTERSECT command to find
the intersection.

35. $\sqrt{3x + 4} = 2 + \sqrt{x}$

$3x + 4 = 4 + 4\sqrt{x} + x$

$2x = 4\sqrt{x}$

$x = 2\sqrt{x}$

$x^2 = 4x$

$x^2 - 4x = 0$

$x(x - 4) = 0$

$x = 0, 4$

Check:

$\sqrt{3(0) + 4} = \sqrt{44} = 2;$ $2 + \sqrt{0} = 4$ √

$\sqrt{3(4) + 4} = \sqrt{16} = 4;$ $2 + \sqrt{4} = 4$ √

Solution: $x = 0, 4$

Graphical check: Graph $y_1 = \sqrt{3x + 4}$ and $y_2 = 2 + \sqrt{x}$ and use the INTERSECT command:

37. $2x^{2/3} + 3x^{1/3} - 2 = 0$
Let $u = x^{1/3}$, then
$$2u^2 + 3u - 2 = 0$$
$$(2u - 1)(u + 2) = 0$$
$$u = \frac{1}{2}, \quad -2$$

$x^{1/3} = \dfrac{1}{2} \qquad x^{1/3} = -2$

$x = \dfrac{1}{8} \qquad\quad x = -8$

Graphical check: Graph $y = 2x^{2/3} + 3x^{1/3} - 2$ and use the INTERSECT command to find the x intercepts.

39. $(m^2 - m)^2 - 4(m^2 - m) = 12$
Let $u = m^2 - m$, then
$$u^2 - 4u = 12$$
$$u^2 - 4u - 12 = 0$$
$$(u - 6)(u + 2) = 0$$
$$u = 6, \quad -2$$

$m^2 - m = 6 \qquad\qquad m^2 - m = -2$

$m^2 - m - 6 = 0 \qquad m^2 - m + 2 = 0$

$(m - 3)(m + 2) = 0 \qquad\qquad m = \dfrac{-(-1) \pm \sqrt{(-1)^2 - 4(1)(2)}}{2(1)}$

$m = 3, \; -2 \qquad\qquad m = \dfrac{1 \pm \sqrt{-7}}{2} \; \text{or} \; \dfrac{1 \pm i\sqrt{7}}{2}$

$$m = \frac{1}{2} \pm \frac{\sqrt{7}}{2} i$$

Graphical check: Graph $y1 = (x^2 - x)^2 - 4(x^2 - x)$ and $y2 = 12$ and use the INTERSECT command to find the intersections.

41.

$$\sqrt{u - 2} = 2 + \sqrt{2u + 3}$$
$$u - 2 = 4 + 4\sqrt{2u + 3} + 2u + 3$$
$$u - 2 = 2u + 7 + 4\sqrt{2u + 3}$$
$$-u - 9 = 4\sqrt{2u + 3}$$
$$u^2 + 18u + 81 = 16(2u + 3)$$
$$u^2 + 18u + 81 = 32u + 48$$
$$u^2 - 14u + 33 = 0$$
$$(u - 3)(u - 11) = 0$$
$$u = 3, \; 11$$

> **Common Error:**
> $u - 2 = 4 + 2u + 3$ is not an equivalent equation to the given equation.
> $(2 + \sqrt{2u + 3})^2 \neq 4 + 2u + 3$

Check: $\sqrt{3 - 2} \overset{?}{=} 2 + \sqrt{2(3) + 3}$

$1 \neq 5$

$\sqrt{11 - 2} \overset{?}{=} 2 + \sqrt{2(11) + 3}$

$3 \neq 7$

No solution.

Graphical check: Graph $y_1 = \sqrt{x - 2}$ and $y_2 = 2 + \sqrt{2x + 3}$:

The graphs do not intersect.

43. $\sqrt{3y - 2} = 3 - \sqrt{3y + 1}$

$3y - 2 = 9 - 6\sqrt{3y + 1} + 3y + 1$

$3y - 2 = 3y + 10 - 6\sqrt{3y + 1}$

$-12 = -6\sqrt{3y + 1}$

$2 = \sqrt{3y + 1}$

$4 = 3y + 1$

$3 = 3y$

$y = 1$

Check: $\sqrt{3(1) - 2} \overset{?}{=} 3 - \sqrt{3(1) + 1}$

$1 \overset{\checkmark}{=} 1$

Solution: $y = 1$

Graphical check: Graph $y1 = \sqrt{3x - 2}$ and $y2 = 3 - \sqrt{3x + 1}$ and use the INTERSECT command to find the intersection.

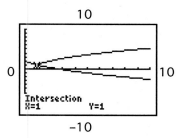

45. $\sqrt{7x - 2} - \sqrt{x + 1} = \sqrt{3}$

$\sqrt{7x - 2} = \sqrt{x + 1} + \sqrt{3}$

$7x - 2 = x + 1 + 2\sqrt{3}\sqrt{x + 1} + 3$

$7x - 2 = x + 4 + 2\sqrt{3(x + 1)}$

$6x - 6 = 2\sqrt{3(x + 1)}$

$3x - 3 = \sqrt{3x + 3}$

$9x^2 - 18x + 9 = 3x + 3$

$9x^2 - 21x + 6 = 0$

$3x^2 - 7x + 2 = 0$

$(3x - 1)(x - 2) = 0$

$x = \frac{1}{3}, \ 2$

Check: $\sqrt{7\left(\frac{1}{3}\right) - 2} - \sqrt{\frac{1}{3} + 1} \overset{?}{=} \sqrt{3}$

$\sqrt{\frac{1}{3}} - \sqrt{\frac{4}{3}} \overset{?}{=} \sqrt{3}$

$-\frac{1}{3}\sqrt{3} \neq \sqrt{3}$

$\sqrt{7(2) - 2} - \sqrt{2 + 1} \overset{?}{=} \sqrt{3}$

$\sqrt{12} - \sqrt{3} \overset{?}{=} \sqrt{3}$

$\sqrt{3} \overset{\checkmark}{=} \sqrt{3}$

Solution: $x = 2$

Graphical check:
Graph $y1 = \sqrt{7x - 2} - \sqrt{x + 1}$ and $y2 = \sqrt{3}$ and use the INTERSECT command to find the intersection.

47. $\sqrt{4x^2 + 12x + 1} - 6x = 9$
$\qquad\qquad x = \dfrac{-b \pm \sqrt{b^2 - 4ac}}{2a}$ $a = 2$ $b = 6$ $c = 5$

$\qquad \sqrt{4x^2 + 12x + 1} = 6x + 9$
$\qquad\qquad x = \dfrac{-6 \pm \sqrt{6^2 - 4(2)(5)}}{2(2)}$

$\qquad 4x^2 + 12x + 1 = 36x^2 + 108x + 81$
$\qquad\qquad x = \dfrac{-6 \pm \sqrt{36 - 40}}{4}$

$\qquad\qquad\qquad 0 = 32x^2 + 96x + 80$
$\qquad\qquad x = \dfrac{-6 \pm \sqrt{-4}}{4}$

$\qquad\qquad\qquad 0 = 2x^2 + 6x + 5$
$\qquad\qquad x = \dfrac{-6 \pm 2i}{4}$

$\qquad\qquad\qquad\qquad x = -\dfrac{3}{2} \pm \dfrac{1}{2}i$

Check: $-\dfrac{3}{2} + \dfrac{1}{2}i$: $\sqrt{4\left(-\dfrac{3}{2} + \dfrac{1}{2}i\right)^2 + 12\left(-\dfrac{3}{2} + \dfrac{1}{2}i\right) + 1} - 6\left(-\dfrac{3}{2} + \dfrac{1}{2}i\right) \overset{?}{=} 9$

$\qquad\qquad\qquad \sqrt{(-3 + i)^2 - 18 + 6i + 1} + 9 - 3i \overset{?}{=} 9$

$\qquad\qquad\qquad \sqrt{9 - 6i + i^2 - 18 + 6i + 1} + 9 - 3i \overset{?}{=} 9$

$\qquad\qquad\qquad \sqrt{9 - 6i - 1 - 18 + 6i + 1} + 9 - 3i \overset{?}{=} 9$

$\qquad\qquad\qquad\qquad \sqrt{-9} + 9 - 3i \overset{?}{=} 9$

$\qquad\qquad\qquad\qquad\quad 3i + 9 - 3i = 9$

$\qquad\qquad\qquad\qquad\qquad\qquad\qquad$ A solution

$-\dfrac{3}{2} - \dfrac{1}{2}i$: $\sqrt{4\left(-\dfrac{3}{2} - \dfrac{1}{2}i\right)^2 + 12\left(-\dfrac{3}{2} - \dfrac{1}{2}i\right) + 1} - 6\left(-\dfrac{3}{2} - \dfrac{1}{2}i\right) \overset{?}{=} 9$

$\qquad\qquad\qquad \sqrt{(-3 - i)^2 - 18 - 6i + 1} + 9 + 3i \overset{?}{=} 9$

$\qquad\qquad\qquad \sqrt{9 + 6i + i^2 - 18 - 6i + 1} + 9 + 3i \overset{?}{=} 9$

$\qquad\qquad\qquad \sqrt{9 + 6i - 1 - 18 - 6i + 1} + 9 + 3i \overset{?}{=} 9$

$\qquad\qquad\qquad\qquad \sqrt{-9} + 9 + 3i \overset{?}{=} 9$

$\qquad\qquad\qquad\qquad\quad 3i + 9 + 3i \overset{?}{=} 9$

$\qquad\qquad\qquad\qquad\qquad 9 + 6i \ne 9$ Not a solution

Solution: $x = -\dfrac{3}{2} + \dfrac{1}{2}i$

Since the solutions are complex, we can't check graphically.

49. $3n^{-2} - 11n^{-1} - 20 = 0$
$\qquad$ Let $u = n^{-1}$, then

Graphical check: Graph $y = 3x^{-2} - 11x^{-1} - 20$ and use the ZERO command to find the zeros.

$\qquad 3u^2 - 11u - 20 = 0$

$\qquad (3u + 4)(u - 5) = 0$

$\qquad\qquad\qquad u = -\dfrac{4}{3}, \ 5$

$\qquad n^{-1} = -\dfrac{4}{3} \qquad n^{-1} = 5$

$\qquad n = -\dfrac{3}{4} \qquad\quad n = \dfrac{1}{5}$

51. $9y^{-4} - 10y^{-2} + 1 = 0$

Let $u = y^{-2}$, then

$9u^2 - 10u + 1 = 0$

$(9u - 1)(u - 1) = 0$

$$u = \frac{1}{9}, \ 1$$

$y^{-2} = \dfrac{1}{9}$ $y^{-2} = 1$

$y^2 = 9$ $y^2 = 1$

$y = \pm 3$ $y = \pm 1$

Graphical check: Graph $y = 9x^{-4} - 10x^{-2} + 1$ and use the ZERO command to find the zeros. (Only the negative zeros are shown here.)

53. $y^{1/2} - 3y^{1/4} + 2 = 0$

Let $u = y^{1/4}$, then

$u^2 - 3u + 2 = 0$

$(u - 1)(u - 2) = 0$

$u = 1, \ 2$

$y^{1/4} = 1$ $y^{1/4} = 2$

$y = 1$ $y = 16$

Common Error:
$y \neq 2^{1/4}$
Both sides must be raised to the fourth power to eliminate the $y^{1/4}$.

Graphical check: Graph $y = x^{1/2} - 3x^{1/4} + 2$ and use the ZERO command to find the zeros.

55. $(m - 5)^4 + 36 = 13(m - 5)^2$

Let $u = (m - 5)^2$, then

$u^2 + 36 = 13u$

$u^2 - 13u + 36 = 0$

$(u - 4)(u - 9) = 0$

$u = 4, \ 9$

$(m - 5)^2 = 4$ $(m - 5)^2 = 9$

$m - 5 = \pm 2$ $m - 5 = \pm 3$

$m = 5 \pm 2$ $m = 5 \pm 3$

$m = 3, \ 7$ $m = 2, \ 8$

Graphical check: Graph $y1 = (x - 5)^4 + 36$ and $y2 = 13(x - 5)^2$ and use the INTERSECT command to find the intersections.
(Only $x = 2$ and $x = 3$ are shown here.)

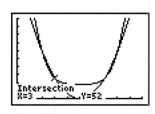

57. The first step was an attempt to square both sides, but it was done incorrectly. The square of the expression $\sqrt{x + 3} + 5$ is NOT $x + 3 + 25$. To solve the equation correctly, you should begin by isolating the root, THEN squaring both sides.

59.
$$\sqrt{5 - 2x} - \sqrt{x + 6} = \sqrt{x + 3}$$
$$\sqrt{5 - 2x} = \sqrt{x + 6} + \sqrt{x + 3}$$
$$5 - 2x = x + 6 + 2\sqrt{x + 6}\sqrt{x + 3} + x + 3$$
$$5 - 2x = 2x + 9 + 2\sqrt{x + 6}\sqrt{x + 3}$$
$$-4x - 4 = 2\sqrt{x + 6}\sqrt{x + 3}$$
$$-2x - 2 = \sqrt{(x + 6)(x + 3)}$$
$$4x^2 + 8x + 4 = (x + 6)(x + 3)$$
$$4x^2 + 8x + 4 = x^2 + 9x + 18$$
$$3x^2 - x - 14 = 0$$
$$(3x - 7)(x + 2) = 0$$
$$x = \frac{7}{3}, -2$$

Check:
$$\sqrt{5 - 2\left(\frac{7}{3}\right)} - \sqrt{\frac{7}{3} + 6} \overset{?}{=} \sqrt{\frac{7}{3} + 3}$$
$$\sqrt{\frac{1}{3}} - \sqrt{\frac{25}{3}} \overset{?}{=} \sqrt{\frac{16}{3}}$$
$$-\frac{4\sqrt{3}}{3} \neq \frac{4\sqrt{3}}{3}$$
$$\sqrt{5 - 2(-2)} - \sqrt{-2 + 6} \overset{?}{=} \sqrt{-2 + 3}$$
$$1 \overset{\checkmark}{=} 1$$

Solution: $x = -2$

Graphical check:
Graph $y1 = \sqrt{5 - 2x} - \sqrt{x + 6}$ and $y2 = \sqrt{x + 3}$ and use the INTERSECT command to find the intersection.

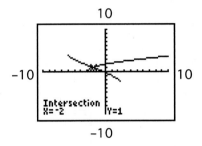

61.
$$2 + 3y^{-4} = 6y^{-2} \qquad \text{Multiply both sides by } y^4 \quad (y \neq 0)$$
$$2y^4 + 3 = 6y^2$$
$$2y^4 - 6y^2 + 3 = 0$$

Let $u = y^2$, then
$$2u^2 - 6u + 3 = 0$$
$$u = \frac{-b \pm \sqrt{b^2 - 4ac}}{2a} \qquad a = 2, \ b = -6, \ c = 3$$
$$u = \frac{-(-6) \pm \sqrt{(-6)^2 - 4(2)(3)}}{2(2)}$$
$$u = \frac{6 \pm \sqrt{36 - 24}}{4} = \frac{6 \pm \sqrt{12}}{4} = \frac{2(3 \pm \sqrt{3})}{4} = \frac{3 \pm \sqrt{3}}{2}$$
$$y^2 = \frac{3 \pm \sqrt{3}}{2}$$
$$y = \pm\sqrt{\frac{3 \pm \sqrt{3}}{2}} \quad \text{(four roots)}$$

Note: $\sqrt{\frac{3 + \sqrt{3}}{2}} \approx 1.538$

$\sqrt{\frac{3 - \sqrt{3}}{2}} \approx 0.7962$

Graphical check:
Graph: $y1 = 2 + 3x^{-4}$ and
$y2 = 6x^{-2}$ and use the
INTERSECT command to find
the intersections. (Only
the positive roots are
shown here.)

63. By squaring: $m - 7\sqrt{m} + 12 = 0$

$$m + 12 = 7\sqrt{m}$$
$$m^2 + 24m + 144 = 49m$$
$$m^2 - 25m + 144 = 0$$
$$(m - 9)(m - 16) = 0$$
$$m = 9, 16$$

Check: $9 - 7\sqrt{9} + 12 \overset{?}{=} 0$ $16 - 7\sqrt{16} + 12 \overset{?}{=} 0$

$0 \overset{\checkmark}{=} 0$ $\phantom{16 - 7\sqrt{16} + 12}0 \overset{\checkmark}{=} 0$

Solution: $m = 9, 16$

By substitution: $m - 7\sqrt{m} + 12 = 0$
Let $u = \sqrt{m}$, then

$$u^2 - 7u + 12 = 0$$
$$(u - 4)(u - 3) = 0$$
$$u = 3, 4$$

$\sqrt{m} = 3$ $\sqrt{m} = 4$
$m = 9$ $m = 16$

These answers have already been checked.

Graphical check:
Graph: $y_1 = x - 7\sqrt{x} + 12$ and use the ZERO command.

65. By squaring: $t - 11\sqrt{t} + 18 = 0$

$$t + 18 = 11\sqrt{t}$$
$$t^2 + 36t + 324 = 121t$$
$$t^2 - 85t + 324 = 0$$
$$(t - 4)(t - 81) = 0$$
$$t = 4, 81$$

Check: $4 - 11\sqrt{4} + 18 \overset{?}{=} 0$

$0 \overset{\checkmark}{=} 0$

$81 - 11\sqrt{81} + 18 \overset{?}{=} 0$

$\phantom{81 - 11\sqrt{81}}0 \overset{\checkmark}{=} 0$

Solution: $t = 4, 81$

By substitution: $t - 11\sqrt{t} + 18 = 0$

Let $u = \sqrt{t}$, then

$u^2 - 11u + 18 = 0$

$(u - 9)(u - 2) = 0$

$\qquad\qquad u = 2, 9$

$\quad u = 2 \qquad u = 9$

$\sqrt{t} = 2 \qquad \sqrt{t} = 9$

$\quad t = 4 \qquad\; t = 81$

These answers have already been checked.

Graphical check:

Graph: $y_1 = x - 11\sqrt{x} + 18$ and use the ZERO command.

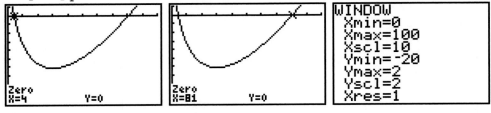

67. Algebraically

$$2\sqrt{x + 5} = 0.01x + 2.04$$
$$200\sqrt{x + 5} = x + 204$$
$$40{,}000(x + 5) = (x + 204)^2$$
$$40{,}000x + 200{,}000 = x^2 + 408x + 41{,}616$$
$$0 = x^2 - 39{,}592x - 158{,}384$$

Although this is factorable it's really hard because of the size of the coefficients, so the quadratic formula is used

$$x = \frac{-b \pm \sqrt{b^2 - 4ac}}{2a} \qquad a = 1$$
$$b = -39{,}592$$
$$c = -158{,}384$$

$$x = \frac{-(-39{,}592) \pm \sqrt{(-39{,}592)^2 - 4(1)(-158{,}384)}}{2(1)}$$

$$x = \frac{39{,}592 \pm 39{,}600}{2}$$

$$x = -4,\ 39{,}596$$

Check: $2\sqrt{-4 + 5} \overset{?}{=} 0.01(-4) + 2.04$

$\qquad\qquad\quad 2 \overset{\checkmark}{=} 2$

$\quad 2\sqrt{39{,}596 + 5} \overset{?}{=} 0.01(39{,}596) + 2.04$

$\qquad\qquad\quad 398 \overset{\checkmark}{=} 398$

Solution: $x = -4,\ 39{,}596$

Graphically:
Graph $y1 = 2\sqrt{x + 5}$ and $y2 = 0.01x + 2.04$ and use the INTERSECT command to find the intersections.

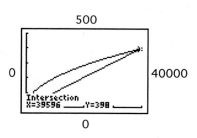

Algebraic method:
   Advantages: straightforward calculation, both solutions arise naturally.
   Disadvantage: messy arithmetic.

Graphical method:
   Advantage: no arithmetic.
   Disadvantages: it is extremely unlikely that one would spot the second solution at $x = 39{,}596$; one must essentially know the answer beforehand; it is not at all obvious what window to choose.

69. Algebraically
$$2x^{-2/5} - 5x^{-1/5} + 1 = 0$$
Let $u = x^{-1/5}$, then
$$2u^2 - 5u + 1 = 0$$

$$u = \frac{-b \pm \sqrt{b^2 - 4ac}}{2a} \qquad \begin{aligned} a &= 2 \\ b &= -5 \\ c &= 1 \end{aligned}$$

$$u = \frac{-(-5) \pm \sqrt{(-5)^2 - 4(2)(1)}}{2(2)}$$

$$u = \frac{5 \pm \sqrt{17}}{4}$$

$$x^{-1/5} = \frac{5 + \sqrt{17}}{4} \qquad x^{-1/5} = \frac{5 - \sqrt{17}}{4}$$

$$x^{1/5} = \frac{4}{5 + \sqrt{17}} \qquad x^{1/5} = \frac{4}{5 - \sqrt{17}}$$

$$x = \left(\frac{4}{5 + \sqrt{17}}\right)^5 \qquad x = \left(\frac{4}{5 - \sqrt{17}}\right)^5$$

$$x \approx 0.016203 \qquad x \approx 1974.98$$

Graphically:
Graph $y1 = 2x^{-2/5} - 5x^{-1/5} + 1$ and use the ZERO command to find the zeros.

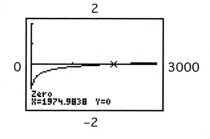

Advantages and disadvantages: See discussion under problem 67.

71.

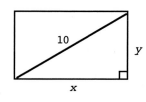

Area = length·width, so $45 = x \cdot y$.
According to the Pythagorean Theorem, $x^2 + y^2 = 10^2$; solve for y:
$$\begin{aligned} x^2 + y^2 &= 100 \\ y^2 &= 100 - x^2 \\ y &= \sqrt{100 - x^2} \\ 45 = x \cdot y &= x\sqrt{100 - x^2} \end{aligned}$$

Graph $y_1 = 45$, $y_2 = x\sqrt{100 - x^2}$ and use the INTERSECT command.

If $x = 5.3$, $y = \sqrt{100 - 5.3^2} = 8.5$.

If $x = 8.5$, $y = \sqrt{100 - 8.5^2} = 5.3$

The only solution is 5.3 in × 8.5 in.

73. As in Example 6 in Section 2-6, we know that the time until the splash is heard can be broken down into two segments.

t_1 = time for stone to reach water

t_2 = time for sound of splash to travel back to surface

We've seen before that the distance covered by a falling object is given by $16t^2$, so if x is the depth of the well, then $x = 16t^2$. From Example 6 we know that the distance traveled by the sound of the splash is $x = 1,100t_2$.

Solve each equation for the time.

$$x = 16t_1^2 \qquad\qquad x = 1,100t_2$$

$$\frac{x}{16} = t_1^2 \qquad\qquad t_2 = \frac{x}{1,100}$$

$$t_1 = \frac{\sqrt{x}}{4}$$

The total time, which is 14 seconds, is the sum of the two time segments.

$$\frac{\sqrt{x}}{4} + \frac{x}{1,100} = 14$$

Graph $y_1 = \dfrac{\sqrt{x}}{4} + \dfrac{x}{1,100}$, $y_2 = 14$ and use the INTERSECT command.

The solution is 2,277, so the well is 2,277 feet deep.

75. (A)

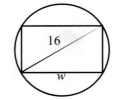

The diameter of the log cross-section is the diagonal of the beam cut out. We can use the Pythagorean Theorem to find the height of the beam.

$$w^2 + height^2 = 16^2$$

$$height^2 = 256 - w^2$$

$$height = \sqrt{256 - w^2}$$

The area is width·height, so $A = w\sqrt{256 - w^2}$. (Note that w must be positive and cannot be 16" or more, so $0 < w < 16$.)

(B) Plug 120 in for A and solve for w:

$$120 = w$$

Graph $y_1 = 120$, $y_2 = x\sqrt{256 - x^2}$ and use the INTERSECT command.

WINDOW
Xmin=0
Xmax=16
Xscl=1
Ymin=0
Ymax=150
Yscl=10
Xres=1

The two solutions are $x = 9.1$ and $x = 13.1$.

If $w = 9.1$, height $= \sqrt{256 - 9.1^2} = 13.1$

If $w = 13.1$, height $= \sqrt{256 - 13.1^2} = 9.1$
The only solution is 13.1 in by 9.1 in.

(C) Graph $y_1 = x\sqrt{256 - x^2}$ and use the MAXIMUM command.

The maximum occurs when the width is 11.3 in; the maximum area is 128 in^2.

77.

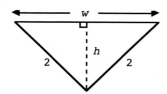

Let w = width of trough
h = altitude of triangular end

Examining the triangular end of the trough sketched above, we see that
$h^2 + \left(\frac{1}{2}w\right)^2 = 2^2$. The area is given by $A = \frac{1}{2}wh$. Since the volume of the trough
V is given by $V = A \cdot 6$, we have

$$9 = 6A$$
$$9 = 6\left(\frac{1}{2}wh\right)$$
$$9 = 3wh$$
$$3 = wh$$

Since $h^2 = 2^2 - \left(\frac{1}{2}w\right)^2$

$$h^2 = 2^2 - \frac{1}{4}w^2$$

$$h = \sqrt{4 - \frac{1}{4}w^2}$$

We solve

$$3 = w\sqrt{4 - \frac{1}{4}w^2}$$

$$9 = w^2\left(4 - \frac{1}{4}w^2\right)$$

$$9 = 4w^2 - \frac{1}{4}(w^2)^2$$

$$36 = 16w^2 - (w^2)^2$$

$$(w^2)^2 - 16w^2 + 36 = 0$$

$$w^2 = \frac{-b \pm \sqrt{b^2 - 4ac}}{2a} \qquad a = 1, \ b = -16, \ c = 36$$

$$w^2 = \frac{-(-16) \pm \sqrt{(-16)^2 - 4(1)(36)}}{2(1)}$$

$$w^2 = \frac{16 \pm \sqrt{256 - 144}}{2}$$

$$w^2 = \frac{16 \pm \sqrt{112}}{2}$$

$$w^2 = 8 \pm 2\sqrt{7}$$

$$w = \sqrt{8 \pm 2\sqrt{7}}$$

$$w = 1.65 \text{ ft or } 3.65 \text{ ft}$$

Section 2-7

1. Answers will vary.

3. Answers will vary

5. $u(x) > 0$ where the graph of u is above the x axis; this occurs when $c < x < f$ or when or (c, f).

7. $v(x) \geq u(x)$ where the graph of v is above or intersects the graph of u; this occurs when $x \leq b$ or $x \geq e$, or $(-\infty, b] \cup [e, \infty)$.

9. $u(x) \leq 0$ where the graph of u is above or intersects the x-axis; this occurs when $x \leq c$ or $x \geq f$, or $(-\infty, c] \cup [f, \infty)$.

11. $v(x) - u(x) > 0$ is equivalent to $v(x) > u(x)$. This occurs where the graph of v is above the graph of u; $x < b$ or $x > e$, or $(-\infty, b) \cup (e, \infty)$.

13. The distance from x to 3 is less than 5.
    $$|x - 3| < 5$$

15. The distance from y to -1 is greater than 6.
    $$|y - (-1)| > 6$$
    $$|y + 1| > 6$$

17. The distance from a to 3 is less than or equal to 5.
    $$|a - 3| \leq 5$$

19. The distance from d to -2 is greater than or equal to 4.
    $$|d - (-2)| \geq 4$$
    $$|d + 2| \geq 4$$

21. The distance from y to the origin is less than or equal to 7.

23. The distance from w to the origin is greater than 7.

25. $7x - 8 < 4x + 7$
    $3x < 15$
    $x < 5$ or $(-\infty, 5)$

27. $-5t < -10$
    $t > 2$ or $(2, \infty)$
    (Note that the direction of the inequality changed.)

29. $3 - m < 4(m - 3)$
    $3 - m < 4m - 12$
    $-5m < -15$
    $m > 3$ or $(3, \infty)$
    (Note that the direction of the inequality changed.)

31. Find zeros of the left side: $(x - 3)(x + 4) = 0$; $x = 3$ or $x = -4$
Choose test numbers and make a table:

Interval	Test Number	Result
$x < -4$	-5	$(-5 - 3)(-5 + 4) = 8$; positive
$-4 < x < 3$	0	$(0 - 3)(0 + 4) = -12$; negative
$x > 3$	4	$(4 - 3)(4 + 4) = 8$; positive

The expression $(x - 3)(x + 4)$ is negative on $-4 < x < 3$, or $(-4, 3)$.

33. Find zeros of the left side: $x(2x - 7) = 0$; $x = 0$ or $2x - 7 = 0$
$$2x = 7$$
$$x = 7/2$$

Choose test numbers and make a table:

Interval	Test Number	Result
$x < 0$	-1	$(-1)(2(-1) - 7) = 9$; positive
$0 < x < 7/2$	1	$(1)(2(1) - 7) = -5$; negative
$x > 7/2$	4	$(4)(2(4) - 7) = 4$; positive

The expression $x(2x - 7)$ is zero or positive on $x \le 0$ or $x \ge 7/2$, or $(-\infty, 0] \cup [7/2, \infty)$.

35. The distance from s to 5 is less than 3, so $s - 5$ is between -3 and 3.
$$|s - 5| < 3$$
$$-3 < s - 5 < 3$$
$$2 < s < 8$$
$$(2, 8)$$

37. The distance from s to 5 is more than 3, so $s - 5$ is either less than -3 or greater than 3.
$$|s - 5| > 3$$
$$s - 5 < -3 \text{ or } s - 5 > 3$$
$$s < 2 \text{ or } s > 8$$
$$(-\infty, 2) \cup (8, \infty)$$

39.
$$-4 < 5t + 6 \le 21$$
$$-10 < 5t \le 15$$
$$-2 < t \le 3 \text{ or } (-2, 3]$$

41.
$$-12 < \frac{3}{4}(2 - x) \le 24$$
$$\frac{4}{3}(-12) < 2 - x \le \frac{4}{3}(24)$$
$$-16 < 2 - x \le 32$$
$$-18 < -x \le 30$$
$$18 > x \ge -30$$
$$-30 \le x < 18 \text{ or } [-30, 18)$$

43.
$$\frac{q}{7} - 3 > \frac{q - 4}{3} + 1$$
$$21\left(\frac{q}{7} - 3\right) > 21\left(\frac{(q - 4)}{3} + 1\right)$$
$$3q - 63 > 7(q - 4) + 21$$
$$3q - 63 > 7q - 28 + 21$$
$$3q - 63 > 7q - 7$$
$$-4q > 56$$
$$q < -14 \text{ or } (-\infty, -14)$$

45. Rearrange so that zero is on the right: $x^2 + 3x - 10 < 0$

Find the zeros of the left side: $x^2 + 3x - 10 = 0$

$$(x + 5)(x - 2) = 0$$

$$x = -5 \quad \text{or} \quad x = 2$$

Choose test numbers and make a table:

Interval	Test Number	Result
$x < -5$	-6	$(-6)^2 + 3(-6) - 10 = 8;$ positive
$-5 < x < 2$	0	$(0)^2 + 3(0) - 10 = -10;$ negative
$x > 2$	3	$(3)^2 + 3(3) - 10 = 8;$ positive

The expression $x^2 + 3x - 10$ is negative on $-5 < x < 2$, or $(-5, 2)$.

47. Rearrange so that zero is on the right: $x^2 - 10x + 21 > 0$

Find the zeros of the left side: $x^2 - 10x + 21 = 0$

$$(x - 7)(x - 3) = 0$$

$$x = 7 \quad \text{or} \quad x = 3$$

Choose test numbers and make a table:

Interval	Test Number	Result
$x < 3$	0	$(0)^2 - 10(0) + 21 = 21;$ positive
$3 < x < 7$	5	$(5)^2 - 10(5) + 21 = -4;$ negative
$x > 7$	8	$(8)^2 - 10(8) + 21 = 5;$ positive

The expression $x^2 - 10x + 21$ is positive on $x < 3$ or $x > 7$, or $(-\infty, 3) \cup (7, \infty)$.

49. Rearrange so that zero is on the right: $x^2 - 8x \leq 0$

Find the zeros of the left side: $x^2 - 8x = 0$

$$x(x - 8) = 0$$

$$x = 0 \quad \text{or} \quad x = 8$$

Choose test numbers and make a table:

Interval	Test Number	Result
$x < 0$	-1	$(-1)^2 - 8(-1) = 9;$ positive
$0 < x < 8$	1	$(1)^2 - 8(1) = -7;$ negative
$x > 8$	10	$(10)^2 - 8(10) = 20;$ positive

The expression $x^2 - 8x$ is zero or negative on $0 \leq x \leq 8$, or $[0, 8]$.

51. Rearrange so that zero is on the right: $x^2 - 2x + 1 < 0$

Find the zeros of the left side: $x^2 - 2x + 1 = 0$

$$(x - 1)(x - 1) = 0$$

$$x = 1$$

Choose test numbers and make a table:

Interval	Test Number	Result
$x < 1$	0	$(0)^2 - 2(0) + 1 = 1;$ positive
$x > 1$	2	$(2)^2 - 2(2) + 1 = 1;$ positive

The expression $x^2 - 2x + 1$ is never negative, so the inequality has no solution.

53. Rearrange so that zero is on the right: $x^2 - 4x - 21 \geq 0$

Find the zeros of the left side: $x^2 - 4x - 21 = 0$
$$(x - 7)(x + 3) = 0$$
$$x = 7 \quad \text{or} \quad x = -3$$

Choose test numbers and make a table:

Interval	Test Number	Result
$x < -3$	-4	$(-4)^2 - 4(-4) - 21 = 11$; positive
$-3 < x < 7$	0	$(0)^2 - 4(0) - 21 = -21$; negative
$x > 7$	8	$(8)^2 - 4(8) - 21 = 11$; positive

The expression $x^2 - 4x - 21$ is zero or positive on $x \leq -3$ or $x \geq 7$, or $(-\infty, -3] \cup [7, \infty)$.

55. Find the zeros of the left side (this requires the quadratic formula):

$x^2 - 5x + 3 = 0 \qquad a = 1, \ b = -5, \ c = 3$

$$x = \frac{-(-5) \pm \sqrt{(-5)^2 - 4(1)(3)}}{2(1)}$$

$$= \frac{5 \pm \sqrt{13}}{2} \approx 4.3, 0.7$$

Choose test numbers and make a table:

Interval	Test Number	Result
$x < 0.7$	0	$(0)^2 - 5(0) + 3 = 3$; positive
$0.7 < x < 4.3$	1	$(1)^2 - 5(1) + 3 = -1$; negative
$x > 4.3$	5	$(5)^2 - 5(5) + 3 = 3$; positive

The expression $x^2 - 5x + 3$ is positive on $x < 0.7$ or $x > 4.3$, or $(-\infty, 0.7) \cup (4.3, \infty)$.

57. Rearrange so that the right side is zero: $-2x^2 + x + 7 \geq 0$
Find the zeros of the left side (this requires the quadratic formula):

$-2x^2 + x + 7 = 0 \qquad a = -2, \ b = 1, \ c = 7$

$$x = \frac{-1 \pm \sqrt{1^2 - 4(-2)(7)}}{2(-2)}$$

$$= \frac{-1 \pm \sqrt{57}}{-4} \approx -1.6, 2.1$$

Choose test numbers and make a table:

Interval	Test Number	Result
$x < -1.6$	-2	$-2(-2)^2 + (-2) + 7 = -3$; negative
$-1.6 < x < 2.1$	0	$-2(0)^2 + (0) + 7 = 7$; positive
$x > 2.1$	3	$-2(3)^2 + (3) + 7 = -8$; negative

The expression $-2x^2 + x + 7$ is zero or positive on $-1.6 \leq x \leq 2.1$, or $[-1.6, 2.1]$.

59. $|3x - 7| \leq 4$

$$-4 \leq 3x - 7 \leq 4$$
$$3 \leq 3x \leq 11$$
$$1 \leq x \leq \frac{11}{3}$$

$$\left[1, \frac{11}{3}\right]$$

61. $|4 - 2t| > 6$

$$4 - 2t > 6 \quad \text{or} \quad 4 - 2t < -6$$
$$-2t > 2 \quad \text{or} \quad -2t < -10$$
$$t < -1 \quad \text{or} \quad t > 5$$
$$(-\infty, -1) \cup (5, \infty)$$

63. $|0.2u + 1.7| \geq 0.5$

$$0.2u + 1.7 \geq 0.5 \quad \text{or} \quad 0.2u + 1.7 \leq -0.5$$
$$0.2u \geq -1.2 \quad \text{or} \qquad 0.2u \leq -2.2$$
$$u \geq -6 \quad \text{or} \qquad u \leq -11$$

$$(-\infty, -11] \cup [-6, \infty)$$

65. If $a - b = 1$, $a = b + 1$ and a is greater than b. $a > b$.

67. If $a < 0$, $b < 0$, then if

$$\frac{b}{a} > 1$$

$$a \cdot \frac{b}{a} < a \cdot 1 \quad \text{(since } a \text{ is negative, multiplying both sides by } a \text{ reverses the inequality)}$$

$$b < a$$
$$a > b$$

69. The distance from x to 3 is between 0 and 0.1. The number x will be within 0.1 of 3 if it is between 2.9 and 3.1. But we also need the distance between x and 3 to be non-zero, so x cannot be 3. So the solution is all numbers between 2.9 and 3.1 except 3, or $(2.9, 3) \cup (3, 3.1)$.

71. The distance from x to c is between 0 and $2c$. The number x will be within $2c$ units of c if it is between $c - 2c$ and $c + 2c$. In other words, x must be between $-c$ and $3c$. But we also need the distance between x and c to be nonzero, so x and c cannot be equal. So the solution is all numbers between $-c$ and $3c$ except c, or $(-c, c) \cup (c, 3c)$.

73. Find the zeros of f: $2x^2 - 5x - 12 = 0$

$$(2x + 3)(x - 4) = 0$$
$$2x + 3 = 0 \quad or \quad x - 4 = 0$$
$$2x = -3 \qquad\qquad x = 4$$
$$x = -3/2$$

Choose test numbers and make a table:

Interval	Test Number	Result
$x < -3/2$	-2	$2(-2)^2 - 5(-2) - 12 = 6;$ positive
$-3/2 < x < 4$	1	$2(1)^2 - 5(1) - 12 = -15;$ negative
$x > 4$	5	$2(5)^2 - 5(5) - 12 = 13;$ positive

The function $f(x) = 2x^2 - 5x - 12$ is positive on $x < -3/2$ or $x > 4$, or $(-\infty, -3/2) \cup (4, \infty)$. It is negative on the interval $(-3/2, 4)$.

75. Find the zeros of h: $x^2 + 9x + 2 = 0$ $a = 1$, $b = 9$, $c = 2$

$$x = \frac{-9 \pm \sqrt{9^2 - 4(1)(2)}}{2(1)}$$

$$= \frac{-9 \pm \sqrt{73}}{2} \approx -0.2, -8.8$$

Choose test numbers and make a table:

Interval	Test Number	Result
$x < -8.8$	-10	$(-10)^2 + 9(-10) + 2 = 12$; positive
$-8.8 < x < -0.2$	-1	$(-1)^2 + 9(-1) + 2 = -6$; negative
$x > -0.2$	0	$(0)^2 + 9(0) + 2 = 2$; positive

The function $h(x) = x^2 + 9x + 2$ is positive on $x < -8.8$ or $x > -0.2$, or $(-\infty, -8.8) \cup (-0.2, \infty)$. It is negative on the interval $(-8.8, -0.2)$.

77. One example is $x^2 \geq 0$. Any inequality of the form $(\,\cdot\,)^2 \geq 0$ will do it, as will any inequality of the form $(\,\cdot\,)^2 > a$, where a is any negative number.

79. The difference between A and 12.436 has an absolute value less than the error of 0.001.
$|A - 12.436| < 0.001$
$-0.001 < A - 12.436 < 0.001$
$12.435 < A < 12.437$ or, in interval notation, $(12.435, 12.437)$

81. Let x = number of calculators sold
Then Revenue = (price per calculator) × (number of calculators sold) = $63x$
    Cost = fixed cost + variable cost
       = 650,000 + (Cost per calculator) × (number sold)
       = 650,000 + 47x

(A) We want Revenue > Cost
          $63x > 650,000 + 47x$
          $16x > 650,000$
           $x > 40,625$
More than 40,625 calculators must be sold for the company to make a profit.

(B) We want Revenue = Cost
          $63x = 650,000 + 47x$
          $16x = 650,000$
           $x = 40,625$

(C) 40,625 calculators sold represents the break-even point, the boundary between profit and loss.

83. (A) The company might try to increase sales and keep the price the same (see part B). It might try to increase the price and keep the sales the same (see part C). Either of these strategies would need further analysis.

(B) Here the cost has been changed to 650,000 + 50.5x, but the revenue is still $63x$.
   Revenue > Cost
     $63x > 650,000 + 50.5x$
     $12.5x > 650,000$
        $x > 52,000$ calculators

(C) Let p = the new price. Here the cost is still $650,000 + 50.5x$ as in part (B) where x is now known to be $40,625$; now cost = $650,000 + 50.5(40,625)$. The revenue is (price per calculator) × (number of calculators) = $p(40,625)$.

Revenue > Cost

$$p(40,625) > 650,000 + 50.5(40,625)$$

$$p > \frac{650,000 + 50.5(40,625)}{40,625}$$

$$p > 66.50$$

The price could be raised by $3.50 to $66.50.

85. The printer will show a profit if revenue exceeds cost; that is, if $R(x) > C(x)$, or

$$10x - 0.05x^2 > 200 + 2.25x$$
$$-200 + 7.75x - 0.05x^2 > 0$$

The solution of this inequality consists of all values of x for which the graph of $y = -200 + 7.75x - 0.05x^2$ is above the x axis. Graphing and using the ZERO command, we get

To the nearest integer, the printer will show a profit for $33 < x < 122$.

87. The Fahrenheit range is described by $60 < F < 80$

Since $F = \frac{9}{5}C + 32$, we write

$$60 < \frac{9}{5}C + 32 < 80$$

$$28 < \frac{9}{5}C < 48$$

$$\frac{5}{9} \cdot 28 < \frac{5}{9} \cdot \frac{9}{5}C < \frac{5}{9} \cdot 48$$

$$16° < C < 27° \text{ (to the nearest degree)}$$

The Celsius range is 16°C to 27°C.

89. (A) Since the quadratic distance function d has a maximum value of 256 feet after 4 seconds, the vertex of the graph is (4, 256) and the vertex form of the function $d(t)$ is

$$d(t) = a(t - 4)^2 + 256$$

To find a we use the fact that $d(0) = 0$ (the object was propelled from ground level).

$$d(0) = a(0 - 4)^2 + 256 = 0$$
$$16a + 256 = 0$$
$$16a = -256$$
$$a = -16$$

The quadratic function is

$$d(t) = -16(t - 4)^2 + 256$$
$$d(t) = -16(t^2 - 8t - 16) + 256$$
$$d(t) = -16t^2 + 128t - 256 + 256$$
$$d(t) = -16t^2 + 128$$

(B) To find when the object is more than 240 feet above the ground, we solve the inequality $-16t^2 + 128t > 240$, or $-16t^2 + 128t - 240 > 0$.

$$-16t^2 + 128t - 240 > 0 \qquad \text{(Divide both sides by -16)}$$
$$t^2 - 8t + 15 < 0 \qquad \text{(Direction of inequality was reversed!)}$$

Find the zeros of the left side: $(t - 5)(t - 3) = 0$
$$t = 5 \text{ or } t = 3$$

Choose test numbers and make a table:

Interval	Test Number	Result
$t < 3$	1	$(1)^2 - 8(1) + 15 = 8$; positive
$3 < t < 5$	4	$(4)^2 - 8(4) + 15 = -1$; negative
$t > 5$	6	$(6)^2 - 8(1) + 15 = 8$; positive

The solution of the inequality is $3 < t < 5$, so the object is above height 240 for times between 3 and 5 seconds after being propelled.

91. The temperature increases at a constant rate so the model will be a linear function. We're given enough information to find two points on the line. The temperature at 1 km is 30°C so (1, 30) is one point. The temperature increases 2.8°C for each 100 meters, so at a depth of 2 km it will have increased 28°C; at 2 km depth, the temperature is 58°C, so (2, 58) is a second point.

$$m = \frac{58 - 30}{2 - 1} = 28$$
$$y - 30 = 28(x - 1)$$
$$y - 30 = 28x - 28$$
$$y = 28x + 2$$

The model is $T(x) = 28x + 2$

We're asked to solve the inequality $150 < 28x + 2 < 200$.
$$150 < 28x + 2 < 200$$
$$148 < 28x < 198$$
$$5.286 < x < 7.071$$

Depths between 5.286 km and 7.071 km will have temperatures between 150°C and 200°C.

93. The independent variable is the altitude so we enter the numbers in the first height column as L_1. The dependent variable is the temperature so we enter the first temperature column as L_2.

(Note that the entire list is not visible on this diagram.)
We use the linear regression command on the STAT CALC menu:

The linear model is $y = -0.00738x + 28.2$.

We're asked to solve the inequality below:
$$-30 < -0.00738x + 28.2 < -10$$
$$-58.2 < -0.00738x < -38.2$$
$$7,890 > \qquad x \qquad > 5,180 \qquad \text{(Note that the direction reversed)}$$

The temperature will be between -10°C and -30°C at altitudes between 5,180 meters and 7,890 meters.

95. The independent variable is altitude so we enter the first height column as L_1. The dependent variable is pressure so we enter the first pressure column as L_2.

(Note that the entire list is not visible on this diagram.)

We use the linear regression command on the STAT CALC menu:

The linear model is $y = -0.0623x + 877$.

We are asked to solve the inequality below:
$$350 < -0.0623x + 877 < 650$$
$$-527 < -0.0623x < -227$$
$$8,460 > \qquad x \qquad > 3,640$$

The pressure will be between 350 hpa and 650 hpa between 3,640 and 8,460 meters.

97. (A) The independent variable is demand so we enter the orange juice demand column as L_1. The dependent variable is price so we enter the orange juice price column as L_2.

We use the linear regression command on the STAT CALC menu.

The linear model is $p = d(x) = -0.000\ 0727x + 3.50$.

To find the domain we first note that x must be at least zero. Also, the price cannot be negative. Using the graph and the ZERO command, we find that price is negative when demand is greater than 48,100 (rounded to 3 significant digits), so the domain is $0 \le x \le 48,100$.

The revenue function is
$$R(x) = xp$$
$$= x(-0.000\ 0727x + 3.50)$$
$$= -0.000\ 0727x^2 + 3.50x$$
The domain is the same as the domain of $d(x)$.

The cost function is
$$C(x) = 20,000 + 0.50x$$
$$\text{Fixed costs} + \$0.50 \text{ per gallon produced}$$

The domain is $x \ge 0$. (There's no stated limitation on how much they can sell.)

(C) The company will break even when revenue = cost so we graph $y_1 = R(x) = -0.000\ 0727x^2 + 3.50x$ and $y_2 = C(x) = 20,000 + 0.50x$ and find the intersection.

The revenue and cost graphs intersect at $x = 8,360$ and $x = 32,900$ so the company will break even if they sell 8,360 gallons or 32,900 gallons. They'll make a profit if they sell between these two levels, and will lose money if they sell less than 8,360 gallons or more than 32,900 gallons.

(D) The profit is calculated by subtracting cost from revenue so the profit function is
$$P(x) = R(x) - C(x) =$$
$$P(x) = -0.000\ 0727x^2 + 3.50x - (20,000 + 0.50x)$$
$$P(x) = -0.000\ 0727x^2 + 3.00x - 20,000$$
Graph the profit function and use the MAXIMUM command

The maximum profit of \$10,900 occurs when 20,600 gallons are sold.

CHAPTER 2 REVIEW

1. The points at two of the vertices of the triangle are (-4, 3) and (1, 1). The line moves 2 units down between these points so the rise is -2. The line moves 5 units to the right between these points so the run is 5. The slope is therefore $m = -\frac{2}{5}$.

$$y - 1 = -\frac{2}{5}(x - 1)$$
$$y - 1 = -\frac{2}{5}x + \frac{2}{5} \quad \text{(Multiply both sides by 5)}$$
$$5y - 5 = -2x + 2$$
$$2x + 5y = 7 \quad \quad (2\text{-}1)$$

2. $3x + 2y = 9$
$$2y = -3x + 9$$
$$y = -\frac{3}{2}x + \frac{9}{2}$$

Slope: $-\frac{3}{2}$

To graph, find a few points by substituting values.

x	y
0	$\frac{9}{2}$
3	0
1	3

$(2\text{-}1)$

3. The line passes through the two given points, (6, 0) and (0, 4), so its slope is given by
$$m = \frac{0 - 4}{6 - 0} = \frac{-4}{6} = -\frac{2}{3}$$
The equation of the line is, therefore, using the point-slope form,
$$y - 0 = -\frac{2}{3}(x - 6)$$
or $\quad 3y = -2(x - 6)$
or $\quad 3y = -2x + 12$.
$$2x + 3y = 12 \quad \quad (2\text{-}1)$$

4. $y = mx + b \quad \quad m = -\frac{2}{3} \quad \quad b = 2$
$$y = -\frac{2}{3}x + 2 \quad \quad (2\text{-}1)$$

5. *vertical:* $x = -3$, slope not defined; *horizontal:* $y = 4$, slope = 0 $\quad \quad (2\text{-}1)$

6. (A) $0.05x + 0.25(30 - x) = 3.3$
$$0.05x + 7.5 - 0.25x = 33$$
$$-0.2x + 7.5 = 3.3$$
$$-0.2x = -4.2$$
$$x = 21$$

(B) $\frac{5x}{3} - \frac{4 + x}{2} = \frac{x - 2}{4} + 1 \quad$ LCD = 12
$$4(5x) - 6(4 + x) = 3(x - 2) + 12$$
$$20x - 24 - 6x = 3x - 6 + 12$$
$$14x - 24 = 3x + 6$$
$$11x = 30$$
$$x = \frac{30}{11} \approx 2.727$$

Check:

Check:

$(2\text{-}2)$

7. (A) $f(x) = -x^2 - 2x + 3$

$\qquad = -(x^2 + 2x + ?) + 3 \qquad \frac{1}{2}(2) = 1 \qquad 1^2 = 1$

$\qquad = -(x^2 + 2x + 1) + 3 + 1$

$\qquad = -(x + 1)^2 + 4$

(B) The graph of the function is the same as the graph of $y = x^2$ reflected in the x axis, shifted left 1 unit, and up 4 units.

(C) Solve $0 = -x^2 - 2x + 3$ Check:

$\qquad\quad 0 = x^2 + 2x - 3$

$\qquad\quad 0 = (x + 3)(x - 1)$

$\qquad x + 3 = 0 \quad$ or $\quad x - 1 = 0$

$\qquad\quad\; x = -3 \qquad\qquad\; x = 1$

x intercepts: -3, 1

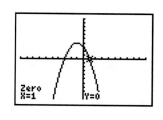

$(2\text{-}3,\ 2\text{-}5)$

8. (A) $f(x) = x^2 - 3x - 2$

$\qquad = (x^2 - 3x + ?) - 2 \qquad \frac{1}{2}(-3) = -\frac{3}{2} \qquad \left(-\frac{3}{2}\right)^2 = \frac{9}{4}$

$\qquad = \left(x^2 - 3x + \frac{9}{4}\right) - 2 - \frac{9}{4}$

$\qquad = \left(x - \frac{3}{2}\right)^2 - \frac{17}{4}$

(B) The graph of the function is the same as the graph of $y = x^2$ shifted right $\frac{3}{2}$ units and down $\frac{17}{4}$ units.

(C) $x^2 - 3x - 2 = 0$ Check:

$\quad x = \dfrac{-b \pm \sqrt{b^2 - 4ac}}{2a} \qquad a = 1,\ b = -3,\ c = -2$

$\quad x = \dfrac{-(-3) \pm \sqrt{(-3)^2 - 4(1)(-2)}}{2(1)}$

$\quad x = \dfrac{3 \pm \sqrt{9 + 8}}{2}$

$\quad x = \dfrac{3 \pm \sqrt{17}}{2}$

x intercepts: $\dfrac{3 + \sqrt{17}}{2} \approx 3.5616$

$\qquad\qquad\quad \dfrac{3 - \sqrt{17}}{2} \approx -0.5616$

$(2\text{-}3,\ 2\text{-}5)$

9. Use the vertex formula:

$\quad x = -\dfrac{b}{2a} = -\dfrac{9}{2(-3)} = -\dfrac{9}{-6} = \dfrac{3}{2}$

To find the second coordinate, plug back into f:

$f\left(\dfrac{3}{2}\right) = -3\left(\dfrac{3}{2}\right)^2 + 9\left(\dfrac{3}{2}\right) - 4 = -3\left(\dfrac{9}{4}\right) + \dfrac{27}{2} - 4 = -\dfrac{27}{4} + \dfrac{54}{4} - \dfrac{16}{4} = \dfrac{11}{4}$

The vertex is $\left(\dfrac{3}{2}, \dfrac{11}{4}\right)$; since $a = -3$ is negative, the parabola opens down and the vertex is a maximum.

$(2\text{-}3)$

10. (A) $(-3 + 2i) + (6 - 8i) = -3 + 2i + 6 - 8i = 3 - 6i$

(B) $(3 - 3i)(2 + 3i) = 6 + 3i - 9i^2 = 6 + 3i + 9 = 15 + 3i$

(C) $\dfrac{13 - i}{5 - 3i} = \dfrac{(13 - i)}{(5 - 3i)}\dfrac{(5 + 3i)}{(5 + 3i)}$

$= \dfrac{65 + 34i - 3i^2}{25 - 9i^2} = \dfrac{65 + 34i + 3}{25 + 9} = \dfrac{68 + 34i}{34} = 2 + i$

(D) $(7 + 6i) - (8 + 9i) = 7 + 6i - 8 - 9i = -1 - 3i$ $\hspace{2cm}$ *(2-4)*

11. $4 - 3(2x + 7) = 5(4 - x) + 11x - 3$

$\qquad 4 - 6x - 21 = 20 - 5x + 11x - 3$

$\qquad\quad -6x - 17 = 17 + 6x$

$\qquad\qquad -12x = 34$

$\qquad\qquad\quad x = -\dfrac{34}{12} = -\dfrac{17}{6} \approx -2.8$

Check: Graph $y_1 = 4 - 3(2x + 7)$ and $y_2 = 5(4 - x) + 11x - 3$, use the INTERSECT command:

$\hspace{11cm}$ *(2-2)*

12. $(2x + 5)^2 = 23$

$\qquad 2x + 5 = \pm\sqrt{23}$

$\qquad\quad 2x = -5 \pm \sqrt{23}$

$\qquad\quad\; x = \dfrac{-5 \pm \sqrt{23}}{2} \approx -0.1, -4.9$

Check: Graph $y_1 = (2x + 5)^2$ and $y_2 = 23$, use the INTERSECT command:

$\hspace{11cm}$ *(2-5)*

13. $x^2 - 2x + 3 = x^2 + 3x - 7$ Check:

$\qquad -2x + 3 = 3x - 7$

$\qquad -5x + 3 = -7$

$\qquad\quad -5x = -10$

$\qquad\qquad x = 2$

$\hspace{9cm}$ *(2-1, 2-5)*

14. $2x^2 - 7 = 0$ Check:

$$2x^2 = 7$$

$$x^2 = \frac{7}{2}$$

$$x = \pm\sqrt{\frac{7}{2}}$$

$$x = \pm\frac{\sqrt{14}}{2} \approx \pm 1.8708$$

(2-5)

15. $2x^2 = 4x$ Check:

$$2x^2 - 4x = 0$$

$$2x(x - 2)$$

$$2x = 0 \quad x - 2 = 0$$

$$x = 0 \qquad x = 2$$

 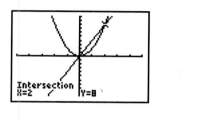

(2-5)

16. $2x^2 = 7x - 3$

$$2x^2 - 7x + 3 = 0$$

$$(2x - 1)(x - 3) = 0$$

$$2x - 1 = 0 \quad x - 3 = 0$$

$$x = \frac{1}{2} \qquad x = 3$$

(2-5)

17. $m^2 + m + 1 = 0$

$$m = \frac{-b \pm \sqrt{b^2 - 4ac}}{2a}$$

$a = 1$

$b = 1$

$c = 1$

$$m = \frac{-1 \pm \sqrt{(1)^2 - 4(1)(1)}}{2(1)}$$

$$m = \frac{-1 \pm \sqrt{-3}}{2}$$

$$m = \frac{-1 \pm i\sqrt{3}}{2}$$

$$m = -\frac{1}{2} \pm \frac{\sqrt{3}}{2}i$$

(2-5)

18.
$$y^2 = \frac{3}{2}(y + 1)$$
$$2y^2 = 3(y + 1)$$
$$2y^2 = 3y + 3$$
$$2y^2 - 3y - 3 = 0$$
$$y = \frac{-b \pm \sqrt{b^2 - 4ac}}{2a} \qquad a = 2$$
$$b = -3$$
$$c = -3$$
$$y = \frac{-(-3) \pm \sqrt{(-3)^2 - 4(2)(-3)}}{2(2)}$$
$$y = \frac{3 \pm \sqrt{33}}{4}$$

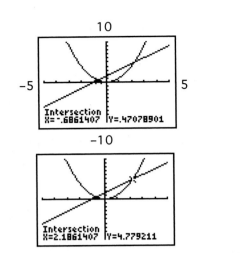

(2-5)

19. $\sqrt{4 + 7x} = 5$
$$4 + 7x = 25$$
$$7x = 21$$
$$x = 3$$

Algebraic check: $\sqrt{4 + 7(3)} = \sqrt{25} = 5$ ✓

Graphical check:

(2-6)

20. $\sqrt{5x - 6} - x = 0$
$$\sqrt{5x - 6} = x$$
$$5x - 6 = x^2$$
$$0 = x^2 - 5x + 6$$
$$x^2 - 5x + 6 = 0$$
$$(x - 3)(x - 2) = 0$$
$$x = 2, \ 3$$

Check: $\sqrt{5(2) - 6} - 2 \overset{?}{=} 0$
$$0 \overset{\checkmark}{=} 0$$
$$\sqrt{5(3) - 6} - 3 \overset{?}{=} 0$$
$$0 \overset{\checkmark}{=} 0$$

Solution: $x = 2, \ 3$

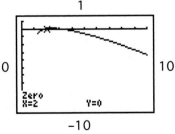

(2-6)

21. $\sqrt{x^2 + 2} = \sqrt{4x + 14}$
$$x^2 + 2 = 4x + 14$$
$$x^2 - 4x - 12 = 0$$
$$(x - 6)(x + 2) = 0$$
$$x = 6 \quad \text{or} \quad x = -2$$

Algebraic check: $\sqrt{6^2 + 2} = \sqrt{38}; \quad \sqrt{4(6) + 14} = \sqrt{38}$ ✓

$$\sqrt{(-2)^2 + 2} = \sqrt{6}; \quad \sqrt{4(-2) + 14} = \sqrt{6}$$ ✓

Graphical check:

(*2-6*)

22. $3(2 - x) - 2 \leq 2x - 1$

$\quad\ 6 - 3x - 2 \leq 2x - 1$

$\quad\quad -3x + 4 \leq 2x - 1$

$\quad\quad\quad\quad -5x \leq -5$

$\quad\quad\quad\quad\quad\ x \geq 1$

$[1, \infty)$

(*2-7*)

23. Rearrange so that zero is on the right: $x^2 + x - 20 < 0$

Find the zeros of the left side: $x^2 + x - 20 = 0$

$$(x + 5)(x - 4) = 0$$

$$x = -5 \quad \text{or} \quad x = 4$$

Choose test numbers and make a table:

Interval	Test Number	Result
$x < -5$	-6	$(-6)^2 + (-6) - 20 = 10;$ positive
$-5 < x < 4$	0	$(0)^2 + (0) - 20 = -20;$ negative
$x > 4$	5	$(5)^2 + (5) - 20 = 10;$ positive

The expression $x^2 + x - 20$ is negative on $-5 < x < 4$, or $(-5, 4)$.   (*2-7*)

24. Rearrange so that zero is on the right: $x^2 - 4x - 12 > 0$

Find zeros of the left side: $x^2 - 4x - 12 = 0$

$$(x - 6)(x + 2) = 0$$

$$x = 6, -2$$

Choose test numbers and make a table:

Interval	Test Number	Result
$x < -2$	-3	$(-3)^2 - 4(-3) - 12 = 9;$ positive
$-2 < x < 6$	0	$(0)^2 - 4(0) - 12 = -12;$ negative
$x > 6$	10	$(10)^2 - 4(10) - 12 = 48;$ positive

The expression $x^2 - 4x - 12$ is positive on $x < -2$ or on $x > 6$, or $(-\infty, -2) \cup (6, \infty)$.

(*2-7*)

25. (A) A line with positive slope is rising as x moves from left to right. It is the graph of an increasing function.

(B) A line with negative slope is falling as x moves from left to right. It is the graph of a decreasing function.

(C) A parabola that opens upward and has vertex at (h, k) is falling to the left of the vertex and rising to the right. It is the graph of a function that is decreasing on $(-\infty, h]$ and increasing on $[h, \infty)$.

(D) A parabola that opens downward and has vertex at (h, k) is rising to the left of the vertex and falling to the right. It is the graph of a function that is increasing on $(-\infty, h]$ and decreasing on $[h, \infty)$. *(2-1, 2-3)*

26. Since two points are given, we find the slope, then apply the point-slope form.

$$m = \frac{-3 - 3}{0 - (-4)} = \frac{-6}{4} = -\frac{3}{2}$$

$$y - 3 = -\frac{3}{2}[x - (-4)]$$
$$2(y - 3) = -3(x + 4)$$
$$2y - 6 = -3x - 12$$
$$3x + 2y = -6$$

(2-1)

27. The line $6x + 3y = 5$, or
$3y = -6x + 5$,

or $y = -2x + \frac{5}{3}$, has slope -2.

(A) We require a line through $(-2, 1)$, with slope -2. Applying the point-slope form, we have

$$y - 1 = -2[x - (-2)]$$
$$y - 1 = -2x - 4$$
$$y = -2x - 3$$

(B) A line perpendicular to one with slope -2 has slope $\frac{1}{2}$, so we need a line

with slope $m = \frac{1}{2}$. Again applying the point-slope form, we have

$$y - 1 = \frac{1}{2}[x - (-2)]$$

$$y - 1 = \frac{1}{2}x + 1$$

$$y = \frac{1}{2}x + 2$$

(2-1)

28. $|y + 9| < 5$
$-5 < y + 9 < 5$
$-14 < y < -4$ *(2-7)*

29. $|2x - 8| \geq 3$
$2x - 8 \geq 3$ or $2x - 8 \leq -3$
$2x \geq 11$ $2x \leq 5$
$x \geq 5.5$ or $x \leq 2.5$ *(2-2)*

30. Rearrange so that zero is on the right: $2x^2 - 7x + 1 \geq 0$
Find zeros of the left side: $2x^2 - 7x + 1 = 0$ $a = 2, b = -7, c = 1$

$$x = \frac{-(-7) \pm \sqrt{(-7)^2 - 4(2)(1)}}{2(2)}$$

$$= \frac{7 \pm \sqrt{41}}{4} \approx 3.4, 0.1$$

Choose test numbers and make a table:

Interval	Test Number	Result
$x < 0.1$	0	$2(0)^2 - 7(0) + 1 = 1$; positive
$0.1 < x < 3.4$	1	$2(1)^2 - 7(1) + 1 = -4$; negative
$x > 3.4$	5	$2(5)^2 - 7(5) + 1 = 16$; positive

The expression $2x^2 - 7x + 1$ is zero or positive on $x \leq 0.1$ or on $x \geq 3.4$. *(2-7)*

31. The distance between y and 5 is no more than 2.
$$|y - 5| \leq 2$$
$$-2 \leq y - 5 \leq 2$$
$$3 \leq y \leq 7 \quad \text{or} \quad [3, 7] \qquad (2\text{-}7)$$

32. The distance between t and -6 is greater than 9.
$$|t - (-6)| > 9, \text{ or } |t + 6| > 9$$
$$t + 6 > 9 \quad \text{or} \quad t + 6 < -9$$
$$t > 3 \quad \text{or} \qquad t < -15$$
$$(-\infty, -15) \cup (3, \infty) \qquad (2\text{-}7)$$

33. $0.1x^2 + x + 1.5 = 0 \quad a = 0.1, \ b = 1, \ c = 1.5$
$b^2 - 4ac = 1^2 - 4(0.1)(1.5) = 1 - 0.6 = 0.4$
The discriminant is positive so there are 2 real zeros. See check below. $(2\text{-}5)$

34. $0.1x^2 + x + 2.5 = 0 \quad a = 0.1, \ b = 1, \ c = 2.5$
$b^2 - 4ac = 1^2 - 4(0.1)(2.5) = 1 - 1 = 0$
The discriminant is zero so there is one real zero. See check below. $(2\text{-}5)$

35. $0.1x^2 + x + 3.5 = 0 \quad a = 0.1, \ b = 1, \ c = 3.5$
$b^2 - 4ac = 1^2 - 4(0.1)(3.5) = 1 - 1.4 = -0.4$
The discriminant is negative so there are two imaginary roots.

Check for questions 30 — 32:

  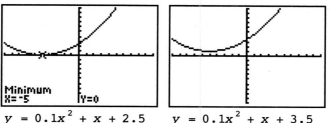

$y = 0.1x^2 + x + 1.5 \qquad y = 0.1x^2 + x + 2.5 \qquad y = 0.1x^2 + x + 3.5 \qquad (2\text{-}5)$

36. (A) $f(x) = 0.5x^2 - 4x + 5$
$$= 0.5(x^2 - 8x) + 5$$
$$= 0.5(x^2 - 8x + 16) + 5 - 0.5(16)$$
$$= 0.5(x - 4)^2 - 3$$

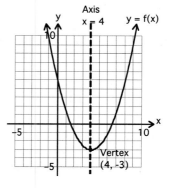

Therefore the line $x = 4$ is the axis of the parabola and $(4, -3)$ is its vertex.

(B) The parabola opens upward, since $0.5 = a > 0$. So, the parabola is decreasing on $(-\infty, 4]$ and increasing on $[4, \infty)$. The range is $[-3, \infty)$.

(C) The minimum value of -3 occurs at $x = 4$. There is no maximum. $(2\text{-}3)$

37. g is a linear function. Its graph passes through $(x_1, y_1) = (-1, 0)$ and $(x_2, y_2) = (1, 4)$. Therefore the slope of the line is
$$m = \frac{y_2 - y_1}{x_2 - x_1} = \frac{4 - 0}{1 - (-1)} = 2. \text{ We use the point-slope form:}$$
$$y - y_1 = m(x - x_1)$$
$$y - 0 = 2[x - (-1)]$$
$$y = 2x + 2$$
The function is given by $g(x) = 2x + 2$.
f is a quadratic function. Its equation must be of the form
$f(x) = y = a(x - h)^2 + k$. The vertex of the parabola is at $(1, 2)$.

Therefore the equation must have form
$$y = a(x - 1)^2 + 2$$
Since the parabola passes through (3, 0), these coordinates must satisfy the equation.
$$0 = a(3 - 1)^2 + 2$$
$$0 = 4a + 2$$
$$-2 = 4a$$
$$a = -0.5$$
The equation is
$$y = f(x) = -0.5(x - 1)^2 + 2$$
$$f(x) = -0.5(x^2 - 2x + 1) + 2$$
$$f(x) = -0.5x^2 + x - 0.5 + 2$$
$$f(x) = -0.5x^2 + x + 1.5 \qquad\qquad (2\text{-}1, \ 2\text{-}3)$$

38. (A) $(3 + i)^2 - 2(3 + i) + 3 = 9 + 6i + i^2 - 6 - 2i + 3$
$$= 9 + 6i - 1 - 6 - 2i + 3 = 5 + 4i$$

(B) $i^{27} = i^{26}i = (i^2)^{13}i = (-1)^{13}i = (-1)i = -i \qquad\qquad (2\text{-}4)$

39. (A) $(2 - \sqrt{-4}) - (3 - \sqrt{-9}) = (2 - i\sqrt{4}) - (3 - i\sqrt{9}) = (2 - 2i) - (3 - 3i)$
$$= 2 - 2i - 3 + 3i = -1 + i$$

(B) $\dfrac{2 - \sqrt{-1}}{3 + \sqrt{-4}} = \dfrac{2 - i\sqrt{1}}{3 + i\sqrt{4}} = \dfrac{2 - i}{3 + 2i} = \dfrac{(2 - i)}{(3 + 2i)}\dfrac{(3 - 2i)}{(3 - 2i)} = \dfrac{6 - 7i + 2i^2}{9 - 4i^2} = \dfrac{6 - 7i - 2}{9 + 4}$

$$= \dfrac{4 - 7i}{13} = \dfrac{4}{13} - \dfrac{7}{13}i$$

(C) $\dfrac{4 + \sqrt{-25}}{\sqrt{-4}} = \dfrac{4 + i\sqrt{25}}{i\sqrt{4}} = \dfrac{4 + 5i}{2i} = \dfrac{4 + 5i}{2i}\dfrac{i}{i} = \dfrac{4i + 5i^2}{2i^2} = \dfrac{4i - 5}{-2} = \dfrac{5}{2} - 2i$

(D) $-\sqrt{-16} \cdot \sqrt{-25} = 4i \cdot 5i = 20i^2 = 20(-1) = -20 \qquad\qquad (2\text{-}4)$

40. $\left(x + \dfrac{5}{2}\right)^2 = \dfrac{5}{4}$ Graphical check: $\dfrac{-5 + \sqrt{5}}{2} \approx -1.382$

$$x + \dfrac{5}{2} = \pm\sqrt{\dfrac{5}{4}} \qquad\qquad\qquad \dfrac{-5 - \sqrt{5}}{2} \approx -3.618$$

$$x + \dfrac{5}{2} = \pm\dfrac{\sqrt{5}}{2}$$ Graph $y_1 = \left(x + \dfrac{5}{2}\right)^2$ and $y_2 = \dfrac{5}{4}$ and find the intersections

$$x = -\dfrac{5}{2} \pm \dfrac{\sqrt{5}}{2}$$ using the INTERSECT command.

$$x = \dfrac{-5 \pm \sqrt{5}}{2}$$

$$(2\text{-}5)$$

41. $1 + \dfrac{3}{u^2} = \dfrac{2}{u}$ Excluded value: $u \neq 0$

$$u^2 + 3 = 2u$$
$$u^2 - 2u = -3$$
$$u^2 - 2u + 1 = -2$$
$$(u - 1)^2 = -2$$
$$u - 1 = \pm\sqrt{-2}$$
$$u = 1 \pm \sqrt{-2}$$
$$u = 1 \pm i\sqrt{2}$$ The solutions are imaginary and cannot be confirmed
graphically. $(2\text{-}6)$

42. $2x + 3\sqrt{4x^2 - 4x + 9} = 1$

$3\sqrt{4x^2 - 4x + 9} = 1 - 2x$

$9(4x^2 - 4x + 9) = 4x^2 - 4x + 1$

$36x^2 - 36x + 81 = 4x^2 - 4x + 1$

$32x^2 - 32x + 80 = 0$

$2x^2 - 2x + 5 = 0$

$$x = \frac{-b \pm \sqrt{b^2 - 4ac}}{2a} \qquad a = 2$$

$b = -2$

$c = 5$

$$x = \frac{-(-2) \pm \sqrt{(-2)^2 - 4(2)(5)}}{2(2)} = \frac{2 \pm \sqrt{-36}}{4} = \frac{2 \pm 6i}{4} = \frac{1 \pm 3i}{2} \quad \text{or} \quad \frac{1}{2} \pm \frac{3}{2}i$$

Check: $\dfrac{1}{2} + \dfrac{3}{2}i$:

$$2\left(\frac{1}{2} + \frac{3}{2}i\right) + 3\sqrt{4\left(\frac{1}{2} + \frac{3}{2}i\right)^2 - 4\left(\frac{1}{2} + \frac{3}{2}i\right) + 9} \overset{?}{=} 1$$

$$1 + 3i + 3\sqrt{1 + 6i + 9i^2 - 2 - 6i + 9} \overset{?}{=} 1$$

$$1 + 3i + 3\sqrt{1 + 6i - 9 - 2 - 6i + 9} \overset{?}{=} 1$$

$$1 + 3i + 3\sqrt{-1} \overset{?}{=} 1$$

$$1 + 3i + 3i \overset{?}{=} 1$$

$$1 + 6i \neq 1$$

Not a solution

Check: $\dfrac{1}{2} - \dfrac{3}{2}i$: $2\left(\dfrac{1}{2} - \dfrac{3}{2}i\right) + 3\sqrt{4\left(\dfrac{1}{2} - \dfrac{3}{2}i\right)^2 - 4\left(\dfrac{1}{2} - \dfrac{3}{2}i\right) + 9} \overset{?}{=} 1$

$$1 - 3i + 3\sqrt{1 - 6i + 9i^2 - 2 + 6i + 9} \overset{?}{=} 1$$

$$1 - 3i + 3\sqrt{1 - 6i - 9 - 2 + 6i + 9} \overset{?}{=} 1$$

$$1 - 3i + 3\sqrt{-1} \overset{?}{=} 1$$

$$1 - 3i + 3i \overset{\checkmark}{=} 1 \quad \text{A solution}$$

Solution: $\dfrac{1}{2} - \dfrac{3}{2}i$ The solutions cannot be confirmed graphically.   *(2-6)*

43. $2x^{2/3} - 5x^{1/3} - 12 = 0$ Graphical check: Graph $y = 2x^{2/3} - 5x^{1/3} - 12$ and
Let $u = x^{1/3}$, then   find the zeros using the ZERO command.

$2u^2 - 5u - 12 = 0$

$(2u + 3)(u - 4) = 0$

$u = -\dfrac{3}{2}, \; 4$

$x^{1/3} = -\dfrac{3}{2} \qquad x^{1/3} = 4$

$x = -\dfrac{27}{8} \qquad x = 64$

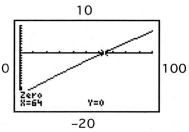

(2-6)

44. $m^4 + 5m^2 - 36 = 0$
Let $u = m^2$, then
$$u^2 + 5u - 36 = 0$$
$$(u + 9)(u - 4) = 0$$
$$u = -9, \; 4$$
$$m^2 = -9 \qquad m^2 = 4$$
$$m = \pm 3i \qquad m = \pm 2$$

Graphical check: Graph $y = x^4 + 5x^2 - 36$ and find the zeros using a built-in routine.

 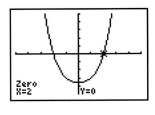

$(2-6)$

45. $\sqrt{y - 2} - \sqrt{5y + 1} = -3$
$$-\sqrt{5y + 1} = -3 - \sqrt{y - 2}$$
$$\left(-\sqrt{5y + 1}\right)^2 = \left(-3 - \sqrt{y - 2}\right)^2$$
$$5y + 1 = 9 + 6\sqrt{y - 2} + y - 2$$
$$5y + 1 = y + 7 + 6\sqrt{y - 2}$$
$$4y - 6 = 6\sqrt{y - 2}$$
$$2y - 3 = 3\sqrt{y - 2}$$
$$4y^2 - 12y + 9 = 9(y - 2)$$
$$4y^2 - 12y + 9 = 9y - 18$$
$$4y^2 - 21y + 27 = 0$$
$$(4y - 9)(y - 3) = 0$$
$$y = \tfrac{9}{4}, \; 3$$

> **Common Error:**
> $y - 2 - 5y + 1 = 9$
> is not equivalent to the equation formed by squaring both members of the given equation.

Check:
$$\sqrt{\tfrac{9}{4} - 2} - \sqrt{5\left(\tfrac{9}{4}\right) + 1} \stackrel{?}{=} -3$$
$$\sqrt{\tfrac{1}{4}} - \sqrt{\tfrac{49}{4}} \stackrel{?}{=} -3$$
$$-3 \stackrel{\checkmark}{=} -3$$
$$\sqrt{3 - 2} - \sqrt{5(3) + 1} \stackrel{?}{=} -3$$
$$-3 \stackrel{\checkmark}{=} -3$$

Solution: $\tfrac{9}{4}$, 3

Graphical check: Graph $y1 = \sqrt{x - 2} - \sqrt{5x + 1}$ and $y2 = -3$ and find the zeros using a built-in routine solutions using the INTERSECT command.

 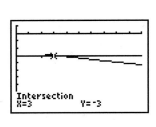

$(2-6)$

46. Enter the x column as L_1 and the y column as L_2 then choose the linear regression command on the STAT CALC menu.

(A)

The regression line is $y = 2x - 5$.

(B)

The regression line is $y = .5x + 2.5$

The graphs are symmetric with respect to the line $y = x$; the two functions are inverses.   *(2-2)*

47. It's possible for a quadratic function to have only imaginary zeros. For example, $f(x) = x^2 + 4$ has zeros $\pm 2i$, $g(x) = x^2 + 4x + 5$ has zeros $-2 \pm i$. If a quadratic function $f(x) = ax^2 + bx + c$ has only imaginary zeros, they will be

$$x = \frac{-b \pm i\sqrt{4ac - b^2}}{2a}$$ and therefore will be a pair of complex conjugate numbers. *(2-5)*

48. The graph of any quadratic function is a parabola. If the function has only imaginary zeros, it still can be graphed. The parabola will simply not intersect the x axis.   *(2-5)*

49. In this problem, $a = 1$, $b = -6$, $c = c$. The discriminant is $b^2 - 4ac =$ $(-6)^2 - 4(1)(c) = 36 - 4c$:

If $36 - 4c > 0$, (in which case $36 > 4c$ or $c < 9$), there are two distinct real roots.

If $36 - 4c = 0$, (in which case $c = 9$), there is one real double root.

If $36 - 4c < 0$, (in which case $36 < 4c$ or $c > 9$), there are two distinct imaginary roots.   *(2-5)*

50.
$$P = M - Mdt$$
$$M - Mdt = P$$
$$M(1 - dt) = P$$
$$M = \frac{P}{1 - dt}$$ *(2-2)*

51.
$$P = EI - RI^2$$
$$RI^2 - EI + P = 0$$
$$I = \frac{-b \pm \sqrt{b^2 - 4ac}}{2a}$$ $a = R$
$$b = -E$$
$$c = P$$
$$I = \frac{-(-E) \pm \sqrt{(-E)^2 - 4(R)(P)}}{2(R)}$$
$$I = \frac{E \pm \sqrt{E^2 - 4PR}}{2R}$$ *(2-5)*

52. The given inequality can be simplified as follows:
$$a + b < b - a$$
$$2a + b < b$$
$$2a < 0$$
$$a < 0$$

This tells us that the truth of the inequality has nothing to do with b; it is true as long as a is negative for any value of b.   *(2-7)*

53. If $a > b$ and b is negative, then $\dfrac{a}{b} < \dfrac{b}{b}$, that is, $\dfrac{a}{b} < 1$, since dividing both sides by b reverses the order of the inequality. $\dfrac{a}{b}$ is less than 1. (*2-7*)

54. The distance from x to 6 is between zero and d; this distance is less than d if x is between $6 - d$ and $6 + d$. But in order for that distance to be nonzero, x cannot be equal to 6. So the solution is all numbers between $6 - d$ and $6 + d$ except 6, or $(6 - d, 6) \cup (6, 6 + d)$. (*2-7*)

55. $(a + bi)\left(\dfrac{a}{a^2 + b^2} - \dfrac{b}{a^2 + b^2}i\right)$

$= \dfrac{(a + bi)}{1}\left(\dfrac{a}{a^2 + b^2} - \dfrac{bi}{a^2 + b^2}\right)$

$= \dfrac{a(a + bi)}{a^2 + b^2} - \dfrac{bi(a + bi)}{a^2 + b^2}$

$= \dfrac{a^2 + abi - abi - b^2 i^2}{a^2 + b^2}$

$= \dfrac{a^2 + b^2}{a^2 + b^2} = 1$ (*2-4*)

56. If $m = 0$, the equations are reduced to $-y = b$ (a horizontal line) and $x = b$ (a vertical line). In this case, the graphs are perpendicular. Otherwise, $m \neq 0$. Solving for y yields

$$mx - y = b \qquad\qquad x + my = b$$
$$mx = y + b \qquad\qquad my = -x + b$$
$$y = mx - b \qquad\qquad y = -\dfrac{1}{m}x + \dfrac{b}{m}$$

The first line has slope m and the second has slope $-\dfrac{1}{m}$. The graphs are perpendicular in this case also. (*2-1*)

57. $2(2 + 3i)^2 - 8(2 + 3i) + 26$

$= 2\left(4 + 12i + 9i^2\right) - 16 - 24i + 26$

$= 8 + 24i - 18 - 16 - 24i + 26$

$= 0$

The imaginary zeros of quadratic equations come in conjugate pairs, so the other one must be the conjugate of $2 + 3i$, or $2 - 3i$. (*2-5*)

58. The other root of the equation must be $5 - 2i$ (conjugate pairs), so the left side of the equation must factor as $(x - (5 + 2i))(x - (5 - 2i))$. Multiply out the parentheses:

$x^2 - (5 + 2i)x - (5 - 2i)x + (5 - 2i)(5 + 2i)$

$= x^2 - 5x - 2ix - 5x + 2ix + 25 - 10i + 10i - 4i^2$

$= x^2 - 10x + 25 - 4(-1)$

$= x^2 - 10x + 29$

So $b = -10$ and $c = 29$. (*2-5*)

59. Algebraically:
$3x^{-2/5} - 4x^{-1/5} + 1 = 0$
Let $u = x^{-1/5}$, then
$$3u^2 - 4u + 1 = 0$$
$$(3u - 1)(u - 1) = 0$$
$$u = \frac{1}{3}, \ 1$$

$x^{-1/5} = \frac{1}{3}$ $x^{-1/5} = 1$

$x^{1/5} = 3$ $x^{1/5} = 1$

$x = 243$ $x = 1$

Graphically: Graph $y = 3x^{-2/5} - 4x^{-1/5} + 1$ and use the ZERO command to find the zeros.

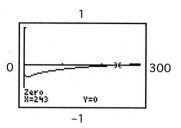

$(2-6)$

60. $x^3 + 1 = 0$
Factor using the sum of cubes formula:
$(x + 1)(x^2 - x + 1) = 0$
$x + 1 = 0$ $x^2 - x + 1 = 0$

$x = -1$ $x = \dfrac{-b \pm \sqrt{b^2 - 4ac}}{2a}$ $a = 1$
$b = -1$
$c = 1$

$$x = \frac{-(-1) \pm \sqrt{(-1)^2 - 4(1)(1)}}{2(1)}$$

$$= \frac{1 \pm \sqrt{-3}}{2}$$

$$= \frac{1 \pm i\sqrt{3}}{2}$$

Solutions: $x = -1, \ \dfrac{1 \pm i\sqrt{3}}{2}$ $(2-5)$

61. Let the three integers be represented by x, $x + 1$, and $x + 2$. Then
$$x + (x + 1) + (x + 2) = 144$$
(The sum of the integers is 144.)
$$3x + 3 = 144$$
$$3x = 141$$
$$x = 47$$
The integers are 47, 48, and 49. $(2-2)$

62. Let the three consecutive integers be represented by x, $x + 2$, and $x + 4$.
(Since all are even, they must be separated by 2.) Then
$$x + 2(x + 2) = 2(x + 4)$$
(First plus two times second is twice the third.)
$$x + 2x + 4 = 2x + 8$$
$$3x + 4 = 2x + 8$$
$$x = 4$$
The integers are 4, 6, and 8. $(2-2)$

63. $b = 5h$ $(2-2)$ **64.** $h = \dfrac{1}{4}b$ $(2-2)$

65. (A) The slope of this linear function is 3.4 (coefficient of x), so the snow is falling at 3.4 inches per hour.

(B) To find the amount of snow at the time it started to snow, we plug in zero for x:
$$S(0) = 3.4(0) + 11.1 = 11.1 \text{ inches}$$
$(2-1)$

66. (A) First, we need to find the rate of growth in terms of months. We were given an increase of 75,000 per day, which corresponds to $75,000 \times 365 = 27,375,000$ per year, or $27,375,000/12 = 2,281,250$ per month. This rate of growth represents the slope of our linear function, and the given initial value (60 million) is the y-intercept. Using the slope-intercept form, we get

$$B(x) = 2,281,250x + 60,000,000$$

(B) We set our function equal to $100,000,000$ and solve for x:
$$2,281,250x + 60,000,000 = 100,000,000$$
$$2,281,250x = 40,000,000$$

$$x = \frac{40,000,000}{2,281,250} \approx 17.5$$

The number of blogs is predicted to hit $100,000,000$ 17.5 months after August 1, 2006, which is mid-January 2008. *(2-1, 2-2)*

67. Find the vertex of the function:
$$x = -\frac{b}{2a} = -\frac{8.54}{2(-0.23)} = 18.6$$

A student should study 18.6 hours to maximize their score. *(2-3)*

68. The area is given by $A = \ell \cdot w$. We can use the Pythagorean Theorem to find a relationship between ℓ and w:
$$\ell^2 + w^2 = 32.5^2$$
$$\ell^2 = 1056.25 - w^2$$
$$\ell = \sqrt{1056.25 - w^2}$$

The area is 375 square inches so
$$375 = \ell \cdot w \quad \sqrt{1056.25 - w^2}$$

Graph $y_1 = 375$, $y_2 = x\sqrt{1056.25 - x^2}$ and use the INTERSECT command.

```
                          WINDOW
                          Xmin=0
                          Xmax=20
                          Xscl=2
                          Ymin=0
                          Ymax=500
                          Yscl=100
Intersection              Xres=1
X=12.5        Y=375
```

The intersection occurs when $x = 12.5$ so the width is 12.5".
$\ell = \sqrt{1056.25 - 12.5^2} = 30$ so the length is 30.0". *(2-5)*

69. The distance in feet covered by a falling object is given by $d(t) = 16t^2$. If the hammer reaches the ground after 3.5 seconds, $d(3.5)$ is the height of the building.
$$d(3.5) = 16(3.5)^2 = 196$$
The building is 196 feet tall. *(2-5)*

70. $C(x) = 0.001x^2 - 9.5x + 30,000$ is a quadratic function.

(A) Graph C and apply the MINIMUM command to find the minimum.

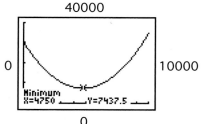

The minimum weekly cost is $7,437.50 when $x = 4,750$ calculators.

(B) Graph $y1 = C(x)$ and $y2 = 12,000$ and apply the INTERSECT command to find the intersections.

The weekly cost of $12,000 is achieved at $x = 2,614$ or $x = 6,886$ calculators.

(C) Since the minimum weekly cost is $7,437.50, a cost of $6000 is impossible. There would be no intersection of $y = C(x)$ and $y = 6,000$ (not shown).   *(2-3)*

71. $C(x) = 0.001x^2 - 9.5x + 30,000$. The revenue function is $R(x) = 3x$. The break-even points occur when $C(x) = R(x)$. Graph $y1 = C(x)$ and $y2 = R(x)$ and apply the INTERSECT command to find the intersections.

The break-even points are $x = 3,240$ or $x = 9,260$ calculators.   *(2-7)*

72. Referring to the graphs in problem 71, a profit will occur when $R(x) > C(x)$, that is, when the line is above the parabola, or $3,240 < x < 9,260$. A loss will occur when $R(x) < C(x)$, that is, when the parabola is above the line, for nonnegative values of x, $0 \le x < 3,240$ or $x > 9,260$ calculators.   *(2-7)*

73. (A) If V is linearly related to t, then we are looking for an equation whose graph passes through $(t_1, V_1) = (0, 12,000)$ and $(t_2, V_2) = (8, 2,000)$. We find the slope, and then we use the point-slope form to find the equation.

$$m = \frac{V_2 - V_1}{t_2 - t_1} = \frac{2,000 - 12,000}{8 - 0} = \frac{-10,000}{8} = -1,250$$

$V - V_1 = m(t - t_1)$
$V - 12,000 = -1,250(t - 0)$
$V - 12,000 = -1,250t$
$\qquad V = -1,250t + 12,000$

(B) We are asked for V when $t = 5$
$v = -1,250(5) + 12,000$
$V = -6,250 + 12,000$
$V = \$5,750$ *(2-1)*

74. (A) If R is linearly related to C, then we are looking for an equation whose graph passes through $(C_1, R_1) = (30, 48)$ and $(C_2, R_2) = (20, 32)$. We find the slope, and then we use the point-slope form to find the equation.

$$m = \frac{R_2 - R_1}{C_2 - C_1} = \frac{32 - 48}{20 - 30} = \frac{-16}{-10} = 1.6$$

$R - R_1 = m(C - C_1)$

$R - 48 = 1.6(C - 30)$

$R - 48 = 1.6C - 48$

$\quad\; R = 1.6C$

(B) We are asked for R when $C = 105$.

$\quad R = 1.6(105)$

$\quad\quad = \$168$ $\hspace{6cm}$ (2-1)

75. If $0 \le x \le 3,000$, $E(x) = 400$

$$\left(\begin{matrix}\text{Base}\\\text{Salary}\end{matrix}\right) + \left(\begin{matrix}\text{Commission on}\\\text{Sales over \$3,000}\end{matrix}\right)$$

If $x > 3,000$, $E(x) = 400 \quad + \quad 0.1(x - 3,000)$

$\hspace{3.8cm} = 400 \quad + \quad 0.1x - 300$

$\hspace{3.8cm} = 0.1x \quad + \quad 100$

Summarizing,

$$E(x) = \begin{cases} 400 & \text{if} \quad 0 \le x \le 3,000 \\ 0.1x + 100 & \text{if} \quad x > 3,000 \end{cases}$$

$E(2,000) = \$400$

$E(5,000) = 0.1(5,000) + 100 = 500 + 100 = \$600$ $\hspace{3cm}$ (2-1)

76. (A) From the figure, we can see that $A = x(y + y) = 2xy$. To get this written only in terms of x, we'll need another way to express variable y. Notice that the fence is made up of four pieces of length y and three pieces of length x, and a total of 120 feet of fence is used, so

$\quad 4y + 3x = 120.$

We can solve this equation for y:

$\quad 4y = 120 - 3x$

$\quad y = 30 - \dfrac{3}{4}x$

Now we can substitute this in for y in our original area expression:

$\quad A = 2x\left(30 - \dfrac{3}{4}x\right)$

$\quad A(x) = 60x - \dfrac{3}{2}x^2$

(B) The domain will be based on the fact that both x and y have to be positive, since they are lengths. One obvious restriction is $x > 0$. Since we also need $y > 0$, we get

$\quad 30 - \dfrac{3}{4}x > 0$

$\quad 30 > \dfrac{3}{4}x$

$\quad \dfrac{4}{3} \cdot 30 > x$

$\quad 40 > x$

So the domain is $0 < x < 40$.

(C) To find the maximum, we find the first coordinate of the vertex of the quadratic function $A(x) = -\frac{3}{2}x^2 + 60x$:

$x = -\frac{b}{2a} = -\frac{60}{2(-3/2)} = 20$; when $x = 20$, $y = 30 - \frac{3}{4}(20) = 30 - 15 = 15$, so the area will be largest when $x = 20$ and $y = 15$.   (2-3)

77. (A) $H = 0.7(220 - A)$

(B) We are asked to find H
when $A = 20$.
$H = 0.7(220 - 20)$
$H = 140$ beats per minute.

(C) We are asked to find A
when $H = 126$.
$126 = 0.7(220 - A)$
$126 = 154 - 0.7A$
$-28 = -0.7A$
$A = 40$ years old.   (2-1)

78. Let $x =$ width of page
$y =$ height of page
Then $xy = 480$, and $y = \frac{480}{x}$.

Since the printed portion is surrounded by margins of 2 cm on each side, we have
$x - 4 =$ width of printed portion
$y - 4 =$ height of printed portion

The printed area is given by $(x - 4)(y - 4)$, and we're told that this area should be 320 sq. cm., so

$(x - 4)(y - 4) = 320$, that is

$(x - 4)\left(\frac{480}{x} - 4\right) = 320$

Solving this, we obtain:

$x\left(\frac{480}{x}\right) - 4x - 4\left(\frac{480}{x}\right) + 16 = 320$

$480 - 4x - \frac{1,920}{x} + 16 = 320$

$-4x - \frac{1,920}{x} = -176$ LCD: x $(x \neq 0)$

$-4x^2 - 1,920 = -176x$

$0 = 4x^2 - 176x + 1,920$

$0 = x^2 - 44x + 480$

$0 = (x - 20)(x - 24)$

$x - 20 = 0$ or $x - 24 = 0$

$\quad\quad x = 20 \quad\quad\quad\quad x = 24$

$\quad\quad \frac{480}{x} = 24 \quad\quad\quad \frac{480}{x} = 20$

The dimensions of the page are 20 cm by 24 cm.   (2-5)

79.

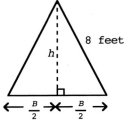

8 feet

In the isosceles triangle we note:

$$\frac{1}{2}Bh = A = 24$$

So

$$h = \frac{48}{B}$$

Applying the Pythagorean theorem, we have

$$h^2 + \left(\frac{B}{2}\right)^2 = 8^2$$

$$\left(\frac{48}{B}\right)^2 + \left(\frac{B}{2}\right)^2 = 8^2$$

$$\frac{2,304}{B^2} + \frac{B^2}{4} = 64$$

$$4B^2\left(\frac{2,304}{B^2}\right) + 4B^2\left(\frac{B^2}{4}\right) = 4B^2(64)$$

$$9,216 + B^4 = 256B^2$$

$$(B^2)^2 - 256B^2 + 9,216 = 0$$

$$B^2 = \frac{-b \pm \sqrt{b^2 - 4ac}}{2a} \qquad a = 1, \ b = -256, \ c = 9,216$$

$$B^2 = \frac{-(-256) \pm \sqrt{(-256)^2 - 4(1)(9,216)}}{2(1)}$$

$$B^2 = \frac{256 \pm \sqrt{65,536 - 36,864}}{2}$$

$$B^2 = \frac{256 \pm \sqrt{28,672}}{2}$$

$$B^2 = 128 \pm 32\sqrt{7}$$

$$B = \sqrt{128 \pm 32\sqrt{7}}$$

$$B = 14.58 \text{ ft or } 6.58 \text{ ft}$$

(*2-6*)

80. Sketch a figure:

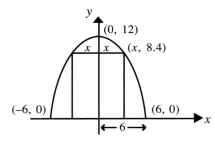

The equation of the parabolic entrance way in the coordinate system with vertex at (0, 12) shown must be of the form $y = ax^2 + 12$. Since (6, 0) is on the parabola, the coordinates of the point must satisfy the equation, so

$$0 = a(6)^2 + 12$$

$$a = -\frac{1}{3}$$

The equation of the parabola is $y = -\frac{1}{3}x^2 + 12$. To find the width of the door that has height 8.4, let $y = 8.4$ and solve for x.

$$8.4 = -\frac{1}{3}x^2 + 12$$

$$-3.6 = -\frac{1}{3}x^2$$

$$x^2 = 10.8$$

$$x = \sqrt{10.8} \approx 3.3 \text{ ft.}$$

The width of the door is $2x = 6.6$ ft.

(*2-3*)

81. (A)

Enter data.   Compute regression equation.

The quadratic regression model is
$$y(x) = 0.0615x^2 - 2.88x + 48.0.$$

(B)

Transfer the regression   Graph the regression equation and
equation to the equation   determine when the level of 26.7
editor.   will be reached a second time.

The model predicts that the percentage of marijuana users will return to the
1979 level when $t \approx 38$, or 2008.   *(2-3)*

82. (A)

Enter data.   Compute regression equation.

The linear regression model is $y = -3.10x + 54.6$.

(B)

Transfer the regression   Graph the regression equation and
equation to the equation   determine the percentage that corresponds
editor.   to the assessed value of $300.

The model predicts a percentage of 45.33% votes for democrats.   *(2-1)*

83. For the first model, the independent variable is supply so enter the supply
column as L_1. The dependent variable is price so enter the price column as L_2.
Then choose linear regression from the STAT CALC menu.

The price-supply model is $y = 0.000\ 0258x + 0.0533$.

For the second model, the independent variable is demand so we enter the demand column as L_1 leaving price as L_2.

The price-demand model is $y = -0.000\ 0221x + 1.63$.

The equilibrium price occurs when supply and demand are equal. We graph the supply and demand curves and find where they intersect.

The equilibrium price is the y-coordinate of the intersection, which is \$0.90. (*2-2*)

84. (A) The revenue function is obtained from multiplying the demand function (price) by pounds sold (x).

$$R(x) = (-0.000\ 0221x + 1.63)x = -0.000\ 0221x^2 + 1.63x$$

The number of pounds must be positive and revenue must be positive also. The graph of $R(x)$ shows that the domain of $R(x)$ in this setting is $0 \leq x \leq 73,800$.

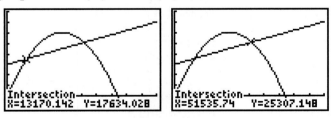

The domain of the cost function is $x \geq 0$: pounds produced cannot be negative and we're not given a limit on production.

The cost is fixed costs (\$15,000) plus \$0.20 times pounds produced (x).

$$C(x) = 15,000 + 0.20x$$

(B) The company will break-even if revenue and cost are equal so we find where the graphs of $R(x)$ and $C(x)$ intersect.

The x-values at the points of intersection are 13,200 and 51,500 (rounded to 3 significant digits), so the company will break-even with production of 13,200 lbs. or 51,500 lbs. They will make a profit if production is between those levels and will lose money if production is less than 13,200 lbs. or greater then 51,500 lbs.

(C) Profit = Revenue - Cost

$P(x) = R(x) - C(x) = -0.0000221x^2 + 1.63x - (15,000 + 0.20x)$

$P(x) = -0.0000221x^2 + 1.43x - 15,000$ (Distribute the negative!)

Graph $y_1 = P(x)$ as above and use the MAXIMUM command.

The maximum profit of $8,130 occurs when 32,400 pounds are sold.

(2-5, 2-7)

85. (A) The independent variable is speed and the dependent variable is mileage, so we enter the speed column as L_1 and the mileage column as L_2. Then we choose quadratic regression from the STAT CALC menu.

The quadratic model is $y = -0.00174x^2 + 0.0865x + 0.980$.

(B) The rental cost is $15 per hour plus $2.55 per gallon for gas. This can be expressed as $C = 15t + 2.55g$ where t is the number of hours and g is gallons of gas used. We can write each of t and g in terms of speed (x) to get a model for cost in terms of speed.

The time of the trip will be the distance divided by the speed since distance = speed times time. The trip is 100 miles so $t = \dfrac{100}{x}$. The gasoline usage can be calculated by dividing the length (100 miles) by the miles per gallon (y). So $g = \dfrac{100}{y}$. Our quadratic model tells us that $y = -0.00174x^2 + 0.0865x + 0.980$.

So, $g = \dfrac{100}{-0.00174x^2 + 0.0865x + 0.980}$

Our rental cost is now $C = 15\left(\dfrac{100}{x}\right) + 2.55\left(\dfrac{100}{-0.00174x^2 + 0.0865x + 0.980}\right)$

$$C = \dfrac{1500}{x} + \dfrac{255}{-0.00174x^2 + 0.0865x + 0.980}$$

We use the MINIMUM command to find when this is minimized.

The minimum cost occurs when the speed is 31.5 mph. Plugging $x = 31.5$ into our quadratic regression model we find that the mileage at 31.5 mph is 1.98 mpg. We use our expressions for t, g, and C to answer the remaining questions.

$t = \dfrac{100}{x} = \dfrac{100}{31.5} = 3.17$ hours

$g = \dfrac{100}{y} = \dfrac{100}{1.98} = 50.5$ gallons

$C = 15t + 2.55g = 15(2.98) + 2.55(52.1) = \$178$

(2-5)

Chapters 1 & 2 Cumulative Review

1. (A)

   (B) Since the x coordinates vary between a minimum of −3 and a maximum of 3, Xmin = −3 and Xmax = 3. Since the y coordinates vary between a minimum of −4 and a maximum of 4, Ymin = −4 and Ymax = 4.

   (C) Not a function (two range values correspond to domain value 1)

   (1-1, 1-2)

2. (A) Use slope-intercept form with $m = \dfrac{6 - 2}{5 - 3} = \dfrac{4}{2} = 2$.

   $y = 2x + b$

   Since (3, 2) is on the line, its coordinates satisfy the equation:

   $2 = 2 \cdot 3 + b$

   $b = -4$

   Equation: $y = 2x - 4$

   (B) A line perpendicular to $y = 2x - 4$, which has slope 2, will have slope satisfying $2m = -1$, or $m = -\dfrac{1}{2}$. We use the point-slope form

   $$y - 6 = -\frac{1}{2}(x - 5)$$

   $$y - 6 = -\frac{1}{2}x + \frac{5}{2}$$

   $$y = -\frac{1}{2}x + \frac{5}{2} + 6$$

   $$y = -\frac{1}{2}x + \frac{17}{2}$$

   (C)

x	$y = 2x - 4$	$y = -\dfrac{1}{2}x + \dfrac{17}{2}$
0	−4	17 / 2
1	−2	8
5	6	6

   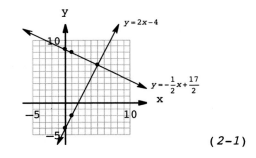

   (2-1)

3. $2x - 3y = 6$

   $-3y = -2x + 6$

   $y = \dfrac{2}{3}x - 2$

   slope: $\dfrac{2}{3}$ y intercept: −2 x intercept: 3 (if $y = 0$, $2x = 6$, and $x = 3$)

x	y
−3	−4
0	−2
0	0

   (2-1)

4. (A) $f(-2) + g(-3) = (-2)^2 - 2(-2) + 5 + 3(-3) - 2 = 4 + 4 + 5 - 9 - 2 = 2$

(B) $f(1) \cdot g(1) = (1^2 - 2 \cdot 1 + 5) \cdot (3 \cdot 1 - 2) = 4 \cdot 1 = 4$

(C) $\dfrac{g(0)}{f(0)} = \dfrac{3 \cdot 0 - 2}{0^2 - 2 \cdot 0 + 5} = \dfrac{-2}{5}$ *(1-2)*

5. (A) Stretched vertically by a factor of 2
(B) Shifted right 2 units
(C) Shifted down 2 units   *(1-4)*

6. Domain: [-2, 3] Range: [-1, 2]   *(1-2)*

7. Notice that $f(2) = 2$, while $f(-2) = -1$. We have found an x value for which $f(x)$ and $f(-x)$ are not equal, so f is not even. We have also found an x value for which $f(x) \neq -f(-x)$, so f is not odd, either.   *(1-4)*

8. (A) The graph of $f(x)$ is shifted left one unit and reflected across the x axis.

(B) The graph of $f(x)$ is stretched vertically by a factor of two and shifted down two units.

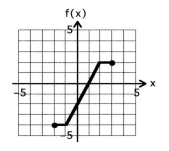

(1-4)

9. $5 - 3(x + 4) = x + 2(11 - 2x)$

$5 - 3x - 4 = x + 22 - 4x$

$-3x + 1 = -3x + 22$

$1 = 22$

This is a contradiction, so there is no solution.

Check:

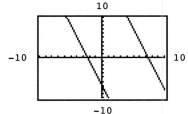

The lines are parallel and never intersect, so there is no solution.   *(2-2)*

10. $(4x + 15)^2 = 9$

$4x + 15 = \pm 3$

$4x = -15 \pm 3 = -18, -12$

$x = -\dfrac{18}{4}, -\dfrac{12}{4}$

$x = -\dfrac{9}{2}, -3$

Check:

11.
$$\frac{7x}{5} - \frac{3 + 2x}{2} = \frac{x - 10}{3} + 2$$
$$30\frac{7x}{5} - 30\frac{(3 + 2x)}{2} = 30\frac{(x - 10)}{3} + 2(30)$$
$$42x - 15(3 + 2x) = 10(x - 10) + 60$$
$$42x - 45 - 30x = 10x - 100 + 60$$
$$12x - 45 = 10x - 40$$
$$2x = 5$$
$$x = \frac{5}{2}$$

Check:

(2-2)

12.
$$3x^2 = -12x$$
$$3x^2 + 12x = 0$$
$$3x(x + 4) = 0$$
$$3x = 0 \quad \text{or} \quad x + 4 = 0$$
$$x = 0 \qquad\qquad x = -4$$

Check:

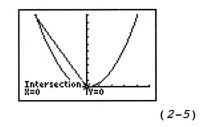

(2-5)

13.
$$4x^2 - 20 = 0$$
$$4x^2 = 20$$
$$x^2 = 5$$
$$x = \pm\sqrt{5}$$

Check: $\sqrt{5} \approx 2.236$

(2-5)

14.
$$x^2 - 6x + 2 = 0$$
$$x^2 - 6x = -2$$
$$x^2 - 6x + 9 = 7$$
$$(x - 3)^2 = 7$$
$$x - 3 = \pm\sqrt{7}$$
$$x = 3 \pm \sqrt{7}$$

Check: $3 + \sqrt{7} \approx 5.646$
$3 - \sqrt{7} \approx 0.354$

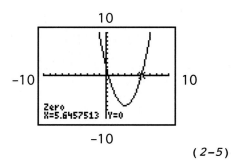

(2-5)

15.
$$x - \sqrt{12 - x} = 0$$
$$x = \sqrt{12 - x}$$
$$x^2 = 12 - x$$
$$x^2 + x - 12 = 0$$
$$(x + 4)(x - 3) = 0$$
$$x = -4, \ 3$$

Check: $-4 - \sqrt{12 - (-4)} = -4 - \sqrt{16} = -8$ Not a solution
$$3 - \sqrt{12 - 3} = 3 - \sqrt{9} = 0 \quad \checkmark$$

Solution: $x = 3$

Check:

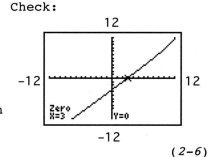

(2-6)

16. $2(3 - y) + 4 \leq 5 - y$

$\qquad 6 - 2y + 4 \leq 5 - y$

$\qquad -2y + 10 \leq 5 - y$

$\qquad\qquad -y \leq -5$

$\qquad\qquad\quad y \geq 5$ $\qquad\qquad\qquad\qquad$ (*2-7*)

17. $-3 \leq 2 - 3x < 11$

$\quad -5 \leq -3x < 9$

$\quad \dfrac{5}{3} \geq x > -3 \qquad$ (Note that the inequalities reversed)

Solution: $\quad -3 < x \leq \dfrac{5}{3}$, or $\left(-3, \dfrac{5}{3}\right]$ $\qquad\qquad\qquad\qquad$ (*2-7*)

18. $|x - 2| < 7$

$\qquad -7 < x - 2 < 7$

$\qquad -5 < x < 9$ $\qquad\qquad\qquad$ (*2-7*)

19. Rearrange so that the right side is zero:

$x^2 + 3x - 10 \geq 0$

Find the zeros of the left side:

$(x + 5)(x - 2) = 0;$ zeros are -5 and 2

Choose test numbers and make a table:

Interval	Test Number	Result
$x < -5$	-6	$(-6)^2 + 3(-6) - 10 = 8;$ positive
$-5 < x < 2$	0	$(0)^2 + 3(0) - 10 = -10;$ negative
$x > 2$	4	$(4)^2 + 3(4) - 10 = 18;$ positive

Solution: $x \leq -5$ or $x \geq 2$, or $(-\infty, -5] \cup [2, \infty)$ $\qquad\qquad\qquad$ (*2-7*)

20. (A) $f(x) = x^2 - 4x - 1$

$\qquad\qquad = (x^2 - 4x \qquad) - 1$

$\qquad\qquad = (x^2 - 4x + 4) - 1 - 4$

$\qquad\qquad = (x - 2)^2 - 5$

(B) The graph of f is the same as the graph of $y = x^2$ shifted to the right 2 units and down 5 units.

(C) Solve $0 = x^2 - 4x - 1$

$\qquad\quad 0 = (x - 2)^2 - 5$

$\qquad\quad (x - 2)^2 = 5$

$\qquad\quad\; x - 2 \;= \pm\sqrt{5}$

$\qquad\qquad\quad x = 2 \pm \sqrt{5}$

Check: $2 + \sqrt{5} \approx 4.236$

$\qquad\quad\; 2 - \sqrt{5} \approx -0.236$

$\qquad\qquad\qquad\qquad\qquad\qquad\qquad\qquad$ (*2-3, 2-5*)

21. (A) $(2 - 3i) - (-5 + 7i) = 2 - 3i + 5 - 7i = 7 - 10i$

(B) $(1 + 4i)(3 - 5i) = 3 + 7i - 20i^2 = 3 + 7i + 20 = 23 + 7i$

(C) $\dfrac{5 + i}{2 + 3i} = \dfrac{(5 + i)(2 - 3i)}{(2 + 3i)(2 - 3i)} = \dfrac{10 - 13i - 3i^2}{4 - 9i^2}$

$\qquad = \dfrac{10 - 13i + 3}{4 + 9} = \dfrac{13 - 13i}{13} = 1 - i$ $\hspace{3cm}$ *(2-4)*

22. (A) All real numbers $(-\infty, \infty)$

(B) From the graph, the possible function values include -2 (only) and all numbers greater than or equal to 1. In set and interval notation: $\{-2\} \cup [1, \infty)$.

(C) $f(-3) = 1$ $f(-2) = 2$ (not -2) $f(2) = -2$ (not 2).
Then $f(-3) + f(-2) + f(2) = 1 + 2 + (-2) = 1$

(D) $[-3, -2]$ and $[2, \infty)$

(E) f is discontinuous at $x = -2$ and at $x = 2$. $\hspace{2cm}$ *(1-4)*

23. $(f \circ g)(x) = f[g(x)] = f\left(\dfrac{x + 3}{x}\right) = \dfrac{1}{\frac{x+3}{x} - 2} = \dfrac{x(1)}{x\left(\frac{x+3}{x} - 2\right)} = \dfrac{x}{x + 3 - 2x} = \dfrac{x}{3 - x}$

The domain of f is all real numbers except 2.
The domain of g is all non-zero real numbers.
The domain of $f \circ g$ is the set of all non-zero real numbers for which $g(x) \neq 2$.

$g(x) = 2$ only if $\dfrac{x + 3}{x} = 2$, that is $x + 3 = 2x$, or $x = 3$.

The domain of $f \circ g$ is the set of all non-zero real numbers except 3, that is, $(-\infty, 0) \cup (0, 3) \cup (3, \infty)$.

> **Common Error:**
> The domain of $f \circ g$ cannot be found by looking at the final form $\frac{x}{3 - x}$. It is not $\{x \mid x \neq 3\}$.

$\hspace{13cm}$ *(1-5)*

24. $f(x) = 2x + 5$
Assume $f(a) = f(b)$
$\qquad 2a + 5 = 2b + 5$
$\qquad\quad 2a = 2b$
$\qquad\quad\ a = b$
f is one-to-one.

Solve $y = f(x)$ for x:
$\qquad y = 2x + 5$
$y - 5 = 2x$
$\qquad x = \dfrac{y - 5}{2} = f^{-1}(y)$

Interchange x and y
$y = f^{-1}(x) = \dfrac{x - 5}{2}$ or $\dfrac{1}{2}x - \dfrac{5}{2}$ Domain: $(-\infty, \infty)$ *(1-6)*

25. $f(x) = \sqrt{x + 4}$
Assume $f(a) = f(b)$
$\qquad \sqrt{a + 4} = \sqrt{b + 4}$
$\qquad\quad a + 4 = b + 4$
$\qquad\qquad a = b$
f is one-to-one.

(A) Solve $y = f(x)$ for x
$\quad y = \sqrt{x + 4}$
$\quad y^2 = x + 4$ $y \geq 0$
$\qquad x = y^2 - 4$ $y \geq 0$ $f^{-1}(y) = y^2 - 4$

Interchange x and y:

$y = f^{-1}(x) = x^2 - 4$ Domain: $x \geq 0$

Check: $f^{-1}[f(x)] = (\sqrt{x + 4})^2 - 4 = x + 4 - 4 = x$

$\qquad f[f^{-1}(x)] = \sqrt{x^2 - 4 + 4} = \sqrt{x^2} = x$ since $x \geq 0$ in the domain of f^{-1}.

(B) Domain of $f = [-4, \infty) =$ Range of f^{-1}

$\qquad$ Range of $f = [0, \infty) =$ Domain of f^{-1}

(C)

$\qquad\qquad\qquad\qquad\qquad\qquad$ (1-6)

26. (A) The graph of f passes the horizontal line test.
f is one-to-one.

(B) Since $g(0) = g(-1) = 0$, g is not one-to-one.

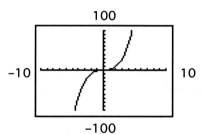

$\qquad\qquad\qquad\qquad\qquad\qquad$ (1-6)

27. The line $3x + 2y = 12$, or

$\qquad 2y = -3x + 12$,

or $y = -\dfrac{3}{2}x + 6$, has slope $-\dfrac{3}{2}$.

(A) We require a line through $(-6, 1)$ with slope $-\dfrac{3}{2}$. Applying the point-slope form, we have $y - 1 = -\dfrac{3}{2}(x + 6)$

$$y - 1 = -\frac{3}{2}x - 9$$

$$y = -\frac{3}{2}x - 8$$

(B) The slope of a line parallel to one with slope $-\dfrac{3}{2}$ will have slope $\dfrac{2}{3}$ (negative reciprocals). Again applying the point-slope form, we have

$$y - 1 = \frac{2}{3}(x + 6)$$

$$y - 1 = \frac{2}{3}x + 4$$

$$y = \frac{2}{3}x + 5 \qquad\qquad\qquad (2-1)$$

28. $f(x) = x^2 - 2x - 8$. Use the vertex formula:

$x = -\dfrac{b}{2a} = -\dfrac{-2}{2(1)} = 1$; $f(1) = 1^2 - 2(1) - 8 = -9$

The vertex is $(1, -9)$ and the axis is
$x = 1$. the parabola opens up since $x > 0$,
so the vertex is a minimum; the minimum
value is -9 and occurs at $x = 1$. The range
is then $[-9, \infty)$.
y intercept: Set $x = 0$, then $f(0) = -8$ is
the y intercept.
x intercepts: Set $f(x) = 0$, then
$0 = x^2 - 2x - 8$
$0 = (x - 4)(x + 2)$
$x = 4$ or $x = -2$ are the x intercepts.

Graph: Locate axis and vertex,
then plot several points on either
side of the axis.

x	$f(x)$
-3	7
-2	0
-1	-5
0	-8
1	-9
2	-8
3	-5
4	0
5	7

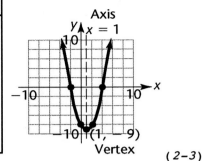

(2-3)

29. $|4x - 9| > 3$

$4x - 9 < -3$ or $4x - 9 > 3$
$\quad\ 4x < 6 \qquad\qquad\ 4x > 12$
$\quad\ \ x < \dfrac{3}{2}$ or $\quad\ \ x > 3$

$\left(-\infty, \dfrac{3}{2}\right) \cup (3, \infty)$ *(2-7)*

30. Multiply both sides by 2 to clear fractions:

$x^2 - 2x - 16 > 0$

Find the zeros of the left side (quadratic formula with $a = 1$, $b = -2$, $c = -16$)

$x = \dfrac{-(-2) \pm \sqrt{(-2)^2 - 4(1)(-16)}}{2(1)}$

$x = \dfrac{2 \pm \sqrt{68}}{2} = \dfrac{2 \pm 2\sqrt{17}}{2} = 1 \pm \sqrt{17} \approx 5.1, -3.1$

Choose test numbers and make a table:

Interval	Test Number	Result
$x < -3.1$	-4	$(-4)^2 - 2(-4) - 16 = 8$; positive
$-3.1 < x < 5.1$	0	$(0)^2 - 2(0) - 16 = -16$; negative
$x > 5.1$	6	$(6)^2 - 2(6) - 16 = 8$; positive

Solution: $x < -3.1$ or $x > 5.1$, or $(-\infty, -3.1) \cup (5.1, \infty)$ *(2-7)*

31. (A) $(2 - 3i)^2 - (4 - 5i)(2 - 3i) - (2 + 10i)$

$= (2)^2 - 2(2)(3i) + (3i)^2 - (8 - 12i - 10i + 15i^2) - 2 - 10i$

$= 4 - 12i + 9i^2 - (8 - 22i + 15i^2) - 2 - 10i$

$= 4 - 12i + 9i^2 - 8 + 22i - 15i^2 - 2 - 10i$

$= 4 - 12i - 9 - 8 + 22i + 15 - 2 - 10i$

$= 0 + 0i$ or 0

(B) $\dfrac{3}{5} + \dfrac{4}{5}i + \dfrac{1}{\frac{3}{5} + \frac{4}{5}i} = \dfrac{3}{5} + \dfrac{4}{5}i + \dfrac{1}{\left(\frac{3}{5} + \frac{4}{5}i\right)}\dfrac{\left(\frac{3}{5} - \frac{4}{5}i\right)}{\left(\frac{3}{5} - \frac{4}{5}i\right)}$

$\qquad\qquad = \dfrac{3}{5} + \dfrac{4}{5}i + \dfrac{\frac{3}{5} - \frac{4}{5}i}{\frac{9}{25} - \frac{16}{25}i^2}$

$\qquad\qquad = \dfrac{3}{5} + \dfrac{4}{5}i + \dfrac{\frac{3}{5} - \frac{4}{5}i}{\frac{9}{25} + \frac{16}{25}}$

$\qquad\qquad = \dfrac{3}{5} + \dfrac{4}{5}i + \dfrac{3}{5} - \dfrac{4}{5}i$

$\qquad\qquad = \dfrac{6}{5}$

(C) $i^{35} = i^{32}i^3 = (i^4)^8(-i) = 1^8(-i) = -i$ $\qquad\qquad\qquad$ (*2-4*)

32. (A) $(5 + 2\sqrt{-9}) - (2 - 3\sqrt{-16}) = (5 + 2i\sqrt{9}) - (2 - 3i\sqrt{16})$

$\qquad\qquad\qquad\qquad\qquad = (5 + 6i) - (2 - 12i)$

$\qquad\qquad\qquad\qquad\qquad = 5 + 6i - 2 + 12i$

$\qquad\qquad\qquad\qquad\qquad = 3 + 18i$

(B) $\dfrac{2 + 7\sqrt{-25}}{3 - \sqrt{-1}} = \dfrac{2 + 7i\sqrt{25}}{3 - i}$

$\qquad\quad = \dfrac{2 + 35i}{3 - i}$

$\qquad\quad = \dfrac{(2 + 35i)(3 + i)}{(3 - i)(3 + i)}$

$\qquad\quad = \dfrac{6 + 107i + 35i^2}{9 - i^2}$

$\qquad\quad = \dfrac{6 + 107i - 35}{9 + 1}$

$\qquad\quad = \dfrac{-29 + 107i}{10}$

$\qquad\quad = -2.9 + 10.7i$

(C) $\dfrac{12 - \sqrt{-64}}{\sqrt{-4}} = \dfrac{12 - i\sqrt{64}}{i\sqrt{4}}$

$\qquad\quad = \dfrac{12 - 8i}{2i}$

$\qquad\quad = \dfrac{12 - 8i}{2i}\dfrac{i}{i}$

$\qquad\quad = \dfrac{12i - 8i^2}{2i^2}$

$\qquad\quad = \dfrac{12i + 8}{-2}$

$\qquad\quad = -4 - 6i$ $\qquad$ (*2-4*)

33.

x	$y = x - 1$		x	$y = x^2 + 1$
-1	-2		0	1
-2	-3		1	2
			2	5

Domain: all real numbers
Range: $(-\infty, -1) \cup [1, \infty)$
Discontinuous at: $x = 0$

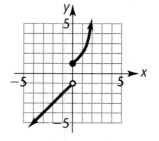

(*1-3*)

34. The graph of $y = |x|$ is vertically shrunk by $\dfrac{1}{2}$, reflected in the x axis, shifted two units to the right and three units up; $y = -\dfrac{1}{2}|x - 2| + 3$. $\qquad$ (*1-4*)

35. The vertex is (-2, -3), so the equation is of the form

$$y = a(x + 2)^2 - 3$$

What remains is to find a. To do so, we can use the other point we know, (-1, -2). This point has to satisfy the equation above, so

$$-2 = a(-1 + 2)^2 - 3 \qquad \text{(Solve for } a\text{)}$$
$$-2 = a - 3$$
$$1 = a$$

The equation is $y = (x + 2)^2 - 3$. *(2-3)*

36. $\qquad 1 + \dfrac{14}{y^2} = \dfrac{6}{y} \qquad$ Excluded value: $y \neq 0$

$$y^2(1) + y^2\frac{14}{y^2} = y^2\frac{6}{y}$$
$$y^2 + 14 = 6y$$
$$y^2 - 6y + 14 = 0$$
$$y^2 - 6y = -14$$
$$y^2 - 6y + 9 = -5$$
$$(y - 3)^2 = -5$$
$$y - 3 = \pm\sqrt{-5}$$
$$y = 3 \pm \sqrt{-5}$$
$$y = 3 \pm i\sqrt{5}$$

(2-6)

37. $4x^{2/3} - 4x^{1/3} - 3 = 0$
Let $u = x^{1/3}$, then
$$4u^2 - 4u - 3 = 0$$
$$(2u - 3)(2u + 1) = 0$$
$$u = \frac{3}{2}, \ -\frac{1}{2}$$

$$x^{1/3} = \frac{3}{2} \qquad x^{1/3} = -\frac{1}{2}$$
$$x = \frac{27}{8} \qquad x = -\frac{1}{8}$$

Check: Graph $y = 4x^{2/3} - 4x^{1/3} - 3$ and use the ZERO command to find the zeros.

 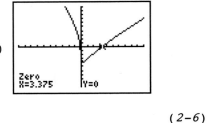

(2-6)

38. $u^4 + u^2 - 12 = 0$
Let $w = u^2$, then
$$w^2 + w - 12 = 0$$
$$(w + 4)(w - 3) = 0$$
$$w = -4, \ 3$$
$$u^2 = -4 \qquad u^2 = 3$$
$$u = \pm 2i \qquad u = \pm\sqrt{3}$$

Check: Graph $y = x^4 + x^2 - 12$ and use the ZERO command to find the zeros.

(2-6)

39.

$$\sqrt{8t - 2} - 2\sqrt{t} = 1$$
$$\sqrt{8t - 2} = 2\sqrt{t} + 1$$
$$8t - 2 = 4t + 4\sqrt{t} + 1$$
$$4t - 3 = 4\sqrt{t}$$
$$16t^2 - 24t + 9 = 16t$$
$$16t^2 - 40t + 9 = 0$$
$$(4t - 1)(4t - 9) = 0$$
$$t = \frac{1}{4}, \frac{9}{4}$$

Check: $\sqrt{8\left(\frac{1}{4}\right) - 2} - 2\sqrt{\frac{1}{4}} = \sqrt{0} - 2\left(\frac{1}{2}\right) = -1 \ne 1$
Not a solution

$\sqrt{8\left(\frac{9}{4}\right) - 2} - 2\sqrt{\frac{9}{4}} = \sqrt{16} - 2\left(\frac{3}{2}\right) = 4 - 3 = 1$ $\checkmark$

Solution: $t = \frac{9}{4}$

Graphical check: Graph $y1 = \sqrt{8x - 2} - 2\sqrt{x}$ and $y2 = 1$ and use the INTERSECT command to find the intersection.

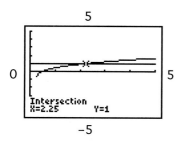

(2-6)

40.

$$6x = \sqrt{9x^2 - 48}$$
$$36x^2 = 9x^2 - 48$$
$$27x^2 = -48$$
$$x^2 = -\frac{48}{27}$$
$$x^2 = -\frac{16}{9}$$
$$x = \pm\sqrt{-\frac{16}{9}}$$
$$x = \pm\frac{4}{3}i$$

Check: $\frac{4}{3}i$:

$$6\left(\frac{4}{3}i\right) \overset{?}{=} \sqrt{9\left(\frac{4}{3}i\right)^2 - 48}$$
$$8i \overset{?}{=} \sqrt{-16 - 48}$$
$$8i \overset{\checkmark}{=} \sqrt{-64}$$
A solution

Check: $-\frac{4}{3}i$:

$$6\left(-\frac{4}{3}i\right) \overset{?}{=} \sqrt{9\left(-\frac{4}{3}i\right)^2 - 48}$$
$$-8i \overset{?}{=} \sqrt{-16 - 48}$$
$$-8i \ne \sqrt{-64}$$
Not a solution

Solution: $\frac{4}{3}i$ *(2-6)*

41. In this problem, $a = 1$, $b = b$, $c = 1$, and the discriminant $b^2 - 4ac = b^2 - 4\cdot1\cdot1 = b^2 - 4$. The number and types of roots depend on the sign of $b^2 - 4 = (b - 2)(b + 2)$. A glance at the graph of $y = x^2 - 4$ (not shown) shows that $b^2 - 4$ is positive if $b > 2$ or $b < -2$, and that there are two distinct real roots of $x^2 + bx + 1 = 0$ in this case. $b^2 - 4$ is equal to zero if $b = -2$ or $b = 2$, and there is one real double root of $x^2 + bx + 1 = 0$ in this case. $b^2 - 4$ is negative if $-2 < b < 2$, and there are two distinct imaginary roots of $x^2 + bx + 1 = 0$ in this case.   *(2-5)*

42. Odd: $f(x) = x$ Even: $f(x) = x^2$
If a function is both even and odd, then $f(x) = f(-x) = -f(x)$; adding $f(x)$ to both sides, we get $2f(x) = 0$ and $f(x) = 0$ for all x in its domain.   *(1-4)*

43. A quadratic equation $ax^2 + bx + c = 0$ with real coefficients will have imaginary roots only if $b^2 - 4ac = $ is negative. In this case it will have two distinct imaginary roots that are conjugates, so it is not possible for it to have one real and one imaginary root or a double imaginary root.   *(2-5)*

44. $\dfrac{g(2 + h) - g(2)}{h} = \dfrac{[-2(2 + h)^2 + 3(2 + h) - 1] - [-2(-2)^2 + 3(-2) - 1]}{h}$

$\qquad\qquad\qquad\quad = \dfrac{-2(4 + 4h + h^2) + 6 + 3h - 1 - [-8 - 6 - 1]}{h}$

$\qquad\qquad\qquad\quad = \dfrac{-8 - 8h - 2h^2 + 6 + 3h - 1 + 8 + 6 + 1}{h}$

$\qquad\qquad\qquad\quad = \dfrac{-5h - 2h^2}{h} = -5 - 2h$ $\hspace{3cm}$ (*1-3*)

45. The graph of $y = x^{1/3}$ has been vertically stretched by a factor of 2 and shifted 1 unit to the left and 1 unit down. $y = 2\sqrt[3]{x + 1} - 1$. $\hspace{2cm}$ (*1-4*)

46. (A) $A = 2\pi r(r + h)$ $\qquad$ (B) $A = 2\pi r(r + h)$
$\qquad\quad A = 2\pi r^2 + 2\pi rh$ $\qquad\qquad\quad A = 2\pi r^2 + 2\pi rh$
$\qquad\quad A - 2\pi r^2 = 2\pi rh$ $\qquad\qquad\qquad r^2 + hr = \dfrac{A}{2\pi}$
$\qquad\qquad\quad h = \dfrac{A - 2\pi r^2}{2\pi r}$ $\qquad\quad r^2 + hr + \dfrac{h^2}{4} = \dfrac{h^2}{4} + \dfrac{A}{2\pi}$

$\qquad\qquad\qquad\qquad\qquad\qquad\qquad\left(r + \dfrac{h}{2}\right)^2 = \dfrac{h^2}{4} + \dfrac{A}{2\pi}$

$\qquad\qquad\qquad\qquad\qquad\qquad\quad r + \dfrac{h}{2} = \pm\sqrt{\dfrac{h^2}{4} + \dfrac{A}{2\pi}}$

$\qquad\qquad\qquad\qquad\qquad\qquad\qquad\quad r = -\dfrac{h}{2} + \sqrt{\dfrac{h^2}{4} + \dfrac{A}{2\pi}}$

The negative root $r = -\dfrac{h}{2} - \sqrt{\dfrac{h^2}{4} + \dfrac{A}{2\pi}}$ is discarded because r, the radius of the cylinder, must be positive. $\hspace{3cm}$ (*2-2, 2-5*)

47. (A) $g(x) = \sqrt{4 - x^2}$

The domain of g is those values of x for which $4 - x^2 \geq 0$.
Graphing $y = 4 - x^2$, we obtain the graph at the right.

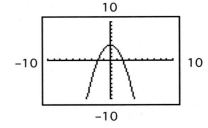

The domain of g is $-2 \leq x \leq 2$, or $[-2, 2]$.

(B) $\left(\dfrac{f}{g}\right)(x) = \dfrac{f(x)}{g(x)} = \dfrac{x^2}{\sqrt{4 - x^2}}$. The domain of $\dfrac{f}{g}$ is the intersection of the domains of f (all real numbers) and g ($[-2, 2]$) with the exclusion of those points (-2 and 2) where $g(x) = 0$. So the domain of $f/g = (-2, 2)$.

(C) $(f \circ g)(x) = f[g(x)] = f(\sqrt{4 - x^2}) = (\sqrt{4 - x^2})^2 = 4 - x^2$.
The domain of $f \circ g$ is the set of all real numbers in the domain of g ($[-2, 2]$) for which $f(x)$ is real, that is, all numbers in $[-2, 2]$. $\hspace{2cm}$ (*1-5*)

48. (A) $f(x) = x^2 - 2x - 3$ $x \geq 1$

f passes the horizontal line test (see graph below, in part C), so f is one-to-one.

Solve $y = f(x)$ for x:
$$
\begin{aligned}
y &= x^2 - 2x - 3 & x &\geq 1 \\
x^2 - 2x &= y + 3 & x &\geq 1 \\
x^2 - 2x + 1 &= y + 4 & x &\geq 1 \\
(x - 1)^2 &= y + 4 & x &\geq 1 \\
x - 1 &= \sqrt{y + 4} & x &\geq 1
\end{aligned}
$$

positive square root only because $x \geq 1$

$x = 1 + \sqrt{y + 4} = f^{-1}(y)$

Interchange x and y:

$y = f^{-1}(x) = 1 + \sqrt{x + 4}$ Domain: $x \geq -4$ (Check omitted)

(B) Domain of $f^{-1} = [-4, \infty)$
    Range of $f^{-1} =$ Domain of $f = [1, \infty)$.

(C)

$(1\text{-}6)$

49.
$$
\begin{aligned}
x^2 - x + 2 &= \left(\frac{1}{2} - \frac{i}{2}\sqrt{7}\right)^2 - \left(\frac{1}{2} - \frac{i}{2}\sqrt{7}\right) + 2 \\
&= \left(\frac{1}{2}\right)^2 - 2\left(\frac{1}{2}\right)\left(\frac{i}{2}\sqrt{7}\right) + \left(\frac{i}{2}\sqrt{7}\right)^2 - \frac{1}{2} + \frac{i}{2}\sqrt{7} + 2 \\
&= \frac{1}{4} - \frac{i}{2}\sqrt{7} + \frac{7i^2}{4} - \frac{1}{2} + \frac{i}{2}\sqrt{7} + 2 \\
&= \frac{1}{4} - \frac{7}{4} - \frac{1}{2} + 2 \\
&= 0 \qquad\qquad\qquad (2\text{-}4)
\end{aligned}
$$

50. Simplify the given inequality:

$\begin{array}{ll} a - b < b - a & \text{Add } a \text{ to both sides} \\ 2a - b < b & \text{Add } b \text{ to both sides} \\ \quad 2a < 2b & \text{Divide both sides by 2} \\ \quad\;\; a < b & \end{array}$

The original inequality is true for any real numbers a and b with $a < b$.

$(2\text{-}2)$

51.
$$
\begin{aligned}
\frac{a + bi}{a - bi} &= \frac{(a + bi)(a + bi)}{(a - bi)(a + bi)} = \frac{(a + bi)^2}{a^2 - (bi)^2} = \frac{a^2 + 2abi + b^2 i^2}{a^2 - b^2 i^2} \\
&= \frac{a^2 + 2abi - b^2}{a^2 + b^2} \\
&= \frac{a^2 - b^2 + 2abi}{a^2 + b^2} \\
&= \frac{a^2 - b^2}{a^2 + b^2} + \frac{2ab}{a^2 + b^2}\, i \qquad (2\text{-}4)
\end{aligned}
$$

52.

$$3x^2 = 2\sqrt{2}\,x - 1$$

$$3x^2 - 2\sqrt{2}\,x + 1 = 0$$

$$x = \frac{-b \pm \sqrt{b^2 - 4ac}}{2a} \qquad a = 3$$

$$b = -2\sqrt{2}$$
$$c = 1$$

$$x = \frac{-(-2\sqrt{2}) \pm \sqrt{(-2\sqrt{2})^2 - 4(3)\,(1)}}{2(3)}$$

$$x = \frac{2\sqrt{2} \pm \sqrt{8 - 12}}{6}$$

$$x = \frac{2\sqrt{2} \pm \sqrt{-4}}{6}$$

$$x = \frac{2\sqrt{2} \pm 2i}{6}$$

$$x = \frac{2(\sqrt{2} \pm i)}{6}$$

$$x = \frac{\sqrt{2} \pm i}{3} \qquad\qquad (2\text{-}5)$$

53.

$$1 + 13x^{-2} + 36x^{-4} = 0$$

$$1 + \frac{13}{x^2} + \frac{36}{x^4} = 0 \qquad \text{LCD} = x^4 \qquad x \neq 0$$

$$x^4 + 13x^2 + 36 = 0$$

$$(x^2 + 4)(x^2 + 9) = 0$$

$$x^2 + 4 = 0 \quad \text{or} \quad x^2 + 9 = 0$$

$$x^2 = -4 \qquad\qquad x^2 = -9$$

$$x = \pm 2i \qquad\qquad x = \pm 3i \qquad\qquad (2\text{-}6)$$

54.

$$\sqrt{16x^2 + 48x + 39} - 2x = 3$$

$$\sqrt{16x^2 + 48x + 39} = 3 + 2x$$

$$16x^2 + 48x + 39 = 9 + 12x + 4x^2$$

$$12x^2 + 36x + 30 = 0$$

$$2x^2 + 6x + 5 = 0$$

$$x = \frac{-b \pm \sqrt{b^2 - 4ac}}{2a} \qquad a = 2$$

$$b = 6$$
$$c = 5$$

$$x = \frac{-6 \pm \sqrt{6^2 - 4(2)\,(5)}}{2(2)}$$

$$x = \frac{-6 \pm \sqrt{36 - 40}}{4}$$

$$x = \frac{-6 \pm \sqrt{-4}}{4}$$

$$x = \frac{-6 \pm 2i}{4}$$

$$x = \frac{-3 \pm i}{2}$$

Check $\dfrac{-3 + i}{2}$:

$$\sqrt{16\left(\tfrac{-3+i}{2}\right)^2 + 48\left(\tfrac{3+i}{2}\right) + 39} - 2\left(\tfrac{-3+i}{2}\right) \overset{?}{=} 3$$

$$\sqrt{4(9 - 6i + i^2) + 24(-3 + i) + 39} + 3 - i \overset{?}{=} 3$$

$$\sqrt{36 - 24i + 4i^2 - 72 + 24i + 39} + 3 - i \overset{?}{=} 3$$

$$\sqrt{36 - 4 - 72 + 39} + 3 - i \overset{?}{=} 3$$

$$\sqrt{-1} + 3 - i \overset{?}{=} 3$$

$$i + 3 - i \overset{?}{=} 3$$

$$i + 3 - i \overset{\checkmark}{=} 3 \quad \text{A solution}$$

Check $\dfrac{-3 - i}{2}$:

$$\sqrt{16\left(\tfrac{-3-i}{2}\right)^2 + 48\left(\tfrac{3-i}{2}\right) + 39} - 2\left(\tfrac{-3-i}{2}\right) \overset{?}{=} 3$$

$$\sqrt{4(9 + 6i + i^2) + 24(-3 - i) + 39} + 3 + i \overset{?}{=} 3$$

$$\sqrt{36 + 24i + 4i^2 - 72 - 24i + 39} + 3 + i \overset{?}{=} 3$$

$$\sqrt{36 - 4 - 72 + 39} + 3 + i \overset{?}{=} 3$$

$$\sqrt{-1} + 3 + i \overset{?}{=} 3$$

$$i + 3 + i \overset{?}{=} 3$$

$$2i + 3 \neq 3 \quad \text{Not a solution}$$

Solution: $x = \dfrac{-3 + i}{2}$ (or $-1.5 + 0.5i$) *(2-6)*

55. $3x^{-2/5} - x^{-1/5} - 1 = 0$

Let $u = x^{-1/5}$, then

$3u^2 - u - 1 = 0$

$$u = \frac{-b \pm \sqrt{b^2 - 4ac}}{2a} \qquad \begin{array}{l} a = 3 \\ b = -1 \\ c = -1 \end{array}$$

$$u = \frac{-(-1) \pm \sqrt{(-1)^2 - 4(3)(-1)}}{2(3)}$$

$$u = \frac{1 \pm \sqrt{13}}{6}$$

$$x^{-1/5} = \frac{1 + \sqrt{13}}{6} \qquad\qquad x^{-1/5} = \frac{1 - \sqrt{13}}{6}$$

$$x^{1/5} = \frac{6}{1 + \sqrt{13}} \qquad\qquad x^{1/5} = \frac{6}{1 - \sqrt{13}}$$

$$x = \left(\frac{6}{1 + \sqrt{13}}\right)^5 \qquad\qquad x = \left(\frac{6}{1 - \sqrt{13}}\right)^5$$

$$x \approx 3.75 \qquad\qquad\qquad x \approx -64.75$$

Check: Graph $y = 3x^{-2/5} - x^{-1/5} - 1$ and use the ZERO command to find the zeros.

56. a is a square root of x if $a^2 = x$. So $5 + i$ is a square root of $24 + 10i$ because $(5 + i)^2 = 25 + 10i + i^2 = 25 + 10i - 1 = 24 + 10i$. $-5 - i$ is also a square root of $24 + 10i$ because $(-5 - i)^2 = 25 + 10i + i^2 = 25 + 10i - 1 = 24 + 10i$. To find the square roots of $24 + 10i$, set $(a + bi)^2 = 24 + 10i$, simplify the left side, and apply the definition of equality for complex numbers. This will yield the system of equations $a^2 - b^2 = 24$, $2ab = 10$, which can be solved for real a and b. *(2-4)*

57. (A)

$$\frac{f(x + h) - f(x)}{h} = \frac{[0.5(x + h)^2 - 3(x + h) - 7] - [0.5x^2 - 3x - 7]}{h}$$

$$= \frac{0.5(x^2 + 2xh + h^2) - 3x - 3h - 7 - 0.5x^2 + 3x + 7}{h}$$

$$= \frac{0.5x^2 + xh + 0.5h^2 - 3x - 3h - 7 - 0.5x^2 + 3x + 7}{h}$$

$$= \frac{xh + 0.5h^2 - 3h}{h}$$

$$= x + 0.5h - 3$$

(B)

$$\frac{f(x) - f(a)}{x - a} = \frac{(0.5x^2 - 3x - 7) - (0.5a^2 - 3a - 7)}{x - a}$$

$$= \frac{0.5x^2 - 3x - 7 - 0.5a^2 + 3a + 7}{x - a}$$

$$= \frac{0.5(x^2 - a^2) - 3(x - a)}{x - a}$$

$$= \frac{(x - a)[0.5(x + a) - 3]}{x - a}$$

$$= 0.5(x + a) - 3$$

$$= 0.5x + 0.5a - 3$$

(1-2)

58. A simple example of a function that satisfies all these conditions, $y = 0.24x^2 - 2.6x + 4$, is graphed as shown.

It would not be possible for a function to have fewer than the two x intercepts shown and still be continuous. Minimum: 2
However, a graph could cross the x axis arbitrarily often (an even number of times) and still satisfy all the stated conditions. There is no maximum number of intercepts.

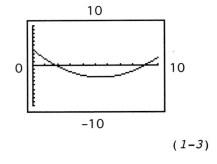

(1-3)

59. $f(x) = |x + 2| + |x - 2|$
If $x < -2$, then $|x + 2| = -(x + 2)$ and $|x - 2| = -(x - 2)$, so
$f(x) = -(x + 2) + -(x - 2)$
$\quad\;\; = -x - 2 - x + 2$
$\quad\;\; = -2x$
If $-2 \le x \le 2$ then $|x + 2| = x + 2$ but $|x - 2| = -(x - 2)$, so
$f(x) = x + 2 + -(x - 2)$
$\quad\;\; = x + 2 - x + 2$
$\quad\;\; = 4$

If $x > 2$, then $|x + 2| = x + 2$ and $|x - 2| = x - 2$, so
$f(x) = x + 2 + x - 2$
$\quad\;\; = 2x$
Domain: $(-\infty, \infty)$

x	$f(x) = -2x$	x	$f(x) = 4$	x	$f(x) = 2x$
-5	10	-2	4	3	6
-3	6	0	4	5	10
		2	4		

Piecewise definition for f:

$$f(x) = \begin{cases} -2x & \text{if} & x < -2 \\ 4 & \text{if} & -2 \le x \le 2 \\ 2x & \text{if} & x > 2 \end{cases}$$

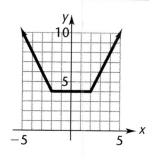

Range: $[4, \infty)$ $(1\text{-}3)$

60. In problems 79 and 81, Section 1-3, we noted that $[\![x/2]\!]$ jumps at intervals of 2 units, and $[\![3x]\!] = [\![x \div \frac{1}{3}]\!]$ jumps at intervals of $\frac{1}{3}$ unit. We expect $[\![2x]\!]$ to jump at intervals of $\frac{1}{2}$ unit.

$f(x) = 2x - [\![2x]\!]$
$f(0) = 2\cdot 0 - [\![2\cdot 0]\!] = 0 - [\![0]\!] = 0 - 0 = 0$
$f(\frac{1}{4}) = 2\cdot\frac{1}{4} - [\![2\cdot\frac{1}{4}]\!] = \frac{1}{2} - [\![\frac{1}{2}]\!] = \frac{1}{2} - 0 = \frac{1}{2}$
$f(\frac{1}{2}) = 2\cdot\frac{1}{2} - [\![2\cdot\frac{1}{2}]\!] = 1 - [\![1]\!] = 1 - 1 = 0$
$f(\frac{3}{4}) = 2\cdot\frac{3}{4} - [\![2\cdot\frac{3}{4}]\!] = \frac{3}{2} - [\![\frac{3}{2}]\!] = \frac{3}{2} - 1 = \frac{1}{2}$
$f(1) = 2\cdot 1 - [\![2\cdot 1]\!] = 2 - [\![2]\!] = 2 - 2 = 0$
$f(x) = 2x$ if $0 \le x \le \frac{1}{2}$, $f(x) = 2x - 1$ if $\frac{1}{2} \le x < 1$

Generalizing, we can write: $f(x) = \begin{cases} \;\vdots & & \vdots \\ 2x + 2 & \text{if} & -1 \le x < -\frac{1}{2} \\ 2x + 1 & \text{if} & -\frac{1}{2} \le x < 0 \\ 2x & \text{if} & 0 \le x < \frac{1}{2} \\ 2x - 1 & \text{if} & \frac{1}{2} \le x < 1 \\ 2x - 2 & \text{if} & 1 \le x < \frac{3}{2} \\ 2x - 3 & \text{if} & \frac{3}{2} \le x < 2 \\ \;\vdots & & \vdots \end{cases}$

Domain: All real numbers
Range: $[0, 1)$
Discontinuous at $x = k/2$, k an integer

$(1\text{-}3)$

61. $x^3 - 8 = 0$
$(x - 2)(x^2 + 2x + 4) = 0$
$x - 2 = 0 \qquad x^2 + 2x + 4 = 0$
$\qquad x = 2 \qquad\;\; x^2 + 2x \qquad\;\; = -4$
$\qquad\qquad\qquad\;\; x^2 + 2x + 1 = -4 + 1$
$\qquad\qquad\qquad\quad\;\; (x + 1)^2 = -3$
$\qquad\qquad\qquad\qquad\;\; x + 1 = \pm\sqrt{-3}$
$\qquad\qquad\qquad\qquad\qquad x = -1 \pm i\sqrt{3}$
Solution: $x = 2, -1 \pm i\sqrt{3}$

$(2\text{-}5)$

62. (A) The rate at which the amount of water is changing (-13,200 gallons per hour, which is the rate for three pumps each working at 4,400 gallons per hour) is the slope of a linear function; the original amount (400,000) is the y intercept. The equation is $y = mx + b = -13,200x + 400,000$

(B) Set the amount of water, represented here by y, equal to zero and solve:
$$0 = -13,200x + 400,000$$
$$13,200x = 400,000$$
$$x = \frac{400,000}{13,200} \approx 30.3$$

It will take 30.3 hours.

(C) $100,000 < -13,2000x + 400,000 < 200,000$
$$-300,000 < -13,200x < -200,000$$
$$22.7 > x > 15.2 \quad \text{(Note change in direction)}$$

The amount of water will be between 100,000 and 200,000 gallons from 15.2 to 22.7 hours after pumping begins. *(2-1, 2-7)*

63. "Break even" means Cost = Revenue
Let x = number of books sold
Revenue = number of books sold × price per book
$$= x(9.65)$$
Cost = Fixed Cost + Variable Cost
$$= 41,800 + \text{number of books} \times \text{cost per book}$$
$$= 41,800 + x(4.90)$$
$$9.65x = 41,800 + 4.90x$$
$$4.75x = 41,800$$
$$x = \frac{41,800}{4.75}$$
$$x = 8,800 \text{ books}$$
(2-2)

64. The distance of p from 200 must be no greater than 10.
$$|p - 200| \le 10$$
(2-2)

65. (A) A profit will result if revenue is greater than cost; that is, if
$$R > C$$
$$15p - 2p^2 > 88 - 12p$$
$$88 + 27p - 2p^2 > 0$$
$$2p^2 - 27p + 88 < 0$$
$$(2p - 11)(p - 8) < 0$$

The solution of this inequality consists of all values of x for which the graph of $y = 2x^2 - 27x + 88$ is below the x axis. The x intercepts of the graph are 5.5 and 8.

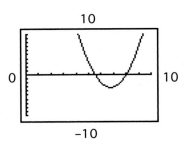

Examining the graph, we see that $2p^2 - 27p + 88 < 0$ and a profit will occur ($R > C$) for $\$5.5 < p < \$8$ or ($\$5.5$, $\$8$).

(B) A loss will result if cost is greater than revenue, that is, if
$$C > R$$
$$88 - 12p > 15p - 2p^2$$
$$2p^2 - 27p + 88 > 0$$

Referring to the graph in part (A), we see that $2p^2 - 27p + 88 > 0$, and a loss will occur ($C > R$), for $p < \$5.5$ or $p > \$8$. Since a negative price doesn't make sense, we delete any number to the left of 0. A loss will occur for $\$0 \le p < \$5.5$ or $p > \$8$. [$\$0$, $\$5.5$) $\cup$ ($\$8$, ∞) *(2-3, 2-7)*

66. (A) If v is linearly related to t, then we are looking for an equation whose graph passes through $(t_1, v_1) = (0, 20000)$ and $(t_2, v_2) = (8, 4000)$. We find the slope, m, and then we use the point-slope form to find the equation.

$$m = \frac{v_2 - v_1}{t_2 - t_1} = \frac{4,000 - 20,000}{8 - 0} = -2,000$$

$$v - v_1 = m(t_2 - t_1)$$
$$v - 20,000 = -2,000(t - 0)$$
$$v - 20,000 = -2,000t$$
$$v = -2,000t + 20,000$$

(B) Solve for t
$$v = -2,000t + 20,000$$
$$v - 20,000 = -2,000t$$
$$t = \frac{v - 20,000}{-2,000}$$
$$t = -0.0005v + 10$$

The inverse function tells us how long it would take to reach a specified value.

(1-6)

67.   Let x = the distance from port A to port B

Then $115 - x$ = the distance from port B to port C

Applying the Pythagorean theorem, we have

$$x^2 + (115 - x)^2 = 85^2$$
$$x^2 + 13,225 - 230x + x^2 = 7,225$$
$$2x^2 - 230x + 6,000 = 0$$
$$x^2 - 115x + 3,000 = 0$$
$$(x - 40)(x - 75) = 0$$
$$x = 40, 75$$
$$115 - x = 75, 40$$

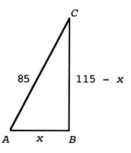

Based on the given information, there are two possible solutions:
    40 miles from A to B and 75 miles from B to C *or*
    75 miles from A to B and 40 miles from B to C (2-5)

68. If x is linearly related to p, then we are looking for an equation whose graph passes through $(p_1, x_1) = (3.79, 1,160)$ and $(p_2, x_2) = (3.59, 1,340)$. We find the slope, and then we use the point-slope form to find the equation.

$$m = \frac{x_2 - x_1}{p_2 - p_1} = \frac{1,340 - 1,160}{3.59 - 3.79} = \frac{180}{-0.2} = -900$$

$$x - x_1 = m(p - p_1)$$
$$x - 1,160 = -900(p - 3.79)$$
$$x - 1,160 = -900p + 3,411$$
$$x = -900p + 4,571$$

If the price is lowered to \$3.29, we are asked for x when $p = 3.29$
$$x = -900(3.29) + 4,571$$
$$= 1,610 \text{ bottles}$$

(2-2)

69. For the first 60 calls: If $0 \le x \le 60$, $C(x) = 0.06x$
For the next 90 calls: If $60 < x \le 150$:

$$\begin{pmatrix} \text{Cost of first} \\ 60 \text{ calls} \end{pmatrix} + \begin{pmatrix} \text{Cost of next } x - 60 \\ \text{calls at 0.05 per call} \end{pmatrix}$$

$$\begin{aligned} C(x) &= 0.06(60) + 0.05(x - 60) \\ &= 3.60 + 0.05x - 3 \\ &= 0.05x + 0.6 \end{aligned}$$

For the next 150 calls, if $150 < x \le 300$,

$$\begin{pmatrix} \text{Cost of first} \\ 150 \text{ calls} \end{pmatrix} + \begin{pmatrix} \text{Cost of next } x - 150 \\ \text{calls at 0.04 per call} \end{pmatrix}$$

$$\begin{aligned} C(x) &= 0.05(150) + 0.6 + 0.04(x - 150) \\ &= 7.5 + 0.6 + 0.04x - 6 \\ &= 0.04x + 2.1 \end{aligned}$$

Finally, if $x > 300$,

$$\begin{pmatrix} \text{Cost of first} \\ 300 \text{ calls} \end{pmatrix} + \begin{pmatrix} \text{Cost of next } x - 300 \\ \text{calls at 0.03 per call} \end{pmatrix}$$

$$\begin{aligned} C(x) &= 0.04(300) + 2.1 + 0.03(x - 300) \\ &= 12 + 2.1 + 0.03x - 9 \\ &= 0.03x + 5.1 \end{aligned}$$

Summarizing,
$$C(x) = \begin{cases} 0.06x & \text{if} & 0 \le x \le 60 \\ 0.05x + 0.6 & \text{if} & 60 < x \le 150 \\ 0.04x + 2.1 & \text{if} & 150 < x \le 300 \\ 0.03x + 5.1 & \text{if} & 300 < x \end{cases}$$

x	$C(x) = 0.06x$	x	$C(x) = 0.05x + 0.6$	x	$C(x) = 0.04x + 2.1$	x	$C(x) = 0.03x + 5.1$
0	0	90	5.1	160	8.5	310	14.4
30	1.8	140	7.6	200	10.1	400	17.1
60	3.6						

(1-3)

70. (A)   Let x = width of pen
Then $2x + \ell = 80$

So $\ell = 80 - 2x$ = length of pen

$A(x) = x\ell = x(80 - 2x) = 80x - 2x^2$

(B) Since all distances must be positive, $x > 0$ and $80 - 2x > 0$, so $80 > 2x$ or $x < 40$. Therefore $0 < x < 40$ or $(0, 40)$ is the domain of $A(x)$.

(C) Note: 0, 40 were excluded from the domain for geometrical reasons; but can be used to help draw the graph since the function $80x - 2x^2$ has domain including these values.

x	$A(x)$
0	0
10	600
20	800
30	600
40	0

Use the vertex formula: $x = -\dfrac{b}{2a} = -\dfrac{80}{2(-2)} = 20$. The maximum value of A occurs when x is 20. The other dimension is $80 - 2(20) = 40$, so the dimensions are 20 feet by 40 feet.   *(1-3, 2-3)*

71. (A) $f(1) = 1 - 2[\![1/2]\!] = 1 - 2(0) = 1$
    $f(2) = 2 - 2[\![2/2]\!] = 2 - 2(1) = 0$
    $f(3) = 3 - 2[\![3/2]\!] = 3 - 2(1) = 1$
    $f(4) = 4 - 2[\![4/2]\!] = 4 - 2(2) = 0$

    (B) If n is an integer, n is either odd or even.
    If n is even, it can be written as $2k$, where k is an integer.
    Then $f(n) = f(2k) = 2k - 2[\![2k/2]\!] = 2k - 2[\![k]\!] = 2k - 2k = 0$.
    Otherwise, n is odd, and n can be written as $2k + 1$, where k is an integer.
    Then $f(n) = f(2k + 1) = 2k + 1 - 2[\![(2k + 1)/2]\!] = 2k + 1 - 2[\![k + \frac{1}{2}]\!]$
    $$= 2k + 1 - 2k$$
    $$= 1$$
    Summarizing, if n is an integer, $f(n) = \begin{cases} 1 \text{ if } n \text{ is an odd integer} \\ 0 \text{ if } n \text{ is an even integer} \end{cases}$ *(1-3)*

72. (A) 30,000 bushels
    (B) The demand decreases by 10,000 bushels to 20,000 bushels.
    (C) The demand increases by 10,000 bushels to 40,000 bushels.
    (D) As the price varies from \$3.00 to \$3.50 per bushel, the demand varies from 15,000 to 50,000 bushels, increasing with decreasing price and decreasing with increasing price.
    (E) From the graph, the following table is constructed:

q	20	25	30	35	40
P	340	332	325	320	315

```
L1      L2     L3     2
20      340    ------
25      332
30      325
35      320
40      315

------
L2(6) =
```

```
QuadReg
y=ax²+bx+c
a=.0228571429
b=-2.611428571
c=383.0285714
```

Enter the data   Compute the regression equation.

The quadratic regression model is $y = -0.0229x^2 - 2.61x + 383$.   *(1-1, 2-1)*

73. (A) The independent variable is number of years since 1970, so we enter 0, 5, 10, 15, 20, 25, 30, and 35 as L_1. The dependent variable is egg consumption, so we enter 309, 276, 271, 255, 233, 234, 252, and 255 as L_2. Then we choose quadratic regression from the STAT CALC menu.

The quadratic regression model is $y = 0.124x^2 - 5.84x + 309$.

(B) We're asked to find x when y is 309 and again when y is 400. Graph $y_1 = 0.124x^2 - 5.84x + 309$, $y_2 = 309$, and $y_3 = 400$, and use the INTERSECT command.

The quadratic model is at height 309 at $x = 47$ so consumption will return to the 1970 level 47 years after 1970 or 2017. The quadratic model is at height 400 when $x = 59$, so consumption will return to the 1945 level 59 years after 1970 or 2029.

(C) Egg consumption declined from 309 eggs per capita in 1970 to a low of 233 in 1990 then began to increase, reaching 255 in 2005. *(2-5)*

74. (A) The independent variable is speed and the dependent variable is length of skid marks, so we enter the speed column as L_1 and the length column as L_2. Then we choose quadratic regression from the STAT CALC menu

The quadratic model is $y = 0.0481x^2 + 0.0690x + 2.21$.

(B) We're asked to find x (speed) when y (length) is 220. Graph $y_1 = 0.0481x^2 + 0.0690x + 2.21$, $y_2 = 220$, and find the intersection.

The speed is 66.6 mph when the length of the skid mark is 220 feet. *(2-5)*

75. (A) The independent variable is speed and the dependent variable is mileage, so we enter the speed column as L1 and the mileage column as L2 then use the quadratic regression command from the STAT CALC menu.

The quadratic model is $y = -0.00149x^2 + 0.0627x + 1.21$.

(B) The rental cost is $10 per hour plus $2.60 per gallon of gas. This can be written as $C = 10t + 2.60g$ where t is the number of hours and g is gallons of gas used. We can write each of t and g in terms of speed (x) to get a model for cost in terms of speed.

The time of the trip is the distance divided by the speed, since distance = speed times time. The trip is 200 miles so $t = \dfrac{200}{x}$.

The gasoline usage can be calculated by dividing the length (200 miles) by the miles per gallon (y); $g = \dfrac{200}{y}$. Substituting in our quadratic model for y:

$$g = \frac{200}{-0.00149x^2 + 0.0627x + 1.21}$$

Our rental cost is

$$C = 10t + 2.60g = 10\left(\frac{200}{x}\right) + 2.60\left(\frac{200}{-0.00149x^2 + 0.0627x + 1.21}\right)$$

$$C = \frac{2,000}{x} + \frac{520}{-0.00149x^2 + 0.0627x + 1.21}$$

We use the MINIMUM command to find the minimum cost.

The minimum cost occurs when the speed is 26.9 mph. Plugging $x = 26.9$ back into our quadratic regression model, we find that the mileage at 26.9 mph is 1.82 mpg. We use our expressions for t, g, and C to answer the remaining questions:

$$t = \frac{200}{x} = \frac{200}{26.9} = 7.43 \text{ hours}$$

$$g = \frac{200}{y} = \frac{200}{1.82} = 110 \text{ gallons}$$

$$C = 10t + 2.60g = 10(7.43) + 2.60(110) = \$360$$

(2-5)

CHAPTER 3   Polynomial and Rational Functions

Section 3-1

1. Answers will vary.

3. Answers will vary.

5. True. The general quadratic function $f(x) = ax^2 + bx + c$ fits the definition of polynomial.

7. False. The coefficients can be imaginary as well.

9. False. The zeros of a polynomial are x intercepts of its graph only if they are real numbers.

11. The degree is odd and the coefficient is positive, so $f(x)$ increases without bound as $x \to \infty$ and decreases without bound as $x \to -\infty$. This matches graph c.

13. The degree is even and the coefficient is positive, so $h(x)$ increases without bound both as $x \to \infty$ and $x \to -\infty$. This matches graph d.

15. The real zeros are the x-intercepts: -1 and 3. The turning point is $(1, 4)$. $P(x) \to -\infty$ as $x \to \infty$ and $P(x) \to -\infty$ as $x \to -\infty$.

17. The real zeros are the x-intercepts: -2 and 1. The turning points are $(-1, 4)$ and $(1, 0)$. $P(x) \to -\infty$ as $x \to -\infty$ and $P(x) \to \infty$ as $x \to \infty$.

19. The graph of a polynomial always increases or decreases without bound as $x \to \infty$ and $x \to -\infty$. This graph does not.

21. The graph of a polynomial always has a finite number of turning points — at most one less than the degree. This graph has infinitely many turning points.

23. This is a polynomial with degree 3.

25. This is not a polynomial because the variable is in the denominator of the last term.

27. This is not a polynomial because the variable is under a radical.

29. This is a polynomial. If the parentheses are multiplied out, it will fit the definition of polynomial, and the degree is 3.

31. Set each factor equal to zero and solve:

$x = 0$	$x^2 - 9 = 0$	$x^2 + 4 = 0$
	$x^2 = 9$	$x^2 = -4$
	$x = 3, -3$	$x = 2i, -2i$

The zeros are 0, 3, -3, $2i$, and $-2i$; of these 0, 3, and -3 are x-intercepts.

33. Set each factor equal to zero and solve:

$x + 5 = 0$	$x^2 + 9 = 0$	$x^2 + 16 = 0$
$x = -5$	$x^2 = -9$	$x^2 = -16$
	$x = 3i, -3i$	$x = 4i, -4i$

The zeros are -5, $3i$, $-3i$, $4i$, and $4i$. Only -5 is an x-intercept.

35. $P(x) = x^3 - 5x^2 + 2x + 6$ $n = 3$ $a_n = 1$

(A) Since $a_n > 0$ and n is odd $P(x) \to \infty$ as $x \to \infty$ and $P(x) \to -\infty$ as $x \to -\infty$. Since the degree of $P(x)$ is 3, $P(x)$ can have a maximum of three x intercepts and two local extrema.

(B) Graphing and applying the ZERO command yields:

The three x intercepts are $x = -0.86$, 1.68, and 4.18. Next, we use the MAXIMUM and MINIMUM commands.

The local maximum is $P(0.21) \approx 6.21$.   The local minimum is $P(3.12) \approx -6.06$.

37. $P(x) = -x^3 + 4x^2 + x + 5$ $n = 3$ $a_n = -1$

(A) Since $a_n < 0$ and n is odd $P(x) \rightarrow -\infty$ as $x \rightarrow \infty$ and $P(x) \rightarrow \infty$ as $x \rightarrow -\infty$. Since the degree of $P(x)$ is 3, $P(x)$ can have a maximum of three x intercepts and two local extrema.

(B) Graphing and applying the ZERO command yields the graph at the right.

There is actually only one x intercept at $x = 4.47$. Next, we use the MAXIMUM and MINIMUM commands.

The local minimum is $P(-0.12) \approx 4.94$ The local maximum is $P(2.79) \approx 17.21$

39. $P(x) = x^4 + x^3 - 5x^2 - 3x + 12$ $n = 4$ $a_n = 1$

(A) Since $a_n > 0$ and n is even $P(x) \rightarrow \infty$ as $x \rightarrow \infty$ and $P(x) \rightarrow \infty$ as $x \rightarrow -\infty$. Since the degree of $P(x)$ is 4, $P(x)$ can have a maximum of four x intercepts and three local extrema.

(B) As seen in the graphs below, there are no x intercepts. Graphing and applying the MAXIMUM and MINIMUM commands yields:

Local minimum at
$P(-1.87) \approx 5.81$

Local maximum at
$P(-0.28) \approx 12.43$

Local minimum at
$P(1.41) \approx 4.59$

41.

43.

45.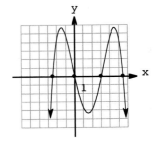

47. $p(x) = x(x - 2)(x + 3)$ (factored form)

$= x(x^2 + x - 6)$

$= x^3 + x^2 - 6x$ (standard form)

49. $p(x) = x(x + \frac{1}{2})(x - 4)$ (to get the specified zeros)

$p(x) = x(2x + 1)(x - 4)$ (multiply $(x + 1/2)$ by 2 to get integer coefficients)

$= x(2x^2 - 7x - 4)$

$= 2x^3 - 7x^2 - 4x$ (standard form)

51. $p(x) = (x - 2)(x - 3)(x - 4)$ (factored form)

$= (x - 2)(x^2 - 7x + 12)$

$= x^3 - 2x^2 - 7x^2 + 14x + 12x - 24$

$= x^3 - 9x^2 + 26x - 24$ (standard form)

53. $p(x) = x(x - i)(x + i)$ (factored form)

$= x(x^2 - i^2)$

$= x(x^2 + 1)$

$= x^3 + x$ (standard form)

55. $P(x) = x^3$ is an example of a third-degree polynomial with one x intercept ($x = 0$).

57. No such polynomial exists; the graph of a third-degree polynomial must cross the x axis at least once.

59. This is the same as the right hand behavior of the function $p(x) = x^2$; even degree, positive leading coefficient, so $\lim_{x \to \infty} x^2 = \infty$.

61. This is the same as the left hand behavior of the function $p(x) = x^3 - x^2 + x$; odd degree, positive leading coefficient, so $\lim_{x \to -\infty} (x^3 - x^2 + x) = -\infty$.

63. This is the same as the right hand behavior of the function $p(x) = 4 - x - 3x^4$; even degree, negative leading coefficient, so $\lim_{x \to \infty} (4 - x - 3x^4) = -\infty$.

65. This is the same as the left hand behavior of the function $p(x) = -x^4 + x^3 + 5x^2$; even degree, negative leading coefficient, so $\lim_{x \to -\infty} (-x^4 + x^3 + 5x^2) = -\infty$

67. Graphing and applying the ZERO command yields:

x intercepts at $x = -12.69, -0.72, 4.41$. Next, we use the MAXIMUM and MINIMUM commands.

Local minimum at
$P(-8.07) \approx -424.07$

Local maximum at
$P(2.07) \approx 96.07$

69. Graphing and applying the ZERO command yields:

x intercepts at $-16.06, 0.50, 15.56$. Next, we use the MAXIMUM and MINIMUM commands.

Local maximum at
$P(-9.13) \approx 65.86$

Local minimum at
$P(9.13) \approx -55.86$

71. Graphing and applying the ZERO commands yields:

x intercepts at $-16.15, -2.53, 1.56, 14.12$. Next, we use the MAXIMUM and MINIMUM commands.

Local minimum at
$P(-11.68) \approx -1,395.99$

Local maximum at
$P(-0.50) \approx 95.72$

Local minimum at
$P(9.92) \approx -1,140.27$

73. Graphing and applying the ZERO command yields:

x intercepts at 1, 1.09. Next, we use the MAXIMUM and MINIMUM commands.

Local minimum at   Local maximum at   Local minimum at
$P(1.05) \approx -0.20$ $P(6.01) \approx 605.01$ $P(10.94) \approx 9.70$

75. (A) If the degree is even the graph has to approach either ∞ or $-\infty$ both to the right and left, so there has to be at least one turning point. $f(x) = x^4$, graphed below, has only 1 turning point.

A degree 4 polynomial can't have any more than 3 turning points.
$f(x) = x^4 - 5x^2$, graphed below, has 3 turning points.

$f(x) = x^4$ $f(x) = x^4 - 5x^2$

(B) An even degree polynomial can have no x-intercepts as illustrated by $f(x) = x^4 + 5$ (graphed below). A polynomial of degree 4 can have at most 4 x-intercepts. For example, $f(x) = x^4 - 5x^2 + 5$, graphed below, has 4 x-intercepts.

$f(x) = x^4 + 5$ $f(x) = x^4 - 5x^2 + 5$

77. No. If $f(x)$ is an even function, then $f(x) = f(-x)$ for any x. But $f(x) = x^2 + x$ is an even degree polynomial and $f(1) = 2$ while $f(-1) = 0$.

79. (A)

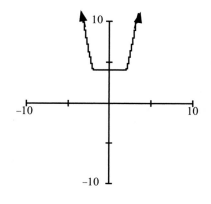

(B) All the points on the horizontal portion of the graph satisfy the definition of local minimum, but none are turning points.
The set $\{c \mid -2 \le x \le 2\}$ represents all such numbers c.

81. It is possible. A third degree polynomial can have at most two turning points, and it has to approach ∞ from either the left or right, and $-\infty$ from the other side; the graph of such a function with real zeros 0 and 2 is shown at the right.

(It is the graph of $p(x) = x^3 - 4x^2 + 4x$.)

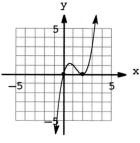

83. (A) The revenue $R(x)$ is given by revenue = (number sold)(price per object)
$$R(x) = xp(x)$$
$$= x(0.0004x^2 - x + 569)$$
$$= 0.0004x^3 - x^2 + 569x$$

(B) Graphing and applying the MAXIMUM command yields the graph at the right.

The number of air conditioners that maximizes the revenue is 364. The maximum revenue is $93,911.

The price is $p(x) = \dfrac{R(x)}{x} = \dfrac{93,911}{364} = \$258.$

85.

(A) The volume of the shielding is equal to the volume of the shielded box (right hand picture) minus the volume of the unshielded box (left hand picture). So

$$\begin{pmatrix}\text{Volume of}\\\text{shielding}\end{pmatrix} = \begin{pmatrix}\text{Volume of}\\\text{shielded box}\end{pmatrix} - \begin{pmatrix}\text{Volume of}\\\text{unshielded box}\end{pmatrix}$$

$$V = (1 + 2x)(2 + 2x)(4 + 2x) - 1 \cdot 2 \cdot 4$$

$$= (2 + 6x + 4x^2)(4 + 2x) - 8$$

$$= 8 + 24x + 16x^2 + 4x + 12x^2 + 8x^3 - 8$$

$$= 28x + 28x^2 + 8x^3$$

(B) Set our expression for volume equal to 3 and solve for x:

$(1 + 2x)(2 + 2x)(4 + 2x) - 8 = 3$

Graphing and applying the INTERSECT command yields the graph at the right.

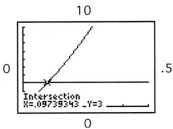

The thickness of the shielding is 0.097 ft.

87. (A) The independent variable is years since 1960, so we enter 0, 10, 20, 30, 35, 40, and 45 as L_1. The dependent variable is total expenditures, so we enter the total expenditures column as L_2. Then we choose the cubic regression command from the STAT CALC menu.

The cubic regression polynomial is $y = 0.0178x^3 + 0.0775x^2 + 3.42x + 23.2$.

(B) 2010 is 50 years after 1960, so we plug 50 into our regression model.

$y = 0.0178(50)^3 + 0.0775(50)^2 + 3.42(50) + 23.2$
$= 2,612.95$

Total expenditures in 2010 will be $2,613.0 billion.

(C) This time, we choose the quadratic regression command from the STAT CALC menu:

The quadratic regression model is $y = 1.29x^2 - 17.0x + 62.8$. The graph shows both the cubic and quadratic models; the cubic model is the one to the left near the top of the screen, which shows that it predicts more rapid long-term increase than the quadratic.

89. (A) The independent variable is years since 1950, so we enter 0, 10, 20, 30, 40, 50, and 55 as L_1. The dependent variable is marriage rate, so we enter the marriage column as L_2. Then we use the cubic regression and quartic regression commands from the STAT CALC menu.

The cubic regression polynomial is $y = -0.000\,151x^3 + 0.0109x^2 - 0.207x + 10.7$, and the quartic is $y = 0.000\,00920x^4 - 0.001\,16x^3 + 0.0449x^2 - 0.558x + 11.0$.

(B) 2008 is 58 years after 1950, so we plug 58 into our regression models.

$y = -0.000\,151(58)^3 + 0.0109(58)^2 - 0.207(58) + 10.7$
$= 5.9$ marriages per 1,000 population

$y = 0.000\,00920(58)^4 - 0.00116(58)^3 + 0.0449(58)^2 - 0.558(58) + 11.0$
$= 7.9$ marriages per 1,000 population

(C) The first graph below shows the cubic model and the scatter plot; the second shows the quartic model:

The quartic model fits the data points better, but the cubic probably provides a more realistic prediction for 2008. After the last data point, the quartic model predicts a sudden increase in the marriage rate that is not at all supported by the data; the trend appears likely to continue downward.

(D) A quadratic model would do a poor job with this data because it can only change direction once, but the data changes direction twice.

Section 3-2

1. $x^2 - 5x + 12 = (x - 2)(x - 3) + 6$. The dividend is $x^2 - 5x + 12$, the divisor is $x - 3$, the quotient is $x - 2$, and the remainder is 6.

3. Answers will vary.

5. Answers will vary.

7.
$$
\begin{array}{r}
2m + 1 \\
2m - 1 \overline{\smash{\big)}\, 4m^2 + 0m - 1} \\
\underline{-\left(4m^2 - 2m\right)} \\
2m - 1 \\
\underline{-(2m - 1)} \\
0
\end{array}
$$
$2m + 1$

9.
$$
\begin{array}{r}
4x - 5 \\
2x + 1 \overline{\smash{\big)}\, 8x^2 - 6x + 6} \\
\underline{-\left(8x^2 + 4x\right)} \\
-10x + 6 \\
\underline{-(-10x - 5)} \\
11
\end{array}
$$
$4x - 5 + \dfrac{11}{2x + 1}$

11.
$$
\begin{array}{r}
x^2 + x + 1 \\
x - 1 \overline{\smash{\big)}\, x^3 - 0x^2 + 0x - 1} \\
\underline{-\left(x^3 - x^2\right)} \\
x^2 + 0x \\
\underline{-\left(x^2 - x\right)} \\
x - 1 \\
\underline{-(x - 1)} \\
0
\end{array}
$$
$x^2 + x + 1$

13.
$$
\begin{array}{r}
2y^2 - 5y + 13 \\
y + 2 \overline{\smash{\big)}\, 2y^3 - y^2 + 3y - 1} \\
\underline{-\left(2y^3 + 4y^2\right)} \\
- 5y^2 + 3y \\
\underline{-\left(-5y^2 - 10y\right)} \\
13y - 1 \\
\underline{-(13y + 26)} \\
- 27
\end{array}
$$
$2y^2 - 5y + 13 - \dfrac{27}{y + 2}$

15.
$$
\begin{array}{r}
\begin{array}{ccc} 1 & 3 & -7 \\ & 2 & 10 \end{array} \\
2 \overline{\big| \begin{array}{ccc} 1 & 5 & 3 \end{array}}
\end{array}
$$
$\dfrac{x^2 + 3x - 7}{x - 2} = x + 5 + \dfrac{3}{x - 2}$

17.
$$
\begin{array}{r}
\begin{array}{ccc} 4 & 10 & -9 \\ & -12 & 6 \end{array} \\
-3 \overline{\big| \begin{array}{ccc} 4 & -2 & -3 \end{array}}
\end{array}
$$
$\dfrac{4x^2 + 10x - 9}{x + 3} = 4x - 2 - \dfrac{3}{x + 3}$

19.

$$
\begin{array}{r}
2\quad 0\quad -3\quad\ 1 \\
\ \ 4\quad\ \ 8\quad 10 \\
\hline
2\ \big|\ 2\quad 4\quad\ \ 5\quad 11
\end{array}
$$

$$\frac{2x^3 - 3x + 1}{x - 2} = 2x^2 + 4x + 5 + \frac{11}{x - 2}$$

21. $P(-2) = 3(-2)^2 - (-2) - 10 = 12 + 2 - 10 = 4$

$$
\begin{array}{r}
3\quad -1\quad -10 \\
\ \ -6\quad\ \ 14 \\
\hline
-2\ \big|\ 3\quad -7\quad\ \ 4
\end{array}
$$

$P(-2) = 4$

23. $P(2) = 2(2)^3 - 5(2)^2 + 7(2) - 7 = 16 - 20 + 14 - 7 = 3$

$$
\begin{array}{r}
2\quad -5\quad\ \ 7\quad -7 \\
\ \ 4\quad -2\quad 10 \\
\hline
2\ \big|\ 2\quad -1\quad\ \ 5\quad\ \ 3
\end{array}
$$

$P(2) = 3$

25. $P(-4) = (-4)^4 - 10(-4)^2 + 25(-4) - 2 = 256 - 160 - 100 - 2 = -6$

$$
\begin{array}{r}
1\quad\ \ 0\quad -10\quad\ \ 25\quad -2 \\
\ \ -4\quad\ 16\quad -24\quad -4 \\
\hline
-4\ \big|\ 1\quad -4\quad\ \ 6\quad\ \ \ 1\quad -6
\end{array}
$$

$P(-4) = -6$

27. $x - 1$ will be a factor of $P(x)$ if $P(1) = 0$. Since $P(x) = x^{18} - 1$, $P(1) = 1^{18} - 1 = 0$. Therefore $x - 1$ is a factor of $x^{18} - 1$.

29. $x + 1$ will be a factor of $P(x)$ if $P(-1) = 0$. Since $P(x) = 3x^3 - 7x^2 - 8x + 2$, $P(-1) = 3(-1)^3 - 7(-1)^2 - 8(-1) + 2 = -3 - 7 + 8 + 2 = 0$. Therefore $x + 1$ is a factor of $3x^3 - 7x^2 - 8x + 2$.

31.

$$
\begin{array}{r}
2x^2 + x + 3 \\
-3x + 4\ \big)\overline{-6x^3 + 5x^2 - 5x + 12} \\
\underline{-\left(-6x^3 + 8x^2\right)} \\
-3x^2 - 5x \\
\underline{-\left(-3x^2 + 4x\right)} \\
-9x + 12 \\
\underline{-\left(-9x + 12\right)} \\
0
\end{array}
$$

$2x^2 + x + 3$

33.

$$
\begin{array}{r}
x^2 - 2x - 5 \\
2x - 1\ \big)\overline{2x^3 - 5x^2 - 8x + 1} \\
\underline{-\left(2x^3 - x^2\right)} \\
-4x^2 - 8x \\
\underline{-\left(-4x^2 + 2x\right)} \\
-10x + 1 \\
\underline{-\left(-10x + 5\right)} \\
-4
\end{array}
$$

$x^2 - 2x - 5 - \dfrac{4}{2x - 1}$

35.

$$\begin{array}{r} x - 4 \\ x^2 + 0x - 2{\overline{\smash{\big)}\,x^3 - 4x^2 - 7x + 1}} \\ -\left(x^3 + 0x^2 - 2x\right) \\ \hline -4x^2 - 5x + 1 \\ -\left(-4x^2 + 0x + 8\right) \\ \hline -5x - 7 \end{array}$$

$$x - 4 + \frac{-5x - 7}{x^2 - 2}$$

37.

$$\begin{array}{r} 3x^2 - 3x - 5 \\ x^2 + x + 2{\overline{\smash{\big)}\,3x^4 + 0x^3 - 2x^2 - 3x + 5}} \\ -\left(3x^4 + 3x^3 + 6x^2\right) \\ \hline -3x^3 - 8x^2 - 3x \\ -\left(-3x^3 - 3x^2 - 6x\right) \\ \hline -5x^2 + 3x + 5 \\ -\left(-5x^2 - 5x - 10\right) \\ \hline 8x + 15 \end{array}$$

$$3x^2 - 3x - 5 + \frac{8x + 15}{x^2 + x + 2}$$

39.

$$\begin{array}{r} \;3\quad\;0\quad\;0\quad -1\quad -4 \\ -3\quad\;3\quad -3\quad\;4 \\ \hline -1{\overline{\smash{\big)}}}\;3\quad -3\quad\;3\quad -4\quad\;0 \end{array}$$

$$3x^3 - 3x^2 + 3x - 4$$

41.

$$\begin{array}{r} \;1\quad\;0\quad\;0\quad\;0\quad\;0\quad\;1 \\ -1\quad\;1\quad -1\quad\;1\quad -1 \\ \hline -1{\overline{\smash{\big)}}}\;1\quad -1\quad\;1\quad -1\quad\;1\quad\;0 \end{array}$$

$$x^4 - x^3 + x^2 - x + 1$$

43.

$$\begin{array}{r} \;3\quad\;2\quad\;0\quad -4\quad -1 \\ -9\quad\;21\quad -63\quad 201 \\ \hline -3{\overline{\smash{\big)}}}\;3\quad -7\quad\;21\quad -67\quad 200 \end{array}$$

$$3x^3 - 7x^2 + 21x - 67 + \frac{200}{x + 3}$$

45.

$$\begin{array}{r} \;2\quad -13\quad\;0\quad\;75\quad\;2\quad\;0\quad -50 \\ 10\quad -15\quad -75\quad\;0\quad 10\quad\;50 \\ \hline 5{\overline{\smash{\big)}}}\;2\quad -3\quad -15\quad\;0\quad\;2\quad 10\quad\;0 \end{array}$$

$$2x^5 - 3x^4 - 15x^3 + 2x + 10$$

47.

$$\begin{array}{r} \phantom{-\frac{1}{2})}\;4\quad\;2\quad -6\quad -5\quad\;1 \\ -2\quad\;0\quad\;3\quad\;1 \\ \hline -\tfrac{1}{2}{\overline{\smash{\big)}}}\;4\quad\;0\quad -6\quad -2\quad\;2 \end{array}$$

$$4x^3 - 6x - 2 + \frac{2}{x + 1/2}$$

49.

$$\begin{array}{r} \phantom{-\frac{3}{2})}\;4\quad\;4\quad -7\quad -6 \\ -6\quad\;3\quad\;6 \\ \hline -\tfrac{3}{2}{\overline{\smash{\big)}}}\;4\quad -2\quad -4\quad\;0 \end{array}$$

$$4x^2 - 2x - 4$$

51.

$$\begin{array}{r} \;3\quad -2\quad\;2\quad -3\quad\;1 \\ 1.2\quad -0.32\quad\;0.672\quad -0.9312 \\ \hline 0.4{\overline{\smash{\big)}}}\;3\quad -0.8\quad\;1.68\quad -2.328\quad\;0.0688 \end{array}$$

$$3x^3 - 0.8x^2 + 1.68x - 2.328 + \frac{0.0688}{x - 0.4}$$

53.

$$\begin{array}{r} \;3\quad\;2\quad\;5\quad\;0\quad -7\quad -3 \\ -2.4\quad\;0.32\quad -4.256\quad\;3.4048\quad\;2.87616 \\ \hline -0.8{\overline{\smash{\big)}}}\;3\quad -0.4\quad\;5.32\quad -4.256\quad -3.5952\quad -0.12384 \end{array}$$

$$3x^4 - 0.4x^3 + 5.32x^2 - 4.256x - 3.5952 - \frac{0.12384}{x + 0.8}$$

55. $P\left(\dfrac{5}{2}\right) = 4\left(\dfrac{5}{2}\right)^3 - 12\left(\dfrac{5}{2}\right)^2 - 7\left(\dfrac{5}{2}\right) + 10 = \dfrac{125}{2} - 75 - \dfrac{35}{2} + 10$

$\qquad = \dfrac{125}{2} - \dfrac{150}{2} - \dfrac{35}{2} + \dfrac{20}{2} = -\dfrac{40}{2} = -20$

$$
\begin{array}{r|rrrr}
 & 4 & -12 & -7 & 10 \\
 &   & 10 & -5 & -30 \\
\hline
5/2 & 4 & -2 & -12 & -20
\end{array}
\qquad P(5/2) = -20
$$

57. $P\left(\dfrac{1}{2}\right) = 3\left(\dfrac{1}{2}\right)^3 + 5\left(\dfrac{1}{2}\right)^2 - \dfrac{1}{2} + 2 = \dfrac{3}{8} + \dfrac{5}{4} - \dfrac{1}{2} + 2$

$\qquad = \dfrac{3}{8} + \dfrac{10}{8} - \dfrac{4}{8} + \dfrac{16}{8} = \dfrac{25}{8}$

$$
\begin{array}{r|rrrr}
 & 3 & 5 & -1 & 2 \\
 &   & 3/2 & 13/4 & 9/8 \\
\hline
1/2 & 3 & 13/2 & 9/4 & 25/8
\end{array}
\qquad P(1/2) = 25/8
$$

59. $P(2i) = (2i)^3 + 1 = 2^3 \cdot i^3 + 1 = 8(-i) + 1 = 1 - 8i$

$$
\begin{array}{r|rrrr}
 & 1 & 0 & 0 & 1 \\
 &   & 2i & -4 & -8i \\
\hline
2i & 1 & 2i & -4 & 1-8i
\end{array}
\qquad P(2i) = 1 - 8i
$$

61.
$$
\begin{array}{r|rrrr}
 & 1 & 0 & -7 & 6 \\
 &   & 1 & 1 & -6 \\
\hline
1 & 1 & 1 & -6 & 0
\end{array}
$$

The remainder is zero, so $x = 1$ is a zero. The quotient is $x^2 + x - 6$, so we need to set this equal to zero and solve.

$x^2 + x - 6 = 0$
$(x + 3)(x - 2) = 0$

$x + 3 = 0 \quad or \quad x - 2 = 0$
$x = -3 \qquad\qquad x = 2$

The three zeros are $x = 1, 2,$ and -3.

63.
$$
\begin{array}{r|rrrr}
 & 3 & -8 & -5 & 6 \\
 &   & 9 & 3 & -6 \\
\hline
3 & 3 & 1 & -2 & 0
\end{array}
$$

The remainder is zero, so $x = 3$ is a zero. The quotient is $3x^2 + x - 2$, so we need to set this equal to zero and solve.

$3x^2 + x - 2 = 0$
$(3x - 2)(x + 1) = 0$

$3x - 2 = 0 \quad or \quad x + 1 = 0$
$3x = 2 \qquad\qquad x = -1$
$x = 2/3$

The three zeros are $x = 3, 2/3,$ and -1.

65.
$$
\begin{array}{r|rrrr}
 & 1 & -5 & 4 & -20 \\
 &   & 5 & 0 & 20 \\
\hline
5 & 1 & 0 & 4 & 0
\end{array}
$$

The remainder is zero, so $x = 5$ is a zero. The quotient is $x^2 + 4$, so we need to set this equal to zero and solve.

$$x^2 + 4 = 0$$
$$x^2 = -4$$
$$x = \pm 2i$$

The three zeros are $x = 5$, $2i$, and $-2i$.

67.

$$
\begin{array}{r}
\ \ 1 \quad -2 \quad -3 \quad\ 10 \\
\ \ \quad\ -2 \quad\ \ 8 \quad -10 \\
\hline
-2\ |\ \ 1 \quad -4 \quad\ \ 5 \quad\ \ 0
\end{array}
$$

The remainder is zero, so $x = -2$ is a zero. The quotient is $x^2 - 4x + 5$, so we need to set this equal to zero and solve.

$$x^2 - 4x + 5 = 0$$

$$x = \frac{-(-4) \pm \sqrt{(-4)^2 - 4(1)(5)}}{2(1)}$$

$$x = \frac{4 \pm \sqrt{-4}}{2} = \frac{4 \pm 2i}{2} = 2 \pm i$$

The three zeros are $x = -2$, $2 + i$, and $2 - i$.

69.

$$
\begin{array}{r}
\ \ 1 \quad\quad -3 \quad\quad\quad\ 1 \quad -3 \\
\ \ \quad\quad\ i \quad\ -3i - 1 \quad\ 3 \\
\hline
i\ |\ 1 \quad -3 + i \quad\quad\quad -3i \quad\ 0
\end{array}
$$

$$x^2 + (-3 + i)x - 3i$$

71. We apply the remainder theorem to evaluate using synthetic division.

(A)
$$
\begin{array}{r}
\ \ 1 \quad\quad 2i \quad -10 \\
\ \ \quad\ 2 - i \quad\ \ 5 \\
\hline
2 - i\ |\ 1 \quad 2 + i \quad -5
\end{array}
$$

$$P(2 - i) = -5$$

(B)
$$
\begin{array}{r}
\ \ 1 \quad\quad\ \ 2i \quad\quad\quad -10 \\
\ \ \quad\quad 5 - 5i \quad\ 10 - 40i \\
\hline
5 - 5i\ |\ 1 \quad 5 - 3i \quad\quad\quad -40i
\end{array}
$$

$$P(5 - 5i) = -40i$$

(C)
$$
\begin{array}{r}
\ \ 1 \quad\quad 2i \quad -10 \\
\ \ \quad\ 3 - i \quad\ 10 \\
\hline
3 - i\ |\ 1 \quad 3 + i \quad\ \ 0
\end{array}
$$

$$P(3 - i) = 0$$

(D)
$$
\begin{array}{r}
\ \ 1 \quad\quad\ \ 2i \quad -10 \\
\ \ \quad\ -3 - i \quad\ 10 \\
\hline
-3 - i\ |\ 1 \quad -3 + i \quad\ \ 0
\end{array}
$$

$$P(-3 - i) = 0$$

73. We will divide by $x - 2$ using synthetic division, and try to figure out what value of k will make the remainder zero (which guarantees that $x = 2$ is a zero).

$$
\begin{array}{r}
\ \ 1 \quad -3 \quad\ \ 4 \quad\quad\ k \\
\ \ \quad\quad\ \ 2 \quad -2 \quad\quad 4 \\
\hline
2\ |\ \ 1 \quad -1 \quad\ \ 2 \quad\ k + 4
\end{array}
$$

If $k = -4$, the remainder is zero, and $x = 2$ is a zero of P.

75. We will use the same approach as in Problem 73.

$$
\begin{array}{r}
\ \ -4 \quad -6 \quad\quad\quad k \quad\quad\quad\quad -6 \\
\ \ \quad\quad\ \ 12 \quad\quad -18 \quad -3k + 54 \\
\hline
-3\ |\ \ -4 \quad\ \ 6 \quad\ k - 18 \quad -3k + 48
\end{array}
$$

Now we need the value of k for which $-3k + 48 = 0$.

$$-3k + 48 = 0 \quad\Rightarrow\quad -3k = -48 \quad\Rightarrow\quad k = 16$$

77. (A) $a_2(x) + (a_1 + a_2 r)$

$$
x - r \overline{)\, a_2 x^2 + a_1 x \qquad\qquad\quad + a_0}
$$

$$
\underline{a_2 x^2 - a_2 r x}
$$

$$
(a_1 + a_2 r)x \qquad\qquad + a_0
$$

$$
\underline{(a_1 + a_2 r)x - r(a_1 + a_2 r)}
$$

$$
a_0 + r(a_1 + a_2 r)
$$

$$
\begin{array}{ccc}
a_2 & a_1 & a_0 \\
 & a_2 r & (a_1 + a_2 r)r \\
\hline
r\,\overline{)\,a_2} & a_1 + a_2 r & a_0 + (a_1 + a_2 r)r
\end{array}
$$

In both cases the coefficient of x is a_2, the constant term $a_2 r + a_1$, and the remainder is $(a_2 r + a_1)r + a_0$.

(B) The remainder expanded is $a_2 r^2 + a_1 r + a_0 = P(r)$.

79.
$$
\begin{aligned}
P(x) &= \{[(2x - 3)x + 2]x - 5\}x + 7 \\
P(-2) &= \{[\{2(-2) - 3\}(-2) + 2](-2) - 5\}(-2) + 7 \\
&= \{[\{-7\}(-2) + 2](-2) - 5\}(-2) + 7 \\
&= \{[16](-2) - 5\}(-2) + 7 \\
&= \{-37\}(-2) + 7 \\
&= 81 \\
P(1.7) &= \{[\{2(1.7) - 3\}(1.7) + 2](1.7) - 5\}(1.7) + 7 \\
&= \{[\{0.4\}(1.7) + 2](1.7) - 5\}(1.7) + 7 \\
&= \{[2.68](1.7) - 5\}(1.7) + 7 \\
&= \{-0.444\}(1.7) + 7 \\
&= 6.2452 \text{ or } 6.2 \text{ to two significant digits.}
\end{aligned}
$$

81. (A) $P(x) = (x - r)Q(x) + R$ where $Q(x)$ is a polynomial with degree one less than the degree of P.

(B) Plug in $x = r$: $\quad P(r) = (r - r)Q(r) + R = 0 + R = R$

Conclusion: $P(r)$ is equal to the remainder when dividing by $x - r$.

83. No. The two remaining zeros could be imaginary, as in $P(x) = x^3 + x = x(x^2 + 1)$. One of the zeros is $x = 0$, and the two remaining zeros are i and $-i$.

Section 3-3

1.

$$
\begin{array}{c|rrrr}
 & 2 & -4 & -18 & 1 \\
 & & 8 & 16 & -8 \\
\hline
4 & 2 & 4 & -2 & -7
\end{array}
$$

The coefficients in the quotient row are not all positive, so $x = 4$ is not an upper bound for the zeros.

$$
\begin{array}{c|rrrr}
 & 2 & -4 & -18 & 1 \\
 & & -6 & 30 & -36 \\
\hline
-3 & 2 & -10 & 12 & -35
\end{array}
$$

The coefficients in the quotient row alternate in sign, so $x = -3$ is a lower bound for the zeros.

3.

```
        1    4   -3  -10
             2   12   18
  2 |   1    6    9    8
```

The coefficients in the quotient row are all positive, so $x = 2$ is an upper bound for the zeros.

```
        1    4   -3  -10
            -4    0   12
 -4 |   1    0   -3    2
```

The coefficients in the quotient row do not alternate in sign, so $x = -4$ is not a lower bound for the zeros.

5.

```
        1    2   -3    4   -5
             1    3    0    4
  1 |   1    3    0    4   -1
```

The coefficients in the quotient row are not all positive, so $x = 1$ is not an upper bound for the zeros.

```
        1    2   -3    4    -5
            -3    3    0   -12
 -3 |   1   -1    0    4   -17
```

The coefficients in the quotient row do not alternate in sign, so $x = -3$ is not a lower bound for the zeros.

7. The degree is 2 so there are at most two zeros. Graph $y = x^2 + 5x - 2$ and use the ZERO command.

The zeros are -5.372 and 0.372.

9. The degree is 3 so there are at most three zeros. Graph $y = 2x^3 - 5x + 2$ and use the ZERO command.

The zeros are -1.752, 0.432, and 1.320.

11. Yes. There are at most three zeros for this third degree polynomial, and the graph has three x intercepts.

13. Yes. The upper and lower bound theorem tells us that $x = 4$ is an upper bound for the zero, which means any number bigger than 4 (10 in particular) is also an upper bound for the zeros.

15. Answers will vary.

17. This inequality is true wherever the graph of $P(x)$ is above the x-axis or intersects the x-axis. This is $[-2, -1] \cup \{1\} \cup [3, \infty)$.

19. This inequality is true only when $P(x)$ is above the x-axis but not where it intersects the x-axis. This is $(-2, -1) \cup (3, \infty)$.

21. See the graph of $P(x) = x^2 + 5x - 2$ in Problem 7. It is positive (above the x-axis) on $(-\infty, -5.372) \cup (0.372, \infty)$.

23. See the graph of $P(x) = 2x^3 - 5x + 2$ in Problem 9. It is negative (below the x-axis) or zero (on the x-axis) on $(-\infty, -1.752] \cup [0.432, 1.320]$.

25. We form a synthetic division table.

	1	0	-3	1	
0	1	0	-3	1	
1	1	1	-2	-1	
2	1	2	1	3	an upper bound
-1	1	-1	-2	3	
-2	1	-2	1	-1	a lower bound

2 is an upper bound; -2 is a lower bound.

27. We form a synthetic division table.

	1	-3	4	2	-9	
0	1	-3	4	2	-9	
1	1	-2	2	4	-5	
2	1	-1	2	6	3	
3	1	0	4	14	33	an upper bound
-1	1	-4	8	-6	-3	
-2	1	-5	14	-26	43	a lower bound

3 is an upper bound; -2 is a lower bound.

29. We form a synthetic division table.

	1	0	-3	3	2	-2	
0	1	0	-3	3	2	-2	
1	1	1	-2	1	3	1	
2	1	2	1	5	12	22	an upper bound
-1	1	-1	-2	5	-3	1	
-2	1	-2	1	1	0	-2	
-3	1	-3	6	-15	47	-143	a lower bound

2 is an upper bound; -3 is a lower bound.

31. (A) We form a synthetic division table.

	1	-2	3	-8
0	1	-2	3	-8
1	1	-1	2	-6
2	1	0	3	-2
3	1	1	6	10
-1	1	-3	6	-14

From the table, 3 is an upper bound and -1 is a lower bound.

(B) The only interval in which a real zero is indicated is $(2, 3)$. The remainders have different signs for $x = 2$ and $x = 3$. We search for this real zero. We organize our calculations in a table.

Sign Change Interval (a, b)	Midpoint m	Sign of P P(a)	P(m)	P(b)
(2, 3)	2.5	−	+	+
(2, 2.5)	2.25	−	+	+
(2, 2.25)	2.125	−	−	+
(2.125, 2.25)	2.1875	−	−	+
(2.1875, 2.25)	2.21875	−	−	+
(2.21875, 2.25)	2.234375	−	−	+
(2.234375, 2.25)	2.2421875	−	−	+
(2.2421875, 2.25)	2.24609375	−	−	+
(2.24609375, 2.25)	We stop here	−		+

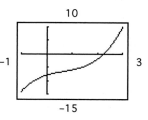

Since each endpoint rounds to 2.25, a real zero lies on this last interval and is given by 2.25 to two decimal place accuracy. A glance at the graph of $P(x)$ on [-1, 3] confirms that this is the only real zero.

33. (A) We form a synthetic division table.

	1	1	-5	7	-22
0	1	1	-5	7	-22
1	1	2	-3	4	-18
2	1	3	1	9	-4
3	1	4	7	28	62
-1	1	0	-5	12	-34
-2	1	-1	-3	13	-48
-3	1	-2	1	4	-34
-4	1	-3	7	-21	62

From the table, 3 is an upper bound and -4 is a lower bound.

(B) There are real zeros in the intervals (2, 3) and (-4, -3) indicated in the table. The remainders have different signs for each pair of x values. We search for the zero in (2, 3). We organize our calculations in a table.

Sign Change Interval (a, b)	Midpoint m	Sign of P P(a)	P(m)	P(b)
(2, 3)	2.5	−	+	+
(2, 2.5)	2.25	−	+	+
(2, 2.25)	2.125	−	+	+
(2, 2.125)	2.0625	−	−	+
(2.0625, 2.125)	2.09375	−	−	+
(2.09375, 2.125)	2.109375	−	−	+
(2.109375, 2.125)	2.1171875	−	−	+
(2.1171875, 2.125)	2.12109375	−	+	+
(2.1171875, 2.12109375)	We stop here	−		+

Since each endpoint rounds to 2.12, a real zero lies on this last interval and is given by 2.12 to two decimal place accuracy. A similar search (details omitted) leads to -3.51 as the other indicated real zero. A glance at the graph of $P(x)$ on [-4, 3] confirms that these are the only zeros.

35. (A) We form a synthetic division table.

	1	0	-3	0	-4	4
0	1	0	-3	0	-4	4
1	1	1	-2	-2	-6	-2
2	1	2	1	2	0	4
-1	1	-1	-2	2	-6	10
-2	1	-2	1	-2	0	4
-3	1	-3	6	-18	50	-146

From the table, 2 is an upper bound and -3 is a lower bound.

(B) There are real zeros in the intervals (0, 1), (1, 2), and (-3, -2) indicated in the table. The remainders have different signs for each pair of x values. We search for the real zero in (0, 1). We organize our calculations in a table.

Sign Change Interval (a, b)	Midpoint m	Sign of P		
		P(a)	P(m)	P(b)
(0, 1)	0.5	+	+	-
(0.5, 1)	0.75	+	-	-
(0.5, 0.75)	0.625	+	+	-
(0.625, 0.75)	0.6875	+	+	-
(0.6875, 0.75)	0.71875	+	+	-
(0.71875, 0.75)	0.734375	+	+	-
(0.734375, 0.75)	0.7421875	+	+	-
(0.7421875, 0.75)	0.74609875	+	+	-
(0.74609875, 0.75)	We stop here	+		-

Since each endpoint rounds to 0.75, a real zero lies on this last interval and is given by 0.75 to two decimal place accuracy. Similar searches (details omitted) lead to -2.09 and 1.88 as the other indicated real zeros. A glance at the graph of $P(x)$ on [-3, 3] confirms that these are the only zeros.

37. (A) We form a synthetic division table.

	1	1	3	1	2	-5
0	1	1	3	1	2	-5
1	1	2	5	6	8	3
-1	1	0	3	-2	4	-9

From the table, 1 is an upper bound and -1 is a lower bound.

(B) There is a real zero in the interval (0, 1) indicated in the table. The remainders have different signs for $x = 0$ and $x = 1$. We search for this real zero. We organize our calculations in a table.

Sign Change Interval (a, b)	Midpoint m	Sign of P		
		P(a)	P(m)	P(b)
(0, 1)	0.5	-	-	+
(0.5, 1)	0.75	-	-	+
(0.75, 1)	0.875	-	+	+
(0.75, 0.875)	0.8125	-	-	+
(0.8125, 0.875)	0.84375	-	+	+
(0.8125, 0.84375)	0.828125	-	-	+
(0.828125, 0.84375)	0.8359375	-	+	+
(0.828125, 0.8359375)	0.83203125	-	-	+
(0.83203125, 0.8359375)	0.833984375	-	-	+
(0.833984375, 0.8359375)	0.8349609375	-	+	+
(0.833984375, 0.8349609375)	We stop here	-		+

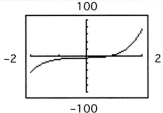

Since each endpoint rounds to 0.83, a real zero lies on this last interval and is given by 0.83 to two decimal place accuracy. A glance at the graph of $P(x)$ on [-2, 2] confirms that this is the only real zero.

39. (A) $P(3) = 3^3 - 2(3)^2 - 5(3) + 4 = -2$ Different signs, so there is
      $P(4) = 4^3 - 2(4)^2 - 5(4) + 4 = 16$ a zero between 3 and 4.

    (B) Midpoint of [3, 4]: 3.5; $P(3.5) = 4.875$; New interval: [3, 3.5]
       Midpoint of [3, 3.5]: 3.25; $P(3.25) = 0.953$; New interval: [3, 3.25]
       Midpoint of [3, 3.25]: 3.125; $P(3.125) = -0.639$;
                           New interval: [3.125, 3.25]
       Midpoint of [3.125, 3.25]: 3.1875; $P(3.1875) = 0.128$;
                           New interval: [3.125, 3.1875]
       Midpoint of [3.125, 3.1875]: 3.15625; $P(3.15625) = -0.263$

The sign is different for $P(3.15625)$ and $P(3.1875)$; both of these x-values
round to 3.2, so 3.2 is the zero to one decimal place accuracy. It took 5
intervals.

> Calculator Note: Here's a convenient way to repeatedly plug different numbers
> into the same function, as we did in Problem 39. Enter the function as y_1 in
> the equation editor and use the "value" command from the calculate menu. All
> you have to do is key in the x-value you want to plug in.

41. (A) $P(-2) = -9$ Different signs, so there's a zero between -2 and -1.
      $P(-1) = 3$

    (B) Midpoint of [-2, -1] = -1.5; $P(-1.5) = -1.375$; New interval: [-1.5, -1]
       Midpoint of [-1.5, -1] = -1.25; $P(-1.25) = 1.17$;
                          New interval: [-1.5, -1.25]
       Midpoint of [-1.5, -1.25] = -1.375; $P(-1.375) = -0.006$;
                         New interval: [-1.375, -1.25]
       Midpoint of [-1.375, -1.25] = -1.3125; $P(-1.3125) = 0.606$;
                         New interval: [-1.375, -1.3125]
       Midpoint of [-1.375, -1.3125] = -1.34375; $P(-1.34375) = 0.306$;
                         New interval: [-1.375, -1.34375]
       Midpoint of [-1.375, -1.34375] = -1.359375; $P(-1.359375) = 0.15$

$P(-1.375)$ and $P(-1.359375)$ have opposite signs and both x-values round to -1.4,
so -1.4 is the zero to one decimal place accuracy. It took 6 intervals.

43. (A) $P(3) = -2$
      $P(4) = 59$ Different signs, so there's a zero between 3 and 4.

    (B) Midpoint of [3, 4] = 3.5; $P(3.5) = 17.1$; New interval: [3, 3.5]
       Midpoint of [3, 3.5] = 3.25; $P(3.25) = 5.2$; New interval: [3, 3.25]
       Midpoint of [3, 3.25] = 3.125; $P(3.125) = 1.1$; New interval: [3, 3.125]
       Midpoint of [3, 3.125] = 3.0625; $P(3.0625) = -0.6$

$P(3.0625)$ and $P(3.125)$ have opposite signs and both x-values round to 3.1, so
3.1 is the zero to one decimal place accuracy. It took 4 intervals.

45. (A) $P(-1) = -3$
      $P(0) = 3$ Different signs, so there's a zero between -1 and 0.

    (B) Midpoint of [-1, 0] = -0.5; $P(-0.5) = .19$; New interval: [-1, -0.5]
       Midpoint of [-1, -0.5] = -0.75; $P(-0.75) = -1.5$;
                         New interval: [-0.75, -0.5]
       Midpoint of [-0.75, -0.5] = -0.625; $P(-0.625) = -0.7$;
                         New interval: [-0.625, -0.5]
       Midpoint of [-0.625, -0.5] = -0.5625; $P(-0.5625) = -0.3$;
                         New interval: [-0.5625, -0.5]
       Midpoint of [-0.5625, -0.5] = -0.53125; $P(-0.53125) = -0.02$

$P(-0.5)$ and $P(-0.53125)$ have opposite signs and both x-values round to -0.5,
so -0.5 is the zero to one decimal place accuracy. It took 5 intervals.

47. Since 1 is a zero of multiplicity 2, $P(x)$ does not change sign at $x = 1$. The bisection method fails. (The TI-84, using a more complicated routine, can find the zero.)

49. Since 3 is a zero of multiplicity 4, $P(x)$ does not change sign at $x = 3$. The bisection method fails. (The TI-84 cannot find this zero.)

51. The graph in a standard window shows two zeros that are turning points. The degree is 4 so there could be more zeros. We'll make a synthetic division table to see if $x = 10$ and $x = -10$ are bounds for the real zeros.

	1	-4	-10	28	49
-10	1	-14	130	-1272	12769
10	1	6	50	528	5329

The signs alternate for -10 so it's a lower bound for the real zeros, and all signs are positive for 10 so it's an upper bound. In other words, the only real zeros are the two we see on the standard screen.

To find the zeros, we use the MINIMUM command:

The zeros are —1.83 and 3.83

53. The graph in a standard window shows three zeros, two of which are turning points. The degree is 5 so there could be more zeros. We'll make a synthetic division table to see if $x = 10$ and $x = -10$ are bounds for the real zeros.

	1	-6	4	24	-16	-32
-10	1	-16	164	-1616	16144	-161472
10	1	4	44	464	4624	46208

The signs alternate for -10 so it's a lower bound for the real zeros, and all signs are positive for 10 so it's an upper bound. In other words, the only real zeros are the two we see on the standard screen.

To find the zero that's not a turning point, we use the ZERO command; and to find the two that are, we use the MAXIMUM and MINIMUM commands:

The zeros are -1.24, 2, and 3.24.

55. On a standard viewing window it's difficult to see what's happening close to $x = 2$, so we examine that area more closely.

Now we can see that there are three zeros, two of which are turning points. The degree is five so there could be other zeros. A synthetic division table will help us check.

	1	-6	11	-4	-3.75	-0.5
-2	1	-8	27	-58	112.25	-225
10	1	4	51	506	5056.25	50562

The signs alternate for -2 so it's a lower bound for the zeros. The signs are all positive for 10 so it's an upper bound for the zeros and we know that we have seen all of them. To find the zero that's not a turning point we use the ZERO command; and to find the two that are, we use the MAXIMUM and MINIMUM commands.

The zeros are -0.22, 2, and 2.22.

57. <u>Algebraic Solution</u>
Find zeros of the left side:
$$(2x - 5)(x + 3)(x + 7) = 0$$
$$x = 5/2, -3, -7$$

Choose test numbers and make a table:

Interval	Test Number	Result
$x < -7$	-10	$(2(-10) - 5)(-10 + 3)(-10 + 7) = -525;$ negative
$-7 < x < -3$	-5	$(2(-5) - 5)(-5 + 3)(-5 + 7) = 60;$ positive
$-3 < x < 5/2$	0	$(2(0) - 5)(0 + 3)(0 + 7) = -105;$ negative
$x > 5/2$	5	$(2(5) - 5)(5 + 3)(5 + 7) = 480;$ positive

Solution: $(-7, -3) \cup (5/2, \infty)$
<u>Graphical Solution</u>

The graph is positive (above the x axis) on $(-7, -3) \cup (5/2, \infty)$.

59. <u>Algebraic Solution</u>
Rearrange so that zero is on the right side:
$$x^2 - 2 > 0$$
Find the zeros of the left side:
$$x^2 - 2 = 0 \quad \Rightarrow x^2 = 2 \quad \Rightarrow x = \pm\sqrt{2} \approx \pm 1.414$$

Interval	Test Number	Result
$x < -1.414$	-2	$(-2)^2 - (-2) = 8;$ positive
$-1.414 < x < 1.414$	0	$0^2 - 2 = -2;$ negative
$x > 1.414$	2	$2^2 - 2 = 2;$ positive

Solution: $(-\infty, -1.414) \cup (1.414, \infty)$

Graphical Solution

The graph is positive (above the x axis) on $(-\infty, -1.414) \cup (1.414, \infty)$.

61. Algebraic Solution
Rearrange so that zero is on the right side:

$x^3 - 9x \leq 0$

Find zeros of the left side:

$x^3 - 9x = 0$
$x(x^2 - 9) = x(x + 3)(x - 3) = 0$

Zeros: $x = 0, -3, 3$

Choose test numbers and make a table:

Interval	Test Number	Result
$x < -3$	-5	$(-5)^3 - 9(-5) = -80;$ negative
$-3 < x < 0$	-1	$(-1)^3 - 9(-1) = 8;$ positive
$0 < x < 3$	1	$(1)^3 - 9(1) = -8;$ negative
$x > 3$	5	$(5)^3 - 9(5) = 80;$ positive

Solution: $(-\infty, -3] \cup [0, 3]$

Graphical Solution

The graph is negative or zero (below or on the x axis) on $(-\infty, -3] \cup [0, 3]$.

63. Algebraic Solution
Rearrange so that zero is on the right side:

$x^3 + 5x^2 + 6x \leq 0$

Find zeros of the left side:

$x^3 + 5x^2 + 6x = 0$
$x(x^2 + 5x + 6) = x(x + 3)(x + 2) = 0$

Zeros: $x = 0, -3, -2$

Choose test numbers and make a table:

Interval	Test Number	Result
$x < -3$	-5	$(-5)^3 + 5(-5)^2 + 6(-5) = -30;$ negative
$-3 < x < -2$	-2.5	$(-2.5)^3 + 5(-2.5)^2 + 6(-2.5) = 0.625;$ positive
$-2 < x < 0$	-1	$(-1)^3 + 5(-1)^2 + 6(-1) = -2;$ negative
$x > 0$	1	$(1)^3 + 5(1)^2 + 6(1) = 12;$ positive

Solution: $(-\infty, -3] \cup [-2, 0]$

Graphical Solution

The graph is negative or zero (below or on the x axis) on $(-\infty, -3] \cup [-2, 0]$.

Problems 65-69 are solved graphically because finding the zeros algebraically is difficult, or maybe impossible.

65. $x^2 + 7x - 3 \leq x^3 + x + 4$

$-x^3 + x^2 + 6x - 7 \leq 0$

Find the zeros of $P(x) = -x^3 + x^2 + 6x - 7$ graphically:

The zeros are -2.507, 1.222, and 2.285. $P(x)$ is negative or zero (below, or on, the x-axis) on $[-2.507, 1.222] \cup [2.285, \infty)$.

67. $x^4 < 8x^3 - 17x^2 + 9x - 2$

$x^4 - 8x^3 + 17x^2 - 9x + 2 < 0$

Find the zeros of $P(x) = x^4 - 8x^3 + 17x^2 - 9x + 2$ graphically:

The zeros are 2.484 and 4.873. $P(x)$ is negative (below the x-axis) on $(2.484, 4.873)$.

69. $(x^2 + 2x - 2)^2 \geq 2$

$(x^2 + 2x - 2)^2 - 2 \geq 0$

Find the zeros of $P(x) = (x^2 + 2x - 2)^2 - 2$ graphically (standard window):

The zeros are -3.101, -2.259, 0.259, and 1.101. $P(x)$ is positive or zero (above, or on, the x-axis) on $(-\infty, -3.101] \cup [-2.259, 0.259] \cup [1.101, \infty)$.

71. (A) There are three real zeros. Because this is a third degree polynomial, there can be at most three zeros, so we can see all of them in this window.

(B) The largest zero looks to be around $x = 2$, so $x = 3$ is an upper bound for the real zeros.

(C)

$$
\begin{array}{r|rrrr}
 & -1 & -1 & 6 & 0 \\
 &   & -3 & -12 & -18 \\
\hline
3 & -1 & -4 & -6 & -18 \\
\end{array}
$$

All coefficients in the quotient row are negative; when the leading coefficient of a polynomial is negative, $x = r$ is an upper bound for the real zeros if dividing by $x - r$ using synthetic division makes all coefficients in the quotient row negative.

73. (A) We form a synthetic division table.

	1	−24	−25	10
0	1	−24	−25	10
10	1	−14	−165	−1640
20	1	−4	−105	−2090
30	1	6	155	4660
−10	1	−34	315	−3140

From the table, 30 is an upper bound and −10 is a lower bound.

(B) Graphing $P(x)$ using the window suggested by part (A), we obtain

The zeros are −1.29, 0.31, and 24.98.

75. (A) We form a synthetic division table.

	1	12	−900	0	5,000
0	1	12	−900	0	5,000
10	1	22	−680	−680	−63,000
20	1	32	−260	−5200	−99,000
30	1	42	360	10,800	329,000
−10	1	2	−920	9,200	−87,000
−20	1	−8	−740	14,800	−291,000
−30	1	−18	−360	10,800	−319,000
−40	1	−28	220	−8,800	357,000

From the table, 30 is an upper bound and −40 is a lower bound.

(B) Graphing $P(x)$ using the window suggested in part (A), we obtain

The zeros are -36.53, -2.33, 2.40, and 24.46.

77. (A) We form a synthetic division table.

	1	0	-100	-1,000	-5,000
0	1	0	-100	-1,000	-5,000
10	1	10	0	-1,000	-15,000
20	1	20	300	5,000	95,000
-10	1	-10	0	-1,000	5,000

From the table, 20 is an upper bound and -10 is a lower bound.

(B) Graphing $P(x)$ using the window suggested in part (A), we obtain

The real zeros are -7.47 and 14.03.

79. (A) We form a synthetic division table.

	4	-40	-1,475	7,875	-10,000
0	4	-40	-1,475	7,875	-10,000
10	4	0	-1,475	-6,875	-78,750
20	4	40	-675	-5,625	-122,500
30	4	80	925	35,625	1,058,750
-10	4	-80	-675	14,625	-156,250
-20	4	-120	925	-10,625	202,500

From the table, 30 is an upper bound and -20 is a lower bound.

(B) Graphing $P(x)$ using the window suggested in part (A), we obtain the graph shown at the right.

It appears that there are zeros between -20 and -10, and between 20 and 30, and there may be a zero of even multiplicity between 0 and 10.

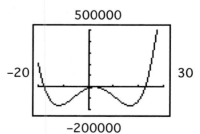

Applying ZERO and MAXIMUM commands, we obtain

-17.66 and 20.66 are simple zeros and 2.50 is probably a double zero. To check that it is really a zero, we note

$$
\begin{array}{r|rrrrr}
 & 4 & -40 & -1475 & 7875 & -10000 \\
 & & 10 & -75 & -3875 & 10000 \\
\hline
2.50 & 4 & -30 & -1550 & 4000 & 0
\end{array}
$$

2.50 is definitely a zero. You can check that it is a double zero using a second synthetic division.

81. (A) We form a synthetic division table.

	0.01	-0.1	-12	0	0	9,000
0	0.01	-0.1	-12	0	0	9,000
10	0.01	0	-12	-120	-1,200	-3,000
20	0.01	0.1	-10	-200	-4,000	-71,000
30	0.01	0.2	-6	-180	-5,400	-153,000
40	0.01	0.3	0	0	0	9,000
-10	0.01	-0.2	-10	100	-1,000	19,000
-20	0.01	-0.3	-6	120	-2,400	57,000
-30	0.01	-0.4	0	0	0	9,000
-40	0.01	-0.5	8	-320	12,800	-503,000

From the table, 40 is an upper bound and -40 is a lower bound.

(B) Graphing $P(x)$ using a window suggested by part (A), we obtain

The real zeros are -30.45, 9.06, and 39.80.

A graph in a smaller window confirms that there are no other real zeros between -10 and 10, and therefore most likely no other real zeros.

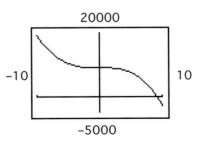

83. Yes. According to the Remainder Theorem, if the remainder is 10 when $P(x)$ is divided by $x + 4$, then $P(-4) = 10$. Similarly, if the remainder is -8 when dividing by $x + 5$, then $P(-5) = -8$. Since $P(-5)$ and $P(-4)$ have opposite signs, there must be a zero between -5 and -4.

85. (A) $P(x) = (x - r)Q(x) + R$ where $Q(x)$ is a polynomial with degree one less than the degree of P, and R is a number. If all numbers in the quotient row are positive after performing the division, it means that all of the coefficients of Q are positive, and so is the number R.

(B) If x is greater than r, then of course $x - r$ is positive. We know that the sign of $Q(x)$ has to be positive because: (1) r is positive, so r raised to any power is positive as well; (2) all of the coefficients of Q are positive; and (3) Q is a polynomial, so it's the sum of some finite number of terms that look like a coefficient times x to a power. All of the powers are positive, and all of the coefficients are positive, so their sum is positive as well.

(C) From (B), we know that both $x - r$ and $Q(x)$ are positive, so $(x - r)Q(x)$ is positive. We also know that R is positive, so from part (A), $P(x)$ is the sum of two positive numbers, and must be positive.

(D) We started by assuming that x is ANY number greater than r, and concluded in part (C) that $P(x)$ cannot be zero. In other words, no number greater than r is a zero for P, so r is an upper bound for the real zeros of P.

87. (A) We need to find the zeros of $P(x)$, which we will do graphically:

The variable x represents the number of toasters produced, so the answer must be a whole number. To break even, the company must make 81 or 460 toasters.

(B) We will add the horizontal line $y = 1,200$ to the graph, and find where the profit function is above this height:

The profit is at or above \$1,200 per week for production levels between 176 and 396 toasters.

89. Let (x, x^2) be a point on the graph of $y = x^2$. Then the distance from $(1, 2)$ to (x, x^2) must equal 1 unit. Applying the distance formula, we have,

$$\sqrt{(x - 1)^2 + (x^2 - 2)^2} = 1$$
$$(x - 1)^2 + (x^2 - 2)^2 = 1$$
$$x^2 - 2x + 1 + x^4 - 4x^2 + 4 = 1$$
$$x^4 - 3x^2 - 2x + 4 = 0$$

Let $P(x) = x^4 - 3x^2 - 2x + 4$

The only rational zero is 1.

$$
\begin{array}{r|rrrrr}
 & 1 & 0 & -3 & -2 & 4 \\
 &   & 1 & 1 & -2 & -4 \\
\hline
1 & 1 & 1 & -2 & -4 & 0 \\
\end{array}
$$

$P(x) = (x - 1)(x^3 + x^2 - 2x - 4)$

We examine $Q(x) = x^3 + x^2 - 2x - 4$.
Graphing $y = Q(x)$ we obtain the graph
shown at the right.

$Q(x)$ (and $P(x)$) has a second zero at
$x = 1.659$. Therefore, the two real zeros
of $P(x)$ are 1 and 1.659. The two required
points are $(1, 1)$ and $(1.659, 1.659^2) =$
$(1.659, 2.752)$.

91.

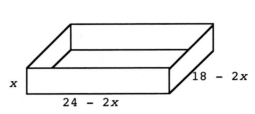

From the above figures it should be clear that
$V = $ length $\times$ width $\times$ height $= (24 - 2x)(18 - 2x)x$ $0 < x < 9$ (So that all the
dimensions are positive)

We solve $(24 - 2x)(18 - 2x)x = 600$

$$432x - 84x^2 + 4x^3 = 600$$
$$4x^3 - 84x^2 + 432x - 600 = 0$$
$$x^3 - 21x^2 + 108x - 150 = 0$$

Let $P(x) = x^3 - 21x^2 + 108x - 150$. Graphing $y = P(x)$ on the interval $0 < x < 9$,
we obtain

The zeros of $P(x)$ are 2.319 and 4.590, $0 < x < 9$.
$x = 2.319$ inches or 4.590 inches.

93.   We note:

$$\begin{pmatrix} \text{Volume} \\ \text{of} \\ \text{tank} \end{pmatrix} = \begin{pmatrix} \text{Volume of} \\ \text{two hemispheres} \\ \text{of radius } x \end{pmatrix} + \begin{pmatrix} \text{Volume of cylinder} \\ \text{with radius } x, \\ \text{height } 10 - 2x \end{pmatrix}$$

$$20\pi = \frac{4}{3}\pi x^3 \qquad\qquad + \pi x^2 (10 - 2x)$$

$$20 = \frac{4}{3}x^3 + 10x^2 - 2x^3$$

$$60 = 4x^3 + 30x^2 - 6x^3$$

$$2x^3 - 30x^2 + 60 = 0$$

$$x^3 - 15x^2 + 30 = 0 \quad \text{From physical considerations, } x > 0 \text{ and } 10 - 2x > 0,$$
so we are interested only in solutions in $(0, 5)$.

Let $P(x) = x^3 - 15x^2 + 30$

Graphing $y = P(x)$ on the interval $0 < x < 5$, we obtain

The only zero of $P(x)$, $0 < x < 5$, is 1.490.

$x = 1.490$ feet.

95. The independent variable is years after 1996, so we enter 0, 2, 4, 6, 7, 8, and 9 as L_1, and the Hispanic row from the table as L_2. Then we use the cubic regression command from the STAT CALC menu:

The cubic model is $y = 0.0142x^3 - 0.219x^2 + 1.55x + 42.7$. The inequality we need to solve is $0.0142x^3 - 0.219x^2 + 1.55x + 42.7 > 60$. To solve, we enter the cubic regression model as y_1 and the horizontal line $y_2 = 60$, and see where the cubic model is above height 60.

The model predicts that the Hispanic homeownership rate will pass 60% about 14 years after 1996, or 2010.

97. The independent variable is years after 1996, so we enter 0, 2, 4, 6, 7, 8, and 9 as L_1, and the White row from the table as L_2. Then we use the linear regression command from the STAT CALC menu:

The linear model is $y = 0.420x + 69.2$. The inequality we need to solve is $0.0142x^3 - 0.219x^2 + 1.55x + 42.7 > 0.420x + 69.2$ (using the model from Problem 95). To solve, we enter the cubic regression model from Problem 95 as as y_1 and the linear model from this problem as y_2, and see where the cubic model is above the linear.

The model predicts that the Hispanic homeownership rate will pass the White rate about 17 years after 1996, or 2013.

99. The independent variable is years after 1970, so we enter 0, 10, 20, 25, 30, and 35 as L_1, and the Males column as L_2. Then we use the cubic regression command from the STAT CALC menu:

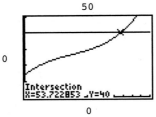

The cubic model is $y = 0.000281x^3 - 0.0220x^2 + 0.853x + 14.1$. The inequality we need to solve is $0.000281x^3 - 0.0220x^2 + 0.853x + 14.1 > 40$. To solve, we enter the cubic model as y_1 and $y_2 = 40$, and see when the cubic model gets above height 40.

The model predicts that the percentage of males in the U.S. that have completed at least four years of college will pass 40% after about 54 years, or in 2024.

Section 3-4

1. -8 (multiplicity 3), 6 (multiplicity 2); degree of $P(x)$ is 5

3. -4 (multiplicity 3), 3 (multiplicity 2); -1; degree of $P(x)$ is 6.

5. $2i$ (multiplicity 3), $-2i$ (multiplicity 4), 2 (multiplicity 5), -2 (multiplicity 5); degree of $P(x) = 17$.

7. The first factor can be factored further:
$$P(x) = x(x + 3)(x - 3)(x^2 + 9)(x + 9)^2$$
0 (multiplicity 1), -3 (multiplicity 1), 3 (multiplicity 1), $3i$ (multiplicity 1), $-3i$ (multiplicity 1), -9 (multiplicity 2); degree of $P(x)$ is 7.

9. Answers will vary. **11.** Answers will vary.

13. False. Explanations will vary.

15. $P(x) = (x - 3)^2(x + 4)$; degree 3

17. $P(x) = (x + 7)^3[x - (-3 + \sqrt{2})][x - (-3 - \sqrt{2})]$; degree 5

19. $P(x) = [x - (2 - 3i)][x - (2 + 3i)](x + 4)^2$; degree 4

21. Since -2, 1, and 3 are zeros, $P(x) = (x + 2)(x - 1)(x - 3)$ is the lowest degree polynomial that has this graph. The degree of $P(x)$ is 3.

23. Since -2 and 1 are zeros, each with multiplicity 2, $P(x) = (x + 2)^2(x - 1)^2$ is the lowest degree polynomial that has this graph. The degree of $P(x)$ is 4.

25. Since -3, -2, 0, 1, and 2 are zeros, $P(x) = (x + 3)(x + 2)x(x - 1)(x - 2)$ is the lowest degree polynomial that has this graph. The degree of $P(x)$ is 5.

27. Possible factors of 6 (the leading coefficient) are ±1, ±2, ±3, ±6. Possible factors of 1 (the constant term) are ±1. To find all possible rational zeros, we form all possible fractions with numerators from the first list and denominators from the second:
$$\pm\frac{1}{1}, \pm\frac{2}{1}, \pm\frac{3}{1}, \pm\frac{6}{1}.$$

The possible rational zeros are ±1, ±2, ±3, ±6.

29. Possible factors of 4 (the leading coefficient) are ±1, ±2, ±4. Possible factors of 3 (the constant term) are ±1, ±3. To find all possible rational zeros, we form all possible fractions with numerators from the first list and denominators from the second:

$$\pm\frac{1}{1},\ \pm\frac{2}{1},\ \pm\frac{4}{1},\ \pm\frac{1}{3},\ \pm\frac{2}{3},\ \pm\frac{4}{3}$$

The possible rational zeros are ±1, ±2, ±4, $\pm\frac{1}{3}$, $\pm\frac{2}{3}$, $\pm\frac{4}{3}$.

31. Possible factors of 3 (the leading coefficient) are ±1, ±3. Possible factors of 12 (the constant term) are ±1, ±2, ±3, ±4, ±6, ±12. To find all possible rational zeros, we form all possible fractions with numerators from the first list and denominators from the second:

$$\pm\frac{1}{1},\ \pm\frac{3}{1},\ \pm\frac{1}{2},\ \pm\frac{3}{2},\ \pm\frac{1}{3},\ \pm\frac{3}{3},\ \pm\frac{1}{4},\ \pm\frac{3}{4},\ \pm\frac{1}{6},\ \pm\frac{3}{6},\ \pm\frac{1}{12},\ \pm\frac{3}{12}$$

Reducing fractions and removing duplicates, we find that the possible rational zeros are ±1, ±3, $\pm\frac{1}{2}$, $\pm\frac{3}{2}$, $\pm\frac{1}{3}$, $\pm\frac{1}{4}$, $\pm\frac{3}{4}$, $\pm\frac{1}{6}$, $\pm\frac{1}{12}$.

33. (A) Let $u = x^2$; then $x^4 + 5x^2 + 4 = u^2 + 5u + 4 = (u + 4)(u + 1)$. Replacing u with x^2, we get:

$$P(x) = (x^2 + 4)(x^2 + 1)$$

(B) The zeros of $x^2 + 4$ are $2i$ and $-2i$, so it factors as $(x - 2i)(x + 2i)$. The zeros of $x^2 + 1$ are i and $-i$, so it factors as $(x - i)(x + i)$.

$$P(x) = (x - 2i)(x + 2i)(x - i)(x + i)$$

35. (A) Factor by grouping:

$$P(x) = (x^3 - x^2) + (25x - 25) = x^2(x - 1) + 25(x - 1) = (x^2 + 25)(x - 1)$$

(B) The zeros of $x^2 + 25$ are $5i$ and $-5i$ so it factors as $(x - 5i)(x + 5i)$.

$$P(x) = (x - 5i)(x + 5i)(x - 1)$$

37. $P(x) = x^3 + 9x^2 + 24x + 16$. Since -1 is a zero of $P(x)$, we can write

```
         1    9    24    16
             -1    -8   -16
   -1 ⌋ 1    8    16     0
```

$$P(x) = (x + 1)(x^2 + 8x + 16)$$

Now we can factor $x^2 + 8x + 16$ as $(x + 4)(x + 4)$, so
$P(x) = (x + 1)(x + 4)(x + 4)$.

39. $P(x) = x^4 - 1$. Since 1 is a zero of $P(x)$, we can write

```
        1    0    0    0    -1
             1    1    1     1
   1 ⌋ 1     1    1    1     0
```

$P(x) = (x - 1)(x^3 + x^2 + x + 1)$. Since -1 is also a zero of $P(x)$, we can write

```
        1    1    1    1
            -1    0   -1
  -1 ⌋ 1    0    1    0
```

$P(x) = (x - 1)(x + 1)(x^2 + 1)$. Since $x^2 + 1 = x^2 - (-1) = x^2 - i^2$, we can write
$P(x) = (x - 1)(x + 1)(x - i)(x + i)$

41. $P(x) = 2x^3 - 17x^2 + 90x - 41$. Since $\frac{1}{2}$ is a zero of $P(x)$, we can write

$$\begin{array}{r}
\phantom{\frac{1}{2}}\ \ 2\quad -17\quad\ \ 90\quad -41\\
1\quad\ \ -8\quad\ \ 41\\
\hline
\tfrac{1}{2}\big|\ 2\quad -16\quad\ \ 82\qquad\ \ 0
\end{array}$$

$$P(x) = \left(x - \frac{1}{2}\right)(2x^2 - 16x + 82) = \left(x - \frac{1}{2}\right)2(x^2 - 8x + 41) = (2x - 1)(x^2 - 8x + 41)$$

Since $x^2 - 8x + 41$ is a quadratic, we can find its zeros by the quadratic formula

$x^2 - 8x + 41 = 0 \qquad a = 1$
$\qquad\qquad\qquad\qquad\ \ b = -8$
$\qquad\qquad\qquad\qquad\ \ c = 41$

$$\begin{aligned}
x &= \frac{-b \pm \sqrt{b^2 - 4ac}}{2a}\\[4pt]
&= \frac{-(-8) \pm \sqrt{(-8)^2 - 4(1)(41)}}{2(1)}\\[4pt]
&= \frac{8 \pm \sqrt{-100}}{2}\\[4pt]
&= \frac{8 \pm 10i}{2}\\[4pt]
&= 4 \pm 5i
\end{aligned}$$

Since the zeros of $x^2 - 8x + 41$ are $4 + 5i$ and $4 - 5i$, by the factor theorem

$\qquad x^2 - 8x + 41 = [x - (4 + 5i)][x - (4 - 5i)]$

and $\quad P(x) = (2x - 1)[x - (4 + 5i)][x - (4 - 5i)]$

43. The possible rational zeros are ±1, ±2, ±3, ±5, ±6, ±10, ±15, ±30.

Examining a portion of the graph of $y = P(x)$ we see that there are possible integer zeros at 2, 3, and -5. We test the likely candidates and find

$$\begin{array}{r}
1\quad\ \ 0\quad -19\quad\ \ 30\\
2\quad\ \ \ 4\quad -30\\
\hline
2\big|\ 1\quad\ \ 2\quad -15\qquad\ \ 0
\end{array}$$

2 is a zero.

So $P(x) = (x - 2)(x^2 + 2x - 15)$
$\qquad\quad\ = (x - 2)(x - 3)(x + 5)$
The zeros of $P(x)$ are 2, 3, -5.

45. $P(x) = x^4 - \dfrac{21}{10}x^3 + \dfrac{2}{5}x = \dfrac{1}{10}(10x^4 - 21x^3 + 4x) = \dfrac{1}{10}x(10x^3 - 21x^2 + 4)$.

Because x is a factor, 0 is a zero.
We examine $Q(x) = 10x^3 - 21x^2 + 4$.
Possible factors of 4 are ±1, ±2, ±4.
Possible factors of 10 are ±1, ±2, ±5, ±10.
The possible rational zeros of

$Q(x)$ are ±1, ±2, ±4, $\pm\dfrac{1}{2}$, $\pm\dfrac{1}{5}$, $\pm\dfrac{2}{5}$, $\pm\dfrac{4}{5}$, $\pm\dfrac{1}{10}$.

Examining the graph of $y = Q(x)$ we see that there are zeros between -1 and 0, and between 0 and 1, and a possible integer zero at 2. Testing 2, we obtain

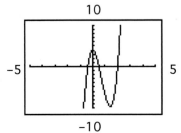

$$\begin{array}{r} 10 \quad -21 \quad\ \ 0 \quad\ \ 4 \\ 20 \quad\ -2 \quad -4 \\ \hline 2\overline{|10 \quad\ -1 \quad -2 \quad\ \ 0} \end{array}$$

2 is a zero.

So $P(x) = \dfrac{1}{10}x(x - 2)(10x^2 - x - 2)$

$\qquad = \dfrac{1}{10}x(x - 2)(5x + 2)(2x - 1)$

The zeros of $P(x)$ are 0, 2, $-\dfrac{2}{5}$, $\dfrac{1}{2}$.

47. $P(x) = x^4 - 5x^3 + \dfrac{15}{2}x^2 - 2x - 2 = \dfrac{1}{2}(2x^4 - 10x^3 + 15x^2 - 4x - 4)$.

Possible factors of -4 are ±1, ±2, ±4.
Possible factors of 2 are ±1, ±2.
The possible rational zeros are

±1, ±2, ±4, $\pm\dfrac{1}{2}$.

Examining the graph of $y = P(x)$ we see that
there are zeros between -1 and 0, and between
1 and 2, and a possible double zero at 2.
Testing 2, we obtain

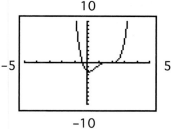

$$\begin{array}{r} 2 \quad -10 \quad\ \ 15 \quad -4 \quad -4 \\ 4 \quad -12 \quad\ \ 6 \quad\ \ 4 \\ \hline 2\overline{|2 \quad\ -6 \quad\ \ \ 3 \quad\ \ 2 \quad\ \ 0} \end{array}$$

2 is a zero. Testing whether it is a double zero, we examine the reduced
polynomial $2x^3 - 6x^2 + 3x + 2$.

$$\begin{array}{r} 2 \quad -6 \quad\ \ 3 \quad\ \ 2 \\ 4 \quad -4 \quad -2 \\ \hline 2\overline{|2 \quad -2 \quad -1 \quad\ \ 0} \end{array}$$

So 2 is a double zero of $P(x)$, and we have
$P(x) = \dfrac{1}{2}(x - 2)^2(2x^2 - 2x - 1)$

To find the remaining zeros, we solve $2x^2 - 2x - 1 = 0$.
Applying the quadratic formula with $a = 2$, $b = -2$, $c = -1$, we obtain

$\qquad x = \dfrac{-(-2) \pm \sqrt{(-2)^2 - 4(2)(-1)}}{2(2)} = \dfrac{2 \pm \sqrt{12}}{4} = \dfrac{1 \pm \sqrt{3}}{2}$ or $\dfrac{1}{2} \pm \dfrac{1}{2}\sqrt{3}$.

So the zeros of $P(x)$ are 2 (double), $\dfrac{1}{2} \pm \dfrac{1}{2}\sqrt{3}$.

49. Let $u = x^2$. Then
$\qquad P(x) = x^4 + 11x^2 + 30 = u^2 + 11u + 30 = (u + 6)(u + 5) = (x^2 + 6)(x^2 + 5)$.
The zeros of $x^2 + 6$ are $i\sqrt{6}$ and $-i\sqrt{6}$; the zeros of $x^2 + 5$ are $i\sqrt{5}$ and $-i\sqrt{5}$.
So the zeros of $P(x)$ are $\pm i\sqrt{6}$, $\pm i\sqrt{5}$.

51. Possible factors of 5 are ±1, ±5. Possible factors of 3 are ±1, ±3. The possible rational zeros are ±1, ±5, $\pm\dfrac{1}{3}$, $\pm\dfrac{5}{3}$.

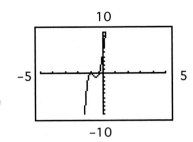

Examining the graph of $y = P(x)$, we see that there is a zero between -1 and 0, and a possible double zero at -1. Testing -1, we obtain

$$
\begin{array}{r|rrrrrr}
 & 3 & -5 & -8 & 16 & 21 & 5 \\
 &   & -3 & 8 & 0 & -16 & -5 \\
\hline
-1 & 3 & -8 & 0 & 16 & 5 & 0 \\
\end{array}
$$

-1 is a zero.
Testing whether it is a double zero, we examine the reduced polynomial $3x^4 - 8x^3 + 16x + 5$.

$$
\begin{array}{r|rrrrr}
 & 3 & -8 & 0 & 16 & 5 \\
 &   & -3 & 11 & -11 & -5 \\
\hline
-1 & 3 & -11 & 11 & 5 & 0 \\
\end{array}
$$

So -1 is a double zero. The only remaining rational candidate between -1 and 0 is $-\dfrac{1}{3}$. Testing $-\dfrac{1}{3}$ in the reduced polynomial $3x^3 - 11x^2 + 11x + 5$, we obtain

$$
\begin{array}{r|rrrr}
 & 3 & -11 & 11 & 5 \\
 &   & -1 & 4 & -5 \\
\hline
-\frac{1}{3} & 3 & -12 & 15 & 0 \\
\end{array}
$$

$-\dfrac{1}{3}$ is a zero.
So $P(x) = (x + 1)^2\left(x + \dfrac{1}{3}\right)(3x^2 - 12x + 15) = (x + 1)^2\left(x + \dfrac{1}{3}\right)3(x^2 - 4x + 5)$

To find the remaining zeros, we solve $x^2 - 4x + 5 = 0$ using the quadratic formula ($a = 1$, $b = -4$, $c = 5$)

$$x = \frac{-(-4) \pm \sqrt{(-4)^2 - 4(1)(5)}}{2(1)}$$

$$x = \frac{4 \pm \sqrt{-4}}{2} = \frac{4 \pm 2i}{2} = 2 \pm i$$

So the zeros of $P(x)$ are -1 (double), $-\dfrac{1}{3}$, $2 \pm i$.

53. Let $P(x) = 2x^3 - 5x^2 + 1$.
Possible rational zeros: ±1, $\pm\dfrac{1}{2}$.

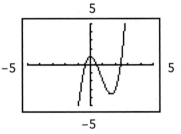

Examining the graph of $y = P(x)$, we see that there is a zero between -1 and 0, between 0 and 1, and between 1 and 2.

We test the likely candidates, $\dfrac{1}{2}$ and $-\dfrac{1}{2}$, and find

$$
\begin{array}{r|rrrr}
 & 2 & -5 & 0 & 1 \\
 &   & 1 & -2 & -1 \\
\hline
\frac{1}{2} & 2 & -4 & -2 & 0 \\
\end{array}
$$

$\dfrac{1}{2}$ is a zero.

So $P(x) = \left(x - \dfrac{1}{2}\right)(2x^2 - 4x - 2) = (2x - 1)(x^2 - 2x - 1)$

To find the remaining zeros, we solve $x^2 - 2x - 1 = 0$, using the quadratic formula:

$x^2 - 2x - 1 \qquad a = 1,\ b = -2,\ c = -1$

$x = \dfrac{-(-2) \pm \sqrt{(-2)^2 - 4(1)(-1)}}{2(1)}$

$= \dfrac{2 \pm \sqrt{8}}{2} = \dfrac{2 \pm 2\sqrt{2}}{2} = 1 \pm \sqrt{2}$

The zeros are $\dfrac{1}{2}$, $1 \pm \sqrt{2}$. These are the roots of the equation.

55. Let $P(x) = x^4 + 4x^3 - x^2 - 20x - 20$. Possible rational zeros: $\pm 1,\ \pm 2,\ \pm 4,\ \pm 5,\ \pm 10,\ \pm 20$.

Examining the graph of $y = P(x)$, we see that there is a zero between 2 and 3, and possibly a double zero near -2. We test -2.

```
     1    4   -1   -20   -20
         -2   -4    10    20
  -2 | 1   2   -5   -10    0
```

-2 is a zero. Testing whether it is a double zero, we examine the reduced polynomial $x^3 + 2x^2 - 5x - 10$.

```
     1    2   -5   -10
         -2    0    10
  -2 | 1   0   -5    0
```

-2 is a double zero.

So $P(x) = (x + 2)^2(x^2 - 5) = (x + 2)^2(x - \sqrt{5})(x + \sqrt{5})$. So the zeros of the polynomial are -2 (double), $\pm\sqrt{5}$. These are the roots of the equation.

57. Let $P(x) = x^4 - 2x^3 - 5x^2 + 8x + 4$. Possible rational zeros: $\pm 1,\ \pm 2,\ \pm 4$.

Examining the graph of $y = P(x)$ we see that there is a zero between -1 and 0 and between 2 and 3. The possible integer zeros are -2 and 2, which we test.

```
     1   -2   -5    8    4
          2    0  -10   -4
   2 | 1   0   -5   -2    0
```

2 is a zero. Next, we examine the reduced polynomial $x^3 - 5x - 2$.

```
     1    0   -5   -2
         -2    4    2
  -2 | 1  -2   -1    0
```

-2 is a zero.

So $P(x) = (x - 2)(x + 2)(x^2 - 2x - 1)$. The zeros of $x^2 - 2x - 1$ are $1 \pm \sqrt{2}$ (see problem 53). So the zeros of the polynomial are ± 2, $1 \pm \sqrt{2}$. These are the roots of the equation.

59. Factoring is the simplest method of solving. Let $u = x^2$.

$$x^4 + 10x^2 + 9 = 0$$
$$u^2 + 10u + 9 = 0$$
$$(u + 9)(u + 1) = 0$$
$$u = -9, \qquad u = -1$$
$$x^2 = -9, \qquad x^2 = -1$$
$$x = 3i, \ -3i, \ i, \ -i$$

61. Let $P(x) = 2x^5 - 3x^4 - 2x + 3$. Possible rational zeros: ± 1, ± 3, $\pm\dfrac{1}{2}$, $\pm\dfrac{3}{2}$.

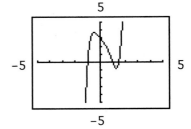

Examining the graph of $y = P(x)$ we see that there is a zero between 1 and 2, and possible zeros at -1 and 1. We test the likely candidates and find

	2	-3	0	0	-2	3
		2	-1	-1	-1	-3
$1\rfloor$	2	-1	-1	-1	-3	0

1 is a zero. We examine the reduced polynomial and find

	2	-1	-1	-1	-3
		-2	3	-2	3
$-1\rfloor$	2	-3	2	-3	0

-1 is a zero. We examine the new reduced polynomial and find

	2	-3	2	-3
		3	0	3
$\dfrac{3}{2}\rfloor$	2	0	2	0

$\dfrac{3}{2}$ is a zero.

So $P(x) = (x - 1)(x + 1)\left(x - \dfrac{3}{2}\right)(2x^2 + 2)$

$$= (x - 1)(x + 1)\left(x - \dfrac{3}{2}\right)2(x^2 + 1)$$

$$= (x - 1)(x + 1)(2x - 3)(x^2 + 1)$$

$$= (x - 1)(x + 1)(2x - 3)(x - i)(x + i)$$

The zeros are ± 1, $\dfrac{3}{2}$, $\pm i$. These are the roots of the equation.

63. The possible rational zeros are ± 1, ± 2, ± 4, $\pm\dfrac{1}{2}$, $\pm\dfrac{1}{3}$, $\pm\dfrac{2}{3}$, $\pm\dfrac{4}{3}$, $\pm\dfrac{1}{6}$.

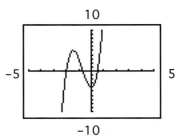

Examining the graph of $y = P(x)$, we see that there is a zero between -1 and 0, between 0 and 1, and a possible integer zero at -2. Testing -2, we obtain

	6	13	0	-4
		-12	-2	4
$-2\rfloor$	6	1	-2	0

-2 is a zero.

$$P(x) = (x + 2)(6x^2 + x - 2) = (x + 2)(3x + 2)(2x - 1)$$

65. The possible rational zeros are ±1, ±2, ±4.
Examining the graph of $y = P(x)$, we see that there
is a zero between -1 and 0, between 2 and 3, and a
possible integer zero at -4. Testing -4, we obtain

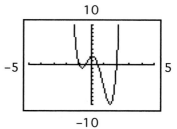

$$
\begin{array}{r|rrrr}
 & 1 & 2 & -9 & -4 \\
 & & -4 & 8 & 4 \\
\hline
-4 & 1 & -2 & -1 & 0 \\
\end{array}
$$

-4 is a zero.

So $P(x) = (x + 4)(x^2 - 2x - 1)$. The zeros of $x^2 - 2x - 1$ are $1 \pm \sqrt{2}$ (see
problem 53). So $P(x) = (x + 4)[x - (1 + \sqrt{2})][x - (1 - \sqrt{2})]$

67. The possible rational zeros are $\pm 1,\ \pm 2,\ \pm\dfrac{1}{2},\ \pm\dfrac{1}{4}$.

Examining the graph of $y = P(x)$, we see that there
is a zero between -1 and 0, between 0 and 1, and
possible integer zeros at -1 and 2. Testing 2, we
obtain

$$
\begin{array}{r|rrrrr}
 & 4 & -4 & -9 & 1 & 2 \\
 & & 8 & 8 & -2 & -2 \\
\hline
2 & 4 & 4 & -1 & -1 & 0 \\
\end{array}
$$

2 is a zero. Testing -1, we examine the reduced polynomial $4x^3 + 4x^2 - x - 1$.

$$
\begin{array}{r|rrrr}
 & 4 & 4 & -1 & -1 \\
 & & -4 & 0 & 1 \\
\hline
-1 & 4 & 0 & -1 & 0 \\
\end{array}
$$

-1 is a zero.

So $P(x) = (x - 2)(x + 1)(4x^2 - 1) = (x - 2)(x + 1)(2x - 1)(2x + 1)$.

69.
$$
\begin{aligned}
[x - (4 - 5i)][x - (4 + 5i)] &= [x - 4 + 5i][x - 4 - 5i] \\
&= [(x - 4) + 5i][(x - 4) - 5i] \\
&= (x - 4)^2 - 25i^2 \\
&= x^2 - 8x + 16 + 25 \\
&= x^2 - 8x + 41
\end{aligned}
$$

71.
$$
\begin{aligned}
[x - (3 + 4i)][x - (3 - 4i)] &= [x - 3 - 4i][x - 3 + 4i] \\
&= [(x - 3) - 4i][(x - 3) + 4i] \\
&= (x - 3)^2 - 16i^2 \\
&= x^2 - 6x + 9 + 16 \\
&= x^2 - 6x + 25
\end{aligned}
$$

73.
$$
\begin{aligned}
[x - (a + bi)][x - (a - bi)] &= [x - a - bi][x - a + bi] \\
&= [(x - a) - bi][(x - a) + bi] \\
&= (x - a)^2 - b^2i^2 \\
&= (x - a)^2 + b^2 \\
&= x^2 - 2ax + a^2 + b^2
\end{aligned}
$$

75. (A) $2 - i$

(B) $\left(x - (2 + i)\right)$ and $\left(x - (2 - i)\right)$

(C)
$$
\begin{aligned}
(x - (2 + i)(x - (2 - i)) &= (x - 2 - i)(x - 2 + i) \\
&= \big((x - 2) - i\big)\big((x - 2) + i\big) \\
&= (x - 2)^2 - i^2 \\
&= x^2 - 4x + 4 - (-1) \\
&= x^2 - 4x + 5
\end{aligned}
$$

(D)

$$\begin{array}{r} x - 3 \\ x^2 - 4x + 5\overline{\smash{\big)}\ x^3 - 7x^2 + 17x - 15} \\ \underline{-\left(x^3 - 4x^2 + 5x\right)} \\ -3x^2 + 12x - 15 \\ \underline{-\left(-3x^2 + 12x - 15\right)} \\ 0 \end{array}$$

The remaining factor is $x - 3$, so the remaining zero is $x = 3$.

77. If $3 - i$ is a zero, then $3 + i$ is a zero. So $Q(x) = [x - (3 - i)][x - (3 + i)]$ divides $P(x)$ evenly. Applying the result of Problem 73, $Q(x) = x^2 - 6x + 9 + 1$ $= x^2 - 6x + 10$. Dividing, we see

$$\begin{array}{r} x\ \ + 1 \\ x^2 - 6x + 10\overline{\smash{\big)}\ x^3 - 5x^2 + \ \ 4x + 10} \\ \underline{x^3 - 6x^2 + 10x} \\ x^2 - \ \ 6x + 10 \\ \underline{x^2 - \ \ 6x + 10} \\ 0 \end{array}$$

So $P(x) = (x + 1)Q(x)$ and the two other zeros are -1 and $3 + i$.

79. If $-5i$ is a zero, then so is $5i$. So $Q(x) = (x - 5i)(x + 5i)$ divides $P(x)$ evenly. $Q(x) = x^2 - 25i^2 = x^2 + 25$. Dividing, we see

$$\begin{array}{r} x - 3 \\ x^2 + 25\overline{\smash{\big)}\ x^3 - 3x^2 + 25x - 75} \\ \underline{x^3 \qquad\quad + 25x} \\ -3x^2 \qquad - 75 \\ \underline{-3x^2 \qquad - 75} \\ 0 \end{array}$$

So $P(x) = (x - 3)Q(x)$ and the two other zeros are $5i$ and 3.

81. If $2 + i$ is a zero, then $2 - i$ is a zero. So $Q(x) = [x - (2 + i)][x - (2 - i)]$ divides $P(x)$ evenly. Applying the result of Problem 73, $Q(x) = x^2 - 4x + 4 + 1$ $= x^2 - 4x + 5$. Dividing, we see

$$\begin{array}{r} x^2 - 2 \\ x^2 - 4x + 5\overline{\smash{\big)}\ x^4 - 4x^3 + 3x^2 + 8x - 10} \\ \underline{x^4 - 4x^3 + 5x^2} \\ -2x^2 + 8x - 10 \\ \underline{-2x^2 + 8x - 10} \\ 0 \end{array}$$

So $P(x) = Q(x)(x^2 - 2)$. $x^2 - 2$ has two zeros: $\sqrt{2}$ and $-\sqrt{2}$.
Summarizing, $P(x)$ has 4 zeros: $2 + i$, $2 - i$, $\sqrt{2}$, $-\sqrt{2}$.

83. $\sqrt{6}$ is a root of $x^2 = 6$ or $x^2 - 6 = 0$. The possible rational roots of this equation are ±1, ±2, ±3, ±6. Since none of them satisfies $x^2 = 6$, there are no rational roots, and $\sqrt{6}$ is not rational.

85. $\sqrt[3]{5}$ is a root of $x^3 = 5$ or $x^3 - 5 = 0$. The possible rational roots of this equation are ±1, ±5. Since none of them satisfies $x^3 = 5$, there are no rational roots, and $\sqrt[3]{5}$ is not rational.

87. The possible rational zeros are ±1, ±2, ±3, ±6. Examining the graph, we can see that there are two real zeros, one between -3 and -2, and the other between 2 and 3. Since none of the possible rational zeros actually are zeros, there must not be any rational zeros.

89. The only possible rational zeros are ±1, but $P(1) = -3$ and $P(-1) = 1$, so there are no rational zeros. A look at the graph in a standard window shows 3 real zeros which is the maximum number possible for a third degree polynomial.

91. Here is a graph of
$P(x) = 3x^3 - 37x^2 + 84x - 24$.

The rational zero theorem gives ±1, ±2, ±3, ±4, ±6, ±8, ±12, ±24, $\pm\frac{1}{3}$, $\pm\frac{2}{3}$, $\pm\frac{4}{3}$, $\pm\frac{8}{3}$ as possible rational zeros. However, the graph suggests that $P(x)$ has no integer zeros; there are no negative zeros, and the positive zeros would appear to lie between 0 and 1, 2 and 3, and 9 and 10. This suggests that we consider only the possible zeros $\frac{1}{3}$, $\frac{2}{3}$, and $\frac{8}{3}$. Testing $\frac{1}{3}$, we obtain

$$
\begin{array}{r}
\phantom{\tfrac{1}{3}}\;\;3 \quad -37 \quad\;\; 84 \quad\; -24 \\
1 \quad\; -12 \quad\;\; 24 \\
\hline
\tfrac{1}{3}\,\lfloor\; 3 \quad -36 \quad\;\; 72 \quad\;\;\;\; 0
\end{array}
$$

$\frac{1}{3}$ is a zero.

So $P(x) = \left(x - \frac{1}{3}\right)(3x^2 - 36x + 72) = \left(x - \frac{1}{3}\right)3(x^2 - 12x + 24)$

To find the remaining zeros, we solve $x^2 - 12x + 24 = 0$ using the quadratic formula:

$x^2 - 12x + 24 = 0 \qquad a = 1, \; b = -12, \; c = 24$

$$x = \frac{-(-12) \pm \sqrt{(-12)^2 - 4(1)(24)}}{2(1)}$$

$$= \frac{12 \pm \sqrt{48}}{2} = \frac{12 \pm 4\sqrt{3}}{2} = 6 \pm 2\sqrt{3}$$

So the zeros of $P(x)$ are $\frac{1}{3}$, $6 \pm 2\sqrt{3}$.

93. Here is a graph of
$$P(x) = 4x^4 + 4x^3 + 49x^2 + 64x - 240.$$

The rational zero theorem gives ± 1, ± 2, ± 3, ± 4, ± 5, ± 6, ± 8, ± 10, ± 12, ± 15, ± 16, ± 20, ± 24, ± 30, ± 40, ± 48, ± 60, ± 80, ± 120, ± 240, $\pm\dfrac{1}{2}$, $\pm\dfrac{3}{2}$, $\pm\dfrac{5}{2}$, $\pm\dfrac{15}{2}$, $\pm\dfrac{1}{4}$, $\pm\dfrac{3}{4}$, $\pm\dfrac{5}{4}$, $\pm\dfrac{15}{4}$ as possible rational zeros. However, the graph suggests that $P(x)$ has no integer zeros, and the zeros appear to lie between -3 and -2, and 1 and 2.

This suggests that we consider only the possible zeros $\dfrac{3}{2}$, $\dfrac{5}{4}$, and $-\dfrac{5}{2}$.

Testing $\dfrac{3}{2}$, we obtain

$$\dfrac{3}{2} \overline{\big)} \begin{array}{ccccc} 4 & 4 & 49 & 64 & -240 \\ & 6 & 15 & 96 & 240 \\ \hline 4 & 10 & 64 & 160 & 0 \end{array}$$

$\dfrac{3}{2}$ is a zero.

So $P(x) = \left(x - \dfrac{3}{2}\right)(4x^3 + 10x^2 + 64x + 160) = \left(x - \dfrac{3}{2}\right)2(2x^3 + 5x^2 + 32x + 80)$

We consider the reduced polynomial $Q(x) = 2x^3 + 5x^2 + 32x + 80$. The graph suggests that the other real zero is negative, so we try $-\dfrac{5}{2}$ next.

$$-\dfrac{5}{2} \overline{\big)} \begin{array}{cccc} 2 & 5 & 32 & 80 \\ & -5 & 0 & -80 \\ \hline 2 & 0 & 32 & 0 \end{array}$$

$-\dfrac{5}{2}$ is a zero.

So $P(x) = \left(x - \dfrac{3}{2}\right)2\left(x + \dfrac{5}{2}\right)(2x^2 + 32) = 4\left(x - \dfrac{3}{2}\right)\left(x + \dfrac{5}{2}\right)(x^2 + 16)$.

The remaining zeros of $P(x)$ are the zeros of $x^2 + 16$, that is, $\pm 4i$. So the zeros of $P(x)$ are $\dfrac{3}{2}$, $-\dfrac{5}{2}$, $\pm 4i$.

95. Here is a graph of
$$P(x) = 4x^4 - 44x^3 + 145x^2 - 192x + 90.$$

The rational zero theorem gives ± 1, ± 2, ± 3, ± 5, ± 6, ± 9, ± 10, ± 15, ± 18, ± 30, ± 45, ± 90, $\pm\dfrac{1}{2}$, $\pm\dfrac{3}{2}$, $\pm\dfrac{5}{2}$, $\pm\dfrac{9}{2}$, $\pm\dfrac{15}{2}$, $\pm\dfrac{45}{2}$, $\pm\dfrac{1}{4}$, $\pm\dfrac{3}{4}$, $\pm\dfrac{5}{4}$, $\pm\dfrac{9}{4}$, $\pm\dfrac{15}{4}$, $\pm\dfrac{45}{4}$ as possible rational zeros. However, the graph suggests that the only possible integer zeros are 1 and 2; there are no negative zeros, and the zeros appear to lie between 0 and 3, and between 6 and 7. This suggests that we consider only the possible zeros 1, 2, $\dfrac{1}{2}$, $\dfrac{3}{2}$, $\dfrac{5}{2}$, $\dfrac{1}{4}$, $\dfrac{3}{4}$, $\dfrac{5}{4}$, $\dfrac{9}{4}$. Testing the possible candidates, we obtain $P(1) = 3$, $P(2) = -2$, and

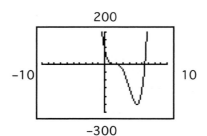

$$\begin{array}{r}
4 \quad -44 \quad 145 \quad -192 \quad \ \ 90 \\
6 \quad -57 \quad \ \ 132 \quad -90 \\
\hline
\end{array}$$

$$\frac{3}{2}\overline{)\begin{array}{r} 4 \quad\ -38 \quad\ \ \ 88 \quad\quad -60 \quad\quad\ \ 0 \end{array}}$$

$\frac{3}{2}$ is a zero.

So $P(x) = \left(x - \frac{3}{2}\right)(4x^3 - 38x^2 + 88x - 60) = \left(x - \frac{3}{2}\right)2(2x^3 - 19x^2 + 44x - 30)$

We consider the reduced polynomial $Q(x) = 2x^3 - 19x^2 + 44x - 30$. The remaining possibilities from the reduced list are $\frac{1}{2}$, $\frac{3}{2}$, $\frac{5}{2}$. Testing for a possible double zero at $\frac{3}{2}$, we obtain

$$\begin{array}{r}
2 \quad -19 \quad\ \ 44 \quad\ -30 \\
3 \quad -24 \quad\ \ \ 30 \\
\hline
\end{array}$$

$$\frac{3}{2}\overline{)\begin{array}{r} 2 \quad\ -16 \quad\quad 20 \quad\quad\ 0 \end{array}}$$

$\frac{3}{2}$ is a double zero.

$$So \ P(x) = \left(x - \frac{3}{2}\right)^2 2(2x^2 - 16x + 20) = \left(x - \frac{3}{2}\right)^2 4(x^2 - 8x + 10)$$

To find the remaining zeros, we solve $x^2 - 8x + 10 = 0$, using the quadratic formula.

$x^2 - 8x + 10 = 0 \qquad a = 1, \ b = -8, \ c = 10$

$$x = \frac{-(-8) \pm \sqrt{(-8)^2 - 4(1)(10)}}{2(1)}$$

$$= \frac{8 \pm \sqrt{24}}{2} = \frac{8 \pm 2\sqrt{6}}{2} = 4 \pm \sqrt{6}$$

So the zeros of $P(x)$ are $\frac{3}{2}$ (double), $4 \pm \sqrt{6}$.

97. (A) Since there are 3 zeros of $x^3 - 1$, there are 3 cube roots of 1.

(B) $x^3 - 1 = (x - 1)(x^2 + x + 1)$. The other cube roots of 1 will be solutions to $x^2 + x + 1 = 0$. Applying the quadratic formula with $a = b = c = 1$, we have

$x = \frac{-1 \pm \sqrt{(1)^2 - 4(1)(1)}}{2(1)} = \frac{-1 \pm \sqrt{-3}}{2} = \frac{-1 \pm i\sqrt{3}}{2}$. The other cube roots of 1 are

$-\frac{1}{2} + \frac{\sqrt{3}}{2}i$ and $-\frac{1}{2} - \frac{\sqrt{3}}{2}i$.

99. $P(x)$ can have at most n and must have at least one real zero. Each zero of $P(x)$ represents a point where $P(x) = y = 0$ so the graph of $P(x)$ will cross the x axis at and only at zeros of $P(x)$. Thus there can be a maximum of n axis crossings and there is a minimum of 1 axis crossing.

101. $P(2 + i) = (2 + i)^2 + 2i(2 + i) - 5$

$\qquad\qquad = 4 + 4i + i^2 + 4i + 2i^2 - 5$

$\qquad\qquad = 4 + 4i - 1 + 4i - 2 - 5$

$\qquad\qquad = -4 + 8i$

So $P(2 + i) \neq 0$ and $2 + i$ is not a zero of $P(x)$. This does not contradict the theorem, since $P(x)$ is not a polynomial with real coefficients (the coefficient of x is the imaginary number $2i$).

103. The right and left behaviors of an odd polynomial are opposite: if the leading coefficient is positive, the graph approaches $-\infty$ to the left, and ∞ to the right, and if the leading coefficient is negative, the graph approaches ∞ to the left and $-\infty$ to the right. In either case, some of the graph is above the x axis and some is below. The graphs of polynomials are always continuous, so the graph has to cross the x axis at least once.

105. If the constant term is zero, then every term has at least one power of the variable. In that case you can factor out the largest power of the variable that appears in every term; this will leave a reduced polynomial with a nonzero constant term, and then the rational zero theorem can be applied as usual. One of the zeros of the polynomial will be zero, and the rest will be the zeros of the reduced polynomial.

107. Let x = the amount of increase.

Then old volume = $1 \times 2 \times 3 = 6$

new volume = $(x + 1)(x + 2)(x + 3) = x^3 + 6x^2 + 11x + 6$

Since (new volume) = $10 \cdot$ (old volume), we must solve

$$x^3 + 6x^2 + 11x + 6 = 10(6)$$
$$x^3 + 6x^2 + 11x + 6 = 60$$
$$P(x) = x^3 + 6x^2 + 11x - 54 = 0$$

The possible rational zeros are ±1, ±2, ±3, ±6, ±9, ±18, ±27, ±54. Testing in order, we obtain $P(1) = -36$ $P(2) = 0$

$$
\begin{array}{r|rrrr}
 & 1 & 6 & 11 & -54 \\
 &   & 2 & 16 & 54 \\
\hline
2 & 1 & 8 & 27 & 0 \\
\end{array}
$$

Now we know that $x^3 + 6x^2 + 11x - 54 = (x - 2)(x^2 + 8x + 27)$. Since $x^2 + 8x + 27 = 0$ has no positive zeros, 2 is the only positive solution of the equation. The increase must equal 2 feet.

 109.

From the figure, we can see that
Volume = $x(5 - 2x)(8 - 2x) = 14$

Since x, $5 - 2x$, and $8 - 2x$ must all be positive, the domain of x is $0 < x < \dfrac{5}{2}$ or $(0, 2.5)$. We solve $x(5 - 2x)(8 - 2x) = 14$, or $4x^3 - 26x^2 + 40x = 14$, for x in this domain.

$$4x^3 - 26x^2 + 40x = 14$$
$$4x^3 - 26x^2 + 40x - 14 = 0$$
$$P(x) = 2x^3 - 13x^2 + 20x - 7 = 0$$

Possible rational zeros: ±1, ±7, $\pm\dfrac{1}{2}$, $\pm\dfrac{7}{2}$. Testing the possibilities, we obtain

$P(1) = 2$ $P(7) = 182$ $P\left(\dfrac{1}{2}\right) = 0$

$$
\begin{array}{r}
2 \quad -13 \quad 20 \quad -7 \\
1 \quad -6 \quad 7 \\
\hline
\frac{1}{2}\lfloor 2 \quad -12 \quad 14 \quad 0
\end{array}
$$

$\frac{1}{2}$ is a zero.

So $2x^3 - 13x^2 + 20x - 7 = \left(x - \frac{1}{2}\right)(2x^2 - 12x + 14) = (2x - 1)(x^2 - 6x + 7)$

To find the remaining zeros, we solve $x^2 - 6x + 7 = 0$, using the quadratic formula:

$x^2 - 6x + 7 = 0$ \qquad $a = 1,\ b = -6,\ c = 7$

$x = \dfrac{-(-6) \pm \sqrt{(-6)^2 - 4(1)(7)}}{2(1)}$

$= \dfrac{6 \pm \sqrt{8}}{2} = \dfrac{6 \pm 2\sqrt{2}}{2} = 3 \pm \sqrt{2}$

The zeros are $\frac{1}{2}$, $3 - \sqrt{2}$, $3 + \sqrt{2}$, or 0.5, 1.59, 4.41 to two significant digits. We discard $3 + \sqrt{2}$ or 4.41, since it is not in the interval (0, 2.5). The square should be 0.5 × 0.5 inches or 1.59 × 1.59 inches.

Section 3-5

1. Answers will vary. \qquad\qquad **3.** Answers will vary.

5. It is not accurate. The graph of a function can never intersect a vertical asymptote, but it can intersect a horizontal asymptote.

7. This graph has a vertical asymptote $x = 2$, and a horizontal asymptote $y = -2$. This corresponds to $g(x)$.

9. This graph has a vertical asymptote $x = 2$, and a horizontal asymptote $y = 2$. This corresponds to $h(x)$.

11. (A) ∞ (B) $-\infty$ (C) 2   (D) 2

13. (A) $-\infty$ (B) ∞ (C) 2   (D) 2

In Problems 15 – 33, we will use the symbol n(x) to represent the numerator of a rational function, and d(x) to represent the denominator.

15. $\dfrac{2x - 4}{x + 1}$ *Domain:* $d(x) = x + 1$ zero: $x = -1$ domain: $(-\infty,\ -1) \cup (-1,\ \infty)$
x intercepts: $n(x) = 2x - 4$ zero: $x = 2$ x intercept: 2

17. $\dfrac{x^2 - 1}{x^2 - 16}$ *Domain:* $d(x) = x^2 - 16$ zeros: $x^2 - 16 = 0$

$$x^2 = 16$$
$$x = \pm 4$$

domain: $(-\infty,\ -4) \cup (-4,\ 4) \cup (4,\ \infty)$

x intercepts: $n(x) = x^2 - 1$ zeros: $x^2 - 1 = 0$

$$x^2 = 1$$
$$x = \pm 1$$

x intercepts: -1, 1

19. $\dfrac{x^2 - x - 6}{x^2 - x - 12}$ *Domain:* $d(x) = x^2 - x - 12$ *zeros:* $x^2 - x - 12 = 0$

$$(x + 3)(x - 4) = 0$$
$$x = -3, \ 4$$

domain: $(-\infty, \ -3) \cup (-3, \ 4) \cup (4, \ \infty)$

x intercepts: $n(x) = x^2 - x - 6$ *zeros:* $x^2 - x - 6 = 0$

$$(x + 2)(x - 3) = 0$$
$$x = -2, \ 3$$

x intercepts: -2, 3

21. $\dfrac{x}{x^2 + 4}$ *Domain:* $d(x) = x^2 + 4$ no real zeros

Domain: all real numbers

23. $\dfrac{4 - x^2}{x}$ *Domain:* $d(x) = x$ zero: 0

Domain: $(-\infty, \ 0) \cup (0, \ \infty)$

x intercepts: $n(x) = 4 - x^2$ *zeros:* $4 - x^2 = 0$

$$(2 + x)(2 - x) = 0$$
$$x = -2, \ 2$$

x intercepts: -2, 2

25. $\dfrac{2x}{x - 4}$ *vertical asymptotes:* $d(x) = x - 4$ zero: $x = 4$

vertical asymptote $x = 4$

> **Common Error:**
> $x = 0$ is not a
> vertical asymptote.

horizontal asymptotes: Since $n(x)$ and $d(x)$ have the same degree, the line $y = 2$ is a horizontal asymptote (The ratio of the leading coefficients is 2/1, or 2).

27. $\dfrac{2x^2 + 3x}{3x^2 - 48}$ *vertical asymptotes:* $d(x) = 3x^2 - 48$ zeros: $3x^2 - 48 = 0$

$$3(x + 4)(x - 4) = 0$$
$$x = -4, \ 4$$

vertical asymptotes: $x = -4$, $x = 4$

horizontal asymptotes: Since $n(x)$ and $d(x)$ have the same degree, the line $y = \dfrac{2}{3}$ is a horizontal asymptote (The ratio of the leading coefficients is 2/3).

29. $\dfrac{2x}{x^4 + 1}$ *vertical asymptote:* $d(x) = x^4 + 1$ No real zeros:

No vertical asymptotes

horizontal asymptotes: Since the degree of $n(x)$ is less than the degree of $d(x)$, the x axis is a horizontal asymptote

horizontal asymptote: $y = 0$

31. $\dfrac{6x^4}{3x^2 - 2x - 5}$ *vertical asymptotes:* $d(x) = 3x^2 - 2x - 5$ zeros: $3x^2 - 2x - 5 = 0$

$$(3x - 5)(x + 1) = 0$$
$$x = \dfrac{5}{3}, \ -1$$

vertical asymptotes: $x = -1$, $x = \dfrac{5}{3}$

horizontal asymptotes: Since the degree of $n(x)$ is greater than the degree of $d(x)$, there are no horizontal asymptotes.

33. $\dfrac{4x^2 + 4x - 24}{x^2 - 2x} = \dfrac{4(x^2 + x - 6)}{x(x - 2)} = \dfrac{4(x + 3)(x - 2)}{x(x - 2)} = \dfrac{4(x + 3)}{x}$, $x \neq 2$

vertical asymptote: $x = 0$
Horizontal asymptotes: Since $n(x)$ and $d(x)$ have the same degree, the horizontal asymptote is $y = 4$ (The ratio of the leading coefficients is 4/1 or 4).

> **Common error:** $x = 2$ is not an asymptote. The graph has a hole at (2, 10) and the function is not defined at $x = 2$, however, the function does not exhibit asymptotic behavior near $x = 2$.

35. The graph has more than one horizontal asymptote.

37. The graph has a sharp corner at (0, 0).

39. $\dfrac{x^2 + 2x}{x} = \dfrac{x(x + 2)}{x} = x + 2$ if $x \neq 0$. The graph of f is the same as the graph of g except that f has a hole at (0, 2).

41. $\dfrac{x + 2}{x^2 + 10x + 16} = \dfrac{x + 2}{(x + 2)(x + 8)} = \dfrac{1}{x + 8}$ if $x \neq -2$. The graph of f is the same as the graph of g except that f has a hole at $\left(-2, \dfrac{1}{6}\right)$.

43. $f(x) = \dfrac{1}{x - 4} = \dfrac{n(x)}{d(x)}$

Intercepts. There are no real zeros of $n(x) = 1$. No x intercept
$f(0) = -\dfrac{1}{4}$ $y = -\dfrac{1}{4}$ y intercept
Vertical asymptotes. $d(x) = x - 4$ zeros: 4 $x = 4$

Horizontal asymptotes. Since the degree of $n(x)$ is less that the degree of $d(x)$, the x axis is a horizontal asymptote.
Complete the sketch. Plot a few points.

x	$f(x)$
4.5	2
5	1
6	.5
3.5	-2
3	-1
2	-.5

> **Common Error:**
> Entering $1/x - 4$ in the calculator.
> Parentheses are needed: $1/(x - 4)$.

45. $f(x) = \dfrac{x}{x + 1} = \dfrac{n(x)}{d(x)}$

Intercepts. Real zeros of $n(x) = x$ $x = 0$ x intercept
$f(0) = 0$ $y = 0$ y intercept
The graph crosses the coordinate axes only at the origin.

Vertical asymptotes. $d(x) = x + 1$ zeros: -1 $x = -1$
Horizontal asymptotes. Since $n(x)$ and $d(x)$ have the same degree, the line $y = 1$ is a horizontal asymptote (The ratio of the leading coefficients is 1/1, or 1).
Complete the sketch. Plot a few additional points.

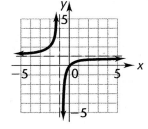

47. $h(x) = \dfrac{x}{2x - 2} = \dfrac{n(x)}{d(x)}$

Intercepts. Real zeros of $n(x) = x$ $x = 0$ x intercept
$h(0) = 0$ $y = 0$ y intercept
The graph crosses the coordinate axes only at the origin.
Vertical asymptotes. $d(x) = 2x - 2$ zeros: 1 $x = 1$

Horizontal asymptotes. Since $n(x)$ and $d(x)$ have the same degree, the line
$y = \dfrac{1}{2}$ is a horizontal asymptote (The ratio of the
leading coefficients is 1/2).
Complete the sketch. Plot a few additional points.

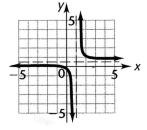

49. $f(x) = \dfrac{2x - 4}{x + 3} = \dfrac{n(x)}{d(x)}$

Intercepts. Real zeros of $n(x) = 2x - 4$ $x = 2$ x intercept
$f(0) = -\dfrac{4}{3}$ y intercept
Vertical asymptotes. $d(x) = x + 3$ zeros: -3 $x = -3$

Horizontal asymptotes. Since $n(x)$ and $d(x)$ have the same degree, the line $y = 2$
is a horizontal asymptote (The ratio of the
leading coefficients is 2/1, or 2).
Complete the sketch. Plot a few additional points.

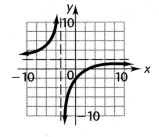

51. $g(x) = \dfrac{1 - x^2}{x^2} = \dfrac{n(x)}{d(x)}$

Intercepts. Real zeros of $n(x) = 1 - x^2$ $1 - x^2 = 0$
$x^2 = 1$
$x = \pm 1$ x intercepts
$g(0)$ is not defined   no y intercepts
Vertical asymptotes. $d(x) = x^2$ zeros: 0 $x = 0$
Horizontal asymptotes. Since $n(x)$ and $d(x)$ have the same degree, the line
$y = -1$ is a horizontal asymptote (The ratio of the
leading coefficients is -1/1, or -1).
Complete the sketch. Plot a few additional points.

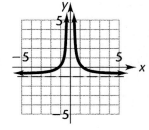

53. $f(x) = \dfrac{9}{x^2 - 9} = \dfrac{n(x)}{d(x)}$

Intercepts. There are no real zeros of $n(x) = 9$. No x intercept
$$f(0) = -1 \quad y \text{ intercept}$$

Vertical asymptotes. $d(x) = x^2 - 9 \quad$ zeros: $x^2 - 9 = 0$
$$x^2 = 9$$
$$x = \pm 3$$

Horizontal asymptotes. Since the degree of $n(x)$ is less than the degree of $d(x)$, the x axis is a horizontal asymptote.

Complete the sketch. Plot a few additional points.

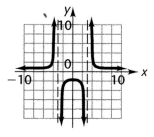

55. $f(x) = \dfrac{x}{x^2 - 1} = \dfrac{n(x)}{d(x)}$

Intercepts. Real zeros of $n(x) = x \quad x = 0 \quad x$ intercept
$$f(0) = 0 \quad y = 0 \quad y \text{ intercept}$$
The graph crosses the coordinate axes only at the origin.

Vertical asymptotes. $d(x) = x^2 - 1 \quad$ zeros: $x^2 - 1 = 0$
$$x^2 = 1$$
$$x = \pm 1$$

Horizontal asymptotes. Since the degree of $n(x)$ is less than the degree of $d(x)$, the x axis is a horizontal asymptote.

Complete the sketch. Plot a few additional points.

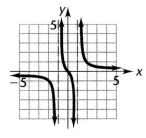

57. $g(x) = \dfrac{2}{x^2 + 1} = \dfrac{n(x)}{d(x)}$

Intercepts. There are no real zeros of $n(x) = 2$.   No x intercept
$$g(0) = 2 \quad y \text{ intercept}$$

Vertical asymptotes. There are no real zeros of $d(x) = x^2 + 1$
        No vertical asymptotes

Horizontal asymptotes. Since the degree of $n(x)$ is less than the degree of $d(x)$, the x axis is a horizontal asymptote.

Complete the sketch. Plot a few additional points.

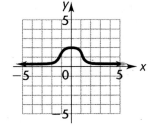

59. $f(x) = \dfrac{12x^2}{(3x + 5)^2} = \dfrac{n(x)}{d(x)}$

Intercepts. Real zeros of $n(x) = 12x^2$ $x = 0$ x intercept
                              $f(0) = 0$ $y = 0$ y intercept
The graph crosses the coordinate axes only at the origin.
Vertical asymptotes. $d(x) = (3x + 5)^2$ zeros: $(3x + 5)^2 = 0$
                                              $3x + 5 = 0$
                                                $x = -\dfrac{5}{3}$

Horizontal asymptotes. Since $n(x)$ and $d(x)$ have the same degree, the line

$y = \dfrac{12}{3^2} = \dfrac{4}{3}$ is a horizontal asymptote (The

denominator in expanded form is $9x^2 + 30x + 25$, so
the ratio of the leading coefficients is 12/9 or
4/3).
Complete the sketch. Plot a few additional points.

61. $f(x) = \dfrac{x^2 - 1}{x^2 + 7x + 10} = \dfrac{n(x)}{d(x)}$

Intercepts. Real zeros of $n(x) = x^2 - 1$ $x^2 - 1 = 0$
                                             $x^2 = 1$
                                       $x = \pm 1$ x intercepts
                     $f(0) = -\dfrac{1}{10}$ y intercept

Vertical asymptotes. Real zeros of $d(x) = x^2 + 7x + 10$ $x^2 + 7x + 10 = 0$
                                         $(x + 2)(x + 5) = 0$
                                               $x = -2, -5$

Horizontal asymptotes. Since $n(x)$ and $d(x)$ have the same degree, the line $y = 1$
is a horizontal asymptote (The ratio of
the leading coefficients is 1/1, or 1).

Complete the sketch. Plot a few
additional points.

63. $f(x) = \dfrac{x^2 - 3x - 10}{x - 5} = \dfrac{(x - 5)(x + 2)}{x - 5} = x + 2$ $x \neq 5$.

The graph is the same as the graph of $y = x + 2$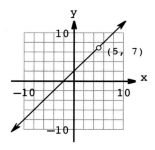
except that f has a hole at $(5, 7)$. There are no
asymptotes. The line has x intercept $x = -2$ and
y intercept $y = 2$.

65. To have zeros -2, -1, 1, and 2, the numerator must have factors $(x + 2)$, $(x + 1)$, $(x - 1)$, and $(x - 2)$. To have a horizontal asymptote at $y = 3$, the degree of the denominator should be the same as the numerator, and the leading coefficient of the numerator should be 3 times as large. To have no vertical asymptote, the denominator should have no real zeros.

$$f(x) = \frac{3(x + 2)(x - 2)(x + 1)(x - 1)}{x^4 + 1} \quad \text{or} \quad f(x) = \frac{3(x^2 - 4)(x^2 - 1)}{x^4 + 1} \quad \text{will work.}$$

67. To get $y = 2x + 5$ as an oblique asymptote, our function should look like

$f(x) = 2x + 5 + \dfrac{r(x)}{q(x)}$ where the degree of $q(x)$ is greater than the degree of

$r(x)$. If $q(x) = x - 10$, we'll have $x = 10$ as vertical asymptote. In that case $r(x)$ will have to be constant so that it's power is less than $q(x)$; $r(x) = 100$ will do.

$$\begin{aligned}
f(x) &= 2x + 5 + \frac{100}{x - 10} \\
&= \frac{(2x + 5)(x - 10)}{x - 10} + \frac{100}{x - 10} \\
&= \frac{(2x + 5)(x - 10) + 100}{x - 10}
\end{aligned}$$

69. $\dfrac{x}{x - 2} \leq 0$

> **Common Error:**
> It is not correct to "multiply both sides" by $x - 2$. The expression $x - 2$ can be either positive or negative, so the direction of the inequality would be uncertain.

Algebraic Solution

Let $f(x) = \dfrac{p(x)}{q(x)} = \dfrac{x}{x - 2}$

The zero of $p(x) = x$ is 0. The zero of $q(x) = x - 2$ is 2. These two zeros partition the x axis into the three intervals shown in the table. A test number is chosen from each interval to determine the sign of $f(x)$.

Interval	Test number x	$f(x)$	Sign of f
$(-\infty, 0)$	-1	$\dfrac{1}{3}$	+
$(0, 2)$	1	-1	-
$(2, \infty)$	3	3	+

The left side is equal to zero at $x = 0$, but not at $x = 2$ (zero denominator). We conclude that the solution set is $[0, 2)$.

Graphical Solution

Graph $y_1 = \dfrac{x}{x - 2}$ and find where the graph is at or below height zero:

The graph is at or below height zero on $[0, 2)$.

71. $\dfrac{x^2 - 16}{5x - 2} > 0$

Algebraic Solution

Let $f(x) = \dfrac{p(x)}{q(x)} = \dfrac{x^2 - 16}{5x - 2}$

The zeros of $p(x) = x^2 - 16 = (x + 4)(x - 4)$ are -4 and 4.

The zero of $q(x) = 5x - 2$ is $\dfrac{2}{5}$.

These three zeros partition the x axis into the four intervals shown in the table. A test number is chosen from each interval to determine the sign of $f(x)$.

Interval	Test number x	$f(x)$	Sign of f
$(-\infty, -4)$	-5	$-\dfrac{1}{3}$	-
$\left(-4, \dfrac{2}{5}\right)$	0	8	+
$\left(\dfrac{2}{5}, 4\right)$	1	-5	-
$(4, \infty)$	5	$\dfrac{9}{23}$	+

We conclude that the solution set is $\left(-4, \dfrac{2}{5}\right) \cup (4, \infty)$.

Graphical Solution

Graph $y_1 = \dfrac{x^2 - 16}{5x - 2}$ and find where the graph above height zero.

The graph is above height zero on $\left(-4, 2/5\right) \cup \left(4, \infty\right)$. (You have to note that the vertical asymptote is at $x = 2/5$; this isn't obvious at all from the graph!)

73. $\dfrac{x^2 + 4x - 20}{3x} \geq 4$

Algebraic Solution

$\dfrac{x^2 + 4x - 20}{3x} - 4 \geq 0$

$\dfrac{x^2 + 4x - 20}{3x} - \dfrac{12x}{3x} \geq 0$

$\dfrac{x^2 - 8x - 20}{3x} \geq 0$

Let $f(x) = \dfrac{p(x)}{q(x)} = \dfrac{x^2 - 8x - 20}{3x}$.

The zeros of $p(x) = x^2 - 8x - 20 = (x + 2)(x - 10)$ are -2 and 10. The zero of $q(x)$ is 0. These three zeros partition the x axis into the four intervals shown in the table. A test number is chosen from each interval to determine the sign of $f(x)$.

Interval	Test number x	$f(x)$	Sign of f
$(-\infty, -2)$	-3	$-\dfrac{13}{9}$	$-$
$(-2, 0)$	-1	$\dfrac{11}{3}$	$+$
$(0, 10)$	1	-9	$-$
$(10, \infty)$	11	$\dfrac{13}{33}$	$+$

The left side is equal to zero at -2 and 10, but not at 0 (zero denominator). We conclude that the solution set is $[-2, 0) \cup [10, \infty)$.

Graphical Solution

Graph $y_1 = \dfrac{x^2 + 4x - 20}{3x}$, $y_2 = 4$ and find where the height of the first graph is at or above the height of the second.

The first graph is at or above the second on $[-2, 0) \cup [10, \infty)$.

75. $\dfrac{5x}{x^2 - 1} < \dfrac{9}{x}$

Algebraic Solution

$$\frac{5x}{x^2 - 1} - \frac{9}{x} < 0$$

$$\frac{5x^2 - 9(x^2 - 1)}{x(x^2 - 1)} < 0$$

$$\frac{-4x^2 + 9}{x(x^2 - 1)} < 0$$

Let $f(x) = \dfrac{p(x)}{q(x)} = \dfrac{-4x^2 + 9}{x(x^2 - 1)}$.

The zeros of $p(x) = -4x^2 + 9 = 9 - 4x^2 = (3 + 2x)(3 - 2x)$ are $-\dfrac{3}{2}$ and $\dfrac{3}{2}$. The zeros of $q(x) = x(x^2 - 1) = x(x + 1)(x - 1)$ are 0, 1, and -1. These five zeros partition the x axis into the six intervals shown in the table. A test number is chosen from each interval to determine the sign of $f(x)$.

Interval	Test number x	$f(x)$	Sign of f
$\left(-\infty, -\dfrac{3}{2}\right)$	-2	$\dfrac{7}{6}$	$+$
$\left(-\dfrac{3}{2}, -1\right)$	$-\dfrac{5}{4}$	$-\dfrac{176}{45}$	$-$
$(-1, 0)$	$-\dfrac{1}{2}$	$\dfrac{64}{3}$	$+$
$(0, 1)$	$\dfrac{1}{2}$	$-\dfrac{64}{3}$	$-$
$\left(1, \dfrac{3}{2}\right)$	$\dfrac{5}{4}$	$\dfrac{176}{45}$	$+$
$\left(\dfrac{3}{2}, \infty\right)$	2	$-\dfrac{7}{6}$	$-$

We conclude that the solution set is $\left(-\dfrac{3}{2}, -1\right) \cup (0,\ 1) \cup \left(\dfrac{3}{2}, \infty\right)$.

Graphical Solution

Graph $y_1 = \dfrac{5x}{x^2 - 1}$, $y_2 = \dfrac{9}{x}$ and find where the height of the first graph is below the height of the second. (The second function is the thicker graph.)

The first graph is below the second on $(-1.5,\ -1)$ (note the vertical asymptote is at $x = -1$), on $(0,\ 1)$ (note the 2nd vertical asymptote is at $x = 1$), and on $(1.5,\ \infty)$, so the solution is $\left(-1.5, -1\right) \cup \left(0, 1\right) \cup \left(1.5, \infty\right)$.

77. $\dfrac{x^2 + 7x + 3}{x + 2} > 0$

Algebraic Solution

zeros of the numerator $x^2 + 7x + 3 = 0$ $a = 1$, $b = 7$, $c = 3$

$$x = \frac{-7 \pm \sqrt{7^2 - 4(1)(3)}}{2(1)} = \frac{-7 \pm \sqrt{37}}{2} = \frac{-7 \pm 6.083}{2} = -6.541,\ -0.459$$

zeros of the denominator: $x = -2$

The three zeros partition the x axis into the four intervals shown in the table below. A test number is chosen from each interval to determine the sign of the rational expression.

Interval	Test number x	$f(x)$	Sign of f
$(-\infty,\ -6.541)$	-10	$-\dfrac{33}{8}$	$-$
$(-6.541,\ -2)$	-5	$\dfrac{7}{3}$	$+$
$(-2,\ -0.459)$	-1	-3	$-$
$(-0.459,\ \infty)$	0	$\dfrac{3}{2}$	$+$

The left side is positive on $(-6.541,\ -2) \cup (-0.459,\ \infty)$. We need to exclude all endpoints, so this is the solution.

Graphical Solution

Graph $y_1 = \dfrac{x^2 + 7x + 3}{x + 2}$ and find where it is above height zero

The graph is above the x axis on $(-6.541,\ -2) \cup (-0.459,\ \infty)$.

79. $\dfrac{5}{x^2} - \dfrac{1}{x+3} < 0$ (Simplify the left side)

Algebraic Solution

$$\dfrac{5(x+3)}{x^2(x+3)} - \dfrac{x^2}{x^2(x+3)} < 0$$

$$\dfrac{5x+15-x^2}{x^2(x+3)} < 0$$

$$\dfrac{-x^2+5x+15}{x^2(x+3)} < 0$$

zeros of the numerator: $-x^2 + 5x + 15 = 0$ $a = -1$, $b = 5$, $c = 15$

$$x = \dfrac{-5 \pm \sqrt{5^2 - 4(-1)(15)}}{2(-1)} = \dfrac{-5 \pm \sqrt{85}}{-2} = \dfrac{-5 \pm 9.220}{-2} = 7.110, -2.110$$

zeros of the denominator: $x^2(x+3) = 0$

$$x = 0, -3$$

The four zeros partition the x axis into the five intervals shown in the table below. A test number is chosen from each interval to determine the sign of the rational expression.

Interval	Test number x	f(x)	Sign of f
$(-\infty, -3)$	-5	$\dfrac{7}{10}$	$+$
$(-3, -2.110)$	-2.5	-1.2	$-$
$(-2.110, 0)$	-1	$\dfrac{9}{2}$	$+$
$(0, 7.110)$	1	$\dfrac{19}{4}$	$+$
$(7.110, \infty)$	8	$-\dfrac{9}{704}$	$-$

The left side is negative on $(-3, -2.110) \cup (7.110, \infty)$. We need to exclude all endpoints so this is the solution.

Graphical Solution

Graph $y_1 = \dfrac{5}{x^2} - \dfrac{1}{x+3}$ and find where it is below height zero

The graph is below the x axis on $(-3, -2.110) \cup (7.110, \infty)$.

81. $\dfrac{9}{x} - \dfrac{5}{x^2} \leq 1$ (Simplify)

<u>Algebraic Solution</u>

$$\dfrac{9}{x} - \dfrac{5}{x^2} - 1 \leq 0$$

$$\dfrac{9x}{x^2} - \dfrac{5}{x^2} - \dfrac{x^2}{x^2} \leq 0$$

$$\dfrac{-x^2 + 9x - 5}{x^2} \leq 0$$

zeros of the numerator: $-x^2 + 9x - 5 = 0$ $a = -1,\ b = 9,\ c = -5$

$$x = \dfrac{-9 \pm \sqrt{9^2 - 4(-1)(-5)}}{2(-1)} = \dfrac{-9 \pm \sqrt{61}}{-2} = \dfrac{-9 \pm 7.810}{-2} = 0.595,\ 8.405$$

zeros of the denominator: $x = 0$

These three zeros partition the x axis into the four intervals shown in the table below. A test number is chosen from each interval to determine the sign of the rational expression.

Interval	Test number x	$f(x)$	Sign of f
$(-\infty, 0)$	-1	-15	$-$
$(0, 0.595)$	0.5	-3	$-$
$(0.595, 8.405)$	1	3	$+$
$(0.8405, \infty)$	10	$-\dfrac{3}{20}$	$-$

The left side is negative on $(-\infty, 0)$, $(0, 0.595)$, and $(8.405, \infty)$. It's equal to zero when $x = 0.595$ and 8.405, so we include those endpoints, but we exclude $x = 0$ since the expression is undefined there. The solution is $(-\infty, 0) \cup (0, 0.595] \cup [8.405, \infty)$.

<u>Graphical Solution</u>

Graph $y_1 = \dfrac{5}{x^2} - \dfrac{1}{x + 3}$, $y_2 = 1$ and find where the first graph is below the second.

The graph of $y_1 = \dfrac{9}{x} - \dfrac{5}{x^2}$ is below the graph of $y_2 = 1$ on $(-\infty, 0)$, $(0, 0.595)$ and $(8.405, \infty)$ and they intersect at $x = 0.595$ and $x = 8.405$.

83. $\dfrac{3x + 2}{x - 5} > 10$ (Simplify)

Algebraic Solution

$$\frac{3x + 2}{x - 5} - 10 > 0$$

$$\frac{3x + 2}{x - 5} - \frac{10(x - 5)}{x - 5} > 0$$

$$\frac{3x + 2 - 10x + 50}{x - 5} > 0$$

$$\frac{-7x + 52}{x - 5} > 0$$

zeros of the numerator: $-7x + 52 = 0$
$$-7x = -52$$
$$x = 7.429$$

zeros of the denominator: $x = 5$

These two zeros partition the x axis into the 3 intervals shown in the table below. A test number is chosen from each interval to determine the sign of the rational expression.

Interval	Test number x	$f(x)$	Sign of f
$(-\infty, 5)$	0	$-\dfrac{52}{5}$	$-$
$(5, 7.429)$	6	10	$+$
$(7.429, \infty)$	8	$-\dfrac{4}{3}$	$-$

The left side is positive on $(5, 7.429)$. We need to exclude the endpoints, so this is the solution.

Graphical Solution

Graph $y_1 = \dfrac{3x + 2}{x - 5}$, $y_2 = 10$ and find where the first graph is above the second.

The graph of $y_1 = \dfrac{3x + 2}{x - 5}$ is above the graph of $y_2 = 10$ on $(5, 7.429)$.

85. $\dfrac{4}{x + 1} \geq \dfrac{7}{x}$ (Simplify)

Algebraic Solution

$$\frac{4}{x + 1} - \frac{7}{x} \geq 0$$

$$\frac{4x}{x(x + 1)} - \frac{7(x + 1)}{x(x + 1)} \geq 0$$

$$\frac{4x - 7x - 7}{x(x + 1)} \geq 0$$

$$\frac{-3x - 7}{x(x + 1)} \geq 0$$

zeros of the numerator: $-3x - 7 = 0$
$$-3x = 7$$
$$x = -2.333$$

zeros of the denominator: $x(x + 1) = 0$
$$x = 0, -1$$

These three zeros partition the x axis into the four intervals shown in the table below. A test number is chosen from each interval to determine the sign of the rational expression.

Interval	Test number x	$f(x)$	Sign of f
$(-\infty, -2.333)$	-3	$\dfrac{1}{3}$	$+$
$(-2.333, -1)$	-2	$-\dfrac{1}{2}$	$-$
$(-1, 0)$	$-\dfrac{1}{2}$	22	$+$
$(0, \infty)$	1	-5	$-$

The left side is positive on $(-\infty, -2.333) \cup (-1, 0)$. We include $x = -2.333$ because it makes the left side zero, but exclude 0 and -1 as they make the expression undefined. The solution is $(-\infty, -2.333] \cup (-1, 0)$.

Graphical Solution
Graph $y_1 = \dfrac{4}{x + 1}$, $y_2 = \dfrac{7}{x}$ and find where the first graph is at or above the second. (The thicker graph is $7/x$.)

 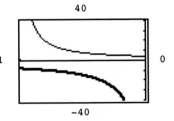

The thinner graph, $y_1 = \dfrac{4}{x + 1}$, is at or above the thicker on $(-\infty, -2.333]$, and again from $x = -1$ to $x = 0$. (That's hard to tell from the first graph; the second shows both graphs on just that interval) So the solution is $(-\infty, -2.333] \cup (-1, 0)$.

87. $f(x) = \dfrac{2x^2}{x - 1} = \dfrac{n(x)}{d(x)}$

Vertical asymptotes. Real zeros of $d(x) = x - 1$ $x = 1$
Horizontal asymptote. Since the degree of $n(x)$ is greater than the degree of $d(x)$, there is no horizontal asymptote.
Oblique asymptote.

$$
\begin{array}{r}
2x + 2 \\
x - 1 \overline{) 2x^2 } \\
\underline{2x^2 - 2x} \\
2x \\
\underline{2x - 2} \\
2
\end{array}
$$

We can write f as $f(x) = 2x + 2 + \dfrac{2}{x - 1}$, so the line $y = 2x + 2$ is an oblique asymptote.

89. $p(x) = \dfrac{x^3}{x^2 + 1} = \dfrac{n(x)}{d(x)}$

Vertical asymptotes. There are no real zeros of $d(x) = x^2 + 1$.
No vertical asymptotes.
Horizontal asymptotes. Since the degree of $n(x)$ is greater than the degree of $d(x)$, there is no horizontal asymptote.
Oblique asymptote:

$$
\begin{array}{r}
x \\
x^2 + 1 \overline{)\, x^3 } \\
\underline{x^3 + x } \\
- x
\end{array}
$$

We can write p as $p(x) = x + \dfrac{-x}{x^2 + 1}$, so the line $y = x$ is an oblique asymptote.

91. $r(x) = \dfrac{2x^2 - 3x + 5}{x} = \dfrac{n(x)}{d(x)}$

Vertical asymptotes. Real zeros of $d(x) = x$ $x = 0$
Horizontal asymptote. Since the degree of $n(x)$ is greater than the degree of $d(x)$, there is no horizontal asymptote.

Oblique asymptote. $\dfrac{2x^2 - 3x + 5}{x} = \dfrac{2x^2}{x} - \dfrac{3x}{x} + \dfrac{5}{x}$

$$= 2x - 3 + \dfrac{5}{x}$$

We can write r as $r(x) = 2x - 3 + \dfrac{5}{x}$, so the line $y = 2x - 3$ is an oblique asymptote.

93. Here is a computer-generated graph of $f(x)$.

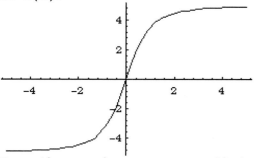

From the graph, we can see that $f(x) \to 5$ as $x \to \infty$ and $f(x) \to -5$ as $x \to -\infty$; the lines $y = 5$ and $y = -5$ are horizontal asymptotes.

95. Here is a computer-generated graph of $f(x)$.

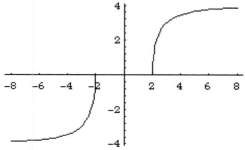

From the graph, we can see that $f(x) \to 4$ as $x \to \infty$ and $f(x) \to 4$ as $x \to -\infty$; the lines $y = 4$ and $y = -4$ are horizontal asymptotes.

97. $f(x) = \dfrac{x^2 + 1}{x} = \dfrac{n(x)}{d(x)}$

Intercepts. There are no real zeros of $n(x) = x^2 + 1$.   No x intercept
 $f(0)$ is not defined   No y intercept
Vertical asymptotes. Real zeros of $d(x) = x$. $x = 0$
Horizontal asymptote. Since the degree of $n(x)$ is greater than the degree of $d(x)$, there is no horizontal asymptote.

Oblique asymptote. $f(x) = \dfrac{x^2 + 1}{x} = \dfrac{x^2}{x} + \dfrac{1}{x} = x + \dfrac{1}{x}$

The line $y = x$ is an oblique asymptote.

Complete the sketch. Plot a few points.

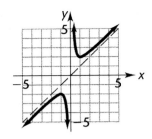

x	$f(x)$
$\frac{1}{2}$	$\frac{5}{2}$
1	2
2	$\frac{5}{2}$
$-\frac{1}{2}$	$-\frac{5}{2}$
-1	-2
-2	$-\frac{5}{2}$

99. $k(x) = \dfrac{x^2 - 4x + 3}{2x - 4} = \dfrac{n(x)}{d(x)}$

Intercepts. Real zeros of $n(x) = x^2 - 4x + 3$

$$x^2 - 4x + 3 = 0$$
$$(x - 1)(x - 3) = 0$$
$$x = 1, 3 \quad x \text{ intercepts}$$

$$k(0) = -\frac{3}{4} \quad y \text{ intercept}$$

Vertical asymptotes. Real zeros of $d(x) = 2x - 4$

$$2x - 4 = 0$$
$$2x = 4$$
$$x = 2$$

Horizontal asymptote. Since the degree of $n(x)$ is greater than the degree of $d(x)$, there is no horizontal asymptote.

Oblique asymptote:

$$
\begin{array}{r}
\frac{1}{2}x - 1 \\
2x - 4 \overline{\smash{\big)}\ x^2 - 4x + 3} \\
\underline{x^2 - 2x} \\
-2x + 3 \\
\underline{-2x + 4} \\
-1
\end{array}
$$

We can write k as $k(x) = \dfrac{1}{2}x - 1 + \dfrac{-1}{2x - 4}$, so the line $y = \dfrac{1}{2}x - 1$ is an oblique asymptote.

Complete the sketch. Plot a few additional points.

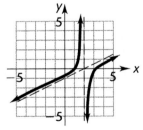

101. $F(x) = \dfrac{8 - x^3}{4x^2} = \dfrac{n(x)}{d(x)}$

Intercepts. Real zeros of $n(x) = 8 - x^3$

$$8 - x^3 = 0$$
$$8 = x^3$$
$$2 = x$$
$$x = 2 \quad x \text{ intercept}$$

$F(0)$ is not defined. No y intercept.

Vertical asymptotes. Real zeros of $d(x) = 4x^2$. $\quad x = 0$

Horizontal asymptote. Since the degree of $n(x)$ is greater than the degree of $d(x)$, there is no horizontal asymptote.

Oblique asymptote. $F(x) = \dfrac{8 - x^3}{4x^2} = \dfrac{8}{4x^2} - \dfrac{x^3}{4x^2} = -\dfrac{1}{4}x + \dfrac{2}{x^2}$.

The line $y = -\dfrac{1}{4}x$ is an oblique asymptote.

Complete the sketch. Plot a few additional points.

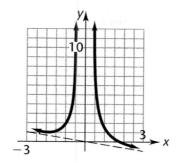

103. $f(x) = \dfrac{x^2 - 4}{x - 2}$. $f(x)$ is not defined if $x - 2 = 0$,
that is, $x = 2$

Domain: $(-\infty, 2) \cup (2, \infty)$

$f(x) = \dfrac{(x - 2)(x + 2)}{(x - 2)}$

$f(x) = x + 2; \; x \ne 2$

The graph is a straight line with slope 1 and y intercept 2, except that the point $(2, 4)$ is not on the graph.

There are no asymptotes.

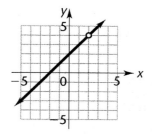

105. $r(x) = \dfrac{x + 2}{x^2 - 4}$. $f(x)$ is not defined if $x^2 - 4 = 0$, that is,

$$x^2 = 4$$
$$x = \pm 2$$

Domain: $(-\infty, -2) \cup (-2, 2) \cup (2, \infty)$

$r(x) = \dfrac{x + 2}{(x + 2)(x - 2)}$

$r(x) = \dfrac{1}{x - 2}$

The graph is the same as the graph of the function $\dfrac{1}{x - 2}$, except that the point $\left(-2, -\dfrac{1}{4}\right)$ is not on the graph.

Intercepts: $y = -\dfrac{1}{2}$. No x intercept.

Vertical asymptote: $x = 2$
Horizontal asymptote: $y = 0$
Complete the sketch. Plot a few additional points.

x intercepts: $n(x) = x$ zero: $x = 0$
x intercept: 0

107. $N(t) = \dfrac{50t}{t + 4}$ $t \geq 0$

Intercepts: Real zeros of $50t$: $t = 0$ $N(0) = 0$
Vertical asymptotes: None, since -4, the only zero of $t + 4$, is not in the domain of N.

Horizontal asymptote: $N = 50$. As $t \to \infty$, $N \to 50$
This number tells us that in the long run, an average employee will be able to assemble 50 components per day.
Complete the sketch. Plot a few additional points.

109. $N(t) = \dfrac{5t + 30}{t}$ $t \geq 1$

Intercepts: Real zeros of $5t + 30$, $t \geq 1$. None, since -6, the only zero of $5t + 30$, is not in the domain of N.
Vertical asymptotes: None, since 0, the only zero of t, is not in the domain of N.

Horizontal asymptote:
$N = 5$. As $t \to \infty$,
$N \to 5$

This tells us that in the long run, the average student will remember 5 of the symbols.
Complete the sketch. Plot a few points.

t	$N(t)$
1	35
3	15
5	11
10	8

111. (A) $\bar{C}(n) = \dfrac{C(n)}{n} = \dfrac{2{,}500 + 175n + 25n^2}{n} = 25n + 175 + \dfrac{2{,}500}{n}$

(B) To find the minimum value, we will graph $\bar{C}(n)$ and use the MINIMUM command:

The minimum occurs after 10 years.

(C) *Intercepts:* Real zeros of $2{,}500 + 175n + 25n^2$.
None. No n intercepts.
0 is not in the domain of n, so there are no $\bar{C}$ intercepts.
Vertical asymptotes: Real zeros of n. The line $n = 0$ is a vertical asymptote.
Sign behavior: $\bar{C}$ is always positive since $n \geq 0$.
Horizontal asymptote: None, since the degree of $C(n)$ is greater than the degree of n.

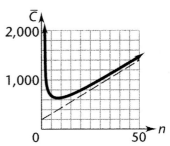

Oblique asymptote: The line $\bar{C} = 25n + 175$ is an oblique asymptote.

113. (A) Begin with a diagram:

The total amount of fence needed is $L = 2x + 2y$, adding up the lengths of the four sides. But we were asked for an expression in terms of x, so we need to eliminate the variable y. We know that the area is 225 sq. ft., and we can see from the diagram that the area is also length × width, or xy. So

$$xy = 225 \quad \Rightarrow \quad y = \frac{225}{x}$$

and we can write the total amount of fence as

$$L(x) = 2x + 2\left(\frac{225}{x}\right) = 2x + \frac{450}{x} \text{ or } \frac{2x^2 + 450}{x}.$$

(B) At the very least, x has to be positive. Also, if the two sides of length x use up all of the fence, there will be no enclosed area, so we need $2x < 225$, or $x < 112.5$. The domain is $0 < x < 112.5$.

(C) To find the minimum, we will graph $L(x)$ on the appropriate domain, and use the MINIMUM command:

The minimum occurs when the width x is 15, in which case the length y is 225/15 = 15. So the dimensions are 15 ft × 15 ft.

(D) *Intercepts:* Real zeros of $2x^2 + 450$. None; no x intercepts. 0 is not in the domain of L, so there are no L intercepts.
Vertical asymptotes: Real zeros of x. The line $x = 0$ is a vertical asymptote.
Horizontal asymptote: None, since the degree of $2x^2 + 450$ is greater than the degree of x.
Oblique asymptote: The line $L = 2x$ is an oblique asymptote.

Section 3-6

1. Answers will vary.   **3.** Answers will vary.

5. Answers will vary.

7. $F = \dfrac{k}{x}$ **9.** $R = kST$ **11.** $L = km^3$ **13.** $A = kc^2d$ **15.** $P = kx$ **17.** $h = \dfrac{k}{\sqrt{s}}$

19. $R = k\dfrac{m}{d^2}$ **21.** $D = k\dfrac{xy^2}{z}$

23. The quantities are inversely proportional, so if F increases, x will decrease.

25. The quantities are directly proportional, so if m increases, L will increase.

27. Write $u = k\sqrt{v}$. Substitute $u = 3$ and $v = 4$ and solve for k.

$$3 = k\sqrt{4}$$
$$3 = 2k$$
$$k = \frac{3}{2}$$

The equation of variation is $u = \frac{3}{2}\sqrt{v}$. When $v = 10$, $u = \frac{3}{2}\sqrt{10}$.

29. Write $L = \frac{k}{M^2}$. Substitute $L = 9$ and $M = 9$ and solve for k.

$$9 = \frac{k}{9^2}$$
$$k = 9^3 = 729$$

The equation of variation is $L = \frac{729}{M^2}$. When $M = 6$, $L = \frac{729}{6^2} = \frac{81}{4}$.

31. Write $Q = k\frac{mn^2}{P}$. Substitute $Q = 2$, $m = 3$, $n = 6$, and $P = 12$ and solve for k.

$$2 = k\frac{3(6)^2}{12}$$
$$2 = 9k$$
$$k = \frac{2}{9}$$

The equation of variation is $Q = \frac{2}{9}\frac{mn^2}{P}$.

When $m = 4$, $n = 18$, and $P = 2$,

$$Q = \frac{2}{9}\frac{4(18)^2}{2} = 144$$

33. The variation is direct. As one quantity increases, the other does as well.

35. The variation is inverse. As one quantity increases, the other decreases.

37. $t = \frac{k}{T}$ **39.** $L = k\frac{wh^2}{x}$ **41.** $N = K\frac{F}{d}$

43. Write $f = kx$. Then $f_1 = kx_1$ and $f_2 = kx_2$. Therefore

$$\frac{kx_1}{kx_2} = \frac{f_1}{f_2}$$
$$\frac{x_1}{x_2} = \frac{f_1}{f_2}$$

45. Write $w = \frac{k}{d^2}$. Substitute $w = 100$ and $d = 4,000$ and solve for k.

$$100 = \frac{k}{(4,000)^2}$$
$$k = 100(4,000)^2 = 1.6 \times 10^9$$

The equation of variation is $w = \frac{1.6 \times 10^9}{d^2}$.

When she is 400 miles above the earth's surface, $d = 4,000 + 400 = 4,400$. Substitute to find

$$w = \frac{1.6 \times 10^9}{(4,400)^2} = 83 \text{ lbs to the nearest pound}$$

47. Write $I = k\dfrac{E}{R}$. Substitute $I = 22$, $E = 110$, and $R = 5$ and solve for k.

$$22 = k\dfrac{110}{5}$$
$$k = 1$$

The equation of variation is $I = \dfrac{E}{R}$. When $E = 220$ and $R = 11$,

$$I = \dfrac{220}{11} = 20 \text{ amperes}$$

49. Write $P = kv^3$. Let $P_1 = kv_1{}^3$ be the original horsepower. To represent doubling the speed, we let $v_2 = 2v_1$; then $P_2 = kv_2{}^3 = k(2v_1)^3 = 8kv_1{}^3 = 8P_1$. The new horsepower ($P_2$) is 8 times the original (P_1), so horsepower must be multiplied by 8 to double the speed of the boat.

51. Write $f = k\dfrac{\sqrt{T}}{L}$. Let $f_1 = k\dfrac{\sqrt{T_1}}{L_1}$ be the original frequency. To represent increasing tension by a factor of 4 and doubling the length, we let $T_2 = 4T_1$, and $L_2 = 2L_1$; then

$$f_2 = k\dfrac{\sqrt{T_2}}{L_2} = k\dfrac{\sqrt{4T_1}}{2L_1} = k\dfrac{\sqrt{T_1}}{L_1} = f_1.$$ The new frequency (f_2) is the same as the original (f_1), so there would be no net effect.

53. Write $t = k\dfrac{r}{v}$. Substitute $t = 1.42$, $r = 4{,}050$, and $v = 18{,}000$ and solve for k.

$$1.42 = k\dfrac{4{,}050}{18{,}000}$$
$$k = 1.42\dfrac{18{,}000}{4{,}050}$$
$$k = 6.311$$

The equation of variation is $t = 6.311\dfrac{r}{v}$. When $r = 4{,}300$ and $v = 18{,}500$,

$$t = 6.311\dfrac{4{,}300}{18{,}500} = 1.47 \text{ hours}$$

55. Write $d = kh$. Substitute $d = 4$ and $h = 500$ and solve for k.

$$4 = k(500)$$
$$k = \dfrac{1}{125}$$

The equation of variation is $d = \dfrac{1}{125}h$. When $h = 2{,}500$,

$$d = \dfrac{1}{125}2{,}500 = 20 \text{ days}.$$

57. Write $L = kv^2$. Let $L_1 = kv_1{}^2$ be the original length. To represent doubling the speed, we let $v_2 = 2v_1$. Then

$$L_2 = kv_2{}^2 = k(2v_1)^2 = 4kv_1{}^2 = 4L_1.$$

The new length (L_2) is 4 times the original length (L_1), so the length would be quadrupled.

59. Write $P = kAv^2$. Substitute $P = 120$, $A = 100$, and $v = 20$ and solve for k.

$$120 = k(100)20^2$$
$$120 = 40,000k$$
$$k = \frac{3}{1000}$$

The equation of variation is $P = \frac{3}{1000}Av^2$. When $A = 200$ and $v = 30$,

$$P = \frac{3}{1000}(200)(30)^2 = 540 \text{ lb}$$

61. (A) $\Delta S = kS$

(B) Substitute $\Delta S = 1$ and $S = 50$ and solve for k.

$$1 = k(50)$$
$$k = \frac{1}{50}$$

The equation of variation is $\Delta S = \frac{1}{50}S$. When $S = 500$, $\Delta S = \frac{1}{50}(500) = 10$ oz.

(C) Substitute $\Delta S = 1$ and $S = 60$ and solve for k.

$$1 = k(60)$$
$$k = \frac{1}{60}$$

The equation of variation is $\Delta S = \frac{1}{60}S$. When $S = 480$, $\Delta S = \frac{1}{60}(480) = 8$ candlepower.

63. Write $V = kr^3$. Let $V_1 = kr_1^3$ be the original volume. To represent doubling the radius, we let $r_2 = 2r_1$; then

$$V_2 = kr_2^3 = k(2r_1)^3 = 8kr_1^3 = 8V_1$$

The new volume (V_1) is 8 times the original volume (V_1).

65. Write frequency $= f$, length $= x$, $f = \frac{k}{x}$. Substitute $f = 16$ and $x = 32$ and solve for k.

$$16 = \frac{k}{32}$$
$$k = 512$$

The equation of variation is $f = \frac{512}{x}$. When $x = 16$, $f = \frac{512}{16} = 32$ times per second.

CHAPTER 3 REVIEW

1. (A) This is not a polynomial since the variable appears in the denominator of the last term.

(B) This is a polynomial with degree 3. *(3-1)*

2. The zeros are -1 and 3; the turning points are $(-1, 0)$, $(1, 2)$, and $(3, 0)$; $P(x) \to \infty$ as $x \to -\infty$ and $P(x) \to \infty$ as $x \to \infty$. *(3-1)*

3.
$$
\begin{array}{rrrr}
2 & 3 & 0 & -1 \\
 & -4 & 2 & -4 \\
\hline
\end{array}
$$
$$-2\overline{)\,2 \quad -1 \quad 2 \quad -5}$$

$2x^3 + 3x^2 - 1$
$= (x + 2)(2x^2 - x + 2) - 5$ *(3-2)*

4.
$$
\begin{array}{rrrrrr}
1 & -4 & 0 & 9 & 0 & -8 \\
 & 3 & -3 & -9 & 0 & 0 \\
\hline
\end{array}
$$
$$3\overline{)\,1 \quad -1 \quad -3 \quad 0 \quad 0 \quad -8}$$

$P(3) = -8$ *(3-2)*

5. 2, -4, -1 *(3-1)*

6. The odd degree and positive leading coefficient tell us that the graph should approach ∞ as $x \to \infty$ (out to the right) and approach $-\infty$ as $x \to -\infty$ (out to the left). The zeros provide the x intercepts of the graph.

(*3-1*)

7. $P(x) = (x - 5)(x + 3)(x - 2)$ is one possible answer. In standard form,

$P(x) = (x - 5)(x^2 + x - 6)$

$= x^3 - 5x^2 + x^2 - 5x - 6x + 30$

$= x^3 - 4x^2 - 11x + 30$ (*3-1*)

8. Since complex zeros come in conjugate pairs, $1 - i$ is a zero.   (*3-4*)

9. (A) Since the graph has x intercepts -2, 0, and 2, these are zeros of $P(x)$. Therefore, $P(x) = (x + 2)x(x - 2) = x^3 - 4x$.

   (B) $P(x) \to \infty$ as $x \to \infty$ and $P(x) \to -\infty$ as $x \to -\infty$.   (*3-1*)

10. We form a synthetic division table:

	1	-4	0	2	
-2	1	-6	12	-22	both are lower bounds, since
-1	1	-5	5	-3	both rows alternate in sign
3	1	-1	-3	-7	
4	1	0	0	2	upper bound

(*3-3*)

11. We investigate $P(1)$ and $P(2)$ by forming a synthetic division table.

	2	-3	1	-5
1	2	-1	0	-5
2	2	1	3	1

   Since $P(1)$ and $P(2)$ have opposite signs, there is at least one real zero between 1 and 2.   (*3-3*)

12. The factors of 6 are ± 1, ± 2, ± 3, ± 6. The leading coefficient is 1, so these are all possible rational zeros.   (*3-4*)

13. Using the possibilities found in problem 12, we form a synthetic division table:

	1	-4	1	6	
1	1	-3	-2	4	
2	1	-2	-3	0	2 is a zero

   Then $x^3 - 4x^2 + x + 6 = (x - 2)(x^2 - 2x - 3) = (x - 2)(x - 3)(x + 1)$. The rational zeros are 2, 3, -1.   (*3-4*)

14. (A) $f(x) = \dfrac{2x - 3}{x + 4} = \dfrac{n(x)}{d(x)}$

   The domain of f is the set of all real numbers x such that $d(x) = x + 4 \neq 0$, that is, $(-\infty, -4) \cup (-4, \infty)$. f has an x intercept where $n(x) = 2x - 3 = 0$, that is, $x = \dfrac{3}{2}$.

(B) $g(x) = \dfrac{3x}{x^2 - x - 6} = \dfrac{n(x)}{d(x)}$

The domain of g is the set of all real numbers x such that
$d(x) = x^2 - x - 6 \neq 0$, that is, $(x + 2)(x - 3) \neq 0$, that is, $x \neq -2, 3$, or
$(-\infty, -2) \cup (-2, 3) \cup (3, \infty)$. g has an x intercept where $n(x) = 3x = 0$,
that is, $x = 0$.

$(3\text{-}5)$

15. (A) Horizontal asymptote: since $n(x)$ and $d(x)$ have the same degree, the line
$y = 2$ is a horizontal asymptote (The ratio of the leading coefficients is 2/1,
or 2). Vertical asymptotes: zeros of $d(x)$

$x + 4 = 0$
$\quad x = -4$

(B) Horizontal asymptote: since the degree of $n(x)$ is less than the degree of
$d(x)$, the line $y = 0$ is a horizontal asymptote.
Vertical asymptotes: zeros of $d(x)$

$x^2 - x - 6 = 0$
$(x + 2)(x - 3) = 0$
$\quad x = -2, \ x = 3$

$(3\text{-}5)$

16. $F = k\sqrt{x}$ $\quad(3\text{-}6)$ $\qquad$ **17.** $G = kxy^2$ $\quad(3\text{-}6)$ $\qquad$ **18.** $H = \dfrac{k}{z^3}$ $\quad(3\text{-}6)$

19. $R = kx^2y^2$ $\quad(3\text{-}6)$ $\qquad$ **20.** $S = \dfrac{k}{u^2}$ $\quad(3\text{-}6)$ $\qquad$ **21.** $T = k\dfrac{v}{w}$ $\quad(3\text{-}6)$

22. When two quantities are directly proportional, an increase in one corresponds
to an increase in the other as well. In this case, the larger surface area
will increase the amount of light captured.

$(3\text{-}6)$

23. When two quantities are inversely proportional, an increase in one corresponds
to a decrease in the other. In this case, an increase in the number of
children will correspond with a decrease in cleanliness.

$(3\text{-}6)$

24. The graph of a polynomial has to approach ∞ or $-\infty$ as $x \to \infty$ and $x \to -\infty$.
This graph approaches zero.

$(3\text{-}1)$

25. $P(x) = x^3 - 3x^2 - 3x + 4$

(A)

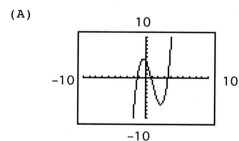

The graph of $P(x)$ has three x
intercepts and two turning points.
$P(x) \to \infty$ as $x \to \infty$ and $P(x) \to -\infty$
as $x \to -\infty$.

(B) Applying the ZERO command yields:

The largest x intercept is 3.53.

$(3\text{-}1)$

26. We use synthetic division:

$$\frac{1}{4} \overline{)\begin{array}{rrrrr} 8 & -14 & -13 & -4 & 7 \\ & 2 & -3 & -4 & -2 \\ \hline 8 & -12 & -16 & -8 & 5 \end{array}}$$

We can write P as

$P(x) = \left(x - \dfrac{1}{4}\right)(8x^3 - 12x^2 - 16x - 8) + 5$

$P\left(\dfrac{1}{4}\right) = 5$

$(3\text{-}2)$

27.

$$\begin{array}{r} x^2 - 9x + 18 \\ x^2 + 2x \overline{\smash{\big)}\ x^4 - 7x^3 + 0x^2 - 5x + 1} \end{array}$$

$$\underline{-\left(x^4 + 2x^3\right)}$$

$$-9x^3 + 0x^2$$

$$\underline{-\left(-9x^3 - 18x^2\right)}$$

$$18x^2 - 5x$$

$$\underline{-\left(18x^2 + 36x\right)}$$

$$-41x + 1$$

$$x^2 - 9x + 18 + \frac{-41x + 1}{x^2 + 2x} \qquad\qquad (3\text{-}2)$$

28.

$$\begin{array}{r} \phantom{-\tfrac{1}{2}\big|}4 \quad -8 \quad -3 \quad -3 \\ \phantom{-\tfrac{1}{2}\big|4}-2 \quad\ \ 5 \quad -1 \\ \hline -\tfrac{1}{2}\big|\ 4 \quad -10 \quad\ 2 \quad -4 \end{array}$$

$$P\left(-\frac{1}{2}\right) = -4 \qquad\qquad (3\text{-}2)$$

29. The quadratic formula tells us that $x^2 - 2x - 1 = 0$ if

$$x = \frac{-b \pm \sqrt{b^2 - 4ac}}{2a} \qquad a = 1,\ b = -2,\ c = -1$$

$$x = \frac{-(-2) \pm \sqrt{(-2)^2 - 4(1)(-1)}}{2(1)}$$

$$= \frac{2 \pm \sqrt{8}}{2}$$

$$= 1 \pm \sqrt{2}$$

Since $1 \pm \sqrt{2}$ are zeros of $x^2 - 2x - 1$, its factors are $x - (1 + \sqrt{2})$ and $x - (1 - \sqrt{2})$, that is, $x^2 - 2x - 1 = [x - (1 + \sqrt{2})][x - (1 - \sqrt{2})]$ $\qquad (3\text{-}2)$

30. $x + 1$ will be a factor of $P(x)$ if $P(-1) = 0$.
$P(-1) = 9(-1)^{26} - 11(-1)^{17} + 8(-1)^{11} - 5(-1)^4 - 7 = 9 + 11 - 8 - 5 - 7 = 0$, so the answer is yes, $x + 1$ is a factor. $\qquad\qquad (3\text{-}2)$

31. First, divide out $x = 2$ using synthetic division:

$$\begin{array}{r} 2 \quad -9 \quad\ \ 3 \quad\ 14 \\ 4 \quad -10 \quad -14 \\ \hline 2\big|\ 2 \quad -5 \quad -7 \quad\ \ 0 \end{array}$$

Now find the zeros of the reduced polynomial, $2x^2 - 5x - 7$ by factoring:

$$2x^2 - 5x - 7 = (2x - 7)(x + 1) = 0$$

$$2x - 7 = 0 \quad\ \text{or} \quad\ x + 1 = 0$$

$$2x = 7 \qquad\qquad\qquad x = -1$$

$$x = 7/2 \qquad\qquad\qquad\qquad\qquad\qquad\qquad (3\text{-}2)$$

32. The possible rational zeros are ± 1, ± 2, ± 4, ± 8, $\pm\dfrac{1}{2}$. We form a synthetic division table:

	2	-3	-18	-8
1	2	-1	-19	-27
2	2	1	-16	-40
4	2	5	2	0

So $2x^3 - 3x^2 - 18x - 8 = (x - 4)(2x^2 + 5x + 2)$
$\qquad\qquad\qquad\qquad\qquad\; = (x - 4)(2x + 1)(x + 2)$

Zeros: 4, $-\dfrac{1}{2}$, -2 *(3-4)*

33. From the given zeros, we get factors of $(x - 4)$, $(x + 1/2)$, and $(x + 2)$. But this can't be the correct factorization because the leading coefficient of the polynomial is 2, and there are no fractional coefficients. We remedy this by multiplying the second factor by 2, making it $(2x + 1)$.

$P(x) = (x - 4)(2x + 1)(x + 2)$ *(3-4)*

34. The possible rational zeros are ± 1, ± 5. We form a synthetic division table:

	1	-3	0	5
1	1	-2	-2	3
5	1	2	10	55
-1	1	-4	4	1
-5	1	-8	40	-195

There are no rational zeros, since all possibilities fail. *(3-4)*

35. $P(x) = 2x^4 - x^3 + 2x - 1$.

Possible rational zeros: ± 1, $\pm\dfrac{1}{2}$

Examining the graph of $y = P(x)$, we see that there is a zero between 0 and 1, and possibly a zero at -1.

We test the likely candidates, and find

-1 is a zero. So $P(x) = (x + 1)(2x^3 - 3x^2 + 3x - 1) = (x + 1)Q(x)$

Testing $\dfrac{1}{2}$ in the reduced polynomial $Q(x)$, we obtain

$$\begin{array}{r} 2 \quad -3 \quad\;\; 3 \quad -1 \\ 1 \quad -1 \quad\;\; 1 \\ \hline \tfrac{1}{2}\,\big|\; 2 \quad -2 \quad\;\; 2 \quad\;\; 0 \end{array}$$

$\dfrac{1}{2}$ is a zero.

So $P(x) = (x + 1)\left(x - \dfrac{1}{2}\right)(2x^2 - 2x + 2) = (x + 1)(2x - 1)(x^2 - x + 1)$

To find the remaining zeros, we solve $x^2 - x + 1 = 0$, by the quadratic formula.
$x^2 - x + 1 = 0$

$$x = \frac{-b \pm \sqrt{b^2 - 4ac}}{2a} \qquad a = 1, \; b = -1, \; c = 1$$

$$x = \frac{-(-1) \pm \sqrt{(-1)^2 - 4(1)(1)}}{2(1)}$$

$$x = \frac{1 \pm \sqrt{-3}}{2}$$

$$x = \frac{1 \pm i\sqrt{3}}{2}$$

The four zeros are -1, $\frac{1}{2}$, and $\frac{1 \pm i\sqrt{3}}{2}$ *(3-4)*

36. $(x + 1)\left(x - \frac{1}{2}\right) 2\left(x - \frac{1 + i\sqrt{3}}{2}\right)\left(x - \frac{1 - i\sqrt{3}}{2}\right) = (x + 1)(2x - 1)\left(x - \frac{1 + i\sqrt{3}}{2}\right)\left(x - \frac{1 - i\sqrt{3}}{2}\right)$

(3-4)

37. The degree is 9. The zeros are 1 (multiplicity 3), -1 (multiplicity 4), i and $-i$. *(3-4)*

38. (A) Let $u = x^2$. Then
$$P(x) = x^4 + 5x^2 - 36$$
$$= u^2 + 5u - 36 = (u + 9)(u - 4) = (x^2 + 9)(x^2 - 4) = (x^2 + 9)(x + 2)(x - 2)$$

(B) $x^2 + 9$ has zeros $-3i$ and $3i$, so it factors as $(x + 3i)(x - 3i)$.
$$P(x) = (x + 3i)(x - 3i)(x + 2)(x - 2)$$ *(3-4)*

39. (A) Examining the graph of $P(x)$, we see that there may be zeros of even multiplicity between -1 and 0, and between 4 and 5, and a possible integer zero at 2.

Testing 2, we obtain

	1	-10	30	-20	-15	-2
		2	-16	28	16	2
2	1	-8	14	8	1	0

2 is a zero. Applying the MAXIMUM and MINIMUM commands, we obtain

There are zeros of even multiplicity at -0.24 and 4.24. Since $P(x)$ is a fifth-degree polynomial, they must be double zeros and 2 must be a simple zero.

(B) -0.24 can be approximated with the MAXIMUM command; 2 can be approximated by the bisection method; 4.24 can be approximated with the MINIMUM command. *(3-3)*

40. (A) We form a synthetic division table.

	1	-2	-30	0	-25	
0	1	-2	-30	0	-25	
1	1	-1	-31	-31	-56	
2	1	0	-30	-60	-145	
3	1	1	-27	-81	-268	
4	1	2	-22	-88	-377	
5	1	3	-15	-75	-400	
6	1	4	-6	-36	-241	
7	1	5	5	35	220	7 is an upper bound
-1	1	-3	-27	27	-52	
-2	1	-4	-22	44	-113	
-3	1	-5	-15	45	-160	
-4	1	-6	-6	24	-121	
-5	1	-7	5	-25	100	-5 is a lower bound

(B) We search for the real zero in (6, 7) indicated in the table. We organize our calculations in a table.

Sign Change Interval (a, b)	Midpoint m	Sign of P		
		$P(a)$	$P(m)$	$P(b)$
(6, 7)	6.5	-	-	+
(6.5, 7)	6.75	-	+	+
(6.5, 6.75)	6.625	-	+	+
(6.5, 6.625)	6.5625	-	-	+
(6.5625, 6.625)	We stop here	-		+

Since each endpoint rounds to 6.6, a real zero lies on this interval and is given by 6.6 to one decimal place accuracy. 4 intervals were required.

(C) Graphing and applying the ZERO command, we obtain:

The zeros are -4.67 and 6.62. (3-3)

41. $f(x) = \dfrac{x - 1}{2x + 2} = \dfrac{n(x)}{d(x)}$

(A) The domain of f is the set of all real numbers x such that $d(x) = 2x + 2 \neq 0$, that is $(-\infty, -1) \cup (-1, \infty)$.

f has an x intercept where $n(x) = x - 1 = 0$, that is, $x = 1$. $f(0) = -\dfrac{1}{2}$, so f has a y intercept at $y = -\dfrac{1}{2}$

(B) Vertical asymptote: $x = -1$. Horizontal asymptote: since $n(x)$ and $d(x)$ have the same degree, the line $y = \dfrac{1}{2}$ is a horizontal asymptote.

(C) Complete the sketch:

Graphing calculator: Hand sketch:

(3–5)

42. Rearrange so that zero is on the left:

$3x^3 - 4x^2 - 15x \geq 0$

Find the zeros of the left side:

$3x^3 - 4x^2 - 15x = 0$

$x\left(3x^2 - 4x - 15\right) = 0$

$x\left(3x + 5\right)\left(x - 3\right) = 0$

$x = 0, \ -5/3, \ 3$

Choose test numbers and make a table:

Interval	Test Number	Result
$x < -5/3$	-2	$3(-2)^3 - 4(-2)^2 - 15(-2) = -10;$ negative
$-5/3 < x < 0$	-1	$3(-1)^3 - 4(-1)^2 - 15(-1) = 8;$ positive
$0 < x < 3$	1	$3(1)^3 - 4(1)^2 - 15(1) = -16;$ negative
$x > 3$	4	$3(4)^3 - 4(4)^2 - 15(4) = 68;$ positive

Solution: $\left[-5/3, 0\right] \cup \left[3, \infty\right)$ (3–3)

43. (A) Find the zeros of $P(x) = x^3 - 5x + 4$ graphically:

The zeros are −2.562, 1, and 1.562. (Two are very close together, so you may need to look closely at that portion of the graph to see that there are two.) $P(x)$ is negative (below the x-axis) on $(-\infty, -2.562) \cup (1, 1.562)$.

(B) Keeping $y_1 = x^3 - 5x + 4$, graph $y_2 = 2$.

The graphs intersect at −2.414, 0.414, and 2. The graph of $x^3 - 5x + 4$ is below the graph of 2 on $(-\infty, -2.414) \cup (0.414, 2)$. (3–3)

44. This is not the graph of a rational function because of what happens at $x = 0$; a rational function can only be discontinuous at an x value because of a vertical asymptote or a hole in the graph. *(3-5)*

45. Write $B = \dfrac{k}{\sqrt{c}}$. Substitute $B = 5$ and $C = 4$ and solve for k.

$$5 = \frac{k}{\sqrt{4}}$$
$$k = 10$$

The equation of variation is $B = \dfrac{10}{\sqrt{c}}$. When $c = 25$,

$$B = \frac{10}{\sqrt{25}} = \frac{10}{5} = 2$$ *(3-6)*

46. Write $D = kxy$. Substitute $D = 10$, $x = 3$, and $y = 2$ and solve for k.
$$10 = k(3)(2)$$
$$k = \frac{5}{3}$$

The equation of variation is $D = \dfrac{5}{3}xy$. When $x = 9$ and $y = 8$,

$$D = \frac{5}{3}(9)(8) = 120$$ *(3-6)*

47.

$$
\begin{array}{c|ccccc}
 & 1 & 0 & 3 & 2 \\
 & & 1+i & 2i & 1+5i \\
\hline
1+i & 1 & 1+i & 3+2i & 3+5i
\end{array}
$$
(See calculations below)

$$(1 + i)^2 = (1 + i)(1 + i) = 1 + 2i + i^2 = 1 + 2i - 1 = 2i$$
$$(1 + i)(3 + 2i) = 3 + 5i + 2i^2 = 3 + 5i - 2 = 1 + 5i$$

$$P(x) = [x^2 + (1 + i)x + (3 + 2i)][x - (1 + i)] + 3 + 5i$$ *(3-2)*

48. $P(x) = \left(x + \dfrac{1}{2}\right)^2 (x + 3)(x - 1)^3$. The degree is 6. *(3-4)*

49. $P(x) = (x + 5)[x - (2 - 3i)][x - (2 + 3i)]$. The degree is 3. *(3-4)*

50. The possible rational zeros are ± 1, ± 2, ± 4, $\pm \dfrac{1}{2}$.

Examining the graph of $y = P(x)$, we see that there are zeros between -1 and 0, between 0 and 1, between 2 and 3, and possible integer zeros at -2 and 2. Testing 2, we obtain

$$
\begin{array}{r|rrrrrr}
 & 2 & -5 & -8 & 21 & 0 & -4 \\
 & & 4 & -2 & -20 & 2 & 4 \\
\hline
2 & 2 & -1 & -10 & 1 & 2 & 0
\end{array}
$$

2 is a zero.
So $P(x) = (x - 2)(2x^4 - x^3 - 10x^2 + x + 2)$.
Testing -2 in the reduced polynomial, we obtain

$$
\begin{array}{r|rrrrr}
 & 2 & -1 & -10 & 1 & 2 \\
 & & -4 & 10 & 0 & -2 \\
\hline
-2 & 2 & -5 & 0 & 1 & 0
\end{array}
$$

-2 is a zero.

$P(x) = (x - 2)(x + 2)(2x^3 - 5x^2 + 1)$. $2x^3 - 5x^2 + 1$ has been shown (see Exercise 3-3, problem 49 for details) to have zeros $\dfrac{1}{2}$, $1 \pm \sqrt{2}$.

Then $P(x)$ has zeros $\dfrac{1}{2}$, ± 2, $1 \pm \sqrt{2}$. *(3-4)*

51. $(x - 2)(x + 2)\left(x - \dfrac{1}{2}\right)2[x - (1 - \sqrt{2})][x - (1 + \sqrt{2})]$

$= (x - 2)(x + 2)(2x - 1)[x - (1 - \sqrt{2})][x - (1 + \sqrt{2})]$ *(3-4)*

52. Graphing $y = P(x)$, we obtain:

Clearly there is a local maximum near −4. Applying the MAXIMUM command, we obtain

To find the other required points, we need different windows. Redrawing and applying the MINIMUM and ZERO commands, we obtain

Summarizing, $P(x)$ has zeros at 0.91 and 1. It has a local minimum $P(-8.94) \approx 9.7$. It has a local maximum $P(-4.01) \approx 605.01$. It has a local minimum $P(0.95) \approx -0.20$. *(3-3)*

53. $P(x)$ changes sign three times. Therefore, it has three zeros and its minimal degree is 3. *(3-1)*

54. Since $1 + 2i$ is a zero, its conjugate $1 - 2i$ is also a zero. Then
$[x - (1 - 2i)][x - (1 + 2i)] = [(x - 1) + 2i][(x - 1) - 2i] = (x - 1)^2 - 4i^2$
$= x^2 - 2x + 5$ is a factor. Since $P(x)$ is a cubic polynomial, it must be of the form $a(x - r)(x^2 - 2x + 5)$. Since the constant term of this polynomial, $-5ar$, must be an integer, r must be a rational number. So there can be no irrational zeros. *(3-4)*

55. (A) The cube roots of 27 are solutions to the equation $x^3 = 27$, or $x^3 - 27 = 0$. The polynomial $x^3 - 27$ is a third degree polynomial and has three zeros, so there must be three cube roots of 27.

(B) Divide out $x = 3$ using synthetic division:

```
        1    0    0  -27
             3    9   27
    3 | 1    3    9    0
```

We can find the zeros of the reduced polynomial $x^2 + 3x + 9$ using the quadratic formula with $a = 1$, $b = 3$, $c = 9$:

$$x = \frac{-3 \pm \sqrt{3^2 - 4(1)(9)}}{2(1)} = \frac{-3 \pm \sqrt{-27}}{2} = \frac{-3 \pm 3\sqrt{3}i}{2} = -\frac{3}{2} \pm \frac{3\sqrt{3}}{2}i$$ *(3-4)*

56. (A) We form a synthetic division table.

	1	2	-500	0	-4,000
0	1	2	-500	0	-4,000
10	1	12	-380	-3,800	-42,000
20	1	22	-60	-1,200	-28,000
30	1	32	460	13,800	410,000
-10	1	-8	-420	4,200	-46,000
-20	1	-18	-140	2,800	-60,000
-30	1	-28	340	-10,200	302,000

From the table, 30 is an upper bound and -30 is a lower bound.

(B) Graphing $P(x)$ in the window suggested by part (A), we obtain

The zeros are -23.54 and 21.57. Examining the graph more closely near $x = 0$, we obtain

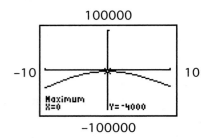

There is no other real zero.   *(3-3)*

57. $f(x) = \dfrac{x^2 + 2x + 3}{x + 1} = \dfrac{n(x)}{d(x)}$

Intercepts. There are no real zeros of $n(x) = x^2 + 2x + 3$. No x intercept
$f(0) = 3$ y intercept
Vertical asymptotes. Real zeros of $d(x) = x + 1$ $x = -1$ is a vertical asymptote
Horizontal asymptotes. Since the degree of $n(x)$ is greater than the degree of $d(x)$, there is no horizontal asymptote.
Oblique asymptote:

$$
\begin{array}{r}
x + 1 \\
x + 1 \overline{)\, x^2 + 2x + 3} \\
\underline{x^2 + x} \\
x + 3 \\
\underline{x + 1} \\
2
\end{array}
$$

$f(x) = x + 1 + \dfrac{2}{x + 1}$.

The line $y = x + 1$ is an oblique asymptote.

Complete the sketch:

Graphing calculator: Hand sketch:

(*3-5*)

58. Graphing $y = f(x)$, we obtain the graph at the right.

From the graph, we can see that $f(x) \to 2$ as $x \to \infty$ and $f(x) \to -2$ as $x \to -\infty$; the lines $y = 2$ and $y = -2$ are horizontal asymptotes.

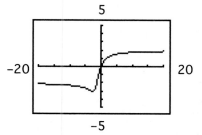

(*3-5*)

59. (A) $\dfrac{x^2 - 3}{x^3 - 3x + 1} \leq 0$

zeros of the numerator: $x^2 - 3 = 0$

$$x^2 = 3$$
$$x = \sqrt{3}, \ -\sqrt{3} \text{ or } 1.732, \ -1.732$$

zeros of the denominator: Find graphically:

The zeros are −1.879, 0.347, 1.532.

Interval	Test Number	Result
$(-\infty, -1.879)$	−2	−1; negative
$(-1.879, -1.732)$	−1.8	0.423; positive
$(-1.732, 0.347)$	0	−3; negative
$(0.347, 1.532)$	1	2; positive
$(1.532, 1.732)$	1.6	−1.49; negative
$(1.732, \infty)$	2	0.333; positive

The expression is negative on $(-\infty, -1.879)$, $(-1.732, 0.347)$, and $(1.532, 1.732)$. We should include −1.732 and 1.732 (since the expression is zero there) and exclude −1.879, 0.347, and 1.532 (since the expression is undefined there.) The solution is $(-\infty, -1.879) \cup [-1.732, 0.347) \cup (1.532, 1.732]$.

Graphical check

The graph is below the x-axis on
$(-\infty, -1.879) \cup (-1.732, 0.347) \cup (1.532, 1.732)$,
and intersects the x-axis at -1.732 and 1.732.

(B) $\dfrac{x^2 - 3}{x^3 - 3x + 1} > \dfrac{5}{x^2}$

$$\dfrac{x^2 - 3}{x^3 - 3x + 1} - \dfrac{5}{x^2} > 0$$

$$\dfrac{x^2(x^2 - 3)}{x^2(x^3 - 3x + 1)} - \dfrac{5(x^3 - 3x + 1)}{x^2(x^3 - 3x + 1)} > 0$$

$$\dfrac{x^4 - 3x^2 - 5x^3 + 15x - 5}{x^2(x^3 - 3x + 1)} > 0$$

zeros of the numerator: Find graphically:

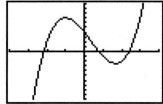

The zeros are -1.843, 0.379, 1.420, and 5.044.

zeros of the denominator:

$x^2(x^3 - 3x + 1 = 0$
$x^2 = 0 \qquad x^3 - 3x + 1 = 0$
$x = 0 \qquad\qquad x = -1.879, 0.347, 1.532 \qquad$ (From part A)

Interval	Test Number	Result
$(-\infty, -1.879)$	-2	-2.25; negative
$(-1.879, -1.843)$	-1.85	0.47; positive
$(-1.843, 0)$	-1	-5.7; negative
$(0, 0.347)$	0.1	-504.3; negative
$(0.347, 0.379)$	0.35	363; positive
$(0.379, 1.420)$	1	-3; negative
$(1.420, 1.532)$	1.5	3.8; positive
$(1.532, 5.044)$	2	-0.92; negative
$(5.044, \infty)$	6	0.03; positive

The expression is positive on $(-1.879, -1.843) \cup (0.347, 0.379) \cup (1.420, 1.532) \cup (5.044, \infty)$

(3-5)

60. The potential rational zeros are ± 1, ± 2, and ± 4. A look at the graph shows 3 zeros, none of which occur at integer values, so none of the candidates for rational zeros are actually zeros.

(3-4)

61. To have real zeros -3, 0, and 2, we should have factors $(x + 3)$, x, and $(x - 2)$ in the numerator. To have vertical asymptotes $x = -1$ and $x = 4$, we should have factors $(x + 1)$ and $(x - 4)$ in the numerator. At the moment, our function is

$$f(x) = \frac{x(x + 3)(x - 2)}{(x + 1)(x - 4)}$$

But the degree of the numerator is greater than the degree of the denominator, so there will be no horizontal asymptote. We can fix that by squaring one of the factors in the denominator. Multiplying the numerator by 5 will then make $y = 5$ a horizontal asymptote.

$$f(x) = \frac{5x(x + 3)(x - 2)}{(x + 1)^2(x - 4)}$$

(There are other possible answers.)   *(3-5)*

62. (A) Graph $P(x)$ on the domain $0 \le x \le 20$ and use the ZERO command:

She should be open 2.7 hours to break even.

(B) Add $y_2 = 325$ to the graph and use the INTERSECT command:

To make a weekly profit of at least \$325, she should be open between 10.9 and 17.5 hours.   *(3-3)*

63. Write $F = \dfrac{k}{r}$. Let $F_1 = \dfrac{k}{r_1}$ be the original force. To represent doubling the radius, we let $r_2 = 2r_1$. Then

$$F_2 = \frac{k}{r_2} = \frac{k}{2r_1} = \frac{1}{2}\frac{k}{r_1} = \frac{1}{2}F_1.$$

The force would be $\dfrac{1}{2}$ the original.   *(3-6)*

64. $v = k\dfrac{\sqrt{T}}{\sqrt{w}}$ *(3-6)*

65. Write $A = kWt$. Substitute $A = A_0$, $W = 10$ and $t = 8$ and solve for k.

$$A_0 = k(10)(8)$$

$$k = \frac{A_0}{80}$$

The equation of variation for this job is $A = \dfrac{A_0}{80}Wt$.

Substitute $A = A_0$ (again) and $W = 4$ and solve for t.

$$A_0 = \frac{A_0}{80}(4)t$$

$$A_0 = \frac{A_0}{20}t$$

$$t = 20 \text{ days}$$ *(3-6)*

66. Write $I = kpr$. Substitute $I = 8$, $p = 100$, and $r = 0.04$ and solve for k.
$$8 = k(100)(0.04)$$
$$k = 2$$
The equation of variation is $I = 2pr$. When $p = 150$ and $r = 0.03$,
$$I = 2(150)(0.03)$$
$$I = \$9.00$$
(3-6)

67. In the given figure, let y = height of door
$$2x = \text{width of door}$$
Then Area of door = $48 = 2xy$
Since (x, y) is a point on the parabola $y = 16 - x^2$, its coordinates satisfy the equation of the parabola.
$$48 = 2x(16 - x^2)$$
$$48 = 32x - 2x^3$$
$$2x^3 - 32x + 48 = 0$$
$$x^3 - 16x + 24 = 0$$
The possible rational solutions of this equation are ±1, ±2, ±3, ±4, ±6, ±8, ±12, ±24. Testing the likely candidates, we obtain

```
      1    0   -16    24
           2    4    -24
   2⎿1    2   -12     0
```

2 is a solution.
The equation can be factored
$$(x - 2)(x^2 + 2x - 12) = 0.$$
To find the remaining zeros, we solve $x^2 + 2x - 12 = 0$, by completing the square.
$$x^2 + 2x = 12$$
$$x^2 + 2x + 1 = 13$$
$$(x + 1)^2 = 13$$
$$x + 1 = \pm\sqrt{13}$$
$$x = -1 + \sqrt{13} \text{ (discarding the negative solution)} \approx 2.61$$
The positive zeros are $x = 2$, 2.61. The dimensions of the door are either $2x = 4$ feet by $16 - x^2 = 12$ feet, or $2x = 5.2$ feet by $16 - x^2 = 9.2$ feet. (3-3)

68. We note:
$$\begin{pmatrix} \text{Volume} \\ \text{of} \\ \text{silo} \end{pmatrix} = \begin{pmatrix} \text{Volume of} \\ \text{hemisphere} \\ \text{of radius } x \end{pmatrix} + \begin{pmatrix} \text{Volume of} \\ \text{cylinder with} \\ \text{radius } x, \text{height 18} \end{pmatrix}$$

$$486\pi = \frac{2}{3}\pi x^3 + \pi x^2 \cdot 18$$

$$486 = \frac{2}{3}x^3 + 18x^2$$

$$0 = x^3 + 27x^2 - 729$$

Examining the graph of
$y = P(x)$, we obtain, for positive x:

The radius is 4.789 feet.

(3-3)

69.

 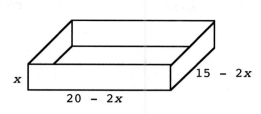

From the above figures we see that
$$V = \text{length} \times \text{width} \times \text{height} = (20 - 2x)(15 - 2x)x$$
To ensure that all dimensions are positive $0 < x < 7.5$
We solve $(20 - 2x)(15 - 2x)x = 300$
$$300x - 70x^2 + 4x^3 = 300$$
$$4x^3 - 70x^2 + 300x - 300 = 0$$
Examining the graph of $y = P(x)$ for $0 < x < 7.5$, we obtain

$x = 1.450$ inches or $x = 4.465$ inches.

$(3\text{-}3)$

70. Let (x, x^2) be a point on the graph of $y = x^2$. Then the distance from $(1, 4)$ to (x, x^2) must equal 3 units. Applying the distance formula, we have,

$$\sqrt{(x - 1)^2 + (x^2 - 4)^2} = 3$$
$$(x - 1)^2 + (x^2 - 4)^2 = 9$$
$$x^2 - 2x + 1 + x^4 - 8x^2 + 16 = 9$$
$$x^4 - 7x^2 - 2x + 8 = 0$$

Let $P(x) = x^4 - 7x^2 - 2x + 8$. The possible rational zeros are ± 1, ± 2, ± 4, ± 8.

Examining the graph of $y = P(x)$, we see that there is a zero between -2 and -1, a zero between 2 and 3, and possible integer zeros at -2 and 1. Testing the likely candidates, we obtain

```
       1    0    -7   -2    8
            1    1    -6   -8
     1|1    1    -6   -8    0
```

1 is a zero. So $P(x) = (x - 1)(x^3 + x^2 - 6x - 8)$. Testing -2 in the reduced polynomial, we obtain

$$
\begin{array}{r}
\ 1 \quad\ 1 \quad -6 \quad -8 \\
-2 \quad\ \ 2 \quad\ \ 8 \\
\hline
-2\,|\ 1 \quad -1 \quad -4 \quad\ \ \ 0
\end{array}
$$

-2 is a zero. The remaining zeros are found by solving $x^2 - x - 4 = 0$. Applying the quadratic formula with $a = 1$, $b = -1$, $c = -4$, we obtain

$$x = \frac{-(-1) \pm \sqrt{(-1)^2 - 4(1)(-4)}}{2(1)} = \frac{1 \pm \sqrt{17}}{2}$$

To three decimal place accuracy, $x = -1.562$ or 2.562. There are four real zeros of $P(x)$, 1, -2, -1.562, and 2.562. There are four points on the graph of $y = x^2$ that are 3 units from $(1, 4)$ and their coordinates are $(1, 1)$, $(-2, 4)$, $(-1.562, 2.440)$, and $(2.562, 6.564)$

(*3-3*)

71. (A)

Enter the data.

Compute the regression equation.

The cubic model is $y = -0.0103x^3 + 0.849x^2 - 2.56x + 222$.

(B)

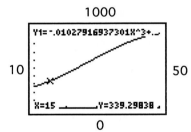

Transfer the regression equation to the equation editor.

Graph the regression equation and determine the value of y corresponding to $x = 15$.

The model predicts that 339 refrigerators would be sold if 15 ads were placed.

(C)

Use the INTERSECT command to determine the value of x that corresponds to $y = 750$.

The model predicts that 750 refrigerators would be sold if 36 ads were placed.

(*3-1*)

72. (A) Since the percentage figures in the second column decrease, then increase, then decrease again with increasing age, this might correspond to a function with both a local minimum and a local maximum. A cubic model can have both.

(B) The independent variable is age and the dependent variable is percentage, so we enter the age column as L_1 and the percentage column as L_2. Then choose cubic regression from the STAT CALC menu.

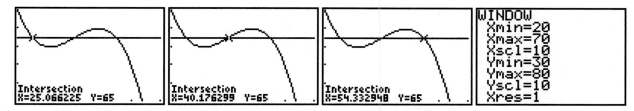

(Note that the entire list is not visible on the graphic.)

The cubic regression model is $y = -0.00424x^3 + 0.507x^2 - 19.3x + 297$.

(C) Graph $y_1 = -0.00424x^3 + 0.507x^2 - 19.3x + 297$, $y_2 = 65$, and use the INTERSECT command.

The percentage is 65 for ages 25, 40, and 54.   (3-1)

CHAPTER 4　　Modeling with Exponential and Logarithmic Functions

Section 4-1

1. Answers will vary.　　　3. Answers will vary.　　　5. Answers will vary.

7. (A) The graph of $y = (0.2)^x$ is decreasing and passes through the point $(-1, 0.2^{-1}) = (-1, 5)$. This corresponds to graph g.

   (B) The graph of $y = 2^x$ is increasing and passes through the point $(1, 2)$. This corresponds to graph n.

   (C) The graph of $y = \left(\dfrac{1}{3}\right)^x$ is decreasing and passes through the point $(-1, 3)$. This corresponds to graph f.

   (D) The graph of $y = 4^x$ is increasing and passes through the point $(1, 4)$. This corresponds to graph m.

9. 16.24　　　　11. 7.524　　　　13. 1.649　　　　15. 4.469

17. Make a table of values:

x	-3	-2	-1	0	1	2
y	$1/27$	$1/9$	$1/3$	0	3	9

19. Make a table of values:

x	-3	-2	-1	0	1	2
y	27	9	3	1	$1/3$	$1/9$

21. $10^{3x-1}10^{4-x} = 10^{(3x-1)+(4-x)} = 10^{2x+3}$

23. $\dfrac{3x}{3^{1-x}} = 3^{x-(1-x)} = 3^{x-1+x} = 3^{2x-1}$

25. $\left(\dfrac{4^x}{5^y}\right)^{3z} = \dfrac{\left(4^x\right)^{3z}}{\left(5^y\right)^{3z}} = \dfrac{4^{3xz}}{5^{3yz}}$

27. $\dfrac{e^{5x}}{e^{2x+1}} = e^{5x-(2x+1)} = e^{5x-2x-1} = e^{3x-1}$

29. (A) Although $1 + \dfrac{1}{x}$ approaches 1, $1 + \dfrac{1}{x}$ is not equal to 1 for any x, so reasoning as if it were 1 is incorrect.

    (B) As $x \to \infty$, $\left(1 + \dfrac{1}{x}\right)^x \to e$.

31. The graph of $y = 2^x$ is shifted 3 units right and 1 unit down.

33. The graph of $y = \left(\dfrac{1}{3}\right)^x$ is shifted 5 units left and 10 units down.

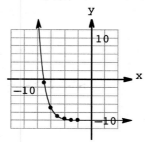

35. The graph of $y = e^x$ is shifted 2 units up.

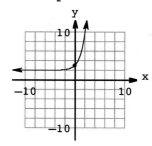

37. The graph of $y = e^x$ is reflected in the x axis, shifted 2 units left, and stretched vertically by a factor of 2.

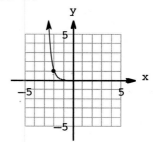

39. $5^{3x} = 5^{4x-2}$ if and only if
$$3x = 4x - 2$$
$$-x = -2$$
$$x = 2$$

41.
$$7^{x^2} = 7^{2x+3} \text{ if and only if}$$
$$x^2 = 2x + 3$$
$$x^2 - 2x - 3 = 0$$
$$(x - 3)(x + 1) = 0$$
$$x = -1, \ 3$$

43. $(1 - x)^5 = (2x - 1)^5$ if and only if
$$1 - x = 2x - 1$$
$$-3x = -2$$
$$x = \dfrac{2}{3}$$

45.
$$9^{x^2} = 3^{3x-1}$$
$$(3^2)^{x^2} = 3^{3x-1}$$
$$3^{2x^2} = 3^{3x-1}$$
if and only if
$$2x^2 = 3x - 1$$
$$2x^2 - 3x + 1 = 0$$
$$(2x - 1)(x - 1) = 0$$
$$x = \dfrac{1}{2}, \ 1$$

47. $4^{x^2} = 8^x$
$$\left(2^2\right)^{x^2} = \left(2^3\right)^x$$
$$2^{2x^2} = 2^{3x} \quad \text{if and only if}$$
$$2x^2 = 3x$$
$$2x^2 - 3x = 0$$
$$x(2x - 3) = 0$$
$$x = 0 \quad or \quad 2x - 3 = 0$$
$$2x = 3$$
$$x = 3/2$$

49. $2xe^{-x} = 0$ if $2x = 0$ or $e^{-x} = 0$.
Since e^{-x} is never 0, the only
solution is $x = 0$.

51. $x^2 e^x - 5xe^x = 0$
$xe^x(x - 5) = 0$
$x = 0$ or $e^x = 0$ or $x - 5 = 0$
never $\qquad\qquad$ $x = 5$
$x = 0, 5$

53. $\qquad a^2 = a^{-2}$
$\qquad a^2 = \dfrac{1}{a^2}$
$\qquad a^4 = 1 \quad (a \neq 0)$
$a^4 - 1 = 0$
$(a - 1)(a + 1)(a^2 + 1) = 0$
$a = 1$ or $a = -1$
This does not violate the exponential property mentioned because $a = 1$ and a
negative are excluded from consideration in the statement of the property.

55. The graph of $y = 1^x$ is a horizontal straight
line since $1^x = 1$ for all real x. This graph
is neither increasing nor decreasing and has
no asymptotes. It is not the graph of an
exponential function.

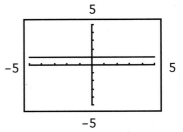

57.

x	−3	−2	−1	0	1	2	3
y	1	1	1	1	1	1	1

The resulting function is simply the constant function $y = 1$.

59. The graph of a polynomial (other than the constant polynomial $P(x) = C$ for any
real number C), doesn't have a horizontal asymptote.

61. $\dfrac{-2x^3 e^{-2x} - 3x^2 e^{-2x}}{x^6} = \dfrac{x^2 e^{-2x}(-2x - 3)}{x^6} = \dfrac{e^{-2x}(-2x - 3)}{x^4}$

63. $(e^x + e^{-x})^2 + (e^x - e^{-x})^2 = (e^x)^2 + 2(e^x)(e^{-x}) + (e^{-x})^2 + (e^x)^2 - 2(e^x)(e^{-x}) + (e^{-x})^2$
$\qquad\qquad\qquad\qquad\qquad = e^{2x} + 2 + e^{-2x} + e^{2x} - 2 + e^{-2x}$
$\qquad\qquad\qquad\qquad\qquad = 2e^{2x} + 2e^{-2x}$

> **Common Errors:**
> $(e^x)^2 \neq e^{x^2}$
> $e^{2x} + e^{2x} \neq e^{4x}$

65. Examining the graph of $y = f(x)$, we obtain

There are no local extrema and no x intercepts. The y intercept is 2.14.
As $x \to -\infty$, $y \to 2$, so the line $y = 2$ is a horizontal asymptote.

67. Examining the graph of $y = m(x)$, we obtain

There is a local minimum at $m(0) = 1$, and 1 is the y intercept. There is no x intercept and no horizontal asymptote.

69. Examining the graph of $y = S(x)$, we obtain

There is a local maximum at $S(0) = 1$, and 1 is the y intercept. There is no x intercept. As $x \to \infty$ or $x \to -\infty$, $y \to 0$, so the line $y = 0$ (the x axis) is a horizontal asymptote.

71. Examining the graph of $y = F(x)$, we obtain

There are no local extrema and no x intercepts.
When $x = 0$, $F(0) = \dfrac{200}{1 + 3e^{-0}} = 50$ is the y intercept. As $x \to -\infty$, $y \to 0$, so the line $y = 0$ (the x axis) is a horizontal asymptote. As $x \to \infty$, $y \to 200$, so the line $y = 200$ is also a horizontal asymptote.

73.

The local maximum is $m(0.91) \approx 2.67$. The x intercept is -0.55, and the y intercept is 2. The line $y = 2$ appears to be a horizontal asymptote. As $x \to -\infty$, $m(x) \to -\infty$.

75.

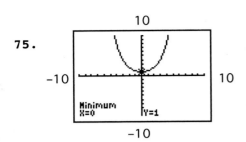

The local minimum is $f(0) = 1$, so zero is the y intercept. There are no x intercepts or horizontal asymptotes; $f(x) \to \infty$ as $x \to \infty$ and $x \to -\infty$.

77. Examining the graph of $y = f(x)$, we obtain

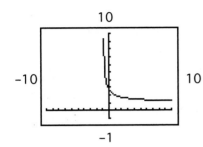

As $x \to 0$, $f(x) = (1 + x)^{1/x}$ seems to approach a value near 3. A table of values near $x = 0$ yields

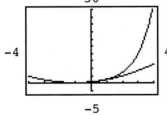

Although $f(0)$ is not defined, as $x \to 0$, $f(x)$ seems to approach a number near 2.718. In fact, it approaches e, since as $x \to 0$, $u = \dfrac{1}{x} \to \infty$, and

$$f(x) = \left(1 + \frac{1}{u}\right)^u \text{ must approach } e \text{ as } u \to \infty.$$

79.

81.

83. Here are graphs of $f_1(x) = \dfrac{x}{e^x}$, $f_2(x) = \dfrac{x^2}{e^x}$, and $f_3(x) = \dfrac{x^3}{e^x}$. In each case as $x \to \infty$, $f_n(x) \to 0$. The line $y = 0$ is a horizontal asymptote.

As $x \to -\infty$, $f_1(x) \to -\infty$ and $f_3(x) \to -\infty$, while $f_2(x) \to \infty$. It appears that as $x \to -\infty$, $f_n(x) \to \infty$ if n is even and $f_n(x) \to -\infty$ if n is odd.

f_1:

f_2:

f_3:

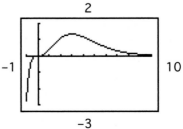

As confirmation of these observations, we show the graph of

$$f_4 = \frac{x^4}{e^x} \text{ (not required)}.$$

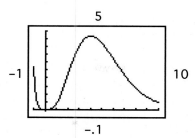

85. Make a table of values, substituting in each requested x value:

x	1.4	1.41	1.414	1.4142	1.41421	1.414214
2^x	2.639016	2.657372	2.664750	2.665119	2.665138	2.665145

The approximate value of $2^{\sqrt{2}}$ is 2.665145 to six decimal places. Using a calculator to compute directly, we get 2.665144.

87. (A) We can use the compound interest formula. Since interest is compounded weekly, m (the number of times per year interest is compounded) is 52. We're given a principal of $4,000 and an interest rate of 11%, which corresponds to $r = 0.11$. After ½ year, interest will have been compounded 26 times.

$$A = P\left(1 + \frac{r}{m}\right)^n ; \quad m = 52; \quad P = 4,000; \quad r = 0.11; \quad n = 26$$

$$A = 4,000\left(1 + \frac{0.11}{52}\right)^{26} = \$4,225.92$$

(B) Everything is the same except n, which in 10 years is $52 \cdot 10 = 520$.

$$A = 4,000\left(1 + \frac{0.11}{365/7}\right)^{520} = \$12,002.71$$

89. We use the Continuous Compound Interest Formula
$A = Pe^{rt}$
$P = 5,250$ $r = 0.1138$ $A = 5,250e^{0.1138t}$
(A) $t = 6.25$ $A = 5,250e^{0.1138(6.25)} = \$10,691.81$
(B) $t = 17$ $A = 5,250e^{0.1138(17)} = \$36,336.69$

91. We use the compound interest formula

$$A = P\left(1 + \frac{r}{m}\right)^n$$

For the first account, $P = 3,000$, $r = 0.08$, $m = 365$. Let y_1 be the amount in the first account; then
$y_1 = 3,000(1 + 0.08/365)^x$ where x is the number of compounding periods (days).
For the second account, $P = 5,000$, $r = 0.05$, $m = 365$. Let y_2 be the amount in the second account; then
$y_2 = 5,000(1 + 0.05/365)^x$

Examining the graphs of y_1 and y_2, we obtain:

X	Y₁	Y₂
6215	11712	11714
6216	11715	11715
6217	11718	11717
6218	11720	11718
6219	11723	11720
6220	11725	11722
6221	11728	11723
X=6218		

The graphs intersect at $x = 6216.15$ days.
Comparing the amounts in the accounts, we see that the first account is worth more than the second for $x \geq 6,217$ days.

93. We use the compound interest formula

$$A = P\left(1 + \frac{r}{m}\right)^n$$

For the first account, $P = 10,000$, $r = 0.089$, $m = 365$. Let y_1 be the amount in the first account, then $y_1 = 10,000(1 + 0.089/365)^x$ where x is the number of compounding periods (days). For the second account, $P = 10,000$, $r = 0.09$, $m = 4$. Let y_2 be the amount in the second account, then $y_2 = 10,000(1 + 0.09/4)^{4x/365}$ where x is the number of days. Examining the graphs of y_1 and y_2, we obtain the graph at the right.

The two graphs are indistinguishable from one another. Examining a table of values, we obtain:

X	Y₁	Y₂
0	10000	10000
91.25	10225	10225
182.5	10455	10455
273.75	10690	10690
365	10931	10931
456.25	11177	11177
547.5	11428	11428

X=0

X	Y₁	Y₂
1186.3	13354	13354
1277.5	13654	13655
1368.8	13961	13962
1460	14275	14276
1551.3	14597	14597
1642.5	14925	14926
1733.75	15261	15262

X=1733.75

The two accounts are extremely close in value, but after 1,277.5 days (14 quarters) the second account begins to be noticeably larger than the first. The first account will never be larger than the second.

95. We use the Continuous Compound Interest Formula and solve for P to obtain a formula for finding the principal.

$A = Pe^{rt}$

$P = \dfrac{A}{e^{rt}}$ or $P = Ae^{-rt}$

$A = 30,000$ $r = 0.09$ $t = 10$

$P = 30,000e^{(-0.09)(10)}$

$P = \$12,197.09$

97. Gill Savings: Use the Continuous Compound Interest Formula

$A = Pe^{rt}$ $P = 1,000$ $r = 0.083$ $t = 2.5$

$A = 1,000e^{(0.083)(2.5)}$

$A = \$1,230.60$

Richardson S & L: Use the Compound Interest Formula

$A = P\left(1 + \dfrac{r}{m}\right)^n$ $P = 1,000$ $r = 0.084$ $n = (4)(2.5)$

$A = 1,000\left(1 + \dfrac{0.084}{4}\right)^{(4)(2.5)}$

$A = \$1,231.00$

U.S.A. Savings: Use the Compound Interest Formula

$A = P\left(1 + \dfrac{r}{m}\right)^n$ $P = 1,000$ $r = 0.0825$ $n = (365)(2.5)$

$A = 1,000\left(1 + \dfrac{0.0825}{365}\right)^{(365)(2.5)}$

$A = \$1,229.03$

99. Use the compound interest formula with $r = 0.0825$, $m = 365$ (daily means interest is compounded 365 times per year), $A = 100,000$ (we're given \$100,000 as the amount in the account after a certain amount of time), and $n = 365(17) = 6,205$ (interest compounded 365 times per year for 17 years).

$$A = P\left(1 + \frac{r}{m}\right)^n \quad \Rightarrow \quad 100,000 = P\left(1 + \frac{0.0825}{365}\right)^{6,205}$$

Now solve for P:

$$P = \frac{100,000}{\left(1 + \dfrac{0.0825}{365}\right)^{6,205}} = \$24,602$$

Section 4-2

1. Answers will vary.   **3.** Answers will vary.

5. Use the doubling time model $A = A_0(2)^{t/d}$ with $A_0 = 200$, $d = 5$.

$A = 200(2)^{t/5}$

7. Use the continuous growth model $A = A_0 e^{rt}$ with $A_0 = 2,000$, $r = 0.02$.

$A = 2,000 e^{0.02t}$

9. Use the half-life model $A = A_0\left(\dfrac{1}{2}\right)^{t/h}$ with $A_0 = 100$, $h = 6$.

$A = 100\left(\dfrac{1}{2}\right)^{t/6}$

11. Use the exponential decay model $A = A_0 e^{-kt}$ with $A_0 = 4$, $k = 0.124$.

$A = 4e^{-0.124t}$

13.

n	L
1	2
2	4
3	8
4	16
5	32
6	64
7	128
8	256
9	512
10	1,024

15. Use the doubling time model $A = A_0\left(\dfrac{1}{2}\right)^{t/h}$ with $A_0 = 10$, $d = 2.4$

$A = 10(2)^{t/2.4}$ (Notice that the doubling time is given in days, so we have to convert any times given to days.)

(A) Plug in $t = 7$ (1 week is 7 days): $A = 10(2)^{7/2.4} = 75.5$
In one week there will be about 76 flies.

(B) Plug in $t = 14$ (2 weeks is 14 days): $A = 10(2)^{14/2.4} = 570.2$
In two weeks there will be about 570 flies.

17. Use the doubling time model $A = A_0\left(\dfrac{1}{2}\right)^{t/h}$ with $A_0 = 2{,}200$, $d = 2$.

$A = 2{,}200(2)^{t/2}$ where t is years after 1970.

(A) For $t = 20$: $A = 2{,}200(2)^{20/2} = 2{,}252{,}800$

(B) For $t = 35$: $A = 2{,}200(2)^{35/2} = 407{,}800{,}360$

19. Use the half-life model $A = A_0\left(\dfrac{1}{2}\right)^{t/h}$ with $A_0 = 25$, $h = 12$

$A = 25\left(\dfrac{1}{2}\right)^{t/12}$

(A) For $t = 5$, $A = 25\left(\dfrac{1}{2}\right)^{5/12} = 19$ pounds

(B) For $t = 20$, $A = 25\left(\dfrac{1}{2}\right)^{20/12} = 7.9$ pounds

21. We use the Unlimited Growth model
$A = A_0 e^{kt}$
$A_0 = 6.5 \times 10^9$ $r = 0.0114$ $t = 10$
$A = 6.5 \times 10^9 e^{(0.0114)(10)}$
$A = 7.3 \times 10^9$, 7.3 billion

23. We use the Unlimited Growth model.
$A = A_0 e^{kt}$
For Russia, $A_0 = 1.43 \times 10^8$, $r = -0.0037$, $A_1 = 1.43 \times 10^8 e^{-0.0037t}$
For Nigeria, $A_0 = 1.29 \times 10^8$, $r = 0.0256$, $A_2 = 1.29 \times 10^8 e^{0.0256t}$
Below is a graph of A_1 and A_2.

From the graph, assuming $t = 0$ in 2005, it appears that the two populations will be equal when t is approximately 3.5, in 2008. After that the population of Nigeria will be greater than that of Russia.

25. A table of values can be generated by a graphing calculator and yields

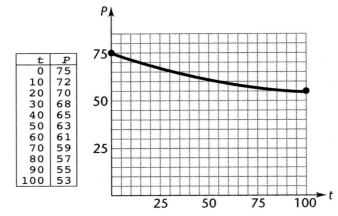

t	P
0	75
10	72
20	70
30	68
40	65
50	63
60	61
70	59
80	57
90	55
100	53

27. $I = I_0 e^{-0.00942d}$

    (A) $d = 50$ $I = I_0 e^{-0.00942(50)} = 0.62I_0$ 62%

    (B) $d = 100$ $I = I_0 e^{-0.00942(100)} = 0.39I_0$ 39%

29. (A) We use the Unlimited Growth model $A = A_0 e^{kt}$ with $A_0 = 39.4$ million,
       $k = 0.032$,
       and $t = 6$ since 2010 is 6 years after 2004.
          $A = 39.4 \times 10^6 e^{0.032(6)}$
          $A = 47.7$ million people living with HIV in 2010.

    (B) Here $A_0 = 39.4$ million, $k = 0.032$, and $t = 11$ since 2015 is 11 years
       after 2004.
          $A = 39.4 \times 10^6 e^{0.032(11)}$
          $A = 56.0$ million people living with HIV in 2015.

31. $T = T_m + (T_0 - T_m) e^{-kt}$
    $T_m = 40°$ $T_0 = 72°$ $k = 0.4$ $t = 3$
    $T = 40 + (72 - 40) e^{-0.4(3)}$
    $T = 50°$

33. As t increases without bound, $e^{-0.2t}$ approaches 0, so $q = 0.0009(1 - e^{-0.2t})$
approaches 0.0009. This tells us that 0.0009 coulomb is the maximum charge on
the capacitor.

35. (A) Examining the graph of $N(t)$, we obtain the graphs below.

After 2 years, 25 deer will be
present. After 6 years, 37 deer will
be present.

    (B) Applying the INTERSECT command, we obtain the graph
    at the right.

    It will take 10 years for the herd to grow to 50 deer.

    (C) As t increases without bound, $e^{-0.14t}$ approaches 0,

    and $N = \dfrac{100}{1 + 4e^{-0.14t}}$ approaches 100. This tells us that

    100 is the number of deer the island can support.

37. The independent variable is years after purchase, so enter 1, 2, 3, 4, 5, and 6
as L_1.   The dependent variable is value, so enter the Value column as L_2.   Then
use the exponential regression command from the STAT CALC menu.

The model gives $y = 14,910(0.8163)^x$.

When $x = 0$, $y = \$14,910$ is the estimated purchase price. Applying the VALUE command, we obtain the graph at the right.

When $x = 10$, the estimated value of the van is $\$1,959$.

39. (A) The independent variable is years since 1980, so enter 0, 5, 10, 15, 18, and 19 as L_1. The dependent variable is power generation in North America, so enter the North America column as L_2. Then use the logistic regression command from the STAT CALC menu.

The model is

$$y = \frac{842}{1 + 2.08e^{-0.187x}}$$

(B) 2010 is 30 years after 1980, so we evaluate our model for $x = 30$ using the VALUE command.

In 2010, approximately 836 billion kilowatt hours will be generated.

Section 4-3

1. Answers will vary.   **3.** Answers will vary.   **5.** Answers will vary.

7. $81 = 3^4$ **9.** $0.001 = 10^{-3}$ **11.** $\dfrac{1}{36} = 6^{-2}$ **13.** $\log_4 8 = \dfrac{3}{2}$

15. $\log_{32} \dfrac{1}{2} = -\dfrac{1}{5}$ **17.** $\log_{2/3} \dfrac{8}{27} = 3$

19. Make a table of values for each function:

x	$f(x) = 3^x$	x	$f^{-1}(x) = \log_3 x$
-3	1 / 27	1 / 27	-3
-2	1 / 9	1 / 9	-2
-1	1 / 3	1 / 3	-1
0	1	1	0
1	3	3	1
2	9	9	2
3	27	27	3

21. Make a table of values for each function:

x	$f(x) = (2/3)^x$		x	$f^{-1}(x) = \log_{2/3} x$
-3	$27/8$		$27/8$	-3
-2	$9/4$		$9/4$	-2
-1	$3/2$		$2/3$	-1
0	1		1	0
1	$2/3$		$2/3$	1
2	$4/9$		$4/9$	2
3	$8/27$		$8/27$	3

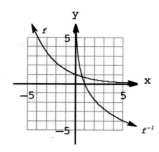

23. 0   **25.** 1   **27.** 4

29. $\log_{10} 0.01 = \log_{10} 10^{-2} = -2$ **31.** $\log_3 27 = \log_3 3^3 = 3$

33. $\log_{1/2} 2 = \log_{1/2} \left(\frac{1}{2}\right)^{-1} = -1$ **35.** 5   **37.** $\log_5 \sqrt[3]{5} = \log_5 5^{1/3} = \frac{1}{3}$

39. 4.6923   **41.** 3.9905

43. $\log_7 13 = \dfrac{\ln 13}{\ln 7} = 1.3181$ using the change of base formula

45. $\log_5 120.24 = \dfrac{\ln 120.24}{\ln 5} = 2.9759$ using the change of base formula

47. $x = 10^{5.3027} = 200,800$ **49.** $x = 10^{-3.1773} = 6.648 \times 10^{-4} = 0.000\ 6648$

51. $x = e^{3.8655} = 47.73$ **53.** $x = e^{-0.3916} = 0.6760$

55. Write $\log_2 x = 2$ in equivalent exponential form. $x = 2^2 = 4$

57. $\log_4 16 = \log_4 4^2 = 2$
$$y = 2$$

59. Write $\log_b 16 = 2$ in equivalent exponential form.
$$16 = b^2$$
$$b^2 = 16$$
$b = 4$ since bases are required to be positive

61. Write $\log_b 1 = 0$ in equivalent exponential form: $1 = b^0$
This statement is true if b is any real number except 0. However, bases are required to be positive and 1 is not allowed, so the original statement is true if b is any positive real number except 1.

63. Write $\log_4 x = \frac{1}{2}$ in equivalent exponential form: $x = 4^{1/2} = 2$

65. $\log_{1/3} 9 = \log_{1/3} 3^2 = \log_{1/3} \dfrac{1}{\left(\frac{1}{3}\right)^2} = \log_{1/3} \left(\frac{1}{3}\right)^{-2} = -2$

67. Write $\log_b 1,000 = \frac{3}{2}$ in equivalent exponential form:
$$1,000 = b^{3/2}$$
$$10^3 = b^{3/2}$$
$(10^3)^{2/3} = (b^{3/2})^{2/3}$ (If two numbers are equal the results are equal if they are raised to the same exponent.)
$$10^{3(2/3)} = b^{3/2(2/3)}$$
$$10^2 = b$$
$$b = 100$$

69. Write $\log_8 x = -\dfrac{4}{3}$ in equivalent exponential form.

$$8^{-4/3} = x$$

$$x = (8^{1/3})^{-4} = 2^{-4} = \dfrac{1}{16}$$

71. Write $\log_{16} 8 = y$ in equivalent exponential form.

$$16^y = 8$$
$$(2^4)^y = 2^3$$
$$2^{4y} = 2^3 \text{ if and only if}$$
$$4y = 3$$
$$y = \dfrac{3}{4}$$

73. 4.959   **75.** 7.861   **77.** 2.280

79. $\log\left(\dfrac{x}{y}\right) = \log x - \log y$

81. $\log\left(x^4 y^3\right) = \log x^4 + \log y^3 = 4 \log x + 3 \log y$

83. $\ln x - \ln y = \ln\left(\dfrac{x}{y}\right)$

85. $2 \ln x + 5 \ln y - \ln z = \ln x^2 + \ln y^5 - \ln z = \ln\left(\dfrac{x^2 y^5}{z}\right)$

87. $\log(xy) = \log x + \log y = -2 + 3 = 1$

89. $\log\left(\dfrac{\sqrt{x}}{y^3}\right) = \log x^{1/2} - \log y^3 = \dfrac{1}{2} \log x - 3 \log y = \dfrac{1}{2}(-2) - 3(3) = -1 - 9 = -10$

91. The graph of g is the same as the graph of f shifted upward 3 units; g is increasing.
Domain: $(0, \infty)$
Vertical asymptote: $x = 0$

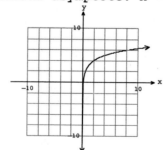

93. The graph of g is the same as the graph of f shifted 2 units to the right; g is decreasing.
Domain: $(2, \infty)$
Vertical asymptote: $x = 2$

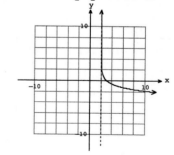

95. The graph of g is the same as the graph of f reflected through the x axis and shifted downward 1 unit; g is decreasing.
Domain: $(0, \infty)$
Vertical asymptote: $x = 0$

97. The graph of g is the same as the graph of f reflected through the x axis, stretched vertically by a factor of 3, and shifted upward 5 units. g is decreasing.
Domain: $(0, \infty)$
Vertical asymptote: $x = 0$

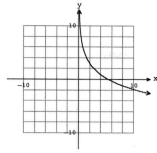

99. Write $y = \log_5 x$
In exponential form:
$\quad 5^y = x$
Interchange x and y:
$\quad 5^x = y$
Therefore $f^{-1}(x) = 5^x$.

101. Write $y = 4 \log_3(x + 3)$
$$\frac{y}{4} = \log_3(x + 3)$$
In exponential form:
$\quad 3^{y/4} = x + 3$
$\quad x = 3^{y/4} - 3$
Interchange x and y:
$\quad y = 3^{x/4} - 3$
Therefore $f^{-1}(x) = 3^{x/4} - 3$

103. (A) Write $y = \log_3(2 - x)$
In exponential form:
$\quad 3^y = 2 - x$
$\quad x = 2 - 3^y$
Interchange x and y:
$\quad y = 2 - 3^x$
Therefore, $f^{-1}(x) = 2 - 3^x$

(B) The graph is the same as the graph of $y = 3^x$ reflected through the x axis and shifted 2 units upward.

(C)

105. The inequality sign in the last step reverses because $\log \frac{1}{3}$ is negative.

107.

109.

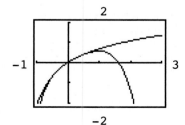

111. Let $u = \log_b M$ and $v = \log_b N$. Changing each equation to exponential form, $b^u = M$ and $b^v = N$. Then we can write M/N as

$$\frac{M}{N} = \frac{b^u}{b^v} = b^{u-v}$$ using a familiar property of exponents. Now change this equation to logarithmic form:

$$\log_b\left(\frac{M}{N}\right) = u - v$$

Finally, recall the way we defined u and v in the first line of our proof:

$$\log_b\left(\frac{M}{N}\right) = \log_b M - \log_b N$$

Section 4-4

1. Answers will vary.

3. Answers will vary.

5. We use the decibel formula
$$D = 10 \log \frac{I}{I_0}$$

   (A) $I = I_0$

   $$D = 10 \log \frac{I_0}{I_0}$$
   $$D = 10 \log 1$$
   $$D = 0 \text{ decibels}$$

   (B) $I_0 = 1.0 \times 10^{-12}$ $I = 1.0$

   $$D = 10 \log \frac{1.0}{1.0 \times 10^{-12}}$$
   $$D = 120 \text{ decibels}$$

7. Use the decibel formula $D = 10 \log \dfrac{I}{I_0}$:

   Let I_1 be the intensity of the first sound, and let I_2 be the intensity of the second sound. Then $I_2 = 1,000 I_1$.

   $$D_1 = 10 \log \frac{I}{I_0}; \quad D_2 = 10 \log \frac{I_2}{I_0}$$

   $$
   \begin{aligned}
   D_2 - D_1 &= 10 \log \frac{I_2}{I_0} - 10 \log \frac{I_1}{I_0} \\
   &= 10 \log\left(\frac{I_2/I_0}{I_1/I_0}\right) \\
   &= 10 \log \frac{I_2}{I_1} \\
   &= 10 \log \frac{1,000 I_1}{I_1} \\
   &= 10 \log 1,000 \\
   &= 30 \text{ decibels}
   \end{aligned}
   $$

9. We use the magnitude formula
$$M = \frac{2}{3} \log \frac{E}{E_0}$$
$$E_0 = 10^{4.40} \quad E = 1.99 \times 10^{17}$$
$$M = \frac{2}{3} \log \frac{1.99 \times 10^{17}}{10^{4.40}}$$
$$= 8.6$$

11. We use the magnitude formula
$$M = \frac{2}{3} \log \frac{E}{E_0}$$

For the Long Beach earthquake,

$$6.3 = \frac{2}{3} \log \frac{E_1}{E_0}$$

$$9.45 = \log \frac{E_1}{E_0}$$

(Change to exponential form)

$$\frac{E_1}{E_0} = 10^{9.45}$$

$$E_1 = E_0 \cdot 10^{9.45}$$

For the Anchorage earthquake,

$$8.3 = \frac{2}{3} \log \frac{E_2}{E_0}$$

$$12.45 = \log \frac{E_2}{E_0}$$

(Change to exponential form)

$$\frac{E_2}{E_0} = 10^{12.45}$$

$$E_2 = E_0 \cdot 10^{12.45}$$

Now we can compare the energy levels by dividing the more powerful (Anchorage) by the less (Long Beach):

$$\frac{E_2}{E_1} = \frac{E_0 \cdot 10^{12.45}}{E_0 \cdot 10^{9.45}} = 10^3$$

$$E_2 = 10^3 E_1, \text{ or } 1,000 \text{ times as powerful}$$

13. Use the magnitude formula $M = \frac{2}{3} \log \frac{E}{E_0}$ with $E = 1.34 \times 10^{14}$, $E_0 = 10^{4.40}$:

$$M = \frac{2}{3} \log \frac{1.34 \times 10^{14}}{10^{4.40}} = 6.5$$

15. Use the magnitude formula $M = \frac{2}{3} \log \frac{E}{E_0}$ with $E = 2.38 \times 10^{21}$, $E_0 = 10^{4.40}$:

$$M = \frac{2}{3} \log \frac{2.38 \times 10^{21}}{10^{4.40}} = 11.3$$

17. We use the rocket equation.

$$v = c \ln \frac{W_t}{W_b}$$

$$v = 2.57 \ln (19.8)$$

$$v = 7.67 \text{ km/s}$$

19. (A) $pH = -\log[H^+] = -\log(4.63 \times 10^{-9}) = 8.3$. Since this is greater than 7, the substance is basic.

(B) $pH = -\log[H^+] = -\log(9.32 \times 10^{-4}) = 3.0$ Since this is less than 7, the substance is acidic.

21. Since $pH = -\log[H^+]$, we have
$$5.2 = -\log[H^+], \text{ or}$$
$$[H^+] = 10^{-5.2} = 6.3 \times 10^{-6} \text{ moles per liter}$$

23. $m = 6 - 2.5 \log \dfrac{L}{L_0}$

(A) We find m when $L = L_0$

$m = 6 - 2.5 \log \dfrac{L_0}{L_0}$

$m = 6 - 2.5 \log 1$

$m = 6$

(B) We compare L_1 for $m = 1$ with L_2 for $m = 6$

$1 = 6 - 2.5 \log \dfrac{L_1}{L_0}$

$-5 = -2.5 \log \dfrac{L_1}{L_0}$

$2 = \log \dfrac{L_1}{L_0}$

$\dfrac{L_1}{L_0} = 10^2$

$L_1 = 100L_0$

$6 = 6 - 2.5 \log \dfrac{L_2}{L_0}$

$0 = -2.5 \log \dfrac{L_2}{L_0}$

$0 = \log \dfrac{L_2}{L_0}$

$\dfrac{L_2}{L_0} = 1$

$L_2 = L_0$

Comparing the two, $\dfrac{L_1}{L_2} = \dfrac{100L_0}{L_0} = 100$. The star of magnitude 1 is 100 times brighter.

25. (A) Enter the x values shown in the table as L_1. Enter the yield values as L_2. Use the logarithmic regression model from the STAT CALC menu.

The model is $y = -525.1353183 + 142.7911807 \ln x$. Evaluating this for $x = 103$ (year 2003) yields 136.7 bushels per acre. Evaluating for $x = 110$ (year 2010) yields 146.0 bushels per acre.

(B) The actual yield was higher than predicted by the model. Entering this as further data would tend to raise the curve slightly and increase the predicted estimate for 2010.

Section 4-5

1. Answers will vary.

3. Answers will vary.

5. Answers will vary.

7. Algebraic Solution:

$10^{-x} = 0.0347$

$-x = \log_{10} 0.0347$

$x = -\log_{10} 0.0347$

$x = 1.46$

Graphical Solution:

9. Algebraic Solution:

$10^{3x+1} = 92$

$3x + 1 = \log_{10} 92$

$3x = \log_{10} 92 - 1$

$x = \dfrac{\log_{10} 92 - 1}{3}$

$x = 0.321$

Graphical Solution:

11. Algebraic Solution:

$e^x = 3.65$

$x = \ln 3.65$

$x = 1.29$

Graphical Solution:

13. <u>Algebraic Solution:</u>
$$e^{2x-1} = 405$$
$$\ln 5^x = \ln 18$$
$$2x - 1 = \ln 405$$
$$2x = 1 + \ln 405$$
$$x = \frac{1 + \ln 405}{2}$$
$$x = 3.50$$
<u>Graphical Solution:</u>

15. <u>Algebraic Solution:</u>
$$5^x = 18$$
$$x \ln 5 = \ln 18$$
$$x = \frac{\ln 18}{\ln 5}$$
$$x = 1.80$$
<u>Graphical Solution:</u>

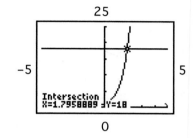

17. <u>Algebraic Solution:</u>
$$2^{-x} = 0.238$$
$$\ln 2^{-x} = \ln 0.238$$
$$-x \ln 2 = \ln 0.238$$
$$-x = \frac{\ln 0.238}{\ln 2}$$
$$x = -\frac{\ln 0.238}{\ln 2}$$
$$x = 2.07$$
<u>Graphical Solution:</u>

19. <u>Algebraic Solution:</u>
$$\log_5(2x - 7) = 2$$

Change to exponential form:
$$5^2 = 2x - 7$$
$$25 = 2x - 7$$
$$32 = 2x$$
$$x = 16$$

<u>Graphical Solution:</u> (Notice that we key in the left side using natural log and the change of base formula)

21. <u>Algebraic Solution:</u>
$$\log_3\left(x^2 - 8x\right) = 2$$

Change to exponential form:
$$3^2 = x^2 - 8x$$
$$0 = x^2 - 8x - 9$$
$$0 = \left(x - 9\right)\left(x + 1\right)$$
$$x = 9, -1$$

<u>Graphical Solution:</u> (Notice that we key in the left side using natural log and the change of base formula)

23. $\log 5 + \log x = 2$
$$\log(5x) = 2$$

Change to exponential form
$$5x = 10^2$$
$$5x = 100$$
$$x = 20$$

25. $\log x + \log(x - 3) = 1$
$$\log[x(x - 3)] = 1$$

Change to exponential form
$$x(x - 3) = 10^1$$
$$x^2 - 3x = 10$$
$$x^2 - 3x - 10 = 0$$
$$(x - 5)(x + 2) = 0$$
$$x = 5 \text{ or } -2$$

> **Common Error:**
> $\log(x - 3) \neq \log x - \log 3$

Check:

$\log 5 + \log(5 - 3) \overset{\checkmark}{=} 1$
$\log(-2) + \log(-2 - 3)$
is not defined.
Solution: $x = 5$

27. $\log(x + 1) - \log(x - 1) = 1$
$$\log \frac{x + 1}{x - 1} = 1$$

Change to exponential form
$$\frac{x + 1}{x - 1} = 10^1$$
$$\frac{x + 1}{x - 1} = 10$$
$$x + 1 = 10(x - 1)$$
$$x + 1 = 10x - 10$$
$$11 = 9x$$
$$x = \frac{11}{9}$$

> **Common Error:**
> $\frac{x + 1}{x - 1} \neq \log 1$

Check:

$\log\left(\frac{11}{9} + 1\right) - \log\left(\frac{11}{9} - 1\right) \overset{?}{=} 1$

$\log \frac{20}{9} - \log \frac{2}{9} \overset{?}{=} 1$

$\log 10 \overset{\checkmark}{=} 1$

29. $\ln(4x - 3) = \ln(x + 1)$
$$4x - 3 = x + 1$$
$$3x = 4$$
$$x = \frac{4}{3}$$

Check: $\ln\left(4\left(\frac{4}{3}\right) - 3\right) = \ln\left(\frac{16}{3} - \frac{9}{3}\right) = \ln\left(\frac{7}{3}\right)$

$\ln\left(\frac{4}{3} + 1\right) = \ln\left(\frac{4}{3} + \frac{3}{3}\right) = \ln\left(\frac{7}{3}\right)$ $\checkmark$

31. $\log_2\left(x^2 - 2x\right) = \log_2(3x - 6)$

$$x^2 - 2x = 3x - 6$$
$$x^2 - 5x + 6 = 0$$
$$(x - 3)(x - 2) = 0$$
$$x = 3, 2$$

Check: $x = 3$: $\log_2\left(3^2 - 2(3)\right) = \log_2 3$

$\log_2\left(3(3) - 6\right) = \log_2 3$ $\checkmark$

$x = 2$: $\log_2\left(2^2 - 2(2)\right) = \log_2 0$ (undefined)

The only solution is $x = 3$.

33. Algebraic Solution:
$$2 = 1.05^x$$
$$\ln 2 = \ln 1.05^x$$
$$\ln 2 = x \ln 1.05$$
$$\frac{\ln 2}{\ln 1.05} = x$$
$$x = 14.2$$

Graphical Solution:

35. <u>Algebraic Solution:</u>

$$e^{-1.4x} = 13$$
$$\ln e^{-1.4x} = \ln 13$$
$$-1.4x = \ln 13$$
$$x = \frac{\ln 13}{-1.4}$$
$$x = -1.83$$

<u>Graphical Solution:</u>

37. <u>Algebraic Solution:</u>

$$5 + 3^x = 10$$
$$3^x = 5$$
$$\ln 3^x = \ln 5$$
$$x \cdot \ln 3 = \ln 5$$
$$x = \frac{\ln 5}{\ln 3} \approx 1.46$$

<u>Graphical Solution:</u>

39. <u>Algebraic Solution:</u>

$$10^{2x+5} - 7 = 13$$
$$10^{2x+5} = 20$$
$$2x + 5 = \log 20$$
$$2x = \log 20 - 5$$
$$x = \frac{\log 20 - 5}{2} \approx -1.85$$

<u>Graphical Solution:</u>

41. <u>Algebraic Solution:</u>

$$123 = 500e^{-0.12x}$$

$$\frac{123}{500} = e^{-0.12x}$$

$$\ln\left(\frac{123}{500}\right) = \ln e^{-0.12x}$$

$$\ln\left(\frac{123}{500}\right) = -0.12x$$

$$\frac{\ln\left(\frac{123}{500}\right)}{-0.12} = x$$

$$x = 11.7$$

<u>Graphical Solution:</u>

43.

$$e^{-x^2} = 0.23$$
$$\ln e^{-x^2} = \ln 0.23$$
$$-x^2 = \ln 0.23$$
$$x^2 = -\ln 0.23$$
$$x = \pm\sqrt{-\ln 0.23}$$
$$x = \pm 1.21$$

Check:

45. $\log x - \log 5 = \log 2 - \log(x - 3)$

$$\log \frac{x}{5} = \log \frac{2}{x - 3}$$

$$\frac{x}{5} = \frac{2}{x - 3}$$

Excluded value: $x \neq 3$

$$5(x - 3)\frac{x}{5} = 5(x - 3)\frac{2}{x - 3}$$

$$(x - 3)x = 10$$

$$x^2 - 3x = 10$$

$$x^2 - 3x - 10 = 0$$
$$(x - 5)(x + 2) = 0$$
$$x = 5, -2$$

Solution: $x = 5$

Check:

$$\log 5 - \log 5 \overset{\checkmark}{=} \log 2 - \log 2$$
$$\log(-2) \text{ is not defined}$$

47. $\ln x = \ln(2x - 1) - \ln(x - 2)$

$$\ln x = \ln \frac{2x - 1}{x - 2}$$

$$x = \frac{2x - 1}{x - 2}$$

Excluded value: $x \neq 2$

$$x(x - 2) = (x - 2)\frac{2x - 1}{x - 2}$$

$$x(x - 2) = 2x - 1$$
$$x^2 - 2x = 2x - 1$$
$$x^2 - 4x + 1 = 0$$

$$x = \frac{-b \pm \sqrt{b^2 - 4ac}}{2a}$$
$$a = 1, \ b = -4, \ c = 1$$

$$x = \frac{-(-4) \pm \sqrt{(-4)^2 - 4(1)(1)}}{2(1)}$$

$$x = \frac{4 \pm \sqrt{12}}{2} = 2 \pm \sqrt{3}$$

Check:

$$\ln(2 + \sqrt{3}) \overset{?}{=} \ln[2(2 + \sqrt{3}) - 1]$$
$$- \ln[(2 + \sqrt{3}) - 2]$$

$$\ln(2 + \sqrt{3}) \overset{?}{=} \ln(3 + 2\sqrt{3}) - \ln\sqrt{3}$$

$$\ln(2 + \sqrt{3}) \overset{?}{=} \ln\left(\frac{3 + 2\sqrt{3}}{\sqrt{3}}\right)$$

$$\ln(2 + \sqrt{3}) \overset{\checkmark}{=} \ln(\sqrt{3} + 2)$$

$\ln(x - 2)$ is not defined if
$x = 2 - \sqrt{3}$
Solution: $x = 2 + \sqrt{3}$

49.

$$\log(2x + 1) = 1 - \log(x - 1)$$
$$\log(2x + 1) + \log(x - 1) = 1$$
$$\log[(2x + 1)(x - 1)] = 1$$
$$(2x + 1)(x - 1) = 10$$
$$2x^2 - x - 1 = 10$$
$$2x^2 - x - 11 = 0$$

$$x = \frac{-b \pm \sqrt{b^2 - 4ac}}{2a}$$
$$a = 2, \ b = -1, \ c = -11$$

$$x = \frac{-(-1) \pm \sqrt{(-1)^2 - 4(2)(-11)}}{2(2)}$$

$$x = \frac{1 \pm \sqrt{89}}{4}$$

Check: $\log\left(2\,\dfrac{1 + \sqrt{89}}{4} + 1\right) \overset{?}{=} 1 - \log\left(\dfrac{1 + \sqrt{89}}{4} - 1\right)$

$$\log\left(\dfrac{1 + \sqrt{89} + 2}{2}\right) \overset{?}{=} 1 - \log\left(\dfrac{1 + \sqrt{89} - 4}{4}\right)$$

$$\log\left(\dfrac{3 + \sqrt{89}}{2}\right) \overset{?}{=} 1 - \log\left(\dfrac{\sqrt{89} - 3}{4}\right)$$

$$\log\left(\dfrac{3 + \sqrt{89}}{2}\right) \overset{?}{=} \log 10 - \log\left(\dfrac{\sqrt{89} - 3}{4}\right)$$

$$\overset{?}{=} \log\left(\dfrac{40}{\sqrt{89} - 3}\right)$$

$$\overset{?}{=} \log\left[\dfrac{40\left(\sqrt{89} + 3\right)}{89 - 9}\right]$$

$$\overset{\surd}{=} \log\left(\dfrac{\sqrt{89} + 3}{2}\right)$$

$\log(x - 1)$ is not defined if $x = \dfrac{1 - \sqrt{89}}{4}$. Solution: $x = \dfrac{1 + \sqrt{89}}{4}$.

51.
$$(\ln x)^3 = \ln x^{\,4}$$
$$(\ln x)^3 = 4 \ln x$$
$$(\ln x)^3 - 4 \ln x = 0$$
$$\ln x[(\ln x)^2 - 4] = 0$$
$$\ln x(\ln x - 2)(\ln x + 2) = 0$$

$$\ln x = 0 \qquad \ln x - 2 = 0 \qquad \ln x + 2 = 0$$
$$x = 1 \qquad\quad \ln x = 2 \qquad\quad \ln x = -2$$
$$x = e^2 \qquad\quad x = e^{-2}$$

Check:

$(\ln 1)^3 \overset{?}{=} \ln 1^4 \qquad (\ln e^2)^3 \overset{?}{=} \ln(e^2)^4 \qquad (\ln e^{-2})^3 \overset{?}{=} \ln(e^{-2})^4$

$0 \overset{\surd}{=} 0 \qquad\qquad 8 \overset{\surd}{=} 8 \qquad\qquad\qquad -8 \overset{\surd}{=} -8$

Solution: $x = 1,\ e^2,\ e^{-2}$

53. $\ln(\ln x) = 1$
Change to
exponential form
$$\ln x = e^1$$
$$\ln x = e$$
$$x = e^e$$

55. $x^{\log x} = 100x$
We start by taking logarithms of both sides.
$$\log(x^{\log x}) = \log 100x$$
$$\log x \log x = \log 100 + \log x$$
$$(\log x)^2 = 2 + \log x$$
$$(\log x)^2 - \log x - 2 = 0$$
$$(\log x - 2)(\log x + 1) = 0$$
$$\log x - 2 = 0 \qquad \log x + 1 = 0$$
$$\log x = 2 \qquad\quad \log x = -1$$
$$x = 10^2 \qquad\quad x = 10^{-1}$$
$$x = 100 \qquad\quad x = 0.1$$

Check: $100^{\log 100} \overset{?}{=} 100 \cdot 100 \qquad (0.1)^{\log(0.1)} \overset{?}{=} 100(0.1)$

$10^4 \overset{\surd}{=} 10^4 \qquad\qquad\qquad 0.1^{-1} \overset{?}{=} 10$

$10 \overset{\surd}{=} 10$

57. (A) In solving an exponential equation, we often can get the variable out of the exponent by using logarithms. If we do that in this case, however, we have made no progress:

$$e^{x/2} = 5 \ln x$$
$$\ln e^{x/2} = \ln(5 \ln x)$$
$$\frac{x}{2} = \ln 5 + \ln(\ln x)$$

This equation is no easier than the original to solve. Other algebraic methods seem to be equally useless.

(B) Examining the graphs of $y = e^{x/2}$ and $y = 5 \ln x$, it appears that there are two solutions of the equation.

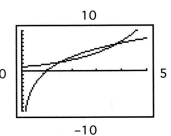

59. (A) In solving an exponential equation, we often can get the variable out of the exponent by using logarithms. If we do that in this case, however, we have made no progress:

$$3^x + 2 = 7 + x - e^{-x}$$
$$3^x = -e^{-x} + x + 5$$
$$\log_3 3^x = \log_3(-e^{-x} + x + 5)$$
$$x = \log_3(-e^{-x} + x + 5)$$

This equation is no easier than the original to solve.

(B) Graphing and applying a built-in routine, we obtain

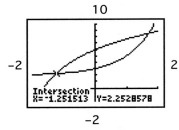

There are two solutions.

61.
$$A = Pe^{rt}$$
$$\frac{A}{P} = e^{rt}$$
$$\ln \frac{A}{P} = rt$$
$$\frac{1}{t} \ln \frac{A}{P} = r$$
$$r = \frac{1}{t} \ln \frac{A}{P}$$

63.
$$D = 10 \log \frac{I}{I_0}$$
$$\frac{D}{10} = \log \frac{I}{I_0}$$
$$\frac{I}{I_0} = 10^{D/10}$$
$$I = I_0(10^{D/10})$$

65.
$$M = 6 - 2.5 \log \frac{I}{I_0}$$
$$6 - M = 2.5 \log \frac{I}{I_0}$$
$$\frac{6 - M}{2.5} = \log \frac{I}{I_0}$$
$$\frac{I}{I_0} = 10^{(6-M)/2.5}$$
$$I = I_0[10^{(6-M)/2.5}]$$

67.

$$I = \frac{E}{R}(1 - e^{-Rt/L})$$

$$RI = E(1 - e^{-Rt/L})$$

$$\frac{RI}{E} = 1 - e^{-Rt/L}$$

$$\frac{RI}{E} - 1 = -e^{-Rt/L}$$

$$-\left(\frac{RI}{E} - 1\right) = e^{-Rt/L}$$

$$-\frac{RI}{E} + 1 = e^{-Rt/L}$$

$$1 - \frac{RI}{E} = e^{-Rt/L}$$

$$\ln\left(1 - \frac{RI}{E}\right) = -\frac{Rt}{L}$$

$$-\frac{L}{R}\ln\left(1 - \frac{RI}{E}\right) = t$$

$$t = -\frac{L}{R}\ln\left(1 - \frac{RI}{E}\right)$$

69.

$$y = \frac{e^x + e^{-x}}{2}$$

$$2y = e^x + e^{-x}$$

$$2y = e^x + \frac{1}{e^x}$$

$$2ye^x = (e^x)^2 + 1$$

$$0 = (e^x)^2 - 2ye^x + 1$$

This equation is quadratic in e^x

$$e^x = \frac{-b \pm \sqrt{b^2 - 4ac}}{2a}$$

$$a = 1, \ b = -2y, \ c = 1$$

$$e^x = \frac{-(-2y) \pm \sqrt{(-2y)^2 - 4(1)(1)}}{2(1)}$$

$$e^x = \frac{2y \pm \sqrt{4y^2 - 4}}{2}$$

$$e^x = \frac{2(y \pm \sqrt{y^2 - 1})}{2}$$

$$e^x = y \pm \sqrt{y^2 - 1}$$

$$x = \ln(y \pm \sqrt{y^2 - 1})$$

71.

$$y = \frac{e^x - e^{-x}}{e^x + e^{-x}}$$

$$y = \frac{e^x - \frac{1}{e^x}}{e^x + \frac{1}{e^x}}$$

$$y = \frac{e^x e^x - \frac{1}{e^x} e^x}{e^x e^x + \frac{1}{e^x} e^x}$$

$$y = \frac{e^{2x} - 1}{e^{2x} + 1}$$

$$y(e^{2x} + 1) = e^{2x} - 1$$

$$ye^{2x} + y = e^{2x} - 1$$

$$1 + y = e^{2x} - ye^{2x}$$

$$1 + y = (1 - y)e^{2x}$$

$$e^{2x} = \frac{1 + y}{1 - y}$$

$$2x = \ln\frac{1 + y}{1 - y}$$

$$x = \frac{1}{2}\ln\frac{1 + y}{1 - y}$$

73. Graphing $y = 2^{-x} - 2x$ and applying the ZERO command, we obtain

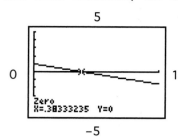

The required solution of $2^{-x} - 2x = 0$, $0 \le x \le 1$, is $x = 0.38$.

75. Graphing $y = x3^x - 1$ and applying the ZERO command, we obtain

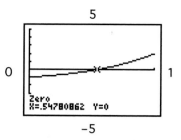

The required solution of $x3^x - 1 = 0$, $0 \le x \le 1$, is $x = 0.55$.

77. Graphing $y = e^{-x} - x$ and applying the ZERO command, we obtain

The required solution of $e^{-x} - x = 0$, $0 \le x \le 1$, is $x = 0.57$.

79. Graphing $y = xe^x - 2$ and applying the ZERO command, we obtain

The required solution of $xe^x - 2 = 0$, $0 \le x \le 1$, is $x = 0.85$.

81. Graphing $y = \ln x + 2x$ and applying the ZERO command, we obtain

The required solution of $\ln x + 2x = 0$, $0 \le x \le 1$, is 0.43.

83. Graphing $y = \ln x + e^x$ and applying the ZERO command, we obtain

The required solution of $\ln x + e^x = 0$, $0 \le x \le 1$, is 0.27.

85. To find the doubling time we replace A in $A = P(1 + 0.15)^n$ with $2P$ and solve for n.

$$2P = P(1.15)^n$$
$$2 = (1.15)^n$$
$$\ln 2 = n \ln 1.15$$
$$n = \frac{\ln 2}{\ln 1.15}$$

$n = 5$ years to the nearest year

87. We solve $A = Pe^{rt}$ for r, with $A = 2{,}500$, $P = 1{,}000$, $t = 10$

$$2{,}500 = 1{,}000e^{r(10)}$$
$$2.5 = e^{10r}$$
$$10r = \ln (2.5)$$
$$r = \frac{1}{10} \ln 2.5$$
$$r = 0.0916 \text{ or } 9.16\%$$

89. (A) We're given $P_0 = 10.5$ (we could use 10.5 million, but if you look carefully at the calculations below, you'll see that the millions will cancel out anyhow), 11.3 for P, and 2 for t (since May 2007 is two years after May 2005).

$$11.3 = 10.5e^{r \cdot 2}$$
$$\frac{11.3}{10.5} = e^{2r}$$
$$\ln\left(\frac{11.3}{10.5}\right) = 2r$$
$$r = \frac{\ln\left(\frac{11.3}{10.5}\right)}{2} \approx 0.0367$$

The annual growth rate is 3.67%.

(B) $P = 10.5e^{0.0367t}$; plug in 20 for P and solve for t.

$$20 = 10.5e^{0.0367t}$$
$$\frac{20}{10.5} = e^{0.0367t}$$
$$\ln\left(\frac{20}{10.5}\right) = 0.0367t$$
$$t = \frac{\ln\left(\frac{20}{10.5}\right)}{0.0367} \approx 17.6$$

The illegal immigrant population is predicted to reach 20 million near the end of 2022, which is 17.6 years after May 2005.

91. We solve $P = P_0 e^{rt}$ for t with $P = 2P_0$, $r = 0.0114$.

$$2P_0 = P_0 e^{0.0114t}$$
$$2 = e^{0.0114t}$$
$$\ln 2 = 0.0114t$$
$$t = \frac{\ln 2}{0.0114}$$

$t = 61$ years to the nearest year

93. We're given $A_0 = 5$, $A = 1$, $t = 6$:

$$A = A_0\left(\frac{1}{2}\right)^{t/h}$$
$$1 = 5\left(\frac{1}{2}\right)^{6/h}$$
$$\frac{1}{5} = \left(\frac{1}{2}\right)^{6/h}$$
$$\ln\frac{1}{5} = \ln\left(\frac{1}{2}\right)^{6/h}$$
$$\ln\frac{1}{5} = \frac{6}{h}\ln\left(\frac{1}{2}\right)$$
$$h \cdot \ln\frac{1}{5} = 6\ln\left(\frac{1}{2}\right)$$
$$h = \frac{6\ln(1/2)}{\ln(1/5)} \approx 2.58$$

The half-life is about 2.58 hours.

95. Let A_0 represent the amount of Carbon-14 originally present. Then the amount left in 2003 was $0.289A_0$. Plug this in for A, and solve for t:

$$0.289A_0 = A_0 e^{-0.000\ 124t}$$

$$0.289 = e^{-.000\ 124t}$$

$$\ln 0.289 = \ln e^{-.000\ 124t}$$
$$\ln 0.289 = -.000\ 124t$$

$$t = \frac{\ln 0.289}{-.000\ 124} \approx 10,010$$

The sample was about 10,010 years old.

97. Let A_0 represent the amount of Carbon-14 originally present. Then the amount left in 2004 was $0.883A_0$. Plug this in for A, and solve for t:

$$0.883A_0 = A_0 e^{-0.000\ 124t}$$

$$0.883 = e^{-.000\ 124t}$$

$$\ln 0.883 = \ln e^{-.000\ 124t}$$
$$\ln 0.883 = -.000\ 124t$$

$$t = \frac{\ln 0.883}{-.000\ 124} \approx 1,003$$

It was 1,003 years old in 2004, so it was made in 1001.

99. We solve $q = 0.0009(1 - e^{-0.2t})$ for t with $q = 0.0007$

$$0.0007 = 0.0009(1 - e^{-0.2t})$$
$$\frac{0.0007}{0.0009} = 1 - e^{-0.2t}$$
$$\frac{7}{9} = 1 - e^{-0.2t}$$
$$-\frac{2}{9} = -e^{-0.2t}$$
$$\frac{2}{9} = e^{-0.2t}$$
$$\ln \frac{2}{9} = -0.2t$$
$$t = \frac{\ln \frac{2}{9}}{-0.2}$$
$$t = 7.52 \text{ seconds}$$

101. First, we solve $T = T_m + (T_0 - T_m)e^{-kt}$ for k, with $T = 61.5°$, $T_m = 40°$, $T_0 = 72°$, $t = 1$

$$61.5 = 40 + (72 - 40)e^{-k(1)}$$
$$21.5 = 32e^{-k}$$
$$\frac{21.5}{32} = e^{-k}$$
$$\ln \frac{21.5}{32} = -k$$
$$k = -\ln \frac{21.5}{32}$$
$$k = 0.40$$

Now we solve $T = T_m + (T_0 - T_m)e^{-0.40t}$ for t, with $T = 50°$, $T_m = 40°$, $T_0 = 72°$

$$50 = 40 + (72 - 40)e^{-0.40t}$$
$$10 = 32e^{-0.40t}$$
$$\frac{10}{32} = e^{-0.40t}$$
$$\ln \frac{10}{32} = -0.40t$$
$$t = \frac{\ln^{10/32}}{-0.40}$$
$$t = 2.9 \text{ hours}$$

103. (A) Plug in $M = 7.0$ and solve for E:

$$7 = \frac{2}{3} \log \frac{E}{10^{4.40}}$$
$$\frac{21}{2} = \log \frac{E}{10^{4.4}}$$
$$10^{21/2} = \frac{E}{10^{4.4}}$$
$$E = 10^{21/2} \cdot 10^{4.4} \approx 7.94 \times 10^{14} \text{ joules}$$

(B) $\dfrac{7.94 \times 10^{14} \text{ joules}}{2.88 \times 10^{14} \text{ joules/day}} = 2.76 \text{ days}$

105. First, find the energy released by one magnitude 7.5 earthquake:

$$7.5 = \frac{2}{3} \log \frac{E}{10^{4.40}}$$

$$11.25 = \log \frac{E}{10^{4.4}}$$

$$10^{11.25} = \frac{E}{10^{4.4}}$$

$$E = 10^{11.25} \cdot 10^{4.4} \approx 4.47 \times 10^{15} \text{ joules}$$

Now multiply by ten to get the energy released by ten such earthquakes:

$$10 \cdot 4.47 \times 10^{15} = 4.47 \times 10^{16}$$

Finally, divide by the energy consumption per year:

$$\frac{4.47 \times 10^{16} \text{ joules}}{1.05 \times 10^{17} \text{ joules/year}} = 0.426$$

So this energy could power the U.S. for 0.426 years, or about 155 days.

CHAPTER 4 REVIEW

1. (A) The graph of $y = \log_2 x$ passes through (1, 0) and (2, 1). This corresponds to graph m.

(B) The graph of $y = 0.5^x$ passes through (0, 1) and (1, 0.5). This corresponds to graph f.

(C) The graph of $y = \log_{0.5} x$ passes through (1, 0) and (0.5, 1). This corresponds to graph n.

(D) The graph of $y = 2^x$ passes through (0, 1) and (1, 2). This corresponds to graph g. *(4-1, 4-3)*

2. $\log m = n$ *(4-3)*   **3.** $\ln x = y$ *(4-3)*   **4.** $x = 10^y$ *(4-3)*   **5.** $y = e^x$ *(4-3)*

6. (A) Make a table of values:

x	−2	−1	0	1	2	3
$\left(\frac{4}{3}\right)^x$	$\frac{9}{16}$	$\frac{3}{4}$	1	$\frac{4}{3}$	$\frac{16}{9}$	$\frac{64}{27}$

(B) The function in part (B) is the inverse of the one graphed in part (A), so its graph is a reflection about the line $y = x$ of the graph to the left. To plot points, just switch the x and y coordinates of the points from the table in part (A).

(4-1)

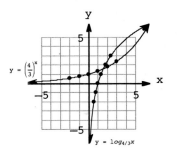

(4-3)

7. $\dfrac{7^{x+2}}{7^{2-x}} = 7^{(x+2)-(2-x)}$

$= 7^{x+2-2+x}$

$= 7^{2x}$ *(4-1)*

8. $\left(\dfrac{e^x}{e^{-x}}\right)^x = [e^{x-(-x)}]^x$

$= (e^{2x})^x = e^{2x \cdot x}$

$= e^{2x^2}$ *(4-1)*

9. $\log_2 x = 3$

$x = 2^3$

$x = 8$ *(4-3)*

10. $\log_x 25 = 2$

$\qquad 25 = x^2$

$\qquad x = 5$

since bases are
restricted positive

$\qquad\qquad (4-3)$

11. $\log_3 27 = x$

$\qquad \log_3 3^3 = x$

$\qquad\qquad x = 3 \qquad (4-3)$

12. $10^x = 17.5$

$\qquad x = \log_{10} 17.5$

$\qquad x = 1.24 \qquad (4-5)$

13. $e^x = 143{,}000$

$\qquad x = \ln 143{,}000$

$\qquad x = 11.9 \qquad (4-5)$

14. $\ln x = -0.01573$

$\qquad x = e^{-0.01573}$

$\qquad x = 0.984 \qquad (4-3)$

15. $\log x = 2.013$

$\qquad x = 10^{2.013}$

$\qquad x = 103 \qquad (4-3)$

16. $1.145 \quad (4-3)$

17. Not defined. ($-e$ is not in the domain of the logarithm function.) $\qquad (4-3)$

18. $2.211 \quad (4-3)$

19. $11.59 \qquad (4-1)$

20. $\log x + 3 \log y - \dfrac{1}{2} \log z$

$\qquad = \log x + \log y^3 - \log z^{1/2}$

$\qquad = \log\left(\dfrac{xy^3}{\sqrt{z}}\right) \qquad (4-3)$

21. $\ln \dfrac{x^3}{y} = \ln x^3 - \ln y = 3 \ln x - \ln y \qquad (4-3)$

22. $\ln(2x - 1) = \ln(x + 3)$

$\qquad 2x - 1 = x + 3$

$\qquad\qquad x = 4$

Check:

$\ln(2\cdot 4 - 1) \overset{?}{=} \ln(4 + 3)$

$\qquad \ln 7 \overset{\checkmark}{=} \ln 7$

$\qquad\qquad (4-5)$

23. $\log(x^2 - 3) = 2 \log(x - 1)$

$\qquad \log(x^2 - 3) = \log(x - 1)^2$

$\qquad\qquad x^2 - 3 = (x - 1)^2$

$\qquad\qquad x^2 - 3 = x^2 - 2x + 1$

$\qquad\qquad\quad -3 = -2x + 1$

$\qquad\qquad\quad -4 = -2x$

$\qquad\qquad\quad\; x = 2$

Check:

$\log(2^2 - 3) \overset{?}{=} 2 \log(2 - 1)$

$\qquad \log 1 \overset{?}{=} 2 \log 1$

$\qquad\quad 0 \overset{\checkmark}{=} 0 \qquad (4-5)$

24. $e^{x^2 - 3} = e^{2x}$

$\qquad x^2 - 3 = 2x$

$\qquad x^2 - 2x - 3 = 0$

$(x - 3)(x + 1) = 0$

$\qquad\qquad x = 3, \; -1$

$\qquad\qquad\qquad (4-5)$

25. $4^{x-1} = 2^{1-x}$

$\qquad (2^2)^{x-1} = 2^{1-x}$

$\qquad 2^{2(x-1)} = 2^{1-x}$

$\qquad 2(x - 1) = 1 - x$

$\qquad 2x - 2 = 1 - x$

$\qquad\quad 3x = 3$

$\qquad\quad\; x = 1 \qquad (4-5)$

26. $4 - 3^x = 2$

$\qquad -3^x = -2$

$\qquad 3^x = 2$

$\qquad \ln 3^x = \ln 2$

$\qquad x \ln 3 = \ln 2$

$\qquad x = \dfrac{\ln 2}{\ln 3} \qquad (4-5)$

27. $5 + \dfrac{1}{2} e^x = \dfrac{17}{2}$

$\qquad \dfrac{1}{2} e^x = \dfrac{17}{2} - 5 = \dfrac{17}{2} - \dfrac{10}{2} = \dfrac{7}{2}$

$\qquad e^x = 2 \cdot \dfrac{7}{2} = 7$

$\qquad x = \ln 7 \qquad (4-5)$

28. $2x^2 e^{-x} = 18 e^{-x}$

$\qquad 2x^2 e^{-x} - 18 e^{-x} = 0$

$\qquad 2 e^{-x}(x^2 - 9) = 0$

$2 e^{-x}(x - 3)(x + 3) = 0$

$2 e^{-x} = 0 \qquad x - 3 = 0 \qquad x + 3 = 0$

never $\qquad\qquad x = 3 \qquad\quad x = -3$

Solution: $x = 3, \; -3 \qquad (4-5)$

29. $\log_{1/4} 16 = x$

$\qquad \log_{1/4} 4^2 = x$

$\qquad \log_{1/4} \left(\dfrac{1}{4}\right)^{-2} = x$

$\qquad\qquad x = -2 \qquad (4-5)$

30. $\log_x 9 = -2$

$\quad\quad x^{-2} = 9$

$\quad\quad \dfrac{1}{x^2} = 9$

$\quad\quad 1 = 9x^2$

$\quad\quad \dfrac{1}{9} = x^2$

$\quad\quad x = \pm\sqrt{\dfrac{1}{9}}$

$\quad\quad x = \dfrac{1}{3}$

since bases are
restricted positive
$\quad\quad (4\text{-}5)$

31. $\log_{16} x = \dfrac{3}{2}$

$\quad\quad 16^{3/2} = x$

$\quad\quad 64 = x$

$\quad\quad x = 64$

$\quad\quad\quad (4\text{-}5)$

32. $\log_x e^5 = 5$

$\quad\quad e^5 = x^5$

$\quad\quad x = e \quad\quad (4\text{-}5)$

33. $10^{\log_{10} x} = 33$

$\quad \log_{10} x = \log_{10} 33$

$\quad\quad\quad x = 33 \quad\quad (4\text{-}5)$

34. $\ln x = 0$

$\quad\quad e^0 = x$

$\quad\quad x = 1 \quad (4\text{-}5)$

35. $x = 2(10^{1.32})$

$\quad x = 41.8 \quad\quad (4\text{-}1)$

36. $x = \log_5 23$

$\quad x = \dfrac{\log 23}{\log 5} \text{ or } \dfrac{\ln 23}{\ln 5}$

$\quad x = 1.95 \quad\quad (4\text{-}3)$

37. $\ln x = -3.218$

$\quad\quad x = e^{-3.218}$

$\quad\quad x = 0.0400 \quad (4\text{-}3)$

38. $x = \log(2.156 \times 10^{-7})$

$\quad x = -6.67 \quad (4\text{-}3)$

39. $x = \dfrac{\ln 4}{\ln 2.31}$

$\quad x = 1.66 \quad\quad (4\text{-}3)$

40. $\quad 25 = 5(2)^x$

$\quad\quad \dfrac{25}{5} = 2^x$

$\quad\quad 5 = 2^x$

$\quad \ln 5 = \ln 2^x$

$\quad \ln 5 = x \ln 2$

$\quad \dfrac{\ln 5}{\ln 2} = x$

$\quad\quad x = 2.32 \quad (4\text{-}5)$

41. $\quad 4{,}000 = 2{,}500 e^{0.12x}$

$\quad\quad \dfrac{4{,}000}{2{,}500} = e^{0.12x}$

$\quad \ln\!\left(\dfrac{4{,}000}{2{,}500}\right) = \ln e^{0.12x}$

$\quad\quad 0.12x = \ln \dfrac{4{,}000}{2{,}500}$

$\quad\quad x = \dfrac{1}{0.12} \ln \dfrac{4{,}000}{2{,}500}$

$\quad\quad x = 3.92 \quad\quad (4\text{-}5)$

42. $\quad 0.01 = e^{-0.05x}$

$\quad \ln 0.01 = \ln e^{-0.05x}$

$\quad -0.05x = \ln 0.01$

$\quad\quad x = \dfrac{\ln 0.01}{-0.05}$

$\quad\quad x = 92.1 \quad\quad\quad (4\text{-}5)$

43. $\quad 5^{2x-3} = 7.08$

$\quad \log 5^{2x-3} = \log 7.08$

$\quad (2x - 3)\log 5 = \log 7.08$

$\quad\quad 2x - 3 = \dfrac{\log 7.08}{\log 5}$

$\quad\quad x = \dfrac{1}{2}\left[3 + \dfrac{\log 7.08}{\log 5}\right]$

$\quad\quad x = 2.11 \quad\quad (4\text{-}5)$

44. $\quad \dfrac{e^x - e^{-x}}{2} = 1$

$\quad\quad e^x - e^{-x} = 2$

$\quad\quad e^x - \dfrac{1}{e^x} = 2$

$\quad e^x e^x - e^x\!\left(\dfrac{1}{e^x}\right) = 2e^x$

$\quad\quad (e^x)^2 - 1 = 2e^x$

$\quad (e^x)^2 - 2e^x - 1 = 0$

This equation is quadratic in e^x:
$$e^x = \frac{-b \pm \sqrt{b^2 - 4ac}}{2a}$$
$$a = 1, \; b = -2, \; c = -1$$

$$e^x = \frac{-(-2) \pm \sqrt{(-2)^2 - 4(1)(-1)}}{2}$$

$$e^x = \frac{2 \pm \sqrt{8}}{2}$$

$$e^x = 1 \pm \sqrt{2}$$

$x = \ln(1 \pm \sqrt{2})$ $1 - \sqrt{2}$ is negative, so not in the domain of the logarithm function.

$x = \ln(1 + \sqrt{2})$
$x = 0.881$ *(4-5)*

45. $\log 3x^2 - \log 9x = 2$

$$\log \frac{3x^2}{9x} = 2$$

$$\frac{3x^2}{9x} = 10^2$$

$$\frac{x}{3} = 100$$

$$x = 300$$

Check:

$$\log(3 \cdot 300^2) - \log(9 \cdot 300) \overset{?}{=} 2$$

$$\log(270{,}000) - \log(2{,}700) \overset{?}{=} 2$$

$$\log \frac{270{,}000}{2{,}700} \overset{?}{=} 2$$

$$\log 100 \overset{\checkmark}{=} 2 \qquad (4\text{-}5)$$

46. $\log x - \log 3 = \log 4 - \log(x + 4)$

$$\log \frac{x}{3} = \log \frac{4}{x + 4}$$

$$\frac{x}{3} = \frac{4}{x + 4} \quad \text{excluded value:}$$

$$\qquad\qquad\qquad\qquad x \neq -4$$

$$3(x + 4)\frac{x}{3} = 3(x + 4)\frac{4}{x + 4}$$

$$(x + 4)x = 12$$

$$x^2 + 4x = 12$$

$$x^2 + 4x - 12 = 0$$

$$(x + 6)(x - 2) = 0$$

$$x = -6 \quad x = 2$$

Check: $\log(-6)$ is not defined

$$\log 2 - \log 3 \overset{?}{=} \log 4 - \log(2 + 4)$$

$$\log \frac{2}{3} \overset{?}{=} \log \frac{4}{6}$$

$$\log \frac{2}{3} \overset{\checkmark}{=} \log \frac{2}{3}$$

Solution: $x = 2$ *(4-5)*

47. $\ln(x + 3) - \ln x = 2 \ln 2$

$$\ln \frac{x + 3}{x} = \ln 2^2$$

$$\frac{x + 3}{x} = 2^2$$

$$\frac{x + 3}{x} = 4$$

$$x + 3 = 4x$$

$$3 = 3x$$

$$x = 1$$

Check:

$$\ln(1 + 3) - \ln 1 \overset{?}{=} 2 \ln 2$$

$$\ln 4 - 0 \overset{?}{=} 2 \ln 2$$

$$\ln 4 \overset{\checkmark}{=} \ln 4 \qquad (4\text{-}5)$$

48. $\ln(2x + 1) - \ln(x - 1) = \ln x$

$$\ln \frac{2x + 1}{x - 1} = \ln x$$

$$\frac{2x + 1}{x - 1} = x \quad \text{Excluded value: } x \neq 1$$

$$(x - 1)\frac{2x + 1}{x - 1} = x(x - 1)$$

$$2x + 1 = x^2 - x$$

$$0 = x^2 - 3x - 1$$

$$x = \frac{-b \pm \sqrt{b^2 - 4ac}}{2a} \quad a = 1, \ b = -3, \ c = -1$$

$$x = \frac{-(-3) \pm \sqrt{(-3)^2 - 4(1)(-1)}}{2(1)}$$

$$x = \frac{3 \pm \sqrt{13}}{2}$$

Check: $\ln\left(\dfrac{3 - \sqrt{13}}{2}\right)$ is not defined

$$\ln\left(2 \cdot \frac{3 + \sqrt{13}}{2} + 1\right) - \ln\left(\frac{3 + \sqrt{13}}{2} - 1\right) \ \overset{?}{=} \ \ln\left(\frac{3 + \sqrt{13}}{2}\right)$$

$$\ln(3 + \sqrt{13} + 1) - \ln\left(\frac{3 + \sqrt{13} - 2}{2}\right) \ \overset{?}{=} \ \ln\left(\frac{3 + \sqrt{13}}{2}\right)$$

$$\ln(4 + \sqrt{13}) - \ln\left(\frac{1 + \sqrt{13}}{2}\right) \ \overset{?}{=} \ \ln\left(\frac{3 + \sqrt{13}}{2}\right)$$

$$\ln\left(\frac{4 + \sqrt{13}}{1} \cdot \frac{2}{1 + \sqrt{13}}\right) \ \overset{?}{=} \ \ln\left(\frac{3 + \sqrt{13}}{2}\right)$$

$$\ln\left(\frac{(4 + \sqrt{13})2}{1 + \sqrt{13}}\right) \ \overset{?}{=} \ \ln\left(\frac{3 + \sqrt{13}}{2}\right)$$

$$\ln\left(\frac{(4 + \sqrt{13})2(1 - \sqrt{13})}{(1 + \sqrt{13})(1 - \sqrt{13})}\right) \ \overset{?}{=} \ \ln\left(\frac{3 + \sqrt{13}}{2}\right)$$

$$\ln\left(\frac{2(4 - 3\sqrt{13} - 13)}{1 - 13}\right) \ \overset{?}{=} \ \ln\left(\frac{3 + \sqrt{13}}{2}\right)$$

$$\ln\left(\frac{-18 - 6\sqrt{13}}{-12}\right) \ \overset{?}{=} \ \ln\left(\frac{3 + \sqrt{13}}{2}\right)$$

$$\ln\left(\frac{3 + \sqrt{13}}{2}\right) \ \overset{\checkmark}{=} \ \ln\left(\frac{3 + \sqrt{13}}{2}\right)$$

Solution: $x = \dfrac{3 + \sqrt{13}}{2}$ (4-5)

49.

$$(\log x)^3 = \log x^9$$
$$(\log x)^3 = 9 \log x$$
$$(\log x)^3 - 9 \log x = 0$$
$$\log x[(\log x)^2 - 9] = 0$$
$$\log x(\log x - 3)(\log x + 3) = 0$$
$$\log x = 0 \quad \log x - 3 = 0 \quad \log x + 3 = 0$$
$$x = 1 \qquad \log x = 3 \qquad \log x = -3$$
$$x = 10^3 \qquad\qquad x = 10^{-3}$$

Check:

$$(\log 1)^3 \ \overset{?}{=} \ \log 1^9$$
$$0 \ \overset{\checkmark}{=} \ 0$$
$$(\log 10^3)^3 \ \overset{?}{=} \ \log(10^3)^9$$
$$27 \ \overset{\checkmark}{=} \ 27$$
$$(\log 10^{-3})^3 \ \overset{?}{=} \ \log(10^{-3})^9$$
$$-27 \ \overset{\checkmark}{=} \ -27$$

Solution: $x = 1, 10^3, 10^{-3}$ (4-5)

50. $\ln(\log x) = 1$
$\qquad \log x = e$
$\qquad\qquad x = 10^e$ (4-5)

51. $(e^x + 1)(e^{-x} - 1) - e^x(e^{-x} - 1) = e^x e^{-x} - e^x + e^{-x} - 1 - e^x e^{-x} + e^x$
$$= 1 - e^x + e^{-x} - 1 - 1 + e^x = e^{-x} - 1 \qquad (4\text{-}1)$$

52. $(e^x + e^{-x})(e^x - e^{-x}) - (e^x - e^{-x})^2 = (e^x)^2 - (e^{-x})^2 - [(e^x)^2 - 2e^x e^{-x} + (e^{-x})^2]$
$$= e^{2x} - e^{-2x} - [e^{2x} - 2 + e^{-2x}]$$
$$= e^{2x} - e^{-2x} - e^{2x} + 2 - e^{-2x}$$
$$= 2 - 2e^{-2x} \qquad (4\text{-}1)$$

53. The graph is shown below, along with a second screen showing the y intercept.

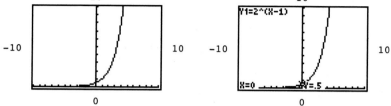

The domain is all real numbers, and the range is $(0, \infty)$. There is no x intercept, and the y intercept is 1/2. There is no vertical asymptote, and the horizontal asymptote is $y = 0$. $\qquad (4\text{-}1)$

54. The graph is shown below, along with a second screen showing the y intercept.

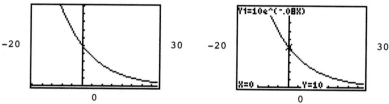

The domain is all real numbers, and the range is $(0, \infty)$. There is no x intercept, and the y intercept is 10. There is no vertical asymptote, and the horizontal asymptote is $y = 0$. $\qquad (4\text{-}1)$

55. The graph is shown below, along with a second screen showing the x intercept.

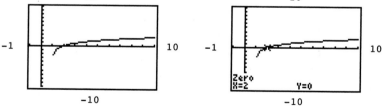

The domain is $[1, \infty)$, and the range is all real numbers. There is no y intercept, and the x intercept is 2. There is no horizontal asymptote, and the vertical asymptote is $x = 1$. $\qquad (4\text{-}3)$

56. The graph is shown below, along with a second screen showing the y intercept.

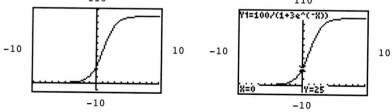

The domain is all real numbers, and the range is $(0, 100)$. There is no x intercept, and the y intercept is 25. There is no vertical asymptote, and the horizontal asymptotes are $y = 0$ and $y = 100$. $\qquad (4\text{-}1)$

57. Reflecting a graph in the line $y = x$ results in the graph of the inverse of the original function. In this case, the result is the inverse of $y = e^x$, which is $y = \ln x$. If the graph of $y = e^x$ is reflected in the x axis, y is replaced by $-y$ and the graph becomes the graph of $-y = e^x$ or $y = -e^x$.

If the graph of $y = e^x$ is reflected in the y axis, x is replaced by $-x$ and the graph becomes the graph of $y = e^{-x}$ or $y = \dfrac{1}{e^x}$ or $y = \left(\dfrac{1}{e}\right)^x$. *(4-1)*

58. Examining the graph of $f(x) = 4 - x^2 + \ln x$, we obtain

The zeros are at 0.018 and 2.187. *(4-5)*

59. Graphing $y_1 = 10^{x-3}$ and $y_2 = 8 \log x$, we obtain

The graphs intersect at (1.003, 0.010) and (3.653, 4.502). *(4-5)*

60.
$$D = 10 \log \frac{I}{I_0}$$
$$\frac{D}{10} = \log \frac{I}{I_0}$$
$$10^{D/10} = \frac{I}{I_0}$$
$$I_0 10^{D/10} = I$$
$$I = I_0(10^{D/10})$$

(4-5)

61.
$$y = \frac{1}{\sqrt{2\pi}} e^{-x^2/2}$$
$$\sqrt{2\pi}\, y = e^{-x^2/2}$$
$$-\frac{x^2}{2} = \ln(\sqrt{2\pi}\, y)$$
$$x^2 = -2 \ln(\sqrt{2\pi}\, y)$$
$$x = \pm\sqrt{-2 \ln(\sqrt{2\pi} y)} \qquad (4\text{-}5)$$

62.
$$x = -\frac{1}{k} \ln \frac{I}{I_0}$$
$$-kx = \ln \frac{I}{I_0}$$
$$\frac{I}{I_0} = e^{-kx}$$
$$I = I_0(e^{-kx})$$

(4-5)

63.

$$r = P \frac{i}{1 - (1 + i)^{-n}}$$

$$\frac{r}{P} = \frac{i}{1 - (1 + i)^{-n}}$$

$$\frac{P}{r} = \frac{1 - (1 + i)^{-n}}{i}$$

$$\frac{Pi}{r} = 1 - (1 + i)^{-n}$$

$$\frac{Pi}{r} - 1 = -(1 + i)^{-n}$$

$$1 - \frac{Pi}{r} = (1 + i)^{-n}$$

$$\ln\left(1 - \frac{Pi}{r}\right) = \ln(1 + i)^{-n}$$

$$\ln\left(1 - \frac{Pi}{r}\right) = -n \ln(1 + i)$$

$$\frac{\ln\left(1 - \frac{Pi}{r}\right)}{-\ln(1 + i)} = n$$

$$n = -\frac{\ln\left(1 - \frac{Pi}{r}\right)}{\ln(1 + i)} \qquad (4\text{-}5)$$

64. (A) For $x > -1$, $y = e^{-x/3}$ decreases from $e^{1/3}$ to 0 while $\ln(x + 1)$ increases from $-\infty$ to ∞. Consequently, the graphs can intersect at exactly one point.

(B) Graphing $y_1 = e^{-x/3}$ and $y_2 = 4 \ln(x + 1)$ we obtain

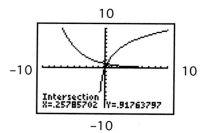

The solution of $e^{-x/3} = 4 \ln(x + 1)$ is $x = 0.258$. $\qquad (4\text{-}5)$

65.

$$\ln y = -5t + \ln c$$

$$\ln y - \ln c = -5t$$

$$\ln\left(\frac{y}{c}\right) = -5t$$

$$\frac{y}{c} = e^{-5t}$$

$$y = ce^{-5t} \qquad (4\text{-}5)$$

66.

x	$y = \log_2 x$	$x = \log_2 y$	y
1	0	0	1
2	1	1	2
4	2	2	4
8	3	3	8

Domain $f = (0, \infty)$ = Range f^{-1}
Range $f = (-\infty, \infty)$ = Domain f^{-1}

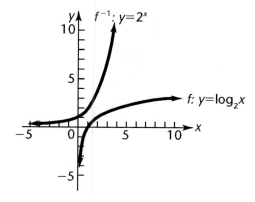

$(4\text{-}3)$

67. If $\log_1 x = y$, then we would have to have $1^y = x$; that is, $1 = x$ for arbitrary positive x, which is impossible. $(4\text{-}3)$

68. Let $u = \log_b M$ and $v = \log_b N$ and then change each equation to exponential form:

$b^u = M$; $b^v = N$

Now multiply the two left sides and use a familiar property of exponents:

$b^u b^v = b^{u+v} = MN$

Next, apply the logarithm base b to both sides:

$\log_b b^{u+v} = \log_b(MN)$

The left side is now just $u + v$, which is $\log_b M + \log_b N$. Switching the order of the equation, we get

$\log_b(MN) = \log_b M + \log_b N$ $(4\text{-}3)$

69. We solve $P = P_0(1.03)^t$ for t, using $P = 2P_0$.

$$2P_0 = P_0(1.03)^t$$
$$2 = (1.03)^t$$
$$\ln 2 = t \ln 1.03$$
$$\frac{\ln 2}{\ln 1.03} = t$$
$$t = 23.4 \text{ years}$$ $(4\text{-}2)$

70. We solve $P = P_0 e^{0.03t}$ for t using $P = 2P_0$.

$$2P_0 = P_0 e^{0.03t}$$
$$2 = e^{0.03t}$$
$$\ln 2 = 0.03t$$
$$\frac{\ln 2}{0.03} = t$$
$$t = 23.1 \text{ years}$$ $(4\text{-}2)$

71. A_0 = original amount
$0.01A_0$ = 1 percent of original amount

We solve $A = A_0 e^{-0.000124t}$ for t, using $A = 0.01A_0$.

$$0.01A_0 = A_0 e^{-0.000124t}$$
$$0.01 = e^{-0.000124t}$$
$$\ln 0.01 = -0.000124t$$
$$\frac{\ln 0.01}{-0.000124} = t$$
$$t = 37{,}100 \text{ years}$$ $(4\text{-}2)$

72. (A) When $t = 0$, $N = 1$. As t increases by $1/2$, N doubles.
$$N = 1 \cdot (2)^{t \div 1/2}$$
$$N = 2^{2t} \text{ (or } N = 4^t)$$

(B) We solve $N = 4^t$ for t, using $N = 10^9$:
$$10^9 = 4^t$$
$$9 = t \log 4$$
$$t = \frac{9}{\log 4}$$
$$t = 15 \text{ days}$$ $(4\text{-}2)$

73. We use $A = Pe^{rt}$ with $P = 1$, $r = 0.03$, and $t = 2010$.
$A = 1e^{0.03(2010)}$
$A = 1.5 \times 10^{26}$ dollars

$(4\text{-}1)$

74. (A)

t	P
0	1,000
5	670
10	449
15	301
20	202
25	135
30	91

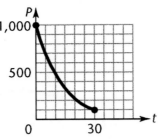

(B) As t tends to infinity, P appears to tend to 0. *(4-1)*

75. $M = \frac{2}{3} \log \frac{E}{E_0}$ $E_0 = 10^{4.40}$

We use $E = 1.99 \times 10^{14}$

$M = \frac{2}{3} \log \frac{1.99 \times 10^{14}}{10^{4.40}}$

$M = \frac{2}{3} \log(1.99 \times 10^{9.6})$

$M = \frac{2}{3}(\log 1.99 + 9.6)$

$M = \frac{2}{3}(0.299 + 9.6)$

$M = 6.6$ *(4-4)*

76. We solve $M = \frac{2}{3} \log \frac{E}{E_0}$ for E, using

$E_0 = 10^{4.40}$, $M = 8.3$

$8.3 = \frac{2}{3} \log \frac{E}{10^{4.40}}$

$\frac{3}{2}(8.3) = \log \frac{E}{10^{4.40}}$

$12.45 = \log \frac{E}{10^{4.40}}$

$\frac{E}{10^{4.40}} = 10^{12.45}$

$E = 10^{4.40} \cdot 10^{12.45}$

$E = 10^{16.85}$ or 7.08×10^{16} joules

(4-4)

77. We use the given formula twice, with

$I_2 = 100,000 I_1$

$D_1 = 10 \log \frac{I_1}{I_0}$ $D_2 = 10 \log \frac{I_2}{I_0}$

$D_2 - D_1 = 10 \log \frac{I_2}{I_0} - 10 \log \frac{I_1}{I_0} = 10 \log \left(\frac{I_2}{I_0} \div \frac{I_1}{I_0} \right) = 10 \log \frac{I_2}{I_1}$

$= 10 \log \frac{100,000 I_1}{I_1} = 10 \log 100,000 = 50$ decibels

The level of the louder sound is 50 decibels more. *(4-4)*

78. $I = I_0 e^{-kd}$

To find k, we solve for k using

$I = \frac{1}{2} I_0$ and $d = 73.6$

$\frac{1}{2} I_0 = I_0 e^{-k(73.6)}$

$\frac{1}{2} = e^{-73.6k}$

$-73.6k = \ln \frac{1}{2}$

$k = \frac{\ln \frac{1}{2}}{-73.6}$

$k = 0.00942$

We now find the depth at which 1% of the surface light remains. We solve

$I = I_0 e^{-0.00942d}$ for d with $I = 0.01 I_0$

$0.01 I_0 = I_0 e^{-0.00942d}$

$0.01 = e^{-0.00942d}$

$-0.00942d = \ln 0.01$

$d = \frac{\ln 0.01}{-0.00942}$

$d = 489$ feet *(4-4)*

79. We solve $N = \dfrac{30}{1 + 29e^{-1.35t}}$ for t with $N = 20$.

$$20 = \frac{30}{1 + 29e^{-1.35t}}$$

$$\frac{1}{20} = \frac{1 + 29e^{-1.35t}}{30}$$

$$1.5 = 1 + 29e^{-1.35t}$$

$$0.5 = 29e^{-1.35t}$$

$$\frac{0.5}{29} = e^{-1.35t}$$

$$-1.35t = \ln\frac{0.5}{29}$$

$$t = \frac{\ln\frac{0.5}{29}}{-1.35}$$

$$t = 3 \text{ years} \qquad\qquad (4\text{-}2)$$

80. (A) The independent variable is years since 1980, so enter 0, 5, 10, 15, 20, and 25 as L_1. The dependent variable is Medicare expenditures, so enter that column as L_2. Then use the exponential regression command on the STAT CALC menu.

The exponential model is $y = 43.3(1.09)^x$.

To find total expenditures in 2010 and 2020, we plug in 30 and 40 for x:

$$y(30) = 43.3(1.09)^{30} = 575; \qquad y(40) = 43.3(1.09)^{40} = 1,360$$

Expenditures are predicted to be \$575 billion in 2010 and \$1,360 billion in 2020.

(B) Graph $y_1 = 43.3(1.09)^x$ and $y_2 = 900$ and use the INTERSECT command:

Expenditures are predicted to reach \$900 billion in 2015. $\qquad (4\text{-}2)$

81. (A)

Enter the data.   Compute the regression   Transfer the data to the
                         equation.   equation editor.

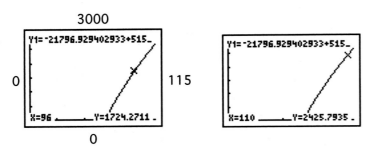

The model predicts a yield of 1,724 million bushels in 1996 and a yield of 2,426 bushels in 2010.

(B) Since this data indicates a lower y value for $x = 96$, it would tend to move the curve lower, and should lead to a lower estimate for 2010. This is indeed what happens:

Enter the new data.

Compute the new regression equation.

Transfer the new data to the equation editor.

The model now predicts a yield of 2,328 million bushels for 2010.   (4-4)

Chapters 3 & 4 Cumulative Review

1. (A) Since the graph has x intercepts -1, 1, and 2, and -1 is at least a double zero (the graph is tangent to the x axis at $x = -1$), the lowest degree equation would be $P(x) = (x + 1)^2(x - 1)(x - 2)$.

   (B) $P(x) \to \infty$ as $x \to \infty$ and as $x \to -\infty$. (3-1)

2. The given zeros are the x intercepts of the graph. The negative leading coefficient tells us that the graph should approach $-\infty$ as x approaches ∞.

(3-1)

3. (A) The graph of $y = \left(\dfrac{3}{4}\right)^x$ passes through $(0, 1)$ and $\left(1, \dfrac{3}{4}\right)$.
   This corresponds to graph m.

   (B) The graph of $y = \left(\dfrac{4}{3}\right)^x$ passes through $(0, 1)$ and $\left(1, \dfrac{4}{3}\right)$.
   This corresponds to graph g.

   (C) The graph of $y = \left(\dfrac{3}{4}\right)^x + \left(\dfrac{4}{3}\right)^x$ passes through $(0, 2)$.
   This corresponds to graph n.

   (D) The graph of $y = \left(\dfrac{4}{3}\right)^x - \left(\dfrac{3}{4}\right)^x$ passes through $(0, 0)$.
   This corresponds to graph f. (4-1)

4.
$$
\begin{array}{r}
3 \quad 5 \quad -18 \quad -3 \\
-9 \quad 12 \quad 18 \\
\hline
-3)\overline{3 \quad -4 \quad -6 \quad 15}
\end{array}
$$

   $3x^3 + 5x^2 - 18x - 3 = (x + 3)(3x^2 - 4x - 6) + 15$ (3-2)

5. -2, 3, 5   (3-1)

6. We investigate $P(1)$ and $P(2)$ by forming a synthetic division table.

	4	-5	-3	-1
1	4	-1	-4	-5
2	4	3	3	5

   Since $P(1)$ and $P(2)$ have opposite signs, there is at least one real zero between 1 and 2. (3-3)

7. The possible rational zeros are ±1, ±2, ±4, ±8. We form a synthetic division table.

	1	1	-10	8	
1	1	2	-8	0	1 is a zero

   $x^3 + x^2 - 10x + 8 = (x - 1)(x^2 + 2x - 8) = (x - 1)(x - 2)(x + 4)$.
   The rational zeros are 1, 2, -4. (3-4)

8. (A) $x = \log_{10} y$ or $x = \log y$ (B) $x = e^y$ (4-3)

9. (A) $(2e^x)^3 = 2^3(e^x)^3 = 8e^{3x}$ (B) $\dfrac{e^{3x}}{e^{-2x}} = e^{3x-(-2x)} = e^{5x}$ *(4-1)*

10. (A) $\log_3 x = 2$ (B) $\log_3 81 = x$ (C) $\log_x 4 = -2$

        $x = 3^2$ $\log_3 3^4 = x$ $x^{-2} = 4$

        $x = 9$ $x = 4$ $\dfrac{1}{x^2} = 4$

                                                         $1 = 4x^2$

                                                   $x^2 = \dfrac{1}{4}$

                                                    $x = \dfrac{1}{2}$

                 since bases are restricted positive   *(4-3)*

11. (A) $10^x = 2.35$ (B) $e^x = 87,500$

        $x = \log 2.35$ $x = \ln 87,500$

        $x = 0.371$ $x = 11.4$

     (C) $\log x = -1.25$ (D) $\ln x = 2.75$

           $x = 10^{-1.25}$ $x = e^{2.75}$

           $x = 0.0562$ $x = 15.6$ *(4-3)*

12. $E = k\dfrac{P}{x^3}$ *(3-6)*   **13.** $F = k\dfrac{q_1 q_2}{r^2}$ *(3-6)*

14. The graph of a nonconstant polynomial cannot have a horizontal asymptote. *(3-1)*

15. The graph of a rational function that has a horizontal asymptote has to approach the horizontal asymptote both as $x \to -\infty$ and $x \to \infty$.   *(3-5)*

16. $f(x) = 3 \ln x - \sqrt{x}$ *(4-3)*

17. The function f multiplies the base e raised to the power of one-half of the domain element by 100 and then subtracts 50.   *(4-1)*

18. $f(x) = \dfrac{2x + 8}{x + 2}$

(A) The domain of f is the set of all real numbers x such that the denominator, $x + 2 \neq 0$; that is $(-\infty, -2) \cup (-2, \infty)$ or $x \neq -2$. f has an x intercept where the numerator, $2x + 8 = 0$; that is, $x = -4$. $f(0) = 4$, so f has a y intercept at $y = 4$.

(B) *Vertical asymptote:* $x = -2$
    Horizontal asymptote: Since the numerator and denominator have the same degree, the line $y = 2$ is a horizontal asymptote.

Complete the sketch:
Graphing calculator: Hand sketch:

                                                   (3-5)

19. $P(x)$ can be factored further:

$$P(x) = x(x^2 + 4)(x + 4)$$

$$x^2 + 4 = 0 \quad \Rightarrow \quad x^2 = -4 \quad \Rightarrow \quad x = \pm\sqrt{-4} = \pm 2i$$

The zeros are 0, $2i$, $-2i$, and -4. Only 0 and -4 are x intercepts.   *(3-1)*

20. <u>Algebraic solution:</u>

The real zeros of $(x^3 + 4x)(x + 4)$ are 0 and -4.

Interval	Test Number	Result
$x < -4$	-5	$\left((-5)^3 + 4(-5)\right)(-5 + 4) = 145;$ positive
$-4 < x < 0$	-1	$\left((-1)^3 + 4(-1)\right)(-1 + 4) = -15;$ negative
$x > 0$	1	$\left((1)^3 + 4(1)\right)(1 + 4) = 25;$ positive

$(x^3 + 4x)(x + 4)$ is negative on $(-4, 0)$ and is equal to zero when $x = -4$ or 0, so the solution is $[-4, 0]$.

<u>Graphical solution:</u>

The graph of $y = (x^3 + 4x)(x + 4)$ is below the x axis or intersects the x axis on $[0, 4]$.   *(3-3)*

21.

$$
\begin{array}{r}
2 \quad -5 \quad\ 3 \quad\ 2 \\
1 \quad -2 \quad \tfrac{1}{2} \\
\hline
\tfrac{1}{2}\, \big|\, 2 \quad -4 \quad\ 1 \quad \tfrac{5}{2}
\end{array}
$$

$$P\left(\frac{1}{2}\right) = \frac{5}{2}$$ *(3-2)*

22. Use synthetic division to divide out $x + 1$:

$$
\begin{array}{r}
3 \quad -7 \quad -18 \quad -8 \\
-3 \quad\ 10 \quad\ 8 \\
\hline
-1\, \big|\, 3 \quad -10 \quad -8 \quad\ 0
\end{array}
$$

The reduced polynomial is $3x^2 - 10x - 8$, which factors as $(3x + 2)(x - 4)$; the remaining zeros are 4 and $-2/3$.   *(3-2)*

23. $x - 1$ will be a factor of $P(x)$ if $P(1) = 0$

$$P(1) = 1^{25} - 1^{20} + 1^{15} + 1^{10} - 1^5 + 1$$

$$= 1 - 1 + 1 + 1 - 1 + 1 = 2 \neq 0,$$

so $x - 1$ is not a factor. $x + 1$ will be a factor of $P(x)$ if $P(-1) = 0$

$$P(-1) = (-1)^{25} - (-1)^{20} + (-1)^{15} + (-1)^{10} - (-1)^5 + 1$$

$$= -1 - 1 - 1 + 1 + 1 + 1 = 0,$$

so $x + 1$ is a factor of $P(x)$.   *(3-2)*

24. (A) We form a synthetic division table:

	1	0	-8	0	3
-3	1	-3	1	-3	12 = $P(-3)$
-2	1	-2	-4	8	-13 = $P(-2)$
-1	1	-1	-7	7	-4 = $P(-1)$
0	1	0	-8	0	3 = $P(0)$
1	1	1	-7	-7	-4 = $P(1)$
2	1	2	-4	-8	-13 = $P(2)$
3	1	3	1	3	12 = $P(3)$

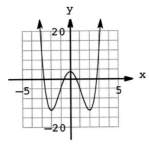

The graph of $P(x)$ has four x intercepts and three turning points; $P(x) \to \infty$ as $x \to \infty$ and as $x \to -\infty$

(B) There are real zeros in the intervals $(-3, -2)$, $(-1, 0)$, $(0, 1)$, and $(2, 3)$ indicated in the table. We search for the real zero in $(2, 3)$. Examining the graph of P in a graphing calculator we obtain the graph at the right.

The largest x intercept is 2.76.

(3-1)

25. Examining the graph of $P(x)$, we see that there may be a zero of even multiplicity between -1 and 0, a zero of odd multiplicity near 2, and a zero of even multiplicity between 2 and 3.

Applying the MAXIMUM command between -1 and 0, the ZERO command near 2, and the MINIMUM command between 2 and 3, we obtain

-0.56 is a zero of even multiplicity, 2 is a zero of odd multiplicity, and 3.56 is a zero of even multiplicity. Since $P(x)$ is a fifth-degree polynomial, -0.56 and 3.56 must be double zeros and 2 must be a simple zero. (3-3)

26. (A) We form a synthetic division table:

	1	2	-20	0	-30
0	1	2	-20	0	-30
1	1	3	-17	-17	-47
2	1	4	-12	-24	-78
3	1	5	-5	-15	-75
4	1	6	4	16	34
-1	1	1	-21	21	-51
-2	1	0	-20	40	-110
-3	1	-1	-17	51	-183
-4	1	-2	-12	48	-222
-5	1	-3	-5	25	-155
-6	1	-4	4	-24	114

From the table, 4 is an upper bound and -6 is a lower bound.

(B) We search for the largest real zero in (3, 4). We organize our calculations in a table.

Sign Change Interval (a, b)	Midpoint m	Sign of P		
		P(a)	P(m)	P(b)
(3, 4)	3.5	−	−	+
(3.5, 4)	3.75	−	−	+
(3.75, 4)	3.875	−	+	+
(3.75, 3.875)	3.8125	−	+	+
(3.75, 3.8125)	We stop here	−	−	+

Since each endpoint rounds to 3.8, a real zero lies on this last interval and is given by 3.8 to one decimal place accuracy. Four further intervals were required.

(C) Examining the graph of $P(x)$, we obtain the graphs at the right.

The real zeros are −5.68 and 3.80.

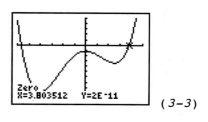

(3-3)

27. The possible rational zeros are ±1, ±3, ±5, ±15, ±$\frac{1}{2}$, ±$\frac{3}{2}$, ±$\frac{5}{2}$, ±$\frac{15}{2}$, ±$\frac{1}{4}$, ±$\frac{3}{4}$, ±$\frac{5}{4}$, ±$\frac{15}{4}$. We form a synthetic division table.

	4	−20	29	−15	
1	4	−16	13	−2	
3	4	−8	5	0	3 is a zero

So $P(x) = (x - 3)(4x^2 - 8x + 5)$

To find the remaining zeros, we solve $4x^2 - 8x + 5 = 0$ by the quadratic formula.

$4x^2 - 8x + 5 = 0$

$$x = \frac{-b \pm \sqrt{b^2 - 4ac}}{2a} \qquad a = 4, \ b = -8, \ c = 5$$

$$x = \frac{-(-8) \pm \sqrt{(-8)^2 - 4(4)(5)}}{2(4)}$$

$$x = \frac{8 \pm \sqrt{64 - 80}}{8}$$

$$x = \frac{8 \pm \sqrt{-16}}{8}$$

$$x = \frac{8 \pm 4i}{8}$$

$$x = 1 \pm \frac{1}{2}i$$

The zeros are 3, $1 \pm \frac{1}{2}i$

(3-4)

28. The possible rational zeros are ±1, ±2, ±3, ±4, ±6, ±12. We form a synthetic division table.

	1	5	1	-15	-12	
1	1	6	7	-8	-20	
2	1	7	15	15	18	a zero between 1 and 2; 2 is an upperbound
-1	1	4	-3	-12	0	-1 is a zero

We now examine $x^3 + 4x^2 - 3x - 12$. This factors by grouping into $(x + 4)(x^2 - 3)$, however, if we don't notice this, we find the remaining possible rational zeros to be -1, -2, -3, -4, -6, -12. We form a synthetic division table.

	1	4	-3	-12	
-1	1	3	-6	-6	not a double zero
-2	1	2	-7	2	a zero between -1 and -2
-3	1	1	-6	6	
-4	1	0	-3	0	-4 is a zero

So $P(x) = (x + 1)(x + 4)(x^2 - 3) = (x + 1)(x + 4)(x - \sqrt{3})(x + \sqrt{3})$. The four zeros are -1, -4, $\pm\sqrt{3}$.

(3-4)

29.
$$2^{x^2} = 4^{x+4}$$
$$2^{x^2} = (2^2)^{x+4}$$
$$2^{x^2} = 2^{2(x+4)}$$
$$x^2 = 2(x + 4)$$
$$x^2 = 2x + 8$$
$$x^2 - 2x - 8 = 0$$
$$(x - 4)(x + 2) = 0$$
$$x - 4 = 0 \qquad x + 2 = 0$$
$$x = 4 \qquad x = -2 \qquad (4\text{-}5)$$

30.
$$\frac{13}{2} - 3^x = \frac{1}{2}$$
$$-3^x = \frac{1}{2} - \frac{13}{2} = -\frac{12}{2} = -6$$
$$3^x = 6$$
$$\ln 3^x = \ln 6$$
$$x \ln 3 = \ln 6$$
$$x = \frac{\ln 6}{\ln 3} \qquad (4\text{-}5)$$

31.
$$2x^2e^{-x} + xe^{-x} = e^{-x}$$
$$2x^2e^{-x} + xe^{-x} - e^{-x} = 0$$
$$e^{-x}(2x^2 + x - 1) = 0$$
$$e^{-x}(2x - 1)(x + 1) = 0$$
$$e^{-x} = 0 \qquad 2x - 1 = 0 \qquad x + 1 = 0$$
$$\text{never} \qquad x = \frac{1}{2} \qquad x = -1$$

Solutions: $\frac{1}{2}$, -1 *(4-5)*

32.
$$e^{\ln x} = 2.5$$
$$x = 2.5 \qquad (4\text{-}5)$$

33.
$$\log_x 10^4 = 4$$
$$x^4 = 10^4$$
$$x = 10 \qquad (4\text{-}5)$$

34.
$$\log_9 x = -\frac{3}{2}$$
$$9^{-3/2} = x$$
$$x = \frac{1}{27} \qquad (4\text{-}5)$$

35. $\ln(x + 4) - \ln(x - 4) = 2 \ln 3$

$\ln \dfrac{x + 4}{x - 4} = \ln 3^2$

$\dfrac{x + 4}{x - 4} = 3^2$

$\dfrac{x + 4}{x - 4} = 9 \qquad x \neq 4$

$x + 4 = 9(x - 4)$

$x + 4 = 9x - 36$

$-8x = -40$

$x = 5$

Check: $\ln(5 + 4) - \ln(5 - 4) \overset{?}{=} 2 \ln 3$

$\ln 9 - \ln 1 \overset{?}{=} 2 \ln 3$

$\ln 9 - 0 \overset{\checkmark}{=} \ln 9$

Solution: $x = 5$ $\qquad\qquad$ (4-5)

---

36. $\ln(2x^2 + 2) = 2 \ln(2x - 4)$

$\ln(2x^2 + 2) = \ln(2x - 4)^2$

$2x^2 + 2 = (2x - 4)^2$

$2x^2 + 2 = 4x^2 - 16x + 16$

$0 = 2x^2 - 16x + 14$

$0 = 2(x - 7)(x - 1)$

$x = 7, 1$

Check: $x = 7$

$\ln(2 \cdot 7^2 + 2) \overset{?}{=} 2 \ln(2 \cdot 7 - 4)$

$\ln 100 \overset{?}{=} 2 \ln 10$

$\ln 100 \overset{\checkmark}{=} \ln 100$

$x = 1$

$\ln(2 \cdot 1^2 + 2) \overset{?}{=} 2 \ln(2 \cdot 1 - 4)$

$\ln(4) \neq 2 \ln(-2)$

Solution: $x = 7$ $\qquad\qquad$ (4-5)

---

37. $\log x + \log(x + 15) = 2$

$\log[x(x + 15)] = 2$

$x(x + 15) = 10^2$

$x^2 + 15x = 100$

$x^2 + 15x - 100 = 0$

$(x - 5)(x + 20) = 0$

$x = 5, -20$

Check: $x = 5$

$\log 5 + \log(5 + 15) \overset{?}{=} 2$

$\log 5 + \log 20 \overset{?}{=} 2$

$\log 100 \overset{\checkmark}{=} 2$

$x = -20$

$\log(-20) + \log(-20 + 15) \neq 2$

Solution: $x = 5$ $\qquad\qquad$ (4-5)

---

38. $\log(\ln x) = -1$

$\ln x = 10^{-1}$

$\ln x = 1/10$

$x = e^{1/10}$ $\qquad$ (4-5)

---

39. $\qquad 4(\ln x)^2 = \ln x^2$

$\qquad 4(\ln x)^2 = 2 \ln x$

$4(\ln x)^2 - 2 \ln x = 0$

$2 \ln x(2 \ln x - 1) = 0$

$2 \ln x = 0 \qquad 2 \ln x - 1 = 0$

$\ln x = 0 \qquad\qquad \ln x = \dfrac{1}{2}$

$x = 1 \qquad\qquad\qquad x = e^{1/2}$

Check: $x = 1$

$4(\ln 1)^2 \overset{?}{=} \ln 1^2$

$0 \overset{\checkmark}{=} 0$

$x = e^{1/2}$

$4(\ln e^{1/2})^2 \overset{?}{=} \ln(e^{1/2})^2$

$4(1/2)^2 \overset{?}{=} \ln(e^{2(1/2)})$

$4(1/4) \overset{?}{=} \ln e$

$1 \overset{\checkmark}{=} 1$

Solution: $1, e^{1/2}$ $\qquad\qquad$ (4-5)

40. $x = \log_3 41$

We use the change of base formula

$$x = \frac{\log 41}{\log 3}$$

$x = 3.38$ *(4-5)*

41. $\ln x = 1.45$

$x = e^{1.45}$

$x = 4.26$ *(4-5)*

42. $4(2^x) = 20$

$2^x = 5$

$x \log 2 = \log 5$

$$x = \frac{\log 5}{\log 2}$$

$x = 2.32$ *(4-5)*

43. $10e^{-0.5x} = 1.6$

$e^{-0.5x} = 0.16$

$-0.5x = \ln 0.16$

$$x = \frac{\ln 0.16}{-0.5}$$

$x = 3.67$ *(4-5)*

44.

$$\frac{e^x - e^{-x}}{e^x + e^{-x}} = \frac{1}{2}$$

$$\frac{e^x - \frac{1}{e^x}}{e^x + \frac{1}{e^x}} = \frac{1}{2}$$

$$\frac{e^x\left(e^x - \frac{1}{e^x}\right)}{e^x\left(e^x + \frac{1}{e^x}\right)} = \frac{1}{2}$$

$$\frac{(e^x)^2 - 1}{(e^x)^2 + 1} = \frac{1}{2}$$

$$\frac{e^{2x} - 1}{e^{2x} + 1} = \frac{1}{2}$$

$$2(e^{2x} + 1)\frac{e^{2x} - 1}{e^{2x} + 1} = 2(e^{2x} + 1)\frac{1}{2}$$

$$2(e^{2x} - 1) = e^{2x} + 1$$

$$2e^{2x} - 2 = e^{2x} + 1$$

$$e^{2x} = 3$$

$$2x = \ln 3$$

$$x = \frac{1}{2}\ln 3$$

$$x = 0.549 \quad (4\text{-}5)$$

45. The graph is shown at the right.

$f(x) = 3^{1-x}$

Domain: $(-\infty, \infty)$

Range: $(0, \infty)$

x intercept: none

y intercept: $f(0) = 3^{1-0} = 3$

Horizontal asymptote:

as $x \to \infty$, $f(x) \to 0$

$y = 0$

(4-1)

46. The graph is shown below, along with commands for finding the intercepts.

$g(x) = \ln(2 - x)$

Domain: $2 - x > 0 \quad (-\infty, 2)$

$\qquad\qquad x < 2$

Range: $(-\infty, \infty)$

y intercept: 0.693 (or $\ln 2$); x intercept: 1

Vertical asymptote: as $x \to 2^-$, $g(x) \to -\infty$; $x = 2$ is a vertical asymptote

(4-3)

47. The graph is shown to the right.
$A(t) = 100e^{-0.3t}$
Domain: $(-\infty, \infty)$
Range: $(0, \infty)$
x intercept: none
y intercept: $A(0) = 100e^{-0.3(0)} = 100$
Horizontal asymptote:
   as $t \to \infty$, $A(t) \to 0$
   $y = 0$

(4-1)

48. The graph is shown below, along with commands for finding the intercepts.
$h(x) = -2e^{-x} + 3$
Domain: $(-\infty, \infty)$
Range: Since $e^{-x} > 0$
$\qquad -2e^{-x} < 0$
$\quad -2e^{-x} + 3 < 3$
$\qquad (-\infty, 3)$
y intercept: 1; x intercept: -0.41
y intercept: $h(0) = -2e^{-0} + 3 = 1$
Horizontal asymptote:
   as $x \to \infty$, $h(x) \to 3$; $y = 3$ is a horizontal asymptote

(4-1)

49. The graph is shown below, along with a window that shows the y intercept.
$$N(t) = \frac{6}{2 + e^{-0.1t}}$$
Domain: $(-\infty, \infty)$
Range: $(0, 3)$
x intercept: none
y intercept: 2

Horizontal asymptotes: As $t \to \infty$, $N(t) \to 3$, so $y = 3$ is a horizontal asymptote.
$\qquad\qquad\qquad\qquad\quad$ As $t \to -\infty$, $N(t) \to 0$, so $y = 0$ is a horizontal asymptote.

 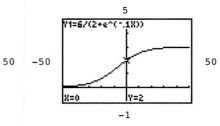

(4-1)

50. Reflecting a graph in the line $y = x$ results in the graph of the inverse
function. In this case, the inverse function of $y = \ln x$ is $y = e^x$. If the
graph of $y = \ln x$ is reflected in the x axis, y is replaced by $-y$ and the graph
becomes the graph of $-y = \ln x$ or $y = -\ln x$.
If the graph of $y = \ln x$ is reflected in the y axis, x is replaced by $-x$ and
the graph becomes the graph of $y = \ln(-x)$.   (4-3)

51. (A) For $x > 0$, $y = e^{-x}$ decreases from 1 to 0 while $\ln x$ increases from $-\infty$ to ∞. Consequently, the graphs can intersect at exactly one point.

(B) Graphing $y1 = e^{-x}$ and $y2 = \ln x$, we obtain

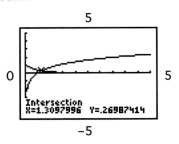

The solution is 1.31.   (4-

52. (A) Let $u = x^2$. Then
$$P(x) = x^4 + 9x^2 + 18$$
$$P(x) = u^2 + 9u + 18$$
$$P(x) = (u + 6)(u + 3)$$
$$P(x) = (x^2 + 6)(x^2 + 3)$$

(B) $x^2 + 6$ has $i\sqrt{6}$ and $-i\sqrt{6}$ as zeros, so it factors as
$$x^2 + 6 = (x + i\sqrt{6})(x - i\sqrt{6}).$$ $x^2 + 3$ has $i\sqrt{3}$ and $-i\sqrt{3}$ as zeros, so it factors as
$$x^2 + 3 = (x + i\sqrt{3})(x - i\sqrt{3}).$$
$$P(x) = (x + i\sqrt{6})(x - i\sqrt{6})(x + i\sqrt{3})(x - i\sqrt{3})$$ (3-4)

53. (A) Let $u = x^2$. Then
$$P(x) = x^4 - 23x^2 - 50$$
$$P(x) = u^2 - 23u - 50$$
$$P(x) = (u - 25)(u + 2)$$
$$P(x) = (x^2 - 25)(x^2 + 2)$$
$$P(x) = (x + 5)(x - 5)(x^2 + 2)$$

(B) $x^2 + 2$ has zeros $i\sqrt{2}$ and $-i\sqrt{2}$ so it factors as
$$x^2 + 2 = (x + i\sqrt{2})(x - i\sqrt{2}).$$
$$P(x) = (x + 5)(x - 5)(x + i\sqrt{2})(x - i\sqrt{2})$$ (3-4)

54. Write $G = kx^2$. Substitute $G = 10$ and $x = 5$ and solve for k.
$$10 = k(5)^2$$
$$10 = 25k$$
$$k = 0.4$$
The equation of variation is $G = 0.4x^2$.
When $x = 7$
$$G = 0.4(7)^2 = 19.6$$ (3-6)

55. Write $H = \dfrac{k}{r^3}$. Substitute $H = 162$ and $r = 2$ and solve for k.
$$162 = \frac{k}{2^3}$$
$$k = 8(162)$$
$$k = 1,296$$

The equation of variation is $H = \dfrac{1,296}{r^3}$.

When $r = 3$
$$H = \frac{1,296}{3^3} = 48$$ (3-6)

56. $f(x) = \dfrac{x^2 + 4x + 8}{x + 2}$

Intercepts. There are no real zeros of the numerator. No x intercept
$\quad\quad\quad f(0) = 4\quad y$ intercept
Vertical asymptotes. The only zero of the denominator is $x = -2$.
Horizontal asymptote. Since the degree of the numerator is greater than the degree of the denominator, there is no horizontal asymptote.
Oblique asymptote.

$$
\begin{array}{r}
x + 2 \\
x + 2 \overline{\big)\; x^2 + 4x + 8} \\
\underline{x^2 + 2x} \\
2x + 8 \\
\underline{2x + 4} \\
4
\end{array}
$$

So, $f(x) = x + 2 + \dfrac{4}{x + 2}$

and the line $y = x + 2$
is an oblique asymptote.

Complete the sketch.
Graphing calculator:

Hand sketch:

(*3-5*)

57. <u>Algebraic solution:</u>
zeros of numerator: $x^3 - x = 0$
$$x(x^2 - 1) = 0$$
$$x(x + 1)(x - 1) = 0$$
$$x = 0, -1, 1$$
zeros of the denominator: $x^3 - 8 = 0$
$$x^3 = 8$$
$$x = 2$$

These four zeros divide the real line into the five intervals shown below.

Interval	Test Number	Result	Sign of f
$(-\infty, -1)$	-2	0.375	$+$
$(-1, 0)$	$-1/2$	-0.05	$-$
$(0, 1)$	$1/2$	0.05	$+$
$(1, 2)$	$3/2$	-0.4	$-$
$(2, \infty)$	3	1.3	$+$

The expression is positive on $(-\infty, -1) \cup (0, 1) \cup (2, \infty)$. It's zero at -1, 0, and 1, so we include them but exclude 2 as the expression is undefined.
Solution: $(-\infty, -1] \cup [0, 1] \cup (2, \infty)$

Graphical solution:

The graph of $\dfrac{x^3 - x}{x^3 - 8}$ is above, or intersects, the x axis on

$(-\infty, -1] \cup [0, 1] \cup (2, \infty)$. $(3\text{-}5)$

58. A preliminary graph of $y = P(x)$ is shown at the right.

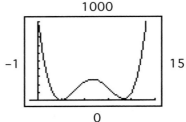

Examining the behavior near the maxima and minima, we obtain

There are zeros at $x = 2.97$ and $x = 3$ and a local minimum at $P(2.98) \approx -0.02$.

There is also a local maximum at $P(7.03) \approx 264.03$ and a local minimum at $P(10.98) \approx 15.98$. $(3\text{-}3)$

59. $[x - (-1)]^2(x - 0)^3[x - (3 + 5i)][x - (3 - 5i)]$

$$= (x + 1)^2 x^3 (x - 3 - 5i)(x - 3 + 5i)$$

degree 7 $(3\text{-}4)$

60. Yes, for example:

$P(x) = (x + i)(x - i)(x + \sqrt{2})(x - \sqrt{2}) = x^4 - x^2 - 2$ has irrational zeros $\sqrt{2}$ and $-\sqrt{2}$.   *(3-4)*

61. (A) We form a synthetic division table:

	1	9	-500	0	20,000
0	1	9	-500	0	20,000
10	1	19	-310	-3,100	-11,000
20	1	29	80	1,600	52,000
-10	1	-1	-490	4,900	-29,000
-20	1	-11	-280	5,600	-92,000
-30	1	-21	130	-3,900	137,000

From the table, 20 is an upper bound and -30 is a lower bound.

(B) Examining the graph of $y = P(x)$, we obtain

The zeros of $P(x)$ are -26.68, -6.22, 7.23, and 16.67.   *(3-3)*

62. The possible rational zeros of $P(x)$ are ±1, ±2, ±3, ±4, ±6, ±12. Examining the graph of $y = P(x)$, it appears that there may be a zero of even multiplicity near -1 and a zero of odd multiplicity near 2. We test the likely candidates and find

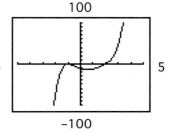

```
      1   -4    3   10  -10  -12
            2   -4   -2   16   12
   2 | 1   -2   -1    8    6    0
```

2 is a zero. $P(x) = (x - 2)(x^4 - 2x^3 - x^2 + 8x + 6)$. Testing -1 in $x^4 - 2x^3 - x^2 + 8x + 6$ we obtain

```
      1   -2   -1    8    6
           -1    3   -2   -6
  -1 | 1   -3    2    6    0
```

-1 is a zero. $P(x) = (x - 2)(x + 1)(x^3 - 3x^2 + 2x + 6)$. Testing -1 again in $x^3 - 3x^2 + 2x + 6$ we obtain

```
      1   -3    2    6
           -1    4   -6
  -1 | 1   -4    6    0
```

-1 is a double zero of $P(x)$. $P(x) = (x - 2)(x + 1)^2(x^2 - 4x + 6)$.

We complete the solution by solving $x^2 - 4x + 6 = 0$ using the quadratic formula with $a = 1$, $b = -4$, and $c = 6$.

$$x = \frac{-(-4) \pm \sqrt{(-4)^2 - 4(1)(6)}}{2(1)} = \frac{4 \pm \sqrt{-8}}{2} = \frac{4 \pm 2i\sqrt{2}\,i}{2} = 2 \pm i\sqrt{2}$$

The zeros of $P(x)$ are 2, -1 (double), and $2 \pm i\sqrt{2}$.

$$P(x) = (x - 2)(x + 1)^2[x - (2 + i\sqrt{2})][x - (2 - i\sqrt{2})]$$
$$= (x - 2)(x + 1)^2(x - 2 - i\sqrt{2})(x - 2 + i\sqrt{2})$$ (3-4)

63. The possible rational zeros of $P(x)$ are ±1, ±2, ±4. Examining the graph of $y = P(x)$, we obtain the graph at the right.

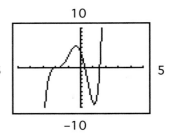

It appears that there may be zeros of unclear multiplicity near -2, as well as zeros between 1 and 2. The only possible rational zero allowed by the graph from the above list is -2. Testing it, we obtain

```
     1    4    1   -11   -8    4
         -2   -4    6    10   -4
 -2 | 1   2   -3   -5    2    0
```

-2 is a zero. Testing whether it is a double zero, we obtain

```
     1    2   -3   -5    2
         -2    0    6   -2
 -2 | 1   0   -3    1    0
```

So $P(x) = (x + 2)^2(x^3 - 3x + 1)$. Examining the graph of $x^3 - 3x + 1$, we obtain

The zeros of $P(x)$ are -2 (double), -1.88, 0.35, and 1.53. (3-4)

64. Since the real zeros are 5 and 8, we should have factors of $x - 5$ and $x - 8$ in the numerator. A factor of $x - 1$ in the denominator provides vertical asymptote 1, but we need the degree of the numerator and denominator to be equal in order to have a non-zero horizontal asymptote. So we'll make the denominator $(x - 1)^2$. Finally, a factor of 3 in the numerator makes $y = 3$ the horizontal asymptote: $f(x) = \dfrac{3(x - 8)(x - 5)}{(x - 1)^2}$. (3-5)

65.
$$A = P\frac{(1 + i)^n - 1}{i}$$

$$Ai = P[(1 + i)^n - 1]$$

$$\frac{Ai}{P} = (1 + i)^n - 1$$

$$1 + \frac{Ai}{P} = (1 + i)^n$$

$$\ln\left(1 + \frac{Ai}{P}\right) = n \ln(1 + i)$$

$$n = \frac{\ln\left(1 + \frac{Ai}{P}\right)}{\ln(1 + i)} \qquad (4\text{-}5)$$

66.
$$\ln y = 5x + \ln A$$
$$\ln y - \ln A = 5x$$
$$\ln\left(\frac{y}{A}\right) = 5x$$
$$\frac{y}{A} = e^{5x}$$
$$y = Ae^{5x} \qquad (4\text{-}5)$$

67.
$$y = \frac{e^x - 2e^{-x}}{2}$$

$$2y = e^x - 2e^{-x}$$

$$2y = e^x - \frac{2}{e^x}$$

$$2ye^x = e^x e^x - e^x\left(\frac{2}{e^x}\right)$$

$$2ye^x = (e^x)^2 - 2$$

$$0 = (e^x)^2 - 2ye^x - 2$$

This equation is quadratic in e^x

$$e^x = \frac{-b \pm \sqrt{b^2 - 4ac}}{2a} \qquad a = 1, \; b = -2y, \; c = -2$$

$$e^x = \frac{-(-2y) \pm \sqrt{(-2y)^2 - 4(1)(-2)}}{2(1)}$$

$$e^x = \frac{2y \pm \sqrt{4y^2 + 8}}{2}$$

$$e^x = y \pm \sqrt{y^2 + 2}$$

Note: Since $0 < 2$, $y^2 < y^2 + 2$, $\sqrt{y^2} < \sqrt{y^2 + 2}$ and $y < \sqrt{y^2 + 2}$ for all real y.
So $y - \sqrt{y^2 + 2}$ is always negative, which means it can't provide a solution.
Also, $y + \sqrt{y^2 + 2}$ is always positive.
$$x = \ln(y + \sqrt{y^2 + 2}) \qquad\qquad (4\text{-}5)$$

68. <u>Algebraic solution:</u>
$$\frac{4x}{x^2 - 1} - 3 < 0$$

$$\frac{4x}{x^2 - 1} - \frac{3(x^2 - 1)}{x^2 - 1} < 0$$

$$\frac{4x - 3x^2 + 3}{x^2 - 1} < 0$$

$$\frac{-3x^2 + 4x + 3}{x^2 - 1} < 0$$

zeros of the numerator: $a = -3$, $b = 4$, $c = 3$
$$x = \frac{-4 \pm \sqrt{16 - 4(-3)(3)}}{2(-3)}$$

$$= \frac{-4 \pm \sqrt{52}}{-6}$$
$$= -0.535, 1.869$$

zeros of the denominator:

$$x^2 - 1 = 0$$
$$x^2 = 1$$
$$x = \pm 1$$

These four zeros divide the real line into the five intervals shown below.

Interval	Test Number	Result	Sign of f
$(-\infty, -1)$	-2	-5.9	$-$
$(-1, -0.535)$	-0.8	5.9	$+$
$(-0.535, 1)$	0.8	-11.9	$-$
$(1, 1.869)$	$3/2$	1.8	$+$
$(1.869, \infty)$	3	-1.5	$-$

The expression is negative on $(-\infty, -1) \cup (-0.535, 1) \cup (1.869, \infty)$.

<u>Graphical solution:</u>

Graph $y_1 = \dfrac{4x}{x^2 - 1}$, $y_2 = 3$ and find where the graph of y_1 is below the graph of y_2.

This occurs on $(-\infty, -1) \cup (-0.535, 1) \cup (1.869, \infty)$. *(3-5)*

69. (A) Graph $y_1 = -4.8x^3 + 47x^2 - 35x - 40$ on the interval $[0,12]$ and use the MAXIMUM command.

The snack bar should be open for 6.1 hours to maximize profit.

(B) Add $y_2 = 300$ to the graph and use the INTERSECT command.

It should be open 4.3 or 7.6 hours to make a profit of $300.

(C) Use the ZERO command.

They will make a profit if open between 1.5 and 8.9 hours.   (*3-1, 3-3*)

70. (A) Evaluate T for $w = 1$: $T(1) = 540 - \dfrac{450(1)}{1 + 2} = 390$ seconds

(B)   The degrees of the numerator and denominator of the fraction are the same, so that term of the function approaches 450/1 (ratio of leading coefficients), or 450.   The entire function approaches $540 - 450 = 90$.   This tells you that in the long run, her best time will approach 90 seconds.

(C) The vertical asymptote occurs at $w = -2$.   Since w represents weeks after learning how to solve the puzzle, negative values are irrelevant.   (*3-5*)

71. We are given y = length of container and x = width of one end; girth is the sum of the four ends, so $4x$ = girth.

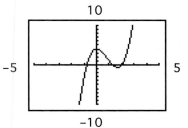

Length + girth = $y + 4x = 10$
         So $y = 10 - 4x$
Since Volume = $8 = x^2 y$, we have
         $8 = x^2(10 - 4x)$
         $8 = 10x^2 - 4x^3$
$4x^3 - 10x^2 + 8 = 0$
$2x^3 - 5x^2 + 4 = 0$
The possible rational solutions of this
equation are ± 1, ± 2, ± 4, $\pm \dfrac{1}{2}$.

Examining the graph of $y = 2x^3 - 5x^2 + 4$,
it appears that there may be an integer zero at 2, a zero between -1 and 0, and a zero between 1 and 2. Testing 2, we obtain

```
   2  -5   0    4
       4  -2   -4
2│ 2  -1  -2    0
```

2 is a zero.
So, the equation can be factored
   $(x - 2)(2x^2 - x - 2) = 0$
To find the remaining zeros, we solve $2x^2 - x - 2 = 0$ by the quadratic formula.
$2x^2 - x - 2 = 0$

$$x = \frac{-b \pm \sqrt{b^2 - 4ac}}{2a} \qquad a = 2,\ b = -1,\ c = -2$$

$$x = \frac{-(-1) \pm \sqrt{(-1)^2 - 4(2)(-2)}}{2(2)}$$

$$x = \frac{1 + \sqrt{17}}{4} \quad \text{(discarding the negative solution)}$$

The positive zeros are $x = 2$, 1.28.
The dimensions of the package are $x = 2$ feet and $y = 2$ feet, or $x = 1.28$ feet and $y = 4.88$ feet.   (*3-4*)

72.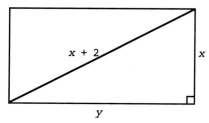

Labeling the rectangle as in the diagram, we have

$$\text{Area} = xy = 6$$

$$\text{So } y = \frac{6}{x}$$

Applying the Pythagorean theorem, we have

$$(x + 2)^2 = x^2 + y^2$$

$$(x + 2)^2 = x^2 + \left(\frac{6}{x}\right)^2$$

$$x^2 + 4x + 4 = x^2 + \frac{36}{x^2}$$

$$4x + 4 = \frac{36}{x^2}$$

$$4x^3 + 4x^2 = 36$$

$$4x^3 + 4x^2 - 36 = 0$$

$$x^3 + x^2 - 9 = 0$$

There are no rational zeros of $P(x) = x^3 + x^2 - 9$. Examining the graph of $y = P(x)$, we see that the only real zero is $x = 1.79$.

Then $y = \dfrac{6}{1.79} = 3.35$. The dimensions of the rectangle are 1.79 feet by 3.35 feet.

(3-4)

73. $t^2 = kd^3$

(3-6)

74. Write $v = \dfrac{k}{\sqrt{w}}$. Substitute $v = 0.3$ and $w = w_1$ and solve for k.

$$0.3 = \frac{k}{\sqrt{w_1}}$$

$$k = 0.3\sqrt{w_1} \text{ where } w_1 = \text{weight of oxygen molecule.}$$

The equation of variation is

$$v = \frac{0.3\sqrt{w_1}}{\sqrt{w_2}}$$

If $w_2 = \dfrac{w_1}{16}$ (weight of hydrogen molecule) then

$$v = \frac{0.3\sqrt{w_1}}{\sqrt{w_1/16}} = \frac{0.3\sqrt{w_1}}{\sqrt{w_1}/4} = 4(0.3) = 1.2 \text{ mile/second}$$

(3-6)

75. We use the Doubling Time Growth Model:

$$P = P_0 2^{t/d}$$

Substituting $P_0 = 60$ million and $d = 23$, we have

$$P = 60(2^{t/23}) \text{ million}$$

(A) For $t = 5$, $P = 60(2^{5/23})$ million
$$= 69.8 \text{ million}$$

(B) For $t = 30$, $P = 60(2^{30/23})$ million
$$= 148 \text{ million}$$

(3-6)

76. We solve $P = P_0(1.07)^t$ for t, using $P = 2P_0$

$$2P_0 = P_0(1.07)^t$$
$$2 = (1.07)^t$$
$$\ln 2 = t \ln 1.07$$
$$\frac{\ln 2}{\ln 1.07} = t$$
$$t = 10.2 \text{ years}$$

$\qquad$ (4-1)

77. We solve $P = P_0 e^{0.07t}$ for t, using $P = 2P_0$

$$2P_0 = P_0 e^{0.07t}$$
$$2 = e^{0.07t}$$
$$\ln 2 = 0.07t$$
$$t = \frac{\ln 2}{0.07}$$
$$t = 9.90 \text{ years}$$

$\qquad$ (4-1)

78. First, we solve $M = \frac{2}{3} \log\left(\frac{E}{E_0}\right)$ for E.

$$M = \frac{2}{3} \log\left(\frac{E}{E_0}\right)$$
$$\frac{3M}{2} = \log \frac{E}{E_0}$$
$$\frac{E}{E_0} = 10^{3m/2}$$
$$E = E_0(10^{3m/2})$$

We now compare E_1 for $M = 8.3$ with E_2 for $M = 7.1$.

$$E_1 = E_0(10^{3 \cdot 8.3/2}) \qquad\qquad E_2 = E_0(10^{3 \cdot 7.1/2})$$
$$E_1 = E_0(10^{12.45}) \qquad\qquad E_2 = E_0(10^{10.65})$$

$$\frac{E_1}{E_2} = \frac{E_0(10^{12.45})}{E_0(10^{10.65})} = 10^{12.45 - 10.65} = 10^{1.8}$$
$$E_1 = 10^{1.8} E_2 \text{ or } 63.1E_2.$$

The 1906 earthquake was 63.1 times as powerful.   (4-4)

79. We solve $D = 10 \log \frac{I}{I_0}$ for I, with $D = 88$, $I_0 = 10^{-12}$

$$88 = 10 \log \frac{I}{10^{-12}}$$
$$8.8 = \log \frac{I}{10^{-12}}$$
$$10^{8.8} = \frac{I}{10^{-12}}$$
$$I = 10^{8.8} \cdot 10^{-12}$$
$$I = 10^{-3.2}$$
$$I = 6.31 \times 10^{-4} \ w/m^2$$

$\qquad$ (4-4)

80. Enter the data (note that the whole list is not visible in this window):

(A) Compute the linear regression equation, graph, and compute the value of the life expectancy that corresponds to $x = 40$ (year 2010).

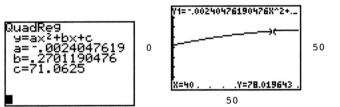

The model predicts a life expectancy of 78.9 years in 2010.

(B) Compute the quadratic regression equation, graph, and compute the value of the life expectancy that corresponds to $x = 40$.

The model predicts a life expectancy of 78.0 years in 2010.

(C) Compute the cubic regression equation, graph, and compute the value of the life expectancy that corresponds to $x = 40$.

The model predicts a life expectancy of 79.1 years in 2010.

(D) Compute the exponential regression equation, graph, and compute the value of the life expectancy that corresponds to $x = 40$

The model predicts a life expectancy of 79.1 years in 2010.   (*3-1, 4-2*)

81. The quadratic model (78.0 years) is closest to the Census Bureau projection.
(*3-1, 4-2*)

CHAPTER 5   Modeling with Systems of Equations and Inequalities

Section 5-1

1. Answers will vary.   3. Answers will vary.   5. Answers will vary.

7. Both lines in the given system are different, but they have the same slope $\left(\frac{1}{2}\right)$ and are therefore parallel. This system corresponds to (b) and has no solution.

9. In slope-intercept form, these equations are $y = 2x - 5$ and $y = -\frac{3}{2}x - \frac{3}{2}$. The first has slope 2 and y intercept -5; the other has slope $-\frac{3}{2}$ and y intercept $-\frac{3}{2}$. This system corresponds to (d) and its solution can be read from the graph as (1, -3). Checking, we see that
$$2x - \ \ y = 2 \cdot 1 - (-3) = 5$$
$$3x + 2y = 3 \cdot 1 + 2(-3) = -3$$

11. $x + y = 7$
$x - y = 3$
If we add, we can eliminate y.
$$\begin{array}{r} x + y = 7 \\ \underline{x - y = 3} \\ 2x \ \ \ \ \ \ = 10 \\ x = 5 \end{array}$$
Now substitute $x = 5$ back into the top equation and solve for y.
$$5 + y = 7$$
$$y = 2$$
(5, 2)

13. $3x - 2y = 12$
$7x + 2y = 8$
If we add, we can eliminate y.
$$\begin{array}{r} 3x - 2y = 12 \\ \underline{7x + 2y = 8} \\ 10x \ \ \ \ \ \ = 20 \\ x = 2 \end{array}$$
Now substitute $x = 2$ back into the bottom equation and solve for y.
$$7(2) + 2y = 8$$
$$2y = -6$$
$$y = -3$$
(2, -3)

15. $3u + 5v = 15$
$6u + 10v = -30$
If we multiply the top equation by -2 and add, we eliminate both u and v.
$$\begin{array}{r} -6u - 10v = -30 \\ \underline{6u + 10v = -30} \\ 0 = -60 \end{array}$$
No solution. The equations represent parallel lines.

17. $y = 2x + 3$
$y = 3x - 5$
Substitute y from the first equation into the second equation to eliminate y.
$$2x + 3 = 3x - 5$$
$$-x + 3 = -5$$
$$-x = -8$$
$$x = 8$$
Now replace x with 8 in the first equation to find y.
$$y = 2 \cdot 8 + 3$$
$$y = 19$$
Solution: $x = 8$, $y = 19$

19. $x - y = 4$
$x + 3y = 12$
Solve the first equation for x in terms of y.
$$x = 4 + y$$
Substitute into the second equation to eliminate x.
$$(4 + y) + 3y = 12$$
$$4y = 8$$
$$y = 2$$
Now replace y with 2 in the first equation to find x.
$$x - 2 = 4$$
$$x = 6$$
Solution: $x = 6$, $y = 2$

21. $3x - y = 7$
$2x + 3y = 1$
Solve the first equation for y in terms of x.
$-y = 7 - 3x$
$y = -7 + 3x$
Substitute into the second equation to eliminate y.
$2x + 3(-7 + 3x) = 1$
$2x - 21 + 9x = 1$
$11x = 22$
$x = 2$

Now replace x with 2 in the first equation to find y.
$3 \cdot 2 - y = 7$
$6 - y = 7$
$y = -1$
Solution: $x = 2$, $y = -1$

23. $4x + 3y = 26$
$3x - 11y = -7$
Solve the second equation for x in terms of y.
$3x = 11y - 7$
$x = \dfrac{11y - 7}{3}$
Substitute into the first equation to eliminate x.
$4\left(\dfrac{11y - 7}{3}\right) + 3y = 26$
$\dfrac{44y - 28}{3} + 3y = 26$
$44y - 28 + 9y = 78$
$53y = 106$
$y = 2$
Now replace y with 2 in the above expression for x:
$x = \dfrac{11 \cdot 2 - 7}{3}$
$x = 5$

Solution: $x = 5$, $y = 2$

25. $7m + 12n = -1$
$5m - 3n = 7$
Solve the first equation for n in terms of m.
$12n = -1 - 7m$
$n = \dfrac{-1 - 7m}{12}$
Substitute into the second equation to eliminate n.
$5m - 3\left(\dfrac{-1 - 7m}{12}\right) = 7$
$5m - \dfrac{-1 - 7m}{4} = 7$
$20m + 1 + 7m = 28$
$27m = 27$
$m = 1$
Now replace m with 1 in the above expression for n:
$n = \dfrac{-1 - 7(1)}{12}$
$n = -\dfrac{2}{3}$
Solution: $m = 1$, $n = -\dfrac{2}{3}$

27. $y = 0.08x$
$y = 100 + 0.04x$
Substitute y from the first equation into the second equation to eliminate y.
$0.08x = 100 + 0.04x$
$0.04x = 100$
$x = 2,500$ $\qquad$ $y = 0.08(2,500)$
$y = 200$

Solution: $x = 2,500$, $y = 200$

29. $0.2u - 0.5v = 0.07$
$0.8u - 0.3v = 0.79$

For convenience, eliminate decimals by multiplying both sides of each equation by 100.

$20u - 50v = 7$ Don't forget to multiply right sides by 100, too!
$80u - 30v = 79$

Solve the first equation for u in terms of v and substitute into the second equation to eliminate u.

$$20u = 50v + 7$$
$$u = \frac{50v + 7}{20}$$
$$80\left(\frac{50v + 7}{20}\right) - 30v = 79$$
$$4(50v + 7) - 30v = 79$$
$$200v + 28 - 30v = 79$$
$$170v = 51$$
$$v = 0.3 \qquad u = \frac{50(0.3) + 7}{20}$$
$$u = 1.1$$

Solution: $u = 1.1$, $v = 0.3$

31. $\dfrac{2}{5}x + \dfrac{3}{2}y = 2$

$\dfrac{7}{3}x - \dfrac{5}{4}y = -5$

Eliminate fractions by multiplying both sides of the first equation by 10 and both sides of the second equation by 12.

$$10\left(\frac{2}{5}x + \frac{3}{2}y\right) = 20$$
$$4x + 15y = 20$$
$$12\left(\frac{7}{3}x - \frac{5}{4}y\right) = -60$$
$$28x - 15y = -60$$

Solve the first equation for y in terms of x and substitute into the second equation to eliminate y.

$$15y = 20 - 4x$$
$$y = \frac{20 - 4x}{15}$$
$$28x - 15\left(\frac{20 - 4x}{15}\right) = -60$$
$$28x - (20 - 4x) = -60$$
$$28x - 20 + 4x = -60$$
$$32x = -40$$
$$x = -\frac{5}{4}$$
$$y = \frac{20 - 4\left(-\frac{5}{4}\right)}{15}$$
$$y = \frac{20 + 5}{15}$$
$$y = \frac{5}{3}$$

Solution: $x = -\dfrac{5}{4}$, $y = \dfrac{5}{3}$

In Problems 33 and 35, a graphing calculator solution is presented.

33. First solve each equation for y.

$$2x - 3y = -5 \qquad\qquad 3x + 4y = 13$$
$$-3y = -2x - 5 \qquad\qquad 4y = -3x + 13$$
$$y = \frac{2}{3}x + \frac{5}{3} \qquad\qquad y = -\frac{3}{4}x + \frac{13}{4}$$

Next enter each equation, graph, and approximate the solution.

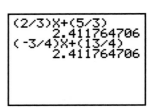

Equation definitions   Intersection point   Check

To two decimal places, the solution is $x = 1.12$, $y = 2.41$ or $(1.12, 2.41)$.

35. First solve each equation for y.

$$3.5x - 2.4y = 0.1 \qquad\qquad 2.6x - 1.7y = -0.2$$
$$-2.4y = -3.5x + 0.1 \qquad\qquad -1.7y = -2.6x - 0.2$$
$$y = \frac{3.5}{2.4}x - \frac{0.1}{2.4} \qquad\qquad y = \frac{2.6x}{1.7} + \frac{0.2}{1.7}$$

Next enter each equation, graph, and approximate the solution.

Equation definitions   Intersection point   Check

To two decimal places, the solution is $x = -2.24$, $y = -3.31$ or $(-2.24, -3.31)$.

37. $x = 2 + p - 2q$
$y = 3 - p + 3q$
Solve the first equation for p in terms of q and x and substitute into the second equation to eliminate p, then solve for q in terms of x and y.

$$p = x - 2 + 2q$$
$$y = 3 - (x - 2 + 2q) + 3q$$
$$y = 3 - x + 2 - 2q + 3q$$
$$y = 5 - x + q$$
$$q = x + y - 5$$

Now substitute this expression for q into $p = x - 2 + 2q$ to find p in terms of x and y.

$$p = x - 2 + 2(x + y - 5)$$
$$p = x - 2 + 2x + 2y - 10$$
$$p = 3x + 2y - 12$$

Solution: $p = 3x + 2y - 12$, $q = x + y - 5$

To check this solution substitute into the original equations to see if true statements result:

$x = 2 + p - 2q$ $\qquad\qquad\qquad\qquad$ $y = 3 - p + 3q$

$x \overset{?}{=} 2 + (3x + 2y - 12) - 2(x + y - 5)$ $\quad$ $y \overset{?}{=} 3 - (3x + 2y - 12) + 3(x + y - 5)$

$x \overset{?}{=} 2 + 3x + 2y - 12 - 2x - 2y + 10$ $\qquad$ $y \overset{?}{=} 3 - 3x - 2y + 12 + 3x + 3y - 15$

$x \overset{\checkmark}{=} x$ $\qquad\qquad\qquad\qquad\qquad\qquad$ $y \overset{\checkmark}{=} y$

39. $ax + by = h$

$cx + dy = k$

Solve the first equation for x in terms of y and the constants.

$ax = h - by$

$x = \dfrac{h - by}{a} \quad (a \neq 0)$

Substitute this expression into the second equation to eliminate x.

$c\left(\dfrac{h - by}{a}\right) + dy = k$ $\qquad\qquad$ Multiply both sides by a

$ac\left(\dfrac{h - by}{a}\right) + ady = ak$

$c(h - by) + ady = ak$

$ch - bcy + ady = ak$

$ady - bcy = ak - ch$ $\qquad\qquad$ Factor y out of the left side

$(ad - bc)y = ak - ch$

$y = \dfrac{ak - ch}{ad - bc} \qquad ad - bc \neq 0$

Similarly, solve the first equation for y in terms of x and the constants.

$by = h - ax$

$y = \dfrac{h - ax}{b} \quad (b \neq 0)$

Substitute this expression into the second equation to eliminate y.

$cx + d\left(\dfrac{h - ax}{b}\right) = k$ $\qquad\qquad$ Multiply both sides by b

$bcx + bd\left(\dfrac{h - ax}{b}\right) = bk$

$bcx + d(h - ax) = bk$

$bcx + dh - adx = bk$

$bcx - adx = bk - dh$ $\qquad\qquad$ Factor x out of the left side

$(bc - ad)x = bk - dh$

$x = \dfrac{bk - dh}{bc - ad} \qquad bc - ad \neq 0$

or, for consistency with the expression for y,

$x = \dfrac{dh - bk}{ad - bc}$

Solution: $x = \dfrac{dh - bk}{ad - bc}$, $\; y = \dfrac{ak - ch}{ad - bc} \qquad ad - bc \neq 0$

41. Let x = airspeed of the plane
$\quad\quad y$ = rate at which wind is blowing
Then $\quad x - y$ = ground speed flying from Atlanta to Los Angeles (head wind)
$\quad\quad\quad\quad x + y$ = ground speed flying from Los Angeles to Atlanta (tail wind)
Then, applying Distance = Rate × Time, we have
$\quad\quad 2,100 = 8.75(x - y)$
$\quad\quad 2,100 = 5(x + y)$
Divide both sides of the first equation by 8.75 and both sides of the second by 5:
$\quad\quad x - y = 240$
$\quad\quad x + y = 420$
Solve the first equation for x in terms of y and substitute into the second equation. $\quad x = 240 + y$
$\quad 240 + y + y = 420$
$\quad\quad\quad\quad\quad 2y = 180$
$\quad\quad\quad\quad\quad\quad y = 90$ mph = wind rate
$\quad\quad x = 240 + y$
$\quad\quad x = 240 + 90$
$\quad\quad x = 330$ mph = airspeed

43. Let x = time rowed upstream
$\quad\quad\quad y$ = time rowed downstream
Then $x + y = \dfrac{1}{4}$ $\quad$ (15 min = $\dfrac{1}{4}$ hr.)
Since $\quad$ rate upstream = 20 - 2 = 18 mph and
$\quad\quad\quad$ rate downstream = 20 + 2 = 22 mph,

applying Distance = Rate × Time to the equal distances upstream and downstream, we have
$\quad\quad 18x = 22y$
Solve the first equation for y in terms of x and substitute into the second equation.
$\quad\quad y = \dfrac{1}{4} - x$
$\quad\quad 18x = 22\left(\dfrac{1}{4} - x\right)$
$\quad\quad 18x = 5.5 - 22x$
$\quad\quad 40x = 5.5$
$\quad\quad\quad x = 0.1375$ hr.
Then the distance rowed upstream is $18x = 18(0.1375) = 2.475$ km.

45. Let x = amount of 50% solution
$\quad\quad\quad y$ = amount of 80% solution
100 milliliters are required, so
$\quad x + y = 100$
68% of the 100 milliliters must be acid, so
$\quad 0.50x + 0.80y = 0.68(100)$
Solve the first equation for y in terms of x and substitute into the second equation.
$\quad\quad y = 100 - x$
$0.50x + 0.80(100 - x) = 0.68(100)$
$\quad\quad 0.5x + 80 - 0.8x = 68$
$\quad\quad\quad\quad\quad\quad -0.3x = -12$
$\quad\quad\quad\quad\quad\quad\quad\quad x = 40$ milliliters of 50% solution
$\quad\quad y = 100 - x$
$\quad\quad y = 100 - 40 = 60$ milliliters of 80% solution

47. "Break even" means Cost = Revenue.
    Let y represent both cost and revenue (since they're equal)
    Let x = number of CDs sold
        y = Revenue = number of CDs sold × price per CD
        $y = x(8.00)$
        y = Cost = Fixed Cost + Variable Cost
            = 17,680 + number of CDs × cost per CD
        $y = 17,680 + x(4.60)$

    Substitute y from the first equation into the second equation to eliminate y.
    $8.00x = 17,680 + 4.60x$
    $3.40x = 17,680$
    $x = \dfrac{17,680}{3.40}$
    $x = 5,200$ CDs

49. Let x = amount invested at 10% $0.1x$ = yield on amount invested at 10%
        y = amount invested at 15% $0.15y$ = yield on amount invested at 15%
    Then
                $x + y = 12,000$ (amount invested)
        $0.1x + 0.15y = 0.12(12,000)$ (total yield)

    Solve the first equation for y in terms of x and substitute into the second equation.
    $y = 12,000 - x$
        $0.1x + 0.15(12,000 - x)$ $= 0.12(12,000)$
        $0.1x + 0.15(12,000) - 0.15x$ $= 0.12(12,000)$
        $0.1x + 1,800 - 0.15x$ $= 1,440$
                    $-0.05x + 1,800 = 1,440$
                    $-0.05x = -360$
                            $x = \$7,200$ invested at 10%
            $y = 12,000 - x = \$4,800$ invested at 15%

51. Let x = number of hours Mexico plant is operated
        y = number of hours Taiwan plant is operated
    Then (Production at Mexico plant) + (Production at Taiwan plant) = (Total Production)
                $40x$ + $20y$ = 4,000 (keyboards)
                $32x$ + $32y$ = 4,000 (screens)

    Solve the first equation for y in terms of x and substitute into the second equation.
            $20y = 4,000 - 40x$
                $y = 200 - 2x$
        $32x + 32(200 - 2x) = 4,000$
        $32x + 6,400 - 64x = 4,000$
                    $-32x = -2,400$
                        $x = 75$ hours Mexico plant
        $y = 200 - 2x = 200 - 2(75)$
            $= 50$ hours Taiwan plant

53. Let x = number of grams of Mix A
        y = number of grams of Mix B
    Then (Nutrition from Mix A) + (Nutrition for Mix B) = (Total Nutrition)
                $0.10x$ + $0.20y$ =   20 (Total protein)
                $0.06x$ + $0.02y$ =   6 (Total fat)

    For convenience, eliminate decimals by multiplying both sides of the first equation by 10 and the second equation by 100.
            $x + 2y = 200$
            $6x + 2y = 600$

Solve the first equation for x in terms of y and substitute into the second equation.

$$x = 200 - 2y$$
$$6(200 - 2y) + 2y = 600$$
$$1200 - 12y + 2y = 600$$
$$-10y = -600$$
$$y = 60 \text{ grams Mix } B$$
$$x = 200 - 2y = 200 - 2(60)$$
$$= 80 \text{ grams Mix } A$$

55. (A) If $p = 4$, then $4 = 0.007q + 3$, $q = \dfrac{1}{0.007} = 143$ T-shirts is the number that suppliers are willing to supply at this price.

$4 = -0.018q + 15$, $q = \dfrac{11}{0.018} = 611$ T-shirts is the number that consumers will purchase.

Demand exceeds supply and the price will rise.

(B) If $p = 8$, then $8 = 0.007q + 3$, $q = \dfrac{5}{0.007} = 714$ T-shirts is the number that suppliers are willing to supply.

$8 = -0.018q + 15$, $q = \dfrac{7}{0.018} = 389$ T-shirts is the number that consumers will purchase at this price.

Supply exceeds demand and the price will fall.

(C) Solve $p = 0.007q + 3$
$\qquad\quad\ p = -0.018q + 15$

(D)

Substitute the expression for p in terms of q from the first equation into the second equation.

$$0.007q + 3 = -0.018q + 15$$
$$0.025q + 3 = 15$$
$$0.025q = 12$$
$$q = 480 \text{ T-shirts is the equilibrium quantity.}$$
$$p = 0.007(480) + 3 = \$6.36 \text{ is the equilibrium price.}$$

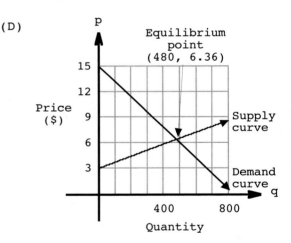

57. (A) Write $p = aq + b$.
Since $p = 0.60$ corresponds to supply $q = 450$,
$$0.60 = 450a + b$$
Since $p = 0.90$ corresponds to supply $q = 750$,
$$0.90 = 750a + b$$

Solve the first equation for b in terms of a and substitute into the second equation.
$$b = 0.60 - 450a$$
$$0.90 = 750a + 0.60 - 450a$$
$$0.30 = 300a$$
$$a = 0.001$$
$$b = 0.60 - 450a = 0.60 - 450(0.001)$$
$$= 0.15$$
The supply equation is $p = 0.001q + 0.15$.

(B) Write $p = cq + d$.
Since $p = 0.60$ corresponds to demand $q = 645$,
$$0.60 = 645c + d$$
Since $p = 0.90$ corresponds to demand $q = 495$,
$$0.90 = 495c + d$$
Solve the first equation for d in terms of c and substitute into the second equation.
$$d = 0.60 - 645c$$
$$0.90 = 495c + 0.60 - 645c$$
$$0.30 = -150c$$
$$c = -0.002$$
$$d = 0.60 - 645c = 0.60 - 645(-0.002)$$
$$= 1.89$$
The demand equation is $p = -0.002q + 1.89$

(C) Solve the system of equations
$$p = 0.001q + 0.15$$
$$p = -0.002q + 1.89$$
Substitute p from the first equation into the second equation to eliminate p.
$$0.001q + 0.15 = -0.002q + 1.89$$
$$0.003q = 1.74$$
$$q = 580 \text{ bushels} = \text{equilibrium quantity}$$
$$p = 0.001q + 0.15 = 0.001(580) + 0.15$$
$$= \$0.73 \text{ equilibrium price}$$

59. $s = a + bt^2$

(A) We are given: When $t = 1$, $s = 180$
When $t = 2$, $s = 132$

Substituting these values in the given equation, we have
$$180 = a + b(1)^2$$
$$132 = a + b(2)^2 \quad \text{or} \quad 180 = a + b$$
$$132 = a + 4b$$
Solve the first equation for a in terms of b and substitute into the second equation to eliminate a.
$$a = 180 - b$$
$$132 = 180 - b + 4b$$
$$132 = 180 + 3b$$
$$-48 = 3b$$
$$b = -16 \qquad 180 = a - 16$$
$$a = 196$$

(B) The height of the building is represented by s, the distance of the object above the ground, when $t = 0$. Since we now know
$$s = 196 - 16t^2$$
from part (A), when $t = 0$, $s = 196$ feet is the height of the building.

(C) The object falls until s, its distance above the ground, is zero. Since
$$s = 196 - 16t^2$$
we substitute $s = 0$ and solve for t.
$$0 = 196 - 16t^2$$
$$16t^2 = 196$$
$$t^2 = \frac{196}{16}$$
$$t = \frac{14}{4} \quad \text{(discarding the negative solution)}$$
$$t = 3.5 \text{ seconds}$$

61. Let p = time of primary wave
s = time for secondary wave
We know
$s - p = 16$ (time difference)
To find a second equation, we have to use Distance = Rate × Time
$5p$ = distance for primary wave
$3s$ = distance for secondary wave
These distances are equal, so
$5p = 3s$

Solve the first equation for s in terms of p and substitute into the second equation to eliminate s.
$s = p + 16$
$5p = 3(p + 16)$
$5p = 3p + 48$
$2p = 48$
$p = 24$ seconds $\qquad s = 24 + 16$
$\qquad\qquad\qquad\qquad\qquad s = 40$ seconds

The distance traveled is $5p = 120$ miles.

Section 5-2

1. Answers will vary.   **3.** Answers will vary.

Note: The checking steps are not shown, but should be performed by the student.

5. $\begin{array}{lll} -2x & = 2 & E_1 \\ x - 3y & = 2 & E_2 \\ -x + 2y + 3z & = -7 & E_3 \end{array}$

Solve E_1 for x.
$\qquad\qquad -2x = 2 \qquad E_1$
$\qquad\qquad\quad x = -1$
Substitute $x = -1$ in E_2 and solve for y.
$\qquad\qquad x - 3y = 2 \qquad E_2$
$\qquad\quad -1 - 3y = 2$
$\qquad\qquad\qquad y = -1$
Substitute $x = -1$ and $y = -1$ in E_3 and solve for z.
$\qquad\quad -x + 2y + 3z = -7 \qquad E_3$
$-(-1) + 2(-1) + 3z = -7$
$\qquad\qquad\qquad\quad z = -2$
$(-1, -1, -2)$

7. $\begin{array}{lll} 2y - z = 2 & E_1 \\ -4y + 2z = 1 & E_2 \\ x - 2y + 3z = 0 & E_3 \end{array}$

Multiply E_1 by 2 and add to E_2
$\begin{array}{ll} 4y - 2z = 4 & 2E_1 \\ \underline{-4y + 2z = 1} & E_2 \\ \quad\quad\;\; 0 = 5 & E_4 \end{array}$

A contradiction. No solution.

9. $\begin{array}{lll} x - 3y & = 2 & E_1 \\ 2y + z & = -1 & E_2 \\ x - y + z & = 1 & E_3 \end{array}$

Multiply E_1 by -1 and add to E_3 to eliminate x.
$\begin{array}{lll} -x + 3y & = -2 & (-1)E_1 \\ \underline{x - y + z = 1} & & E_3 \\ \quad\;\; 2y + z = -1 & & E_4 \end{array}$

Equivalent system:
$\begin{array}{lll} x - 3y & = 2 & E_1 \\ 2y + z & = -1 & E_2 \\ 2y + z & = -1 & E_4 \end{array}$

If E_2 is multiplied by -1 and added to E_4, $0 = 0$ results. The system is dependent and equivalent to

$$x - 3y = 2$$
$$2y + z = -1$$

Let $y = s$. Then

$$2s + z = -1$$
$$z = -2s - 1$$
$$x - 3s = 2$$
$$x = 3s + 2$$

Solutions: $\{(3s + 2, \; s, \; -2s - 1)|\; s$ is any real number$\}$

11.
$$4y - z = -13 \quad E_1$$
$$3y + 2z = 4 \quad E_2$$
$$6x - 5y - 2z = -18 \quad E_3$$

Multiply E_1 by 2 and add to E_2 to eliminate z.

$$8y - 2z = -26 \qquad 2E_1$$
$$\underline{3y + 2z = 4} \qquad E_2$$
$$11y \quad = -22 \qquad E_4$$
$$y = -2$$

Substitute $y = -2$ into E_2 and solve for z.

$$3y + 2z = 4 \quad E_2$$
$$3(-2) + 2z = 4$$
$$z = 5$$

Substitute $y = -2$ and $z = 5$ into E_3 and solve for x.

$$6x - 5(-2) - 2(5) = -18$$
$$x = -3$$

$(-3, -2, 5)$

13.
$$2x + y - z = 5 \qquad E_1$$
$$x - 2y - 2z = 4 \qquad E_2$$
$$3x + 4y + 3z = 3 \qquad E_3$$

Multiply E_2 by -2 and add to E_1 to eliminate x. Also multiply E_2 by -3 and add to E_3 to eliminate x.

$$2x + y - z = 5 \qquad E_1$$
$$\underline{-2x + 4y + 4z = -8} \qquad (-2)E_2$$
$$5y + 3z = -3 \qquad E_4$$

$$3x + 4y + 3z = 3 \qquad E_3$$
$$\underline{-3x + 6y + 6z = -12} \qquad (-3)E_2$$
$$10y + 9z = -9 \qquad E_5$$

Equivalent system:

$$x - 2y - 2z = 4 \qquad E_2$$
$$5y + 3z = -3 \qquad E_4$$
$$10y + 9z = -9 \qquad E_5$$

Multiply E_4 by -2 and add to E_5 to eliminate y.

$$-10y - 6z = 6 \qquad (-2)E_4$$
$$\underline{10y + 9z = -9} \qquad E_5$$
$$3z = -3$$
$$z = -1$$

Substitute $z = -1$ into E_4 and solve for y.

$$5y + 3z = -3 \qquad E_4$$
$$5y + 3(-1) = -3$$
$$y = 0$$

Substitute $y = 0$ and $z = -1$ into E_2 and solve for x.

$$x - 2y - 2z = 4 \quad E_2$$
$$x - 2(0) - 2(-1) = 4$$
$$x = 2$$
$$(2, \ 0, \ -1)$$

15. $\quad x - y + \ \ z = \ \ 1 \quad E_1$
$$2x + y + \ \ z = \ \ 6 \quad E_2$$
$$7x - y + 5z = 15 \quad E_3$$

Multiply E_1 by -2 and add to E_2 to eliminate x. Also multiply E_1 by -7 and add to E_3 to eliminate x.

$$-2x + 2y - 2z = -2 \quad (-2)E_1$$
$$\underline{\ 2x + \ \ y + \ \ z = \ \ 6} \quad E_2$$
$$3y - \ \ z = \ \ 4 \quad E_4$$

$$-7x + 7y - 7z = -7 \quad (-7)E_1$$
$$\underline{\ 7x - \ \ y + 5z = 15} \quad E_3$$
$$6y - 2z = \ \ 8 \quad E_5$$

Equivalent system:
$$x - y + z = 1 \quad E_1$$
$$3y - z = 4 \quad E_4$$
$$6y - 2z = 8 \quad E_5$$

If E_4 is multiplied by -2 and added to E_5, $0 = 0$ results. The system is dependent and equivalent to
$$x - y + z = 1$$
$$3y - z = 4$$

Let $y = s$. Then $\quad 3s - z = 4$
$$z = 3s - 4$$
$$x - s + (3s - 4) = 1$$
$$x = -2s + 5$$

Solutions: $\{(-2s + 5, \ s, \ 3s - 4) \mid s$ is any real number$\}$

17. $\quad 2a + 4b + 3c = \ \ -6 \quad E_1$
$$a - 3b + 2c = -15 \quad E_2$$
$$a + 2b - \ \ c = \ \ \ 9 \quad E_3$$

Add E_2 to E_3 to eliminate a. Also multiply E_2 by -2 and add to E_1 to eliminate a.

$$a - 3b + 2c = -15 \quad E_2$$
$$\underline{-a + 2b - \ \ c = \ \ \ 9} \quad E_3$$
$$-b + \ \ c = \ \ -6 \quad E_4$$

$$-2a + 6b - 4c = \ \ 30 \quad (-2)E_2$$
$$\underline{\ 2a + 4b + 3c = \ \ -6} \quad E_1$$
$$10b - \ \ c = \ \ 24 \quad E_5$$

Equivalent system:
$$a - 3b + 2c = -15 \quad E_2$$
$$-b + \ \ c = \ \ -6 \quad E_4$$
$$10b - \ \ c = \ \ 24 \quad E_5$$

Add E_4 to E_5 to eliminate c
$$-b + \ \ c = \ \ -6 \quad E_4$$
$$\underline{10b - \ \ c = \ \ 24} \quad E_5$$
$$9b \ \ \ \ \ \ \ = \ \ 18$$
$$b = \ \ \ 2$$

Substitute $b = 2$ into E_4 and solve for c.

$$-b + c = -6 \qquad E_4$$
$$-2 + c = -6$$
$$c = -4$$

Substitute $b = 2$ and $c = -4$ into E_2 and solve for a.

$$a - 3b + 2c = -15 \qquad E_2$$
$$a - 3(2) + 2(-4) = -15$$
$$a = -1$$

$(-1,\ 2,\ -4)$

19.
$$2x - 3y + 3z = -5 \qquad E_1$$
$$3x + 2y - 5z = 34 \qquad E_2$$
$$5x - 4y - 2z = 23 \qquad E_3$$

Multiply E_1 by $-\dfrac{3}{2}$ and add to E_2 to eliminate x. Also multiply E_1 by $-\dfrac{5}{2}$ and add to E_3 to eliminate x.

$$-3x + \frac{9}{2}y - \frac{9}{2}z = \frac{15}{2} \qquad \left(-\frac{3}{2}\right)E_1$$
$$\underline{3x + 2y - 5z = 34} \qquad E_2$$
$$\frac{13}{2}y - \frac{19}{2}z = \frac{83}{2} \qquad E_4$$

$$-5x + \frac{15}{2}y - \frac{15}{2}z = \frac{25}{2} \qquad \left(-\frac{5}{2}\right)E_1$$
$$\underline{5x - 4y - 2z = 23} \qquad E_3$$
$$\frac{7}{2}y - \frac{19}{2}z = \frac{71}{2} \qquad E_5$$

Equivalent system:
$$2x - 3y + 3z = -5 \qquad E_1$$
$$\frac{13}{2}y - \frac{19}{2}z = \frac{83}{2} \qquad E_4$$
$$\frac{7}{2}y - \frac{19}{2}z = \frac{71}{2} \qquad E_5$$

Multiply E_4 by -1 and add to E_5 to eliminate z.

$$-\frac{13}{2}y + \frac{19}{2}z = -\frac{83}{2} \qquad (-1)E_4$$
$$\underline{\frac{7}{2}y - \frac{19}{2}z = \frac{71}{2}} \qquad E_5$$
$$-3y = -6 \qquad E_6$$

Solve E_6 for y to obtain $y = 2$. Substitute $y = 2$ into E_4 and solve for z.

$$\frac{13}{2}y - \frac{19}{2}z = \frac{83}{2} \qquad E_4$$
$$\frac{13}{2}(2) - \frac{19}{2}z = \frac{83}{2}$$
$$-\frac{19}{2}z = \frac{57}{2}$$
$$z = -3$$

Substitute $y = 2$ and $z = -3$ into E_1 and solve for x.

$$2x - 3y + 3z = -5 \qquad E_1$$
$$2x - 3(2) + 3(-3) = -5$$
$$x = 5$$

$(5,\ 2,\ -3)$

21.
$$x + 2y + z = 2 \quad E_1$$
$$-2x + 3y - 2z = -3 \quad E_2$$
$$x - 5y + z = 2 \quad E_3$$

Multiply E_1 by 2 and add to E_2 to eliminate x. Also multiply E_1 by -1 and add to E_3 to eliminate x.

$$\begin{array}{rl} 2x + 4y + 2z = 4 & 2E_1 \\ \underline{-2x + 3y - 2z = -3} & E_2 \\ 7y = 1 & E_4 \end{array}$$

$$\begin{array}{rl} -x - 2y - z = -2 & (-1)E_1 \\ \underline{x - 5y + z = 2} & E_3 \\ -7y = 0 & E_5 \end{array}$$

Equivalent system
$$\begin{array}{rl} x + 2y + z = 2 & E_1 \\ 7y = 1 & E_4 \\ -7y = 0 & E_5 \end{array}$$

y cannot simultaneously satisfy E_4 and E_5; this gives us a contradiction (You can also obtain a contradiction by adding E_4 and E_5). There is no solution.

23.
$$-x + 2y - z = -4 \quad E_1$$
$$2x + 5y - 4z = -16 \quad E_2$$
$$x + y - z = -4 \quad E_3$$

Multiply E_1 by 2 and add to E_2 to eliminate x. Also add E_1 to E_3 to eliminate x.

$$\begin{array}{rl} -2x + 4y - 2z = -8 & 2E_1 \\ \underline{2x + 5y - 4z = -16} & E_2 \\ 9y - 6z = -24 & E_4 \end{array}$$

$$\begin{array}{rl} -x + 2y - z = -4 & E_1 \\ \underline{x + y - z = -4} & E_3 \\ 3y - 2z = -8 & E_5 \end{array}$$

Equivalent system:
$$\begin{array}{rl} -x + 2y - z = -4 & E_1 \\ 9y - 6z = -24 & E_4 \\ 3y - 2z = -8 & E_5 \end{array}$$

If E_5 is multiplied by -3 and added to E_4, $0 = 0$ results. The system is dependent and equivalent to
$$-x + 2y - z = -4$$
$$3y - 2z = -8$$

Let $z = s$. Then
$$3y - 2s = -8$$
$$y = \frac{2s - 8}{3} \text{ or } \frac{2}{3}s - \frac{8}{3}$$

$$-x + 2\left(\frac{2}{3}s - \frac{8}{3}\right) - s = -4$$

$$-x = -\frac{4}{3}s + \frac{16}{3} + s - 4$$

$$x = \frac{1}{3}s - \frac{4}{3}$$

Solutions: $\left\{\left(\frac{1}{3}s - \frac{4}{3}, \frac{2}{3}s - \frac{8}{3}, s\right) \,\middle|\, s \text{ is any real number}\right\}$

25. $\begin{array}{lll} 2x + 3y + 2z = 2 & \quad & E_1 \\ 5x + 8y + 4z = 3 & \quad & E_2 \\ 4x + 7y + 2z = 0 & \quad & E_3 \end{array}$

Multiply E_1 by $-\dfrac{5}{2}$ and add to E_2 to eliminate x. Also multiply E_1 by -2 and add to E_3 to eliminate x.

$\begin{array}{lll} -5x - \dfrac{15}{2}y - 5z = -5 & \quad & \left(-\dfrac{5}{2}\right)E_1 \\ \underline{5x + 8y + 4z = 3} & \quad & E_2 \\ \dfrac{1}{2}y - z = -2 & \quad & E_4 \\ -4x - 6y - 4z = -4 & \quad & (-2)E_1 \\ \underline{4x + 7y + 2z = 0} & \quad & E_3 \\ y - 2z = -4 & \quad & E_5 \end{array}$

Equivalent system:

$\begin{array}{lll} 2x + 3y + 2z = 2 & \quad & E_1 \\ \dfrac{1}{2}y - z = -2 & \quad & E_4 \\ y - 2z = -4 & \quad & E_5 \end{array}$

If E_4 is multiplied by -2 and added to E_5, $0 = 0$ results. The system is dependent and equivalent to

$\begin{array}{l} 2x + 3y + 2z = 2 \\ y - 2z = -4 \end{array}$

Let $z = s$. Then

$\begin{array}{l} y - 2s = -4 \\ y = 2s - 4 \\ 2x + 3(2s - 4) + 2s = 2 \\ 2x = -6s + 12 - 2s + 2 \\ x = -4s + 7 \end{array}$

Solutions: $\{(-4s + 7,\ 2s - 4,\ s) \mid s \text{ is any real number}\}$

27. If the points are on the graph, the coordinates satisfy the equation. Substituting, in turn, $(-2, 9)$, $(1, -9)$ and $(4, 9)$ into $ax^2 + bx + c = y$ yields

$\begin{array}{lll} 4a - 2b + c = 9 & \quad & E_1 \\ a + b + c = -9 & \quad & E_2 \\ 16a + 4b + c = 9 & \quad & E_3 \end{array}$

Multiply E_2 by -1 and add to E_1 to eliminate c. Also multiply E_2 by -1 and add to E_3 to eliminate c.

$\begin{array}{lll} -a - b - c = 9 & \quad & (-1)E_2 \\ \underline{4a - 2b + c = 9} & \quad & E_1 \\ 3a - 3b = 18 & \quad & E_4 \\ -a - b - c = 9 & \quad & (-1)E_2 \\ \underline{16a + 4b + c = 9} & \quad & E_3 \\ 15a + 3b = 18 & \quad & E_5 \end{array}$

Equivalent system

$\begin{array}{lll} a + b + c = -9 & \quad & E_2 \\ 3a - 3b = 18 & \quad & E_4 \\ 15a + 3b = 18 & \quad & E_5 \end{array}$

Add E_4 to E_5 to eliminate b.

$\begin{array}{l} 3a - 3b = 18 \\ \underline{15a + 3b = 18} \\ 18a = 36 \\ a = 2 \end{array}$

Substitute $a = 2$ into E_5 and solve for b.

$$15a \quad + 3b = 18 \qquad E_5$$
$$15(2) + 3b = 18$$
$$b = -4$$

Substitute $a = 2$ and $b - 4$ into E_2 and solve for c.

$$a + b + c = -9 \qquad E_2$$
$$2 + (-4) + c = -9$$
$$c = -7$$

$a = 2$, $b = -4$, $c = -7$

29. If the points are on the graph, the coordinates satisfy the equation. Substituting, in turn, $(-1, 1)$, $(5, 1)$, and $(6, -6)$ into $x^2 + y^2 + ax + by + c = 0$ yields

$$1 + 1 - \quad a + b + c = 0$$
$$25 + 1 + 5a + b + c = 0$$
$$36 + 36 + 6a - 6b + c = 0$$

Simplifying yields

$$-a + \quad b + c = -2 \qquad E_1$$
$$5a + \quad b + c = -26 \qquad E_2$$
$$6a - 6b + c = -72 \qquad E_3$$

Multiply E_1 by -1 and add to E_2 to eliminate c (and b).

Multiply E_1 by -1 and add to E_3 to eliminate c.

$$a - b - c = \quad 2 \qquad (-1)E_1$$
$$\underline{5a + b + c = -26} \qquad E_2$$
$$6a \qquad\qquad = -24 \qquad E_4$$

$$a - \quad b - c = \quad 2 \qquad (-1)E_1$$
$$\underline{6a - 6b + c = -72} \qquad E_3$$
$$7a - 7b \qquad = -70 \qquad E_5$$

Equivalent system:

$$-a + \quad b + c = -2 \qquad E_1$$
$$6a \qquad\qquad = -24 \qquad E_4$$
$$7a - 7b \qquad = -70 \qquad E_5$$

Solve E_4 for a to obtain $a = -4$. Substitute $a = -4$ into E_5 and solve for b.

$$7a - 7b = -70 \qquad E_5$$
$$7(-4) - 7b = -70$$
$$b = \quad 6$$

Substitute $a = -4$ and $b = 6$ into E_1 and solve for c.

$$-a + b + c = \quad -2 \qquad E_1$$
$$-(-4) + 6 + c = \quad -2$$
$$c = -12$$

$a = -4$, $b = 6$, $c = -12$

31. Let

x = number of lawn mowers manufactured each week
y = number of snowblowers manufactured each week
z = number of chain saws manufactured each week

Then

E_1 $\quad 20x + 30y + 45z = 35,000 \qquad$ Labor
E_2 $\quad 35x + 50y + 40z = 50,000 \qquad$ Materials
E_3 $\quad 15x + 25y + 10z = 20,000 \qquad$ Shipping

Multiply E_3 by -4.5 and add to E_1 to eliminate z. Also multiply E_3 by -4 and add to E_2 to eliminate z.

$$20x \ + 30y \ \ \ \ \ + 45z = \ \ 35{,}000 \qquad E_1$$
$$\underline{-67.5x - 112.5y - 45z = -90{,}000} \qquad (-4.5)E_3$$
$$-47.5x - \ 82.5y \ \ \ \ \ \ \ \ \ = -55{,}000 \qquad E_4$$

$$35x + \ 50y \ \ \ + 40z = \ \ 50{,}000 \qquad E_2$$
$$\underline{-60x - 100y \ \ \ \ - 40z = -80{,}000} \qquad (-4)E_3$$
$$-25x - \ 50y \ \ \ \ \ \ \ \ = -30{,}000 \qquad E_5$$

Equivalent system:
$$15x \ \ + 25y + 10z = \ \ 20{,}000 \qquad E_3$$
$$-47.5x \ \ - 82.5y \ \ \ \ \ \ \ = -55{,}000 \qquad E_4$$
$$-25x \ \ - 50y \ \ \ \ \ \ \ = -30{,}000 \qquad E_5$$

Multiply E_5 by -1.9 and add to E_4 to eliminate x.
$$47.5x + 95y \ \ \ = \ \ 57{,}000 \qquad (-1.9)E_5$$
$$\underline{47.5x - 82.5y = -55{,}000} \qquad E_4$$
$$12.5y = \ \ 2{,}000$$
$$y = \ \ \ \ \ 160$$

Substitute $y = 160$ into E_5 and solve for x.
$$-25x - 50y \ \ \ \ \ = -30{,}000 \qquad E_5$$
$$-25x - 50(160) = -30{,}000$$
$$x = \ \ \ \ \ \ 880$$

Substitute $x = 880$ and $y = 160$ into E_3 and solve for z
$$15x \ \ + 25y \ \ \ \ + 10z = 20{,}000 \qquad E_3$$
$$15(880) + 25(160) + 10z = 20{,}000$$
$$z = 280$$

880 lawn mowers, 160 snowblowers, 280 chain saws.

33. Let
x = number of days operating the Michigan plant
y = number of days operating the New York plant
z = number of days operating the Ohio plant
Then

E_1	$10x + 70y + 60z = 2{,}150$	Notebooks
E_2	$20x + 50y + 80z = 2{,}300$	Desktops
E_3	$40x + 30y + 90z = 2{,}500$	Servers

Multiply E_1 by -2 and add to E_2 to eliminate x. Also multiply E_1 by -4 and add to E_3 to eliminate x.

$$-20x - 140y - 120z = -4{,}300 \qquad (-2)E_1$$
$$\underline{20x + \ \ 50y + \ \ 80z = \ \ 2{,}300} \qquad E_2$$
$$-90y - \ \ 40z = -2{,}000 \qquad E_4$$

$$-40x - 280y - 240z = -8{,}600 \qquad (-4)E_1$$
$$\underline{40x + \ \ 30y + \ \ 90z = \ \ 2{,}500} \qquad E_3$$
$$-250y - 150z = -6{,}100 \qquad E_5$$

Equivalent system:
$$10x + 70y + \ \ 60z = \ \ 2{,}150 \qquad E_1$$
$$-90y - \ \ 40z = -2{,}000 \qquad E_4$$
$$-250y - 150z = -6{,}100 \qquad E_5$$

Multiply E_5 by -0.36 and add to E_4 to eliminate y.
$$-90y - 40z = -2{,}000 \qquad E_4$$
$$\underline{90y + 54z = \ \ 2{,}196} \qquad (0.36)E_5$$
$$14z = \ \ \ \ \ 196$$
$$z = \ \ \ \ \ \ \ 14$$

Substitute $z = 14$ into E_4 and solve for y

$$-90y - 40z = -2,000 \quad E_4$$
$$-90y - 40(14) = -2,000$$
$$y = 16$$

Substitute $y = 16$ and $z = 14$ into E_1 and solve for x.

$$10x + 70y + 60z = 2,150 \quad E_1$$
$$10x + 70(16) + 60(14) = 2,150$$
$$x = 19$$

19 days Michigan plant, 16 days New York plant, 14 days Ohio plant.

35. Let x = amount invested in treasury bonds
y = amount invested in municipal bonds
z = amount invested in corporate bonds
Then

$$x + y + z = 100,000 \quad \text{total investment}$$
$$.04x + .035y + .05z = 4,400 \quad \text{total income}$$
$$x + y = z \quad \text{risk control factor}$$

Clearing of decimals and changing to standard form yields

$$x + y + z = 100,000 \quad E_1$$
$$40x + 35y + 50z = 4,400,000 \quad E_2$$
$$x + y - z = 0 \quad E_3$$

Multiply E_3 by -1 and add to E_1 to eliminate x and y.

$$x + y + z = 100,000 \quad E_1$$
$$\underline{-x - y + z = 0 \quad (-1)E_3}$$
$$2z = 100,000$$
$$z = 50,000$$

Substituting $z = 50,000$ into E_1 and into E_2 yields

$$x + y + z = 100,000 \quad E_1$$
$$x + y + 50,000 = 100,000$$
$$40x + 35y + 50z = 4,400,000 \quad E_2$$
$$40x + 35y + 50(50,000) = 4,400,000$$

Simplifying yields

$$x + y = 50,000 \quad E_4$$
$$40x + 35y = 1,900,000 \quad E_5$$

Multiply E_4 by -35 and add to E_5 to eliminate y.

$$-35x - 35y = -1,750,000 \quad (-35)E_4$$
$$\underline{40x + 35y = 1,900,000 \quad E_5}$$
$$5x = 150,000$$
$$x = 30,000$$

Substituting $x = 30,000$ into E_4 yields $y = 20,000$. $30,000 in treasury bonds, $20,000 in municipal bonds, $50,000 in corporate bonds.

37. Let x = number of pounds of peanuts
y = number of pounds of cashews
z = number of pounds of walnuts
Then

$$x + y + z = 1,500 \quad \text{total weight}$$
$$1.4x + 4y + 2.9z = 3,500 \quad \text{total value}$$
$$x = y + z \quad \text{blend is half peanuts}$$

Clearing of decimals and changing to standard form yields

$$x + y + z = 1,500 \quad E_1$$
$$14x + 40y + 29z = 35,000 \quad E_2$$
$$x - y - z = 0 \quad E_3$$

Adding equations E_1 and E_3 eliminates y and z.

$$\begin{array}{rcl} x + y + z &=& 1{,}500 \quad E_1 \\ x - y - z &=& 0 \quad E_3 \\ \hline 2x &=& 1{,}500 \\ x &=& 750 \end{array}$$

Substitute $x = 750$ into E_1 and E_2.

$$\begin{array}{rcl} 750 + y + z &=& 1{,}500 \\ 14(750) + 40y + 29z &=& 35{,}000 \end{array}$$

Simplifying yields

$$\begin{array}{rcl} y + z &=& 750 \quad E_4 \\ 40y + 29z &=& 24{,}500 \quad E_5 \end{array}$$

Multiply E_4 by -40 and add to E_5 to eliminate y

$$\begin{array}{rcl} -40y - 40z &=& -30{,}000 \quad (-40)E_4 \\ 40y + 29z &=& 24{,}500 \quad E_5 \\ \hline -11z &=& -5{,}500 \\ z &=& 500 \end{array}$$

Substituting $z = 500$ into E_4 yields $y = 250$.
750 lbs. peanuts, 250 lbs. cashews, 500 lbs. walnuts.

39. Let $f(x) = ax^2 + bx + c$ be the quadratic model and evaluate f at each of the 3 data points.

$$\begin{array}{lrcl} x = 0: & c &=& 75 \quad E_1 \\ x = 50: & 2{,}500a + 50b + c &=& 150 \quad E_2 \\ x = 100: & 10{,}000a + 100b + c &=& 275 \quad E_3 \end{array}$$

Substitute $c = 75$ into E_2 and E_3.

$$\begin{array}{rcl} 2{,}500a + 50b + 75 &=& 150 \\ 10{,}000a + 100b + 75 &=& 275 \end{array}$$

Simplifying yields

$$\begin{array}{rcl} 2{,}500a + 50b &=& 75 \quad E_4 \\ 10{,}000a + 100b &=& 200 \quad E_5 \end{array}$$

Multiply E_4 by -2 and add to E_5 to eliminate b.

$$\begin{array}{rcl} -5{,}000a - 100b &=& -150 \quad (-2)E_4 \\ 10{,}000a + 100b &=& 200 \quad E_5 \\ \hline 5{,}000a &=& 50 \\ a &=& 0.01 \end{array}$$

Substitute $a = 0.01$ into E_5 and solve for b.

$$\begin{array}{rcl} 10{,}000a + 100b &=& 200 \\ 10{,}000(0.01) + 100b &=& 200 \\ b &=& 1 \end{array}$$

The model is $f(x) = 0.01x^2 + x + 75$.
To find the predicted population in 2050 calculate $f(150)$.

$$f(150) = 0.01(150)^2 + 150 + 75 = 450 \text{ million}$$

41. Let $f(x) = ax^2 + bx + c$ be the quadratic model and evaluate f at each of the three data points.

$$\begin{array}{lrcl} x = 0: & c &=& 77.6 \quad E_1 \\ x = 5: & 25a + 5b + c &=& 78 \quad E_2 \\ x = 10: & 100a + 10b + c &=& 78.6 \quad E_3 \end{array}$$

Substitute $c = 77.6$ into E_2 and E_3

$$\begin{array}{rcl} 25a + 5b + 77.6 &=& 78 \\ 100a + 10b + 77.6 &=& 78.6 \end{array}$$

Simplifying yields
$$25a + 5b = 0.4 \qquad E_4$$
$$100a + 10b = 1 \qquad E_5$$

Multiply E_4 by -2 and add to E_5 to eliminate b.

$$\begin{array}{r} -50a - 10b = -0.8 \\ \underline{100a + 10b = 1} \\ 50a = 0.2 \\ a = 0.004 \end{array}$$

Substitute $a = 0.004$ into E_5 and solve for b.
$$100a + 10b = 1 \qquad E_5$$
$$100(0.004) + 10b = 1$$
$$b = 0.06$$

The model is $f(x) = 0.004x^2 + 0.06x + 77.6$.
Predicted life expectancy for females born 1995-2000:
$$f(15) = 0.004(15)^2 + 0.06(15) + 77.6 = 79.4 \text{ yrs.}$$
Predicted life expectancy for females born 2000-2005:
$$f(20) = 0.004(20)^2 + 0.06(20) + 77.6 = 80.4 \text{ yrs.}$$

43. The independent variable is given by the first year in the interval minus the starting year 1980, so enter 0, 5, 10, 15, 20 as L_1. The dependent variable is life expectancy, so enter the data 77.6, 78, 78.6, 79.1, 79.7 as L_2. Then use the quadratic regression command from the STAT CALC menu and plot using the STAT PLOT screen.

  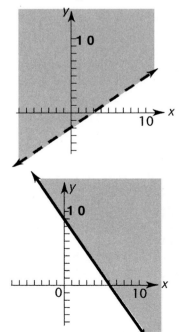

Section 5-3

1. Answers will vary.

3. Answers will vary.

5. Graph $2x - 3y = 6$ as a dashed line, since equality is not included in the original statement. The origin is a suitable test point.
$$2x - 3y < 6$$
$$2(0) - 3(0) = 0 < 6$$
So $(0, 0)$ is in the solution set. The solution region is the set of points (x, y) that are above the line $2x - 3y = 6$.

7. Graph $3x + 2y = 18$ as a solid line, since equality is included in the original statement. The origin is a suitable test point.
$$3(0) + 2(0) = 0 \not\geq 18$$
So $(0, 0)$ is not in the solution set. The solution region is the set of points (x, y) that are on or above the line $3x + 2y = 18$.

9. Graph $y = \dfrac{2}{3}x + 5$ as a solid line, since
equality is included in the original
statement. The origin is a suitable test
point.
$$0 \overset{?}{\le} (0) + 5$$
$$0 \le 5$$
So $(0, 0)$ is in the solution set. The
solution region is the set of points (x, y)
that are on or below the line $y = \dfrac{2}{3}x + 5$.

11. Graph $y = 8$ as a dashed line, since
equality is not included in the original
statement. The solution region consists of
all points whose y-coordinates are less
than 8, that is, the set of points (x, y)
that are below the line $y = 8$.

13. This system is equivalent to the system
$$y \ge -3$$
$$y < 2$$
and its graph is the intersection of the
graphs of these inequalities. The solution
region is the set of points (x, y) that are
below the horizontal line $y = 2$, and on or
above the horizontal line $y = -3$.

15. $x + 2y \le 8$
$3x - 2y \ge 0$
Choose a suitable test point that lies on neither line, for example, $(2, 0)$.
   $2 + 2(0) = 2 \le 8$ So $(2, 0)$ is in the solution region for the first
inequality, and the solution region is *below* the graph of $x + 2y = 8$.
$3(2) - 2(0) = 6 \ge 0$ So $(2, 0)$ is also in the solution region of the second
inequality, and the solution region is *below* the graph of $3x - 2y = 0$.
The region in the diagram that is below both lines is region IV.

17. $x + 2y \ge 8$
$3x - 2y \ge 0$
Choose a suitable test point that lies on neither line, for example, $(2, 0)$.
   $2 + 2(0) = 2 \ne 8$ So $(2, 0)$ is not in the solution region of the first
inequality, and the solution region is *above* the graph of $x + 2y = 8$.
$3(2) - 2(0) = 6 \ge 0$ So $(2, 0)$ is in the solution region of the second
inequality, and the solution region is *below* the graph of $3x - 2y = 0$.
The region in the diagram that is above $x + 2y = 8$ and below $3x - 2y = 0$ is
Region I.

19.

21.

23.

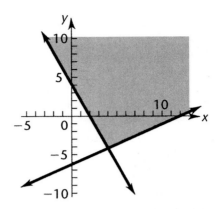

25. (A) Solution region is the double-shaded region.

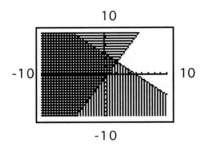

(B) Solution region is the unshaded region.

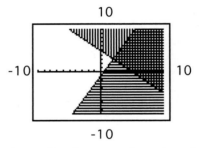

27. (A) Solution region is the double-shaded region.

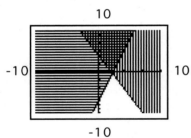

(B) Solution region is the unshaded region.

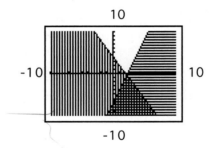

29. Choose a suitable test point that lies on none of the lines, say (5, 1).

$5 + 3(1) = 8 \le 18$ So (5, 1) is in the solution region of the first inequality, and the solution region is *below* the graph of $x + 3y = 18$.

$2(5) + 1 = 11 \not\le 16$ So (5, 1) is not in the solution region of the first inequality, and the solution region is *above* the graph of $2x + y = 16$.

From $x \ge 0$ and $y \ge 0$, we conclude that the solution region is in Quadrant I. So the solution region is region IV in the diagram. The corner points are the labeled points (6, 4), (8, 0), and (18, 0).

31. Choose a suitable test point that lies on none of the lines, say (5, 1).

$5 + 3(1) = 8 \not\ge 18$ So (5, 1) is not in the solution region of the first inequality, and the solution region is *above* the graph of $x + 3y = 18$.

$2(5) + 1 = 11 \not\ge 16$ So (5, 1) is not in the solution region of the first inequality, and the solution region is *above* the graph of $2x + y = 16$.

From $x \ge 0$ and $y \ge 0$, we conclude that the solution region is in Quadrant I. So the solution region is region I in the diagram. The corner points are the labeled points (0, 16), (6, 4), and (18, 0).

33. The solution region is bounded by the 2 axes and the line $2x + 3y = 6$. The corner points are obvious from the graph:
$(0, 0)$, $(0, 2)$, $(3, 0)$.

Check:

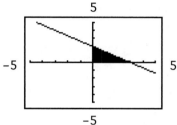

35. The solution region is unbounded. The corner points are obvious from the graph: $(0, 4)$ and $(5, 0)$.

Check:

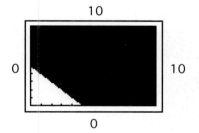

37. The solution region is bounded. Three corner points are obvious from the graph: $(0, 4)$, $(0, 0)$, $(4, 0)$. The fourth corner point is obtained by solving the system
$x + 3y = 12$

$2x + y = 8$ to obtain $\left(\dfrac{12}{5}, \dfrac{16}{5}\right)$.

(Solution left to the student.)

Check:

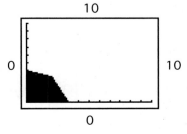

39. The solution region is unbounded. Two corner points are obvious from the graph: $(9, 0)$ and $(0, 8)$. The third corner point is obtained by solving the system
$4x + 3y = 24$

$2x + 3y = 18$ to obtain $(3, 4)$.
(Solution left to the student.)

Check:

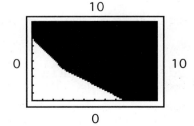

41. The solution region is bounded. Three corner points are obvious from the graph: (6, 0), (0, 0), and (0, 5). The other corner points are obtained by solving:

$2x + y = 12$ and $x + y = 7$
$x + y = 7$ $x + 2y = 10$
to obtain to obtain
$(5, 2)$ $(4, 3)$
(Solutions left to the student.)

Check:

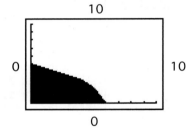

43. The solution region is unbounded. Two of the corner points are obvious from the graph: (16, 0) and (0, 14). The other corner points are obtained by solving:

$x + 2y = 16$ and $x + y = 12$
$x + y = 12$ $2x + y = 14$
to obtain to obtain
$(8, 4)$ $(2, 10)$
(Solutions left to the student.)

Check:

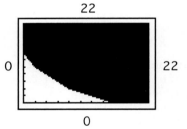

45. The solution region is bounded. The corner points are obtained by solving:

$\begin{cases} x + y = 11 \\ 5x + y = 15 \end{cases}$ to obtain (1, 10)

$\begin{cases} 5x + y = 15 \\ x + 2y = 12 \end{cases}$ to obtain (2, 5), and

$\begin{cases} x + y = 11 \\ x + 2y = 12 \end{cases}$ to obtain (10, 1)
(Solutions left to the student.)

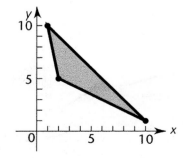

47. The lighter shaded regions are the solution regions to $3x + 2y \geq 24$ and $3x + y \leq 15$; the darker is the intersection of the two. We can see that no portion of the darker region lies to the right of $x = 4$, so there are no points that satisfy both of the first two inequalities and $x > 4$. The solution region is empty.

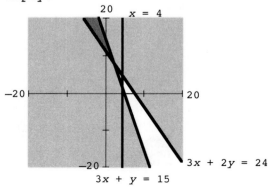

49. The solution region is bounded. The corner points are obtained by solving:

$$\begin{cases} x + y = 10 \\ 3x - 2y = 15 \end{cases}$$ to obtain (7, 3)

$$\begin{cases} 3x - 2y = 15 \\ 3x + 5y = 15 \end{cases}$$ to obtain (5, 0),

$$\begin{cases} 3x + 5y = 15 \\ -5x + 2y = 6 \end{cases}$$ to obtain (0, 3), and

$$\begin{cases} -5x + 2y = 6 \\ x + y = 10 \end{cases}$$ to obtain (2, 8)

(Solutions left to the student.)

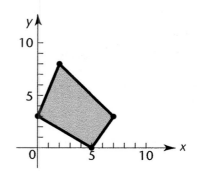

51.   The solution region is bounded. The corner points are obtained by solving:

$$\begin{cases} 16x + 13y = 119 \\ 12x + 16y = 101 \end{cases}$$ to obtain (5.91, 1.88)

$$\begin{cases} 16x + 13y = 119 \\ -4x + 3y = 11 \end{cases}$$ to obtain (2.14, 6.52)

and

$$\begin{cases} 12x + 16y = 101 \\ -4x + 3y = 11 \end{cases}$$ to obtain (1.27, 5.36)

(Solutions left to the student.)

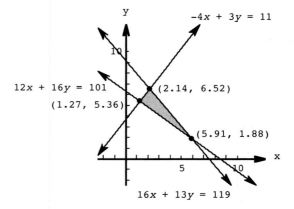

53. Let x = number of trick skis produced per day.
    y = number of slalom skis produced per day

Clearly x and y must be non-negative.
So $x \geq 0$ (1)
   $y \geq 0$ (2)
To fabricate x trick skis requires $6x$ hours.
To fabricate y slalom skis requires $4y$ hours.
108 hours are available for fabricating; so
$6x + 4y \leq 108$ (3)
To finish x trick skis requires $1x$ hours.
To finish y slalom skis requires $1y$ hours.
24 hours are available for finishing, so
$x + y \leq 24$ (4)

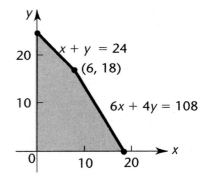

Graphing the inequality system (1), (2), (3), (4), we get the above diagram.

55. (A) All production schedules in the feasible region that are on the graph
        of $50x + 60y = 1{,}100$ will result in a profit of \$1,100.

   (B) There are many possible choices. For example, producing 5 trick and 15
       slalom skis will produce a profit of \$1,150. The graph of the line
       $50x + 60y = 1{,}150$ includes all the production schedules in the feasible
       region that result in a profit of \$1,150.

   (C) A graphical approach would involve drawing other lines of the type
       $50x + 60y = A$. The graphs of these lines include all production schedules
       that will result in a profit of A. Increase A until the line either
       intersects the feasible region only in 1 corner point or contains an edge
       of the feasible region. This value of A will be the maximum profit
       possible. For more details, see Section 5-4 of the text.

57. Clearly x and y must be non-negative.

So $x \geq 0$ (1)

  $=y \geq 0$ (2)

x cubic yards of mix A contains $20x$ pounds of phosphoric acid.
y cubic yards of mix B contains $10y$ pounds of phosphoric acid.

At least 460 pounds of phosphoric acid are required, so
$20x + 10y \geq 460$ (3)
x cubic yards of mix A contains $30x$ pounds of nitrogen.
y cubic yards of mix B contains $30y$ pounds of nitrogen.
At least 960 pounds of nitrogen are required, so
$30x + 30y \geq 960$ (4)

x cubic yards of mix A contains $5x$ pounds of potash.
y cubic yards of mix B contains $10y$ pounds of potash.
At least 220 pounds of potash are required, so
$5x + 10y \geq 220$ (5)

Graphing the inequality system (1), (2), (3), (4), (5), we get the following diagram:

59. Clearly x and y must be non-negative.

So $x \geq 0$ (1)

   $y \geq 0$ (2)

Each sociologist will spend 10 hours collecting data: $10x$ hours.
Each research assistant will spend 30 hours collecting data: $30y$ hours.
At least 280 hours must be spent collecting data; so
$10x + 30y \geq 280$ (3)
Each sociologist will spend 30 hours analyzing data: $30x$ hours.
Each research assistant will spend 10 hours analyzing data: $10y$ hours.
At least 360 hours must be spent analyzing data; so
$30x + 10y \geq 360$ (4)
Graphing the inequality system (1), (2), (3), (4), we get the above diagram.

Section 5-4

1. Answers will vary.   **3.** Answers will vary.   **5.** Answers will vary.

7. Answers will vary.

9.

Corner Point (x, y)	Objective Function $z = x + y$	
(0, 12)	12	
(7, 9)	16	Maximum value
(10, 0)	10	
(0, 0)	0	

The maximum value of z on S is 16 at (7, 9).

11.

Corner Point $(x,\ y)$	Objective Function $z = 3x + 7y$		
(0, 12)	84	Minimum value	Multiple optimal solutions
(7, 9)	84	Minimum value	
(10, 0)	30		
(0, 0)	0		

The maximum value of z on S is 84 at both (0, 12) and (7, 9).

13. Plugging in zero for x, we get $z = 2y$ or $y = \dfrac{z}{2}$.

Plugging in zero for y, we get $z = x$. So the intercepts of the line $z = x + 2y$ are $\left(0, \dfrac{z}{2}\right)$ and $(z,\ 0)$. The feasible region is shown at the right with several constant z lines drawn in.

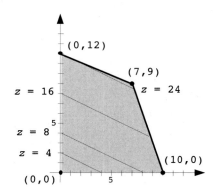

Sliding a straight edge parallel to these constant z lines in the direction of increasing z, we can see that the point in the feasible region that will intersect the constant z line for largest possible z is (7, 9). When $x = 7$ and $y = 9$, $z = 25$.

Check:

Corner Point $(x,\ y)$	Objective Function $z = x + 2y$	
(0, 12)	24	
(7, 9)	25	Maximum Value
(10, 0)	10	
(0, 0)	0	

The maximum value of z on S is 25 at (7, 9).

15. Plugging in zero for x, we get $z = 2y$ or $y = \dfrac{z}{2}$.

Plugging in zero for y, we get $z = 7x$ or $x = \dfrac{z}{7}$. So the intercepts of the line $z = 7x + 2y$ are $\left(0, \dfrac{z}{2}\right)$ and $\left(\dfrac{z}{7}, 0\right)$. The feasible region is shown at the right with several constant z lines drawn in.

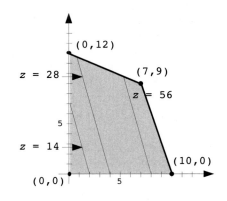

If we slide a straight edge parallel to the constant z lines as z increases, the last point in S that will intersect our lines is (10, 0). When $x = 10$ and $y = 0$, $z = 70$.

Check:

Corner Point $(x,\ y)$	Objective Function $z = 7x + 2y$	
(0, 12)	24	
(7, 9)	67	
(10, 0)	70	Maximum Value
(0, 0)	0	

The maximum value of z on S is 70 at (10, 0).

17.

Corner Point (x, y)	Objective Function $z = 7x + 4y$	
(0, 12)	48	
(12, 0)	84	
(4, 3)	40	
(0, 8)	32	Minimum value

The minimum value of z on S is 32 at (0, 8).

19.

Corner Point (x, y)	Objective Function $z = 3x + 8y$	
(0, 12)	96	
(12, 0)	36	Minimum value ⎫ Multiple optimal solutions
(4, 3)	36	Minimum value ⎭
(0, 8)	64	

The minimum value of z on S is 36 at both (12, 0) and (4, 3).

21. If $x = 5$ and $y = 5$, $z = 5 + 2(5) = 15$, so the constant-value line we need is $x + 2y = 15$. The feasible region is shown at the right with the constant-value line.

If we slide a straightedge parallel to this constant-value line in the direction of decreasing z (downward), the last point in T that will intersect our line is (4, 3). When $x = 4$ and $y = 3$, $z = 10$.

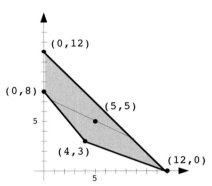

Check:

Corner Point (x, y)	Objective Function $z = x + 2y$	
(0, 12)	24	
(12, 0)	12	
(0, 8)	16	
(4, 3)	10	Minimum Value

The minimum value of z on T is 10 at (4, 3).

23. If $x = 5$ and $y = 5$, $z = 5(5) + 4(5) = 45$, so the constant-value line we need is $5x + 4y = 45$. The feasible region is shown at the right with the constant-value line.

The constant-value line appears to be parallel to the edge connecting (0, 8) and (4, 3). So the minimum could occur at either (0, 8) or (4, 3) when $x = 0$ and $y = 8$, $z = 32$. When $x = 4$ and $y = 3$, $z = 32$ also.

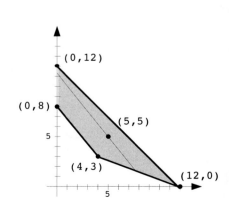

Check:

Corner Point (x, y)	Objective Function z = 5x + 4y	
(0, 12)	48	
(12, 0)	60	
(0, 8)	32	Minimum Value
(4, 3)	32	Minimum Value

The minimum value of 32 occurs at both (0, 8) and (4, 3).

25. The feasible region is graphed as follows:
The corner points (0, 5), (5, 0) and (0, 0) are obvious from the graph. The corner point (4, 3) is obtained by solving the system
$x + 2y = 10$
$3x + y = 15$
We now evaluate the objective function at each corner point.

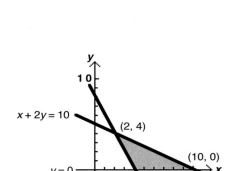

Corner Point (x, y)	Objective Function z = 3x + 2y	
(0, 5)	10	
(0, 0)	0	
(5, 0)	15	
(4, 3)	18	Maximum value

The maximum value of z on S is 18 at (4, 3).

27. The feasible region is graphed as follows:
The corner points (4, 0) and (10, 0) are obvious from the graph. The corner point (2, 4) is obtained by solving the system
$x + 2y = 10$
$2x + y = 8$
We now evaluate the objective function at each corner point.

Corner Point (x, y)	Objective Function z = 3x + 4y	
(4, 0)	12	Minimum value
(10, 0)	30	
(2, 4)	22	

The minimum value of z on S is 12 at (4, 0).

29. The feasible region is graphed below. The corner points (0, 12), (0, 0), and (12, 0) are obvious from the graph. The other corner points are obtained by solving:
$x + 2y = 24$ and $x + y = 14$
 $x + y = 14$ to obtain (4, 10) $2x + y = 24$ to obtain (10, 4)
We now evaluate the objective function at each corner point.

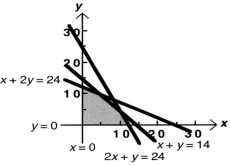

Corner Point (x, y)	Objective Function z = 3x + 4y	
(0, 12)	48	
(0, 0)	0	
(12, 0)	36	
(10, 4)	46	
(4, 10)	52	Maximum value

The maximum value of z on S is 52 at (4, 10).

31. The feasible region is graphed as follows:
The corner points (0, 20) and (20, 0) are obvious
from the graph. The third corner point is obtained
by solving:
$x + 4y = 20$
$4x + y = 20$ to obtain (4, 4)
We now evaluate the objective function at each
corner point.

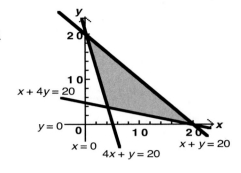

Corner Point (x, y)	Objective Function z = 5x + 6y	
(0, 20)	120	
(20, 0)	100	
(4, 4)	44	Minimum value

The minimum value of z on S is 44 at (4, 4).

33. The feasible region is graphed as follows:
The corner points (60, 0) and (120, 0) are
obvious from the graph. The other corner
points are obtained by solving:

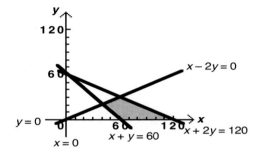

$x + y = 60$ and $x + 2y = 120$
$x - 2y = 0$ $x - 2y = 0$
to obtain (40, 20)   to obtain (60, 30)

We now evaluate the objective function at
each corner point.

Corner Point (x, y)	Objective Function 25x + 50y		
(60, 0)	1,500	Minimum value	
(40, 20)	2,000		
(60, 30)	3,000	Maximum value	Multiple optimal solutions
(120, 0)	3,000	Maximum value	

The minimum value of z on S is 1,500 at (60, 0). The maximum value of z on S is
3,000 at (60, 30) and (120, 0).

35. The feasible region is graphed as follows:
The corner points (0, 45), (0, 20), (25, 0), and (60, 0) are obvious from the
graph shown below. The other corner points are obtained by solving:

$3x + 4y = 240$ and $3x + 4y = 240$
       $y = 45$ to obtain (20, 45) $x = 60$ to obtain (60, 15)
We now evaluate the objective function at each corner point.

Corner Point (x, y)	Objective Function 25x + 15y	
(0, 45)	675	
(0, 20)	300	Minimum value
(25, 0)	625	
(60, 0)	1,500	
(60, 15)	1,725	Maximum value
(20, 45)	1,175	

The minimum value of z on S is 300 at $(0, 20)$. The maximum value of z on S is 1,725 at $(60, 15)$.

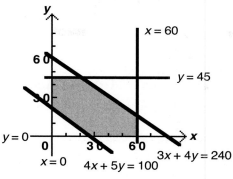

37. The feasible region is graphed as shown at the right.

    The corner point $(0, 0)$ is obvious from the graph. The other corner points are obtained by solving:

$x_1 = 0$	$350x_1 + 340x_2 = 3,762$
$275x_1 + 322x_2 = 3,381$	$275x_1 + 322x_2 = 3,381$
to obtain $(0, 10.5)$	to obtain $(3.22, 7.75)$

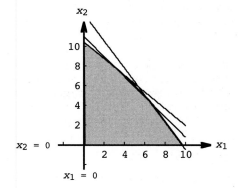

$350x_1 + 340x_2 = 3,762$	$425x_1 + 306x_2 = 4,114$
$425x_1 + 306x_2 = 4,114$	$x_2 = 0$
to obtain $(6.62, 4.25)$	to obtain $(9.68, 0)$

We now evaluate the objective function at each corner point.

Corner Point (x_1, x_2)	Objective Function $525x_1 + 478x_2$	
$(0, 0)$	0	Minimum value
$(0, 10.5)$	5,019	
$(3.22, 7.75)$	5,395	
$(6.62, 4.25)$	5,507	Maximum value
$(9.68, 0)$	5,082	

The maximum value of P is 5,507 at the corner point $(6.62, 4.25)$.

39. The feasible region is graphed as follows:
    Consider the objective function $x + y$. We can see that it takes on the value 2 along $x + y = 2$, the value 4 along $x + y = 4$, the value 7 along $x + y = 7$, and so on. The maximum value of the objective function, then, on S, is 7, which occurs at B. Graphically this occurs when the line $x + y = c$ coincides with the boundary of S. So to answer questions (A)-(E) we must determine values of a and b so that the appropriate line $ax + by = c$ coincides with the boundary of S only at the specified points.

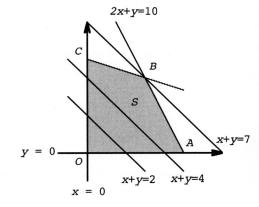

    (A) The line $ax + by = c$ must have slope negative, but greater in absolute value than that of line segment AB, $2x + y = 10$. Therefore $a > 2b$.

(B) The line $ax + by = c$ must have slope negative but between that of $x + 3y = 15$ and $2x + y = 10$. Therefore $\frac{1}{3}b < a < 2b$.

(C) The line $ax + by = c$ must have slope greater than that of line segment BC, $x + 3y = 15$. Therefore $a < \frac{1}{3}b$ or $b > 3a$

(D) The line $ax + by = c$ must be parallel to line segment AB, therefore $a = 2b$.

(E) The line $ax + by = c$ must be parallel to line segment BC, therefore $b = 3a$.

41. We let x = the number of trick skis
y = the number of slalom skis
The problem constraints were
$6x + 4y \leq 108$
$x + y \leq 24$
The non-negative constraints were
$x \geq 0$
$y \geq 0$
The feasible region was graphed there.
(A) We note now: the linear objective function
$P = 40x + 30y$ represents the profit.

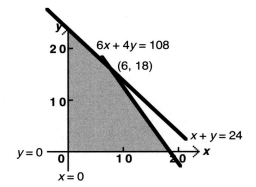

Three of the corner points are obvious from the graph: (0, 24), (0, 0), and (18, 0). The fourth corner point is obtained by solving:
$6x + 4y = 108$
$x + y = 24$ to obtain (6, 18).
Summarizing: the mathematical model for this problem is:
Maximize $P = 40x + 30y$
subject to: $6x + 4y \leq 108$
$x + y \leq 24$
$x, y \geq 0$
We now evaluate the objective function $40x + 30y$ at each corner point.

Corner Point (x, y)	Objective Function $40x + 30y$	
(0, 0)	0	
(18, 0)	720	
(6, 18)	780	Maximum value
(0, 24)	720	

The optimal value is 780 at the corner point (6, 18). So 6 trick skis and 18 slalom skis should be manufactured to obtain the maximum profit of $780.

(B) The objective function now becomes $40x + 25y$. We evaluate this at each corner point.

Corner Point (x, y)	Objective Function $40x + 25y$	
(0, 0)	0	
(18, 0)	720	Maximum value
(6, 18)	690	
(0, 24)	600	

The optimal value is now 720 at the corner point (18, 0). So 18 trick skis and no slalom skis should be produced to obtain a maximum profit of $720.

(C) The objective function now becomes $40x + 45y$. We evaluate this at each corner point.

Corner Point (x, y)	Objective Function $40x + 45y$	
(0, 0)	0	
(18, 0)	720	
(6, 18)	1,050	
(0, 24)	1,080	Maximum value

The optimal value is now 1,080 at the corner point (0, 24). So no trick skis and 24 slalom skis should be produced to obtain a maximum profit of $1,080.

43. Let x = number of model A trucks
   y = number of model B trucks
We form the linear objective function
$C = 15{,}000x + 24{,}000y$
We wish to minimize C, the cost of buying
x trucks @ $15,000 and y trucks @ $24,000,
subject to the constraints.
   $x + y \leq 15$ maximum number of trucks constraint
$2x + 3y \geq 36$ capacity constraint
   $x, y \geq 0$ non-negative constraints.

Solving the system of constraint inequalities graphically, we obtain the feasible region S shown in the diagram.
Next we evaluate the objective function at each corner point.

Corner Point (x, y)	Objective Function $C = 15{,}000x + 24{,}000y$	
(0, 12)	288,000	
(0, 15)	360,000	
(9, 6)	279,000	Minimum value

The optimal value is $279,000 at the corner point (9, 6). So the company should purchase 9 model A trucks and 6 model B trucks to realize the minimum cost of $279,000.

45. (A) Let x = number of tables
        y = number of chairs
We form the linear objective function
$P = 90x + 25y$
We wish to maximize P, the profit from x tables @ $90 and y chairs @ $25, subject to the constraints
$8x + 2y \leq 400$ assembly department constraint
 $2x + y \leq 120$ finishing department constraint
   $x, y \geq 0$ non-negative constraints

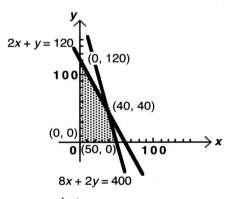

Solving the system of constraint inequalities graphically, we obtain the feasible region S shown in the diagram.
Next we evaluate the objective function at each corner point.

Corner Point (x, y)	Objective Function $P = 90x + 25y$	
(0, 0)	0	
(50, 0)	4,500	
(40, 40)	4,600	Maximum value
(0, 120)	3,000	

The optimal value is 4,600 at the corner point (40, 40). So the company should manufacture 40 tables and 40 chairs for a maximum profit of $4,600.

(B) We are faced with the further condition that $y \geq 4x$. We wish, then, to maximize $P = 90x + 25y$ under the constraints

$$8x + 2y \leq 400$$
$$2x + y \leq 120$$
$$y \geq 4x$$
$$x, y \geq 0$$

The feasible region is now S' as graphed.

Note that the new condition has the effect of excluding $(40, 40)$ from the feasible region.

We now evaluate the objective function at the new corner points.

Corner Point (x, y)	Objective Function $P = 90x + 25y$	
(0, 120)	3,000	
(0, 0)	0	
(20, 80)	3,800	Maximum value

The optimal value is now 3,800 at the corner point (20, 80). So the company should manufacture 20 tables and 80 chairs for a maximum profit of $3,800.

47. Let x = number of gallons produced using the old process
y = number of gallons produced using the new process

We form the linear objective function
$$P = 0.6x + 0.2y$$

(A) We wish to maximize P, the profit from x gallons using the old process and y gallons using the new process, subject to the constraints

$20x + 5y \leq 16,000$	sulfur dioxide constraint
$40x + 20y \leq 30,000$	particulate matter constraint
$x, y \geq 0$	non-negative constraints

Solving the system of constraint inequalities graphically, we obtain the feasible region S shown in the diagram. Note that no corner points are determined by the sulfur dioxide constraint.

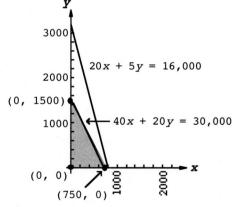

We evaluate the objective function at each corner point.

Corner Point (x, y)	Objective Function $P = 0.6x + 0.2y$	
(0, 0)	0	
(0, 1500)	300	
(750, 0)	450	Maximum value

The optimal value is 450 at the corner point (750, 0). So the company should manufacture 750 gallons by the old process exclusively, for a profit of $450.

(B) The sulfur dioxide constraint is now $20x + 5y \leq 11{,}500$. We now wish to maximize P subject to the constraints

$$20x + 5y \leq 11{,}500$$
$$40x + 20y \leq 30{,}000$$
$$x,\ y \geq 0$$

The feasible region is now S_1 as shown.

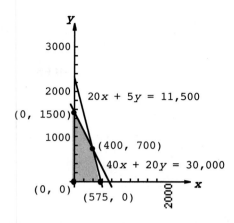

We evaluate the objective function at the new corner points.

Corner Point $(x,\ y)$	Objective Function $P = 0.6x + 0.2y$	
(0, 0)	0	
(575, 0)	345	
(400, 700)	380	Maximum value
(0, 1,500)	300	

The optimal value is now 380 at the corner point (400, 700). So the company should manufacture 400 gallons by the old process and 700 gallons by the new process, for a profit of $380.

(C) The sulfur dioxide constraint is now $20x + 5y \leq 7{,}200$. We now wish to maximize P subject to the constraints

$$20x + 5y \leq 7{,}200$$
$$40x + 20y \leq 30{,}000$$
$$x,\ y \geq 0$$

The feasible region is now S_2 as shown.

Note that now no corner points are determined by the particulate matter constraint.

We evaluate the objective function at the new corner points.

Corner Point $(x,\ y)$	Objective Function $P = 0.6x + 0.2y$	
(0, 0)	0	
(360, 0)	216	
(0, 1,440)	288	Maximum value

The optimal value is now 288 at the corner point (0, 1440). So the company should manufacture 1,440 gallons by the new process exclusively, for a profit of $288.

49. (A) Let x = number of bags of Brand A
y = number of bags of Brand B
We form the objective function
$N = 6x + 7y$
N represents the amount of nitrogen in x bags @ 6 pounds per bag and y bags @ 7 pounds per bag.

We wish to optimize N subject to the constraints

$2x + 4y \geq 480$ phosphoric acid constraint
$6x + 3y \geq 540$ potash constraint
$3x + 4y \leq 620$ chlorine constraint
$x, y \geq 0$ non-negative contraints

Solving the system of constraint inequalities graphically, we obtain the feasible region S shown in the diagram.

Next we evaluate the objective function at the corner points.

Corner Point (x, y)	Objective Function $N = 6x + 7y$	
(20, 140)	1,100	
(40, 100)	940	Minimum value
(140, 50)	1,190	Maximum value

So the nitrogen will range from a minimum of 940 pounds when 40 bags of brand A and 100 bags of Brand B are used to a maximum of 1,190 pounds when 140 bags of brand A and 50 bags of brand B are used.

Chapter 5 Review

1. $y = 4x - 9$
$y = -x + 6$
Substitute y from the first equation into the second equation to eliminate y.
$4x - 9 = -x + 6$
$5x - 9 = 6$
$5x = 15$
$x = 3$
$y = -x + 6 = -3 + 6 = 3$

$x = 3, y = 3$ $\qquad$ (5–1)

2. $3x + 2y = 5$
$4x - y = 14$
Solve the first equation for y in terms of x and substitute into the second equation
$$2y = 5 - 3x$$
$$y = \frac{5 - 3x}{2}$$
$$4x - \left(\frac{5 - 3x}{2}\right) = 14$$
$$8x - \frac{2}{1}\left(\frac{5 - 3x}{2}\right) = 28$$
$$8x - (5 - 3x) = 28$$
$$8x - 5 + 3x = 28$$
$$11x = 33$$
$$x = 3$$
$$y = \frac{5 - 3x}{2} = \frac{5 - 3 \cdot 3}{2}$$
$$y = -2$$
$x = 3, y = -2$ $\qquad$ (5–1)

3. Graph both lines on the same set of axes and find the intersection point:

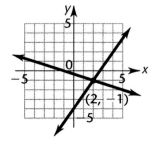

The solution is $x = 2, y = -1$.

(5–1)

4. First solve each equation for y.

$$2x + y = 4 \qquad\qquad -2x + 7y = 9$$
$$y = 4 - 2x \qquad\qquad 7y = 2x + 9$$
$$y = \frac{2}{7}x + \frac{9}{7}$$

Next, enter each equation, graph, and approximate the solution.

Equation definitions

Intersection point

Check

The solution is $x = 1.1875$, $y = 1.625$. $\hfill$ (5–1)

5. Graph $3x - 4y = 24$ as a solid line since equality is included in the original statement. $(0, 0)$ is a suitable test point.
$3(0) - 4(0) = 0 \neq 24$
Therefore $(0, 0)$ is not in the solution set. The line $3x - 4y = 24$ and the half-plane not containing $(0, 0)$ form the graph.

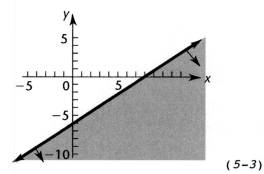

(5–3)

6. Graph both $2x + y = 2$ and $x + 2y = -2$ as solid lines since equality is included in each. $(0, 0)$ is a suitable test point.

$2(0) + 0 = 0 \leq 2$ so $(0, 0)$ satisfies the first inequality, and the solution region is below the line $2x + y = 2$.

$0 + 2(0) = 0 \geq -2$ so $(0, 0)$ satisfies the second inequality, and the solution region is above the line $x + 2y = -2$. The solution region is graphed at the right.

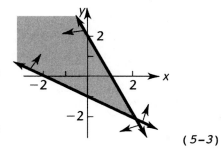

(5–3)

7. $2x + y = 7$
$3x - 2y = 0$
We multiply the top equation by 2 and add.

$4x + 2y = 14$
$\underline{3x - 2y = 0}$
$7x = 14$
$x = 2$

Substituting $x = 2$ in the top equation, we have
$2(2) + y = 7$
$ y = 3$
Solution: $(2, 3)$ $\hfill$ (5–1)

8. $3x - 6y = 5$
$-2x + 4y = 1$
We multiply the top equation by 2, the bottom by 3, and add.

$6x - 12y = 10$
$\underline{-6x + 12y = 3}$
$0 = 13$

No solution $\hfill$ (5–1)

9. $4x - 3y = -8$

$-2x + \dfrac{3}{2}y = 4$

We multiply the bottom equation by 2 and add.

$$
\begin{array}{r}
4x - 3y = -8 \\
-4x + 3y = 8 \\
\hline
0 = 0
\end{array}
$$

There are infinitely many solutions. For any real number t, $4t - 3y = -8$, and
$-3y = -4t - 8$

$y = \dfrac{4t + 8}{3}$

Then $\left(t, \dfrac{4t + 8}{3}\right)$ is a solution for any real number t.   (5-1)

10. $x - 3y + z = 4 \quad E_1$

$-x + 4y - 4z = 1 \quad E_2$

$2x - y + 5z = -3 \quad E_3$

Add E_1 to E_2 to eliminate x.
Also multiply E_1 by -2 and add to E_3 to eliminate x.

$$
\begin{array}{rl}
x - 3y + z = 4 & E_1 \\
-x + 4y - 4z = 1 & E_2 \\
\hline
y - 3z = 5 & E_4
\end{array}
$$

$$
\begin{array}{rl}
-2x + 6y - 2z = -8 & (-2)E_1 \\
2x - y + 5z = -3 & E_3 \\
\hline
5y + 3z = -11 & E_5
\end{array}
$$

Equivalent system:

$$
\begin{array}{rl}
x - 3y + z = 4 & E_1 \\
y - 3z = 5 & E_4 \\
5y + 3z = -11 & E_5
\end{array}
$$

Add E_4 to E_5 to eliminate z

$$
\begin{array}{rl}
y - 3z = 5 & E_4 \\
5y + 3z = -11 & E_5 \\
\hline
6y = -6 & \\
y = -1 &
\end{array}
$$

Substitute $y = -1$ into E_5 and solve for z.

$5y + 3z = -11 \quad E_5$

$5(-1) + 3z = -11$

$z = -2$

Substitute $y = -1$ and $z = -2$ into E_1 and solve for x.

$x - 3y + z = 4 \quad E_1$

$x - 3(-1) + (-2) = 4$

$x = 3$

$(3, -1, -2)$ (5-2)

11.
$$2x + y - z = 5 \quad E_1$$
$$x - 2y - 2z = 4 \quad E_2$$
$$3x + 4y + 3z = 3 \quad E_3$$

Multiply E_2 by -2 and add to E_1 to eliminate x. Also multiply E_2 by -3 and add to E_3 to eliminate x.

$$
\begin{array}{ll}
2x + y - z = 5 & E_1 \\
\underline{-2x + 4y + 4z = -8} & (-2)E_2 \\
 5y + 3z = -3 & E_4 \\
3x + 4y + 3z = 3 & E_3 \\
\underline{-3x + 6y + 6z = -12} & (-3)E_2 \\
 10y + 9z = -9 & E_5
\end{array}
$$

Equivalent system:
$$
\begin{array}{ll}
x - 2y - 2z = 4 & E_2 \\
5y + 3z = -3 & E_4 \\
10y + 9z = -9 & E_5
\end{array}
$$

Multiply E_4 by -2 and add to E_5 to eliminate y.
$$
\begin{array}{ll}
-10y - 6z = 6 & (-2)E_4 \\
\underline{10y + 9z = -9} & E_5 \\
 3z = -3 & \\
z = -1 &
\end{array}
$$

Substitute $z = -1$ into E_4 and solve for y.
$$
\begin{array}{ll}
5y + 3z = -3 & E_4 \\
5y + 3(-1) = -3 & \\
y = 0 &
\end{array}
$$

Substitute $y = 0$ for $z = -1$ into E_2 and solve for x.
$$
\begin{array}{l}
x - 2y - 2z = 4 \quad E_2 \\
x - 2(0) - 2(-1) = 4 \\
x = 2
\end{array}
$$
$(2, 0, -1)$ *(5-2)*

12. The boundary line goes through $(0, -4)$ and $(6, 0)$:
$$m = \frac{0 - (-4)}{6 - 0} = \frac{4}{6} = \frac{2}{3}$$
$$y - 0 = \frac{2}{3}(x - 6)$$
Multiply both sides by 3
$$3y = 2(x - 6)$$
$$3y = 2x - 12$$
$$2x - 3y = 12$$

Now note that $(0, 0)$ is in the shaded region, and $2(0) - 3(0) = 0 < 12$, so the inequality is $2x - 3y \leq 12$ (equality included since the boundary line is solid).

(5-3)

13. The boundary line goes through $(0, 8)$ and $(2, 0)$:
$$m = \frac{0 - 8}{2 - 0} = \frac{-8}{2} = -4$$
$$y - 0 = -4(x - 2)$$
$$y = -4x + 8$$
$$4x + y = 8$$

Now note that $(4, 0)$ is in the shaded region, and $4(4) - 3(0) = 16 > 8$, so the inequality is $4x + y \geq 8$ (equality included since the boundary line is solid).

(5-3)

14.

Corner Point (x, y)	Objective Function $z = 5x + 3y$	
$(0, 10)$	30	
$(0, 6)$	18	Minimum value
$(4, 2)$	26	
$(6, 4)$	42	Maximum value

The maximum value of z on S is 42 at $(6, 4)$. The minimum value of z on S is 18 at $(0, 6)$.

(5-4)

15. $x + 2y + 3z = 1$ E_1
$2x + 3y + 4z = 3$ E_2
$x + 2y + z = 3$ E_3

Multiply E_1 by -1 and add to E_3 to eliminate x and y:

$-x - 2y - 3z = -1$ $(-1)E_1$
$\underline{x + 2y + z = 3}$ E_2
$\qquad\quad -2z = 2$
$\qquad\qquad z = -1$

Substitute $z = -1$ into E_1 and E_2 to obtain a simpler system:

$x + 2y + 3(-1) = 1$
$2x + 3y + 4(-1) = 3$
or
$x + 2y = 4$ E_4
$2x + 3y = 7$ E_5

Multiply E_4 by -2 and add to E_5 to eliminate x.

$-2x - 4y = -8$ $(-2)E_4$
$\underline{2x + 3y = 7}$ E_5
$\qquad -y = -1$
$\qquad\quad y = 1$

Substitute $y = 1$ into E_4 to find x.

$x + 2(1) = 4$
$\qquad x = 2$
$(2, 1, -1)$ (5-2)

16. $x + 2y - z = 2$ E_1
$2x + 3y + z = -3$ E_2
$3x + 5y = -1$ E_3

Add E_1 and E_2 to eliminate z.

$x + 2y - z = 2$ E_1
$\underline{2x + 3y + z = -3}$ E_2
$3x + 5y = -1$ E_4

Multiply E_3 by -1 and add to E_4.

$-3x - 5y = 1$ $(-1)E_3$
$\underline{3x + 5y = -1}$ E_4
$\qquad 0 = 0$

There are infinitely many solutions; the system is dependent and equivalent to:

$x + 2y - z = 2$ E_1
$3x + 5y = -1$ E_3

Multiply E_1 by -3 and add to E_3.

$-3x - 6y + 3z = -6$ $(-3)E_1$
$\underline{3x + 5y = -1}$ E_3
$\qquad -y + 3z = -7$
$\qquad\quad y = 3z + 7$

Let $z = t$. Then $y = 3t + 7$. Substitute this into E_3.

$3x + 5(3t + 7) = -1$
$3x + 15t + 35 = -1$
$\qquad\quad 3x = -15t - 36$
$\qquad\quad\; x = -5t - 12$

Solutions: $\{(-5t - 12, 3t + 7, t) \mid t$ is any real number$\}$ (5-2)

17.

$$\begin{array}{ll} x - y = 1 & E_1 \\ 2x - y = 0 & E_2 \\ x - 3y = -2 & E_3 \end{array}$$

Multiply E_1 by -2 and add to E_2 to eliminate x.

$$\begin{array}{ll} -2x + 2y = -2 & (-2)E_1 \\ \underline{2x - y = 0} & E_2 \\ y = -2 \end{array}$$

Substitute $y = -2$ back into E_1 and E_2.

E_1 : $x - (-2) = 1$; $x = -1$

E_2 : $2x - (-2) = 0$; $x = 1$

Since x cannot be both 1 and -1, the system is inconsistent, and there is no solution.

(5-2)

18.

$$\begin{array}{ll} x + 2y - z = 2 & E_1 \\ 3x - y + 2z = -3 & E_2 \\ 4x + y + z = 7 & E_3 \end{array}$$

Add E_2 and E_3 to eliminate y.

$$\begin{array}{ll} 3x - y + 2z = -3 & E_2 \\ \underline{4x + y + z = 7} & E_3 \\ 7x + 3z = 4 & E_4 \end{array}$$

Multiply E_2 by 2 and add to E_1 to eliminate y again:

$$\begin{array}{ll} 6x - 2y + 4z = -6 & 2E_2 \\ \underline{x + 2y - z = 2} & E_1 \\ 7x + 3z = -4 & E_5 \end{array}$$

Now multiply E_4 by -1 and add to E_5.

$$\begin{array}{ll} -7x - 3z = -4 & (-1)E_4 \\ \underline{7x + 3z = -4} & E_5 \\ 0 = -8 \end{array}$$

The system is inconsistent, and there is no solution.

(5-2)

19. The solution region is bounded. Three corner points are obvious from the graph: (0, 4), (0, 0), and (4, 0). The fourth corner point is obtained by solving the system
$2x + y = 8$
$2x + 3y = 12$ to obtain (3, 2).

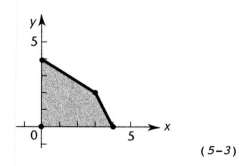

(5-3)

20. The solution region is unbounded. Two corner points are obvious from the graph: (0, 8) and (12, 0). The third corner point is obtained by solving the system
$2x + y = 8$
$x + 3y = 12$
to obtain $\left(\dfrac{12}{5}, \dfrac{16}{5} \right)$.

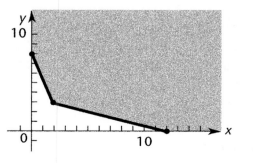

(5-3)

21. The solution region is bounded. The corner point (20, 0) is obvious from the graph. The other corner points are found by solving the system
$$\begin{cases} x + y = 20 \\ x - y = 0 \end{cases} \text{ to obtain } (10, 10)$$
and
$$\begin{cases} x - y = 0 \\ x + 4y = 20 \end{cases} \text{ to obtain } (4, 4)$$

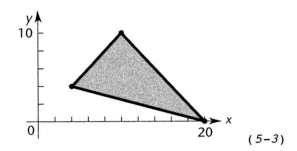

(5-3)

22. The feasible region is graphed as follows. The corner points (0, 4), (0, 0) and (5, 0) are obvious from the graph. The corner point (4, 2) is obtained by solving the system
$$x + 2y = 8$$
$$2x + y = 10$$
We now evaluate the objective function at each corner point.

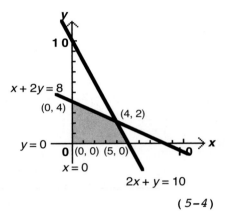

Corner Point (x, y)	Objective Function $z = 7x + 9y$	
(0, 4)	36	
(0, 0)	0	
(5, 0)	35	
(4, 2)	46	Maximum value

The maximum value of z on S is 46 at (4, 2).

(5-4)

23. The feasible region is graphed as follows. The corner points (0, 20), (0, 15), (15, 0) and (20, 0) are obvious from the graph. The corner point (3, 6) is obtained by solving the system
$$3x + y = 15$$
$$x + 2y = 15$$
We now evaluate the objective function at each corner point.

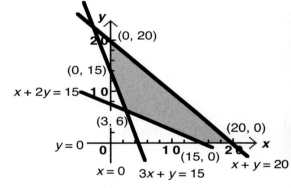

Corner Point (x, y)	Objective Function $z = 5x + 10y$	
(0, 20)	200	
(0, 15)	150	
(3, 6)	75	Minimum value ⎫
(15, 0)	75	Minimum value ⎬ Multiple optimal solutions
(20, 0)	100	

The minimum value of z on S is 75 at (3, 6) and (15, 0).

(5-4)

24. The feasible region is graphed below. The corner points $(0, 10)$ and $(0, 7)$ are obvious from the graph. The other corner points are obtained by solving the systems:

$x + 2y = 20$ and $3x + y = 15$
$3x + y = 15$ to obtain $(2, 9)$ $x + y = 7$ to obtain $(4, 3)$

We now evaluate the objective function at each corner point.

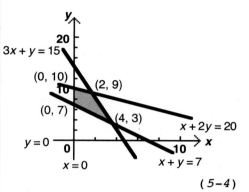

Corner Point (x, y)	Objective Function $z = 5x + 8y$	
$(0, 10)$	80	
$(0, 7)$	56	
$(4, 3)$	44	Minimum value
$(2, 9)$	82	Maximum value

The minimum value of z on S is 44 at $(4, 3)$.
The maximum value of z on S is 82 at $(2, 9)$.

$(5-4)$

25.
$$x_1 + x_2 + x_3 = 7{,}000 \quad E_1$$
$$0.04x_1 + 0.05x_2 + 0.06x_3 = 360 \quad E_2$$
$$0.04x_1 + 0.05x_2 - 0.06x_3 = 120 \quad E_3$$

Multiply E_2 by -1 and add to E_3 to eliminate x_1 and x_2.

$$-0.04x_1 - 0.05x_2 - 0.06x_3 = -360 \qquad (-1)E_2$$
$$\underline{0.04x_1 + 0.05x_2 - 0.06x_3 = 120} \qquad E_3$$
$$-0.12x_3 = -240$$
$$x_3 = 2{,}000$$

Substitute $x_3 = 2{,}000$ back into E_1 and E_2 to obtain a simpler system.
$$x_1 + x_2 + 2{,}000 = 7{,}000$$
$$0.04x_1 + 0.05x_2 + 0.06(2{,}000) = 360$$
or
$$x_1 + x_2 = 5{,}000 \qquad E_4$$
$$0.04x_1 + 0.05x_2 = 240 \qquad E_5$$

Multiply E_4 by -0.04 and add to E_5.
$$-0.04x_1 - 0.04x_2 = -200 \qquad (-0.04)E_4$$
$$\underline{0.04x_1 + 0.05x_2 = 240} \qquad E_5$$
$$0.01x_2 = 40$$
$$x_2 = 4{,}000$$

Substitute $x_2 = 4{,}000$ back into E_4.
$$x_1 + 4{,}000 = 5{,}000$$
$$x_1 = 1{,}000$$
$(1{,}000, 4{,}000, 2{,}000)$

$(5-2)$

26. For convenience we rewrite the constraint conditions:
$1.2x + 0.6y \leq 960$ becomes $2x + y \leq 1{,}600$
$0.04x + 0.03y \leq 36$ becomes $4x + 3y \leq 3{,}600$
$0.2x + 0.3y \leq 270$ becomes $2x + 3y \leq 2{,}700$
       $x, y \geq 0$ are unaltered.

The feasible region is graphed in the diagram. The corner points $(0, 900)$, $(0, 0)$, and $(800, 0)$ are obvious from the graph. The other corner points are obtained by solving the system:

$2x + y = 1{,}600$ and $4x + 3y = 3{,}600$
$4x + 3y = 3{,}600$ to obtain $(600, 400)$ $2x + 3y = 2{,}700$ to obtain $(450, 600)$

We now evaluate the objective function at each corner point.

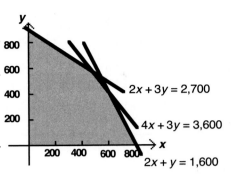

Corner Point (x, y)	Objective Function $z = 30x + 20y$	
(0, 900)	18,000	
(0, 0)	0	
(800, 0)	24,000	
(600, 400)	26,000	Maximum value
(450, 600)	25,500	

The maximum value of z on S is 26,000 at (600, 400).

(5-4)

27. Let x = number of $\frac{1}{2}$-pound packages

   y = number of $\frac{1}{3}$-pound packages

There are 120 packages. So

$x + y = 120$ (1)

Since x $\frac{1}{2}$-pound packages weigh $\frac{1}{2}x$ pounds and y $\frac{1}{3}$-pound packages weigh $\frac{1}{3}y$ pounds, we have

$\frac{1}{2}x + \frac{1}{3}y = 48$ (2)

We solve the system (1), (2) using elimination by addition. We multiply the second equation by -3 and add.

$x + y = 120$

$-\frac{3}{2}x - y = -144$

$-\frac{1}{2}x = -24$

$x = 48$

Substituting into equation (1), we have

$48 + y = 120$

$y = 72$

48 $\frac{1}{2}$-pound packages and 72 $\frac{1}{3}$-pound packages.

(5-1)

28. Using the formulas for perimeter and area of a rectangle, we have

$2a + 2b = 28$

$ab = 48$

Solving the first equation for b and substituting into the second equation, we have

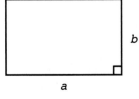

$2b = 28 - 2a$

$b = 14 - a$

$a(14 - a) = 48$

$-a^2 + 14a = 48$

$a^2 - 14a + 48 = 0$

$(a - 6)(a - 8) = 0$

$a = 6, 8$

For $a = 6$ For $a = 8$

$b = 14 - 6$ $b = 14 - 8$

$= 8$ $= 6$

Dimensions: 6 meters by 8 meters.

(5-1)

29. Let x = the number of grams of mix A
y = the number of grams of mix B
z = the number of grams of mix C

Then using the percentages in the table, we get
$0.30x + 0.20y + 0.10z = 27$ (protein)
$0.03x + 0.05y + 0.04z = 5.4$ (fat)
$0.10x + 0.20y + 0.10z = 19$ (moisture)

We can clear decimals easily by multiplying the 1st and 3rd equation by 10, and the 2nd by 100:
$3x + 2y + z = 270$ E_1
$3x + 5y + 4z = 540$ E_2
$x + 2y + z = 190$ E_3

Multiply E_1 by -4 and add to E_2.

$$
\begin{array}{lll}
-12x - 8y - 4z = -1{,}080 & \quad (-4)E_1 \\
\underline{3x + 5y + 4z = \phantom{-1{,}0}540} & \quad E_2 \\
-9x - 3y = \phantom{-1{,}0}-540 & \quad E_4
\end{array}
$$

Multiply E_1 by -1 and add to E_3.

$$
\begin{array}{lll}
-3x - 2y - z = -270 & \quad (-1)E_1 \\
\underline{x + 2y + z = 190} & \quad E_3 \\
-2x = -80 \\
x = 40
\end{array}
$$

Substitute $x = 40$ back into E_4.
$-9(40) - 3y = -540$
$-3y = -180$
$y = 60$

Substitute $x = 40$, $y = 60$ back into E_1.
$3(40) + 2(60) + z = 270$
$240 + z = 270$
$z = 30$

They should use 40 grams of mix A, 60 grams of mix B, and 30 grams of mix C.

$(5\text{-}2)$

30. We begin by writing an expression for the amount of time needed in each machine. Let x = the number of calculator boards produced, and let y = the number of toaster boards produced. Now we can write an expression for the total minutes in each machine for each type of board:

Selective machine: Calculator boards $4x$ (4 minutes per board times x boards), toaster boards $3y$.

Wave machine: Calculator boards $2x$, toaster boards y

Total minutes in selective machine: $4x + 3y$

Total minutes in wave machine: $2x + y$

(A) The total minutes in the selective machine must be no more than 300: $4x + 3y \leq 300$. Both x and y must be non-negative: $x \geq 0$, $y \geq 0$.

The graph is shown at the right.

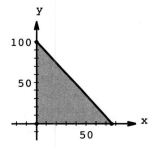

(B) The total minutes in the wave machine must be no more than 120: $2x + y \le 120$. Both x and y must be non-negative: $x \ge 0$, $y \ge 0$.

The graph is shown at the right.

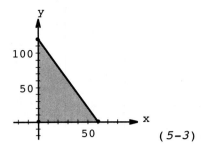

(5-3)

31. Let x = number of regular sails
y = number of competition sails

(A) We form the linear objective function
$P = 60x + 100y$
We wish to maximize P, the profit from x regular sails @ $60 and y competition sails @ $100, subject to the constraints

$x + 2y \le 140$ cutting department constraint
$3x + 4y \le 360$ sewing department constraint
$x, y \ge 0$ non-negative constraints

Solving the system of constraint inequalities graphically, we obtain the feasible region S shown.

Next we evaluate the objective function at each corner point.

Corner Point (x, y)	Objective Function $P = 60x + 100y$	
(0, 0)	0	
(0, 70)	7,000	
(80, 30)	7,800	Maximum value
(120, 0)	7,200	

The optimal value is 7,800 at the corner point (80, 30). So the company should manufacture 80 regular sails and 30 competition sails for a maximum profit of $7,800.

(B) The objective function now becomes $60x + 125y$. We evaluate this at each corner point.

Corner Point (x, y)	Objective Function $P = 60x + 125y$	
(0, 0)	0	
(0, 70)	8,750	Maximum value
(80, 30)	8,550	
(120, 0)	7,200	

The optimal value is now 8,750 at the corner point (0, 70). So the company should manufacture 70 competition sails and no regular sails for a maximum profit of $8,750.

(C) The objective function now becomes $60x + 75y$. We evaluate this at each corner point.

Corner Point (x, y)	Objective Function $P = 60x + 75y$	
(0, 0)	0	
(0, 70)	5,250	
(80, 30)	7,050	
(120, 0)	7,200	Maximum value

The optimal value is now 7,200 at the corner point (120, 0). So the company should manufacture 120 regular sails and no competition sails for a maximum profit of $7,200.

(5-4)

32. Let x = number of grams of mix A
   y = number of grams of mix B

(A) We form the objective function
   $C = 0.07x + 0.04y$

C represents the cost of x grams at \$0.07 per gram and y grams at \$0.04 per gram. We wish to minimize C subject to

$5x + 2y \geq 800$	vitamin constraint
$2x + 4y \geq 800$	mineral constraint
$4x + 4y \leq 1300$	calorie constraint
$x, y \geq 0$	non-negative constraints

Solving the system of constraint inequalities graphically, we obtain the feasible region S shown in the diagram.

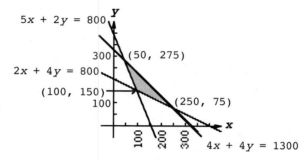

Next we evaluate the objective function at each corner point.

Corner Point (x, y)	Objective Function $C = 0.07x + 0.04y$	
(100, 150)	13	Minimum value
(250, 75)	20.5	
(50, 275)	14.5	

The optimal value is 13 at the corner point (100, 150). So 100 grams of mix A and 150 grams of mix B should be used for a cost of \$13.

(B) The objective function now becomes $0.07x + 0.02y$. We evaluate this at each corner point.

Corner Point (x, y)	Objective Function $C = 0.07x + 0.02y$	
(100, 150)	10	
(250, 75)	19	
(50, 275)	9	Minimum value

The optimal value is now 9 at the corner point (50, 275). So 50 grams of mix A and 275 grams of mix B should be used for a cost of \$9.

(C) The objective function now becomes $0.07x + 0.15y$. We evaluate this at each corner point.

Corner Point (x, y)	Objective Function $C = 0.07x + 0.15y$	
(100, 150)	29.5	
(250, 75)	28.75	Minimum value
(50, 275)	44.75	

The optimal value is now 28.75 at the corner point (250, 75). So 250 grams of mix A and 75 grams of mix B should be used for a cost of \$28.75.

(5-4)

CHAPTER 6   Matrices & Determinants

Section 6-1

1. Answers will vary.   3. Answers will vary.   5. Answers will vary.

7. Answers will vary.   9. Answers will vary.

11. No. Condition 2 is violated: The first nonzero entry in row 2 is not 1.

13. Yes

15. No. Condition 4 is violated: The first nonzero entry in row 3 is to the left of the first nonzero entry in row 2.

17. Yes

19. x_1 $= -2$
       x_2 $= 3$
          $x_3 = 0$

The system is already solved.

21. x_1 $- 2x_3 = 3$
       $x_2 +$ $x_3 = -5$

Solution:
$x_3 = t$
$x_2 = -5 - x_3 = -5 - t$
$x_1 = 3 + 2x_3 = 3 + 2t$
So the solution is $x_1 = 2t + 3$,
$x_2 = -t - 5$,
$x_3 = t$ for any real number t.

23. x_1 $= 0$
       x_2 $= 0$
          $0 = 1$

The system has no solution.

25. $x_1 - 2x_2$ $- 3x_4 = -5$
                $x_3 + 3x_4 = 2$

Solution:
$x_4 = t$
$x_3 = 2 - 3x_4 = 2 - 3t$
$x_2 = s$
$x_1 = -5 + 2x_2 + 3x_4 = -5 + 2s + 3t$
So the solution is $x_1 = 2s + 3t - 5$,
$x_2 = s$, $x_3 = -3t + 2$, $x_4 = t$ for any real numbers s and t.

27. $\begin{bmatrix} 1 & 2 & | & -1 \\ 0 & 1 & | & 3 \end{bmatrix}$ $(-2)R_2 + R_1 \rightarrow R_1$

Need a 0 here

$\sim \begin{bmatrix} 1 & 0 & | & -7 \\ 0 & 1 & | & 3 \end{bmatrix}$

29. $\begin{bmatrix} 1 & 0 & -3 & | & 1 \\ 0 & 1 & 2 & | & 0 \\ 0 & 0 & 3 & | & -6 \end{bmatrix} \frac{1}{3} R_3 \rightarrow R_3$

Need a 1 here
Need 0's here

$\sim \begin{bmatrix} 1 & 0 & -3 & | & 1 \\ 0 & 1 & 2 & | & 0 \\ 0 & 0 & 1 & | & -2 \end{bmatrix} \begin{matrix} 3R_3 + R_1 \rightarrow R_1 \\ (-2)R_3 + R_2 \rightarrow R_2 \end{matrix}$

$\sim \begin{bmatrix} 1 & 0 & 0 & | & -5 \\ 0 & 1 & 0 & | & 4 \\ 0 & 0 & 1 & | & -2 \end{bmatrix}$

31.
$$\begin{bmatrix} 1 & 2 & -2 & | & -1 \\ 0 & 3 & -6 & | & 1 \\ 0 & -1 & 2 & | & -\frac{1}{3} \end{bmatrix} \frac{1}{3} R_2 \rightarrow R_2$$

Need a 1 here

$$\sim \begin{bmatrix} 1 & 2 & -2 & | & -1 \\ 0 & 1 & -2 & | & \frac{1}{3} \\ 0 & -1 & 2 & | & -\frac{1}{3} \end{bmatrix} \begin{matrix} (-2)R_2 + R_1 \rightarrow R_1 \\ \\ R_3 + R_2 \rightarrow R_3 \end{matrix}$$

Need 0's here

$$\sim \begin{bmatrix} 1 & 0 & 2 & | & -\frac{5}{3} \\ 0 & 1 & -2 & | & \frac{1}{3} \\ 0 & 0 & 0 & | & 0 \end{bmatrix}$$

33.
$$\begin{bmatrix} 2 & 4 & -10 & | & -2 \\ 3 & 9 & -21 & | & 0 \\ 1 & 5 & -12 & | & 1 \end{bmatrix} R_1 \leftrightarrow R_3$$

Need a 1 here

$$\sim \begin{bmatrix} 1 & 5 & -12 & | & 1 \\ 3 & 9 & -21 & | & 0 \\ -2 & 4 & -10 & | & -2 \end{bmatrix} \begin{matrix} (-3)R_1 + R_2 \rightarrow R_2 \\ (-2)R_1 + R_3 \rightarrow R_3 \end{matrix}$$

Need 0's here

$$\sim \begin{bmatrix} 1 & 5 & -12 & | & 1 \\ 0 & -6 & 15 & | & -3 \\ 0 & 6 & 14 & | & -4 \end{bmatrix} -\frac{1}{6} R_2 \rightarrow R_2$$

Need a 1 here

$$\sim \begin{bmatrix} 1 & 5 & -12 & | & 1 \\ 0 & 1 & -\frac{5}{2} & | & \frac{1}{2} \\ 0 & -6 & 14 & | & -4 \end{bmatrix} \begin{matrix} (-5)R_2 + R_1 \rightarrow R_1 \\ \\ 6R_2 + R_3 \rightarrow R_3 \end{matrix}$$

Need 0's here

$$\sim \begin{bmatrix} 1 & 0 & \frac{1}{2} & | & -\frac{3}{2} \\ 0 & 1 & -\frac{5}{2} & | & \frac{1}{2} \\ 0 & 0 & -1 & | & -1 \end{bmatrix} -R_3 \rightarrow R_3$$

Need a 1 here

$$\sim \begin{bmatrix} 1 & 0 & \frac{1}{2} & | & -\frac{3}{2} \\ 0 & 1 & \frac{5}{2} & | & \frac{1}{2} \\ 0 & 0 & 1 & | & 1 \end{bmatrix} \begin{matrix} (-\frac{1}{2})R_3 + R_1 \rightarrow R_1 \\ \frac{5}{2} R_3 + R_2 \rightarrow R_2 \end{matrix}$$

Need 0's here

$$\sim \begin{bmatrix} 1 & 0 & 0 & | & -2 \\ 0 & 1 & 0 & | & 3 \\ 0 & 0 & 1 & | & 1 \end{bmatrix}$$

Solution: $x_1 = -2$, $x_2 = 3$, and $x_3 = 1$.

35.
$$\begin{bmatrix} 3 & 8 & -1 & | & -18 \\ 2 & 1 & 5 & | & 8 \\ 2 & 4 & 2 & | & -4 \end{bmatrix} \begin{matrix} \frac{1}{2} R_3 \rightarrow R_3 \\ R_3 \leftrightarrow R_1 \end{matrix}$$

Need a 1 here

$$\sim \begin{bmatrix} 1 & 2 & 1 & | & -2 \\ 2 & 1 & 5 & | & 8 \\ 3 & 8 & -1 & | & -18 \end{bmatrix} \begin{matrix} (-2)R_1 + R_2 \rightarrow R_2 \\ (-3)R_1 + R_3 \rightarrow R_3 \end{matrix}$$

Need 0's here

$$\sim \begin{bmatrix} 1 & 2 & 1 & | & -2 \\ 0 & -3 & 3 & | & 12 \\ 0 & 2 & -4 & | & -12 \end{bmatrix} -\frac{1}{3} R_2 \rightarrow R_2$$

Need a 1 here

$$\sim \begin{bmatrix} 1 & 2 & 1 & | & -2 \\ 0 & 1 & -1 & | & -4 \\ 0 & 2 & -4 & | & -12 \end{bmatrix} \begin{matrix} (-2)R_2 + R_1 \rightarrow R_1 \\ (-2)R_2 + R_3 \rightarrow R_3 \end{matrix}$$

Need 0's here

$$\sim \begin{bmatrix} 1 & 0 & 3 & | & 6 \\ 0 & 1 & -1 & | & -4 \\ 0 & 0 & -2 & | & -4 \end{bmatrix} -\frac{1}{2} R_3 \rightarrow R_3$$

Need a 1 here

$$\sim \begin{bmatrix} 1 & 0 & 3 & | & 6 \\ 0 & 1 & -1 & | & -4 \\ 0 & 0 & 1 & | & 2 \end{bmatrix} \begin{matrix} (-3)R_3 + R_1 \rightarrow R_1 \\ R_3 + R_2 \rightarrow R_2 \end{matrix}$$

Need 0's here

$$\begin{bmatrix} 1 & 0 & 0 & | & 0 \\ 0 & 1 & 0 & | & -2 \\ 0 & 0 & 1 & | & 2 \end{bmatrix}$$ Solution: $x_1 = 0$, $x_2 = -2$, and $x_3 = 2$.

37. $\begin{bmatrix} 2 & -1 & -3 & | & 8 \\ 1 & -2 & 0 & | & 7 \end{bmatrix}$ $R_1 \leftrightarrow R_2$

$\sim \begin{bmatrix} 1 & -2 & 0 & | & 7 \\ 2 & -1 & -3 & | & 8 \end{bmatrix}$ $(-2)R_1 + R_2 \rightarrow R_2$

$\sim \begin{bmatrix} 1 & -2 & 0 & | & 7 \\ 0 & 3 & -3 & | & -6 \end{bmatrix}$ $\frac{1}{3}R_2 \rightarrow R_2$

$\sim \begin{bmatrix} 1 & -2 & 0 & | & 7 \\ 0 & 1 & -1 & | & -2 \end{bmatrix}$ $2R_2 + R_1 \rightarrow R_1$

$\sim \begin{bmatrix} 1 & 0 & -2 & | & 3 \\ 0 & 1 & -1 & | & -2 \end{bmatrix}$

Let $x_3 = t$. Then

$x_2 - x_3 = -2$

$\quad x_2 = x_3 - 2 = t - 2$

$x_1 - 2x_3 = 3$

$\quad x_1 = 2x_3 + 3 = 2t + 3$

Solution: $x_1 = 2t + 3$, $x_2 = t - 2$, $x_3 = t$, t any real number.

39. $\begin{bmatrix} 2 & -1 & | & 0 \\ 3 & 2 & | & 7 \\ 1 & -1 & | & -1 \end{bmatrix}$ $R_1 \leftrightarrow R_3$

$\sim \begin{bmatrix} 1 & -1 & | & -1 \\ 3 & 2 & | & 7 \\ 2 & -1 & | & 0 \end{bmatrix}$ $\begin{matrix}(-3)R_1 + R_2 \rightarrow R_2 \\ (-2)R_1 + R_3 \rightarrow R_3\end{matrix}$

$\sim \begin{bmatrix} 1 & -1 & | & -1 \\ 0 & 5 & | & 10 \\ 0 & 1 & | & 2 \end{bmatrix}$ $R_2 \leftrightarrow R_3$

$\sim \begin{bmatrix} 1 & -1 & | & -1 \\ 0 & 1 & | & 2 \\ 0 & 5 & | & 10 \end{bmatrix}$ $\begin{matrix}R_2 + R_1 \rightarrow R_1 \\ (-5)R_2 + R_3 \rightarrow R_3\end{matrix}$

$\sim \begin{bmatrix} 1 & 0 & | & 1 \\ 0 & 1 & | & 2 \\ 0 & 0 & | & 0 \end{bmatrix}$

Solution: $x_1 = 1$ and $x_2 = 2$.

41. $\begin{bmatrix} 3 & -4 & -1 & | & 1 \\ 2 & -3 & 1 & | & 1 \\ 1 & -2 & 3 & | & 2 \end{bmatrix}$ $R_1 \leftrightarrow R_3$

$\sim \begin{bmatrix} 1 & -2 & 3 & | & 2 \\ 2 & -3 & 1 & | & 1 \\ 3 & -4 & -1 & | & 1 \end{bmatrix}$ $\begin{matrix}(-2)R_1 + R_2 \rightarrow R_2 \\ (-3)R_1 + R_3 \rightarrow R_3\end{matrix}$

$\sim \begin{bmatrix} 1 & -2 & 3 & | & 2 \\ 0 & 1 & -5 & | & -3 \\ 0 & 2 & -10 & | & -5 \end{bmatrix}$ $(-2)R_2 + R_3 \rightarrow R_3$

$\sim \begin{bmatrix} 1 & -2 & 3 & | & 2 \\ 0 & 1 & -5 & | & -3 \\ 0 & 0 & 0 & | & 1 \end{bmatrix}$

Since the last row corresponds to the equation $0x_1 + 0x_2 + 0x_3 = 1$, there is no solution.

43. $\begin{bmatrix} -2 & 1 & 3 & | & -7 \\ 1 & -4 & 2 & | & 0 \\ 1 & -3 & 1 & | & 1 \end{bmatrix}$ $R_1 \leftrightarrow R_2$

$\sim \begin{bmatrix} 1 & -4 & 2 & | & 0 \\ -2 & 1 & 3 & | & -7 \\ 1 & -3 & 1 & | & 1 \end{bmatrix}$ $\begin{matrix}2R_1 + R_2 \rightarrow R_2 \\ (-1)R_1 + R_3 \rightarrow R_3\end{matrix}$

$\sim \begin{bmatrix} 1 & -4 & 2 & | & 0 \\ 0 & -7 & 7 & | & -7 \\ 0 & 1 & -1 & | & 1 \end{bmatrix}$ $R_2 \leftrightarrow R_3$

$\sim \begin{bmatrix} 1 & -4 & 2 & | & 0 \\ 0 & 1 & -1 & | & 1 \\ 0 & -7 & 7 & | & -7 \end{bmatrix}$ $\begin{matrix}4R_2 + R_1 \rightarrow R_1 \\ 7R_2 + R_3 \rightarrow R_3\end{matrix}$

$\sim \begin{bmatrix} 1 & 0 & -2 & | & 4 \\ 0 & 1 & -1 & | & 1 \\ 0 & 0 & 0 & | & 0 \end{bmatrix}$

Let $x_3 = t$. Then

$\quad x_2 - x_3 = 1$

$\qquad x_2 = x_3 + 1 = t + 1$

$\quad x_1 - 2x_3 = 4$

$x_1 = 2x_3 + 4 = 2t + 4$

Solution: $x_1 = 2t + 4$, $x_2 = t + 1$, $x_3 = t$, t any real number.

45. $\begin{bmatrix} 2 & -2 & -4 & | & -2 \\ -3 & 3 & 6 & | & 3 \end{bmatrix}$ $\begin{matrix}\frac{1}{2}R_1 \rightarrow R_1 \\ \frac{1}{3}R_2 \rightarrow R_2\end{matrix}$

$\sim \begin{bmatrix} 1 & -1 & -2 & | & -1 \\ -1 & 1 & 2 & | & 1 \end{bmatrix}$ $R_1 + R_2 \rightarrow R_2$

$\sim \begin{bmatrix} 1 & -1 & -2 & | & -1 \\ 0 & 0 & 0 & | & 0 \end{bmatrix}$

Let $x_3 = t$, $x_2 = s$. Then

$\quad x_1 - x_2 - 2x_3 = -1$

$\quad x_1 \qquad\qquad = x_2 + 2x_3 - 1$

$\qquad\qquad\qquad = s + 2t - 1$

Solution: $x_1 = s + 2t - 1$, $x_2 = s$, $s_3 = t$, s and t any real numbers.

47.
$$\begin{bmatrix} 4 & -1 & 2 & | & 3 \\ -4 & 1 & -3 & | & -10 \\ 8 & -2 & 9 & | & -1 \end{bmatrix} \begin{matrix} \\ R_1 + R_2 \rightarrow R_2 \\ (-2)R_1 + R_3 \rightarrow R_3 \end{matrix}$$

$$\sim \begin{bmatrix} 4 & -1 & 2 & | & 3 \\ 0 & 0 & -1 & | & -7 \\ 0 & 0 & 5 & | & -7 \end{bmatrix} \begin{matrix} \\ \\ 5R_2 + R_3 \rightarrow R_3 \end{matrix}$$

$$\sim \begin{bmatrix} 4 & -1 & 2 & | & 3 \\ 0 & 0 & -1 & | & -7 \\ 0 & 0 & 0 & | & -42 \end{bmatrix}$$

Since the last row corresponds to the equation $0x_1 + 0x_2 + 0x_3 = -42$, there is no solution.

49.
$$\begin{bmatrix} 2 & -5 & -3 & | & 7 \\ -4 & 10 & 2 & | & 6 \\ 6 & -15 & -1 & | & -19 \end{bmatrix} \begin{matrix} \\ 2R_1 + R_2 \rightarrow R_2 \\ (-3)R_1 + R_3 \rightarrow R_3 \end{matrix}$$

$$\sim \begin{bmatrix} 2 & -5 & -3 & | & 7 \\ 0 & 0 & -4 & | & 20 \\ 0 & 0 & 8 & | & -40 \end{bmatrix} \begin{matrix} \frac{1}{2} R_1 \rightarrow R_1 \\ -\frac{1}{4} R_2 \rightarrow R_2 \\ \\ \end{matrix}$$

$$\sim \begin{bmatrix} 1 & -2.5 & -1.5 & | & 3.5 \\ 0 & 0 & 1 & | & -5 \\ 0 & 0 & 8 & | & -40 \end{bmatrix} \begin{matrix} 1.5R_2 + R_1 \rightarrow R_1 \\ \\ (-8)R_2 + R_3 \rightarrow R_3 \end{matrix}$$

$$\sim \begin{bmatrix} 1 & -2.5 & 0 & | & -4 \\ 0 & 0 & 1 & | & -5 \\ 0 & 0 & 0 & | & 0 \end{bmatrix}$$

Let $x_2 = t$. Then $x_3 = -5$ and
$$x_1 - 2.5x_2 = -4$$
$$x_1 = 2.5x_2 - 4$$
$$x_1 = 2.5t - 4$$
Solution: $x_1 = 2.5t - 4$, $x_2 = t$, $x_3 = -5$, t any real number.

51.
$$\begin{bmatrix} 5 & -3 & 2 & | & 13 \\ 2 & -1 & -3 & | & 1 \\ 4 & -2 & 4 & | & 12 \end{bmatrix} \frac{1}{4} R_3 \rightarrow R_3$$

$$\sim \begin{bmatrix} 5 & -3 & 2 & | & 13 \\ 2 & -1 & -3 & | & 1 \\ 1 & -\frac{1}{2} & 1 & | & 3 \end{bmatrix} R_1 \leftrightarrow R_3$$

$$\sim \begin{bmatrix} 1 & -\frac{1}{2} & 1 & | & 3 \\ 2 & -1 & -3 & | & 1 \\ 5 & -3 & 2 & | & 13 \end{bmatrix} \begin{matrix} \\ (-2)R_1 + R_2 \rightarrow R_2 \\ (-5)R_1 + R_3 \rightarrow R_3 \end{matrix}$$

$$\sim \begin{bmatrix} 1 & -\frac{1}{2} & 1 & | & 3 \\ 0 & 0 & -5 & | & -5 \\ 0 & -\frac{1}{2} & -3 & | & -2 \end{bmatrix} \begin{matrix} (-1)R_3 + R_1 \rightarrow R_1 \\ -\frac{1}{5} R_2 \rightarrow R_2 \\ \\ \end{matrix}$$

$$\sim \begin{bmatrix} 1 & 0 & 4 & | & 5 \\ 0 & 0 & 1 & | & 1 \\ 0 & -\frac{1}{2} & -3 & | & -2 \end{bmatrix} \begin{matrix} (-4)R_2 + R_1 \rightarrow R_1 \\ 3R_2 + R_3 \rightarrow R_3 \end{matrix}$$

$$\sim \begin{bmatrix} 1 & 0 & 0 & | & 1 \\ 0 & 0 & 1 & | & 1 \\ 0 & -\frac{1}{2} & 0 & | & 1 \end{bmatrix} (-2)R_3 \rightarrow R_3$$

$$\sim \begin{bmatrix} 1 & 0 & 0 & | & 1 \\ 0 & 0 & 1 & | & 1 \\ 0 & 1 & 0 & | & -2 \end{bmatrix} R_2 \leftrightarrow R_3$$

$$\sim \begin{bmatrix} 1 & 0 & 0 & | & 1 \\ 0 & 1 & 0 & | & -2 \\ 0 & 0 & 1 & | & 1 \end{bmatrix}$$

Solution: $x_1 = 1$, $x_2 = -2$, $x_3 = 1$.

53. (A) The reduced form matrix will have the form

$$\begin{bmatrix} 1 & a & b & | & c \\ 0 & 0 & 0 & | & 0 \\ 0 & 0 & 0 & | & 0 \end{bmatrix}$$

So the system is equivalent to

$$x_1 + ax_2 + bx_3 = c$$
$$0 = 0$$
$$0 = 0$$

The system is dependent, and x_2 and x_3 can assume any real values. There are two parameters in the solution.

(B) The reduced form matrix will have the form

$$\begin{bmatrix} 1 & 0 & a & | & b \\ 0 & 1 & c & | & d \\ 0 & 0 & 0 & | & 0 \end{bmatrix}$$

So the system is equivalent to

$$x_1 + ax_3 = b$$
$$x_2 + cx_3 = d$$
$$0 = 0$$

The system is dependent, with a solution for any real value of x_3. There is one parameter in the solution.

(C) The reduced form matrix will have the form $\begin{bmatrix} 1 & 0 & 0 & | & a \\ 0 & 1 & 0 & | & b \\ 0 & 0 & 1 & | & c \end{bmatrix}$

So there is only one solution, $x_1 = a$, $x_2 = b$, $x_3 = c$, and the system is independent.

(D) This is impossible; there are only 3 equations.

55. The augmented matrix of the system is

$$\begin{bmatrix} 1 & -1 & | & 4 \\ 3 & k & | & 7 \end{bmatrix} (-3)R_1 + R_2 \rightarrow R_2$$

$$\sim \begin{bmatrix} 1 & -1 & | & 4 \\ 0 & k+3 & | & -5 \end{bmatrix}$$

If $k = -3$, this is $\begin{bmatrix} 1 & -1 & | & 4 \\ 0 & 0 & | & -5 \end{bmatrix}$ and the system has no solutions.

If $k \neq -3$, the second matrix shown can be reduced further to

$$\sim \begin{bmatrix} 1 & 0 & | & 4 - \frac{5}{k+3} \\ 0 & 1 & | & \frac{-5}{k+3} \end{bmatrix}$$

(steps not shown). The system has one solution: $x_1 = 4 - \dfrac{5}{k+3}$, $x_2 = \dfrac{-5}{k+3}$.

57. The augmented matrix of the system is

$$\begin{bmatrix} 1 & k & | & 3 \\ 2 & 6 & | & 6 \end{bmatrix} (-2)R_1 + R_2 \rightarrow R_2$$

$$\sim \begin{bmatrix} 1 & k & | & 3 \\ 0 & 6-2k & | & 0 \end{bmatrix}$$

If $k = 3$, this is $\begin{bmatrix} 1 & 3 & | & 3 \\ 0 & 0 & | & 0 \end{bmatrix}$ and the system has infinite solutions: Let $x_2 = s$.
Then

$$x_1 + 3s = 3$$
$$x_1 = 3 - 3s$$

The solutions are $x_1 = 3 - 3s$, $x_2 = s$, s any real number.
If $k \neq 3$, the second matrix shown can be reduced further to

$$\sim \begin{bmatrix} 1 & 0 & | & 3 \\ 0 & 1 & | & 0 \end{bmatrix}$$

(steps not shown). The system has one solution: $x_1 = 3$, $x_2 = 0$.

59. $\begin{bmatrix} 1 & 2 & -4 & -1 & | & 7 \\ 2 & 5 & -9 & -4 & | & 16 \\ 1 & 5 & -7 & -7 & | & 13 \end{bmatrix} \begin{matrix} \\ (-2)R_1 + R_2 \rightarrow R_2 \\ (-1)R_1 + R_3 \rightarrow R_3 \end{matrix}$

$\sim \begin{bmatrix} 1 & 2 & -4 & -1 & | & 7 \\ 0 & 1 & -1 & -2 & | & 2 \\ 0 & 3 & -3 & -6 & | & 6 \end{bmatrix} \begin{matrix} (-2)R_2 + R_1 \rightarrow R_1 \\ \\ (-3)R_2 + R_3 \rightarrow R_3 \end{matrix}$

$\sim \begin{bmatrix} 1 & 0 & -2 & 3 & | & 3 \\ 0 & 1 & -1 & -2 & | & 2 \\ 0 & 0 & 0 & 0 & | & 0 \end{bmatrix}$

Let $x_4 = t$, $x_3 = s$. Then
$$x_2 - x_3 - 2x_4 = 2$$
$$x_2 = s + 2t + 2 + 2$$
$$x_1 - 2x_3 + 3x_4 = 3$$
$$x_1 = 2x_3 - 3x_4 + 3$$
Solution: $x_1 = 2s - 3t + 3$,
$x_2 = s + 2t + 2$, $x_3 = s$, $x_4 = t$,
s and t any real numbers.

61. $\begin{bmatrix} 1 & -1 & 3 & -2 & | & 1 \\ -2 & 4 & -3 & 1 & | & 0.5 \\ 3 & -1 & 10 & -4 & | & 2.9 \\ 4 & -3 & 8 & -2 & | & 0.6 \end{bmatrix} \begin{matrix} \\ 2R_1 + R_2 \rightarrow R_2 \\ (-3)R_1 + R_3 \rightarrow R_3 \\ (-4)R_1 + R_4 \rightarrow R_4 \end{matrix}$

$\sim \begin{bmatrix} 1 & -1 & 3 & -2 & | & 1 \\ 0 & 2 & 3 & -3 & | & 2.5 \\ 0 & 2 & 1 & 2 & | & -0.1 \\ 0 & 1 & -4 & 6 & | & -3.4 \end{bmatrix} \begin{matrix} \\ \\ \\ R_4 \leftrightarrow R_2 \end{matrix}$

$\sim \begin{bmatrix} 1 & -1 & 3 & -2 & | & 1 \\ 0 & 1 & -4 & 6 & | & -3.4 \\ 0 & 2 & 1 & 2 & | & -0.1 \\ 0 & 2 & 3 & -3 & | & 2.5 \end{bmatrix} \begin{matrix} R_2 + R_1 \rightarrow R_1 \\ \\ (-2)R_2 + R_3 \rightarrow R_3 \\ (-2)R_2 + R_4 \rightarrow R_4 \end{matrix}$

$\sim \begin{bmatrix} 1 & 0 & -1 & 4 & | & -2.4 \\ 0 & 1 & -4 & 6 & | & -3.4 \\ 0 & 0 & 9 & -10 & | & 6.7 \\ 0 & 0 & 11 & -15 & | & 9.3 \end{bmatrix} \begin{matrix} \\ \\ (-1)R_4 + R_3 \rightarrow R_3 \end{matrix}$

$\sim \begin{bmatrix} 1 & 0 & -1 & 4 & | & -2.4 \\ 0 & 1 & -4 & 6 & | & -3.4 \\ 0 & 0 & -2 & 5 & | & -2.6 \\ 0 & 0 & 11 & -15 & | & 9.3 \end{bmatrix} \begin{matrix} \\ \\ -\frac{1}{2}R_3 \leftrightarrow R_3 \end{matrix}$

$\sim \begin{bmatrix} 1 & 0 & -1 & 4 & | & -2.4 \\ 0 & 1 & -4 & 6 & | & -3.4 \\ 0 & 0 & 1 & -2.5 & | & 1.3 \\ 0 & 0 & 11 & -15 & | & 9.3 \end{bmatrix} \begin{matrix} R_3 + R_1 \rightarrow R_1 \\ 4R_3 + R_2 \rightarrow R_2 \\ \\ (-11)R_3 + R_4 \rightarrow R_4 \end{matrix}$

$$\sim \begin{bmatrix} 1 & 0 & 0 & 1.5 & | & -1.1 \\ 0 & 1 & 0 & -4 & | & 1.8 \\ 0 & 0 & 1 & -2.5 & | & 1.3 \\ 0 & 0 & 0 & 12.5 & | & -5 \end{bmatrix} \frac{1}{12.5} R_4 \rightarrow R_4$$

$$\sim \begin{bmatrix} 1 & 0 & 0 & 1.5 & | & -1.1 \\ 0 & 1 & 0 & -4 & | & 1.8 \\ 0 & 0 & 1 & -2.5 & | & 1.3 \\ 0 & 0 & 0 & 1 & | & -0.4 \end{bmatrix} \begin{matrix} (-1.5)R_4 + R_1 \rightarrow R_1 \\ 4R_4 + R_2 \rightarrow R_2 \\ 2.5R_4 + R_3 \rightarrow R_3 \end{matrix}$$

$$\sim \begin{bmatrix} 1 & 0 & 0 & 0 & | & -0.5 \\ 0 & 1 & 0 & 0 & | & 0.2 \\ 0 & 0 & 1 & 0 & | & 0.3 \\ 0 & 0 & 0 & 1 & | & -0.4 \end{bmatrix}$$ Solution: $x_1 = -0.5$, $x_2 = 0.2$, $x_3 = 0.3$, $x_4 = -0.4$

63.
$$\begin{bmatrix} 1 & -2 & 1 & 1 & 2 & | & 2 \\ -2 & 4 & 2 & 2 & -2 & | & 0 \\ 3 & -6 & 1 & 1 & 5 & | & 4 \\ -1 & 2 & 3 & 1 & 1 & | & 3 \end{bmatrix} \begin{matrix} 2R_1 + R_2 \rightarrow R_2 \\ (-3)R_1 + R_3 \rightarrow R_3 \\ R_1 + R_4 \rightarrow R_4 \end{matrix}$$

$$\sim \begin{bmatrix} 1 & -2 & 1 & 1 & 2 & | & 2 \\ -2 & 4 & 2 & 2 & -2 & | & 0 \\ 3 & -6 & 1 & 1 & 5 & | & 4 \\ -1 & 2 & 3 & 1 & 1 & | & 3 \end{bmatrix} \frac{1}{4} R_2 \rightarrow R_2$$

$$\sim \begin{bmatrix} 1 & -2 & 1 & 1 & 2 & | & 2 \\ 0 & 0 & 1 & 1 & 0.5 & | & 1 \\ 0 & 0 & -2 & -2 & -1 & | & -2 \\ 0 & 0 & 4 & 2 & 3 & | & 5 \end{bmatrix} \begin{matrix} (-1)R_2 + R_1 \rightarrow R_1 \\ \\ 2R_2 + R_3 \rightarrow R_3 \\ (-4)R_2 + R_4 \rightarrow R_4 \end{matrix}$$

$$\sim \begin{bmatrix} 1 & -2 & 0 & 0 & 1.5 & | & 1 \\ 0 & 0 & 1 & 1 & 0.5 & | & 1 \\ 0 & 0 & 0 & 0 & 0 & | & 0 \\ 0 & 0 & 0 & -2 & 1 & | & 1 \end{bmatrix} R_3 \leftrightarrow R_4$$

$$\sim \begin{bmatrix} 1 & -2 & 0 & 0 & 1.5 & | & 1 \\ 0 & 0 & 1 & 1 & 0.5 & | & 1 \\ 0 & 0 & 0 & -2 & 1 & | & 1 \\ 0 & 0 & 0 & 0 & 0 & | & 0 \end{bmatrix} \left(-\frac{1}{2}\right)R_3 \rightarrow R_3$$

$$\sim \begin{bmatrix} 1 & -2 & 0 & 0 & 1.5 & | & 1 \\ 0 & 0 & 1 & 1 & 0.5 & | & 1 \\ 0 & 0 & 0 & 1 & -0.5 & | & -0.5 \\ 0 & 0 & 0 & 0 & 0 & | & 0 \end{bmatrix} (-1)R_3 + R_2 \rightarrow R_2$$

$$\sim \begin{bmatrix} 1 & -2 & 0 & 0 & 1.5 & | & 1 \\ 0 & 0 & 1 & 0 & 1 & | & 1.5 \\ 0 & 0 & 0 & 1 & -0.5 & | & -0.5 \\ 0 & 0 & 0 & 0 & 0 & | & 0 \end{bmatrix}$$

Let $x_5 = t$. Then

$$x_4 - 0.5x_5 = -0.5$$
$$x_4 = 0.5x_5 - 0.5 = 0.5t - 0.5$$
$$x_3 + x_5 = 1.5$$
$$x_3 = -x_5 + 1.5 = -t + 1.5$$

Let $x_2 = s$. Then

$$x_1 - 2x_2 + 1.5x_5 = 1$$
$$x_1 = 2x_2 - 1.5x_5 + 1 = 2s - 1.5t + 1$$

Solution: $x_1 = 2s - 1.5t + 1$, $x_2 = s$, $x_3 = -t + 1.15$, $x_4 = 0.5t - 0.5$, $x_5 = t$, s and t any real numbers.

65. Let x_1 = number of 15-cent stamps

x_2 = number of 20-cent stamps

x_3 = number of 35-cent stamps

Then $x_1 + x_2 + x_3 = 45$ (total number of stamps)

$15x_1 + 20x_2 + 35x_3 = 1400$ (total value of stamps)

We write the augmented matrix and solve by Gauss-Jordan elimination.

$$\begin{bmatrix} 1 & 1 & 1 & | & 45 \\ 15 & 20 & 35 & | & 1,400 \end{bmatrix} \quad (-15)R_1 + R_2 \to R_2$$

$$\sim \begin{bmatrix} 1 & 1 & 1 & | & 45 \\ 0 & 5 & 20 & | & 725 \end{bmatrix} \quad \tfrac{1}{5}R_2 \to R_2$$

$$\sim \begin{bmatrix} 1 & 1 & 1 & | & 45 \\ 0 & 1 & 4 & | & 145 \end{bmatrix} \quad (-1)R_2 + R_1 \to R_1$$

$$\sim \begin{bmatrix} 1 & 0 & -3 & | & -100 \\ 0 & 1 & 4 & | & 145 \end{bmatrix}$$

This augmented matrix is in reduced form. It corresponds to the system:

$x_1 - 3x_3 = -100$

$x_2 + 4x_3 = 145$

Let $x_3 = t$. Then $x_2 = -4x_3 + 145$

$= -4t + 145$

$x_1 = 3x_3 - 100$

$= 3t - 100$

A solution is achieved, not for every real value of t, but for integer values of t that give rise to non-negative x_1, x_2, x_3.

$x_1 \geq 0$ means $3t - 100 \geq 0$ or $t \geq 33\tfrac{1}{3}$

$x_2 \geq 0$ means $-4t + 145 \geq 0$ or $t \leq 36\tfrac{1}{4}$

The only integer values of t that satisfy these conditions are 34, 35, 36. So we have the solutions:

$x_1 = (3t - 100)$ 15-cent stamps

$x_2 = (145 - 4t)$ 20-cent stamps

$x_3 = t$ 35-cent stamps

where t = 34, 35, or 36.

67. Let x_1 = number of 500-cc containers of 10% solution

x_2 = number of 500-cc containers of 20% solution

x_3 = number of 1,000-cc containers of 50% solution

Then $500x_1 + 500x_2 + 1,000x_3 = 12,000$ (total number of cc)

$0.10(500x_1) + 0.20(500x_2) + 0.50(1,000x_3) = 0.30(12,000)$

(total amount of ingredient in solution.)

After dividing the first equation by 500 and the second by 50, we get

$x_1 + x_2 + 2x_3 = 24$

$x_1 + 2x_2 + 10x_3 = 72$

We write the augmented matrix and solve by Gauss-Jordan elimination:

$$\begin{bmatrix} 1 & 1 & 2 & | & 24 \\ 1 & 2 & 10 & | & 72 \end{bmatrix} (-1)R_1 + R_2 \to R_2$$

$$\sim \begin{bmatrix} 1 & 1 & 2 & | & 24 \\ 0 & 1 & 8 & | & 48 \end{bmatrix} (-1)R_2 + R_1 \to R_1$$

$$\sim \begin{bmatrix} 1 & 0 & -6 & | & -24 \\ 0 & 1 & 8 & | & 48 \end{bmatrix}$$

This augmented matrix is in reduced form. It corresponds to the system:

$x_1 - 6x_3 = -24$

$x_2 + 8x_3 = 48$

Let $x_3 = t$. Then

$x_2 = -8x_3 + 48$

$\quad = -8t + 48$

$x_1 = 6x_3 - 24$

$\quad = 6t - 24$

A solution is achieved, not for every real value of t, but for integer values of t that give rise to non-negative x_1, x_2, x_3.

$x_1 \geq 0$ means $6t - 24 \geq 0$ or $t \geq 4$

$x_2 \geq 0$ means $-8t + 48 \geq 0$ or $t \leq 6$

So we have the solution:

$x_1 = (6t - 24)$ 500-cc containers of 10% solution

$x_2 = (48 - 8t)$ 500-cc containers of 20% solution

$x_3 = t$ 1000-cc containers of 50% solution

where $t = 4$, 5, or 6.

69. If the curve passes through a point, the coordinates of the point satisfy the equation of the curve. So

$3 = a + b(-2) + c(-2)^2$

$2 = a + b(-1) + c(-1)^2$

$6 = a + b(1) + c(1)^2$

After simplification, we have

$a - 2b + 4c = 3$

$a - b + c = 2$

$a + b + c = 6$

We write the augmented matrix and solve by Gauss-Jordan elimination.

$$\begin{bmatrix} 1 & -2 & 4 & | & 3 \\ 1 & -1 & 1 & | & 2 \\ 1 & 1 & 1 & | & 6 \end{bmatrix} \begin{matrix} \\ (-1)R_1 + R_2 \to R_2 \\ (-1)R_1 + R_3 \to R_3 \end{matrix}$$

$$\sim \begin{bmatrix} 1 & -2 & 4 & | & 3 \\ 0 & 1 & -3 & | & -1 \\ 0 & 3 & -3 & | & 3 \end{bmatrix} \begin{matrix} 2R_2 + R_1 \to R_1 \\ \\ (-3)R_2 + R_3 \to R_3 \end{matrix}$$

$$\sim \begin{bmatrix} 1 & 0 & -2 & | & 1 \\ 0 & 1 & -3 & | & -1 \\ 0 & 0 & 6 & | & 6 \end{bmatrix} \frac{1}{6} R_3 \to R_3$$

$$\sim \begin{bmatrix} 1 & 0 & -2 & | & 1 \\ 0 & 1 & -3 & | & -1 \\ 0 & 0 & 1 & | & 1 \end{bmatrix} \begin{matrix} 2R_3 + R_1 \to R_1 \\ 3R_3 + R_2 \to R_2 \\ \\ \end{matrix}$$

$$\sim \begin{bmatrix} 1 & 0 & 0 & | & 3 \\ 0 & 1 & 0 & | & 2 \\ 0 & 0 & 1 & | & 1 \end{bmatrix}$$

Solution: $a = 3$, $b = 2$, $c = 1$.

71. If the curve passes through a point, the coordinates of the point satisfy the equation of the curve. So

$$6^2 + 2^2 + a(6) + b(2) + c = 0$$
$$4^2 + 6^2 + a(4) + b(6) + c = 0$$
$$(-3)^2 + (-1)^2 + a(-3) + b(-1) + c = 0$$

After simplification, we have

$$6a + 2b + c = -40$$
$$4a + 6b + c = -52$$
$$-3a - b + c = -10$$

We write the augmented matrix and solve by Gauss-Jordan elimination.

$$\begin{bmatrix} 6 & 2 & 1 & -40 \\ 4 & 6 & 1 & -52 \\ -3 & -1 & 1 & -10 \end{bmatrix} \frac{1}{6} R_1 \to R_1$$

$$\sim \begin{bmatrix} 1 & \frac{1}{3} & \frac{1}{6} & -\frac{20}{3} \\ 4 & 6 & 1 & -52 \\ -3 & -1 & 1 & -10 \end{bmatrix} \begin{array}{l}(-4)R_1 + R_2 \to R_2 \\ 3R_1 + R_3 \to R_3\end{array}$$

$$\sim \begin{bmatrix} 1 & \frac{1}{3} & \frac{1}{6} & -\frac{20}{3} \\ 0 & \frac{14}{3} & \frac{1}{3} & -\frac{76}{3} \\ -3 & 0 & \frac{3}{2} & -30 \end{bmatrix} \frac{2}{3} R_3 \to R_3$$

$$\sim \begin{bmatrix} 1 & \frac{1}{3} & \frac{1}{6} & -\frac{20}{3} \\ 0 & \frac{14}{3} & \frac{1}{3} & -\frac{76}{3} \\ 0 & 0 & 1 & -20 \end{bmatrix} \begin{array}{l}(-\frac{1}{6})R_3 + R_1 \to R_1 \\ (-\frac{1}{3})R_3 + R_2 \to R_2\end{array}$$

$$\sim \begin{bmatrix} 1 & \frac{1}{3} & 0 & -\frac{10}{3} \\ 0 & \frac{14}{3} & 0 & -\frac{56}{3} \\ 0 & 0 & 1 & -20 \end{bmatrix} \frac{3}{14} R_2 \to R_2$$

$$\sim \begin{bmatrix} 1 & \frac{1}{3} & 0 & -\frac{10}{3} \\ 0 & 1 & 0 & -4 \\ 0 & 0 & 1 & -20 \end{bmatrix} (-\frac{1}{3})R_2 + R_1 \to R_1$$

$$\begin{bmatrix} 1 & 0 & 0 & -2 \\ 0 & 1 & 0 & -4 \\ 0 & 0 & 1 & -20 \end{bmatrix}$$

Solution: $a = -2$, $b = -4$, and $c = -20$.

73. Let x_1 = number of one-person boats
x_2 = number of two-person boats
x_3 = number of four-person boats

(A) We have

$0.5x_1 + 1.0x_2 + 1.5x_3 = 380$ cutting department
$0.6x_1 + 0.9x_2 + 1.2x_3 = 330$ assembly department
$0.2x_1 + 0.3x_2 + 0.5x_3 = 120$ packing department

> **Common Error:**
> The facts in this problem do not justify the equation
> $0.5x_1 + 0.6x_2 + 0.2x_3 = 380$

Clearing of decimals for convenience:

$x_1 + 2x_2 + 3x_3 = 760$ (First equation multiplied by 2)
$6x_1 + 9x_2 + 12x_3 = 3,300$ (Second equation multiplied by 10)
$2x_1 + 3x_2 + 5x_3 = 1,200$ (Third equation multiplied by 10)

We write the augmented matrix and solve by Gauss-Jordan elimination:

$$\begin{bmatrix} 1 & 2 & 3 & | & 760 \\ 6 & 9 & 12 & | & 3,300 \\ 2 & 3 & 5 & | & 1,200 \end{bmatrix} \begin{matrix} \\ (-6)R_1 + R_2 \rightarrow R_2 \\ (-2)R_1 + R_3 \rightarrow R_3 \end{matrix}$$

$$\sim \begin{bmatrix} 1 & 2 & 3 & | & 760 \\ 0 & -3 & -6 & | & -1,260 \\ 0 & -1 & -1 & | & -320 \end{bmatrix} -\tfrac{1}{3}R_2 \rightarrow R_2$$

$$\sim \begin{bmatrix} 1 & 2 & 3 & | & 760 \\ 0 & 1 & 2 & | & 420 \\ 0 & -1 & -1 & | & -320 \end{bmatrix} \begin{matrix} (-2)R_2 + R_1 \rightarrow R_1 \\ \\ R_2 + R_3 \rightarrow R_3 \end{matrix}$$

$$\sim \begin{bmatrix} 1 & 0 & -1 & | & -80 \\ 0 & 1 & 2 & | & 420 \\ 0 & 0 & 1 & | & 100 \end{bmatrix} \begin{matrix} R_3 + R_1 \rightarrow R_1 \\ (-2)R_3 + R_2 \rightarrow R_2 \\ \end{matrix}$$

$$\begin{bmatrix} 1 & 0 & 0 & | & 20 \\ 0 & 1 & 0 & | & 220 \\ 0 & 0 & 1 & | & 100 \end{bmatrix}$$

Therefore

$x_1 = 20$ one-person boats

$x_2 = 220$ two-person boats

$x_3 = 100$ four-person boats

(B) This assumption discards the third equation. The system, cleared of decimals, reads

$$x_1 + 2x_2 + 3x_3 = 760$$
$$6x_1 + 9x_2 + 12x_3 = 3,300$$

The augmented matrix becomes

$$\begin{bmatrix} 1 & 2 & 3 & | & 760 \\ 6 & 9 & 12 & | & 3,300 \end{bmatrix}$$

We solve by Gauss-Jordan elimination. We start by introducing a 0 into the lower left corner using $(-6)R_1 + R_2$ as in the previous problem:

$$\sim \begin{bmatrix} 1 & 2 & 3 & | & 760 \\ 0 & -3 & -6 & | & -1,260 \end{bmatrix} -\tfrac{1}{3}R_2 \rightarrow R_2$$

$$\sim \begin{bmatrix} 1 & 2 & 3 & | & 760 \\ 0 & 1 & 2 & | & 420 \end{bmatrix} (-2)R_2 + R_1 \rightarrow R_1$$

$$\sim \begin{bmatrix} 1 & 0 & -1 & | & -80 \\ 0 & 1 & 2 & | & 420 \end{bmatrix}$$

This augmented matrix is in reduced form. It corresponds to the system:

$$x_1 - x_3 = -80$$
$$x_2 + 2x_3 = 420$$

Let $x_3 = t$. Then $x_2 = -2x_3 + 420$

$$= -2t + 420$$
$$x_1 = x_3 - 80$$
$$= t - 80$$

A solution is achieved, not for every real value of t, but for integer values of t that give rise to non-negative x_1, x_2, x_3.

$x_1 \geq 0$ means $t - 80 \geq 0$ or $t \geq 80$

$x_2 \geq 0$ means $-2t + 420 \geq 0$ or $210 \geq t$

So we have the solution

$x_1 = (t - 80)$ one-person boats

$x_2 = (-2t + 420)$ two-person boats

$x_3 = t$ four-person boats

$80 \leq t \leq 210$, t an integer

(C) In this case we have $x_3 = 0$ from the beginning. The three equations of part (A), cleared of decimals, read:

$$x_1 + 2x_2 = 760$$
$$6x_1 + 9x_2 = 3,300$$
$$2x_1 + 3x_2 = 1,200$$

The augmented matrix becomes:

$$\begin{bmatrix} 1 & 2 & | & 760 \\ 6 & 9 & | & 3,300 \\ 2 & 3 & | & 1,200 \end{bmatrix}$$

Notice that the row operation

$$(-3)R_3 + R_2 \rightarrow R_2$$

transforms this into the equivalent augmented matrix: $\begin{bmatrix} 1 & 2 & | & 760 \\ 0 & 0 & | & -300 \\ 2 & 3 & | & 1,200 \end{bmatrix}$.

Therefore, since the second row corresponds to the equation

$$0x_1 + 0x_2 = -300$$

there is no solution.

No production schedule will use all the work-hours in all departments.

75. Let x_1 = number of ounces of food A.
    x_2 = number of ounces of food B.
    x_3 = number of ounces of food C.

> **Common Error:**
> The facts in this problem do not justify the equation
> $30x_1 + 10x_2 + 10x_3 = 340$

(A) Then

$$30x_1 + 10x_2 + 20x_3 = 340 \text{ (calcium)}$$
$$10x_1 + 10x_2 + 20x_3 = 180 \text{ (iron)}$$
$$10x_1 + 30x_2 + 20x_3 = 220 \text{ (vitamin A)}$$

or

$$3x_1 + x_2 + 2x_3 = 34$$
$$x_1 + x_2 + 2x_3 = 18$$
$$x_1 + 3x_2 + 2x_3 = 22$$

is the system to be solved. We form the augmented matrix and solve by Gauss-Jordan elimination.

$$\begin{bmatrix} 3 & 1 & 2 & | & 34 \\ 1 & 1 & 2 & | & 18 \\ 1 & 3 & 2 & | & 22 \end{bmatrix} \quad R_1 \leftrightarrow R_2$$

$$\sim \begin{bmatrix} 1 & 1 & 2 & | & 18 \\ 3 & 1 & 2 & | & 34 \\ 1 & 3 & 2 & | & 22 \end{bmatrix} \quad \begin{matrix} (-3)R_1 + R_2 \rightarrow R_2 \\ (-1)R_1 + R_3 \rightarrow R_3 \end{matrix}$$

$$\sim \begin{bmatrix} 1 & 1 & 2 & | & 18 \\ 0 & -2 & -4 & | & -20 \\ 0 & 2 & 0 & | & 4 \end{bmatrix} \quad -\tfrac{1}{2}R_2 \rightarrow R_2$$

$$\sim \begin{bmatrix} 1 & 1 & 2 & | & 18 \\ 0 & 1 & 2 & | & 10 \\ 0 & 2 & 0 & | & 4 \end{bmatrix} \quad \begin{matrix} (-1)R_2 + R_1 \rightarrow R_1 \\ (-2)R_2 + R_3 \rightarrow R_3 \end{matrix}$$

$$\sim \begin{bmatrix} 1 & 0 & 0 & | & 8 \\ 0 & 1 & 2 & | & 10 \\ 0 & 0 & -4 & | & -16 \end{bmatrix} \quad -\tfrac{1}{4}R_3 \rightarrow R_3$$

$$\sim \begin{bmatrix} 1 & 0 & 0 & | & 8 \\ 0 & 1 & 2 & | & 10 \\ 0 & 0 & 1 & | & 4 \end{bmatrix} (-2)R_3 + R_2 \rightarrow R_2$$

$$\sim \begin{bmatrix} 1 & 0 & 0 & | & 8 \\ 0 & 1 & 0 & | & 2 \\ 0 & 0 & 1 & | & 4 \end{bmatrix}$$

Solution: $x_1 = 8$ ounces food A

$x_2 = 2$ ounces food B

$x_3 = 4$ ounces food C

(B) In this case we have $x_3 = 0$ from the beginning. The three equations of part (A) become $30x_1 + 10x_2 = 340$

$$10x_1 + 10x_2 = 180$$

$$10x_1 + 30x_2 = 220$$

or

$$3x_1 + x_2 = 34$$

$$x_1 + x_2 = 18$$

$$x_1 + 3x_2 = 22$$

The augmented matrix becomes $\begin{bmatrix} 3 & 1 & | & 34 \\ 1 & 1 & | & 18 \\ 1 & 3 & | & 22 \end{bmatrix}$

We solve by Gauss-Jordan elimination, starting by the row operation $R_1 \leftrightarrow R_2$

$$\begin{bmatrix} 1 & 1 & | & 18 \\ 3 & 1 & | & 34 \\ 1 & 3 & | & 22 \end{bmatrix} \begin{matrix} \\ (-3)R_1 + R_2 \rightarrow R_2 \\ (-1)R_1 + R_3 \rightarrow R_3 \end{matrix}$$

$$\sim \begin{bmatrix} 1 & 1 & | & 18 \\ 0 & -2 & | & -20 \\ 0 & 2 & | & 4 \end{bmatrix} R_2 + R_3 \rightarrow R_3$$

$$\sim \begin{bmatrix} 1 & 1 & | & 18 \\ 0 & -2 & | & -20 \\ 0 & 0 & | & -16 \end{bmatrix}$$

Since the third row corresponds to the equation

$$0x_1 + 0x_2 = -16$$

there is no solution.

(C) In this case we discard the third equation. The system becomes

$$30x_1 + 10x_2 + 20x_3 = 340 \quad \text{or} \quad 3x_1 + x_2 + 2x_3 = 34$$

$$10x_1 + 10x_2 + 20x_3 = 180 \qquad\qquad x_1 + x_2 + 2x_3 = 18$$

The augmented matrix becomes

$$\begin{bmatrix} 3 & 1 & 2 & | & 34 \\ 1 & 1 & 2 & | & 18 \end{bmatrix}$$

We solve by Gauss-Jordan elimination, starting by the row operation $R_1 \leftrightarrow R_2$.

$$\begin{bmatrix} 1 & 1 & 2 & | & 18 \\ 3 & 1 & 2 & | & 34 \end{bmatrix} (-3)R_1 + R_2 \rightarrow R_2$$

$$\sim \begin{bmatrix} 1 & 1 & 2 & | & 18 \\ 0 & -2 & -4 & | & -20 \end{bmatrix} -\tfrac{1}{2}R_2 \rightarrow R_2$$

$$\sim \begin{bmatrix} 1 & 1 & 2 & | & 18 \\ 0 & 1 & 2 & | & 10 \end{bmatrix} (-1)R_2 + R_1 \rightarrow R_1$$

$$\sim \begin{bmatrix} 1 & 0 & 0 & | & 8 \\ 0 & 1 & 2 & | & 10 \end{bmatrix}$$

This augmented matrix is in reduced form. It corresponds to the system

$$x_1 = 8$$
$$x_2 + 2x_3 = 10$$

Let $x_3 = t$

Then $x_2 = -2x_3 + 10$
$$= -2t + 10$$

A solution is achieved, not for every real value t, but for values of t that give rise to non-negative x_2, x_3.

$x_3 \geq 0$ means $t \geq 0$

$x_2 \geq 0$ means $-2t + 10 \geq 0$, $5 \geq t$

So we have the solution

$x_1 = 8$ ounces food A

$x_2 = -2t + 10$ ounces food B

$x_3 = t$ ounces food C

$0 \leq t \leq 5$

77. Let x_1 = number of barrels of Mix A

$\qquad x_2$ = number of barrels of Mix B

$\qquad x_3$ = number of barrels of Mix C

$\qquad x_4$ = number of barrels of Mix D

Then $\quad 30x_1 + 30x_2 + 30x_3 + 60x_4 = 900$ (phosphoric acid)

$\qquad\quad 50x_1 + 75x_2 + 25x_3 + 25x_4 = 750$ (nitrogen)

$\qquad\quad 30x_1 + 20x_2 + 20x_3 + 50x_4 = 700$ (potash)

or $\qquad x_1 + x_2 + x_3 + 2x_4 = 30$

$\qquad\quad 2x_1 + 3x_2 + x_3 + x_4 = 30$

$\qquad\quad 3x_1 + 2x_2 + 2x_3 + 5x_4 = 70$

is the system to be solved. We form the augmented matrix and solve by Gauss-Jordan elimination.

$$\begin{bmatrix} 1 & 1 & 1 & 2 & | & 30 \\ 2 & 3 & 1 & 1 & | & 30 \\ 3 & 2 & 2 & 5 & | & 70 \end{bmatrix} \begin{matrix} \\ (-2)R_1 + R_2 \rightarrow R_2 \\ (-3)R_1 + R_3 \rightarrow R_3 \end{matrix}$$

$$\sim \begin{bmatrix} 1 & 1 & 1 & 2 & | & 30 \\ 0 & 1 & -1 & -3 & | & -30 \\ 0 & -1 & -1 & -1 & | & -20 \end{bmatrix} \begin{matrix} (-1)R_2 + R_1 \rightarrow R_1 \\ \\ R_2 + R_3 \rightarrow R_3 \end{matrix}$$

$$\sim \begin{bmatrix} 1 & 0 & 2 & 5 & | & 60 \\ 0 & 1 & -1 & -3 & | & -30 \\ 0 & 0 & -2 & -4 & | & -50 \end{bmatrix} -\tfrac{1}{2}R_3 \rightarrow R_3$$

$$\sim \begin{bmatrix} 1 & 0 & 2 & 5 & | & 60 \\ 0 & 1 & -1 & -3 & | & -30 \\ 0 & 0 & 1 & 2 & | & 25 \end{bmatrix} \begin{matrix} (-2)R_3 + R_1 \rightarrow R_1 \\ R_3 + R_2 \rightarrow R_2 \end{matrix}$$

$$\sim \begin{bmatrix} 1 & 0 & 0 & 1 & | & 10 \\ 0 & 1 & 0 & -1 & | & -5 \\ 0 & 0 & 1 & 2 & | & 25 \end{bmatrix}$$

This augmented matrix is in reduced form. It corresponds to the system

$$x_1 \qquad\quad + x_4 = 10$$
$$x_2 \quad - x_4 = -5$$
$$x_3 + 2x_4 = 25$$

Let $x_4 = t.$ Then $x_3 = -2x_4 + 25$

$$= -2t + 25$$
$$x_2 = x_4 - 5$$
$$= t - 5$$
$$x_1 = -x_4 + 10$$
$$= -t + 10$$

A solution is achieved, not for every real value t, but for values of t that give rise to non-negative x_1, x_2, x_3, x_4.

$x_4 \geq 0$ means $t \geq 0$

$x_3 \geq 0$ means $-2t + 25 \geq 0$ $t \leq 12.5$

$x_2 \geq 0$ means $t - 5 \geq 0$ $t \geq 5$

$x_1 \geq 0$ means $-t + 10 \geq 0$ $t \leq 10$

So we have the solution: $x_1 = 10 - t$ barrels of mix A,

$x_2 = t - 5$ barrels of mix B,

$x_3 = 25 - 2t$ barrels of mix C,

$x_4 = t$ barrels of mix D

where $5 \leq t \leq 10$.

Section 6-2

1. Answers will vary.   **3.** Answers will vary.   **5.** Answers will vary.

7. Answers will vary.   **9.** Answers will vary.

11. $\begin{bmatrix} 5 & -2 \\ 3 & 0 \end{bmatrix} + \begin{bmatrix} -3 & 7 \\ 1 & -6 \end{bmatrix} = \begin{bmatrix} 5 + (-3) & (-2) + 7 \\ 3 + 1 & 0 + (-6) \end{bmatrix} = \begin{bmatrix} 2 & 5 \\ 4 & -6 \end{bmatrix}$

13. $\begin{bmatrix} 4 & 0 \\ -2 & 3 \\ 8 & 1 \end{bmatrix} + \begin{bmatrix} -1 & 2 \\ 0 & 5 \\ 4 & -6 \end{bmatrix} = \begin{bmatrix} 4 + (-1) & 0 + 2 \\ (-2) + 0 & 3 + 5 \\ 8 + 4 & 1 + (-6) \end{bmatrix} = \begin{bmatrix} 3 & 2 \\ -2 & 8 \\ 12 & -5 \end{bmatrix}$

15. These matrices have different sizes, so the sum is not defined.

17. $\begin{bmatrix} 5 & -1 & 0 \\ 4 & 6 & 3 \end{bmatrix} - \begin{bmatrix} 2 & 4 & -6 \\ 3 & 5 & -5 \end{bmatrix} = \begin{bmatrix} 5 - 2 & (-1) - 4 & 0 - (-6) \\ 4 - 3 & 6 - 5 & 3 - (-5) \end{bmatrix} = \begin{bmatrix} 3 & -5 & 6 \\ 1 & 1 & 8 \end{bmatrix}$

19. $\begin{bmatrix} 12 & -16 & 28 \\ -8 & 36 & 20 \end{bmatrix}$ **21.** $[5 \quad 3]\begin{bmatrix} 4 \\ 7 \end{bmatrix} = [5 \cdot 4 + 3 \cdot 7] = [41]$

23. $\begin{bmatrix} -6 & 3 \\ 2 & -5 \end{bmatrix}\begin{bmatrix} 1 \\ 3 \end{bmatrix} = \begin{bmatrix} (-6)1 + 3 \cdot 3 \\ 2 \cdot 1 + (-5)3 \end{bmatrix} = \begin{bmatrix} 3 \\ -13 \end{bmatrix}$

25. $\begin{bmatrix} 5 & 1 \\ 4 & 6 \end{bmatrix}\begin{bmatrix} 2 & 0 \\ 3 & 8 \end{bmatrix} = \begin{bmatrix} 5 \cdot 2 + 1 \cdot 3 & 5 \cdot 0 + 1 \cdot 8 \\ 4 \cdot 2 + 6 \cdot 3 & 4 \cdot 0 + 6 \cdot 8 \end{bmatrix} = \begin{bmatrix} 13 & 8 \\ 26 & 48 \end{bmatrix}$

27. $\begin{bmatrix} 8 & -3 \\ -5 & 3 \end{bmatrix}\begin{bmatrix} 2 & 0 \\ 0 & 6 \end{bmatrix} = \begin{bmatrix} 8 \cdot 2 + (-3)0 & 8 \cdot 0 + (-3)6 \\ (-5)2 + 3 \cdot 0 & (-5)0 + 3 \cdot 6 \end{bmatrix} = \begin{bmatrix} 16 & -18 \\ -10 & 18 \end{bmatrix}$

29. $[4 \quad -2]\begin{bmatrix} -5 \\ -3 \end{bmatrix} = [4(-5) + (-2)(-3)] = [-14]$

31. $\begin{bmatrix} -5 \\ -3 \end{bmatrix}[4 \quad -2] = \begin{bmatrix} (-5)4 & (-5)(-2) \\ (-3)4 & (-3)(-2) \end{bmatrix} = \begin{bmatrix} -20 & 10 \\ -12 & 6 \end{bmatrix}$

33. $[3 \quad -2 \quad -4]\begin{bmatrix} 1 \\ 2 \\ -3 \end{bmatrix} = [3 \cdot 1 + (-2)2 + (-4)(-3)] = [11]$

35. $\begin{bmatrix} 1 \\ 2 \\ -3 \end{bmatrix}[3 \quad -2 \quad -4] = \begin{bmatrix} 1 \cdot 3 & 1(-2) & 1(-4) \\ 2 \cdot 3 & 2(-2) & 2(-4) \\ (-3)3 & (-3)(-2) & (-3)(-4) \end{bmatrix} = \begin{bmatrix} 3 & -2 & -4 \\ 6 & -4 & -8 \\ -9 & 6 & 12 \end{bmatrix}$

37. C has 3 columns. A has 2 rows. Therefore, CA is not defined.

39. $BA = \begin{bmatrix} -3 & 1 \\ 2 & 5 \end{bmatrix}\begin{bmatrix} 2 & -1 & 3 \\ 0 & 4 & -2 \end{bmatrix} = \begin{bmatrix} (-3)2 + 1 \cdot 0 & (-3)(-1) + 1 \cdot 4 & (-3)3 + 1(-2) \\ 2 \cdot 2 + 5 \cdot 0 & 2(-1) + 5 \cdot 4 & 2 \cdot 3 + 5(-2) \end{bmatrix}$

$= \begin{bmatrix} -6 & 7 & -11 \\ 4 & 18 & -4 \end{bmatrix}$

41. $C^2 = \begin{bmatrix} -1 & 0 & 2 \\ 4 & -3 & 1 \\ -2 & 3 & 5 \end{bmatrix}\begin{bmatrix} -1 & 0 & 2 \\ 4 & -3 & 1 \\ -2 & 3 & 5 \end{bmatrix}$

$= \begin{bmatrix} (-1)(-1) + 0 \cdot 4 + 2(-2) & (-1)0 + 0(-3) + 2 \cdot 3 & (-1)2 + 0 \cdot 1 + 2 \cdot 5 \\ 4(-1) + (-3)4 + 1(-2) & 4 \cdot 0 + (-3)(-3) + 1 \cdot 3 & 4 \cdot 2 + (-3)1 + 1 \cdot 5 \\ (-2)(-1) + 3 \cdot 4 + 5(-2) & (-2)0 + 3(-3) + 5 \cdot 3 & (-2)2 + 3 \cdot 1 + 5 \cdot 5 \end{bmatrix}$

$= \begin{bmatrix} -3 & 6 & 8 \\ -18 & 12 & 10 \\ 4 & 6 & 24 \end{bmatrix}$

43. $DA = \begin{bmatrix} 3 & -2 \\ 0 & -1 \\ 1 & 2 \end{bmatrix}\begin{bmatrix} 2 & -1 & 3 \\ 0 & 4 & -2 \end{bmatrix}$

$= \begin{bmatrix} 3 \cdot 2 + (-2)0 & 3(-1) + (-2)4 & 3 \cdot 3 + (-2)(-2) \\ 0 \cdot 2 + (-1)0 & 0(-1) + (-1)4 & 0 \cdot 3 + (-1)(-2) \\ 1 \cdot 2 + 2 \cdot 0 & 1(-1) + 2 \cdot 4 & 1 \cdot 3 + 2(-2) \end{bmatrix} = \begin{bmatrix} 6 & -11 & 13 \\ 0 & -4 & 2 \\ 2 & 7 & -1 \end{bmatrix}$

$C + DA = \begin{bmatrix} -1 & 0 & 2 \\ 4 & -3 & 1 \\ -2 & 3 & 5 \end{bmatrix} + \begin{bmatrix} 6 & -11 & 13 \\ 0 & -4 & 2 \\ 2 & 7 & -1 \end{bmatrix} = \begin{bmatrix} 5 & -11 & 15 \\ 4 & -7 & 3 \\ 0 & 10 & 4 \end{bmatrix}$

45. $0.2CD = 0.2\begin{bmatrix} -1 & 0 & 2 \\ 4 & -3 & 1 \\ -2 & 3 & 5 \end{bmatrix}\begin{bmatrix} 3 & -2 \\ 0 & -1 \\ 1 & 2 \end{bmatrix}$

$= 0.2\begin{bmatrix} (-1)3 + 0 \cdot 0 + 2 \cdot 1 & (-1)(-2) + 0(-1) + 2 \cdot 2 \\ 4 \cdot 3 + (-3)0 + 1 \cdot 1 & 4(-2) + (-3)(-1) + 1 \cdot 2 \\ (-2)3 + 3 \cdot 0 + 5 \cdot 1 & (-2)(-2) + 3(-1) + 5 \cdot 2 \end{bmatrix}$

$= 0.2\begin{bmatrix} -1 & 6 \\ 13 & -3 \\ -1 & 11 \end{bmatrix} = \begin{bmatrix} -0.2 & 1.2 \\ 2.6 & -0.6 \\ -0.2 & 2.2 \end{bmatrix}$

47. $DB = \begin{bmatrix} 3 & -2 \\ 0 & -1 \\ 1 & 2 \end{bmatrix}\begin{bmatrix} -3 & 1 \\ 2 & 5 \end{bmatrix} = \begin{bmatrix} 3(-3) + (-2)2 & 3 \cdot 1 + (-2)5 \\ 0(-3) + (-1)2 & 0 \cdot 1 + (-1)5 \\ 1(-3) + 2 \cdot 2 & 1 \cdot 1 + 2 \cdot 5 \end{bmatrix} = \begin{bmatrix} -13 & -7 \\ -2 & -5 \\ 1 & 11 \end{bmatrix}$

$CD = \begin{bmatrix} -1 & 6 \\ 13 & -3 \\ -1 & 11 \end{bmatrix}$ (see problem 45)

$2DB + 5CD = 2\begin{bmatrix} -13 & -7 \\ -2 & -5 \\ 1 & 11 \end{bmatrix} + 5\begin{bmatrix} -1 & 6 \\ 13 & -3 \\ -1 & 11 \end{bmatrix} = \begin{bmatrix} -26 & -14 \\ -4 & -10 \\ 2 & 22 \end{bmatrix} + \begin{bmatrix} -5 & 30 \\ 65 & -15 \\ -5 & 55 \end{bmatrix} = \begin{bmatrix} -31 & 16 \\ 61 & -25 \\ -3 & 77 \end{bmatrix}$

49. $(-1)AC$ is a matrix of size 2×3. $3DB$ is a matrix of size 3×2. So, $(-1)AC + 3DB$ is not defined.

51. $CD = \begin{bmatrix} -1 & 6 \\ 13 & -3 \\ -1 & 11 \end{bmatrix}$ (see problem 45)

$CDA = \begin{bmatrix} -1 & 6 \\ 13 & -3 \\ -1 & 11 \end{bmatrix} \begin{bmatrix} 2 & -1 & 3 \\ 0 & 4 & -2 \end{bmatrix}$

$= \begin{bmatrix} (-1)2 + 6 \cdot 0 & (-1)(-1) + 6 \cdot 4 & (-1)3 + 6(-2) \\ 13 \cdot 2 + (-3)0 & 13(-1) + (-3)4 & 13 \cdot 3 + (-3)(-2) \\ (-1)2 + 11 \cdot 0 & (-1)(-1) + 11 \cdot 4 & (-1)3 + 11(-2) \end{bmatrix} = \begin{bmatrix} -2 & 25 & -15 \\ 26 & -25 & 45 \\ -2 & 45 & -25 \end{bmatrix}$

53. $DB = \begin{bmatrix} -13 & -7 \\ -2 & -5 \\ 1 & 11 \end{bmatrix}$ (see problem 47)

$DBA = \begin{bmatrix} -13 & -7 \\ -2 & -5 \\ 1 & 11 \end{bmatrix} \begin{bmatrix} 2 & -1 & 3 \\ 0 & 4 & -2 \end{bmatrix}$

$= \begin{bmatrix} (-13)2 + (-7)0 & (-13)(-1) + (-7)4 & (-13)3 + (-7)(-2) \\ (-2)2 + (-5)0 & (-2)(-1) + (-5)4 & (-2)3 + (-5)(-2) \\ 1 \cdot 2 + 11 \cdot 0 & 1(-1) + 11 \cdot 4 & 1 \cdot 3 + 11(-2) \end{bmatrix} = \begin{bmatrix} -26 & -15 & -25 \\ -4 & -18 & 4 \\ 2 & 43 & -19 \end{bmatrix}$

55. Entering matrix B in a graphing calculator, we obtain the results

```
[B]²
      [[.28 .72]
       [.24 .76]]
[B]³
      [[.256 .744]
       [.248 .752]]
```

```
[B]^4
      [[.2512 .7488]
       [.2496 .7504]]
[B]^5
      [[.25024 .74976…
       [.24992 .75008…
```

```
[B]^6
[[.250048 .7499…
 [.249984 .7500…
[B]^7
[[.2500096 .749…
 [.2499968 .750…
```

It appears that $B^n \rightarrow \begin{bmatrix} 0.25 & 0.75 \\ 0.25 & 0.75 \end{bmatrix}$

We calculate AB, AB^2, AB^3, … and obtain the results

```
[A]*[B]
        [[.26 .74]]
[A]*[B]²
        [[.252 .748]]
[A]*[B]³
        [[.2504 .7496]]
```

```
[A]*[B]^4
[[.25008 .74992…
[A]*[B]^5
[[.250016 .7499…
[A]*[B]^6
[[.2500032 .749…
```

It appears that $AB^n \rightarrow [0.25 \quad 0.75]$.

57. $\begin{bmatrix} a & b \\ c & d \end{bmatrix} + \begin{bmatrix} 2 & -3 \\ 0 & 1 \end{bmatrix} = \begin{bmatrix} a + 2 & b - 3 \\ c & d + 1 \end{bmatrix}$

This matrix will be equal to $\begin{bmatrix} 1 & -2 \\ 3 & -4 \end{bmatrix}$ if all of the corresponding elements are equal.

$\begin{array}{llll} a + 2 = 1 & b - 3 = -2 & c = 3 & d + 1 = -4 \\ \quad a = -1 & \quad b = 1 & c = 3 & \quad d = -5 \end{array}$

59. $\begin{bmatrix} 3x & 5 \\ -1 & 4x \end{bmatrix} + \begin{bmatrix} 2y & -3 \\ -6 & -y \end{bmatrix} = \begin{bmatrix} 3x + 2y & 2 \\ -7 & 4x - y \end{bmatrix}$

This matrix will be equal to $\begin{bmatrix} 7 & 2 \\ -7 & 2 \end{bmatrix}$ if all of the corresponding elements are equal.

$$3x + 2y = 7 \qquad\qquad 2 \overset{\vee}{=} 2 \Big\}$$
$$-7 \overset{\vee}{=} -7 \quad 4x - y = 2 \Big\} \quad \text{Two conditions are already met.}$$

To find x and y, we solve the system:

$3x + 2y = 7$

$4x - y = 2$ to obtain $x = 1$, $y = 2$. (Solution left to the student.)

61. Compute the square of matrix A:

$$A^2 = \begin{bmatrix} a & b \\ c & -a \end{bmatrix}\begin{bmatrix} a & b \\ c & -a \end{bmatrix} = \begin{bmatrix} a^2 + bc & ab + (-ab) \\ ac + (-ac) & cb + a^2 \end{bmatrix} = \begin{bmatrix} a^2 + bc & 0 \\ 0 & a^2 + bc \end{bmatrix}$$

Two of the entries are already zero and the other two are both $a^2 + bc$. So if $a^2 + bc = 0$, then $A^2 = 0$.

If $a = 1$, $b = 1$, $c = -1$, then $a^2 + bc = 0$, so the matrix $A = \begin{bmatrix} 1 & 1 \\ -1 & -1 \end{bmatrix}$

will have $A^2 = 0$. If $a = 2$, $b = 4$, $c = -1$, then $a^2 + bc = 0$, so the matrix

$A = \begin{bmatrix} 2 & 4 \\ -1 & -2 \end{bmatrix}$ will have $A^2 = 0$. (There are many other possible examples.)

63. Compute the product AB:

$$AB = \begin{bmatrix} a & b \\ c & d \end{bmatrix}\begin{bmatrix} 1 & 1 \\ 1 & -1 \end{bmatrix} = \begin{bmatrix} a + b & a + b \\ c + d & c + d \end{bmatrix}$$

Two of the entries are $a + b$ and the other two are $c + d$, so if $a = -b$ and $c = -d$, then $AB = 0$.

The following are a couple of examples of matrices A that will satisfy $AB = 0$:

$$\begin{bmatrix} 2 & -2 \\ 4 & -4 \end{bmatrix} \quad \begin{bmatrix} -5 & 5 \\ 1 & -1 \end{bmatrix}$$

65. $\begin{bmatrix} 1 & 3 \\ -2 & -2 \end{bmatrix}\begin{bmatrix} x & 1 \\ 3 & 2 \end{bmatrix} = \begin{bmatrix} x + 9 & 7 \\ -2x - 6 & -6 \end{bmatrix}$

This matrix will be equal to $\begin{bmatrix} y & 7 \\ y & -6 \end{bmatrix}$ if all of the corresponding elements are equal.

$$x + 9 = y \qquad 7 \overset{\vee}{=} 7 \Big\}$$
$$-2x - 6 = y \quad -6 \overset{\vee}{=} -6 \Big\} \quad \text{Two conditions are already met.}$$

To find x and y, we solve the system:

$x + 9 = y$

$-2x - 6 = y$ to obtain $x = -5$, $y = 4$. (Solution left to the student.)

67. $\begin{bmatrix} 1 & 3 \\ 1 & 4 \end{bmatrix}\begin{bmatrix} a & b \\ c & d \end{bmatrix} = \begin{bmatrix} a + 3c & b + 3d \\ a + 4c & b + 4d \end{bmatrix}$

This matrix will be equal to $\begin{bmatrix} 6 & -5 \\ 7 & -7 \end{bmatrix}$ if all of the corresponding elements are equal.

$a + 3c = 6 \qquad b + 3d = -5$

$a + 4c = 7 \qquad b + 4d = -7$

Solving these systems we obtain $a = 3$, $b = 1$, $c = 1$, $d = -2$.

(Solution left to the student.)

69. (A) Since $\begin{bmatrix} a_1 & 0 \\ 0 & d_1 \end{bmatrix} + \begin{bmatrix} a_2 & 0 \\ 0 & d_2 \end{bmatrix} = \begin{bmatrix} a_1 + a_2 & 0 \\ 0 & d_1 + d_2 \end{bmatrix}$, the statement is true.

(B) $A + B = B + A$ is true for any matrices for which $A + B$ is defined, as it is in this case.

(C) Since $\begin{bmatrix} a_1 & 0 \\ 0 & d_1 \end{bmatrix}\begin{bmatrix} a_2 & 0 \\ 0 & d_2 \end{bmatrix} = \begin{bmatrix} a_1 a_2 & 0 \\ 0 & d_1 d_2 \end{bmatrix}$, the statement is true.

(D) Since $\begin{bmatrix} a_1 & 0 \\ 0 & d_1 \end{bmatrix} \begin{bmatrix} a_2 & 0 \\ 0 & d_2 \end{bmatrix} = \begin{bmatrix} a_1 a_2 & 0 \\ 0 & d_1 d_2 \end{bmatrix} = \begin{bmatrix} a_2 a_1 & 0 \\ 0 & d_2 d_1 \end{bmatrix} = \begin{bmatrix} a_2 & 0 \\ 0 & d_2 \end{bmatrix} \begin{bmatrix} a_1 & 0 \\ 0 & d_2 \end{bmatrix}$,
the statement is true.

71. $\frac{1}{2}(A + B) = \frac{1}{2}\left(\begin{bmatrix} 30 & 25 \\ 60 & 80 \end{bmatrix} + \begin{bmatrix} 36 & 27 \\ 54 & 74 \end{bmatrix} \right) = \frac{1}{2} \begin{bmatrix} 66 & 52 \\ 114 & 154 \end{bmatrix} = \begin{matrix} \\ \end{matrix}\begin{bmatrix} 33 & 26 \\ 57 & 77 \end{bmatrix}\begin{matrix} \text{Guitar Banjo} \\ \text{Materials} \\ \text{Labor} \end{matrix}$

73. If a quantity is increased by 15%, the result is a multiplication by 1.15.
If a quantity is increased by 10%, the result is a multiplication by 1.1.
So we must calculate $1.1N - 1.15M$. The mark-up matrix is:

$1.1N - 1.15M = 1.1\begin{bmatrix} 13,900 & 783 & 263 & 215 \\ 15,000 & 838 & 395 & 236 \\ 18,300 & 967 & 573 & 248 \end{bmatrix} - 1.15\begin{bmatrix} 10,400 & 682 & 215 & 182 \\ 12,500 & 721 & 295 & 182 \\ 16,400 & 827 & 443 & 192 \end{bmatrix}$

$= \begin{bmatrix} 15,290 & 861.3 & 289.3 & 236.5 \\ 16,500 & 921.8 & 434.5 & 259.6 \\ 20,130 & 1,063.7 & 630.3 & 272.8 \end{bmatrix} - \begin{bmatrix} 11,960 & 784.3 & 247.25 & 209.3 \\ 14,375 & 829.15 & 339.25 & 209.3 \\ 18,860 & 951.05 & 509.45 & 220.8 \end{bmatrix}$

$= \begin{matrix} \text{Model } A \\ \text{Model } B \\ \text{Model } C \end{matrix}\begin{bmatrix} \$3,330 & \$77 & \$42 & \$27 \\ \$2,125 & \$93 & \$95 & \$50 \\ \$1,270 & \$113 & \$121 & \$52 \end{bmatrix} = \text{Mark up}$

Basic Car, Air, AM/FM Radio, Cruise Control

75. (A) $[0.6 \quad 0.6 \quad 0.2]\begin{bmatrix} 8 \\ 10 \\ 5 \end{bmatrix} = (0.6)8 + (0.6)10 + (0.2)5 = 11.80$ dollars per boat

(B) $[1.5 \quad 1.2 \quad 0.4]\begin{bmatrix} 9 \\ 12 \\ 6 \end{bmatrix} = (1.5)9 + (1.2)12 + (0.4)6 = 30.30$ dollars per boat

(C) The matrix NM has no clear meaning, but the matrix MN gives the labor costs per boat at each plant.

(D) $MN = \begin{bmatrix} 0.6 & 0.6 & 0.2 \\ 1.0 & 0.9 & 0.3 \\ 1.5 & 1.2 & 0.4 \end{bmatrix}\begin{bmatrix} 8 & 9 \\ 10 & 12 \\ 5 & 6 \end{bmatrix}$

$= \begin{bmatrix} (0.6)8 + (0.6)10 + (0.2)5 & (0.6)9 + (0.6)12 + (0.2)6 \\ (1.0)8 + (0.9)10 + (0.3)5 & (1.0)9 + (0.9)12 + (0.3)6 \\ (1.5)8 + (1.2)10 + (0.4)5 & (1.5)9 + (1.2)12 + (0.4)6 \end{bmatrix}$

$= \begin{bmatrix} \$11.80 & \$13.80 \\ \$18.50 & \$21.60 \\ \$26.00 & \$30.30 \end{bmatrix}\begin{matrix} \text{One-person boat} \\ \text{Two-person boat} \\ \text{Four-person boat} \end{matrix}$

with column headings Plant I, Plant II.

This matrix gives the labor costs for each type of boat at each plant.

77. (A) $A^2 = AA = \begin{bmatrix} 0 & 1 & 0 & 1 & 0 \\ 0 & 0 & 1 & 0 & 0 \\ 1 & 0 & 0 & 0 & 1 \\ 0 & 0 & 1 & 0 & 0 \\ 0 & 0 & 0 & 1 & 0 \end{bmatrix}\begin{bmatrix} 0 & 1 & 0 & 1 & 0 \\ 0 & 0 & 1 & 0 & 0 \\ 1 & 0 & 0 & 0 & 1 \\ 0 & 0 & 1 & 0 & 0 \\ 0 & 0 & 0 & 1 & 0 \end{bmatrix} = \begin{bmatrix} 0 & 0 & 2 & 0 & 0 \\ 1 & 0 & 0 & 0 & 1 \\ 0 & 1 & 0 & 2 & 0 \\ 1 & 0 & 0 & 0 & 1 \\ 0 & 0 & 1 & 0 & 0 \end{bmatrix}$

The 1 in row 2 and column 1 of A^2 indicates that there is one way to travel from Baltimore to Atlanta with one intermediate connection. The 2 in row 1 and column 3 indicates that there are two ways to travel from Atlanta to Chicago with one intermediate connection. In general, the elements in A^2 indicate the number of different ways to travel from the ith city to the jth city with one intermediate connection.

(B) $A^3 = A^2 A = \begin{bmatrix} 0 & 0 & 2 & 0 & 0 \\ 1 & 0 & 0 & 0 & 1 \\ 0 & 1 & 0 & 2 & 0 \\ 1 & 0 & 0 & 0 & 1 \\ 0 & 0 & 1 & 0 & 0 \end{bmatrix}\begin{bmatrix} 0 & 1 & 0 & 1 & 0 \\ 0 & 0 & 1 & 0 & 0 \\ 1 & 0 & 0 & 0 & 1 \\ 0 & 0 & 1 & 0 & 0 \\ 0 & 0 & 0 & 1 & 0 \end{bmatrix} = \begin{bmatrix} 2 & 0 & 0 & 0 & 2 \\ 0 & 1 & 0 & 2 & 0 \\ 0 & 0 & 3 & 0 & 0 \\ 0 & 1 & 0 & 2 & 0 \\ 1 & 0 & 0 & 0 & 1 \end{bmatrix}$

The 1 in row 4 and column 2 of A^3 indicates that there is one way to travel from Denver to Baltimore with two intermediate connections. The 2 in row 1 and column 5 indicates that there are two ways to travel from Atlanta to El Paso with two intermediate connections. In general, the elements in A^3 indicate the number of different ways to travel from the ith city to the the jth city with two intermediate connections.

(C) A is given above.

$A + A^2 = \begin{bmatrix} 0 & 1 & 0 & 1 & 0 \\ 0 & 0 & 1 & 0 & 0 \\ 1 & 0 & 0 & 0 & 1 \\ 0 & 0 & 1 & 0 & 0 \\ 0 & 0 & 0 & 1 & 0 \end{bmatrix} + \begin{bmatrix} 0 & 0 & 2 & 0 & 0 \\ 1 & 0 & 0 & 0 & 1 \\ 0 & 1 & 0 & 2 & 0 \\ 1 & 0 & 0 & 0 & 1 \\ 0 & 0 & 1 & 0 & 0 \end{bmatrix} = \begin{bmatrix} 0 & 1 & 2 & 1 & 0 \\ 1 & 0 & 1 & 0 & 1 \\ 1 & 1 & 0 & 2 & 1 \\ 1 & 0 & 1 & 0 & 1 \\ 0 & 0 & 1 & 1 & 0 \end{bmatrix}$

$A + A^2 + A^3 = \begin{bmatrix} 0 & 1 & 2 & 1 & 0 \\ 1 & 0 & 1 & 0 & 1 \\ 1 & 1 & 0 & 2 & 1 \\ 1 & 0 & 1 & 0 & 1 \\ 0 & 0 & 1 & 1 & 0 \end{bmatrix} + \begin{bmatrix} 2 & 0 & 0 & 0 & 2 \\ 0 & 1 & 0 & 2 & 0 \\ 0 & 0 & 3 & 0 & 0 \\ 0 & 1 & 0 & 2 & 0 \\ 1 & 0 & 0 & 0 & 1 \end{bmatrix} = \begin{bmatrix} 2 & 1 & 2 & 1 & 2 \\ 1 & 1 & 1 & 2 & 1 \\ 1 & 1 & 3 & 2 & 1 \\ 1 & 1 & 1 & 2 & 1 \\ 1 & 0 & 1 & 1 & 1 \end{bmatrix}$

A zero element remains, so we must compute A^4.

$A^4 = A^3 A = \begin{bmatrix} 2 & 0 & 0 & 0 & 2 \\ 0 & 1 & 0 & 2 & 0 \\ 0 & 0 & 3 & 0 & 0 \\ 0 & 1 & 0 & 2 & 0 \\ 1 & 0 & 0 & 0 & 1 \end{bmatrix}\begin{bmatrix} 0 & 1 & 0 & 1 & 0 \\ 0 & 0 & 1 & 0 & 0 \\ 1 & 0 & 0 & 0 & 1 \\ 0 & 0 & 1 & 0 & 0 \\ 0 & 0 & 0 & 1 & 0 \end{bmatrix} = \begin{bmatrix} 0 & 2 & 0 & 4 & 0 \\ 0 & 0 & 3 & 0 & 0 \\ 3 & 0 & 0 & 0 & 3 \\ 0 & 0 & 3 & 0 & 0 \\ 0 & 1 & 0 & 2 & 0 \end{bmatrix}$

Then $A + A^2 + A^3 + A^4 = \begin{bmatrix} 2 & 1 & 2 & 1 & 2 \\ 1 & 1 & 1 & 2 & 1 \\ 1 & 1 & 3 & 2 & 1 \\ 1 & 1 & 1 & 2 & 1 \\ 1 & 0 & 1 & 1 & 1 \end{bmatrix} + \begin{bmatrix} 0 & 2 & 0 & 4 & 0 \\ 0 & 0 & 3 & 0 & 0 \\ 3 & 0 & 0 & 0 & 3 \\ 0 & 0 & 3 & 0 & 0 \\ 0 & 1 & 0 & 2 & 0 \end{bmatrix} = \begin{bmatrix} 2 & 3 & 2 & 5 & 2 \\ 1 & 1 & 4 & 2 & 1 \\ 4 & 1 & 3 & 2 & 4 \\ 1 & 1 & 4 & 2 & 1 \\ 1 & 1 & 1 & 3 & 1 \end{bmatrix}$

This matrix indicates that it is possible to travel from any origin to any destination with at most 3 intermediate connections.

79. (A) $[1,000 \quad 500 \quad 5,000]\begin{bmatrix} \$0.80 \\ \$1.50 \\ \$0.40 \end{bmatrix} = \begin{matrix} 1,000(\$0.80) + 500(\$1.50) + 5,000(\$0.40), \\ = \$3,550 \end{matrix}$

(B) $[2,000 \quad 800 \quad 8,000]\begin{bmatrix} \$0.80 \\ \$1.50 \\ \$0.40 \end{bmatrix} = \begin{matrix} 2,000(\$0.80) + 800(\$1.50) + 8,000(\$0.40) \\ = \$6,000 \end{matrix}$

(C) The matrix MN has no obvious interpretations, but the matrix NM represents the total cost of all contacts in each town.

(D) $NM = \begin{bmatrix} 1,000 & 500 & 5,000 \\ 2,000 & 800 & 8,000 \end{bmatrix}\begin{bmatrix} \$0.80 \\ \$1.50 \\ \$0.40 \end{bmatrix} = \begin{bmatrix} 1,000(0.80) + 500(1.50) + 5,000(0.40) \\ 2,000(0.80) + 800(1.50) + 8,000(0.40) \end{bmatrix}$

$= \begin{bmatrix} \$3,550 \\ \$6,000 \end{bmatrix}\begin{matrix} \text{Berkeley} \\ \text{Oakland} \end{matrix} = \text{cost of all contacts in each town.}$

(E) The matrix $[1 \quad 1]N$ can be used to find the total number of each of the three types of contact:

$[1 \quad 1]\begin{bmatrix} 1,000 & 500 & 5,000 \\ 2,000 & 800 & 8,000 \end{bmatrix} = [1,000 + 2,000 \quad 500 + 800 \quad 5,000 + 8,000]$

$[\text{Telephone House Letter}] = [3,000 \quad 1,300 \quad 13,000]$

(F) The matrix $N\begin{bmatrix}1\\1\\1\end{bmatrix}$ can be used to find the total number of contacts in each town:

$$\begin{bmatrix}1,000 & 500 & 5,000\\2,000 & 800 & 8,000\end{bmatrix}\begin{bmatrix}1\\1\\1\end{bmatrix}$$

$$=\begin{bmatrix}1,000+500+5,000\\2,000+800+8,000\end{bmatrix}=\begin{bmatrix}6,500\\10,800\end{bmatrix}=\begin{bmatrix}\text{Berkeley contacts}\\\text{Oakland contacts}\end{bmatrix}$$

81. (A) Since player 1 did not defeat player 1, a 0 is placed in row 1, column 1.
Since player 1 did not defeat player 2, a 0 is placed in row 1, column 2.
Since player 1 defeated player 3, a 1 is placed in row 1, column 3.
Since player 1 defeated player 4, a 1 is placed in row 1, column 4.
Since player 1 defeated player 5, a 1 is placed in row 1, column 5.
Since player 1 did not defeat player 6, a 0 is placed in row 1, column 6.
Proceeding in this manner, we obtain

$$A=\begin{bmatrix}0 & 0 & 1 & 1 & 1 & 0\\1 & 0 & 0 & 1 & 1 & 0\\0 & 1 & 0 & 1 & 0 & 0\\0 & 0 & 0 & 0 & 0 & 1\\0 & 0 & 1 & 1 & 0 & 1\\1 & 1 & 1 & 0 & 0 & 0\end{bmatrix}$$

(B) $A^2=\begin{bmatrix}0 & 0 & 1 & 1 & 1 & 0\\1 & 0 & 0 & 1 & 1 & 0\\0 & 1 & 0 & 1 & 0 & 0\\0 & 0 & 0 & 0 & 0 & 1\\0 & 0 & 1 & 1 & 0 & 1\\1 & 1 & 1 & 0 & 0 & 0\end{bmatrix}+\begin{bmatrix}0 & 0 & 1 & 1 & 1 & 0\\1 & 0 & 0 & 1 & 1 & 0\\0 & 1 & 0 & 1 & 0 & 0\\0 & 0 & 0 & 0 & 0 & 1\\0 & 0 & 1 & 1 & 0 & 1\\1 & 1 & 1 & 0 & 0 & 0\end{bmatrix}=\begin{bmatrix}0 & 1 & 1 & 2 & 0 & 2\\0 & 0 & 2 & 2 & 1 & 2\\1 & 0 & 0 & 1 & 1 & 1\\1 & 1 & 1 & 0 & 0 & 0\\1 & 2 & 1 & 1 & 0 & 1\\1 & 1 & 1 & 3 & 2 & 0\end{bmatrix}$

$$A+A^2=\begin{bmatrix}0 & 1 & 2 & 3 & 1 & 2\\1 & 0 & 2 & 3 & 2 & 2\\1 & 1 & 0 & 2 & 1 & 1\\1 & 1 & 1 & 0 & 0 & 1\\1 & 2 & 2 & 2 & 0 & 2\\2 & 2 & 2 & 3 & 2 & 0\end{bmatrix}=B$$

Check (using graphing calculator):

```
[A]
 [[0 0 1 1 1 0]
  [1 0 0 1 1 0]
  [0 1 0 1 0 0]
  [0 0 0 0 0 1]
  [0 0 1 1 0 1]
  [1 1 1 0 0 0]]
```
```
[A]²
 [[0 1 1 2 0 2]
  [0 0 2 2 1 2]
  [1 0 0 1 1 1]
  [1 1 1 0 0 0]
  [1 2 1 1 0 1]
  [1 1 1 3 2 0]]
```
```
[A]+[A]²→[B]
 [[0 1 2 3 1 2]
  [1 0 2 3 2 2]
  [1 1 0 2 1 1]
  [1 1 1 0 0 1]
  [1 2 2 2 0 2]
  [2 2 2 3 2 0]]
```

(C) To add up the rows of AB, we can multiply by a column matrix of all ones: that is, if

$$C=\begin{bmatrix}1\\1\\1\\1\\1\\1\end{bmatrix}\quad\text{Then }BC=\begin{bmatrix}0 & 1 & 2 & 3 & 1 & 2\\1 & 0 & 2 & 3 & 2 & 2\\1 & 1 & 0 & 2 & 1 & 1\\1 & 1 & 1 & 0 & 0 & 1\\1 & 2 & 2 & 2 & 0 & 2\\2 & 2 & 2 & 3 & 2 & 0\end{bmatrix}\begin{bmatrix}1\\1\\1\\1\\1\\1\end{bmatrix}=\begin{bmatrix}9\\10\\6\\4\\9\\11\end{bmatrix}$$

(D) BC measures the relative strength of the players, with the larger numbers representing greater strength. Player 6 is the strongest and player 4 the weakest; ranking:

Frank, Bart, Aaron & Elvis (tie), Charles, Dan.

Section 6-3

1. Answers will vary.   3. Answers will vary.   5. Answers will vary.

7. $\begin{bmatrix} 2 & -3 \\ 4 & 5 \end{bmatrix}$ 9. $\begin{bmatrix} 2 & -3 \\ 4 & 5 \end{bmatrix}$ 11. $\begin{bmatrix} -2 & 1 & 3 \\ 2 & 4 & -2 \\ 5 & 1 & 0 \end{bmatrix}$ 13. $\begin{bmatrix} -2 & 1 & 3 \\ 2 & 4 & -2 \\ 5 & 1 & 0 \end{bmatrix}$

15. $\begin{bmatrix} 3 & -4 \\ -2 & 3 \end{bmatrix}\begin{bmatrix} 3 & 4 \\ 2 & 3 \end{bmatrix} = \begin{bmatrix} 3\cdot 3 + (-4)2 & 3\cdot 4 + (-4)3 \\ (-2)3 + 3\cdot 2 & (-2)4 + 3\cdot 3 \end{bmatrix} = \begin{bmatrix} 1 & 0 \\ 0 & 1 \end{bmatrix}$
These two matrices are inverses of each other.

17. $\begin{bmatrix} 2 & 2 \\ -1 & -1 \end{bmatrix}\begin{bmatrix} 1 & 1 \\ -1 & -1 \end{bmatrix} = \begin{bmatrix} 2\cdot 1 + 2(-1) & 2\cdot 1 + 2(-1) \\ (-1)1 + (-1)(-1) & (-1)1 + (-1)(-1) \end{bmatrix} = \begin{bmatrix} 0 & 0 \\ 0 & 0 \end{bmatrix}$
These two matrices are not inverses of each other.

19. $\begin{bmatrix} -5 & 2 \\ -8 & 3 \end{bmatrix}\begin{bmatrix} 3 & -2 \\ 8 & -5 \end{bmatrix} = \begin{bmatrix} (-5)3 + 2\cdot 8 & (-5)(-2) + 2(-5) \\ (-8)3 + 3\cdot 8 & (-8)(-2) + 3(-5) \end{bmatrix} = \begin{bmatrix} 1 & 0 \\ 0 & 1 \end{bmatrix}$
These two matrices are inverses of each other.

21. $\begin{bmatrix} 1 & 2 & 0 \\ 0 & 1 & 0 \\ -1 & -1 & 1 \end{bmatrix}\begin{bmatrix} 1 & -2 & 0 \\ 0 & 1 & 0 \\ 1 & -1 & 0 \end{bmatrix} =$

$\begin{bmatrix} 1\cdot 1 + 2\cdot 0 + 0\cdot 1 & 1(-2) + 2\cdot 1 + 0(-1) & 1\cdot 0 + 2\cdot 0 + 0\cdot 0 \\ 0\cdot 1 + 1\cdot 1 + 0(-1) & 0(-2) + 1\cdot 1 + 0(-1) & 0\cdot 0 + 1\cdot 0 + 0\cdot 0 \\ (-1)1 + (-1)1 + 1\cdot 1 & (-1)(-2) + (-1)1 + 1(-1) & (-1)0 + (-1)0 + 0\cdot 0 \end{bmatrix}$

$= \begin{bmatrix} 1 & 0 & 0 \\ 1 & 1 & 0 \\ -1 & 0 & 0 \end{bmatrix}$

These two matrices are not inverses of each other.

23. $\begin{bmatrix} 1 & -1 & 1 \\ 0 & 2 & -1 \\ 2 & 3 & 0 \end{bmatrix}\begin{bmatrix} 3 & 3 & -1 \\ -2 & -2 & 1 \\ -4 & -5 & 2 \end{bmatrix}$

$= \begin{bmatrix} 1\cdot 3 + (-1)(-2) + 1(-4) & 1\cdot 3 + (-1)(-2) + 1(-5) & 1\cdot(-1) + (-1)\cdot 1 + 1\cdot 2 \\ 0\cdot 3 + 2(-2) + (-1)(-4) & 0\cdot 3 + 2(-2) + (-1)(-5) & 0(-1) + 2\cdot 1 + (-1)\cdot 2 \\ 2\cdot 3 + 3(-2) + 0(-4) & 2\cdot 3 + 3(-2) + 0(-5) & 2(-1) + 3\cdot 1 + 0\cdot 2 \end{bmatrix}$

$= \begin{bmatrix} 1 & 0 & 0 \\ 0 & 1 & 0 \\ 0 & 0 & 1 \end{bmatrix}$

These two matrices are inverses of each other.

25. $\begin{bmatrix} 1 & 9 & | & 1 & 0 \\ 0 & 1 & | & 0 & 1 \end{bmatrix}$ $(-9)R_2 + R_1 \rightarrow R_1$

$\sim \begin{bmatrix} 1 & 0 & | & 1 & -9 \\ 0 & 1 & | & 0 & 1 \end{bmatrix}$

The inverse is $\begin{bmatrix} 1 & -9 \\ 0 & 1 \end{bmatrix}$

Check: $M^{-1}M = \begin{bmatrix} 1 & -9 \\ 0 & 1 \end{bmatrix}\begin{bmatrix} 1 & 9 \\ 0 & 1 \end{bmatrix} = \begin{bmatrix} 1\cdot 1 + (-9)0 & 1\cdot 9 + (-9)1 \\ 0\cdot 1 + 1\cdot 0 & 0\cdot 9 + 1\cdot 1 \end{bmatrix} = \begin{bmatrix} 1 & 0 \\ 0 & 1 \end{bmatrix}$

27. $\begin{bmatrix} -1 & -2 & | & 1 & 0 \\ 2 & 5 & | & 0 & 1 \end{bmatrix} 2R_1 + R_2 \rightarrow R_2$

$\sim \begin{bmatrix} -1 & -2 & | & 1 & 0 \\ 0 & 1 & | & 2 & 1 \end{bmatrix} 2R_2 + R_1 \rightarrow R_1$

$\sim \begin{bmatrix} -1 & -2 & | & 1 & 0 \\ 0 & 1 & | & 2 & 1 \end{bmatrix} 2R_2 + R_1 \rightarrow R_1$

$\sim \begin{bmatrix} -1 & 0 & | & 5 & 2 \\ 0 & 1 & | & 2 & 1 \end{bmatrix} (-1)R_1 \rightarrow R_1$

$\sim \begin{bmatrix} 1 & 0 & | & -5 & -2 \\ 0 & 1 & | & 2 & 1 \end{bmatrix}$

The inverse is $\begin{bmatrix} -5 & -2 \\ 2 & 1 \end{bmatrix}$

Check: $M^{-1}M = \begin{bmatrix} -5 & -2 \\ 2 & 1 \end{bmatrix} \begin{bmatrix} -1 & -2 \\ 2 & 5 \end{bmatrix} = \begin{bmatrix} (-5)(-1) + (-2)2 & (-5)(-2) + (-2)5 \\ 2(-1) + 1 \cdot 2 & 2(-2) + 1 \cdot 5 \end{bmatrix}$

$= \begin{bmatrix} 1 & 0 \\ 0 & 1 \end{bmatrix}$

29. $\begin{bmatrix} -5 & 7 & | & 1 & 0 \\ 2 & -3 & | & 0 & 1 \end{bmatrix} 3R_2 + R_1 \rightarrow R_1$

$\sim \begin{bmatrix} 1 & -2 & | & 1 & 3 \\ 2 & -3 & | & 0 & 1 \end{bmatrix} (-2)R_1 + R_2 \rightarrow R_2$

$\sim \begin{bmatrix} 1 & -2 & | & 1 & 3 \\ 0 & 1 & | & -2 & -5 \end{bmatrix} 2R_2 + R_1 \rightarrow R_1$

$\sim \begin{bmatrix} 1 & 0 & | & -3 & -7 \\ 0 & 1 & | & -2 & -5 \end{bmatrix}$

The inverse is $\begin{bmatrix} -3 & -7 \\ -2 & -5 \end{bmatrix}$

Check: $M^{-1}M = \begin{bmatrix} -3 & -7 \\ -2 & -5 \end{bmatrix} \begin{bmatrix} -5 & 7 \\ 2 & -3 \end{bmatrix} = \begin{bmatrix} (-3)(-5) + (-7)2 & (-3)7 + (-7)(-3) \\ (-2)(-5) + (-5)2 & (-2)7 + (-5)(-3) \end{bmatrix} = \begin{bmatrix} 1 & 0 \\ 0 & 1 \end{bmatrix}$

31. If the inverse existed we would find it by row operations on the following matrix:

$= \begin{bmatrix} 3 & 9 & | & 1 & 0 \\ 2 & 6 & | & 0 & 1 \end{bmatrix}$

But consider what happens if we perform $(-\frac{2}{3})R_1 + R_2 \rightarrow R_2$

$= \begin{bmatrix} 3 & 9 & | & 1 & 0 \\ 0 & 0 & | & -\frac{2}{3} & 1 \end{bmatrix}$

Since a row of zeros results to the left of the vertical line, no inverse exists.

33. $\begin{bmatrix} 2 & 3 & | & 1 & 0 \\ 3 & 5 & | & 0 & 1 \end{bmatrix} \frac{1}{2}R_1 \rightarrow R_1$

$\sim \begin{bmatrix} 1 & 1.5 & | & 0.5 & 0 \\ 3 & 5 & | & 0 & 1 \end{bmatrix} (-3)R_1 + R_2 \rightarrow R_2$

$\sim \begin{bmatrix} 1 & 1.5 & | & 0.5 & 0 \\ 0 & 0.5 & | & -1.5 & 1 \end{bmatrix} 2R_2 \rightarrow R_2$

$\sim \begin{bmatrix} 1 & 1.5 & | & 0.5 & 0 \\ 0 & 1 & | & -3 & 2 \end{bmatrix} (-1.5)R_2 + R_1 \rightarrow R_1$

$$\sim \begin{bmatrix} 1 & 0 & | & 5 & -3 \\ 0 & 1 & | & -3 & 2 \end{bmatrix}$$

The inverse is $\begin{bmatrix} 5 & -3 \\ -3 & 2 \end{bmatrix}$

The checking steps are left to the student in this and some subsequent problems.

35. $\begin{bmatrix} 1 & -1 & 0 & | & 1 & 0 & 0 \\ -1 & 1 & -1 & | & 0 & 1 & 0 \\ 0 & -1 & 1 & | & 0 & 0 & 1 \end{bmatrix}$ $R_1 + R_2 \rightarrow R_2$

$$\sim \begin{bmatrix} 1 & -1 & 0 & | & 1 & 0 & 0 \\ 0 & 0 & -1 & | & 1 & 1 & 0 \\ 0 & -1 & 1 & | & 0 & 0 & 1 \end{bmatrix}$$ $R_2 \leftrightarrow R_3$

$$\sim \begin{bmatrix} 1 & -1 & 0 & | & 1 & 0 & 0 \\ 0 & -1 & 1 & | & 0 & 0 & 1 \\ 0 & 0 & -1 & | & 1 & 1 & 0 \end{bmatrix}$$ $(-1)R_2 + R_1 \rightarrow R_1$

$$\sim \begin{bmatrix} 1 & 0 & -1 & | & 1 & 0 & -1 \\ 0 & -1 & 1 & | & 0 & 0 & 1 \\ 0 & 0 & -1 & | & 1 & 1 & 0 \end{bmatrix}$$ $\begin{array}{l}(-1)R_3 + R_1 \rightarrow R_1 \\ R_3 + R_2 \rightarrow R_2\end{array}$

$$\sim \begin{bmatrix} 1 & 0 & 0 & | & 0 & -1 & -1 \\ 0 & -1 & 0 & | & 1 & 1 & 1 \\ 0 & 0 & -1 & | & 1 & 1 & 0 \end{bmatrix}$$ $\begin{array}{l}(-1)R_2 \rightarrow R_2 \\ (-1)R_3 \rightarrow R_3\end{array}$

$$\sim \begin{bmatrix} 1 & 0 & 0 & | & 0 & -1 & -1 \\ 0 & 1 & 0 & | & -1 & -1 & -1 \\ 0 & 0 & 1 & | & -1 & -1 & 0 \end{bmatrix}$$

The inverse is $\begin{bmatrix} 0 & -1 & -1 \\ -1 & 1 & -1 \\ -1 & -1 & 0 \end{bmatrix}$

Check: $M^{-1}M = \begin{bmatrix} 0 & -1 & -1 \\ -1 & 1 & -1 \\ -1 & -1 & 0 \end{bmatrix}\begin{bmatrix} 1 & -1 & 0 \\ -1 & 1 & -1 \\ 0 & -1 & 1 \end{bmatrix}$

$$= \begin{bmatrix} 0\cdot 1 + (-1)(-1) + (-1)0 & 0(-1) + (-1)1 + (-1)(-1) & 0\cdot 0 + (-1)(-1) + (-1)1 \\ (-1)1 + (-1)(-1) + (-1)0 & (-1)(-1) + (-1)1 + (-1)(-1) & (-1)0 + (-1)(-1) + (-1)1 \\ (-1)1 + (-1)(-1) + 0\cdot 0 & (-1)(-1) + (-1)1 + 0(-1) & (-1)0 + (-1)(-1) + 0\cdot 1 \end{bmatrix}$$

$$= \begin{bmatrix} 1 & 0 & 0 \\ 0 & 1 & 0 \\ 0 & 0 & 1 \end{bmatrix}$$

37. $\begin{bmatrix} 1 & 2 & 5 & | & 1 & 0 & 0 \\ 3 & 5 & 9 & | & 0 & 1 & 0 \\ 1 & 1 & -2 & | & 0 & 0 & 1 \end{bmatrix}$ $\begin{array}{l}(-3)R_1 + R_2 \rightarrow R_2 \\ (-1)R_1 + R_3 \rightarrow R_3\end{array}$

$$\sim \begin{bmatrix} 1 & 2 & 5 & | & 1 & 0 & 0 \\ 0 & -1 & -6 & | & -3 & 1 & 0 \\ 0 & -1 & -7 & | & -1 & 0 & 1 \end{bmatrix}$$ $\begin{array}{l}2R_2 + R_1 \rightarrow R_1 \\ \\ (-1)R_2 + R_3 \rightarrow R_3\end{array}$

$$\sim \begin{bmatrix} 1 & 0 & -7 & | & -5 & 2 & 0 \\ 0 & -1 & -6 & | & -3 & 1 & 0 \\ 0 & 0 & -1 & | & 2 & -1 & 1 \end{bmatrix}$$ $\begin{array}{l}(-7)R_3 + R_1 \rightarrow R_1 \\ (-6)R_3 + R_2 \rightarrow R_2\end{array}$

$$\sim \begin{bmatrix} 1 & 0 & 0 & | & -19 & 9 & -7 \\ 0 & -1 & 0 & | & -15 & 7 & -6 \\ 0 & 0 & -1 & | & 2 & -1 & 1 \end{bmatrix} \begin{matrix} (-1)R_2 \rightarrow R_2 \\ (-1)R_3 \rightarrow R_3 \end{matrix}$$

$$\sim \begin{bmatrix} 1 & 0 & 0 & | & -19 & 9 & -7 \\ 0 & 1 & 0 & | & 15 & -7 & 6 \\ 0 & 0 & 1 & | & -2 & 1 & -1 \end{bmatrix}$$

The inverse is $\begin{bmatrix} -19 & 9 & -7 \\ 15 & -7 & 6 \\ -2 & 1 & -1 \end{bmatrix}$

Check: $M^{-1}M = \begin{bmatrix} -19 & 9 & -7 \\ 15 & -7 & 6 \\ -2 & 1 & -1 \end{bmatrix}\begin{bmatrix} 1 & 2 & 5 \\ 3 & 5 & 9 \\ 1 & 1 & -2 \end{bmatrix}$

$$= \begin{bmatrix} (-19)1 + 9 \cdot 3 + (-7)1 & (-19)2 + 9 \cdot 5 + (-7)1 & (-19)5 + 9 \cdot 9 + (-7)(-2) \\ 15 \cdot 1 + (-7)3 + 6 \cdot 1 & 15 \cdot 2 + (-7)5 + 6 \cdot 1 & 15 \cdot 5 + (-7)9 + 6(-2) \\ (-2)1 + 1 \cdot 3 + (-1)1 & (-2)2 + 1 \cdot 5 + (-1)1 & (-2)5 + 1 \cdot 9 + (-1)(-2) \end{bmatrix}$$

$$= \begin{bmatrix} 1 & 0 & 0 \\ 0 & 1 & 0 \\ 0 & 0 & 1 \end{bmatrix}$$

39. $\begin{bmatrix} 2 & 2 & -1 & | & 1 & 0 & 0 \\ 0 & 4 & -1 & | & 0 & 1 & 0 \\ -1 & -2 & 1 & | & 0 & 0 & 1 \end{bmatrix} R_1 \leftrightarrow R_3$

$$\sim \begin{bmatrix} -1 & -2 & 1 & | & 0 & 0 & 1 \\ 0 & 4 & -1 & | & 0 & 1 & 0 \\ 2 & 2 & -1 & | & 1 & 0 & 0 \end{bmatrix} 2R_1 + R_3 \rightarrow R_3$$

$$\sim \begin{bmatrix} -1 & -2 & 1 & | & 0 & 0 & 1 \\ 0 & 4 & -1 & | & 0 & 1 & 0 \\ 0 & -2 & 1 & | & 1 & 0 & 2 \end{bmatrix} \begin{matrix} (-1)R_3 + R_1 \rightarrow R_1 \\ 2R_3 + R_2 \rightarrow R_2 \end{matrix}$$

$$\sim \begin{bmatrix} -1 & 0 & 0 & | & -1 & 0 & -1 \\ 0 & 0 & 1 & | & 2 & 1 & 4 \\ 0 & -2 & 1 & | & -1 & 0 & 2 \end{bmatrix} (-1)R_2 + R_3 \rightarrow R_3$$

$$\sim \begin{bmatrix} -1 & 0 & 0 & | & -1 & 0 & -1 \\ 0 & 0 & 1 & | & 2 & 1 & 4 \\ 0 & -2 & 0 & | & -1 & -1 & -2 \end{bmatrix} R_2 \leftrightarrow R_3$$

$$\sim \begin{bmatrix} -1 & 0 & 0 & | & -1 & 0 & -1 \\ 0 & -2 & 0 & | & -1 & -1 & -2 \\ 0 & 0 & 1 & | & 2 & 1 & 4 \end{bmatrix} \begin{matrix} (-1)R_1 \rightarrow R_1 \\ (-\frac{1}{2})R_2 \rightarrow R_2 \end{matrix}$$

$$\sim \begin{bmatrix} 1 & 0 & 0 & | & 1 & 0 & 1 \\ 0 & 1 & 0 & | & \frac{1}{2} & \frac{1}{2} & 1 \\ 0 & 0 & 1 & | & 2 & 1 & 4 \end{bmatrix}$$

The inverse is $\begin{bmatrix} 1 & 0 & 1 \\ \frac{1}{2} & \frac{1}{2} & 1 \\ 2 & 1 & 4 \end{bmatrix}$

41. If the inverse existed we would find it by row operations on the following matrix:

$$\begin{bmatrix} 2 & 1 & 1 & | & 1 & 0 & 0 \\ 1 & 1 & 0 & | & 0 & 1 & 0 \\ -1 & -1 & 0 & | & 0 & 0 & 1 \end{bmatrix}$$

But consider what happens if we perform $R_2 + R_3 \to R_2$

$$\begin{bmatrix} 2 & 1 & 1 & | & 1 & 0 & 0 \\ 0 & 0 & 0 & | & 0 & 1 & 1 \\ -1 & -1 & 0 & | & 0 & 0 & 1 \end{bmatrix}$$

Since a row of zeros results to the left of the vertical line, no inverse exists.

43.
$$\begin{bmatrix} 1 & 5 & 10 & | & 1 & 0 & 0 \\ 0 & 1 & 4 & | & 0 & 1 & 0 \\ 1 & 6 & 15 & | & 0 & 0 & 1 \end{bmatrix} (-1)R_1 + R_3 \to R_3$$

$$\sim \begin{bmatrix} 1 & 5 & 10 & | & 1 & 0 & 0 \\ 0 & 1 & 4 & | & 0 & 1 & 0 \\ 0 & 1 & 5 & | & -1 & 0 & 1 \end{bmatrix} \begin{matrix} (-5)R_2 + R_1 \to R_1 \\ \\ (-1)R_2 + R_3 \to R_3 \end{matrix}$$

$$\sim \begin{bmatrix} 1 & 0 & -10 & | & 1 & -5 & 0 \\ 0 & 1 & 4 & | & 0 & 1 & 0 \\ 0 & 0 & 1 & | & -1 & -1 & 1 \end{bmatrix} \begin{matrix} 10R_3 + R_1 \to R_1 \\ (-4)R_3 + R_2 \to R_2 \end{matrix}$$

$$\sim \begin{bmatrix} 1 & 0 & 0 & | & -9 & -15 & 10 \\ 0 & 1 & 0 & | & 4 & 5 & -4 \\ 0 & 0 & 1 & | & -1 & -1 & 1 \end{bmatrix}$$

The inverse is $\begin{bmatrix} -9 & -15 & 10 \\ 4 & 5 & -4 \\ -1 & -1 & 1 \end{bmatrix}$

45. Try to find A^{-1}:

$$\begin{bmatrix} a & 0 & | & 1 & 0 \\ 0 & d & | & 0 & 1 \end{bmatrix} \frac{1}{a}R_1 \to R_1$$

$$\sim \begin{bmatrix} 1 & 0 & | & \frac{1}{a} & 0 \\ 0 & d & | & 0 & 1 \end{bmatrix} \frac{1}{d}R_2 \to R_2$$

$$\sim \begin{bmatrix} 1 & 0 & | & \frac{1}{a} & 0 \\ 0 & 1 & | & 0 & \frac{1}{d} \end{bmatrix} \quad \text{The inverse is } A^{-1} = \begin{bmatrix} \frac{1}{a} & 0 \\ 0 & \frac{1}{d} \end{bmatrix}$$

This will exist unless either a or d is zero, which would make $\frac{1}{a}$ or $\frac{1}{d}$ undefined. So A^{-1} exists exactly when both a and d are non-zero.

47. (A) A^{-1}:

$$\begin{bmatrix} 3 & 2 & | & 1 & 0 \\ -4 & -3 & | & 0 & 1 \end{bmatrix} R_2 + R_1 \to R_1$$

$$\sim \begin{bmatrix} -1 & -1 & | & 1 & 1 \\ -4 & -3 & | & 0 & 1 \end{bmatrix} (-1)R_1 \to R_1$$

$$\sim \begin{bmatrix} 1 & 1 & | & -1 & -1 \\ -4 & -3 & | & 0 & 1 \end{bmatrix} 4R_1 + R_2 \to R_2$$

$$\sim \begin{bmatrix} 1 & 1 & | & -1 & -1 \\ 0 & 1 & | & -4 & 1-3 \end{bmatrix} -1R_2 + R_1 \to R_1$$

$$\sim \begin{bmatrix} 1 & 0 & | & 3 & 2 \\ 0 & 1 & | & -4 & -3 \end{bmatrix} \quad A^{-1} = \begin{bmatrix} 3 & 2 \\ -4 & -3 \end{bmatrix}$$

A^2:

$$\begin{bmatrix} 3 & 2 \\ -4 & -3 \end{bmatrix} \begin{bmatrix} 3 & 2 \\ -4 & -3 \end{bmatrix}$$

$$= \begin{bmatrix} 3 \cdot 3 + 2(-4) & 3 \cdot 2 + 2(-3) \\ (-4)3 + (-3)(-4) & (-4)2 + (-3)(-3) \end{bmatrix}$$

$$= \begin{bmatrix} 1 & 0 \\ 0 & 1 \end{bmatrix}$$

(B) A^{-1}:

$$\begin{bmatrix} -2 & -1 & | & 1 & 0 \\ 3 & 2 & | & 0 & 1 \end{bmatrix} R_2 + R_1 \rightarrow R_1$$

$$\sim \begin{bmatrix} 1 & 1 & | & 1 & 1 \\ 3 & 2 & | & 0 & 1 \end{bmatrix} -3R_1 + R_2 \rightarrow R_2$$

$$\sim \begin{bmatrix} 1 & 1 & | & 1 & 1 \\ 0 & -1 & | & -3 & -2 \end{bmatrix} R_2 + R_1 \rightarrow R_1$$

$$\sim \begin{bmatrix} 1 & 0 & | & -2 & -1 \\ 0 & -1 & | & -3 & -2 \end{bmatrix} -1R_2 \rightarrow R_2$$

$$\sim \begin{bmatrix} 1 & 0 & | & -2 & -1 \\ 0 & 1 & | & 3 & 2 \end{bmatrix} \quad A^{-1} = \begin{bmatrix} -2 & -1 \\ 3 & 2 \end{bmatrix}$$

A^2:

$$\begin{bmatrix} -2 & -1 \\ 3 & 2 \end{bmatrix} \begin{bmatrix} -2 & -1 \\ 3 & 2 \end{bmatrix}$$

$$= \begin{bmatrix} (-2)(-2) + (-1)3 & (-2)(-1) + (-1)2 \\ 3(-2) + 2 \cdot 3 & 3(-1) + 2 \cdot 2 \end{bmatrix}$$

$$= \begin{bmatrix} 1 & 0 \\ 0 & 1 \end{bmatrix}$$

Note that in both cases $A^{-1} = A$ and $A^2 = I$.

49. (A) We calculate A^{-1} by row operations on

$$\begin{bmatrix} 4 & 2 & | & 1 & 0 \\ 1 & 3 & | & 0 & 1 \end{bmatrix} R_1 \leftrightarrow R_2$$

$$\sim \begin{bmatrix} 1 & 3 & | & 0 & 1 \\ 4 & 2 & | & 1 & 0 \end{bmatrix} (-4)R_1 + R_2 \rightarrow R_2$$

$$\sim \begin{bmatrix} 1 & 3 & | & 0 & 1 \\ 0 & -10 & | & 1 & -4 \end{bmatrix} \left(-\frac{1}{10}\right)R_2 \rightarrow R_2$$

$$\sim \begin{bmatrix} 1 & 3 & | & 0 & 1 \\ 0 & 1 & | & -\frac{1}{10} & \frac{2}{5} \end{bmatrix} (-3)R_2 + R_1 \rightarrow R_1$$

$$\sim \begin{bmatrix} 1 & 0 & | & \frac{3}{10} & -\frac{1}{5} \\ 0 & 1 & | & -\frac{1}{10} & \frac{2}{5} \end{bmatrix}$$

$$A^{-1} = \begin{bmatrix} \frac{3}{10} & -\frac{1}{5} \\ -\frac{1}{10} & \frac{2}{5} \end{bmatrix}$$

We calculate $(A^{-1})^{-1}$ by row operations on

$$\begin{bmatrix} \frac{3}{10} & -\frac{1}{5} & | & 1 & 0 \\ -\frac{1}{10} & \frac{2}{5} & | & 0 & 1 \end{bmatrix} \frac{10}{3} R_1 \rightarrow R_1$$

$$\sim \begin{bmatrix} 1 & -\frac{2}{3} & | & \frac{10}{3} & 0 \\ -\frac{1}{10} & \frac{2}{5} & | & 0 & 1 \end{bmatrix} \frac{1}{10} R_1 + R_2 \rightarrow R_2$$

$$\sim \begin{bmatrix} 1 & -\frac{2}{3} & | & \frac{10}{3} & 0 \\ 0 & \frac{1}{3} & | & \frac{1}{3} & 1 \end{bmatrix} 2R_2 + R_1 \rightarrow R_1$$

$$\sim \begin{bmatrix} 1 & 0 & | & 4 & 2 \\ 0 & \frac{1}{3} & | & \frac{1}{3} & 1 \end{bmatrix} 3R_2 \rightarrow R_2$$

$$\sim \begin{bmatrix} 1 & 0 & | & 4 & 2 \\ 0 & 1 & | & 1 & 3 \end{bmatrix}$$

$$(A^{-1})^{-1} = \begin{bmatrix} 4 & 2 \\ 1 & 3 \end{bmatrix}$$

(B) We calculate A^{-1} by row operations on

$$\begin{bmatrix} 5 & 5 & | & 1 & 0 \\ -1 & 3 & | & 0 & 1 \end{bmatrix} R_1 \leftrightarrow R_2$$

$$\sim \begin{bmatrix} -1 & 3 & | & 0 & 1 \\ 5 & 5 & | & 1 & 0 \end{bmatrix} 5R_1 + R_2 \rightarrow R_2$$

$$\sim \begin{bmatrix} -1 & 3 & | & 0 & 1 \\ 0 & 20 & | & 1 & 5 \end{bmatrix} \frac{1}{20} R_2 \rightarrow R_2$$

$$\sim \begin{bmatrix} -1 & 3 & | & 0 & 1 \\ 0 & 1 & | & \frac{1}{20} & \frac{1}{4} \end{bmatrix} (-3)R_2 + R_1 \rightarrow R_1$$

$$\sim \begin{bmatrix} -1 & 0 & | & -\frac{3}{20} & \frac{1}{4} \\ 0 & 1 & | & \frac{1}{20} & \frac{1}{4} \end{bmatrix} (-1)R_1 \rightarrow R_1$$

$$\sim \begin{bmatrix} 1 & 0 & | & \frac{3}{20} & -\frac{1}{4} \\ 0 & 1 & | & \frac{1}{20} & \frac{1}{4} \end{bmatrix}$$

$$A^{-1} = \begin{bmatrix} \frac{3}{20} & -\frac{1}{4} \\ \frac{1}{20} & \frac{1}{4} \end{bmatrix}$$

We calculate A^{-1} by row operations on

$$\begin{bmatrix} \frac{3}{20} & -\frac{1}{4} & | & 1 & 0 \\ \frac{1}{20} & \frac{1}{4} & | & 0 & 1 \end{bmatrix} \frac{20}{3} R_1 \rightarrow R_1$$

$$\sim \begin{bmatrix} 1 & -\frac{5}{3} & | & \frac{20}{3} & 0 \\ \frac{1}{20} & \frac{1}{4} & | & 0 & 1 \end{bmatrix} -\frac{1}{20} R_1 + R_2 \rightarrow R_2$$

$$\sim \begin{bmatrix} 1 & -\frac{5}{3} & | & \frac{20}{3} & 0 \\ 0 & \frac{1}{3} & | & -\frac{1}{3} & 1 \end{bmatrix} 5R_2 + R_1 \rightarrow R_1$$

$$\sim \begin{bmatrix} 1 & 0 & | & 5 & 5 \\ 0 & \frac{1}{3} & | & -\frac{1}{3} & 1 \end{bmatrix} 3R_2 \rightarrow R_2$$

$$\sim \begin{bmatrix} 1 & 0 & | & 5 & 5 \\ 0 & 1 & | & -1 & 3 \end{bmatrix}$$

$$(A^{-1})^{-1} = \begin{bmatrix} 5 & 5 \\ -1 & 3 \end{bmatrix}$$

Note that in both cases the inverse of the inverse works out to be the original matrix.

51. (A) Using a graphing calculator, we enter
A and B and calculate A^{-1} and B^{-1}

Then $(AB)^{-1}$ and
$A^{-1}B^{-1}$ are
calculated as

Finally $B^{-1}A^{-1}$ is calculated as

(B) Using a graphing calculator, we enter
A and B and calculate A^{-1} and B^{-1}

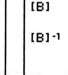

Then $(AB)^{-1}$ and
$A^{-1}B^{-1}$ are
calculated as

Finally $B^{-1}A^{-1}$ is calculated as

Notice that in each case $(AB)^{-1} = B^{-1}A^{-1}$ but $(AB)^{-1} \neq A^{-1}B^{-1}$.

53. Using the assignation numbers 1 to 27 with the letters of the alphabet and a blank as in the text, write

```
C  A  T       I  N     T   H  E     H  A  T
3  1  20  27  9  14  27  20  8  5  27  8  1  20
```

and calculate

$$\begin{bmatrix} 3 & 5 \\ 1 & 2 \end{bmatrix}\begin{bmatrix} 3 & 20 & 9 & 27 & 8 & 27 & 1 \\ 1 & 27 & 14 & 20 & 5 & 8 & 20 \end{bmatrix} = \begin{bmatrix} 14 & 195 & 97 & 181 & 49 & 121 & 103 \\ 5 & 74 & 37 & 67 & 18 & 43 & 41 \end{bmatrix}$$

The encoded message is
14 5 195 74 97 37 181 67 49 18 121 43 103 41

55. The inverse of matrix A is easily calculated to be

$$A^{-1} = \begin{bmatrix} 2 & -5 \\ -1 & 3 \end{bmatrix} \text{ (Details omitted)}$$

Putting the coded message into matrix form and multiplying by A^{-1} yields:

$$\begin{bmatrix} 2 & -5 \\ -1 & 3 \end{bmatrix}\begin{bmatrix} 111 & 40 & 177 & 50 & 116 & 86 & 62 & 121 & 68 \\ 43 & 15 & 68 & 19 & 45 & 29 & 22 & 43 & 27 \end{bmatrix}$$

$$= \begin{bmatrix} 7 & 5 & 14 & 5 & 7 & 27 & 14 & 27 & 1 \\ 18 & 5 & 27 & 7 & 19 & 1 & 4 & 8 & 13 \end{bmatrix}$$

This decodes to   7 18 5 5 14 27 5 7 7 19 27 1 14 4 27 8 1 13
                    G R E E N   E G G S   A N D   H A M

57. Using the assignation of numbers 1 to 27 with the letters of the alphabet and a blank as in the text, write

```
D  W  I  G  H  T     D  A  V  I  D     E  I  S  E  N  H  O  W  E  R
4 23  9  7  8 20 27  4  1 22  9  4 27  5  9 19  5 14  8 15 23  5 18 27 27
```

and calculate

$$\begin{bmatrix} 1 & 0 & 1 & 0 & 1 \\ 0 & 1 & 1 & 0 & 3 \\ 2 & 1 & 1 & 1 & 1 \\ 0 & 0 & 1 & 0 & 2 \\ 1 & 1 & 1 & 2 & 1 \end{bmatrix} \begin{bmatrix} 4 & 20 & 9 & 19 & 23 \\ 23 & 27 & 4 & 5 & 5 \\ 9 & 4 & 27 & 14 & 18 \\ 7 & 1 & 5 & 8 & 27 \\ 8 & 22 & 9 & 15 & 27 \end{bmatrix} = \begin{bmatrix} 21 & 46 & 45 & 48 & 68 \\ 56 & 97 & 58 & 64 & 104 \\ 55 & 94 & 63 & 80 & 123 \\ 25 & 48 & 45 & 44 & 72 \\ 58 & 75 & 59 & 69 & 127 \end{bmatrix}$$

The encoded message is

```
21  56  55  25  58   46   97  94  48   75   45  58   63  45   59  48  64
80  44  69  68  104  123  72  127
```

59. The inverse of B is calculated to be

$$B^{-1} = \begin{bmatrix} -2 & -1 & 2 & 2 & -1 \\ 3 & 2 & -2 & -4 & 1 \\ 6 & 2 & -4 & -5 & 2 \\ -2 & -1 & 1 & 2 & 0 \\ -3 & -1 & 2 & 3 & -1 \end{bmatrix} \quad \text{(Details omitted)}$$

Putting the coded message into matrix form and multiplying by B^{-1} yields

$$\begin{bmatrix} -2 & -1 & 2 & 2 & -1 \\ 3 & 2 & -2 & -4 & 1 \\ 6 & 2 & -4 & -5 & 2 \\ -2 & -1 & 1 & 2 & 0 \\ -3 & -1 & 2 & 3 & -1 \end{bmatrix} \begin{bmatrix} 41 & 25 & 43 & 44 & 68 \\ 84 & 56 & 54 & 67 & 135 \\ 82 & 67 & 89 & 86 & 136 \\ 44 & 20 & 39 & 44 & 81 \\ 74 & 54 & 102 & 90 & 149 \end{bmatrix} = \begin{bmatrix} 12 & 14 & 14 & 15 & 14 \\ 25 & 27 & 5 & 8 & 27 \\ 14 & 2 & 19 & 14 & 27 \\ 4 & 1 & 27 & 19 & 27 \\ 15 & 9 & 10 & 15 & 27 \end{bmatrix}$$

This decodes to

```
12 25 14 4 15 14 27 2 1 9 14 5 19 27 10 15 8 14 19 15 14 27 27 27
 L  Y  N  D  O  N     B  A  I  N  E  S     J  O  H  N  S  O  N
```

Section 6-4

1. Answers will vary.   **3.** Answers will vary.   **5.** Answers will vary.

7. Answers will vary.   **9.** Answers will vary.

11. $2x_1 - x_2 = 3$
$x_1 + 3x_2 = -2$

13. $-2x_1 \quad\quad + x_3 = 3$
$x_1 + 2x_2 + x_3 = -4$
$x_2 - x_3 = 2$

15. $\begin{bmatrix} 4 & -3 \\ 1 & 2 \end{bmatrix} \begin{bmatrix} x_1 \\ x_2 \end{bmatrix} = \begin{bmatrix} 2 \\ 1 \end{bmatrix}$

17. $\begin{bmatrix} 1 & -2 & 1 \\ -1 & 1 & 0 \\ 2 & 3 & 1 \end{bmatrix} \begin{bmatrix} x_1 \\ x_2 \\ x_3 \end{bmatrix} = \begin{bmatrix} -1 \\ 2 \\ -3 \end{bmatrix}$

19. Since $\begin{bmatrix} 3 & -2 \\ 1 & 4 \end{bmatrix} \begin{bmatrix} -2 \\ 1 \end{bmatrix} = \begin{bmatrix} 3(-2) + (-2)1 \\ 1(-2) + 4 \cdot 1 \end{bmatrix}$

$= \begin{bmatrix} -8 \\ 2 \end{bmatrix}$, $\begin{bmatrix} x_1 \\ x_2 \end{bmatrix} = \begin{bmatrix} -8 \\ 2 \end{bmatrix}$ if and only if $x_1 = -8$ and $x_2 = 2$.

21. Since $\begin{bmatrix} -2 & 3 \\ 2 & -1 \end{bmatrix} \begin{bmatrix} 3 \\ 2 \end{bmatrix} = \begin{bmatrix} (-2)3 + 3 \cdot 2 \\ 2 \cdot 3 + (-1)2 \end{bmatrix} = \begin{bmatrix} 0 \\ 4 \end{bmatrix}$

$\begin{bmatrix} x_1 \\ x_2 \end{bmatrix} = \begin{bmatrix} 0 \\ 4 \end{bmatrix}$ if and only if $x_1 = 0$ and $x_2 = 4$.

23. $\begin{bmatrix} 1 & -1 \\ 1 & -2 \end{bmatrix} \begin{bmatrix} x_1 \\ x_2 \end{bmatrix} = \begin{bmatrix} 5 \\ 7 \end{bmatrix}$

$\quad\quad A \quad\quad\quad X \quad = \quad B$

$AX = B$ has solution $X = A^{-1}B$.
To find A^{-1}, we perform row operations on

$\begin{bmatrix} 1 & -1 & | & 1 & 0 \\ 1 & -2 & | & 0 & 1 \end{bmatrix} (-1)R_1 + R_2 \to R_2$

$\sim \begin{bmatrix} 1 & -1 & | & 1 & 0 \\ 0 & -1 & | & -1 & 1 \end{bmatrix} (-1)R_2 + R_1 \to R_1$

$\sim \begin{bmatrix} 1 & 0 & | & 2 & -1 \\ 0 & -1 & | & -1 & 1 \end{bmatrix} (-1)R_2 \to R_2$

$\sim \begin{bmatrix} 1 & 0 & | & 2 & -1 \\ 0 & 1 & | & 1 & -1 \end{bmatrix}$

$A^{-1} = \begin{bmatrix} 2 & -1 \\ 1 & -1 \end{bmatrix}$

$X = A^{-1}B = \begin{bmatrix} 2 & -1 \\ 1 & -1 \end{bmatrix}\begin{bmatrix} 5 \\ 7 \end{bmatrix} = \begin{bmatrix} 2 \cdot 5 + (-1)(7) \\ 1 \cdot 5 + (-1)(7) \end{bmatrix} = \begin{bmatrix} 3 \\ -2 \end{bmatrix}$

So $x_1 = 3$, $x_2 = -2$

25. $\begin{bmatrix} 1 & 1 \\ 2 & -3 \end{bmatrix} \begin{bmatrix} x_1 \\ x_2 \end{bmatrix} = \begin{bmatrix} 15 \\ 10 \end{bmatrix}$

$\quad\quad A \quad\quad\quad X \quad = \quad B$

$AX = B$ has solution $X = A^{-1}B$.
To find A^{-1}, we perform row operations on

$\begin{bmatrix} 1 & 1 & | & 1 & 0 \\ 2 & -3 & | & 0 & 1 \end{bmatrix} (-2)R_1 + R_2 \to R_2$

$\sim \begin{bmatrix} 1 & 1 & | & 1 & 0 \\ 0 & -5 & | & -2 & 1 \end{bmatrix} \left(-\frac{1}{5}\right)R_2 \to R_2$

$\sim \begin{bmatrix} 1 & 1 & | & 1 & 0 \\ 0 & 1 & | & \frac{2}{5} & -\frac{1}{5} \end{bmatrix} (-1)R_2 + R_1 \to R_1$

$\sim \begin{bmatrix} 1 & 0 & | & \frac{3}{5} & \frac{1}{5} \\ 0 & 1 & | & \frac{2}{5} & -\frac{1}{5} \end{bmatrix}$

$A^{-1} = \begin{bmatrix} \frac{3}{5} & \frac{1}{5} \\ \frac{2}{5} & -\frac{1}{5} \end{bmatrix}$

$X = A^{-1}B = \begin{bmatrix} \frac{3}{5} & \frac{1}{5} \\ \frac{2}{5} & -\frac{1}{5} \end{bmatrix}\begin{bmatrix} 15 \\ 10 \end{bmatrix} = \begin{bmatrix} \frac{3}{5} \cdot 15 + \frac{1}{5} \cdot 10 \\ \frac{2}{5} \cdot 15 + \left(-\frac{1}{5}\right)10 \end{bmatrix} = \begin{bmatrix} 11 \\ 4 \end{bmatrix}$

So $x_1 = 11$, $x_2 = 4$

27. $\begin{bmatrix} -1 & -2 \\ 2 & 5 \end{bmatrix} \begin{bmatrix} x_1 \\ x_2 \end{bmatrix} = \begin{bmatrix} k_1 \\ k_2 \end{bmatrix}$

$AX = K$ has solution $X = A^{-1}K$.

We find $A^{-1}K$ for each given K. From problem 27 in Section 6-3, $A^{-1} = \begin{bmatrix} -5 & -2 \\ 2 & 1 \end{bmatrix}$

(A) $K = \begin{bmatrix} 2 \\ 5 \end{bmatrix}$ $A^{-1}K = \begin{bmatrix} -5 & -2 \\ 2 & 1 \end{bmatrix}\begin{bmatrix} 2 \\ 5 \end{bmatrix} = \begin{bmatrix} -20 \\ 9 \end{bmatrix}$ $x_1 = -20$, $x_2 = 9$

(B) $K = \begin{bmatrix} -4 \\ 1 \end{bmatrix}$ $A^{-1}K = \begin{bmatrix} -5 & -2 \\ 2 & 1 \end{bmatrix}\begin{bmatrix} -4 \\ 1 \end{bmatrix} = \begin{bmatrix} 18 \\ -7 \end{bmatrix}$ $x_1 = 18$, $x_2 = -7$

(C) $K = \begin{bmatrix} -3 \\ -2 \end{bmatrix}$ $A^{-1}K = \begin{bmatrix} -5 & -2 \\ 2 & 1 \end{bmatrix}\begin{bmatrix} -3 \\ -2 \end{bmatrix} = \begin{bmatrix} 19 \\ -8 \end{bmatrix}$ $x_1 = 19$, $x_2 = -8$

29. $\begin{bmatrix} -5 & 7 \\ 2 & -3 \end{bmatrix} \begin{bmatrix} x_1 \\ x_2 \end{bmatrix} = \begin{bmatrix} k_1 \\ k_2 \end{bmatrix}$

$AX = K$ has solution $X = A^{-1}K$.

We find $A^{-1}K$ for each given K. From problem 29 in Section 6-3, $A^{-1} = \begin{bmatrix} -3 & -7 \\ -2 & -5 \end{bmatrix}$

(A) $K = \begin{bmatrix} -5 \\ 1 \end{bmatrix}$ $A^{-1}K = \begin{bmatrix} -3 & -7 \\ -2 & -5 \end{bmatrix}\begin{bmatrix} -5 \\ 1 \end{bmatrix} = \begin{bmatrix} 8 \\ 5 \end{bmatrix}$ $x_1 = 8$, $x_2 = 5$

(B) $K = \begin{bmatrix} 8 \\ -4 \end{bmatrix}$ $A^{-1}K = \begin{bmatrix} -3 & -7 \\ -2 & -5 \end{bmatrix}\begin{bmatrix} 8 \\ -4 \end{bmatrix} = \begin{bmatrix} 4 \\ 4 \end{bmatrix}$ $x_1 = 4$, $x_2 = 4$

(C) $K = \begin{bmatrix} 6 \\ 0 \end{bmatrix}$ $A^{-1}K = \begin{bmatrix} -3 & -7 \\ -2 & -5 \end{bmatrix}\begin{bmatrix} 6 \\ 0 \end{bmatrix} = \begin{bmatrix} -18 \\ -12 \end{bmatrix}$ $x_1 = -18$, $x_2 = -12$

31. $\begin{bmatrix} 1 & -1 & 0 \\ -1 & 1 & -1 \\ 0 & -1 & 1 \end{bmatrix} \begin{bmatrix} x_1 \\ x_2 \\ x_3 \end{bmatrix} = \begin{bmatrix} k_1 \\ k_2 \\ k_3 \end{bmatrix}$

$AX = K$ has solution $X = A^{-1}K$.

We find $A^{-1}K$ for each given K. From problem 35 in Section 6-3,

$A^{-1} = \begin{bmatrix} 0 & -1 & -1 \\ -1 & -1 & -1 \\ -1 & -1 & 0 \end{bmatrix}$

(A) $K = \begin{bmatrix} 1 \\ 1 \\ 2 \end{bmatrix}$ $\qquad A^{-1}K = \begin{bmatrix} 0 & -1 & -1 \\ -1 & -1 & -1 \\ -1 & -1 & 0 \end{bmatrix} \begin{bmatrix} 1 \\ 1 \\ 2 \end{bmatrix} = \begin{bmatrix} -3 \\ -4 \\ -2 \end{bmatrix}$ $\qquad x_1 = -3,\ x_2 = -4,\ x_3 = -2$

(B) $K = \begin{bmatrix} -1 \\ 0 \\ -4 \end{bmatrix}$ $\qquad A^{-1}K = \begin{bmatrix} 0 & -1 & -1 \\ -1 & -1 & -1 \\ -1 & -1 & 0 \end{bmatrix} \begin{bmatrix} -1 \\ 0 \\ -4 \end{bmatrix} = \begin{bmatrix} 4 \\ 5 \\ 1 \end{bmatrix}$ $\qquad x_1 = 4,\ x_2 = 5,\ x_3 = 1$

(C) $K = \begin{bmatrix} 3 \\ -2 \\ 0 \end{bmatrix}$ $\qquad A^{-1}K = \begin{bmatrix} 0 & -1 & -1 \\ -1 & -1 & -1 \\ -1 & -1 & 0 \end{bmatrix} \begin{bmatrix} 3 \\ -2 \\ 0 \end{bmatrix} = \begin{bmatrix} 2 \\ -1 \\ -1 \end{bmatrix}$ $\qquad x_1 = 2,\ x_2 = -1,\ x_3 = -1$

33. $\begin{bmatrix} 1 & 2 & 5 \\ 3 & 5 & 9 \\ 1 & 1 & -2 \end{bmatrix} \begin{bmatrix} x_1 \\ x_2 \\ x_3 \end{bmatrix} = \begin{bmatrix} k_1 \\ k_2 \\ k_3 \end{bmatrix}$

$AX = K$ has solution $X = A^{-1}K$.

We find $A^{-1}K$ for each given K. From problem 37 in Section 6-3,

$A^{-1} = \begin{bmatrix} -19 & 9 & -7 \\ 15 & -7 & 6 \\ -2 & 1 & -1 \end{bmatrix}$

(A) $K = \begin{bmatrix} 0 \\ 1 \\ 4 \end{bmatrix}$ $\qquad A^{-1}K = \begin{bmatrix} -19 & 9 & -7 \\ 15 & -7 & 6 \\ -2 & 1 & -1 \end{bmatrix} \begin{bmatrix} 0 \\ 1 \\ 4 \end{bmatrix} = \begin{bmatrix} -19 \\ 17 \\ -3 \end{bmatrix}$ $\qquad x_1 = -19,\ x_2 = 17,\ x_3 = -3$

(B) $K = \begin{bmatrix} 5 \\ -1 \\ 0 \end{bmatrix}$ $\qquad A^{-1}K = \begin{bmatrix} -19 & 9 & -7 \\ 15 & -7 & 6 \\ -2 & 1 & -1 \end{bmatrix} \begin{bmatrix} 5 \\ -1 \\ 0 \end{bmatrix} = \begin{bmatrix} -104 \\ 82 \\ -11 \end{bmatrix}$ $\qquad x_1 = -104,\ x_2 = 82,\ x_3 = -11$

(C) $K = \begin{bmatrix} -6 \\ 0 \\ 2 \end{bmatrix}$ $\qquad A^{-1}K = \begin{bmatrix} -19 & 9 & -7 \\ 15 & -7 & 6 \\ -2 & 1 & -1 \end{bmatrix} \begin{bmatrix} -6 \\ 0 \\ 2 \end{bmatrix} = \begin{bmatrix} 100 \\ -78 \\ 10 \end{bmatrix}$ $\qquad x_1 = 100,\ x_2 = -78,\ x_3 = 10$

35. The system cannot be solved by matrix inverse methods because the matrix of the system, $\begin{bmatrix} -2 & 4 \\ 6 & -12 \end{bmatrix}$, is singular. (The second row is -3 times the first, so row-reduction will result in a row of zeros.) It can be solved by Gauss-Jordan elimination, as follows.

$\begin{bmatrix} -2 & 4 & | & -5 \\ 6 & -12 & | & 15 \end{bmatrix} \quad 3R_1 + R_2 \to R_2$

$\sim \begin{bmatrix} -2 & 4 & | & -5 \\ 0 & 0 & | & 0 \end{bmatrix} \quad (-0.5)R_1 \to R_1$

$\sim \begin{bmatrix} 1 & -2 & | & 2.5 \\ 0 & 0 & | & 0 \end{bmatrix}$

53. If the graph of $f(x) = ax^2 + bx + c$ passes through a point, the coordinates of the point must satisfy the equation of the graph. Then

$k_1 = a(1)^2 + b(1) + c$

$k_2 = a(2)^2 + b(2) + c$

$k_3 = a(3)^2 + b(3) + c$

After simplification, we obtain:

$a + b + c = k_1$

$4a + 2b + c = k_2$

$9a + 3b + c = k_3$

We solve this system, for arbitrary k_1, k_2, k_3, by writing it as a matrix equation.

$$\begin{matrix} A & X & B \end{matrix}$$

$$\begin{bmatrix} 1 & 1 & 1 \\ 4 & 2 & 1 \\ 9 & 3 & 1 \end{bmatrix} \begin{bmatrix} a \\ b \\ c \end{bmatrix} = \begin{bmatrix} k_1 \\ k_2 \\ k_3 \end{bmatrix}$$

If A^{-1} exists, then $X = A^{-1}B$. To find A^{-1} we perform row operations on

$$\left[\begin{array}{ccc|ccc} 1 & 1 & 1 & 1 & 0 & 0 \\ 4 & 2 & 1 & 0 & 1 & 0 \\ 9 & 3 & 1 & 0 & 0 & 1 \end{array}\right] \begin{array}{l} \\ (-4)R_1 + R_2 \to R_2 \\ (-9)R_1 + R_3 \to R_3 \end{array}$$

$$\sim \left[\begin{array}{ccc|ccc} 1 & 1 & 1 & 1 & 0 & 0 \\ 0 & -2 & -3 & -4 & 1 & 0 \\ 0 & -6 & -8 & -9 & 0 & 1 \end{array}\right] -\tfrac{1}{2}R_2 \to R_2$$

$$\sim \left[\begin{array}{ccc|ccc} 1 & 1 & 1 & 1 & 0 & 0 \\ 0 & 1 & \frac{3}{2} & 2 & -\frac{1}{2} & 0 \\ 0 & -6 & -8 & -9 & 0 & 1 \end{array}\right] \begin{array}{l} (-1)R_2 + R_1 \to R_1 \\ \\ 6R_2 + R_3 \to R_3 \end{array}$$

$$\sim \left[\begin{array}{ccc|ccc} 1 & 0 & -\frac{1}{2} & -1 & \frac{1}{2} & 0 \\ 0 & 1 & \frac{3}{2} & 2 & -\frac{1}{2} & 0 \\ 0 & 0 & 1 & 3 & -3 & 1 \end{array}\right] \begin{array}{l} \frac{1}{2}R_3 + R_1 \to R_1 \\ (-\frac{3}{2})R_3 + R_2 \to R_2 \\ \end{array}$$

$$\sim \left[\begin{array}{ccc|ccc} 1 & 0 & 0 & \frac{1}{2} & -1 & \frac{1}{2} \\ 0 & 1 & 0 & -\frac{5}{2} & 4 & -\frac{3}{2} \\ 0 & 0 & 1 & 3 & -3 & 1 \end{array}\right]$$

$$A^{-1} = \frac{1}{2}\begin{bmatrix} 1 & -2 & 1 \\ -5 & 8 & -3 \\ 6 & -6 & 2 \end{bmatrix}$$

Check: $A^{-1}A = \frac{1}{2}\begin{bmatrix} 1 & -2 & 1 \\ -5 & 8 & -3 \\ 6 & -6 & 2 \end{bmatrix}\begin{bmatrix} 1 & 1 & 1 \\ 4 & 2 & 1 \\ 9 & 3 & 1 \end{bmatrix} = \begin{bmatrix} 1 & 0 & 0 \\ 0 & 1 & 0 \\ 0 & 0 & 1 \end{bmatrix}$

We can now solve the system as

$$\begin{matrix} X & A^{-1} & B \end{matrix}$$

$$\begin{bmatrix} a \\ b \\ c \end{bmatrix} = \frac{1}{2}\begin{bmatrix} 1 & -2 & 1 \\ -5 & 8 & -3 \\ 6 & -6 & 2 \end{bmatrix}\begin{bmatrix} k_1 \\ k_2 \\ k_3 \end{bmatrix}$$

(A) $\begin{bmatrix} a \\ b \\ c \end{bmatrix} = \frac{1}{2}\begin{bmatrix} 1 & -2 & 1 \\ -5 & 8 & -3 \\ 6 & -6 & 2 \end{bmatrix}\begin{bmatrix} -2 \\ 1 \\ 6 \end{bmatrix} = \begin{bmatrix} 1 \\ 0 \\ -3 \end{bmatrix}$ $a = 1$, $b = 0$, $c = -3$

(B) $\begin{bmatrix} a \\ b \\ c \end{bmatrix} = \frac{1}{2} \begin{bmatrix} 1 & -2 & 1 \\ -5 & 8 & -3 \\ 6 & -6 & 2 \end{bmatrix} \begin{bmatrix} 4 \\ 3 \\ -2 \end{bmatrix} = \begin{bmatrix} -2 \\ 5 \\ 1 \end{bmatrix}$ $a = -2, \; b = 5, \; c = 1$

(C) $\begin{bmatrix} a \\ b \\ c \end{bmatrix} = \frac{1}{2} \begin{bmatrix} 1 & -2 & 1 \\ -5 & 8 & -3 \\ 6 & -6 & 2 \end{bmatrix} \begin{bmatrix} 8 \\ -5 \\ 4 \end{bmatrix} = \begin{bmatrix} 11 \\ -46 \\ 43 \end{bmatrix}$ $a = 11, \; b = -46, \; c = 43$

Using Quadratic Regression

Enter 1, 2, 3 as L_1.

(A) Enter -2, 1, 6 as L_2 and use the quadratic regression command from the STAT CALC menu.

(B) Enter 4, 3, -2 as L_2 and repeat.

(C) Enter 8, -5, 4 as L_2 and repeat.

55. The system to be solved, for an arbitrary diet, is derived as follows:
Let x_1 = amount of mix A
$\quad\;\; x_2$ = amount of mix B
Then $0.20x_1 + 0.10x_2 = k_1$ (k_1 = amount of protein)
$\quad\;\;\; 0.02x_1 + 0.06x_2 = k_2$ (k_2 = amount of fat)
We solve the system by writing it as a matrix equation.

$$\overset{A}{\begin{bmatrix} 0.20 & 0.10 \\ 0.02 & 0.06 \end{bmatrix}} \overset{X}{\begin{bmatrix} x_1 \\ x_2 \end{bmatrix}} = \overset{B}{\begin{bmatrix} k_1 \\ k_2 \end{bmatrix}}$$

If A^{-1} exists, then $X = A^{-1}B$. To find A^{-1}, we perform row operations on

$$\begin{bmatrix} 0.20 & 0.10 & | & 1 & 0 \\ 0.02 & 0.06 & | & 0 & 1 \end{bmatrix} \begin{matrix} 5R_1 \to R_1 \\ 50R_2 \to R_2 \end{matrix}$$

$$\sim \begin{bmatrix} 1 & 0.5 & | & 5 & 0 \\ 1 & 3 & | & 0 & 50 \end{bmatrix} (-1)R_1 + R_2 \to R_2$$

$$\sim \begin{bmatrix} 1 & 0.5 & | & 5 & 0 \\ 0 & 2.5 & | & -5 & 50 \end{bmatrix} 0.4R_2 \to R_2$$

$$\sim \begin{bmatrix} 1 & 0.5 & | & 5 & 0 \\ 0 & 1 & | & -2 & 20 \end{bmatrix} (-0.5)R_2 + R_1 \to R_1$$

$$\sim \begin{bmatrix} 1 & 0 & | & 6 & -10 \\ 0 & 1 & | & -2 & 20 \end{bmatrix}$$

$$A^{-1} = \begin{bmatrix} 6 & -10 \\ -2 & 20 \end{bmatrix}$$

Check: $A^{-1}A = \begin{bmatrix} 6 & -10 \\ -2 & 20 \end{bmatrix}\begin{bmatrix} 0.20 & 0.10 \\ 0.02 & 0.06 \end{bmatrix} = \begin{bmatrix} 1 & 0 \\ 0 & 1 \end{bmatrix}$

We can now solve the system as

$$\begin{matrix} X & A^{-1} & B \end{matrix}$$

$$\begin{bmatrix} x_1 \\ x_2 \end{bmatrix} = \begin{bmatrix} 6 & -10 \\ -2 & 20 \end{bmatrix}\begin{bmatrix} k_1 \\ k_2 \end{bmatrix}$$

For Diet 1, $k_1 = 20$ and $k_2 = 6$

$$\begin{bmatrix} x_1 \\ x_2 \end{bmatrix} = \begin{bmatrix} 6 & -10 \\ -2 & 20 \end{bmatrix}\begin{bmatrix} 20 \\ 6 \end{bmatrix} = \begin{bmatrix} 60 \\ 80 \end{bmatrix}$$ Diet 1: 60 ounces Mix A and 80 ounces Mix B

For Diet 2, $k_1 = 10$ and $k_2 = 4$

$$\begin{bmatrix} x_1 \\ x_2 \end{bmatrix} = \begin{bmatrix} 6 & -10 \\ -2 & 20 \end{bmatrix}\begin{bmatrix} 10 \\ 4 \end{bmatrix} = \begin{bmatrix} 20 \\ 60 \end{bmatrix}$$ Diet 2: 20 ounces Mix A and 60 ounces Mix B

For Diet 3, $k_1 = 10$ and $k_2 = 6$

$$\begin{bmatrix} x_1 \\ x_2 \end{bmatrix} = \begin{bmatrix} 6 & -10 \\ -2 & 20 \end{bmatrix}\begin{bmatrix} 10 \\ 6 \end{bmatrix} = \begin{bmatrix} 0 \\ 100 \end{bmatrix}$$ Diet 3: 0 ounces Mix A and 100 ounces Mix B

Section 6-5

1. Answers will vary.   3. Answers will vary.   5. Answers will vary.

7. $\begin{vmatrix} 5 & 4 \\ 2 & 3 \end{vmatrix} = 5 \cdot 3 - 2 \cdot 4 = 7$ 9. $\begin{vmatrix} 3 & -7 \\ -5 & 6 \end{vmatrix} = 3 \cdot 6 - (-5)(-7) = -17$

11. $\begin{vmatrix} 4.3 & -1.2 \\ -5.1 & 3.7 \end{vmatrix} = (4.3)(3.7) - (-5.1)(-1.2) = 9.79$

13. $\begin{vmatrix} 5 & 1 & 3 \\ 3 & 4 & 6 \\ 0 & -2 & 8 \end{vmatrix} = \begin{vmatrix} 4 & 6 \\ -2 & 8 \end{vmatrix}$ 15. $\begin{vmatrix} 5 & -1 & -3 \\ 3 & 4 & 6 \\ 0 & -2 & 8 \end{vmatrix} = \begin{vmatrix} 5 & -1 \\ 0 & -2 \end{vmatrix}$

17. $(-1)^{1+1}\begin{vmatrix} 4 & 6 \\ -2 & 8 \end{vmatrix} = (-1)^2[4 \cdot 8 - (-2)6] = 44$

19. $(-1)^{2+3}\begin{vmatrix} 5 & -1 \\ 0 & -2 \end{vmatrix} = (-1)^5[5(-2) - 0(-1)] = 10$

21. We expand by row 1

$$\begin{vmatrix} 1 & 0 & 0 \\ -2 & 4 & 3 \\ 5 & -2 & 1 \end{vmatrix} = a_{11}(\text{cofactor of } a_{11}) + a_{12}(\text{cofactor of } a_{12}) + a_{13}(\text{cofactor of } a_{13})$$

$$= 1(-1)^{1+1}\begin{vmatrix} 4 & 3 \\ -2 & 1 \end{vmatrix} + 0(\diagup) + 0(\diagdown)$$

It is unnecessary to evaluate these since they are multiplied by 0.

$$= (-1)^2[4 \cdot 1 - (-2)3]$$
$$= 10$$

23. We expand by column 1

$$\begin{vmatrix} 0 & 1 & 5 \\ 3 & -7 & 6 \\ 0 & -2 & -3 \end{vmatrix} = a_{11}(\text{cofactor of } a_{11}) + a_{21}(\text{cofactor of } a_{21}) + a_{31}(\text{cofactor of } a_{31})$$

$$= 0() + 3(-1)^{2+1}\begin{vmatrix} 1 & 5 \\ -2 & -3 \end{vmatrix} + 0()$$

It is unnecessary to evaluate these since they are multiplied by 0.

$$= 3(-1)^3[1(-3) - (-2)5]$$
$$= -21$$

Common Error: Neglecting the sign of the cofactor. The cofactor is often called the "signed" minor.

---

25. We expand by column 2

$$\begin{vmatrix} -1 & 2 & -3 \\ -2 & 0 & -6 \\ 4 & -3 & 2 \end{vmatrix} = a_{12}(\text{cofactor of } a_{12}) + a_{22}(\text{cofactor of } a_{22}) + a_{32}(\text{cofactor of } a_{32})$$

$$= 2(-1)^{1+2}\begin{vmatrix} -2 & -6 \\ 4 & 2 \end{vmatrix} + 0() + (-3)(-1)^{3+2}\begin{vmatrix} -1 & -3 \\ -2 & -6 \end{vmatrix}$$

It is unnecessary to evaluate this since it's multiplied by zero.

$$= 2(-1)^3[(-2)2 - 4(-6)] + (-3)(-1)^5[(-1)(-6) - (-2)(-3)]$$
$$= (-2)(20) + 3(0) = -40$$

27. $(-1)^{1+1}\begin{vmatrix} a_{11} & a_{12} & a_{13} & a_{14} \\ a_{21} & a_{22} & a_{23} & a_{24} \\ a_{31} & a_{32} & a_{33} & a_{34} \\ a_{41} & a_{42} & a_{43} & a_{44} \end{vmatrix} = (-1)^{1+1}\begin{vmatrix} a_{22} & a_{23} & a_{24} \\ a_{32} & a_{33} & a_{34} \\ a_{42} & a_{43} & a_{44} \end{vmatrix}$

29. $(-1)^{4+3}\begin{vmatrix} a_{11} & a_{12} & a_{13} & a_{14} \\ a_{21} & a_{22} & a_{23} & a_{24} \\ a_{31} & a_{32} & a_{33} & a_{34} \\ a_{41} & a_{42} & a_{43} & a_{44} \end{vmatrix} = (-1)^{4+3}\begin{vmatrix} a_{11} & a_{12} & a_{14} \\ a_{21} & a_{22} & a_{24} \\ a_{31} & a_{32} & a_{34} \end{vmatrix}$

31. We expand by the second column

$$\begin{vmatrix} 3 & -2 & -8 \\ -2 & 0 & -3 \\ 1 & 0 & -4 \end{vmatrix} = a_{12}(\text{cofactor of } a_{12}) + a_{22}(\text{cofactor of } a_{22}) + a_{32}(\text{cofactor of } a_{32})$$

$$= (-2)(-1)^{1+2}\begin{vmatrix} -2 & -3 \\ 1 & -4 \end{vmatrix} + 0 + 0$$

$$= (-2)(-1)^3[(-2)(-4) - 1(-3)]$$
$$= 2(11)$$
$$= 22$$

33. We expand by the first row

$$\begin{vmatrix} 1 & 4 & 1 \\ 1 & 1 & -2 \\ 2 & 1 & -1 \end{vmatrix} = a_{11}(\text{cofactor of } a_{11}) + a_{12}(\text{cofactor of } a_{12}) + a_{13}(\text{cofactor of } a_{13})$$

$$= 1(-1)^{1+1}\begin{vmatrix} 1 & -2 \\ 1 & -1 \end{vmatrix} + 4(-1)^{1+2}\begin{vmatrix} 1 & -2 \\ 2 & -1 \end{vmatrix} + 1(-1)^{1+3}\begin{vmatrix} 1 & 1 \\ 2 & 1 \end{vmatrix}$$

$$= (-1)^2[1(-1) - 1(-2)] + 4(-1)^3[1(-1) - 2(-2)] + (-1)^4[1\cdot1 - 2\cdot1]$$
$$= 1 + (-12) + (-1)$$
$$= -12$$

35. We expand by the first row

$$\begin{vmatrix} 1 & 4 & 3 \\ 2 & 1 & 6 \\ 3 & -2 & 9 \end{vmatrix} = a_{11}(\text{cofactor of } a_{11}) + a_{12}(\text{cofactor of } a_{12}) + a_{13}(\text{cofactor of } a_{13})$$

$$= 1(-1)^{1+1}\begin{vmatrix} 1 & 6 \\ -2 & 9 \end{vmatrix} + 4(-1)^{1+2}\begin{vmatrix} 2 & 6 \\ 3 & 9 \end{vmatrix} + 3(-1)^{1+3}\begin{vmatrix} 2 & 1 \\ 3 & -2 \end{vmatrix}$$
$$= (-1)^2[1\cdot 9 - (-2)6] + 4(-1)^3[2\cdot 9 - 3\cdot 6] + 3(-1)^4[2(-2) - 1\cdot 3]$$
$$= 21 + 0 - 21$$
$$= 0$$

37. We expand by the second row. Clearly the only non-zero term will be a_{22} (cofactor of a_{22}), which is

$$3(-1)^{2+2}\begin{vmatrix} 2 & 1 & 7 \\ 3 & 2 & 5 \\ 0 & 0 & 2 \end{vmatrix}$$

The order 3 determinant is expanded by the third row. Again there is only one non-zero term, a_{33} (cofactor of a_{33}). So the original determinant is reduced to

$$3(-1)^{2+2} \ 2(-1)^{3+3}\begin{vmatrix} 2 & 1 \\ 3 & 2 \end{vmatrix} = 6(-1)^{10}(2\cdot 2 - 3\cdot 1) = 6$$

39. In every case, we'll be expanding by the first row.

$$\begin{vmatrix} -2 & 0 & 0 & 0 & 0 \\ 9 & -1 & 0 & 0 & 0 \\ 2 & 1 & 3 & 0 & 0 \\ -1 & 4 & 2 & 2 & 0 \\ 7 & -2 & 3 & 5 & 5 \end{vmatrix} = (-2)(-1)^{1+1}\begin{vmatrix} -1 & 0 & 0 & 0 \\ 1 & 3 & 0 & 0 \\ 4 & 2 & 2 & 0 \\ -2 & 3 & 5 & 5 \end{vmatrix} + 0 \text{ terms}$$

$$= -2\begin{vmatrix} -1 & 0 & 0 & 0 \\ 1 & 3 & 0 & 0 \\ 4 & 2 & 2 & 0 \\ -2 & 3 & 5 & 5 \end{vmatrix} = (-2)\left[(-1)(-1)^{1+1}\begin{vmatrix} 3 & 0 & 0 \\ 2 & 2 & 0 \\ 3 & 5 & 5 \end{vmatrix} + 0 \text{ terms}\right]$$

$$= (-2)(-1)\begin{vmatrix} 3 & 0 & 0 \\ 2 & 2 & 0 \\ 3 & 5 & 5 \end{vmatrix} = (-2)(-1)\left[3(-1)^{1+1}\begin{vmatrix} 2 & 0 \\ 5 & 5 \end{vmatrix} + 0 \text{ terms}\right]$$

$$= (-2)(-1)3\begin{vmatrix} 2 & 0 \\ 5 & 5 \end{vmatrix}$$

$$= (-2)(-1)(3)[2\cdot 5 - 5\cdot 0]$$
$$= (-2)(-1)(3)(2)(5)$$
$$= 60$$

41.

$$\begin{vmatrix} 2 & 6 & -1 & 2 & 6 \\ 5 & 3 & -7 & 5 & 3 \\ -4 & -2 & 1 & -4 & -2 \end{vmatrix}$$

$$2\cdot 3\cdot 1 + 6(-7)(-4) + (-1)(5)(-2) - (-4)(3)(-1) - (-2)(-7)2 - 6\cdot 5\cdot 1$$
$$= 6 + 168 + 10 - 12 - 28 - 30 = 114$$

43. False. $\begin{vmatrix} 10 & 10 \\ 0 & 0 \end{vmatrix}$ is a counterexample.

45. True. Expanding $\begin{vmatrix} a_{11} & a_{12} & a_{13} & a_{14} \\ 0 & a_{22} & a_{23} & a_{24} \\ 0 & 0 & a_{33} & a_{34} \\ 0 & 0 & 0 & a_{44} \end{vmatrix}$ by the first column, we obtain successively

$$a_{11} \begin{vmatrix} a_{22} & a_{23} & a_{24} \\ 0 & a_{33} & a_{34} \\ 0 & 0 & a_{44} \end{vmatrix} = a_{11}a_{22} \begin{vmatrix} a_{33} & a_{34} \\ 0 & a_{44} \end{vmatrix} = a_{11}a_{22}a_{33}a_{44}.$$

Similarly for the determinant of an $n \times n$ upper triangular matrix, we would obtain $a_{11}a_{22}a_{33} \cdot \ldots \cdot a_{nn}$ as proposed.

47. $\begin{vmatrix} a & b \\ c & d \end{vmatrix} = ad - bc$ $\qquad \begin{vmatrix} c & d \\ a & b \end{vmatrix} = cb - ad = -(ad - bc)$

We can see that $\begin{vmatrix} a & b \\ c & d \end{vmatrix} = -\begin{vmatrix} c & d \\ a & b \end{vmatrix}$; interchanging the rows of this determinant changes its sign.

49. $\begin{vmatrix} a & b \\ c & d \end{vmatrix} = ad - bc$ $\qquad \begin{vmatrix} ka & b \\ kc & d \end{vmatrix} = kad - kcb = k(ad - bc)$

We can see that $\begin{vmatrix} ka & b \\ kc & d \end{vmatrix} = k \begin{vmatrix} a & b \\ c & d \end{vmatrix}$; multiplying a column of this determinant by a number k multiplies the value of the determinant by k.

51. $\begin{vmatrix} a & b \\ c & d \end{vmatrix} = ad - bc$ $\qquad \begin{vmatrix} kc + a & kd + b \\ c & d \end{vmatrix} = (kc + a)d - (kd + b)c$

$$= kcd + ad - kdc - bc$$
$$= ad - bc$$

We can see that $\begin{vmatrix} kc + a & kd + b \\ c & d \end{vmatrix} = \begin{vmatrix} a & b \\ c & d \end{vmatrix}$; adding a multiple of one row to the other row does not change the value of this determinant.

53. Expanding by the first column

$\begin{vmatrix} a_{11} & a_{12} & a_{13} \\ a_{21} & a_{22} & a_{23} \\ a_{31} & a_{32} & a_{33} \end{vmatrix}$

$= a_{11}(-1)^{1+1} \begin{vmatrix} a_{22} & a_{23} \\ a_{32} & a_{33} \end{vmatrix} + a_{21}(-1)^{2+1} \begin{vmatrix} a_{12} & a_{13} \\ a_{32} & a_{33} \end{vmatrix} + a_{31}(-1)^{3+1} \begin{vmatrix} a_{12} & a_{13} \\ a_{22} & a_{23} \end{vmatrix}$

$= a_{11} \begin{vmatrix} a_{22} & a_{23} \\ a_{32} & a_{33} \end{vmatrix} - a_{21} \begin{vmatrix} a_{12} & a_{13} \\ a_{32} & a_{33} \end{vmatrix} + a_{31} \begin{vmatrix} a_{12} & a_{13} \\ a_{22} & a_{23} \end{vmatrix}$

$= a_{11}(a_{22}a_{33} - a_{32}a_{23}) - a_{21}(a_{12}a_{33} - a_{32}a_{13}) + a_{31}(a_{12}a_{23} - a_{22}a_{13})$

$= a_{11}a_{22}a_{33} - a_{11}a_{32}a_{23} - a_{21}a_{12}a_{33} + a_{21}a_{32}a_{13} + a_{31}a_{12}a_{23} - a_{31}a_{22}a_{13}$
$\qquad\quad ① \qquad\qquad ② \qquad\qquad ③ \qquad\qquad ④ \qquad\qquad ⑤ \qquad\qquad ⑥$

Expanding by the third row

$\begin{vmatrix} a_{11} & a_{12} & a_{13} \\ a_{21} & a_{22} & a_{23} \\ a_{31} & a_{32} & a_{33} \end{vmatrix}$

$= a_{31}(-1)^{3+1} \begin{vmatrix} a_{12} & a_{13} \\ a_{22} & a_{23} \end{vmatrix} + a_{32}(-1)^{3+2} \begin{vmatrix} a_{11} & a_{13} \\ a_{21} & a_{23} \end{vmatrix} + a_{33}(-1)^{3+3} \begin{vmatrix} a_{11} & a_{12} \\ a_{21} & a_{22} \end{vmatrix}$

$$= a_{31} \begin{vmatrix} a_{12} & a_{13} \\ a_{22} & a_{23} \end{vmatrix} - a_{32} \begin{vmatrix} a_{11} & a_{13} \\ a_{21} & a_{23} \end{vmatrix} + a_{33} \begin{vmatrix} a_{11} & a_{12} \\ a_{21} & a_{22} \end{vmatrix}$$

$$= a_{31}(a_{12}a_{23} - a_{13}a_{22}) - a_{32}(a_{11}a_{23} - a_{13}a_{21}) + a_{33}(a_{11}a_{22} - a_{12}a_{21})$$

$$= a_{31}a_{12}a_{23} - a_{31}a_{13}a_{22} - a_{32}a_{11}a_{23} + a_{32}a_{13}a_{21} + a_{33}a_{11}a_{22} - a_{33}a_{12}a_{21}$$

⑤   ⑥   ②   ④   ①   ③

Comparing the two expressions, with the aid of the numbers under the terms, shows that the expressions are the same.

55. $A = \begin{bmatrix} 2 & 3 \\ 1 & -2 \end{bmatrix}$ $B = \begin{bmatrix} -1 & 3 \\ 2 & 1 \end{bmatrix}$

We calculate $AB = \begin{bmatrix} 2 & 3 \\ 1 & -2 \end{bmatrix}\begin{bmatrix} -1 & 3 \\ 2 & 1 \end{bmatrix} = \begin{bmatrix} 2(-1) + 3 \cdot 2 & 2 \cdot 3 + 3 \cdot 1 \\ 1(-1) + (-2)2 & 1 \cdot 3 + (-2) \cdot 1 \end{bmatrix}$

$$= \begin{bmatrix} 4 & 9 \\ -5 & 1 \end{bmatrix}$$

$$\det (AB) = \begin{vmatrix} 4 & 9 \\ -5 & 1 \end{vmatrix} = 4 \cdot 1 - (-5)9 = 49$$

$$\det A = \begin{vmatrix} 2 & 3 \\ 1 & -2 \end{vmatrix} = 2(-2) - 1 \cdot 3 = -7$$

$$\det B = \begin{vmatrix} -1 & 3 \\ 2 & 1 \end{vmatrix} = (-1)1 - 2 \cdot 3 = -7$$

57. The matrix $xI - A$ is calculated as

$$x\begin{bmatrix} 1 & 0 \\ 0 & 1 \end{bmatrix} - \begin{bmatrix} 5 & -4 \\ 2 & -1 \end{bmatrix} = \begin{bmatrix} x & 0 \\ 0 & x \end{bmatrix} - \begin{bmatrix} 5 & -4 \\ 2 & -1 \end{bmatrix} = \begin{bmatrix} x - 5 & 4 \\ -2 & x + 1 \end{bmatrix}$$

The characteristic polynomial is the determinant of this matrix:

$$\begin{bmatrix} x - 5 & 4 \\ -2 & x + 1 \end{bmatrix} = (x - 5)(x + 1) - (4)(-2) = x^2 - 4x - 5 + 8 = x^2 - 4x + 3$$

The zeros of this polynomial are the solutions of $x^2 - 4x + 3 = 0$

$$x^2 - 4x + 3 = 0$$
$$(x - 1)(x - 3) = 0$$
$$x = 1, 3$$

The eigenvalues of this matrix are 1 and 3.

59. The matrix $xI - A$ is calculated as

$$x\begin{bmatrix} 1 & 0 & 0 \\ 0 & 1 & 0 \\ 0 & 0 & 1 \end{bmatrix} - \begin{bmatrix} 4 & -4 & 0 \\ 2 & -2 & 0 \\ 4 & -8 & -4 \end{bmatrix} = \begin{bmatrix} x & 0 & 0 \\ 0 & x & 0 \\ 0 & 0 & x \end{bmatrix} - \begin{bmatrix} 4 & -4 & 0 \\ 2 & -2 & 0 \\ 4 & -8 & -4 \end{bmatrix} = \begin{bmatrix} x - 4 & 4 & 0 \\ -2 & x + 2 & 0 \\ -4 & 8 & x + 4 \end{bmatrix}$$

The characteristic polynomial is the determinant of this matrix.

$$\begin{bmatrix} x - 4 & 4 & 0 \\ -2 & x + 2 & 0 \\ -4 & 8 & x + 4 \end{bmatrix} = (x + 4)(-1)^{3+3}\begin{bmatrix} x - 4 & 4 \\ -2 & x + 2 \end{bmatrix} \text{ expanding by the third column}$$

$$= (x + 4)(1)[(x - 4)(x + 2) - 4(-2)]$$
$$= (x + 4)(x^2 - 2x - 8 + 8)$$
$$= (x + 4)(x^2 - 2x)$$
$$= x^3 + 2x^2 - 8x$$

The zeros of this polynomial are the solutions of

$$x^3 + 2x^2 - 8x = 0$$
$$x(x^2 + 2x - 8) = 0$$
$$x(x + 4)(x - 2) = 0$$
$$x = 0, -4, 2$$

The eigenvalues of this matrix are 0, -4, and 2.

Section 6-6

1. Answers will vary.

3. Theorem 1   **5.** Theorem 1   **7.** Theorem 2   **9.** Theorem 3   **11.** Theorem 5

13. $3C_1 + C_2 \rightarrow C_2$ has been used, $x = 3(-1) + 3 = 0$

15. $3C_1 + C_3 \rightarrow C_3$ has been used, $x = 3(1) + 2 = 5$

17. Interchanging two rows of a determinant changes the sign of the determinant (Theorem 3).

$$\begin{vmatrix} c & d \\ a & b \end{vmatrix} = -10$$

19. Adding a multiple (in this case a multiple by 1) of a row to another row does not change the value of the determinant (Theorem 5).

$$\begin{vmatrix} a+c & b+d \\ c & d \end{vmatrix} = 10$$

21. Adding a multiple of a column to another column does not change the value of the determinant (Theorem 5).

$$\begin{vmatrix} a & a-b \\ c & c-d \end{vmatrix} = \begin{vmatrix} a & -b \\ c & -d \end{vmatrix}$$

If every element of a column of a determinant is multiplied by -1, the value of the determinant is -1 times the original.

$$\begin{vmatrix} a & a-b \\ c & c-d \end{vmatrix} = \begin{vmatrix} a & -b \\ c & -d \end{vmatrix} = (-1)\begin{vmatrix} a & b \\ c & d \end{vmatrix} = -1(10) = -10$$

23. $\begin{vmatrix} -1 & 0 & 3 \\ 2 & 5 & 4 \\ 1 & 5 & 2 \end{vmatrix} = \begin{vmatrix} -1 & 0 & 3 \\ 1 & 0 & 2 \\ 1 & 5 & 2 \end{vmatrix}$ $-1R_3 + R_2 \rightarrow R_2$

$$= 5(-1)^{3+2} \begin{vmatrix} -1 & 3 \\ 1 & 2 \end{vmatrix} = -5[(-1)2 - 1(3)] = 25$$

25. $\begin{vmatrix} 3 & 5 & 0 \\ 1 & 1 & -2 \\ 2 & 1 & -1 \end{vmatrix} = \begin{vmatrix} 3 & 5 & 0 \\ -3 & -1 & 0 \\ 2 & 1 & -1 \end{vmatrix}$ $-2R_3 + R_2 \rightarrow R_2$

$$= (-1)(-1)^{3+3} \begin{vmatrix} 3 & 5 \\ -3 & -1 \end{vmatrix} = (-1)[3(-1) - (-3)5] = -12$$

27. Theorem 1   **29.** Theorem 2   **31.** Theorem 5

33. $2C_3 + C_1 \rightarrow C_1$ has been used, $x = 2 \cdot 1 + 3 = 5$
$C_3 + C_2 \rightarrow C_2$ has been used, $y = (-2) + 2 = 0$

35. $(-4)R_2 + R_1 \rightarrow R_1$ has been used, $x = (-4)3 + 9 = -3$
$2R_2 + R_3 \rightarrow R_3$ has been used, $y = 2 \cdot 3 + 4 = 10$

37. We will generate zeros in the first column by row operations.

$$\begin{vmatrix} 1 & 5 & 3 \\ 4 & 2 & 1 \\ 3 & 1 & 2 \end{vmatrix} = \begin{vmatrix} 1 & 5 & 3 \\ 0 & -18 & -11 \\ 0 & -14 & -7 \end{vmatrix} \begin{matrix} (-4)R_1 + R_2 \rightarrow R_2 \\ (-3)R_1 + R_3 \rightarrow R_3 \end{matrix}$$

$$= 1(-1)^{1+1}\begin{vmatrix} -18 & -11 \\ -14 & -7 \end{vmatrix} + 0 + 0$$

$$= (-1)^2[(-18)(-7) - (-14)(-11)] = -28$$

39. We will generate zeros in the second row by column operations.

$$\begin{vmatrix} 5 & 2 & -3 \\ -2 & 4 & 4 \\ 1 & -1 & 3 \end{vmatrix} = \begin{vmatrix} 5 & 12 & 7 \\ -2 & 0 & 0 \\ 1 & 1 & 5 \end{vmatrix} \begin{matrix} 2C_1 + C_2 \rightarrow C_2 \\ 2C_1 + C_3 \rightarrow C_3 \end{matrix}$$

$$= (-2)(-1)^{2+1}\begin{vmatrix} 12 & 7 \\ 1 & 5 \end{vmatrix} + 0 + 0$$

$$= (-2)(-1)^3[12 \cdot 5 - 1 \cdot 7] = 106$$

41. The column operation $(-3)C_3 + C_1 \rightarrow C_1$ transforms this determinant into

$$\begin{vmatrix} 0 & -4 & 1 \\ 0 & -1 & 2 \\ 0 & 2 & 3 \end{vmatrix}$$

By Theorem 2, the value of this determinant is 0.

43. We start by generating one more zero in the first row.

$$\begin{vmatrix} 0 & 1 & 0 & 1 \\ 1 & -2 & 4 & 3 \\ 2 & 1 & 5 & 4 \\ 1 & 2 & 1 & 2 \end{vmatrix} = \begin{vmatrix} 0 & 0 & 0 & 1 \\ 1 & -5 & 4 & 3 \\ 2 & -3 & 5 & 4 \\ 1 & 0 & 1 & 2 \end{vmatrix} (-1)C_4 + C_2 \rightarrow C_2$$

$$= 1(-1)^{1+4}\begin{vmatrix} 1 & -5 & 4 \\ 2 & -3 & 5 \\ 1 & 0 & 1 \end{vmatrix} + 0 + 0 + 0$$

$$= (-1)\begin{vmatrix} 1 & -5 & 4 \\ 2 & -3 & 5 \\ 1 & 0 & 1 \end{vmatrix}$$

$$= \begin{vmatrix} 1 & 5 & 4 \\ 2 & 3 & 5 \\ 1 & 0 & 1 \end{vmatrix} \text{ by Theorem 1}$$

We now generate one more zero in the third row.

$$\begin{vmatrix} 1 & 5 & 4 \\ 2 & 3 & 5 \\ 1 & 0 & 1 \end{vmatrix} = \begin{vmatrix} 1 & 5 & 3 \\ 2 & 3 & 3 \\ 1 & 0 & 0 \end{vmatrix} (-1)C_1 + C_3 \rightarrow C_3$$

$$= 1(-1)^{3+1}\begin{vmatrix} 5 & 3 \\ 3 & 3 \end{vmatrix} = (-1)^4[5 \cdot 3 - 3 \cdot 3] = 6$$

45. We start by generating zeros in the third row.

$$\begin{vmatrix} 3 & 2 & 3 & 1 \\ 3 & -2 & 8 & 5 \\ 2 & 1 & 3 & 1 \\ 4 & 5 & 4 & -3 \end{vmatrix} = \begin{vmatrix} 1 & 1 & 0 & 1 \\ -7 & -7 & -7 & 5 \\ 0 & 0 & 0 & 1 \\ 10 & 8 & 13 & -3 \end{vmatrix} \begin{matrix} (-2)C_4 + C_1 \rightarrow C_1 \\ (-1)C_4 + C_2 \rightarrow C_2 \\ (-3)C_4 + C_3 \rightarrow C_3 \end{matrix}$$

$$= 1(-1)^{3+4}\begin{vmatrix} 1 & 1 & 0 \\ -7 & -7 & -7 \\ 10 & 8 & 13 \end{vmatrix} + 0 + 0 + 0$$

$$= (-1) \begin{vmatrix} 1 & 1 & 0 \\ -7 & -7 & -7 \\ 10 & 8 & 13 \end{vmatrix}$$

$$= \begin{vmatrix} 1 & 1 & 0 \\ 7 & 7 & 7 \\ 10 & 8 & 13 \end{vmatrix} \text{ by Theorem 1}$$

We now generate one more zero in the first row.

$$\begin{vmatrix} 1 & 1 & 0 \\ 7 & 7 & 7 \\ 10 & 8 & 13 \end{vmatrix} = \begin{vmatrix} 1 & 0 & 0 \\ 7 & 0 & 7 \\ 10 & -2 & 13 \end{vmatrix} \quad (-1)C_1 + C_2 \rightarrow C_2$$

$$= (-2)(-1)^{3+2} \begin{vmatrix} 1 & 0 \\ 7 & 7 \end{vmatrix} + 0 + 0$$

$$= (-2)(-1)^5 [1 \cdot 7 - 0 \cdot 7] = 14$$

47. Expand, for example, by the first column.

$$\begin{vmatrix} a & b & a \\ d & e & d \\ g & h & g \end{vmatrix} = a(-1)^{1+1} \begin{vmatrix} e & d \\ h & g \end{vmatrix} + d(-1)^{2+1} \begin{vmatrix} b & a \\ h & g \end{vmatrix} + g(-1)^{3+1} \begin{vmatrix} b & a \\ e & d \end{vmatrix}$$

$$= a \begin{vmatrix} e & d \\ h & g \end{vmatrix} - d \begin{vmatrix} b & a \\ h & g \end{vmatrix} + g \begin{vmatrix} b & a \\ e & d \end{vmatrix}$$

$$= a(eg - hd) - d(bg - ha) + g(bd - ea)$$

$$= aeg - adh - bdg + adh + bdg - aeg$$

$$= 0$$

49. We expand the left side by the first column, and the right side by the second column.

$$\begin{vmatrix} a_1 & b_1 & c_1 \\ a_2 & b_2 & c_2 \\ a_3 & b_3 & c_3 \end{vmatrix} = a_1(-1)^{1+1} \begin{vmatrix} b_2 & c_2 \\ b_3 & c_3 \end{vmatrix} + a_2(-1)^{2+1} \begin{vmatrix} b_1 & c_1 \\ b_3 & c_3 \end{vmatrix} + a_3(-1)^{3+1} \begin{vmatrix} b_1 & c_1 \\ b_2 & c_2 \end{vmatrix}$$

$$= a_1 \begin{vmatrix} b_2 & c_2 \\ b_3 & c_3 \end{vmatrix} - a_2 \begin{vmatrix} b_1 & c_1 \\ b_3 & c_3 \end{vmatrix} + a_3 \begin{vmatrix} b_1 & c_1 \\ b_2 & c_2 \end{vmatrix}$$

$$-\begin{vmatrix} b_1 & a_1 & c_1 \\ b_2 & a_2 & c_2 \\ b_3 & a_3 & c_3 \end{vmatrix} = -\left[a_1(-1)^{1+2} \begin{vmatrix} b_2 & c_2 \\ b_3 & c_3 \end{vmatrix} + a_2(-1)^{2+2} \begin{vmatrix} b_1 & c_1 \\ b_3 & c_3 \end{vmatrix} + a_3(-1)^{3+2} \begin{vmatrix} b_1 & c_1 \\ b_2 & c_3 \end{vmatrix} \right]$$

$$= -\left[-a_1 \begin{vmatrix} b_2 & c_2 \\ b_3 & c_3 \end{vmatrix} + a_2 \begin{vmatrix} b_1 & c_1 \\ b_3 & c_3 \end{vmatrix} - a_3 \begin{vmatrix} b_1 & c_1 \\ b_2 & c_2 \end{vmatrix} \right]$$

$$= a_1 \begin{vmatrix} b_2 & c_2 \\ b_3 & c_3 \end{vmatrix} - a_2 \begin{vmatrix} b_1 & c_1 \\ b_3 & c_3 \end{vmatrix} + a_3 \begin{vmatrix} b_1 & c_1 \\ b_2 & c_2 \end{vmatrix}$$

The two original expressions are equal.

51. The statements: (2, 5) satisfies the equation $\begin{vmatrix} x & y & 1 \\ 2 & 5 & 1 \\ -3 & 4 & 1 \end{vmatrix} = 0$

and (-3, 4) satisfies the equation $\begin{vmatrix} x & y & 1 \\ 2 & 5 & 1 \\ -3 & 4 & 1 \end{vmatrix} = 0$

are equivalent to the statements $\begin{vmatrix} 2 & 5 & 1 \\ 2 & 5 & 1 \\ -3 & 4 & 1 \end{vmatrix} = 0$ and $\begin{vmatrix} -3 & 4 & 1 \\ 2 & 5 & 1 \\ -3 & 4 & 1 \end{vmatrix} = 0.$

The latter statements are true by Theorem 4.

53. The statement $\begin{vmatrix} x & y & 1 \\ x_1 & y_1 & 1 \\ x_2 & y_2 & 1 \end{vmatrix} = 0$ is the equation of a line because, expanding by the

first row, we have $x(-1)^{1+1}\begin{vmatrix} y_1 & 1 \\ y_2 & 1 \end{vmatrix} + y(-1)^{1+2}\begin{vmatrix} x_1 & 1 \\ x_2 & 1 \end{vmatrix} + 1(-1)^{1+3}\begin{vmatrix} x_1 & y_1 \\ x_2 & y_2 \end{vmatrix} = 0$

This is in the standard form for the equation of a line
$Ax + By + C = 0$

To show that the line passes through (x_1, y_1) and (x_2, y_2), we note that (x_1, y_1) and (x_2, y_2) satisfy the equation because

$\begin{vmatrix} x_1 & y_1 & 1 \\ x_2 & y_2 & 1 \\ x_3 & y_3 & 1 \end{vmatrix} = 0$ and $\begin{vmatrix} x_2 & y_2 & 1 \\ x_1 & y_1 & 1 \\ x_2 & y_2 & 1 \end{vmatrix} = 0$ are true by Theorem 4.

55. Using the result stated in problem 54, we have
$\begin{vmatrix} x_1 & y_1 & 1 \\ x_2 & y_2 & 1 \\ x_3 & y_3 & 1 \end{vmatrix} = 2 \times$ (area of triangle formed by the three points).

If the determinant is 0, then the area of the triangle formed by the three points is zero. The only way this can happen is if the three points are on the same line; that is, the points are collinear.

Section 6-7

1. Answers will vary.

3. No. If the coefficient matrix is not square, it does not have a determinant, and Cramer's rule cannot be used.

5. $D = \begin{vmatrix} 1 & 2 \\ 1 & 3 \end{vmatrix} = 1;$ $x = \dfrac{\begin{vmatrix} 1 & 2 \\ -1 & 3 \end{vmatrix}}{D} = \dfrac{5}{1} = 5;$ $y = \dfrac{\begin{vmatrix} 1 & 1 \\ 1 & -1 \end{vmatrix}}{D} = \dfrac{-2}{1} = -2$

7. $D = \begin{vmatrix} 2 & 1 \\ 5 & 3 \end{vmatrix} = 1;$ $x = \dfrac{\begin{vmatrix} 1 & 1 \\ 2 & 3 \end{vmatrix}}{D} = \dfrac{1}{1} = 1;$ $y = \dfrac{\begin{vmatrix} 2 & 1 \\ 5 & 2 \end{vmatrix}}{D} = \dfrac{-1}{1} = -1$

9. $D = \begin{vmatrix} 2 & -1 \\ -1 & 3 \end{vmatrix} = 5;$ $x = \dfrac{\begin{vmatrix} -3 & -1 \\ 3 & 3 \end{vmatrix}}{D} = \dfrac{-6}{5} = -\dfrac{6}{5};$ $y = \dfrac{\begin{vmatrix} 2 & -3 \\ -1 & 3 \end{vmatrix}}{D} = \dfrac{3}{5}$

11. $D = \begin{vmatrix} 4 & -3 \\ 3 & 2 \end{vmatrix} = 17;$ $x = \dfrac{\begin{vmatrix} 4 & -3 \\ -2 & 2 \end{vmatrix}}{D} = \dfrac{2}{17};$ $y = \dfrac{\begin{vmatrix} 4 & 4 \\ 3 & -2 \end{vmatrix}}{D} = \dfrac{-20}{17} = -\dfrac{20}{17}$

13. $D = \begin{vmatrix} 0.9925 & -0.9659 \\ 0.1219 & 0.2588 \end{vmatrix} = 0.37460$

$x = \dfrac{\begin{vmatrix} 0 & -0.9659 \\ 2,500 & 0.2588 \end{vmatrix}}{D} = \dfrac{2,414.75}{0.37460} = 6,400$ to two significant digits

$y = \dfrac{\begin{vmatrix} 0.9925 & 0 \\ 0.1219 & 2,500 \end{vmatrix}}{D} = \dfrac{2,481.25}{0.37460} = 6,600$ to two significant digits

15. $D = \begin{vmatrix} 0.9954 & -0.9942 \\ 0.0958 & 0.1080 \end{vmatrix} = 0.20275$

$x = \dfrac{\begin{vmatrix} 0 & -0.9942 \\ 155 & 0.1080 \end{vmatrix}}{D} = \dfrac{154.10}{0.20275} = 760$ to two significant digits

$y = \dfrac{\begin{vmatrix} 0.9954 & 0 \\ 0.0958 & 155 \end{vmatrix}}{D} = \dfrac{154.29}{0.20275} = 760$ to two significant digits

17. $D = \begin{vmatrix} 1 & 1 & 0 \\ 0 & 2 & 1 \\ -1 & 0 & 1 \end{vmatrix} = 1$; $x = \dfrac{\begin{vmatrix} 0 & 1 & 0 \\ -5 & 2 & 1 \\ -3 & 0 & 1 \end{vmatrix}}{D} = \dfrac{2}{1} = 2$; $y = \dfrac{\begin{vmatrix} 1 & 0 & 0 \\ 0 & -5 & 1 \\ -1 & -3 & 1 \end{vmatrix}}{D} = \dfrac{-2}{1} = -2$

$z = \dfrac{\begin{vmatrix} 1 & 1 & 0 \\ 0 & 2 & -5 \\ -1 & 0 & -3 \end{vmatrix}}{D} = \dfrac{-1}{1} = -1$

19. $D = \begin{vmatrix} 1 & 1 & 0 \\ 0 & 2 & 1 \\ 0 & -1 & 1 \end{vmatrix} = 3$; $x = \dfrac{\begin{vmatrix} 1 & 1 & 0 \\ 0 & 2 & 1 \\ 1 & -1 & 1 \end{vmatrix}}{D} = \dfrac{4}{3}$; $y = \dfrac{\begin{vmatrix} 1 & 1 & 0 \\ 0 & 0 & 1 \\ 0 & 1 & 1 \end{vmatrix}}{D} = \dfrac{-1}{3} = -\dfrac{1}{3}$

$z = \dfrac{\begin{vmatrix} 1 & 1 & 0 \\ 0 & 2 & 0 \\ 0 & -1 & 1 \end{vmatrix}}{D} = \dfrac{2}{3}$

21. $D = \begin{vmatrix} 0 & 3 & 1 \\ 1 & 0 & 2 \\ 1 & -3 & 0 \end{vmatrix} = 3$; $x = \dfrac{\begin{vmatrix} -1 & 3 & 1 \\ 3 & 0 & 2 \\ -2 & -3 & 0 \end{vmatrix}}{D} = \dfrac{-27}{3} = -9$; $y = \dfrac{\begin{vmatrix} 0 & -1 & 1 \\ 1 & 3 & 2 \\ 1 & -2 & 0 \end{vmatrix}}{D} = \dfrac{-7}{3} = -\dfrac{7}{3}$

$z = \dfrac{\begin{vmatrix} 0 & 3 & -1 \\ 1 & 0 & 3 \\ 1 & -3 & -2 \end{vmatrix}}{D} = \dfrac{18}{3} = 6$

23. $D = \begin{vmatrix} 0 & 2 & -1 \\ 1 & -1 & -1 \\ 1 & -1 & 2 \end{vmatrix} = -6$; $x = \dfrac{\begin{vmatrix} -3 & 2 & -1 \\ 2 & -1 & -1 \\ 4 & -1 & 2 \end{vmatrix}}{D} = \dfrac{-9}{-6} = \dfrac{3}{2}$; $y = \dfrac{\begin{vmatrix} 0 & -3 & -1 \\ 1 & 2 & -1 \\ 1 & 4 & 2 \end{vmatrix}}{D} = \dfrac{7}{-6} = -\dfrac{7}{6}$

$z = \dfrac{\begin{vmatrix} 0 & 2 & -3 \\ 1 & -1 & 2 \\ 1 & -1 & 4 \end{vmatrix}}{D} = \dfrac{-4}{-6} = \dfrac{2}{3}$

25. Compute the coefficient determinant:

$D = \begin{vmatrix} a & 3 \\ 2 & 4 \end{vmatrix} = 4a - 3(2) = 4a - 6$

If $D \neq 0$, there is a unique solution:
$$4a - 6 = 0$$
$$4a = 6$$
$$a = \dfrac{6}{4} = \dfrac{3}{2}$$

So if $a \neq \dfrac{3}{2}$ there is one solution. If $a = \dfrac{3}{2}$ we need to use Gauss-Jordan elimination to determine the nature of the solutions after plugging in $\frac{3}{2}$ for a.

$$\begin{bmatrix} \frac{3}{2} & 3 & \bigm| & b \\ 2 & 4 & \bigm| & 5 \end{bmatrix} \frac{2}{3} R_1 \to R_1$$

$$\sim \begin{bmatrix} 1 & 2 & \bigm| & \frac{2b}{3} \\ 2 & 4 & \bigm| & 5 \end{bmatrix} - 2R_1 + R_2 \to R_2$$

$$\sim \begin{bmatrix} 1 & 2 & \bigm| & \frac{2b}{3} \\ 0 & 0 & \bigm| & -\frac{4b}{3} + 5 \end{bmatrix}$$

If the bottom row is all zeros, there are infinitely many solutions. If the bottom row has zero in the third position, there is no solution. We need to know when $\frac{-4b}{3} + 5 = 0$.

$$\frac{-4b}{3} + 5 = 0$$
$$-4b + 15 = 0$$
$$-4b = -15$$
$$b = \frac{15}{4}$$

So there are infinitely many solutions if $a = \frac{3}{2}$ and $b = \frac{15}{4}$, and no solutions if $a = \frac{3}{2}$ and $b \neq \frac{15}{4}$.

27. $x = \dfrac{\begin{vmatrix} -3 & -3 & 1 \\ -11 & 3 & 2 \\ 3 & -1 & -1 \end{vmatrix}}{\begin{vmatrix} 2 & -3 & 1 \\ -4 & 3 & 2 \\ 1 & -1 & -1 \end{vmatrix}} = \dfrac{20}{5} = 4$

29. $y = \dfrac{\begin{vmatrix} 12 & 5 & 11 \\ 15 & -13 & -9 \\ 5 & 0 & 2 \end{vmatrix}}{\begin{vmatrix} 12 & -14 & 11 \\ 15 & 7 & -9 \\ 5 & -3 & 2 \end{vmatrix}} = \dfrac{28}{14} = 2$

31. $z = \dfrac{\begin{vmatrix} 3 & -4 & 18 \\ -9 & 8 & -13 \\ 5 & -7 & 33 \end{vmatrix}}{\begin{vmatrix} 3 & -4 & 5 \\ -9 & 8 & 7 \\ 5 & -7 & 10 \end{vmatrix}} = \dfrac{5}{2}$

33. $D = \begin{vmatrix} 1 & -4 & 9 \\ 4 & -1 & 6 \\ 1 & -1 & 3 \end{vmatrix} = \begin{vmatrix} -2 & -1 & 0 \\ 2 & 1 & 0 \\ 1 & -1 & 3 \end{vmatrix} \begin{array}{l} (-3)R_3 + R_1 \to R_1 \\ (-2)R_3 + R_2 \to R_2 \end{array} = -\begin{vmatrix} 2 & 1 & 0 \\ 2 & 1 & 0 \\ 1 & -1 & 3 \end{vmatrix}$ by Theorem 1

$\qquad\qquad\qquad\qquad\qquad\qquad\qquad\qquad\qquad\qquad\qquad = 0$ by Theorem 4

Since $D = 0$, the system either has no solution or infinitely many. Since $x = 0$, $y = 0$, $z = 0$ is a solution, the second case must hold.

35. We start with
$$a_{11}x + a_{12}y = k_1$$
$$a_{21}x + a_{22}y = k_2$$

To solve for y, we will eliminate x. We multiply the top equation by $-a_{21}$ and the bottom equation by a_{11}, then add.

$$\begin{array}{rcl} -a_{11}a_{21}x - a_{21}a_{12}y & = & -k_1a_{21} \\ a_{11}a_{21}x + a_{11}a_{22}y & = & k_2a_{11} \\ \hline 0x + a_{11}a_{22}y + a_{21}a_{12}y & = & k_2a_{11} - k_1a_{21} \\ (a_{11}a_{22} - a_{21}a_{12})y & = & a_{11}k_2 - a_{21}k_1 \end{array}$$

$$y = \frac{a_{11}k_2 - a_{21}k_1}{a_{11}a_{22} - a_{21}a_{12}} = \frac{\begin{vmatrix} a_{11} & k_1 \\ a_{21} & k_2 \end{vmatrix}}{\begin{vmatrix} a_{11} & a_{12} \\ a_{21} & a_{22} \end{vmatrix}}$$

37. **(A)** $R = xp + yq = (200 - 6p + 4q)p + (300 + 2p - 3q)q$

$= 200p - 6p^2 + 4pq + 300q + 2pq - 3q^2 = 200p + 300q - 6p^2 + 6pq - 3q^2$

(B) Rewrite the demand equations as

$6p - 4q = 200 - x$

$-2p + 3q = 300 - y$

Apply Cramer's rule: $D = \begin{vmatrix} 6 & -4 \\ -2 & 3 \end{vmatrix} = 10$

$$p = \dfrac{\begin{vmatrix} 200 - x & -4 \\ 300 - y & 3 \end{vmatrix}}{D} = \dfrac{1800 - 3x - 4y}{10} = -0.3x - 0.4y + 180$$

$$q = \dfrac{\begin{vmatrix} 6 & 200 - x \\ -2 & 300 - y \end{vmatrix}}{D} = \dfrac{2200 - 2x - 6y}{10} = -0.2x - 0.6y + 220$$

Then

$R = xp + yq = x(-0.3x - 0.4y + 180) + y(-0.2x - 0.6y + 220)$

$= -0.3x^2 - 0.4xy + 180x - 0.2xy - 0.6y^2 + 220y$

$= 180x + 220y - 0.3x^2 - 0.6xy - 0.6y^2$

CHAPTER 6 REVIEW

1. $R_1 \leftrightarrow R_2$ means interchange Rows 1 and 2.

$\begin{bmatrix} 3 & -6 & | & 12 \\ 1 & -4 & | & 5 \end{bmatrix}$ *(6-1)*

2. $\frac{1}{3}R_2 \rightarrow R_2$ means multiply Row 2 by $\frac{1}{3}$.

$\begin{bmatrix} 1 & -4 & | & 5 \\ 1 & -2 & | & 4 \end{bmatrix}$ *(6-1)*

3. $(-3)R_1 + R_2 \rightarrow R_2$ means replace Row 2 by itself plus -3 times Row 1.

$\begin{bmatrix} 1 & -4 & | & 5 \\ 0 & 6 & | & -3 \end{bmatrix}$ *(6-1)*

4. $x_1 = 4$

$x_2 = -7$

The solution is $(4, -7)$ *(6-1)*

5. $x_1 - x_2 = 4$

$\qquad\quad 0 = 1$

No solution   *(6-1)*

6. $x_1 - x_2 = 4$

$\qquad\quad 0 = 0$

Solution:

$x_2 = t$

$x_1 = x_2 + 4 = t + 4$

The solution is $x_1 = t + 4$, $x_2 = t$ for any real number t.   *(6-1)*

7. $AB = \begin{bmatrix} 4 & -2 \\ 0 & 3 \end{bmatrix}\begin{bmatrix} -1 & 5 \\ -4 & 6 \end{bmatrix} = \begin{bmatrix} 4(-1) + (-2)(-4) & 4 \cdot 5 + (-2)6 \\ 0(-1) + 3(-4) & 0 \cdot 5 + 3 \cdot 6 \end{bmatrix} = \begin{bmatrix} 4 & 8 \\ -12 & 18 \end{bmatrix}$ *(6-2)*

8. $CD = \begin{bmatrix} -1 & 4 \end{bmatrix}\begin{bmatrix} 3 \\ -2 \end{bmatrix} = [(-1)3 + 4(-2)] = [-11]$ *(6-2)*

9. $CB = \begin{bmatrix} -1 & 4 \end{bmatrix}\begin{bmatrix} -1 & 5 \\ -4 & 6 \end{bmatrix} = [(-1)(-1) + 4(-4) \quad (-1)5 + 4 \cdot 6] = [-15 \quad 19]$ *(6-2)*

10. $AD = \begin{bmatrix} 4 & -2 \\ 0 & 3 \end{bmatrix}\begin{bmatrix} 3 \\ -2 \end{bmatrix} = \begin{bmatrix} 4 \cdot 3 + (-2)(-2) \\ 0 \cdot 3 + 3(-2) \end{bmatrix} = \begin{bmatrix} 16 \\ -6 \end{bmatrix}$ *(6-2)*

11. $A + B = \begin{bmatrix} 4 & -2 \\ 0 & 3 \end{bmatrix} + \begin{bmatrix} -1 & 5 \\ -4 & 6 \end{bmatrix} = \begin{bmatrix} 4 + (-1) & (-2) + 5 \\ 0 + (-4) & 3 + 6 \end{bmatrix} = \begin{bmatrix} 3 & 3 \\ -4 & 9 \end{bmatrix}$ *(6-2)*

12. $C + D$ is not defined   *(6-2)*   **13.** $A + C$ is not defined   *(6-2)*

14. $2A - 5B = 2\begin{bmatrix} 4 & -2 \\ 0 & 3 \end{bmatrix} - 5\begin{bmatrix} -1 & 5 \\ -4 & 6 \end{bmatrix} = \begin{bmatrix} 8 & -4 \\ 0 & 6 \end{bmatrix} - \begin{bmatrix} -5 & 25 \\ -20 & 30 \end{bmatrix} = \begin{bmatrix} 13 & -29 \\ 20 & -24 \end{bmatrix}$ (6-2)

15. $CA + C = \begin{bmatrix} -1 & 4 \end{bmatrix}\begin{bmatrix} 4 & -2 \\ 0 & 3 \end{bmatrix} + \begin{bmatrix} -1 & 4 \end{bmatrix} = \begin{bmatrix} (-1)4 + 4 \cdot 0 & (-1)(-2) + 4 \cdot 3 \end{bmatrix} + \begin{bmatrix} -1 & 4 \end{bmatrix}$

$\qquad\qquad = \begin{bmatrix} -4 & 14 \end{bmatrix} + \begin{bmatrix} -1 & 4 \end{bmatrix} = \begin{bmatrix} -5 & 18 \end{bmatrix}$ (6-2)

16. $\begin{bmatrix} 4 & 7 & | & 1 & 0 \\ -1 & -2 & | & 0 & 1 \end{bmatrix}$ $R_1 \leftrightarrow R_2$

$\sim \begin{bmatrix} -1 & -2 & | & 1 & 0 \\ 4 & 7 & | & 0 & 1 \end{bmatrix}$ $4R_1 + R_2 \rightarrow R_2$

$\sim \begin{bmatrix} -1 & -2 & | & 0 & 1 \\ 0 & -1 & | & 1 & 4 \end{bmatrix}$ $(-2)R_2 + R_1 \rightarrow R_1$

$\sim \begin{bmatrix} -1 & 0 & | & -2 & -7 \\ 0 & -1 & | & 1 & 4 \end{bmatrix}$ $\begin{matrix}(-1)R_1 \rightarrow R_1 \\ (-1)R_2 \rightarrow R_2\end{matrix}$

$\sim \begin{bmatrix} 1 & 0 & | & 2 & 7 \\ 0 & 1 & | & -1 & -4 \end{bmatrix}$

The inverse is $\begin{bmatrix} 2 & 7 \\ -1 & -4 \end{bmatrix}$

Check: $A^{-1}A = \begin{bmatrix} 2 & 7 \\ -1 & -4 \end{bmatrix}\begin{bmatrix} 4 & 7 \\ -1 & -2 \end{bmatrix} = \begin{bmatrix} 2 \cdot 4 + 7(-1) & 2 \cdot 7 + 7(-2) \\ (-1)4 + (-4)(-1) & (-1)7 + (-4)(-2) \end{bmatrix}$

$\qquad\qquad = \begin{bmatrix} 1 & 0 \\ 0 & 1 \end{bmatrix} = I$ (6-3)

17. As a matrix equation the system becomes

$\qquad\quad A \qquad X \qquad B$
$\begin{bmatrix} 3 & 2 \\ 4 & 3 \end{bmatrix}\begin{bmatrix} x_1 \\ x_2 \end{bmatrix} = \begin{bmatrix} k_1 \\ k_2 \end{bmatrix}$

The solution of $AX = B$ is $X = A^{-1}B$.

Applying matrix reduction methods, we obtain $A^{-1} = \begin{bmatrix} 3 & -2 \\ -4 & 3 \end{bmatrix}$ (details omitted).

Applying this inverse, we have

$X = \begin{bmatrix} 3 & -2 \\ -4 & 3 \end{bmatrix}\begin{bmatrix} k_1 \\ k_2 \end{bmatrix}$

(A) $\begin{bmatrix} x_1 \\ x_2 \end{bmatrix} = \begin{bmatrix} 3 & -2 \\ -4 & 3 \end{bmatrix}\begin{bmatrix} 3 \\ 5 \end{bmatrix} = \begin{bmatrix} 3 \cdot 3 + (-2)5 \\ (-4)3 + 3 \cdot 5 \end{bmatrix} = \begin{bmatrix} -1 \\ 3 \end{bmatrix}$ $x_1 = -1,\ x_2 = 3$

(B) $\begin{bmatrix} x_1 \\ x_2 \end{bmatrix} = \begin{bmatrix} 3 & -2 \\ -4 & 3 \end{bmatrix}\begin{bmatrix} 7 \\ 10 \end{bmatrix} = \begin{bmatrix} 3 \cdot 7 + (-2)10 \\ (-4)7 + 3 \cdot 10 \end{bmatrix} = \begin{bmatrix} 1 \\ 2 \end{bmatrix}$ $x_1 = 1,\ x_2 = 2$

(C) $\begin{bmatrix} x_1 \\ x_2 \end{bmatrix} = \begin{bmatrix} 3 & -2 \\ -4 & 3 \end{bmatrix}\begin{bmatrix} 4 \\ 2 \end{bmatrix} = \begin{bmatrix} 3 \cdot 4 + (-2)2 \\ (-4)4 + 3 \cdot 2 \end{bmatrix} = \begin{bmatrix} 8 \\ -10 \end{bmatrix}$ $x_1 = 8,\ x_2 = -10$ (6-4)

18. $\begin{vmatrix} 2 & -3 \\ -5 & -1 \end{vmatrix} = 2(-1) - (-5)(-3) = -17$ (6-5)

19. $\begin{vmatrix} 2 & 3 & -4 \\ 0 & 5 & 0 \\ 1 & -4 & -2 \end{vmatrix} = 0 + 5(-1)^{2+2}\begin{vmatrix} 2 & -4 \\ 1 & -2 \end{vmatrix} + 0 = 5(-1)^4[2(-2) - 1(-4)] = 0$ (6-5, 6-6)

20. $D = \begin{vmatrix} 3 & -2 \\ 1 & 3 \end{vmatrix} = 11$

$x = \dfrac{\begin{vmatrix} 8 & -2 \\ -1 & 3 \end{vmatrix}}{D} = \dfrac{22}{11} = 2 \qquad y = \dfrac{\begin{vmatrix} 3 & 8 \\ 1 & -1 \end{vmatrix}}{D} = \dfrac{-11}{11} = -1$ *(6-7)*

21. (A) Interchanging two rows of a determinant changes the sign of the determinant (Theorem 3, Section 6-6).

$\begin{vmatrix} g & h & i \\ d & e & f \\ a & b & c \end{vmatrix} = -2$

(B) If every element of a column of a determinant is multiplied by 3, (Theorem 1, Section 6-6), the value of the determinant is 3 times the original.

$\begin{vmatrix} a & 3b & c \\ d & 3e & f \\ g & 3h & i \end{vmatrix} = 3 \cdot 2 = 6$

(C) Adding a multiple of a column to another column does not change the value of the determinant (Theorem 5, Section 6-6).

$\begin{vmatrix} a & b & a+b+c \\ d & e & d+e+f \\ g & h & g+h+i \end{vmatrix} = \begin{vmatrix} a & b & b+c \\ d & e & e+f \\ g & h & h+i \end{vmatrix} = \begin{vmatrix} a & b & c \\ d & e & f \\ g & h & i \end{vmatrix} = 2$ *(6-6)*

22. We write the augmented matrix:

$\begin{bmatrix} 1 & -1 & | & 4 \\ 2 & 1 & | & 2 \\ -2 & 2 & -8 \end{bmatrix}$ $(-2)R_1 + R_2 \rightarrow R_2$

   Need a 0 here

$\sim \begin{bmatrix} 1 & -1 & | & 4 \\ 0 & 3 & | & -6 \end{bmatrix}$ $\frac{1}{3}R_2 \rightarrow R_2$ corresponds to the linear system $\begin{aligned} x_1 - x_2 &= 4 \\ 3x_2 &= -6 \end{aligned}$

Need a 1 here
Need a 0 here

$\sim \begin{bmatrix} 1 & -1 & | & 4 \\ 0 & 1 & | & -2 \end{bmatrix}$ $R_2 + R_1 \rightarrow R_1$ corresponds to the linear system $\begin{aligned} x_1 - x_2 &= 4 \\ x_2 &= -2 \end{aligned}$

$\sim \begin{bmatrix} 1 & 0 & | & 2 \\ 0 & 1 & | & -2 \end{bmatrix}$ corresponds to the linear system $\begin{aligned} x_1 &= 2 \\ x_2 &= -2 \end{aligned}$

The solution is $x_1 = 2$, $x_2 = -2$. Each pair of lines graphed below has the same intersection point, (2, -2).

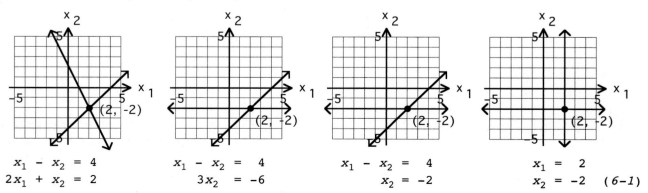

$\begin{aligned} x_1 - x_2 &= 4 \\ 2x_1 + x_2 &= 2 \end{aligned}$ $\qquad$ $\begin{aligned} x_1 - x_2 &= 4 \\ 3x_2 &= -6 \end{aligned}$ $\qquad$ $\begin{aligned} x_1 - x_2 &= 4 \\ x_2 &= -2 \end{aligned}$ $\qquad$ $\begin{aligned} x_1 &= 2 \\ x_2 &= -2 \end{aligned}$ *(6-1)*

23. $\begin{bmatrix} 3 & 2 & | & 3 \\ 1 & 3 & | & 8 \end{bmatrix} R_1 \leftrightarrow R_2$

$\sim \begin{bmatrix} 1 & 3 & | & 8 \\ 3 & 2 & | & 3 \end{bmatrix} (-3)R_1 + R_2 \rightarrow R_2$

$ \underbrace{-3 \quad -9 \quad -24}_{\nwarrow}$

$\sim \begin{bmatrix} 1 & 3 & | & 8 \\ 0 & -7 & | & -21 \end{bmatrix} -\frac{1}{7} R_2 \rightarrow R_2$

$\sim \begin{bmatrix} 1 & 3 & | & 8 \\ 0 & 1 & | & 3 \end{bmatrix} (-3)R_2 + R_1 \rightarrow R_1$

$ \underbrace{0 \quad -3 \quad -9}_{\nwarrow}$

$\sim \begin{bmatrix} 1 & 0 & | & -1 \\ 0 & 1 & | & 3 \end{bmatrix}$

Solution: $x_1 = -1$, $x_2 = 3$ (6-1)

24. $\begin{bmatrix} 1 & 1 & 0 & | & 1 \\ 1 & 0 & -1 & | & -2 \\ 0 & 1 & 2 & | & 4 \end{bmatrix} (-1)R_1 + R_2 \rightarrow R_2$

$\sim \begin{bmatrix} 1 & 1 & 0 & | & 1 \\ 0 & -1 & -1 & | & -3 \\ 0 & 1 & 2 & | & 4 \end{bmatrix} (-1)R_2 \rightarrow R_2$

$\sim \begin{bmatrix} 1 & 1 & 0 & | & 1 \\ 0 & 1 & 1 & | & 3 \\ 0 & 1 & 2 & | & 4 \end{bmatrix} \begin{matrix} (-1)R_2 + R_1 \rightarrow R_1 \\ (-1)R_2 + R_3 \rightarrow R_3 \end{matrix}$

$\sim \begin{bmatrix} 1 & 0 & -1 & | & -2 \\ 0 & 1 & 1 & | & 3 \\ 0 & 0 & 1 & | & 1 \end{bmatrix} \begin{matrix} R_3 + R_1 \rightarrow R_1 \\ (-1)R_3 + R_2 \rightarrow R_2 \end{matrix}$

$\sim \begin{bmatrix} 1 & 0 & 0 & | & -1 \\ 0 & 1 & 0 & | & 2 \\ 0 & 0 & 1 & | & 1 \end{bmatrix}$

Solution: $x_1 = -1$, $x_2 = 2$, $x_3 = 1$
    (6-1)

25. $\begin{bmatrix} 1 & 2 & 3 & | & 1 \\ 2 & 3 & 4 & | & 3 \\ 1 & 2 & 1 & | & 3 \end{bmatrix} \begin{matrix} (-2)R_1 + R_3 \rightarrow R_3 \\ (-1)R_1 + R_3 \rightarrow R_3 \end{matrix}$

$\sim \begin{bmatrix} 1 & 2 & 3 & | & 1 \\ 0 & -1 & -2 & | & 1 \\ 0 & 0 & -2 & | & 2 \end{bmatrix} \begin{matrix} (-1)R_2 \rightarrow R_2 \\ -\frac{1}{2} R_3 \rightarrow R_3 \end{matrix}$

$\sim \begin{bmatrix} 1 & 2 & 3 & | & 1 \\ 0 & 1 & 2 & | & -1 \\ 0 & 0 & 1 & | & -1 \end{bmatrix} (-2)R_2 + R_1 \rightarrow R_1$

$\sim \begin{bmatrix} 1 & 0 & -1 & | & 3 \\ 0 & 1 & 2 & | & -1 \\ 0 & 0 & 1 & | & -1 \end{bmatrix} \begin{matrix} R_3 + R_1 \rightarrow R_1 \\ (-2)R_3 + R_2 \rightarrow R_2 \end{matrix}$

$\sim \begin{bmatrix} 1 & 0 & 0 & | & 2 \\ 0 & 1 & 0 & | & 1 \\ 0 & 0 & 1 & | & -1 \end{bmatrix}$

Solution: $x_1 = 2$, $x_2 = 1$, $x_3 = -1$
    (6-1)

26. $\begin{bmatrix} 1 & 2 & -1 & | & 2 \\ 2 & 3 & 1 & | & -3 \\ 3 & 5 & 0 & | & -1 \end{bmatrix} \begin{matrix} (-2)R_1 + R_2 \rightarrow R_2 \\ (-3)R_1 + R_3 \rightarrow R_3 \end{matrix}$

$\sim \begin{bmatrix} 1 & 2 & -1 & | & 2 \\ 0 & -1 & 3 & | & -7 \\ 0 & -1 & 3 & | & -7 \end{bmatrix} (-1)R_2 + R_3 \rightarrow R_3$

$\sim \begin{bmatrix} 1 & 2 & -1 & | & 2 \\ 0 & -1 & 3 & | & -7 \\ 0 & 0 & 0 & | & 0 \end{bmatrix} 2R_2 + R_1 \rightarrow R_1$

$\sim \begin{bmatrix} 1 & 0 & 5 & | & -12 \\ 0 & -1 & 3 & | & -7 \\ 0 & 0 & 0 & | & 0 \end{bmatrix} (-1)R_2 \rightarrow R_2$

$\sim \begin{bmatrix} 1 & 0 & 5 & | & -12 \\ 0 & 1 & -3 & | & 7 \\ 0 & 0 & 0 & | & 0 \end{bmatrix}$

This corresponds to the system
$x_1 + 5x_3 = -12$
$ x_2 - 3x_3 = 7$

Let $x_3 = t$
Then $x_2 = 3x_3 + 7$
$ = 3t + 7$
$ x_1 = -5x_3 - 12$
$ = -5t - 12$
Solution: $x_1 = -5t - 12$,
$x_2 = 3t + 7$, $x_3 = t$ for every real
number t. There are infinitely many
solutions.   (6-1)

27.
$$\begin{bmatrix} 1 & -2 & | & 1 \\ 2 & -1 & | & 0 \\ 1 & -3 & | & -2 \end{bmatrix} \begin{array}{l} (-2)R_1 + R_2 \rightarrow R_2 \\ (-1)R_1 + R_3 \rightarrow R_3 \end{array}$$

$$\sim \begin{bmatrix} 1 & -2 & | & 1 \\ 0 & 3 & | & -2 \\ 0 & -1 & | & -3 \end{bmatrix} 3R_3 + R_2 \rightarrow R_2$$

$$\sim \begin{bmatrix} 1 & -2 & | & 1 \\ 0 & 0 & | & -11 \\ 0 & -1 & | & -3 \end{bmatrix}$$

The second row corresponds to the equation

$$0x_1 + 0x_2 = -11,$$

so there is no solution. (6-1)

28.
$$\begin{bmatrix} 1 & 2 & -1 & | & 2 \\ 3 & -1 & 2 & | & -3 \end{bmatrix} (-3)R_1 + R_2 \rightarrow R_2$$

$$\sim \begin{bmatrix} 1 & 2 & -1 & | & 2 \\ 0 & -7 & 5 & | & -9 \end{bmatrix} -\tfrac{1}{7}R_2 \rightarrow R_2$$

$$\sim \begin{bmatrix} 1 & 2 & -1 & | & 2 \\ 0 & 1 & -\tfrac{5}{7} & | & \tfrac{9}{7} \end{bmatrix} (-2)R_2 + R_1 \rightarrow R_1$$

$$\sim \begin{bmatrix} 1 & 0 & \tfrac{3}{7} & | & -\tfrac{4}{7} \\ 0 & 1 & -\tfrac{5}{7} & | & \tfrac{9}{7} \end{bmatrix}$$

This corresponds to the system
$$x_1 + \tfrac{3}{7}x_3 = -\tfrac{4}{7}$$
$$x_2 - \tfrac{5}{7}x_3 = \tfrac{9}{7}$$
Let $x_3 = t$
Then $x_2 = \tfrac{5}{7}x_3 + \tfrac{9}{7}$
$$= \tfrac{5}{7}t + \tfrac{9}{7}$$
$$x_1 = -\tfrac{3}{7}x_3 - \tfrac{4}{7}$$
$$= -\tfrac{3}{7}t - \tfrac{4}{7}$$

Solution: $x_1 = -\tfrac{3}{7}t - \tfrac{4}{7}$, $x_2 = \tfrac{5}{7}t + \tfrac{9}{7}$, $x_3 = t$ for every real number t. There are infinitely many solutions. (6-1)

29. $AD = \begin{bmatrix} 1 & 2 \\ 4 & 5 \\ -3 & -1 \end{bmatrix} \begin{bmatrix} 7 & 0 & -5 \\ 0 & 8 & -2 \end{bmatrix} = \begin{bmatrix} 1 \cdot 7 + 2 \cdot 0 & 1 \cdot 0 + 2 \cdot 8 & 1(-5) + 2(-2) \\ 4 \cdot 7 + 5 \cdot 0 & 4 \cdot 0 + 5 \cdot 8 & 4(-5) + 5(-2) \\ (-3)7 + (-1)0 & (-3)0 + (-1)8 & (-3)(-5) + (-1)(-2) \end{bmatrix}$

$$= \begin{bmatrix} 7 & 16 & -9 \\ 28 & 40 & -30 \\ -21 & -8 & 17 \end{bmatrix} \qquad (6-2)$$

30. $DA = \begin{bmatrix} 7 & 0 & -5 \\ 0 & 8 & -2 \end{bmatrix} \begin{bmatrix} 1 & 2 \\ 4 & 5 \\ -3 & -1 \end{bmatrix} = \begin{bmatrix} 7 \cdot 1 + 0 \cdot 4 + (-5)(-3) & 7 \cdot 2 + 0 \cdot 5 + (-5)(-1) \\ 0 \cdot 1 + 8 \cdot 4 + (-2)(-3) & 0 \cdot 2 + 8 \cdot 5 + (-2)(-1) \end{bmatrix}$

$$= \begin{bmatrix} 22 & 19 \\ 38 & 42 \end{bmatrix} \qquad (6-2)$$

31. $BC = \begin{bmatrix} 6 \\ 0 \\ -4 \end{bmatrix} [2 \quad 4 \quad -1] = \begin{bmatrix} 6 \cdot 2 & 6 \cdot 4 & 6(-1) \\ 0 \cdot 2 & 0 \cdot 4 & 0(-1) \\ (-4)2 & (-4)4 & (-4)(-1) \end{bmatrix} = \begin{bmatrix} 12 & 24 & -6 \\ 0 & 0 & 0 \\ -8 & -16 & 4 \end{bmatrix}$ *(6-2)*

32. $CB = [2 \quad 4 \quad -1] \begin{bmatrix} 6 \\ 0 \\ -4 \end{bmatrix} = [2 \cdot 6 + 4 \cdot 0 + (-1)(-4)] = [16]$ *(6-2)*

33. Since D has 3 columns and E has 2 rows, DE is not defined. *(6-2)*

34. $ED = \begin{bmatrix} 9 & -3 \\ -6 & 2 \end{bmatrix} \begin{bmatrix} 7 & 0 & -5 \\ 0 & 8 & -2 \end{bmatrix}$

$= \begin{bmatrix} 9 \cdot 7 + (-3)0 & 9 \cdot 0 + (-3)8 & 9(-5) + (-3)(-2) \\ (-6)7 + 2 \cdot 0 & (-6)0 + 2 \cdot 8 & (-6)(-5) + 2(-2) \end{bmatrix} = \begin{bmatrix} 63 & -24 & -39 \\ -42 & 16 & 26 \end{bmatrix}$ *(6-2)*

35. $\begin{bmatrix} 1 & 0 & 4 & | & 1 & 0 & 0 \\ -2 & 1 & 0 & | & 0 & 1 & 0 \\ 4 & -1 & 4 & | & 0 & 0 & 1 \end{bmatrix}$ $\begin{array}{l} 2R_1 + R_1 \to R_2 \\ (-4)R_1 + R_3 \to R_3 \end{array}$

$\sim \begin{bmatrix} 1 & 0 & 4 & | & 1 & 0 & 0 \\ 0 & 1 & 8 & | & 2 & 1 & 0 \\ 0 & -1 & -12 & | & -4 & 0 & 1 \end{bmatrix}$ $R_2 + R_3 \to R_3$

$\sim \begin{bmatrix} 1 & 0 & 4 & | & 1 & 0 & 0 \\ 0 & 1 & 8 & | & 2 & 1 & 0 \\ 0 & 0 & -4 & | & -2 & 1 & 1 \end{bmatrix}$ $\begin{array}{l} R_3 + R_1 \to R_1 \\ 2R_3 + R_2 \to R_2 \end{array}$

$\sim \begin{bmatrix} 1 & 0 & 0 & | & -1 & 1 & 1 \\ 0 & 1 & 0 & | & -2 & 3 & 2 \\ 0 & 0 & -4 & | & -2 & 1 & 1 \end{bmatrix}$ $\left(-\frac{1}{4}\right)R_3 \to R_3$

$\sim \begin{bmatrix} 1 & 0 & 0 & | & -1 & 1 & 1 \\ 0 & 1 & 0 & | & -2 & 3 & 2 \\ 0 & 0 & 1 & | & \frac{1}{2} & -\frac{1}{4} & -\frac{1}{4} \end{bmatrix}$

The inverse is $\begin{bmatrix} -1 & 1 & 1 \\ -2 & 3 & 2 \\ \frac{1}{2} & -\frac{1}{4} & -\frac{1}{4} \end{bmatrix}$

Check:

$A^{-1}A = \begin{bmatrix} -1 & 1 & 1 \\ -2 & 3 & 2 \\ \frac{1}{2} & -\frac{1}{4} & -\frac{1}{4} \end{bmatrix} \begin{bmatrix} 1 & 0 & 4 \\ -2 & 1 & 0 \\ 4 & -1 & 4 \end{bmatrix}$

$= \begin{bmatrix} (-1)1 + 1(-2) + 1 \cdot 4 & (-1)0 + 1 \cdot 1 + 1(-1) & (-1)4 + 1 \cdot 0 + 1 \cdot 4 \\ (-2)1 + 3(-2) + 2 \cdot 4 & (-2)0 + 3 \cdot 1 + 2(-1) & (-2)4 + 3 \cdot 0 + 2 \cdot 4 \\ (\frac{1}{2})1 + (-\frac{1}{4})(-2) + (-\frac{1}{4})4 & (\frac{1}{2})0 + (-\frac{1}{4})1 + (-\frac{1}{4})(-1) & (\frac{1}{2})4 + (-\frac{1}{4})0 + (-\frac{1}{4})4 \end{bmatrix}$

$= \begin{bmatrix} 1 & 0 & 0 \\ 0 & 1 & 0 \\ 0 & 0 & 1 \end{bmatrix}$ *(6-3)*

36.
$\overset{A}{\begin{bmatrix} 1 & 2 & 3 \\ 2 & 3 & 4 \\ 1 & 2 & 1 \end{bmatrix}} \overset{X}{\begin{bmatrix} x_1 \\ x_2 \\ x_3 \end{bmatrix}} = \overset{B}{\begin{bmatrix} k_1 \\ k_2 \\ k_3 \end{bmatrix}}$

The solution to $AX = B$ is $X = A^{-1}B$.

Applying reduction methods, we obtain $A^{-1} = \begin{bmatrix} -\frac{5}{2} & 2 & -\frac{1}{2} \\ 1 & -1 & 1 \\ \frac{1}{2} & 0 & -\frac{1}{2} \end{bmatrix}$ (details omitted).

Applying the inverse, we get

(A) $B = \begin{bmatrix} 1 \\ 3 \\ 3 \end{bmatrix}$ $X = \begin{bmatrix} x_1 \\ x_2 \\ x_3 \end{bmatrix} = \begin{bmatrix} -\frac{5}{2} & 2 & -\frac{1}{2} \\ 1 & -1 & 1 \\ \frac{1}{2} & 0 & -\frac{1}{2} \end{bmatrix} \begin{bmatrix} 1 \\ 3 \\ 3 \end{bmatrix} = \begin{bmatrix} 2 \\ 1 \\ -1 \end{bmatrix}$ $x_1 = 2$, $x_2 = 1$, $x_3 = -1$

(B) $B = \begin{bmatrix} 0 \\ 0 \\ -2 \end{bmatrix}$ $X = \begin{bmatrix} x_1 \\ x_2 \\ x_3 \end{bmatrix} = \begin{bmatrix} -\frac{5}{2} & 2 & -\frac{1}{2} \\ 1 & -1 & 1 \\ \frac{1}{2} & 0 & -\frac{1}{2} \end{bmatrix} \begin{bmatrix} 0 \\ 0 \\ -2 \end{bmatrix} = \begin{bmatrix} 1 \\ -2 \\ 1 \end{bmatrix}$ $x_1 = 1$, $x_2 = -2$, $x_3 = 1$

(C) $B = \begin{bmatrix} -3 \\ -4 \\ 1 \end{bmatrix}$ $X = \begin{bmatrix} x_1 \\ x_2 \\ x_3 \end{bmatrix} = \begin{bmatrix} -\frac{5}{2} & 2 & -\frac{1}{2} \\ 1 & -1 & 1 \\ \frac{1}{2} & 0 & -\frac{1}{2} \end{bmatrix} \begin{bmatrix} -3 \\ -4 \\ 1 \end{bmatrix} = \begin{bmatrix} -1 \\ 2 \\ -2 \end{bmatrix}$ $x_1 = -1$, $x_2 = 2$, $x_3 = -2$ (6-4)

37. $\begin{vmatrix} -\frac{1}{4} & \frac{3}{2} \\ \frac{1}{2} & \frac{2}{3} \end{vmatrix} = \left(-\frac{1}{4}\right)\left(\frac{2}{3}\right) - \left(\frac{1}{2}\right)\left(\frac{3}{2}\right) = -\frac{1}{6} - \frac{3}{4} = -\frac{11}{12}$ (6-4)

38. $\begin{vmatrix} 2 & -1 & 1 \\ -3 & 5 & 2 \\ 1 & -2 & 4 \end{vmatrix} = \begin{vmatrix} 0 & 0 & 1 \\ -7 & 7 & 2 \\ -7 & 2 & 4 \end{vmatrix}$ $\begin{array}{l} (-2)C_3 + C_1 \rightarrow C_1 \\ C_3 + C_2 \rightarrow C_2 \end{array}$

$= 0 + 0 + 1(-1)^{1+3}\begin{vmatrix} -7 & 7 \\ -7 & 2 \end{vmatrix} = (-1)^4[(-7)2 - (-7)7] = 35$ (6-5, 6-6)

39. $y = \dfrac{\begin{vmatrix} 1 & -6 & 1 \\ 0 & 4 & -1 \\ 2 & 2 & 1 \end{vmatrix}}{\begin{vmatrix} 1 & -2 & 1 \\ 0 & 1 & -1 \\ 2 & 2 & 1 \end{vmatrix}} = \dfrac{\begin{vmatrix} 1 & -6 & 1 \\ 0 & 4 & -1 \\ 0 & 14 & -1 \end{vmatrix}}{\begin{vmatrix} 1 & -2 & -1 \\ 0 & 1 & 0 \\ 2 & 2 & 3 \end{vmatrix}} = \dfrac{1(-1)^{1+1}\begin{vmatrix} 4 & -1 \\ 14 & -1 \end{vmatrix}}{1(-1)^{2+2}\begin{vmatrix} 1 & -1 \\ 2 & 3 \end{vmatrix}} = \dfrac{(-1)^2[4(-1) - 14(-1)]}{(-1)^4[1\cdot3 - 2(-1)]}$

$= \dfrac{10}{5} = 2$ (6-7)

40. (A) If the coefficient matrix has an inverse, then the system can be written as $AX = B$ and its solution can be written $X = A^{-1}B$. The system has one solution.

(B) If the coefficient matrix does not have an inverse, then the system can be solved by Gauss-Jordan elimination, but it will not have exactly one solution. The other possibilities are that the system has no solution or an infinite number of solutions, and either possibility may occur. (6-4)

41. If we assume that A is a non-zero matrix with an inverse A^{-1}, then if $A^2 = 0$ we can write $A^{-1}A^2 = A^{-1}0$ or $A^{-1}AA = 0$ or $IA = 0$ or $A = 0$. But A was assumed non-zero, so there is a contradiction. Therefore A^{-1} cannot exist for such a matrix. (6-4)

42. **(A)** The system is independent. There is one solution.

(B) The matrix is then

$$\begin{bmatrix} 1 & 0 & -3 & | & 4 \\ 0 & 1 & 2 & | & 5 \\ 0 & 0 & 0 & | & n \end{bmatrix}$$

The third row corresponds to the equation $0x_1 + 0x_2 + 0x_3 = n$. This is impossible. The system has no solution.

(C) The matrix is then

$$\begin{bmatrix} 1 & 0 & -3 & | & 4 \\ 0 & 1 & 2 & | & 5 \\ 0 & 0 & 0 & | & 0 \end{bmatrix}$$

There are an infinite number of solutions ($x_3 = t$, $x_2 = 5 - 2t$, $x_1 = 4 + 3t$, for t any real number.) *(6-4)*

43.

$$\begin{aligned} AX - B &= CX \\ AX - B + B - CX &= CX - CX + B \qquad &\text{Addition property} \\ AX + 0 - CX &= 0 + B \qquad &M + (-M) = 0 \\ AX - CX &= B \qquad &M + 0 = M \\ (A - C)X &= B \qquad &\text{Right distributive property} \end{aligned}$$

(applied to subtraction of matrices; see problem 39, Exercise 6-4)

$$\begin{aligned} (A - C)^{-1}[(A - C)X] &= (A - C)^{-1}B \qquad &\text{Left multiplication property} \\ [(A - C)^{-1}(A - C)]X &= (A - C)^{-1}B \qquad &\text{Associative property} \\ IX &= (A - C)^{-1}B \qquad &A^{-1}A = I \\ X &= (A - C)^{-1}B \qquad &IX = X \end{aligned}$$

Common Errors: $(A - C)^{-1}B \neq B(A - C)^{-1}$
$(A - C)^{-1} \neq A^{-1} - C^{-1}$

(6-4)

44. $\begin{bmatrix} 4 & 5 & 6 & | & 1 & 0 & 0 \\ 4 & 5 & -6 & | & 0 & 1 & 0 \\ 1 & 1 & 1 & | & 0 & 0 & 1 \end{bmatrix}$ $(-1)R_1 + R_2 \rightarrow R_2$

$\sim \begin{bmatrix} 4 & 5 & 6 & | & 1 & 0 & 0 \\ 0 & 0 & -12 & | & -1 & 1 & 0 \\ 1 & 1 & 1 & | & 0 & 0 & 1 \end{bmatrix}$ $R_1 \leftrightarrow R_3$

$\sim \begin{bmatrix} 1 & 1 & 1 & | & 0 & 0 & 1 \\ 0 & 0 & -12 & | & -1 & 1 & 0 \\ 4 & 5 & 6 & | & 1 & 0 & 0 \end{bmatrix}$ $(-4)R_1 + R_3 \rightarrow R_3$

$\sim \begin{bmatrix} 1 & 1 & 1 & | & 0 & 0 & 1 \\ 0 & 0 & -12 & | & -1 & 1 & 0 \\ 0 & 1 & 2 & | & 1 & 0 & -4 \end{bmatrix}$ $R_2 \leftrightarrow R_3$

$\sim \begin{bmatrix} 1 & 1 & 1 & | & 0 & 0 & 1 \\ 0 & 1 & 2 & | & 1 & 0 & -4 \\ 0 & 0 & -12 & | & -1 & 1 & 0 \end{bmatrix}$ $(-1)R_2 + R_1 \rightarrow R_1$

$\sim \begin{bmatrix} 1 & 0 & -1 & | & -1 & 0 & 5 \\ 0 & 1 & 2 & | & 1 & 0 & -4 \\ 0 & 0 & -12 & | & -1 & 1 & 0 \end{bmatrix}$ $-\frac{1}{12}R_3 \rightarrow R_3$

$\sim \begin{bmatrix} 1 & 0 & -1 & | & -1 & 0 & 5 \\ 0 & 1 & 2 & | & 1 & 0 & -4 \\ 0 & 0 & 1 & | & \frac{1}{12} & -\frac{1}{12} & 0 \end{bmatrix}$ $\begin{array}{l} R_3 + R_1 \rightarrow R_1 \\ (-2)R_3 + R_2 \rightarrow R_2 \end{array}$

$$\sim \begin{bmatrix} 1 & 0 & 0 & \bigm| & -\frac{11}{12} & -\frac{1}{12} & 5 \\ 0 & 1 & 0 & \bigm| & \frac{10}{12} & \frac{2}{12} & -4 \\ 0 & 0 & 1 & \bigm| & \frac{1}{12} & -\frac{1}{12} & 0 \end{bmatrix}$$

The inverse is $\begin{bmatrix} -\frac{11}{12} & -\frac{1}{12} & 5 \\ \frac{10}{12} & \frac{2}{12} & -4 \\ \frac{1}{12} & -\frac{1}{12} & 0 \end{bmatrix}$ or $\frac{1}{12}\begin{bmatrix} -11 & -1 & 60 \\ 10 & 2 & -48 \\ 1 & -1 & 0 \end{bmatrix}$

Check: $A^{-1}A = \frac{1}{12}\begin{bmatrix} -11 & -1 & 60 \\ 10 & 2 & -48 \\ 1 & -1 & 0 \end{bmatrix}\begin{bmatrix} 4 & 5 & 6 \\ 4 & 5 & -6 \\ 1 & 1 & 1 \end{bmatrix}$

$$= \frac{1}{12}\begin{bmatrix} (-11)4 + (-1)4 + 60\cdot1 & (-11)5 + (-1)5 + 60\cdot1 & (-11)6 + (-1)(-6) + 60\cdot1 \\ 10\cdot4 + 2\cdot4 + (-48)1 & 10\cdot5 + 2\cdot5 + (-48)1 & 10\cdot6 + 2(-6) + (-48)1 \\ 1\cdot4 + (-1)4 + 0\cdot1 & 1\cdot5 + (-1)5 + 0\cdot1 & 1\cdot6 + (-1)(-6) + 0\cdot1 \end{bmatrix}$$

$$= \frac{1}{12}\begin{bmatrix} 12 & 0 & 0 \\ 0 & 12 & 0 \\ 0 & 0 & 12 \end{bmatrix} = \begin{bmatrix} 1 & 0 & 0 \\ 0 & 1 & 0 \\ 0 & 0 & 1 \end{bmatrix} = I \qquad\qquad (6\text{-}3)$$

45. Multiplying the first two equations by 100, the system becomes

$4x_1 + 5x_2 + 6x_3 = 36,000$

$4x_1 + 5x_2 - 6x_3 = 12,000$

$x_1 + x_2 + x_3 = 7,000$

As a matrix equation, we have

$$\begin{matrix} A & & X & & B \end{matrix}$$
$$\begin{bmatrix} 4 & 5 & 6 \\ 4 & 5 & -6 \\ 1 & 1 & 1 \end{bmatrix}\begin{bmatrix} x_1 \\ x_2 \\ x_3 \end{bmatrix} = \begin{bmatrix} 36,000 \\ 12,000 \\ 7,000 \end{bmatrix}$$

The solution to $AX = B$ is $X = A^{-1}B$. Using A^{-1} from problem 44, we have

$$X = \begin{bmatrix} x_1 \\ x_2 \\ x_3 \end{bmatrix} = \frac{1}{12}\begin{bmatrix} -11 & -1 & 60 \\ 10 & 2 & -48 \\ 1 & -1 & 0 \end{bmatrix}\begin{bmatrix} 36,000 \\ 12,000 \\ 7,000 \end{bmatrix}$$

$$= \frac{1}{12}\begin{bmatrix} (-11)(36,000) + (-1)(12,000) + (60)(7,000) \\ (10)(36,000) + (2)(12,000) + (-48)(7,000) \\ 1(36,000) + (-1)(12,000) + (0)(7,000) \end{bmatrix} = \frac{1}{12}\begin{bmatrix} 12,000 \\ 48,000 \\ 24,000 \end{bmatrix} = \begin{bmatrix} 1,000 \\ 4,000 \\ 2,000 \end{bmatrix}$$

Solution: $x_1 = 1,000$, $x_2 = 4,000$, $x_3 = 2,000$ \qquad\qquad (6\text{-}4)

46. $\begin{vmatrix} -1 & 4 & 1 & 1 \\ 5 & -1 & 2 & -1 \\ 2 & -1 & 0 & 3 \\ -3 & 3 & 0 & 3 \end{vmatrix} = \begin{vmatrix} -1 & 4 & 1 & 1 \\ 7 & -9 & 0 & -3 \\ 2 & -1 & 0 & 3 \\ -3 & 3 & 0 & 3 \end{vmatrix}$ $(-2)R_1 + R_2 \to R_2$

$$= 1(-1)^{1+3}\begin{vmatrix} 7 & -9 & -3 \\ 2 & -1 & 3 \\ -3 & 3 & 3 \end{vmatrix} + 0 + 0 + 0$$

$$= \begin{vmatrix} 7 & -9 & -3 \\ 2 & -1 & 3 \\ -3 & 3 & 3 \end{vmatrix} \quad \begin{matrix} R_1 + R_2 \to R_2 \\ R_1 + R_3 \to R_3 \end{matrix}$$

$$= \begin{vmatrix} 7 & -9 & -3 \\ 9 & -10 & 0 \\ 4 & -6 & 0 \end{vmatrix} = (-3)(-1)^{1+3}\begin{vmatrix} 9 & -10 \\ 4 & -6 \end{vmatrix} + 0 + 0$$

$$= (-3)(-1)^4[9(-6) - 4(-10)]$$
$$= (-3)(-14) = 42 \qquad\qquad (6\text{-}6)$$

47. $\begin{vmatrix} u + kv & v \\ w + kx & x \end{vmatrix} = (u + kv)x - (w + kx)v = ux + kvx - wv - kvx = ux - wv$

$$= \begin{vmatrix} u & v \\ w & x \end{vmatrix}$$

$(6\text{-}6)$

48. The statements: $(1, 2)$ satisfies the equation $\begin{vmatrix} x & y & 1 \\ 1 & 2 & 1 \\ -1 & 5 & 1 \end{vmatrix} = 0$ and $(-1, 5)$

satisfies the equation $\begin{vmatrix} x & y & 1 \\ 1 & 2 & 1 \\ -1 & 5 & 1 \end{vmatrix} = 0$ are equivalent to the statements

$\begin{vmatrix} 1 & 2 & 1 \\ 1 & 2 & 1 \\ -1 & 5 & 1 \end{vmatrix} = 0$ and $\begin{vmatrix} -1 & 5 & 1 \\ 1 & 2 & 1 \\ -1 & 5 & 1 \end{vmatrix} = 0$. The latter statements are true by Theorem 4,

Section 6-6. All other points on the line through the given points will also
satisfy the equation. $(6\text{-}6)$

49. Let x_1 = number of grams of mix A

   x_2 = number of grams of mix B

   x_3 = number of grams of mix C

We have

$0.30x_1 + 0.20x_2 + 0.10x_3 = 27$ (protein)

$0.03x_1 + 0.05x_2 + 0.04x_3 = 5.4$ (fat)

$0.10x_1 + 0.20x_2 + 0.10x_3 = 19$ (moisture)

Multiply the first and third equations by 10, and the second equation by 100 to
clear decimals:

$3x_1 + 2x_2 + x_3 = 270$

$3x_1 + 5x_2 + 4x_3 = 540$

$x_1 + 2x_2 + x_3 = 190$

Form the augmented matrix and solve by Gauss-Jordan elimination.

$\begin{bmatrix} 3 & 2 & 1 & | & 270 \\ 3 & 5 & 4 & | & 540 \\ 1 & 2 & 1 & | & 190 \end{bmatrix} R_3 \leftrightarrow R_1$

$\sim \begin{bmatrix} 1 & 2 & 1 & | & 190 \\ 3 & 5 & 4 & | & 540 \\ 3 & 2 & 1 & | & 270 \end{bmatrix} \begin{matrix} \\ (-3)R_1 + R_2 \rightarrow R_2 \\ (-3)R_1 + R_3 \rightarrow R_3 \end{matrix}$

$\sim \begin{bmatrix} 1 & 2 & 1 & | & 190 \\ 0 & -1 & 1 & | & -30 \\ 0 & -4 & -2 & | & -300 \end{bmatrix} (-1)R_2 \rightarrow R_2$

$\sim \begin{bmatrix} 1 & 2 & 1 & | & 190 \\ 0 & 1 & -1 & | & 30 \\ 0 & -4 & -2 & | & -300 \end{bmatrix} \begin{matrix} (-2)R_2 + R_1 \rightarrow R_1 \\ \\ 4R_2 + R_3 \rightarrow R_3 \end{matrix}$

$\sim \begin{bmatrix} 1 & 0 & 3 & | & 130 \\ 0 & 1 & -1 & | & 30 \\ 0 & 0 & -6 & | & -180 \end{bmatrix} -\tfrac{1}{6} R_3 \rightarrow R_3$

$\sim \begin{bmatrix} 1 & 0 & 3 & | & 130 \\ 0 & 1 & -1 & | & 30 \\ 0 & 0 & 1 & | & 30 \end{bmatrix} \begin{matrix} (-3)R_3 + R_1 \rightarrow R_1 \\ R_3 + R_2 \rightarrow R_2 \\ \\ \end{matrix}$

$$\sim \begin{bmatrix} 1 & 0 & 0 & | & 40 \\ 0 & 1 & 0 & | & 60 \\ 0 & 0 & 1 & | & 30 \end{bmatrix}$$

Therefore

$x_1 = 40$ grams Mix A
$x_2 = 60$ grams Mix B
$x_3 = 30$ grams Mix C (6-1)

50. (A) Let $x_1 =$ number of nickels
              $x_2 =$ number of dimes
         Then $x_1 + x_2 = 30$ (total number of coins)
              $5x_1 + 10x_2 = 190$ (total value of coins)
We form the augmented matrix and solve by Gauss-Jordan elimination.

$$\begin{bmatrix} 1 & 1 & | & 30 \\ 5 & 10 & | & 190 \end{bmatrix} (-5)R_1 + R_2 \rightarrow R_2$$

$$\sim \begin{bmatrix} 1 & 1 & | & 30 \\ 0 & 5 & | & 40 \end{bmatrix} \tfrac{1}{5}R_2 \rightarrow R_2$$

$$\sim \begin{bmatrix} 1 & 1 & | & 30 \\ 0 & 1 & | & 8 \end{bmatrix} (-1)R_2 + R_1 \rightarrow R_1$$

$$\sim \begin{bmatrix} 1 & 0 & | & 22 \\ 0 & 1 & | & 8 \end{bmatrix}$$

The augmented matrix is in reduced form. It corresponds to the system
$x_1 = 22$ nickels
$x_2 = 8$ dimes

(B) Let $x_1 =$ number of nickels
          $x_2 =$ number of dimes
          $x_3 =$ number of quarters
      Then $x_1 + x_2 + x_3 = 30$ (total number of coins)
          $5x_1 + 10x_2 + 25x_3 = 190$ (total value of coins)

We form the augmented matrix and solve by Gauss-Jordan elimination

$$\begin{bmatrix} 1 & 1 & 1 & | & 30 \\ 5 & 10 & 25 & | & 190 \end{bmatrix} (-5)R_1 + R_2 \rightarrow R_2$$

$$\sim \begin{bmatrix} 1 & 1 & 1 & | & 30 \\ 0 & 5 & 20 & | & 40 \end{bmatrix} \tfrac{1}{5}R_2 \rightarrow R_2$$

$$\sim \begin{bmatrix} 1 & 1 & 1 & | & 30 \\ 0 & 1 & 4 & | & 8 \end{bmatrix} (-1)R_2 + R_1 \rightarrow R_1$$

$$\sim \begin{bmatrix} 1 & 0 & -3 & | & 22 \\ 0 & 1 & 4 & | & 8 \end{bmatrix}$$

The augmented matrix is in reduced form. It corresponds to the system:
$x_1 - \qquad 3x_3 = 22$
$\qquad x_2 + 4x_3 = 8$

Let $x_3 = t$. Then $x_2 = -4x_3 + 8$
$$= -4t + 8$$
$$x_1 = 3x_3 + 22$$
$$= 3t + 22$$

A solution is achieved, not for every real value of t, but for integer values of t that give rise to non-negative x_1, x_2, x_3.

$x_1 \geq 0$ means $3t + 22 \geq 0$ or $t \geq -7\frac{1}{3}$
$x_2 \geq 0$ means $-4t + 8 \geq 0$ or $t \leq 2$
$x_3 \geq 0$ means $t \geq 0$

The only integer values of t that satisfy these conditions are 0, 1, 2. So we have the solutions

$x_1 = 3t + 22$ nickels

$x_2 = 8 - 4t$ dimes

$x_3 = t$ quarters

where $t = 0$, 1, or 2   *(6-1)*

51. Let x_1 = number of tons at Big Bend

x_2 = number of tons at Saw Pit

Then

$0.05x_1 + 0.03x_2$ = number of tons of nickel at both mines = k_1

$0.07x_1 + 0.04x_2$ = number of tons of copper at both mines = k_2

We solve

$0.05x_1 + 0.03x_2 = k_1$

$0.07x_1 + 0.04x_4 = k_2$,

For arbitrary k_1 and k_2, by writing the system as a matrix equation.

$$\overset{A}{\begin{bmatrix} 0.05 & 0.03 \\ 0.07 & 0.04 \end{bmatrix}} \overset{X}{\begin{bmatrix} x_1 \\ x_2 \end{bmatrix}} = \overset{B}{\begin{bmatrix} k_1 \\ k_2 \end{bmatrix}}$$

If A^{-1} exists, then $X = A^{-1}B$. To find A^{-1}, we perform row operations on

$$\begin{bmatrix} 0.05 & 0.03 & | & 1 & 0 \\ 0.07 & 0.04 & | & 0 & 1 \end{bmatrix} 20R_1 \rightarrow R_1$$

$$\sim \begin{bmatrix} 1 & 0.6 & | & 20 & 0 \\ 0.07 & 0.04 & | & 0 & 1 \end{bmatrix} -0.07R_1 + R_2 \rightarrow R_2$$

$$\sim \begin{bmatrix} 1 & 0.6 & | & 20 & 0 \\ 0 & -0.002 & | & -1.4 & 1 \end{bmatrix} -500R_2 \rightarrow R_2$$

$$\sim \begin{bmatrix} 1 & 0.6 & | & 20 & 0 \\ 0 & 1 & | & 700 & -500 \end{bmatrix} -0.6R_2 + R_1 \rightarrow R_1$$

$$\sim \begin{bmatrix} 1 & 0 & | & -400 & 300 \\ 0 & 1 & | & 700 & -500 \end{bmatrix}$$

$$A^{-1} = \begin{bmatrix} -400 & 300 \\ 700 & -500 \end{bmatrix}$$

Check: $A^{-1}A = \begin{bmatrix} -400 & 300 \\ 700 & -500 \end{bmatrix} \begin{bmatrix} 0.05 & 0.03 \\ 0.07 & 0.04 \end{bmatrix} = \begin{bmatrix} 1 & 0 \\ 0 & 1 \end{bmatrix}$

We can now solve the system as:

$$\overset{X}{\begin{bmatrix} x_1 \\ x_2 \end{bmatrix}} = \overset{A^{-1}}{\begin{bmatrix} -400 & 300 \\ 700 & -500 \end{bmatrix}} \overset{B}{\begin{bmatrix} k_1 \\ k_2 \end{bmatrix}}$$

(A) If $k_1 = 3.6$, $k_2 = 5$,

$$\begin{bmatrix} x_1 \\ x_2 \end{bmatrix} = \begin{bmatrix} -400 & 300 \\ 700 & -500 \end{bmatrix} \begin{bmatrix} 3.6 \\ 5 \end{bmatrix} = \begin{bmatrix} 60 \\ 20 \end{bmatrix}$$

60 tons of ore must be produced at Big Bend, 20 tons of ore at Saw Pit.

(B) If $k_1 = 3$, $k_2 = 4.1$,

$$\begin{bmatrix} x_1 \\ x_2 \end{bmatrix} = \begin{bmatrix} -400 & 300 \\ 700 & -500 \end{bmatrix} \begin{bmatrix} 3 \\ 4.1 \end{bmatrix} = \begin{bmatrix} 30 \\ 50 \end{bmatrix}$$

30 tons of ore must be produced at Big Bend, 50 tons of ore at Saw Pit.

(C) If $k_1 = 3.2$, $k_2 = 4.4$,

$$\begin{bmatrix} x_1 \\ x_2 \end{bmatrix} = \begin{bmatrix} -400 & 300 \\ 700 & -500 \end{bmatrix} \begin{bmatrix} 3.2 \\ 4.4 \end{bmatrix} = \begin{bmatrix} 40 \\ 40 \end{bmatrix}$$

40 tons of ore must be produced at Big Bend, 40 tons of ore at Saw Pit.   *(6-4)*

52. (A) The labor cost of producing one printer stand at the South Carolina plant is the product of the stand row of L with South Carolina column of H.

$$[0.9 \quad 1.8 \quad 0.6] \begin{bmatrix} 10.00 \\ 8.50 \\ 4.50 \end{bmatrix} = 27 \text{ dollars}$$

(B) The matrix HL has no clear meaning, but the matrix LH represents the total labor costs for each item at each plant.

(C) $LH = \begin{bmatrix} 1.7 & 2.4 & 0.8 \\ 0.9 & 1.8 & 0.6 \end{bmatrix} \begin{bmatrix} 11.50 & 10.00 \\ 9.50 & 8.50 \\ 5.00 & 4.50 \end{bmatrix}$

$$= \begin{bmatrix} 1.7 \cdot 11.50 + 2.4 \cdot 9.50 + 0.8 \cdot 5.00 & 1.7 \cdot 10.00 + 2.4 \cdot 8.50 + 0.8 \cdot 4.50 \\ 0.9 \cdot 11.50 + 1.8 \cdot 9.50 + 0.6 \cdot 5.00 & 0.9 \cdot 10.00 + 1.8 \cdot 8.50 + 0.6 \cdot 4.50 \end{bmatrix}$$

$$\begin{array}{cc} \text{N.C.} & \text{S.C.} \\ \begin{bmatrix} \$46.35 & \$41.00 \\ \$30.45 & \$27.00 \end{bmatrix} & \begin{array}{l} \text{Desks} \\ \text{Stands} \end{array} \end{array}$$

(6-2)

53. (A) The average monthly production for the months of January and February is represented by the matrix $\frac{1}{2}(J + F)$

$$\frac{1}{2}(J + F) = \frac{1}{2}\left(\begin{bmatrix} 1,500 & 1,650 \\ 850 & 700 \end{bmatrix} + \begin{bmatrix} 1,700 & 1,810 \\ 850 & 740 \end{bmatrix} \right) = \frac{1}{2}\begin{bmatrix} 3,200 & 3,460 \\ 1,780 & 1,440 \end{bmatrix}$$

$$\begin{array}{cc} \text{N.C.} & \text{S.C.} \\ = \begin{bmatrix} 1,600 & 1,730 \\ 890 & 720 \end{bmatrix} & \begin{array}{l} \text{Desks} \\ \text{Stands} \end{array} \end{array}$$

(B) The increase in production from January to February is represented by the matrix $F - J$.

$$F - J = \begin{bmatrix} 1,700 & 1,810 \\ 930 & 740 \end{bmatrix} - \begin{bmatrix} 1,500 & 1,650 \\ 850 & 700 \end{bmatrix}$$

$$\begin{array}{cc} \text{N.C.} & \text{S.C.} \\ = \begin{bmatrix} 200 & 160 \\ 80 & 40 \end{bmatrix} & \begin{array}{l} \text{Desks} \\ \text{Stands} \end{array} \end{array}$$

(C) $J\begin{bmatrix} 1 \\ 1 \end{bmatrix} = \begin{bmatrix} 1,500 & 1,650 \\ 850 & 700 \end{bmatrix}\begin{bmatrix} 1 \\ 1 \end{bmatrix} = \begin{bmatrix} 3,150 \\ 1,550 \end{bmatrix} \begin{array}{l} \text{Desks} \\ \text{Stands} \end{array}$

This matrix represents the total production of each item in January.

(6-2)

54. The inverse of matrix B is calculated to be

$$B^{-1} = \begin{bmatrix} 1 & 1 & -1 \\ 0 & -1 & 1 \\ -1 & 0 & 1 \end{bmatrix}$$

Putting the coded message into matrix form and multiplying by B^{-1} yields

$$\begin{bmatrix} 1 & 1 & -1 \\ 0 & -1 & 1 \\ -1 & 0 & 1 \end{bmatrix} \begin{bmatrix} 25 & 24 & 21 & 41 & 21 & 52 \\ 8 & 25 & 41 & 30 & 32 & 52 \\ 26 & 33 & 48 & 50 & 41 & 79 \end{bmatrix} = \begin{bmatrix} 7 & 16 & 14 & 21 & 12 & 25 \\ 18 & 8 & 7 & 20 & 9 & 27 \\ 1 & 9 & 27 & 9 & 20 & 27 \end{bmatrix}$$

This decodes to

7 18 1 16 8 9 14 7 27 21 20 9 12 9 20 25 27 27

G R A P H I N G   U T I L I T Y

(*6-3*)

Chapters 5 & 6 Cumulative Review

1. We choose elimination by addition.
We multiply the top equation by 3,
the bottom by 5, and add.

$$9x - 15y = 33$$
$$10x + 15y = 5$$
$$\overline{19x = 38}$$
$$x = 2$$

Substituting $x = 2$ in the bottom
equation, we have
$$2(2) + 3y = 1$$
$$4 + 3y = 1$$
$$3y = -3$$
$$y = -1$$
Solution: $(2, -1)$ $(5-1)$

2.

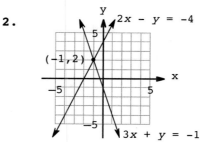

$(-1, 2)$ $(5-1)$

3. $-6x + 3y = 2$
   $2x - y = 1$
Solve the second equation for y
and substitute into the first.
$$2x - y = 1$$
$$2x - 1 = y$$
$$-6x + 3(2x - 1) = 2$$
$$-6x + 6x - 3 = 2$$
$$-3 = 2$$
This is a contradiction, so the
system has no solution. $(5-1)$

4.

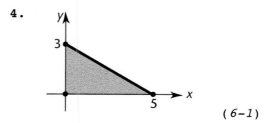

$(6-1)$

5.

Corner Point (x, y)	Objective Function $z = 2x + 3y$	
$(0, 4)$	12	
$(5, 0)$	10	Minimum value
$(6, 7)$	33	Maximum value
$(0, 10)$	30	

The minimum value of z on S is 10 at $(5, 0)$.
The maximum value of z on S is 33 at $(6, 7)$. $(5-5)$

6. (A) $M - 2N = \begin{bmatrix} 2 & 1 \\ 1 & -3 \end{bmatrix} - 2\begin{bmatrix} 1 & 2 \\ -1 & 3 \end{bmatrix} = \begin{bmatrix} 2 & 1 \\ 1 & -3 \end{bmatrix} - \begin{bmatrix} 2 & 4 \\ -2 & 6 \end{bmatrix} = \begin{bmatrix} 0 & -3 \\ 3 & -9 \end{bmatrix}$

   (B) $P + Q$ is not defined

   (C) $PQ = \begin{bmatrix} 1 & 2 \end{bmatrix}\begin{bmatrix} -1 \\ 2 \end{bmatrix} = [1(-1) + 2 \cdot 2] = [3]$

   (D) $MN = \begin{bmatrix} 2 & 1 \\ 1 & -3 \end{bmatrix}\begin{bmatrix} 1 & 2 \\ -1 & 3 \end{bmatrix} = \begin{bmatrix} 2 \cdot 1 + 1(-1) & 2 \cdot 2 + 1 \cdot 3 \\ 1 \cdot 1 + (-3)(-1) & 1 \cdot 2 + (-3)3 \end{bmatrix} = \begin{bmatrix} 1 & 7 \\ 4 & -7 \end{bmatrix}$

   (E) $PN = \begin{bmatrix} 1 & 2 \end{bmatrix}\begin{bmatrix} 1 & 2 \\ -1 & 3 \end{bmatrix} = [1 \cdot 1 + 2(-1) \quad 1 \cdot 2 + 2 \cdot 3] = \begin{bmatrix} -1 & 8 \end{bmatrix}$

   (F) QM is not defined $(6-2)$

7. $\begin{vmatrix} 0 & 2 & 0 \\ 1 & 3 & 2 \\ -1 & 4 & 3 \end{vmatrix} = 0 + 2(-1)^{1+2}\begin{vmatrix} 1 & 2 \\ -1 & 3 \end{vmatrix} + 0 = 2(-1)^3[1 \cdot 3 - (-1)2] = -10$ $(6-5)$

8. (A) $x_1 = 3$

$\quad\quad x_2 = -4$

(B) $x_1 - 2x_2 = 3$

$\quad\quad$ Let $x_2 = t$

$\quad\quad$ Then $x_1 = 2x_2 + 3$

$\quad\quad\quad\quad\quad = 2t + 3$

Solution: $x_1 = 2t + 3$, $x_2 = t$ for every real number t.

(C) $\quad x_1 - 2x_2 = 3$

$\quad\quad 0x_1 + 0x_2 = 1$

No solution. $\hspace{6cm}$ *(6-1)*

9. (A) $\begin{bmatrix} 1 & 1 & | & 3 \\ -1 & 1 & | & 5 \end{bmatrix}$

(B) $\begin{bmatrix} 1 & 1 & | & 3 \\ -1 & 1 & | & 5 \end{bmatrix} R_1 + R_2 \to R_2$

$\sim \begin{bmatrix} 1 & 1 & | & 3 \\ 0 & 2 & | & 8 \end{bmatrix} \frac{1}{2}R_2 \to R_2$

$\sim \begin{bmatrix} 1 & 1 & | & 3 \\ 0 & 1 & | & 4 \end{bmatrix} (-1)R_2 + R_1 \to R_1$

$\sim \begin{bmatrix} 1 & 0 & | & -1 \\ 0 & 1 & | & 4 \end{bmatrix}$

(C) Solution: $x_1 = -1$, $x_2 = 4$

$\hspace{5cm}$ *(6-1, 6-2)*

10. (A) As a matrix equation the system becomes

$\quad\quad \underset{A}{\begin{bmatrix} 1 & -3 \\ 2 & -5 \end{bmatrix}} \underset{X}{\begin{bmatrix} x_1 \\ x_2 \end{bmatrix}} = \underset{B}{\begin{bmatrix} k_1 \\ k_2 \end{bmatrix}}$

(B) To find A^{-1}, we perform row operations on

$\begin{bmatrix} 1 & -3 & | & 1 & 0 \\ 2 & -5 & | & 0 & 1 \end{bmatrix} (-2)R_1 + R_2 \to R_2$

$\sim \begin{bmatrix} 1 & -3 & | & 1 & 0 \\ 0 & 1 & | & -2 & 1 \end{bmatrix} 3R_2 + R_1 \to R_1$

$\sim \begin{bmatrix} 1 & 0 & | & -5 & 3 \\ 0 & 1 & | & -2 & 1 \end{bmatrix}$

$A^{-1} = \begin{bmatrix} -5 & 3 \\ -2 & 1 \end{bmatrix}$

Check: $A^{-1}A = \begin{bmatrix} -5 & 3 \\ -2 & 1 \end{bmatrix}\begin{bmatrix} 1 & -3 \\ 2 & -5 \end{bmatrix} = \begin{bmatrix} (-5)1 + 3\cdot2 & (-5)(-3)+3(-5) \\ (-2)1 + 1\cdot2 & (-2)(-3)+1(-5) \end{bmatrix} = \begin{bmatrix} 1 & 0 \\ 0 & 1 \end{bmatrix}$

(C) The solution of $AX = B$ is $X = A^{-1}B$. Using the result of (B), we get

$\quad X = \begin{bmatrix} -5 & 3 \\ -2 & 1 \end{bmatrix}\begin{bmatrix} k_1 \\ k_2 \end{bmatrix}$

If $k_1 = -2$, $k_2 = 1$, we get

$\begin{bmatrix} x_1 \\ x_2 \end{bmatrix} = \begin{bmatrix} -5 & 3 \\ -2 & 1 \end{bmatrix}\begin{bmatrix} -2 \\ 1 \end{bmatrix} = \begin{bmatrix} (-5)(-2)+3\cdot1 \\ (-2)(-2)+1\cdot1 \end{bmatrix} = \begin{bmatrix} 13 \\ 5 \end{bmatrix} \quad x_1 = 13, \ x_2 = 5$

(D) If $k_1 = 1$, $k_2 = -2$, reasoning as in (C) we get

$\begin{bmatrix} x_1 \\ x_2 \end{bmatrix} = \begin{bmatrix} -5 & 3 \\ -2 & 1 \end{bmatrix}\begin{bmatrix} 1 \\ -2 \end{bmatrix} = \begin{bmatrix} (-5)1 + 3(-2) \\ (-2)1 + 1(-2) \end{bmatrix} = \begin{bmatrix} -11 \\ -4 \end{bmatrix} \quad x_1 = -11, \ x_2 = -4 \hspace{2cm}$ *(6-4)*

11. (A) $D = \begin{vmatrix} 2 & -3 \\ 4 & -5 \end{vmatrix} = 2(-5) - 4(-3) = 2$

(B) $x = \dfrac{\begin{vmatrix} 1 & -3 \\ 2 & -5 \end{vmatrix}}{D} = \dfrac{1}{2} \qquad y = \dfrac{\begin{vmatrix} 2 & 1 \\ 4 & 2 \end{vmatrix}}{D} = \dfrac{0}{2} = 0 \hspace{4cm}$ *(6-7)*

12. We write the augmented matrix:

$$\begin{bmatrix} 1 & 3 & | & 10 \\ 2 & -1 & | & -1 \end{bmatrix} \underbrace{(-2)R_1 + R_2 \to R_2}$$

-2   -6   -20

↑

Need a 0 here

$$\sim \begin{bmatrix} 1 & 3 & | & 10 \\ 0 & -7 & | & -21 \end{bmatrix} -\frac{1}{7} R_2 \to R_2 \qquad \text{corresponds to the linear system} \qquad \begin{array}{rcl} x_1 + 3x_2 &=& 10 \\ -7x_2 &=& -21 \end{array}$$

↑

Need a 1 here

Need a 0 here

↓

$$\begin{bmatrix} 1 & 3 & | & 10 \\ 0 & 1 & | & 3 \end{bmatrix} \underbrace{(-3)R_2 + R_1 \to R_1} \qquad \text{corresponds to the linear system} \qquad \begin{array}{rcl} x_1 + 3x_2 &=& 10 \\ x_2 &=& 3 \end{array}$$

0   -3   -9

$$\sim \begin{bmatrix} 1 & 0 & | & 1 \\ 0 & 1 & | & 3 \end{bmatrix} \qquad \text{corresponds to the linear system} \qquad \begin{array}{rcl} x_1 &=& 1 \\ x_2 &=& 3 \end{array}$$

The solution is $x_1 = 1$, $x_2 = 3$. Each pair of lines graphed below has the same intersection point, (1, 3).

  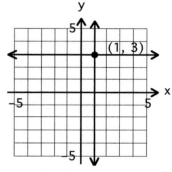

$$\begin{array}{rcl} x_1 + 3x_2 &=& 10 \\ 2x_1 - x_2 &=& -1 \end{array} \qquad \begin{array}{rcl} x_1 + 3x_2 &=& 10 \\ -7x_2 &=& -21 \end{array} \qquad \begin{array}{rcl} x_1 + 3x_2 &=& 10 \\ x_2 &=& 3 \end{array} \qquad \begin{array}{rcl} x_1 &=& 1 \\ x_2 &=& 3 \quad (5\text{-}1) \end{array}$$

13. Here is a computer-generated graph of the system, entered as

$$y = \frac{7 + 2x}{3}$$

$$y = \frac{18 - 3x}{4}$$

After zooming in (graph not shown) the intersection point is located at (1.53, 3.35) to two decimal places.

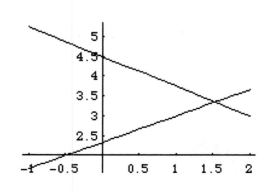

(5-1)

14. $\begin{bmatrix} 1 & 2 & -1 & | & 3 \\ 0 & 1 & 1 & | & -2 \\ 2 & 3 & 1 & | & 0 \end{bmatrix}$ $(-2)R_1 + R_3 \rightarrow R_3$

$\sim \begin{bmatrix} 1 & 2 & -1 & | & 3 \\ 0 & 1 & 1 & | & -2 \\ 0 & -1 & 3 & | & -6 \end{bmatrix}$ $\begin{array}{l} (-2)R_2 + R_1 \rightarrow R_1 \\ \\ R_2 + R_3 \rightarrow R_3 \end{array}$

$\sim \begin{bmatrix} 1 & 0 & -3 & | & 7 \\ 0 & 1 & 1 & | & -2 \\ 0 & 0 & 4 & | & -8 \end{bmatrix}$ $\frac{1}{4} R_3 \rightarrow R_3$

$\sim \begin{bmatrix} 1 & 0 & -3 & | & 7 \\ 0 & 1 & 1 & | & -2 \\ 0 & 0 & 1 & | & -2 \end{bmatrix}$ $\begin{array}{l} 3R_3 + R_1 \rightarrow R_1 \\ (-1)R_3 + R_2 \rightarrow R_2 \end{array}$

$\sim \begin{bmatrix} 1 & 0 & 0 & | & 1 \\ 0 & 1 & 0 & | & 0 \\ 0 & 0 & 1 & | & -2 \end{bmatrix}$

Solution: $x_1 = 1$, $x_2 = 0$, $x_3 = -2$

(6-1)

15. $\begin{bmatrix} 1 & 1 & -2 & | & 2 \\ 0 & 4 & 6 & | & -1 \\ 0 & 6 & 9 & | & 0 \end{bmatrix}$ $(-\frac{2}{3})R_2 + R_3 \rightarrow R_3$

$\sim \begin{bmatrix} 1 & 1 & -2 & | & 2 \\ 0 & 4 & 6 & | & -1 \\ 0 & 0 & 0 & | & \frac{2}{3} \end{bmatrix}$

The last row corresponds to the equation
$$0x_1 + 0x_2 + 0x_3 = \frac{2}{3},$$
so there is no solution. (6-1)

16. $\begin{bmatrix} 1 & -2 & 1 & | & 1 \\ 3 & -2 & -1 & | & -5 \end{bmatrix}$ $(-3)R_1 + R_2 \rightarrow R_2$

$\sim \begin{bmatrix} 1 & -2 & 1 & | & 1 \\ 0 & 4 & -4 & | & -8 \end{bmatrix}$ $\frac{1}{4} R_2 \rightarrow R_2$

$\sim \begin{bmatrix} 1 & -2 & 1 & | & 1 \\ 0 & 1 & -1 & | & -2 \end{bmatrix}$ $2R_2 + R_1 \rightarrow R_1$

$\sim \begin{bmatrix} 1 & 0 & -1 & | & -3 \\ 0 & 1 & -1 & | & -2 \end{bmatrix}$

This corresponds to the system

$x_1 \quad\quad - x_3 = -3$

$\quad\quad x_2 - x_3 = -2$

Let $x_3 = t$

Then $x_2 = x_3 - 2$

$\quad\quad = t - 2$

$\quad x_1 = x_3 - 3$

$\quad\quad = t - 3$

Solution: $x_1 = t - 3$, $x_2 = t - 2$, $x_3 = t$ for every real number t. (6-1)

17. (A) $MN = \begin{bmatrix} 1 & 2 & -1 \end{bmatrix} \begin{bmatrix} 1 \\ -1 \\ 2 \end{bmatrix} = \begin{bmatrix} 1 \cdot 1 + 2(-1) + (-1)2 \end{bmatrix} = \begin{bmatrix} -3 \end{bmatrix}$

(B) $NM = \begin{bmatrix} 1 \\ -1 \\ 2 \end{bmatrix} \begin{bmatrix} 1 & 2 & -1 \end{bmatrix} = \begin{bmatrix} 1 \cdot 1 & 1 \cdot 2 & 1(-1) \\ (-1)1 & (-1)2 & (-1)(-1) \\ 2 \cdot 1 & 2 \cdot 2 & 2(-1) \end{bmatrix} = \begin{bmatrix} 1 & 2 & -1 \\ -1 & -2 & 1 \\ 2 & 4 & -2 \end{bmatrix}$ (6-2)

18. (A) $LM = \begin{bmatrix} 2 & -1 & 0 \\ 1 & 2 & 1 \end{bmatrix} \begin{bmatrix} 1 & 2 \\ -1 & 0 \\ 1 & 1 \end{bmatrix} = \begin{bmatrix} 2 \cdot 1 + (-1)(-1) + 0 \cdot 1 & 2 \cdot 2 + (-1)0 + 0 \cdot 1 \\ 1 \cdot 1 + 2(-1) + 1 \cdot 1 & 1 \cdot 2 + 2 \cdot 0 + 1 \cdot 1 \end{bmatrix}$

$= \begin{bmatrix} 3 & 4 \\ 0 & 3 \end{bmatrix}$

$LM - 2N = \begin{bmatrix} 3 & 4 \\ 0 & 3 \end{bmatrix} - 2\begin{bmatrix} 2 & 1 \\ -1 & 0 \end{bmatrix} = \begin{bmatrix} 3 & 4 \\ 0 & 3 \end{bmatrix} - \begin{bmatrix} 4 & 2 \\ -2 & 0 \end{bmatrix} = \begin{bmatrix} -1 & 2 \\ 2 & 3 \end{bmatrix}$

(B) Since ML is a 3×3 matrix and N is a 2×2 matrix, $ML + N$ is not defined. (6-2)

19. The solution region is unbounded. Two corner points are obvious from the graph: (0, 6) and (8, 0). The third corner point is obtained by solving the system

$3x + 2y = 12$
$x + 2y = 8$ to obtain (2, 3).

(Solution left to the student.)

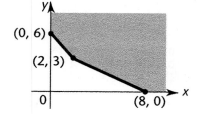

(5-4)

20. The feasible region is graphed as follows: The corner points (0, 7), (0, 0) and (8, 0) are obvious from the graph. The corner point (6, 4) is obtained by solving the system

$x + 2y = 14$
$2x + y = 16$ (Solution left to the student.)

We now evaluate the objective function at each corner point.

Corner Point $(x,\ y)$	Objective Function $z = 4x + 9y$	
(0, 7)	63	Maximum value
(0, 0)	0	
(8, 0)	32	
(6, 4)	60	

(5-5)

21. (A) As a matrix equation the system becomes

$$\underset{A}{\begin{bmatrix} 1 & 4 & 2 \\ 2 & 6 & 3 \\ 2 & 5 & 2 \end{bmatrix}} \underset{X}{\begin{bmatrix} x_1 \\ x_2 \\ x_3 \end{bmatrix}} = \underset{B}{\begin{bmatrix} k_1 \\ k_2 \\ k_3 \end{bmatrix}}$$

(B) To find A^{-1}, we perform row operations on

$$\begin{bmatrix} 1 & 4 & 2 & | & 1 & 0 & 0 \\ 2 & 6 & 3 & | & 0 & 1 & 0 \\ 2 & 5 & 2 & | & 0 & 0 & 1 \end{bmatrix} \begin{matrix} \\ (-2)R_1 + R_2 \rightarrow R_2 \\ (-2)R_1 + R_3 \rightarrow R_3 \end{matrix}$$

$$\sim \begin{bmatrix} 1 & 4 & 2 & | & 1 & 0 & 0 \\ 0 & -2 & -1 & | & -2 & 1 & 0 \\ 0 & -3 & -2 & | & -2 & 0 & 1 \end{bmatrix} -\frac{1}{2}R_2 \rightarrow R_2$$

$$\sim \begin{bmatrix} 1 & 4 & 2 & | & 1 & 0 & 0 \\ 0 & 1 & \frac{1}{2} & | & 1 & -\frac{1}{2} & 0 \\ 0 & -3 & -2 & | & -2 & 0 & 1 \end{bmatrix} (-4)R_2 + R_1 \rightarrow R_1$$

$$\sim \begin{bmatrix} 1 & 0 & 0 & | & -3 & 2 & 0 \\ 0 & 1 & \frac{1}{2} & | & 1 & -\frac{1}{2} & 0 \\ 0 & 0 & -\frac{1}{2} & | & 1 & -\frac{3}{2} & 1 \end{bmatrix} R_3 + R_2 \rightarrow R_2$$

$$\sim \begin{bmatrix} 1 & 0 & 0 & | & -3 & 2 & 0 \\ 0 & 1 & 0 & | & 2 & -2 & 1 \\ 0 & 0 & -\frac{1}{2} & | & 1 & -\frac{3}{2} & 1 \end{bmatrix} -2R_3 \rightarrow R_3$$

$$\sim \begin{bmatrix} 1 & 0 & 0 & | & -3 & 2 & 0 \\ 0 & 1 & 0 & | & 2 & -2 & 1 \\ 0 & 0 & 1 & | & -2 & 3 & -2 \end{bmatrix}$$

$$A^{-1} = \begin{bmatrix} -3 & 2 & 0 \\ 2 & -2 & 1 \\ -2 & 3 & -2 \end{bmatrix}$$

Check: $A^{-1}A = \begin{bmatrix} -3 & 2 & 0 \\ 2 & -2 & 1 \\ -2 & 3 & -2 \end{bmatrix} \begin{bmatrix} 1 & 4 & 2 \\ 2 & 6 & 3 \\ 2 & 5 & 2 \end{bmatrix}$

$$= \begin{bmatrix} (-3)1 + 2 \cdot 2 + 0 \cdot 2 & (-3)4 + 2 \cdot 6 + 0 \cdot 5 & -3 \cdot 2 + 2 \cdot 3 + 0 \cdot 2 \\ 2 \cdot 1 + (-2)2 + 1 \cdot 2 & 2 \cdot 4 + (-2)6 + 1 \cdot 5 & 2 \cdot 2 + (-2)3 + 1 \cdot 2 \\ (-2)1 + 3 \cdot 2 + (-2)2 & (-2)4 + 3 \cdot 6 + (-2)5 & (-2)2 + 3 \cdot 3 + (-2)2 \end{bmatrix}$$

$$= \begin{bmatrix} 1 & 0 & 0 \\ 0 & 1 & 0 \\ 0 & 0 & 1 \end{bmatrix}$$

(C) The solution of $AX = B$ is $X = A^{-1}B$. Using the result of (B), we get

$$X = \begin{bmatrix} -3 & 2 & 0 \\ 2 & -2 & 1 \\ -2 & 3 & -2 \end{bmatrix} \begin{bmatrix} k_1 \\ k_2 \\ k_3 \end{bmatrix}$$

If $k_1 = -1$, $k_2 = 2$, $k_3 = 1$, we get

$$\begin{bmatrix} x_1 \\ x_2 \\ x_3 \end{bmatrix} = \begin{bmatrix} -3 & 2 & 0 \\ 2 & -2 & 1 \\ -2 & 3 & -2 \end{bmatrix} \begin{bmatrix} -1 \\ 2 \\ 1 \end{bmatrix} = \begin{bmatrix} 7 \\ -5 \\ 6 \end{bmatrix} \qquad x_1 = 7, \; x_2 = -5, \; x_3 = 6$$

(D) If $k_1 = 2$, $k_2 = 0$, $k_3 = -1$, reasoning as in (C) we get

$$\begin{bmatrix} x_1 \\ x_2 \\ x_3 \end{bmatrix} = \begin{bmatrix} -3 & 2 & 0 \\ 2 & -2 & 1 \\ -2 & 3 & -2 \end{bmatrix} \begin{bmatrix} 2 \\ 0 \\ -1 \end{bmatrix} = \begin{bmatrix} -6 \\ 3 \\ 2 \end{bmatrix} \qquad x_1 = -6, \; x_2 = 3, \; x_3 = -2 \qquad (6-4)$$

22. (A) $D = \begin{vmatrix} 1 & 2 & -1 \\ 2 & 8 & 1 \\ -1 & 3 & 5 \end{vmatrix} = 1(-1)^{1+1}\begin{vmatrix} 8 & 1 \\ 3 & 5 \end{vmatrix} + 2(-1)^{1+2}\begin{vmatrix} 2 & 1 \\ -1 & 5 \end{vmatrix} + (-1)(-1)^{1+3}\begin{vmatrix} 2 & 8 \\ -1 & 3 \end{vmatrix}$

$$= (-1)^2(8 \cdot 5 - 3 \cdot 1) + 2(-1)^3[2 \cdot 5 - (-1)1] + (-1)^5[2 \cdot 3 - (-1)8]$$

$$= 37 - 22 - 14 = 1$$

(B) $z = \dfrac{\begin{vmatrix} 1 & 2 & 1 \\ 2 & 8 & -2 \\ -1 & 3 & 2 \end{vmatrix}}{D} = \dfrac{1(-1)^{1+1}\begin{vmatrix} 8 & 1 \\ 3 & 5 \end{vmatrix} + 2(-1)^{1+2}\begin{vmatrix} 2 & -2 \\ -1 & 2 \end{vmatrix} + 1(-1)^{1+3}\begin{vmatrix} 2 & 8 \\ -1 & 3 \end{vmatrix}}{1}$

$$= 32$$

$$(6-6, \; 6-7)$$

23. (A) If m and n are both zero, the matrix is

$$\begin{bmatrix} 1 & 0 & -5 & | & 2 \\ 0 & 1 & 3 & | & 6 \\ 0 & 0 & 0 & | & 0 \end{bmatrix}$$

Then there are an infinite number of solutions ($x_3 = t$, $x_2 = 6 - 3t$, $x_1 = 2 + 5t$, for t any real number.)

(B) If $m = 0$, the matrix is

$$\begin{bmatrix} 1 & 0 & -5 & | & 2 \\ 0 & 1 & 3 & | & 6 \\ 0 & 0 & 0 & | & n \end{bmatrix}$$

The third row corresponds to the equation $0x_1 + 0x_2 + 0x_3 = n$. This is impossible (since $n \neq 0$). The system has no solution.

(C) The system is independent. There is one solution. *(6-1)*

24. If $A^2 = A$ and A^{-1} exists, then

$A^{-1}A^2 = A^{-1}A$

$A^{-1}AA = I$ (since $A^{-1}A$ is the identity matrix for any A)

  $IA = I$ (same reason applied to the left side)

   $A = I$ (since the identity matrix times A is just A)

This shows that A must be equal to the identity matrix of the same size. *(6-4)*

25. L, M, and P are in reduced form.
N is not in reduced form; it violates condition 1 of the definition of reduced matrix: there is a row of 0's above rows having non-zero elements. *(6-1)*

26. $k\begin{vmatrix} a & b \\ c & d \end{vmatrix} = k(ad - bc) = kad - kbc = kad - kcb = \begin{vmatrix} ka & b \\ kc & d \end{vmatrix}$ *(6-6)*

27. $\begin{vmatrix} a & b \\ c & d \end{vmatrix} = ad - bc = ad + akb - akb - bc = ad + akb - (bc + bka)$

$= a(d + kb) - b(c + ka) = \begin{vmatrix} a & b \\ c + ka & d + kb \end{vmatrix}$ *(6-6)*

28. We assume that $\det M = ad - bc \neq 0$ and find M^{-1} by applying row operations to

$\begin{bmatrix} a & b & | & 1 & 0 \\ c & d & | & 0 & 1 \end{bmatrix}$ Case I: $a \neq 0$
$\quad (-\frac{c}{a})R_1 + R_2 \rightarrow R_2$

$\sim \begin{bmatrix} a & b & | & 1 & 0 \\ 0 & \frac{ad-bc}{a} & | & -\frac{c}{a} & 1 \end{bmatrix} \frac{a}{ad-bc} R_2 \rightarrow R_2$

$\sim \begin{bmatrix} a & b & | & 1 & 0 \\ 0 & 1 & | & \frac{-c}{ad-bc} & \frac{a}{ad-bc} \end{bmatrix} (-b)R_2 + R_1 \rightarrow R_1$

$\sim \begin{bmatrix} a & 0 & | & 1 + \frac{bc}{ad-bc} & \frac{-ab}{ad-bc} \\ 0 & 1 & | & \frac{-c}{ad-bc} & \frac{a}{ad-bc} \end{bmatrix}$

$\sim \begin{bmatrix} a & 0 & | & \frac{ad}{ad-bc} & \frac{-ab}{ad-bc} \\ 0 & 1 & | & \frac{-c}{ad-bc} & \frac{a}{ad-bc} \end{bmatrix} \frac{1}{a}R_1 \rightarrow R_1$

$\sim \begin{bmatrix} 1 & 0 & | & \frac{d}{ad-bc} & \frac{-b}{ad-bc} \\ 0 & 1 & | & \frac{-c}{ad-bc} & \frac{a}{ad-bc} \end{bmatrix}$

Case II: $a = 0$ (b, $c \neq 0$, since $b = 0$ or $c = 0$ would imply det $M = ad - bc = 0$)

$$\begin{bmatrix} 0 & b & | & 1 & 0 \\ c & d & | & 0 & 1 \end{bmatrix} \quad R_1 \leftrightarrow R_2$$

$$\sim \begin{bmatrix} c & d & | & 0 & 1 \\ 0 & b & | & 1 & 0 \end{bmatrix} \quad \tfrac{1}{c} R_1 \rightarrow R_1$$

$$\sim \begin{bmatrix} 1 & \frac{d}{c} & | & 0 & \frac{1}{c} \\ 0 & b & | & 1 & 0 \end{bmatrix} \quad (-\tfrac{d}{bc}) R_2 + R_1 \rightarrow R_1$$

$$\sim \begin{bmatrix} 1 & 0 & | & -\frac{d}{bc} & \frac{1}{c} \\ 0 & b & | & 1 & 0 \end{bmatrix} \quad \tfrac{1}{b} R_2 \rightarrow R_2$$

$$\sim \begin{bmatrix} 1 & 0 & | & -\frac{d}{bc} & \frac{1}{c} \\ 0 & 1 & | & \frac{1}{b} & 0 \end{bmatrix}$$

$$\sim \begin{bmatrix} 1 & 0 & | & \frac{d}{ad-bc} & \frac{-b}{ad-bc} \\ 0 & 1 & | & \frac{-c}{ad-bc} & \frac{a}{ad-bc} \end{bmatrix} \quad \text{since } a = 0.$$

In either case,

$$M^{-1} = \begin{bmatrix} \frac{d}{ad-bc} & \frac{-b}{ad-bc} \\ \frac{-c}{ad-bc} & \frac{a}{ad-bc} \end{bmatrix} = \frac{1}{ad-bc}\begin{bmatrix} d & -b \\ -c & a \end{bmatrix} = \frac{1}{\det M}\begin{bmatrix} d & -b \\ -c & a \end{bmatrix} \tag{6-6}$$

29. True. For example,

$$\begin{bmatrix} a_{11} & a_{12} & a_{13} \\ 0 & a_{22} & a_{23} \\ 0 & 0 & a_{33} \end{bmatrix} + \begin{bmatrix} b_{11} & b_{12} & b_{13} \\ 0 & b_{22} & b_{23} \\ 0 & 0 & b_{33} \end{bmatrix} = \begin{bmatrix} a_{11} + b_{11} & a_{12} + b_{12} & a_{13} + b_{13} \\ 0 & a_{22} + b_{22} & a_{23} + b_{23} \\ 0 & 0 & a_{33} + b_{33} \end{bmatrix}$$

proves the statement for 3×3 matrices. A similar proof can be given for any order matrix. (6-2)

30. True. For example, $\begin{bmatrix} a_{11} & 0 \\ a_{21} & a_{22} \end{bmatrix}\begin{bmatrix} b_{11} & 0 \\ b_{21} & b_{22} \end{bmatrix} = \begin{bmatrix} a_{11}b_{11} & 0 \\ a_{21}b_{11} + a_{22}b_{21} & a_{22}b_{22} \end{bmatrix}$ proves the statement for 2×2 matrices. A similar proof can be given for any order matrix. (6-1)

31. False. For example, $\begin{bmatrix} 1 & 1 \\ 0 & 1 \end{bmatrix} + \begin{bmatrix} 1 & 0 \\ 1 & 1 \end{bmatrix} = \begin{bmatrix} 2 & 1 \\ 1 & 2 \end{bmatrix}$ which is not a diagonal matrix. (6-2)

32. False. For example, $\begin{bmatrix} 1 & 1 \\ 0 & 1 \end{bmatrix}\begin{bmatrix} 1 & 0 \\ 1 & 1 \end{bmatrix} = \begin{bmatrix} 2 & 1 \\ 1 & 1 \end{bmatrix}$ which is not a diagonal matrix. (6-2)

33. True. If all elements above and all elements below the principal diagonal are zero, then all elements not on the principal diagonal are zero. (6-3)

34. True. For example, the inverse of $\begin{bmatrix} a_{11} & 0 & 0 \\ 0 & a_{22} & 0 \\ 0 & 0 & a_{33} \end{bmatrix}$ is $\begin{bmatrix} \frac{1}{a_{11}} & 0 & 0 \\ 0 & \frac{1}{a_{22}} & 0 \\ 0 & 0 & \frac{1}{a_{33}} \end{bmatrix}$.

This is defined if none of elements on the principal diagonal are zero. A similar situation holds for any order matrix. (6-3)

35. True. For example, $\begin{vmatrix} a_{11} & 0 & 0 & 0 \\ 0 & a_{22} & 0 & 0 \\ 0 & 0 & a_{33} & 0 \\ 0 & 0 & 0 & a_{44} \end{vmatrix}$ can be expanded by the first column to

yield, successively, $a_{11}\begin{vmatrix} a_{22} & 0 & 0 \\ 0 & a_{33} & 0 \\ 0 & 0 & a_{44} \end{vmatrix} = a_{11}a_{22}\begin{vmatrix} a_{33} & 0 \\ 0 & a_{44} \end{vmatrix} = a_{11}a_{12}a_{33}a_{44}.$

A similar result holds for any order matrix. $(6-5)$

36. True. For example, $\begin{vmatrix} a_{11} & 0 & 0 & 0 \\ a_{22} & a_{22} & 0 & 0 \\ a_{31} & a_{32} & a_{33} & 0 \\ a_{41} & a_{42} & a_{43} & a_{44} \end{vmatrix}$ can be expanded by the first row to

yield, successively, $a_{11}\begin{vmatrix} a_{22} & 0 & 0 \\ a_{32} & a_{33} & 0 \\ a_{42} & a_{43} & a_{44} \end{vmatrix} = a_{11}a_{22}\begin{vmatrix} a_{33} & 0 \\ a_{43} & a_{44} \end{vmatrix} = a_{11}a_{22}a_{33}a_{44}.$

A similar result holds for any order matrix. $(6-5)$

37. Let x = amount invested at 8%
    y = amount invested at 14%
Then $x + y = 12,000$ (total amount invested)   (1)
$0.08x + 0.14y = 0.10(12,000)$ (total yield on investment)   (2)
We solve the system of equations (1), (2) using elimination by addition.

$-0.08x - 0.08y = -0.08(12,000) - 0.08$ [equation (1)]
$0.08x + 0.14y = 0.10(12,000)$ equation (2)

$0.06y = -0.08(12,000) + 0.10(12,000)$

$0.06y = 240$
$y = 4,000$
$x + y = 12,000$
$x = 8,000$

$8,000 at 8% and $4,000 at 14%. $(5-1)$

38. Let x_1 = number of ounces of mix A.
    x_2 = number of ounces of mix B.
    x_3 = number of ounces of mix C.
Then
$0.2x_1 + 0.1x_2 + 0.15x_3 = 23$ (protein)
$0.02x_1 + 0.06x_2 + 0.05x_3 = 6.2$ (fat)
$0.15x_1 + 0.1x_2 + 0.05x_3 = 16$ (moisture)
or
$4x_1 + 2x_2 + 3x_3 = 460$ (1st equation multiplied by 20)
$2x_1 + 6x_2 + 5x_3 = 620$ (2nd equation multiplied by 100)
$3x_1 + 2x_2 + x_3 = 320$ (3rd equation multiplied by 20)
is the system to be solved. We form the augmented matrix and solve by Gauss-Jordan elimination.

$\begin{bmatrix} 4 & 2 & 3 & | & 460 \\ 2 & 6 & 5 & | & 620 \\ 3 & 2 & 1 & | & 320 \end{bmatrix} \frac{1}{4}R_1 \rightarrow R_1$

$\sim \begin{bmatrix} 1 & \frac{1}{2} & \frac{3}{4} & | & 115 \\ 2 & 6 & 5 & | & 620 \\ 3 & 2 & 1 & | & 320 \end{bmatrix} \begin{matrix} \\ (-2)R_1 + R_2 \rightarrow R_2 \\ (-3)R_1 + R_3 \rightarrow R_3 \end{matrix}$

$$\sim \begin{bmatrix} 1 & \frac{1}{2} & \frac{3}{4} & | & 115 \\ 0 & 5 & \frac{7}{2} & | & 390 \\ 0 & \frac{1}{2} & -\frac{5}{4} & | & -25 \end{bmatrix} R_2 \leftrightarrow R_3$$

$$\sim \begin{bmatrix} 1 & \frac{1}{2} & \frac{3}{4} & | & 115 \\ 2 & \frac{1}{2} & -\frac{5}{4} & | & -25 \\ 3 & 5 & \frac{7}{2} & | & 390 \end{bmatrix} \begin{matrix} (-1)R_2 + R_1 \to R_1 \\ \\ (-10)R_2 + R_3 \to R_3 \end{matrix}$$

$$\sim \begin{bmatrix} 1 & 0 & 2 & | & 140 \\ 0 & \frac{1}{2} & -\frac{5}{4} & | & -25 \\ 0 & 0 & 16 & | & 640 \end{bmatrix} \begin{matrix} 2R_2 \to R_2 \\ \frac{1}{16} R_3 \to R_3 \end{matrix}$$

$$\sim \begin{bmatrix} 1 & 0 & 2 & | & 140 \\ 0 & 1 & -\frac{5}{2} & | & -50 \\ 0 & 0 & 1 & | & 40 \end{bmatrix} \begin{matrix} (-2)R_3 + R_1 \to R_1 \\ \left(-\frac{5}{2}\right)R_3 + R_2 \to R_2 \end{matrix}$$

$$\sim \begin{bmatrix} 1 & 0 & 0 & | & 60 \\ 0 & 1 & 0 & | & 50 \\ 0 & 0 & 1 & | & 40 \end{bmatrix}$$

Solution: $x_1 = 60$g Mix A,
$x_2 = 50$g Mix B, $x_3 = 40$g Mix C (6-1)

39. Let x_1 = number of model A trucks
     x_2 = number of model B trucks
     x_3 = number of model C trucks
Then $x_1 + x_2 + x_3 = 12$ (total number of trucks)
   $18{,}000x_1 + 22{,}000x_2 + 30{,}000x_3 = 300{,}000$ (total funds needed)

We form the augmented matrix and solve by Gauss-Jordan elimination.

$$\begin{bmatrix} 1 & 1 & 1 & | & 12 \\ 18{,}000 & 22{,}000 & 30{,}000 & | & 300{,}000 \end{bmatrix} \frac{1}{2{,}000} R_2 \to R_2$$

$$\sim \begin{bmatrix} 1 & 1 & 1 & | & 12 \\ 9 & 11 & 15 & | & 150 \end{bmatrix} (-9)R_1 + R_2 \to R_2$$

$$\sim \begin{bmatrix} 1 & 1 & 1 & | & 12 \\ 0 & 2 & 6 & | & 42 \end{bmatrix} -\frac{1}{2}R_2 + R_1 \to R_1$$

$$\sim \begin{bmatrix} 1 & 0 & -2 & | & -9 \\ 0 & 2 & 6 & | & 42 \end{bmatrix} \frac{1}{2}R_2 \to R_2$$

$$\sim \begin{bmatrix} 1 & 0 & -2 & | & -9 \\ 0 & 1 & 3 & | & 21 \end{bmatrix}$$

Let $x_3 = t$. Then $x_2 = -3x_3 + 21$
                    $= -3t + 21$
             $x_1 = 2x_3 - 9$
                    $= 2t - 9$

A solution is achieved, not for every value of t, but for integer values of t that give rise to non-negative x_1, x_2, x_3.

   $x_1 \geq 0$ means $2t - 9 \geq 0$ or $t \geq 4\frac{1}{2}$
   $x_2 \geq 0$ means $-3t + 21 \geq 0$ or $t \leq 7$

The only integer values of t that satisfy these conditions are 5, 6, 7. So we get the solutions
   $x_1 = 2t - 9$ model A trucks
   $x_2 = -3t + 21$ model B trucks
   $x_3 = t$ model C trucks

where t = 5, 6, or 7. The distributor can purchase
   1 model A truck, 6 model B trucks, and 5 model C trucks, or
   3 model A trucks, 3 model B trucks, and 6 model C trucks, or
   5 model A trucks and 7 model C trucks.   (6-1)

40. Let x = number of standard day packs
    y = number of deluxe day packs

(A) We form the linear objective function
    $P = 8x + 12y$
    We wish to maximize P, the profit from x standard packs @ \$8 and y deluxe packs @ \$12, subject to the constraints
    $0.5x + 0.5y \leq 300$ fabricating constraint
    $0.3x + 0.6y \leq 240$ sewing constraint
    $x, y \geq 0$ non-negative constraints

Solving the system of constraint inequalities graphically, we obtain the feasible region S shown in the diagram.

The three corner points $(0, 400)$, $(0, 0)$, and $(600, 0)$ are obvious from the diagram. The corner point $(400, 200)$ is found by solving the system
    $0.5x + 0.5y = 300$
    $0.3x + 0.6y = 240$ (Left to the student.)

Next we evaluate the objective function at each corner point.

Corner Point (x, y)	Objective Function $P = 8x + 12y$	
$(0, 0)$	0	
$(600, 0)$	4,800	
$(400, 200)$	5,600	Maximum value
$(0, 400)$	4,800	

The optimal value is 5,600 at the corner point $(400, 200)$. So the company should manufacture 400 standard packs and 200 deluxe packs for a maximum profit of \$5,600.

(B) The objective function now becomes $5x + 15y$. We evaluate this at each corner point.

Corner Point (x, y)	Objective Function $P = 5x + 15y$	
$(0, 0)$	0	
$(600, 0)$	3,000	
$(400, 200)$	5,000	
$(0, 400)$	6,000	Maximum value

The optimal value is now 6,000 at the corner point $(0, 400)$. So the company should manufacture no standard packs and 400 deluxe packs for a maximum profit of \$6,000.

(C) The objective function now becomes $11x + 9y$. We evaluate this at each corner point.

Corner Point (x, y)	Objective Function $P = 11x + 9y$	
$(0, 0)$	0	
$(600, 0)$	6,600	Maximum value
$(400, 200)$	6,200	
$(0, 400)$	3,600	

The optimal value is now 6,600 at the corner point $(600, 0)$. So the company should manufacture 600 standard packs and no deluxe packs for a maximum profit of \$6,600.

                                                            (5-5)

41. (A) $M\begin{bmatrix}0.25\\0.25\\0.25\\0.25\end{bmatrix} = \begin{bmatrix}78 & 84 & 81 & 86\\91 & 65 & 84 & 92\\95 & 90 & 92 & 91\\75 & 82 & 87 & 91\\83 & 88 & 81 & 76\end{bmatrix}\begin{bmatrix}0.25\\0.25\\0.25\\0.25\end{bmatrix}$

$= \begin{bmatrix}0.25(78 + 84 + 81 + 86)\\0.25(91 + 65 + 84 + 92)\\0.25(95 + 90 + 92 + 91)\\0.25(75 + 82 + 87 + 91)\\0.25(83 + 88 + 81 + 76)\end{bmatrix} = \begin{bmatrix}82.25\\83\\92\\83.75\\82\end{bmatrix}\begin{matrix}\text{Ann}\\\text{Bob}\\\text{Carol}\\\text{Dan}\\\text{Eric}\end{matrix}$

(B) $M\begin{bmatrix}0.2\\0.2\\0.2\\0.4\end{bmatrix} = \begin{bmatrix}78 & 84 & 81 & 86\\91 & 65 & 84 & 92\\95 & 90 & 92 & 91\\75 & 82 & 87 & 91\\83 & 88 & 81 & 76\end{bmatrix}\begin{bmatrix}0.2\\0.2\\0.2\\0.4\end{bmatrix}$

$= \begin{bmatrix}0.2(78 + 84 + 81) + 0.4 \cdot 86\\0.2(91 + 65 + 84) + 0.4 \cdot 92\\0.2(95 + 90 + 92) + 0.4 \cdot 91\\0.2(75 + 82 + 87) + 0.4 \cdot 91\\0.2(83 + 88 + 81) + 0.4 \cdot 76\end{bmatrix} = \begin{bmatrix}83\\84.8\\91.8\\85.2\\80.8\end{bmatrix}\begin{matrix}\text{Ann}\\\text{Bob}\\\text{Carol}\\\text{Dan}\\\text{Eric}\end{matrix}$

(C) $[0.2 \quad 0.2 \quad 0.2 \quad 0.2 \quad 0.2]M = [0.2 \quad 0.2 \quad 0.2 \quad 0.2 \quad 0.2]\begin{bmatrix}78 & 84 & 81 & 86\\91 & 65 & 84 & 92\\95 & 90 & 92 & 91\\75 & 82 & 87 & 91\\83 & 88 & 81 & 76\end{bmatrix}$

$= [0.2(78 + 91 + 95 + 75 + 83) \quad 0.2(84 + 65 + 90 + 82 + 88) \quad 0.2(81 + 84 + 92 + 87 + 81) \quad 0.2(86 + 92 + 91 + 91 + 76)]$

$\begin{matrix}\text{Test 1} & \text{Test 2} & \text{Test 3} & \text{Test 4}\\= [84.4 & 81.8 & 85 & 87.2]\end{matrix}$

(6-2)

CHAPTER 7   Sequences, Induction, and Probability

Section 7-1

1. Answers will vary.

3. Answers will vary.

5. $a_1 = 1 - 2 = -1$; $a_2 = 2 - 2 = 0$; $a_3 = 3 - 2 = 1$; $a_4 = 4 - 2 = 2$

7. $a_1 = \dfrac{1-1}{1+1} = 0$; $a_2 = \dfrac{2-1}{2+1} = \dfrac{1}{3}$; $a_3 = \dfrac{3-1}{3+1} = \dfrac{2}{4} = \dfrac{1}{2}$; $a_4 = \dfrac{4-1}{4+1} = \dfrac{3}{5}$

9. $a_1 = (-2)^{1+1} = (-2)^2 = 4$; $a_2 = (-2)^{2+1} = (-2)^3 = -8$
$a_3 = (-2)^{3+1} = (-2)^4 = 16$; $a_4 = (-2)^{4+1} = (-2)^5 = -32$

11. $a_8 = 8 - 2 = 6$

13. $a_{100} = \dfrac{100-1}{100+1} = \dfrac{99}{101}$

15. $S_5 = 1 + 2 + 3 + 4 + 5$

17. $S_3 = \dfrac{1}{10^1} + \dfrac{1}{10^2} + \dfrac{1}{10^3} = \dfrac{1}{10} + \dfrac{1}{100} + \dfrac{1}{1,000}$

19. $S_4 = (-1)^1 + (-1)^2 + (-1)^3 + (-1)^4 = (-1) + 1 + (-1) + 1 = -1 + 1 - 1 + 1$

21. $a_1 = (-1)^{1+1}\, 1^2 = 1$ $a_2 = (-1)^{2+1}\, 2^2 = -4$ $a_3 = (-1)^{3+1}\, 3^2 = 9$
$a_4 = (-1)^{4+1}\, 4^2 = -16$ $a_5 = (-1)^{5+1}\, 5^2 = 25$

23.
$$a_1 = \dfrac{1}{3}\left[1 - \dfrac{1}{10^1}\right] = \dfrac{1}{3}\cdot\dfrac{9}{10} = \dfrac{3}{10} = 0.3$$
$$a_2 = \dfrac{1}{3}\left[1 - \dfrac{1}{10^2}\right] = \dfrac{1}{3}\cdot\dfrac{99}{100} = \dfrac{33}{100} = 0.33$$
$$a_3 = \dfrac{1}{3}\left[1 - \dfrac{1}{10^3}\right] = \dfrac{1}{3}\cdot\dfrac{999}{1,000} = \dfrac{333}{1,000} = 0.333$$
$$a_4 = \dfrac{1}{3}\left[1 - \dfrac{1}{10^4}\right] = \dfrac{1}{3}\cdot\dfrac{9,999}{10,000} = \dfrac{3,333}{10,000} = 0.3333$$
$$a_5 = \dfrac{1}{3}\left[1 - \dfrac{1}{10^5}\right] = \dfrac{1}{3}\cdot\dfrac{99,999}{100,000} = \dfrac{33,333}{100,000} = 0.33333$$

25. $a_1 = \left(-\dfrac{1}{2}\right)^{1-1} = \left(-\dfrac{1}{2}\right)^0 = 1$
$a_2 = \left(-\dfrac{1}{2}\right)^{2-1} = -\dfrac{1}{2}$
$a_3 = \left(-\dfrac{1}{2}\right)^{3-1} = \dfrac{1}{4}$
$a_4 = \left(-\dfrac{1}{2}\right)^{4-1} = -\dfrac{1}{8}$
$a_5 = \left(-\dfrac{1}{2}\right)^{5-1} = \dfrac{1}{16}$

27. $a_1 = 7$
$a_2 = a_{2-1} - 4 = a_1 - 4 = 7 - 4 = 3$
$a_3 = a_{3-1} - 4 = a_2 - 4 = 3 - 4 = -1$
$a_4 = a_{4-1} - 4 = a_3 - 4 = -1 - 4 = -5$
$a_5 = a_{5-1} - 4 = a_4 - 4 = -5 - 4 = -9$

> **Common Error:**
> $a_2 \neq a_2 - 1 - 4$. The 1 is in the subscript: a_{2-1}; that is, a_1

29. $a_1 = 4$

$a_2 = \dfrac{1}{4} a_{2-1} = \dfrac{1}{4} a_1 = \dfrac{1}{4} \cdot 4 = 1$

$a_3 = \dfrac{1}{4} a_{3-1} = \dfrac{1}{4} a_2 = \dfrac{1}{4} \cdot 1 = \dfrac{1}{4}$

$a_4 = \dfrac{1}{4} a_{4-1} = \dfrac{1}{4} a_3 = \dfrac{1}{4} \cdot \dfrac{1}{4} = \dfrac{1}{16}$

$a_5 = \dfrac{1}{4} a_{5-1} = \dfrac{1}{4} a_4 = \dfrac{1}{4} \cdot \dfrac{1}{16} = \dfrac{1}{64}$

31. $a_1 = 1$

$a_2 = 2$

$a_3 = a_1 + 2a_2 = 1 + 2(2) = 5$

$a_4 = a_2 + 2a_3 = 2 + 2(5) = 12$

$a_5 = a_3 + 2a_4 = 5 + 2(12) = 29$

$a_6 = a_4 + 2a_5 = 12 + 2(29) = 70$

$a_7 = a_5 + 2a_6 = 29 + 2(70) = 169$

The graphic demonstrates use of the RECUR program on a TI-84.

```
prgmRECUR
FIRST TERM? 1
COEF OF 1ST TERM
1
SECOND TERM? 2
COEF OF 2ND TERM
2
NBR OF TERMS? 7
{1 2 5 12 29 70 169}
                  Done
```

33. $a_1 = -1$

$a_2 = 2$

$a_3 = 2a_1 + a_2 = 2(-1) + 2 = 0$

$a_4 = 2a_2 + a_3 = 2(2) + 0 = 4$

$a_5 = 2a_3 + a_4 = 2(0) + 4 = 4$

$a_6 = 2a_4 + a_5 = 2(4) + 4 = 12$

$a_7 = 2a_5 + a_6 = 2(4) + 12 = 20$

The graphic demonstrates use of the RECUR program on a TI-84.

```
prgmRECUR
FIRST TERM? -1
COEF OF 1ST TERM
2
SECOND TERM? 2
COEF OF 2ND TERM
1
NBR OF TERMS? 7
{-1 2 0 4 4 12 20}
                  Done
```

35. a_n: 4, 5, 6, 7, ⋯

$n = 1, 2, 3, 4, ⋯$

Comparing a_n with n, we see that
$a_n = n + 3$

37. a_n: 3, 6, 9, 12, ⋯

$n = 1, 2, 3, 4, ⋯$

Comparing a_n with n, we see that
$a_n = 3n$

39. a_n: $\dfrac{1}{2}$, $\dfrac{2}{3}$, $\dfrac{3}{4}$, $\dfrac{4}{5}$, ⋯

$n = 1, 2, 3, 4, ⋯$

Comparing a_n with n, we see that

$a_n = \dfrac{n}{n + 1}$

41. a_n: 1, −1, 1, −1, ⋯

$n = 1, 2, 3, 4, ⋯$
Comparing a_n with n, we see that a_n involves (−1) to successively even and odd powers, so to a power that depends on n. We could write
$a_n = (-1)^{n-1}$ or $a_n = (-1)^{n+1}$
or other choices. $a_n = (-1)^{n+1}$ is one of many correct answers.

43. a_n: -2, 4, -8, 16, $\cdots$

$n = 1$, 2, 3, 4, $\cdots$

Comparing a_n with n, we see that a_n involves -1 and 2 to successively higher powers, so to powers that depend on n. We write

$a_n = (-1)^n(2)^n$ or $a_n = (-2)^n$

45. a_n: x, $\dfrac{x^2}{2}$, $\dfrac{x^3}{3}$, $\dfrac{x^4}{4}$, $\cdots$ or $\dfrac{x^1}{1}$, $\dfrac{x^2}{2}$, $\dfrac{x^3}{3}$, $\dfrac{x^4}{4}$, $\cdots$

Comparing a_n with n, we see that $a_n = \dfrac{x^n}{n}$

47. The most obvious formula is $a_n = 2^n$ since $a_1 = 1$ (or 2^0), $a_2 = 2$ (or 2^1) and $a_3 = 4$ (or 2^2). We'll use quadratic regression to find another, treating n as independent variable and a_n as dependent variable. We enter 1, 2, 3 as L_1 and 1, 2, 4 as L_2 then use the quadratic regression command.

The quadratic regression is $a_n = 0.5n^2 - 0.5n + 1$. Using this formula, $a_1 = 1$, $a_2 = 2$, $a_3 = 4$, so this fits the desired criteria.

49. The first three terms are the cubes of 1, 2, and 3, so one choice is $a_n = n^3$. We'll use quadratic regression to find another, treating n as independent variable and a_n as dependent variable. We enter 1, 2, 3 as L_1 and 1, 8, 27 as L_2, then use the quadratic regression command.

The quadratic regression is $a_n = 6n^2 - 11n + 6$. Using this formula: $a_1 = 1$, $a_2 = 8$, $a_3 = 27$, so this fits the desired criteria.

51. Compare to problem 37: this sequence is the same as $\{3n\}$ with alternating signs. We accomplish this by including $(-1)^n$, so one choice is $a_n = (-1)^n \cdot 3n$. We'll use quadratic regression to find another, treating n as independent variable and a_n as dependent variable. We enter 1, 2, 3 as L_1 and 1, 8, 27 as L_2, then use the quadratic regression command.

The quadratic regression is $a_n = -12n^2 + 45n - 36$. Using this formula, $a_1 = -3$, $a_2 = 6$, $a_3 = -9$, so this fits the desired criteria.

53.

55.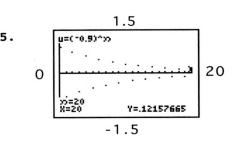

57. $S_4 = \dfrac{(-2)^{1+1}}{1} + \dfrac{(-2)^{2+1}}{2} + \dfrac{(-2)^{3+1}}{3} + \dfrac{(-2)^{4+1}}{4}$

$= \dfrac{4}{1} - \dfrac{8}{2} + \dfrac{16}{3} - \dfrac{32}{4}$

59. $S_3 = \dfrac{1}{1} x^{1+1} + \dfrac{1}{2} x^{2+1} + \dfrac{1}{3} x^{3+1}$

$= x^2 + \dfrac{x^3}{2} + \dfrac{x^4}{3}$

61. $S_5 = \dfrac{(-1)^{1+1}}{1} x^1 + \dfrac{(-1)^{2+1}}{2} x^2 + \dfrac{(-1)^{3+1}}{3} x^3 + \dfrac{(-1)^{4+1}}{4} x^4 + \dfrac{(-1)^{5+1}}{5} x^5$

$= x - \dfrac{x^2}{2} + \dfrac{x^3}{3} - \dfrac{x^4}{4} + \dfrac{x^5}{5}$

63. $S_4 = 1^2 + 2^2 + 3^2 + 4^2$

$k = 1, 2, 3, 4, \cdots$

$a_k = k^2, \ k = 1, 2, 3, 4$

$S_4 = \sum\limits_{k=1}^{4} k^2$

65. $S_5 = \dfrac{1}{2^1} + \dfrac{1}{2^2} + \dfrac{1}{2^3} + \dfrac{1}{2^4} + \dfrac{1}{2^5}$

$k = 1, 2, 3, 4, \cdots$

$a_k = \dfrac{1}{2^k}, \ k = 1, 2, 3, 4, 5$

$S_5 = \sum\limits_{k=1}^{5} \dfrac{1}{2^k}$

67. $S_n = 1 + \dfrac{1}{2^2} + \dfrac{1}{3^2} + \cdots + \dfrac{1}{n^2}$

$k = 1, 2, 3, \cdots, n$

$a_k = \dfrac{1}{k^2}, \ k = 1, 2, 3, \cdots, n$

$S_n = \sum\limits_{k=1}^{n} \dfrac{1}{k^2}$

69. $S_n = 1 - 4 + 9 + \cdots + (-1)^{n+1} n^2$

$k = 1, 2, 3, \cdots, n$

$a_k = (-1)^{k+1} k^2, \ k = 1, 2, 3, \cdots, n$

$S_n = \sum\limits_{k=1}^{n} (-1)^{k+1} k^2$

71. (A) $a_1 = 3$

$a_2 = \dfrac{a_{2-1}^2 + 2}{2a_{2-1}} = \dfrac{a_1^2 + 2}{2a_1} = \dfrac{3^2 + 2}{2 \cdot 3} \approx 1.83$

$a_3 = \dfrac{a_{3-1}^2 + 2}{2a_{3-1}} = \dfrac{a_2^2 + 2}{2a_2} = \dfrac{(1.83)^2 + 2}{2(1.83)} \approx 1.46$

$a_4 = \dfrac{a_{4-1}^2 + 2}{2a_{4-1}} = \dfrac{a_3^2 + 2}{2a_3} = \dfrac{(1.46)^2 + 2}{2(1.46)} \approx 1.415$

(B) Calculator $\sqrt{2} = 1.4142135\ldots$

(C) $a_1 = 1$

$a_2 = \dfrac{a_1^2 + 2}{2a_1} = \dfrac{1^2 + 2}{2 \cdot 1} = 1.5$

$a_3 = \dfrac{a_2^2 + 2}{2a_2} = \dfrac{(1.5)^2 + 2}{2(1.5)} \approx 1.417$

$a_4 = \dfrac{a_3^2 + 2}{2a_3} = \dfrac{(1.417)^2 + 2}{2(1.417)} \approx 1.414$

73. The first ten terms of the Fibonacci sequence a_n are

1, 1, 2, 3, 5, 8, 13, 21, 34, 55

The first ten terms of b_n are

1, 3, 4, 7, 11, 18, 29, 47, 76, 123

The first ten terms of $c_n = \dfrac{b_n}{a_n}$ are

$$\frac{1}{1}, \frac{3}{1}, \frac{4}{2}, \frac{7}{3}, \frac{11}{5}, \frac{18}{8}, \frac{29}{13}, \frac{47}{21}, \frac{76}{34}, \frac{123}{55}$$

In decimal notation this becomes

1, 3, 2, 2.33…, 2.2, 2.25, 2.23…, 2.238…, 2.235…, 2.236…

75. $e^{0.2} = 1 + \dfrac{0.2}{1!} + \dfrac{(0.2)^2}{2!} + \dfrac{(0.2)^3}{3!} + \dfrac{(0.2)^4}{4!}$

$= 1.2214000$

$e^{0.2} = 1.2214028$ (calculator—direct evaluation)

77. $\displaystyle\sum_{k=1}^{n} ca_k = ca_1 + ca_2 + ca_3 + \cdots + ca_n = c(a_1 + a_2 + a_3 + \cdots a_n) = c\sum_{k=1}^{n} a_k$

79. (A) The height of the fifth bounce is a_5:

$$a_5 = 10\left(\frac{1}{2}\right)^{5-1} = 10\left(\frac{1}{2}\right)^4 = 10\left(\frac{1}{16}\right) = \frac{10}{16} = \frac{5}{8} = 0.625 \text{ feet}$$

The height of the tenth bounce is a_{10}:

$$a_{10} = 10\left(\frac{1}{2}\right)^{10-1} = 10\left(\frac{1}{2}\right)^9 = 10\left(\frac{1}{512}\right) = \frac{10}{512} = \frac{5}{256} = 0.01953 \text{ feet}$$

(B) $\displaystyle\sum_{n=1}^{10} 10\left(\frac{1}{2}\right)^{n-1} = 10\left(\frac{1}{2}\right)^{1-1} + 10\left(\frac{1}{2}\right)^{2-1} + 10\left(\frac{1}{2}\right)^{3-1} + 10\left(\frac{1}{2}\right)^{4-1} + 10\left(\frac{1}{2}\right)^{5-1} + 10\left(\frac{1}{2}\right)^{6-1} + 10\left(\frac{1}{2}\right)^{7-1}$

$$+ 10\left(\frac{1}{2}\right)^{8-1} + 10\left(\frac{1}{2}\right)^{9-1} + 10\left(\frac{1}{2}\right)^{10-1}$$

$= 10 + 5 + 2.5 + 1.25 + 0.625 + 0.3125 + 0.1563 + 0.0781 + 0.0391 + 0.0195$

$= 91.98$ feet

This is the sum of the heights of all ten bounces.

81. (A) First year: $\$40,000 = a_1$

Second year: $\$40,000 + 0.04(\$40,000) = \$41,600 = a_2$

Third year: $\$41,600 + 0.04(\$41,600) = \$43,264 = a_3$

Fourth year: $\$43,264 + 0.04(\$43,264) = \$44,995 = a_4$

Fifth year: $\$44,995 + 0.04(\$44,995) = \$46,795 = a_5$

Sixth year: $\$46,795 + 0.04(\$46,795) = \$48,667 = a_6$

(B) The salary for each year is calculated by multiplying the previous year's salary by 0.04 and adding to the previous year's salary. This can easily be written as a recursive sequence: $a_n = a_{n-1} + 0.04a_{n-1}$, which can be simplified a bit as $a_n = 1.04a_{n-1}$. But we can also write as a non-recursive formula: at each stage, we multiply by one more factor of 1.04, so we can write

$a_n = 40,000(1.04)^{n-1}$.

(C) $\displaystyle\sum_{n=1}^{6} a_n = 40,000 + 41,600 + 43,264 + 44,995 + 46,795 + 48,667 = \$265,321$

This is the total amount earned in the first six years.

Section 7-2

1. Answers will vary.

3. $(3 + 5)^1 = 3^1 + 5^1$ True
   $(3 + 5)^2 = 3^2 + 5^2$ False
   Fails at $n = 2$

5. $1^2 = 3 \cdot 1 - 2$ True
   $2^2 = 3 \cdot 2 - 2$ True
   $3^2 = 3 \cdot 3 - 2$ False
   Fails at $n = 3$

7. (A) P_1: $2 = 2 \cdot 1^2$ $\quad 2 = 2$

   P_2: $2 + 6 = 2 \cdot 2^2$ $\quad 8 = 8$

   P_3: $2 + 6 + 10 = 2 \cdot 3^2$ $\quad 18 = 18$

   (B) P_k: $2 + 6 + 10 + \cdots + (4k - 2) = 2k^2$

   P_{k+1}: $2 + 6 + 10 + \cdots + (4k - 2) + [4(k + 1) - 2] = 2(k + 1)^2$

   or $2 + 6 + 10 + \cdots + (4k - 2) + (4k + 2) = 2(k + 1)^2$

   (C) Prove: $2 + 6 + 10 + \cdots + (4n - 2) = 2n^2$, $n \in N$

   Write: P_n: $2 + 6 + 10 + \cdots + (4n - 2) = 2n^2$

   Part 1: Show that P_1 is true:
   P_1: $2 = 2 \cdot 1^2$
   $\quad\quad = 2 \quad\quad P_1$ is true

   Part 2: Show that if P_k is true, then P_{k+1} is true:
   Write out P_k and P_{k+1}.

   $\quad P_k$: $2 + 6 + 10 + \cdots + (4k - 2) = 2k^2$

   $\quad P_{k+1}$: $2 + 6 + 10 + \cdots + (4k - 2) + (4k + 2) = 2(k + 1)^2$

   We start with P_k:
   $\quad 2 + 6 + 10 + \cdots + (4k - 2) = 2k^2$

   Adding $(4k + 2)$ to both sides, we get
   $\quad 2 + 6 + 10 + \cdots + (4k - 2) + (4k + 2) = 2k^2 + 4k + 2$
   $$= 2(k^2 + 2k + 1)$$
   $$= 2(k + 1)^2$$

   We have shown that if P_k is true, then P_{k+1} is true.

   Conclusion: P_n is true for all positive integers n.

Common Errors:
P_1 is not the nonsensical statement
$2 + 6 + 10 + \cdots + (4 \cdot 1 - 2) = 2 \cdot 1^2$
achieved by "substitution" of 1 for n. The left side of P_n has n terms; the left side of P_1 must have 1 term. Also, $P_{k+1} \neq P_k + 1$
Nor is P_{k+1} simply its $k + 1$st term. $4k + 2 \neq 2(k + 1)^2$

9. (A) P_1: $a^5 a^1 = a^{5+}$ $\quad\quad a^6 = a^6$

   P_2: $a^5 a^2 = a^{5+2}$ $\quad a^5 a^2 = a^5(a^1 a) = (a^5 a)a = a^6 a = a^7 = a^{5+2}$

   P_3: $a^5 a^3 = a^{5+3}$ $\quad a^5 a^3 = a^5(a^2 a) = a^5(a^1 a)a = [(a^5 a)a]a = a^8 = a^{5+3}$

(B) P_k: $a^5 a^k = a^{5+k}$

P_{k+1}: $a^5 a^{k+1} = a^{5+k+1}$

(C) Prove: $a^5 a^n = a^{5+n}$, $n \in N$
Write: P_n: $a^5 a^n = a^{5+n}$

Part 1: Show that P_1 is true:
P_1: $a^5 a^1 = a^{5+1}$

$a^5 a = a^{5+1}$. True by the recursive definition of a^n.

Part 2: Show that if P_k is true, then P_{k+1} is true:
Write out P_k and P_{k+1}.

P_k: $a^5 a^k = a^{5+k}$
P_{k+1}: $a^5 a^{k+1} = a^{5+k+1}$

We start with P_k:
$a^5 a^k = a^{5+k}$

Multiply both sides by a:
$a^5 a^k a = a^{5+k} a$
$a^5 a^{k+1} = a^{5+k+1}$ by the recursive definition of a^n.

We have shown that if P_k is true, then P_{k+1} is true.

Conclusion: P_n is true for all positive integers n.

11. (A) P_1: $9^1 - 1 = 8$ is divisible by 4

   P_2: $9^2 - 1 = 80$ is divisible by 4

   P_3: $9^3 - 1 = 728$ is divisible by 4

   (B) P_k: $9^k - 1 = 4r$ for some integer r

   P_{k+1}: $9^{k+1} - 1 = 4s$ for some integer s

   (C) Prove: $9^n - 1$ is divisible by 4.
      Write: P_n: $9^n - 1$ is divisible by 4

Part 1: Show that P_1 is true:
P_1: $9^1 - 1 = 8 = 4 \cdot 2$ is divisible by 4. Clearly true.

Part 2: Show that if P_k is true, then P_{k+1} is true:
Write out P_k and P_{k+1}.

P_k: $9^k - 1 = 4r$ for some integer r
P_{k+1}: $9^{k+1} - 1 = 4s$ for some integer s

We start with P_k:
$9^k - 1 = 4r$ for some integer r

Now, $9^{k+1} - 1 = 9^{k+1} - 9^k + 9^k - 1$
$\qquad\qquad = 9^k(9 - 1) + 9^k - 1$
$\qquad\qquad = 9^k \cdot 8 + 9^k - 1$
$\qquad\qquad = 4 \cdot 9^k \cdot 2 + 4r$
$\qquad\qquad = 4(9^k \cdot 2 + r)$

Therefore,

$9^{k+1} - 1 = 4s$ for some integer s $(= 2 \cdot 9^k + r)$
$9^{k+1} - 1$ is divisible by 4.

We have shown that if P_k is true, then P_{k+1} is true.

Conclusion: P_n is true for all positive integers n.

13. The polynomial $x^4 + 1$ is a counterexample, since it has degree 4, and no real zeros.

15. 23 is a counterexample, since there are no prime numbers p such that $23 < p < 29$.

17. P_n: $2 + 2^2 + 2^3 + \cdots + 2^n = 2^{n+1} - 2$

Part 1: Show that P_1 is true:
$$P_1: \ 2 = 2^{1+1} - 2$$
$$= 2^2 - 2$$
$$= 4 - 2$$
$$= 2 \qquad P_1 \text{ is true}$$

Part 2: Show that if P_k is true, then P_{k+1} is true:
Write out P_k and P_{k+1}.

$$P_k: \ 2 + 2^2 + 2^3 + \cdots + 2^k = 2^{k+1} - 2$$
$$P_{k+1}: \ 2 + 2^2 + 2^3 + \cdots + 2^k + 2^{k+1} = 2^{k+2} - 2$$

We start with P_k:
$$2 + 2^2 + 2^3 + \cdots + 2^k = 2^{k+1} - 2$$

Adding 2^{k+1} to both sides,
$$2 + 2^2 + 2^3 + \cdots + 2^k + 2^{k+1} = 2^{k+1} - 2 + 2^{k+1}$$
$$= 2^{k+1} + 2^{k+1} - 2$$
$$= 2 \cdot 2^{k+1} - 2$$
$$= 2^1 \cdot 2^{k+1} - 2$$
$$= 2^{k+2} - 2$$

We have shown that if P_k is true, then P_{k+1} is true.

Conclusion: P_n is true for all positive integers n.

19. P_n: $1^2 + 3^2 + 5^2 + \cdots + (2n - 1)^2 = \dfrac{1}{3}(4n^3 - n)$

Part 1: Show that P_1 is true:
$$P_1: \ 1^2 = \dfrac{1}{3}(4 \cdot 1^3 - 1)$$
$$= \dfrac{1}{3}(4 - 1)$$
$$= 1 \qquad \text{True.}$$

Part 2: Show that if P_k is true, then P_{k+1} is true:
Write out P_k and P_{k+1}.

$$P_k: \ 1^2 + 3^2 + 5^2 + \cdots + (2k - 1)^2 = \dfrac{1}{3}(4k^3 - k)$$

$$P_{k+1}: \ 1^2 + 3^2 + 5^2 + \cdots + (2k - 1)^2 + (2k + 1)^2 = \dfrac{1}{3}[4(k + 1)^3 - (k + 1)]$$

We start with P_k:

$$1^2 + 3^2 + 5^2 + \cdots + (2k - 1)^2 = \frac{1}{3}(4k^3 - k)$$

Adding $(2k + 1)^2$ to both sides,

$$1^2 + 3^2 + 5^2 + \cdots + (2k - 1)^2 + (2k + 1)^2 = \frac{1}{3}(4k^3 - k) + (2k + 1)^2$$

$$= \frac{1}{3}[4k^3 - k + 3(2k + 1)^2]$$

$$= \frac{1}{3}[4k^3 - k + 3(4k^2 + 4k + 1)]$$

$$= \frac{1}{3}[4k^3 - k + 12k^2 + 12k + 3]$$

$$= \frac{1}{3}[4k^3 + 12k^2 + 12k - k + 3]$$

$$= \frac{1}{3}[4k^3 + 12k^2 + 12k + 4 - k - 1]$$

$$= \frac{1}{3}[4(k^3 + 3k^2 + 3k + 1) - (k + 1)]$$

$$= \frac{1}{3}[4(k + 1)^3 - (k + 1)]$$

We have shown that if P_k is true, then P_{k+1} is true.

Conclusion: P_n is true for all positive integers n.

21. P_n: $1^2 + 2^2 + 3^2 + \cdots + n^2 = \dfrac{n(n + 1)(2n + 1)}{6}$

Part 1: Show that P_1 is true:

$$P_1: \quad 1^2 = \frac{1(1 + 1)(2 \cdot 1 + 1)}{6}$$

$$= \frac{1 \cdot 2 \cdot 3}{6}$$

$$= 1 \qquad P_1 \text{ is true.}$$

Part 2: Show that if P_k is true, then P_{k+1} is true:

Write out P_k and P_{k+1}.

$$P_k: \quad 1^2 + 2^2 + 3^2 + \cdots + k^2 = \frac{k(k + 1)(2k + 1)}{6}$$

$$P_{k+1}: \quad 1^2 + 2^2 + 3^2 + \cdots + k^2 + (k + 1)^2 = \frac{(k + 1)(k + 2)(2k + 3)}{6}$$

We start with P_k:

$$1^2 + 2^2 + 3^2 + \cdots + k^2 = \frac{k(k + 1)(2k + 1)}{6}$$

Adding $(k + 1)^2$ to both sides,

$$1^2 + 2^2 + 3^2 + \cdots + k^2 + (k + 1)^2 = \frac{k(k + 1)(2k + 1)}{6} + (k + 1)^2$$

$$= \frac{k(k + 1)(2k + 1)}{6} + \frac{6(k + 1)^2}{6}$$

$$= \frac{k(k + 1)(2k + 1) + 6(k + 1)^2}{6}$$

$$= \frac{(k + 1)[k(2k + 1) + 6(k + 1)]}{6}$$

$$= \frac{(k + 1)[2k^2 + 7k + 6]}{6}$$

$$= \frac{(k + 1)(k + 2)(2k + 3)}{6}$$

We have shown that if P_k is true, then P_{k+1} is true.

Conclusion: P_n is true for all positive integers n.

23. P_n: $\dfrac{a^n}{a^3} = a^{n-3}$ $\quad n > 3$

Part 1: Show that P_4 is true:

P_4: $\dfrac{a^4}{a^3} = a^{4-3}$

But $\dfrac{a^4}{a^3} = \dfrac{a^3 a}{a^3} = a = a^1 = a^{4-3}$ $\quad$ So, P_4 is true.

Part 2: Show that if P_k is true, then P_{k+1} is true:
Write out P_k and P_{k+1}.

$\qquad P_k$: $\dfrac{a^k}{a^3} = a^{k-3}$

$\qquad P_{k+1}$: $\dfrac{a^{k+1}}{a^3} = a^{k-2}$

We start with P_k:

$\qquad \dfrac{a^k}{a^3} = a^{k-3}$

Multiply both sides by a:

$\qquad \dfrac{a^k}{a^3} a = a^{k-3} a$

$\qquad \dfrac{a^{k+1}}{a^3} = a^{k-3+1}$ by the recursive definition of a^n

$\qquad \dfrac{a^{k+1}}{a^3} = a^{k-2}$

We have shown that if P_k is true, then P_{k+1} is true.

Conclusion: P_n is true for all integers $n > 3$.

25. Write: P_n: $a^m a^n = a^{m+n}$ $\quad m$ an arbitrary element of N.

Part 1: Show that P_1 is true:

P_1: $a^m a^1 = a^{m+1}$ This is true by the recursive definition of a^n.

Part 2: Show that if P_k is true, then P_{k+1} is true:
Write out P_k and P_{k+1}.

$\qquad P_k$: $a^m a^k = a^{m+k}$

$\qquad P_{k+1}$: $a^m a^{k+1} = a^{m+k+1}$

We start with P_k:

$\qquad a^m a^k = a^{m+k}$

Multiplying both sides by a:

$\qquad a^m a^k a = a^{m+k} a$

$\qquad a^m a^{k+1} = a^{m+k+1}$ by the recursive definition of a^n.

We have shown that if P_k is true, then P_{k+1} is true.

Conclusion: $a^m a^n = a^{m+n}$ is true for arbitrary m, n positive integers.

27. Write: P_n: $x^n - 1 = (x - 1)Q_n(x)$ for some polynomial $Q_n(x)$.

Part 1: Show that P_1 is true:

P_1: $x^1 - 1 = (x - 1)Q_1(x)$ for some $Q_1(x)$. This is true since we can choose
$Q_1(x) = 1$.

Part 2: Show that if P_k is true, then P_{k+1} is true:
Write out P_k and P_{k+1}.

P_k: $x^k - 1 = (x - 1)Q_k(x)$ for some $Q_k(x)$

P_{k+1}: $x^{k+1} - 1 = (x - 1)Q_{k+1}(x)$ for some $Q_{k+1}(x)$

We start with P_k:

$x^k - 1 = (x - 1)Q_k(x)$ for some $Q_k(x)$

Now, $x^{k+1} - 1 = x^{k+1} - x^k + x^k - 1$

$$= x^k(x - 1) + x^k - 1$$

$$= x^k(x - 1) + (x - 1)Q_k(x)$$

$$= (x - 1)(x^k + Q_k(x))$$

$$= (x - 1)Q_{k+1}(x), \text{ where } Q_{k+1}(x) = x^k + Q_k(x)$$

Conclusion: $x^n - 1 = (x - 1)Q_n(x)$ for all positive integers n. Therefore $x^n - 1$ is divisible by $x - 1$ for all positive integers n.

29. Write: P_n: $x^{2n} - 1 = (x - 1)Q_n(x)$ for some polynomial $Q_n(x)$.

Part 1: Show that P_1 is true:

P_1: $x^2 - 1 = (x - 1)Q_1(x)$ for some polynomial $Q_1(x)$. This is true since we can choose $Q_1(x) = x + 1$.

Part 2: Show that if P_k is true, then P_{k+1} is true:
Write out P_k and P_{k+1}.

P_k: $x^{2k} - 1 = (x - 1)Q_k(x)$ for some $Q_k(x)$

P_{k+1}: $x^{2k+2} - 1 = (x - 1)Q_{k+1}(x)$ for some $Q_{k+1}(x)$

Now, $x^{2k+2} - 1 = x^{2k+2} - x^{2k} + x^{2k} - 1$

$$= x^{2k}(x^2 - 1) + x^{2k} - 1$$

$$= x^{2k}(x + 1)(x - 1) + (x - 1)Q_k(x)$$

$$= (x - 1)[x^{2k}(x + 1) + Q_k(x)]$$

$$= (x - 1)Q_{k+1}(x), \text{ where } Q_{k+1}(x) = x^{2k}(x + 1) + Q_k(x)$$

Conclusion: $x^{2n} - 1 = (x - 1)Q_n(x)$ for all positive integers n. Therefore $x^{2n} - 1$ is divisible by $x - 1$ for all positive integers n.

31. P_n: $1^3 + 2^3 + 3^3 + \cdots + n^3 = (1 + 2 + 3 + \cdots + n)^2$

Part 1: Show that P_1 is true:

P_1: $1^3 = 1^2$ True

Part 2: Show that if P_k is true, then P_{k+1} is true:
Write out P_k and P_{k+1}.

P_k: $1^3 + 2^3 + 3^3 + \cdots + k^3 = (1 + 2 + 3 + \cdots + k)^2$

P_{k+1}: $1^3 + 2^3 + 3^3 + \cdots + k^3 + (k + 1)^3 = (1 + 2 + 3 + \cdots + k + k + 1)^2$

We take it as proved in text Matched Problem 2 that

$$1 + 2 + 3 + \cdots + n = \frac{n(n + 1)}{2} \quad \text{for} \quad n \in N$$

So $1 + 2 + 3 + \cdots + k = \dfrac{k(k + 1)}{2}$

$1 + 2 + 3 + \cdots + k + (k + 1) = \dfrac{(k + 1)(k + 2)}{2}$ are known

We start with P_k:

$1^3 + 2^3 + 3^3 + \cdots + k^3 = (1 + 2 + 3 + \cdots + k)^2$

Adding $(k + 1)^3$ to both sides:

$1^3 + 2^3 + 3^3 + \cdots + k^3 + (k + 1)^3 = (1 + 2 + 3 + \cdots + k)^2 + (k + 1)^3$

$= \left[\dfrac{k(k + 1)}{2}\right]^2 + (k + 1)^3$

$= (k + 1)^2\left[\left(\dfrac{k}{2}\right)^2 + (k + 1)\right]$

$= (k + 1)^2\left[\dfrac{k^2}{4} + \dfrac{4(k + 1)}{4}\right]$

$= (k + 1)^2\left[\dfrac{k^2 + 4k + 4}{4}\right]$

$= \dfrac{(k + 1)^2(k + 2)^2}{4}$

$= \left[\dfrac{(k + 1)(k + 2)}{2}\right]^2$

$= [1 + 2 + 3 + \cdots + k + (k + 1)]^2$

Conclusion: P_n is true for all positive integers n.

33. We note:

$$\begin{array}{rcccl}
2 &=& 2 &= 1 \cdot 2 & n = 1 \\
2 + 4 &=& 6 &= 2 \cdot 3 & n = 2 \\
2 + 4 + 6 &=& 12 &= 3 \cdot 4 & n = 3 \\
2 + 4 + 6 + 8 &=& 20 &= 4 \cdot 5 & n = 4
\end{array}$$

Hypothesis: P_n: $2 + 4 + 6 + \cdots + 2n = n(n + 1)$

Proof: Part 1: Show P_1 is true

P_1: $2 = 1 \cdot 2$ True

Part 2: Show that if P_k is true, then P_{k+1} is true:

Write out P_k and P_{k+1}.

P_k: $2 + 4 + 6 + \cdots + 2k = k(k + 1)$

P_{k+1}: $2 + 4 + 6 + \cdots + 2k + (2k + 2) = (k + 1)(k + 2)$

We start with P_k:

$2 + 4 + 6 + \cdots + 2k = k(k + 1)$

Adding $2k + 2$ to both sides:

$2 + 4 + 6 + \cdots + 2k + (2k + 2) = k(k + 1) + 2k + 2$

$= (k + 1)k + (k + 1)2$

$= (k + 1)(k + 2)$

Conclusion: The hypothesis P_n is true for all positive integers n.

35. $n = 1$: no line is determined.
$n = 2$: one line is determined.

$n = 3$: three lines are determined.

$n = 4$: six lines are determined.

$n = 5$: ten lines are determined.

Hypothesis: P_n: n points (no three collinear) determine
$1 + 2 + 3 + \cdots + (n - 1) = \dfrac{n(n - 1)}{2}$ lines, $n \geq 2$.

Proof: Part 1: Show that P_2 is true.

P_2: 2 points determine one line:

$$1 = \frac{2(2 - 1)}{2}$$
$$= \frac{2 \cdot 1}{2}$$
$$= 1 \text{ is true}$$

Part 2: Show that if P_k is true, then P_{k+1} is true:
Write out P_k and P_{k+1}.

P_k: k points determine $1 + 2 + 3 + \cdots + (k - 1) = \dfrac{k(k - 1)}{2}$ lines

P_{k+1}: $k + 1$ points determine $1 + 2 + 3 + \cdots + (k - 1) + k = \dfrac{(k + 1)k}{2}$ lines

We start with P_k:

k points determine $1 + 2 + 3 + \cdots + (k - 1) = \dfrac{k(k - 1)}{2}$ lines

Now, the $k + 1$st point will determine a total of k new lines, one with each of the previously existing k points. These k new lines will be added to the previously existing lines. Therefore, $k + 1$ points determine

$$1 + 2 + 3 + \cdots + (k - 1) + k = \frac{k(k - 1)}{2} + k \text{ lines}$$
$$= k\left[\frac{k - 1}{2} + 1\right] \text{ lines}$$
$$= k\left[\frac{k - 1 + 2}{2}\right] \text{ lines}$$
$$= \frac{k(k + 1)}{2} \text{ lines}$$

Conclusion: The hypothesis P_n is true for $n \geq 2$.

37. P_n: $a > 1 \Rightarrow a^n > 1$, $n \in N$.

Part 1: Show that P_1 is true:

P_1: $a > 1 \Rightarrow a^1 > 1$. This is automatically true.

Part 2: Show that if P_k is true, then P_{k+1} is true:

Write out P_k and P_{k+1}.

  P_k: $a > 1 \Rightarrow a^k > 1$

  P_{k+1}: $a > 1 \Rightarrow a^{k+1} > 1$

We start with P_k. Further, assume $a > 1$ and try to derive $a^{k+1} > 1$. If this succeeds, we have proved P_{k+1}.

Assume $a > 1$.

From P_k we know that $a^k > 1$, also $1 > 0$, so $a > 0$. We may therefore multiply both sides of the inequality $a^k > 1$ by a without changing the direction of the inequality.

  $a^k a > 1a$

  $a^{k+1} > a$

But, $a > 1$ by assumption, so $a^{k+1} > 1$. We have derived this from P_k and $a > 1$. So if P_k is true, then P_{k+1} is true.

Conclusion: P_n is true for all $n \in N$.

39. P_n: $n^2 > 2n$, $n \geq 3$.

Part 1: Show that P_3 is true:

P_3: $3^2 > 2 \cdot 3$

  $9 > 6$ True

Part 2: Show that if P_k is true, then P_{k+1} is true:

Write out P_k and P_{k+1}.

  P_k: $k^2 > 2k$

  P_{k+1}: $(k + 1)^2 > 2(k + 1)$

We start with P_k:

  $k^2 > 2k$

Adding $2k + 1$ to both sides: $k^2 + 2k + 1 > 2k + 2k + 1$

  $(k + 1)^2 > 2k + 2 + 2k - 1$

Now, $2k - 1 > 0$, $k \in N$, so $2k + 2 + 2k - 1 > 2k + 2$.

Therefore, $(k + 1)^2 > 2k + 2$

  $(k + 1)^2 > 2(k + 1)$

So if P_k is true, then P_{k+1} is true.

Conclusion: P_n is true for all integers $n \geq 3$.

41. $3^4 + 4^4 + 5^4 + 6^4 = 2,258$

  $7^4 = 2,401$

$3^4 + 4^4 + 5^4 + 6^4 \neq 7^4$

There is no true obvious generalization of the given facts.

43. To prove $a_n = b_n$, $n \in N$, write:
P_n: $a_n = b_n$.

Proof: Part 1: Show P_1 is true.
P_1: $a_1 = b_1$; $a_1 = 1$; $b_1 = 2 \cdot 1 - 1 = 1$
So $a_1 = b_1$

Part 2: Show that if P_k is true, then P_{k+1} is true:
Write out P_k and P_{k+1}.
   P_k: $a_k = b_k$
   P_{k+1}: $a_{k+1} = b_{k+1}$

We start with P_k:
   $a_k = b_k$

Now, $a_{k+1} = a_{k+1-1} + 2 = a_k + 2 = b_k + 2 = 2k - 1 + 2 = 2k + 1$
$$= 2(k + 1) - 1 = b_{k+1}$$

Therefore, $a_{k+1} = b_{k+1}$

So if P_k is true, then P_{k+1} is true.

Conclusion: P_n is true for all $n \in N$, and $\{a_n\} = \{b_n\}$.

45. To prove: $a_n = b_n$, $n \in N$, write:
P_n: $a_n = b_n$.

Proof: Part 1: Show P_1 is true.
P_1: $a_1 = b_1$; $a_1 = 2$; $b_1 = 2^{2 \cdot 1 - 1} = 2^1 = 2$
so $a_1 = b_1$

Part 2: Show that if P_k is true, then P_{k+1} is true:
Write out P_k and P_{k+1}.
   P_k: $a_k = b_k$
   P_{k+1}: $a_{k+1} = b_{k+1}$

We start with P_k:
   $a_k = b_k$

Now, $a_{k+1} = 2^2 a_{k+1-1} = 2^2 a_k = 2^2 b_k = 2^2 2^{2k-1} = 2^{2+2k-1} = 2^{2(k+1)-1} = b_{k+1}$

Therefore, $a_{k+1} = b_{k+1}$

So if P_k is true, then P_{k+1} is true.

Conclusion: P_n is true for all $n \in N$, and $\{a_n\} = \{b_n\}$.

Section 7-3

1. Answers will vary.   **3.** Answers will vary.

5. (A) Since $(-16) - (-11) = -5$ and $(-21) - (-16) = -5$, the given terms can start an arithmetic sequence with $d = -5$. Then next terms are then $-21 + (-5) = -26$, and $(-26) + (-5) = -31$.

(B) Since $(-4) - 2 = -6$ and $8 - (-4) = 12$, there is no common difference. Since $(-4) \div 2 = -2$ and $8 \div (-4) = -2$, the given terms can start a geometric sequence with $r = -2$. The next terms are then $8 \cdot (-2) = -16$, and $(-16) \cdot (-2) = 32$.

(C) Since $4 - 1 \neq 9 - 4$, there is no common difference, so the sequence is not an arithmetic sequence. Since $4 \div 1 \neq 9 \div 4$, there is no common ratio, so the sequence is not geometric either.

(D) Since $\frac{1}{6} - \frac{1}{2} = -\frac{1}{3}$ and $\frac{1}{18} - \frac{1}{6} = -\frac{1}{9}$, there is no common difference. Since $\frac{1}{6} \div \frac{1}{2} = \frac{1}{3}$ and $\frac{1}{18} \div \frac{1}{6} = \frac{1}{3}$, the given terms can start a geometric sequence with $r = \frac{1}{3}$. The next terms are then $\frac{1}{18} \cdot \frac{1}{3} = \frac{1}{54}$ and $\frac{1}{54} \cdot \frac{1}{3} = \frac{1}{162}$.

7. $a_2 = a_1 + d = -5 + 4 = -1$
$a_3 = a_2 + d = -1 + 4 = 3$
$a_4 = a_3 + d = 3 + 4 = 7$
$a_n = a_1 + (n - 1)d$

9. $a_{15} = a_1 + 14d = -3 + 14 \cdot 5 = 67$
$$S_n = \frac{n}{2}[2a_1 + (n - 1)d]$$
$$S_{11} = \frac{11}{2}[2(-3) + (11 - 1)5]$$
$$= \frac{11}{2}(44)$$
$$= 242$$

11. $a_2 - a_1 = 5 - 1 = d = 4$
$$S_n = \frac{n}{2}[2a_1 + (n - 1)d]$$
$$S_{21} = \frac{21}{2}[2 \cdot 1 + (21 - 1)4]$$
$$= \frac{21}{2}(82)$$
$$= 861$$

13. $a_2 - a_1 = 5 - 7 = -2 = d$
$a_n = a_1 + (n - 1)d$
$a_{15} = 7 + (15 - 1)(-2)$
$= -21$

15. $a_2 = a_1 r = (-6)\left(-\frac{1}{2}\right) = 3$
$a_3 = a_2 r = 3\left(-\frac{1}{2}\right) = -\frac{3}{2}$
$a_4 = a_3 r = \left(-\frac{3}{2}\right)\left(-\frac{1}{2}\right) = \frac{3}{4}$

17. $a_n = a_1 r^{n-1}$
$a_{10} = 81\left(\frac{1}{3}\right)^{10-1}$
$= \frac{1}{243}$

19. $S_n = \frac{a_1 - ra_n}{1 - r}$
$S_7 = \frac{3 - 3(2,187)}{1 - 3}$
$= 3,279$

21. $a_n = a_1 + (n - 1)d$
$a_{20} = a_1 + 19d$
$117 = 3 + 19d$
$114 = 19d$
So, $d = 6$. Therefore
$a_{101} = a_1 + (100)d$
$= 3 + 100(6) = 603$

23. $S_n = \frac{n}{2}(a_1 + a_n)$
$S_{40} = \frac{40}{2}(-12 + 22)$
$= 200$

25. $a_2 - a_1 = d$
$\frac{1}{2} - \frac{1}{3} = \frac{1}{6} = d$
$a_n = a_1 + (n - 1)d$
$a_{11} = \frac{1}{3} + (11 - 1)\frac{1}{6}$
$= 2$
$S_n = \frac{n}{2}(a_1 + a_n)$
$S_{11} = \frac{11}{2}\left(\frac{1}{3} + 2\right) = \frac{77}{6}$

27. $a_n = a_1 + (n - 1)d$
$a_{10} = a_1 + 9d$
$a_3 = a_1 + 2d$
Eliminating d between these two statements by addition, we have
$2a_{10} = 2a_1 + 18d$
$-9a_3 = -9a_1 - 18d$
$\overline{2a_{10} - 9a_3 = -7a_1}$

$a_1 = \frac{2a_{10} - 9a_3}{-7} = \frac{2(55) - 9(13)}{-7} = 1$

29.
$$a_n = a_1 r^{n-1}$$
$$1 = 100 r^{6-1}$$
$$\frac{1}{100} = r^5$$
$$r = \sqrt[5]{0.01} = 10^{-2/5}$$
$$r = 0.398$$

31.
$$S_n = \frac{a_1 - a_1 r^n}{1 - r}$$
$$S_{10} = \frac{5 - 5(-2)^{10}}{1 - (-2)}$$
$$= \frac{-5,115}{3}$$
$$= -1,705$$

33. First find r:
$$a_n = a_1 r^{n-1}$$
$$\frac{8}{3} = 9 r^{4-1}$$
$$\frac{8}{27} = r^3$$
$$r = \frac{2}{3}$$
$$a_2 = r a_1 = \frac{2}{3}(9) = 6$$
$$a_3 = r a_2 = \frac{2}{3}(6) = 4$$

35.
$$a_n = a_1 + (n - 1)d$$
$$d = a_2 - a_1$$
$$= (3 \cdot 2 + 3) - (3 \cdot 1 + 3) = 3$$
$$a_{51} = a_1 + (51 - 1)d$$
$$= (3 \cdot 1 + 3) + 50 \cdot 3$$
$$= 156$$
$$S_n = \frac{n}{2}(a_1 + a_n)$$
$$S_{51} = \frac{51}{2}(6 + 156)$$
$$= \frac{51}{2}[162]$$
$$= 4,131$$

37.
$$S_n = \frac{a_1 - a_1 r^n}{1 - r}$$
First, note $a_1 = (-3)^{1-1} = 1$
$$r = \frac{a_2}{a_1} = \frac{(-3)^{2-1}}{(-3)^{1-1}} = -3$$
$$S_7 = \frac{1 - 1(-3)^7}{1 - (-3)}$$
$$= \frac{2,188}{4}$$
$$= 547$$

39.
$$g(t) = 5 - t$$
$$g(1) = 5 - 1 = 4$$
$$g(51) = 5 - 51 = -46$$
$$g(1) + g(2) + g(3) + \cdots + g(51) = S_{51}$$
$$S_n = \frac{n}{2}(a_1 + a_n)$$
$$S_{51} = \frac{51}{2}(g(1) + g(51))$$
$$= \frac{51}{2}[4 + (-46)]$$
$$= -1,071$$

41. $g(1) + g(2) + \cdots + g(10)$ is a geometric series, with
$$g(1) = a_1 = \left(\frac{1}{2}\right)^1 = \frac{1}{2}$$
$$r = \frac{g(2)}{g(1)} = \frac{\left(\frac{1}{2}\right)^2}{\left(\frac{1}{2}\right)^1} = \frac{1}{2}$$
$$S_n = \frac{a_1 - a_1 r^n}{1 - r}$$
$$S_{10} = \frac{\frac{1}{2} - \frac{1}{2}\left(\frac{1}{2}\right)^{10}}{1 - \frac{1}{2}}$$
$$= \frac{1 - \left(\frac{1}{2}\right)^{10}}{2 - 1}$$
$$= \frac{1,023}{1,024}$$

43. First, find n:
$$a_n = a_1 + (n - 1)d$$
$$134 = 22 + (n - 1)2$$
$$n = 57$$
Now, find S_{57}
$$S_n = \frac{n}{2}(a_1 + a_n)$$
$$S_{57} = \frac{57}{2}(22 + 134)$$
$$= 4,446$$

45. To prove:
$$1 + 3 + 5 + \cdots + (2n - 1) = n^2$$
The sequence 1, 3, 5, $\cdots$ is an arithmetic sequence, with $d = 2$. We are to find S_n.

But, $S_n = \dfrac{n}{2}(a_1 + a_n)$

$\qquad = \dfrac{n}{2}[1 + (2n - 1)]$

$\qquad = \dfrac{n}{2} \cdot 2n$

So, $S_n = n^2$

47. $\dfrac{a_2}{a_1} = \dfrac{a_3}{a_2} = r$ for $a_1 + a_2 + a_3$ to be a geometric series. Then,

$\dfrac{x}{-2} = \dfrac{-6}{x}$

$x^2 = 12$

$\quad x = 2\sqrt{3}$,

since x is specified positive.

49. Note that $a_n - a_{n-1} = 3$, so this is an arithmetic sequence, with $d = 3$. We are to find a_n.

But $a_n = a_1 + (n - 1)d$

So $a_n = -3 + (n - 1)3$ or $3n - 6$

51. Here's what an arithmetic sequence with $d = 0$ would look like:

$\quad a_1, a_1, a_1, a_1, \cdots$

So the sum S_n is given by $S_n = a_1 + a_1 + \cdots + a_1$ where there are n terms in the sum.   But this is exactly how we define multiplication!   A simple form is $S_n = n \cdot a_1$.

53.

Plot1 Plot2 Plot3 nMin=1 \u(n)🗐5+8n u(nMin)🗐 \v(n)🗐(1.1)^n v(nMin)🗐 \w(n)= w(nMin)=	

Enter the sequences in the sequence editor.

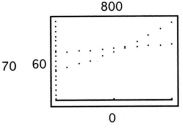

Graph the sequences.

n	u(n)	v(n)
60	485	304.48
61	493	334.93
62	501	368.42
63	509	405.27
64	517	445.79
65	525	490.37
66	533	539.41

u(n)🗐5+8n

Display the points (in the appropriate interval) in a table.

From the second graph, 66 is the least positive integer n such that $a_n < b_n$. The table confirms this.

55.

Plot1 Plot2 Plot3 nMin=1 \u(n)🗐1000(.99)^ n u(nMin)🗐 \v(n)🗐2n+1 v(nMin)🗐 \w(n)=	

Enter the sequences in the sequence editor.

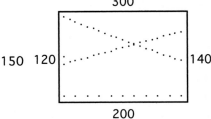

Graph the sequences.

n	u(n)	v(n)
130	270.75	261
131	268.05	263
132	265.37	265
133	262.71	267
134	260.09	269
135	257.48	271
136	254.91	273

u(n)🗐1000(.99)^n

Display the points (in the appropriate interval) in a table.

From the second graph, 133 is the least positive integer n such that $a_n < b_n$. The table confirms this.

57. $\dfrac{a_2}{a_1} = 1 \div 3 = \dfrac{1}{3} = r. \; |r| < 1$

Therefore, this infinite geometric series has a sum.

$S_\infty = \dfrac{a_1}{1 - r}$

$\quad = \dfrac{3}{1 - \frac{1}{3}} = \dfrac{9}{2}$

59. $\dfrac{a_2}{a_1} = \dfrac{4}{2} = 2 = r \geq 1$

Therefore, this infinite geometric series has no sum.

61. $\dfrac{a_2}{a_1} = \left(-\dfrac{1}{2}\right) \div 2 = -\dfrac{1}{4} = r. \; |r| < 1$

Therefore, this infinite geometric series has a sum.

$S_\infty = \dfrac{a_1}{1 - r}$

$\quad = \dfrac{2}{1 - \left(-\frac{1}{4}\right)}$

$\quad = \dfrac{8}{5}$

63. $0.\overline{7} = 0.777\ldots = 0.7 + 0.07 + 0.007$
$\qquad\qquad\qquad\qquad\quad + 0.0007 + \ldots$

This is an infinite geometric series with
$a_1 = 0.7$ and $r = 0.1$.
So

$S_\infty = \dfrac{a_1}{1 - r} = \dfrac{0.7}{1 - 0.1} = \dfrac{0.7}{0.9} = \dfrac{7}{9}$

65. $0.\overline{54} \ldots = 0.54 + 0.0054 + 0.000054 + \ldots$
This is an infinite geometric series with $a_1 = 0.54$ and $r = 0.01$. So

$S_\infty = \dfrac{a_1}{1 - r} = \dfrac{0.54}{1 - 0.01} = \dfrac{0.54}{0.99} = \dfrac{6}{11}$

67. $3.\overline{216} = 3.216216216\ldots = 3 + 0.216 + 0.000216 + 0.000000216 + \ldots$
Therefore, we note: $0.216 + 0.000216 + 0.000000216 + \ldots$ is an infinite geometric series with $a_1 = 0.216$ and $r = 0.001$. So

$3.\overline{216} = 3 + S_\infty = 3 + \dfrac{a_1}{1 - r} = 3 + \dfrac{0.216}{1 - 0.001} = 3 + \dfrac{0.216}{0.999} = 3\dfrac{8}{37}$ or $\dfrac{119}{37}$

69. Write: P_n: $a_n = a_1 + (n - 1)d$

Proof: Part 1: Show that P_1 is true:
P_1: $a_1 = a_1 + (1 - 1)d = a_1$ P_1 is true.

Part 2: Show that if P_k is true, then P_{k+1} is true:
Write out P_k and P_{k+1}.

$\quad P_k$: $a_k = a_1 + (k - 1)d$
$\quad P_{k+1}$: $a_{k+1} = a_1 + kd$
We start with P_k:
$\quad a_k = a_1 + (k - 1)d$
Adding d to both sides:
$\quad a_k + d = a_1 + (k - 1)d + d$
$\qquad a_{k+1} = a_1 + d[(k - 1) + 1]$
$\qquad a_{k+1} = a_1 + kd$
So if P_k is true, then P_{k+1} is true.

Conclusion: P_n is true for all positive integers n.

71. This is a geometric sequence with ratio $\dfrac{a_n}{a_{n-1}} = -3$.

$$a_n = a_1 r^{n-1}$$
$$= (-2)(-3)^{n-1}$$

73. Assume x, y, z are consecutive terms of an arithmetic progression.
Then, $y - x = d$, $z - y = d$
To show $a = x^2 + xy + y^2$, $b = z^2 + xz + x^2$, $c = y^2 + yz + z^2$, are consecutive terms of an arithmetic progression, we need only show that
$b - a = c - b$

$$
\begin{aligned}
b - a &= z^2 + xz + x^2 - (x^2 + xy + y^2) \\
&= z^2 + xz - xy - y^2 \\
&= z^2 + x(z - y) - y^2 \\
&= z^2 - y^2 + x(z - y) \\
&= (z - y)(z + y + x) \\
&= d(x + y + z)
\end{aligned}
$$

$$
\begin{aligned}
c - b &= y^2 + yz + z^2 - (z^2 + xz + x^2) \\
&= y^2 + yz - xz - x^2 \\
&= y^2 - x^2 + yz - xz \\
&= y^2 - x^2 + z(y - x) \\
&= (y - x)[y + x + z] \\
&= d(x + y + z)
\end{aligned}
$$

Therefore, a, b, c, that is, $x^2 + xy + y^2$, $z^2 + xz + x^2$, $y^2 + yz + z^2$ are consecutive terms of an arithmetic progression.

75. Write: P_n: $a_n = a_1 r^{n-1}$, $n \in N$

Proof: Part 1: Show that P_1 is true:

$$
\begin{aligned}
P_1: a_1 &= a_1 r^{1-1} \\
&= a_1 r^0 \\
&= a_1 \quad \text{True.}
\end{aligned}
$$

Part 2: Show that if P_k is true, then P_{k+1} is true:
Write out P_k and P_{k+1}.

$$P_k: a_k = a_1 r^{k-1}$$
$$P_{k+1}: a_{k+1} = a_1 r^k$$

We start with P_k:

$$a_k = a_1 r^{k-1}$$

Multiply both sides by r:

$$a_k\, r = a_1 r^{k-1} r$$
$$a_k r = a_1 r^k$$

But, we are given that $a_{k+1} = r a_k = a_k r$.
Then $a_{k+1} = a_1 r^k$
So if P_k is true, then P_{k+1} is true.

Conclusion: P_n is true for all $n \in N$.

77. Given a, b, c, d, e, f is an arithmetic progression, let D be the common difference. Then, assuming $D \neq 0$:

$$b = a + D, \ c = a + 2D, \ d = a + 3D, \ e = a + 4D, \ f = a + 5D$$

From the solution to Exercise 5-1, problem 39, we know that there will be a solution of

$$ax + by = c$$
$$dx + ey = f$$

if $\ ae - bd \neq 0$

In this case, $ae - bd = a(a + 4D) - (a + 3D)(a + D)$

$$= a^2 + 4aD - (a^2 + 4aD + 3D^2)$$

$$= -3D^2 \neq 0$$

Then, from the same problem, we have

$$x = \frac{ce - bf}{ae - bd} = \frac{(a + 2D)(a + 4D) - (a + D)(a + 5D)}{-3D^2}$$

$$= \frac{a^2 + 6aD + 8D^2 - (a^2 + 6aD + 5D^2)}{-3D^2}$$

$$= \frac{3D^2}{-3D^2}$$

$$= -1$$

$$y = \frac{af - dc}{ae - bd} = \frac{a(a + 5D) - (a + 3D)(a + 2D)}{-3D^2}$$

$$= \frac{a^2 + 5aD - (a^2 + 5aD + 6D^2)}{-3D^2}$$

$$= \frac{-6D^2}{-3D^2}$$

$$= 2$$

79. We can use the formula for a sum from Theorem 5.

$$S_5 = \frac{1 - 2^5}{1 - 2} = \frac{-31}{-1} = 31$$

$$S_{10} = \frac{1 - 2^{10}}{1 - 2} = \frac{-1,023}{-1} = 1,023$$

$$S_{15} = \frac{1 - 2^{15}}{1 - 2} = \frac{-32,767}{-1} = 32,767$$

$$S_{20} = \frac{1 - 2^{20}}{1 - 2} = \frac{-1,048,575}{-1} = 1,048,575$$

It looks like S_n continues to get larger and approaches ∞ as n approaches ∞.

81. With each firm, the salaries form an arithmetic sequence. We are asked for the sum of fifteen terms, or S_{15}, given a_1 and d.

$$S_n = \frac{n}{2}[2a_1 + (n - 1)d\ ]$$

$$S_{15} = \frac{15}{2}[2a_1 + 14d\ ] = 15(a_1 + 7d)$$

Firm A: $a_1 = 25,000 \quad d = 1,200 \quad S_{15} = 15(25,000 + 7 \cdot 1,200) = \$501,000$

Firm B: $a_1 = 28,000 \quad d = 800 \quad S_{15} = 15(28,000 + 7 \cdot 800) = \$504,000$

83. We are asked for the sum of an infinite geometric series.

$a_1 = \$800,000$

$r = 0.8 \quad |r| \le 1,$

so the series has a sum,

$S_\infty = \dfrac{a_1}{1 - r}$

$\quad = \dfrac{\$800,000}{1 - 0.8}$

$\quad = \dfrac{\$800,000}{0.2}$

$\quad = \$4,000,000$

85. After one year, $P(1 + r)$ is present, so the geometric sequence has $a_1 = P(1 + r)$. The ratio is given as $(1 + r)$, so

$a_n = a_1(1 + r)^{n-1}$

$\quad = P(1 + r)(1 + r)^{n-1}$

$\quad = P(1 + r)^n$ is the amount present after n years.

The time taken for P to double is represented by n years. We set $A = 2P$, $r = 0.06$, then solve

$2P = P(1 + 0.06)^n$

$2 = (1.06)^n$

$\log 2 = n \log 1.06$

$n = \dfrac{\log 2}{\log 1.06}$

$\quad \approx 12$ years

87. This involves an arithmetic sequence. Let d = increase in earnings each year.

$a_1 = 7,000$

$a_{11} = 14,000$

$a_n = a_1 + (n - 1)d$

$14,000 = 7,000 + (11 - 1)d$

$d = \$700$

The amount of money received over the 11 years is

$S_{11} = a_1 + a_2 + \cdots + a_{11}$

$S_n = \dfrac{n}{2}(a_1 + a_n)$

$S_{11} = \dfrac{11}{2}(7,000 + 14,000) = \$115,500$

89. We are asked for the sum of an infinite geometric series.

a_1 = number of revolutions in the first minute = 300

$r = \dfrac{2}{3} \quad |r| < 1,$ so the series has a sum,

$S_\infty = \dfrac{a_1}{1 - r} = \dfrac{300}{1 - \frac{2}{3}} = 900$ revolutions.

91. This involves a geometric sequence. Let $a_n = 2,000$ calories. There are five stages: $n = 5$. We require a_1 on the assumption that $r = 20\% = \dfrac{1}{5}$.

$$a_n = a_1 r^{n-1}$$
$$2000 = a_1 \left(\frac{1}{5}\right)^{5-1}$$
$$a_1 = 2{,}000 \cdot 5^4$$
$$= 1{,}250{,}000 \text{ calories}$$

93. We have an arithmetic sequence, 16, 48, 80, $\cdots$, with $a_1 = 16$, $d = 32$.

(A) This requires a_{11}: $\quad a_n = a_1 + (n - 1)d$
$$a_{11} = 16 + (11 - 1)32$$
$$= 336 \text{ feet}$$

(B) This requires s_{11}: $\quad s_n = \dfrac{n}{2}(a_1 + a_n)$
$$s_{11} = \frac{11}{2}(16 + 336)$$
$$= 1{,}936 \text{ feet}$$

(C) This requires s_t: $\quad s_n = \dfrac{n}{2}[2a_1 + (n - 1)d]$
$$s_t = \frac{t}{2}[2a_1 + (t - 1)32]$$
$$= \frac{t}{2}[32 + (t - 1)32]$$
$$= \frac{t}{2} \cdot 32t$$
$$= 16t^2 \text{ feet}$$

95. This involves a geometric sequence. Let $a_1 = 2A_0$ = number present after 1 half-hour period. In t hours, $2t$ half-hours will have elapsed, so $n = 2t$; $r = 2$, since the number of bacteria doubles in each period.

$$a_n = a_1 r^{n-1}$$
$$a_{2t} = 2A_0 2^{2t-1}$$
$$= A_0 2^1 2^{2t-1}$$
$$= A_0 2^{2t}$$

97. If b_n is the brightness of an nth-magnitude star, we find r for the geometric progression $b_1, b_2, b_3, \cdots$, given $b_1 = 100b_6$.

$$b_n = b_1 r^{n-1}$$
$$b_6 = b_1 r^{6-1}$$
$$b_6 = b_1 r^5$$
$$b_6 = 100b_6 r^5$$
$$\frac{1}{100} = r^5$$
$$r = \sqrt[5]{0.01} = 10^{-0.4} = 0.398$$

99. This involves a geometric sequence. a_1 = amount of money on first square; $a_1 = \$0.01$, $r = 2$; a_n = amount of money on nth square.

$$a_n = a_1 r^{\,n-1}$$
$$a_{64} = 0.01 \cdot 2^{64-1}$$
$$= 9.22 \times 10^{16} \text{ dollars}$$

The amount of money on the whole board $= a_1 + a_2 + a_3 + \cdots + a_{64} = S_{64}$.

$$S_n = \frac{a_1 - r a_n}{1 - r}$$
$$S_{64} = \frac{0.01 - 2(9.223 \times 10^{16})}{1 - 2}$$
$$= 1.845 \times 10^{17} \text{ dollars}$$

101. This involves a geometric sequence. Let $a_1 = 15$ = pressure at sea level, $r = \dfrac{1}{10}$ = factor of decrease. We require a_5 = pressure after four 10-mile increases in altitude.

$$a_5 = 15\left(\frac{1}{10}\right)^{5-1}$$

$$= 0.0015 \text{ pounds per square inch.}$$

103. This involves an infinite geometric series. a_1 = perimeter of first triangle = 1, $r = \dfrac{1}{2}$, $|r| < 1$, so the series has a sum.

$$S_\infty = \frac{a_1}{1 - r} = \frac{1}{1 - \frac{1}{2}} = 2$$

105.

From the figure, and ordinary induction, we can see that the interior angles of an $n + 2$-sided polygon, $n = 1, 2, 3, \cdots$, are those of n triangles, that is, $180°n$. So for the sequence of interior angles $\{a_n\}$,

$a_n - a_{n-1} = 180°n - 180°(n - 1) = 180° = d.$
A detailed proof by mathematical induction is omitted.

For a 21-sided polygon, we use $a_n = a_1 + (n - 1)d$ with $a_1 = 180°$, $d = 180°$, and $n + 2 = 21$, in which case $n = 19$.
$a_{19} = 180° + (19 - 1)180°$
$\qquad = 3,420°$

Section 7-4

1. Answers will vary.   **3.** Answers will vary.   **5.** Answers will vary.

7. $9! = 362,880$ **9.** $11! = 39,916,800$

11. $\dfrac{11!}{8!} = \dfrac{11 \cdot 10 \cdot 9 \cdot 8!}{8!} = 990$ **13.** $\dfrac{5!}{2!\,3!} = \dfrac{5 \cdot 4 \cdot 3!}{2 \cdot 1 \cdot 3!} = 10$

15. $\dfrac{7!}{4!\,(7 - 4)!} = \dfrac{7!}{4!\,3!} = \dfrac{7 \cdot 6 \cdot 5 \cdot 4!}{4!\,3 \cdot 2 \cdot 1} = 35$ **17.** $\dfrac{7!}{7!\,(7 - 7)!} = \dfrac{7!}{7!\,0!} = \dfrac{7!}{7!\,(1)} = 1$

19. The alleged solution for number 11 is incorrect. The function "11 n Pr8" represents the permutations of 11 objects taken 8 at a time, which is calculated by
$$\frac{11!}{(11 - 8)!} = \frac{11!}{3!}$$
This is different from problem 11, which is $11!/8!$.
The alleged solution for number 13 is incorrect also. The calculator interprets the input $5!/2!3!$ as $\dfrac{5!}{2!} \cdot 3!$. The correct calculation is $5!/(2!3!)$.
The solution for number 15 is correct.

21. $P_{5,3} = \dfrac{5!}{(5-3)!} = \dfrac{5!}{2!} = \dfrac{5 \cdot 4 \cdot 3 \cdot 2!}{2!} = 60$

23. $P_{52,4} = \dfrac{52!}{(52-4)!} = \dfrac{52!}{48!} = \dfrac{52 \cdot 51 \cdot 50 \cdot 49 \cdot 48!}{48!} = 6{,}497{,}400$

25. $C_{5,3} = \dfrac{5!}{3!(5-3)!} = \dfrac{5!}{3!\,2!} = 10$

(See problem 13)

27. $C_{52,4} = \dfrac{52!}{4!(52-4)!} = \dfrac{P_{52,4}}{4!} = \dfrac{6{,}497{,}400}{4 \cdot 3 \cdot 2 \cdot 1} = 270{,}725$

29. (A) Order is important. The selection is a permutation.

(B) Order is not important. The selection is a combination.

31. Order is important here. We use permutations, selecting, in order, three horses out of ten:

$P_{10,3} = 10 \cdot 9 \cdot 8 = 720$ different finishes

33. For the subcommittee, order is not important. We use combinations, selecting three persons out of seven:

$C_{7,3} = \dfrac{7!}{3!(7-3)!} = \dfrac{7!}{3!\,4!} = \dfrac{7 \cdot 6 \cdot 5 \cdot 4!}{3 \cdot 2 \cdot 1 \cdot 4!} = 35$ subcommittees

In choosing a president, a vice-president, and a secretary, we can use permutations, or apply the multiplication principle.

O_1: Selecting the president N_1: 7 ways

O_2: Selecting the vice-president N_2: 6 ways (the president is not considered)

O_3: Selecting the secretary N_3: 5 ways (the president and vice-president are not considered)

There are $N_1 \cdot N_2 \cdot N_3 = 7 \cdot 6 \cdot 5 \; (= P_{7,3}) = 210$ ways.

35. For each game, we are selecting two teams out of ten to be opponents. Since the order of the opponents does not matter (this has nothing to do with the order in which the games might be played, which is not under discussion here), we use combinations.

$C_{10,2} = \dfrac{10!}{2!(10-2)!} = \dfrac{10!}{2!\,8!} = \dfrac{10 \cdot 9 \cdot 8!}{2 \cdot 1 \cdot 8!} = 45$ games

37.

	No letter can be repeated	Allowing letters to repeat
O_1: Selecting first letter	N_1: 6 ways	N_1: 6 ways
O_2: Selecting second letter	N_2: 5 ways	N_2: 6 ways
O_3: Selecting third letter	N_3: 4 ways	N_3: 6 ways
O_4: Selecting fourth letter	N_4: 3 ways	N_4: 6 ways
	$P_{6,4} = 6 \cdot 5 \cdot 4 \cdot 3 = 360$ possible code words	$6 \cdot 6 \cdot 6 \cdot 6 = 1{,}296$ possible code words

39.

	No digit can be repeated	Allowing digits to repeat
O_1: Selecting first digit	N_1: 10 ways	N_1: 10 ways
O_2: Selecting second digit	N_2: 9 ways	N_2: 10 ways
⋮	⋮	⋮
O_5: Selecting fifth digit	N_5: 6 ways	N_5: 10 ways

$P_{10,5} = 10 \cdot 9 \cdot 8 \cdot 7 \cdot 6 =$ $10 \cdot 10 \cdot 10 \cdot 10 \cdot 10 =$

$100,000$

$30,240$ lock combinations   lock combinations

41. We are selecting five cards out of the 13 hearts in the deck. The order is not important, so we use combinations.

$$C_{13,5} = \frac{13!}{5!(13-5)!} = \frac{13!}{5!\,8!} = \frac{13 \cdot 12 \cdot 11 \cdot 10 \cdot 9 \cdot 8!}{5 \cdot 4 \cdot 3 \cdot 2 \cdot 1 \cdot 8!} = 1,287$$

43.

	Repeats allowed	No repeats allowed
O_1: Selecting first letter	N_1: 26 ways	N_1: 26 ways
O_2: Selecting second letter	N_2: 26 ways	N_2: 25 ways
O_3: Selecting third letter	N_3: 26 ways	N_3: 24 ways
O_4: Selecting first digit	N_4: 10 ways	N_4: 10 ways
O_5: Selecting second digit	N_5: 10 ways	N_5: 9 ways
O_6: Selecting third digit	N_6: 10 ways	N_6: 8 ways

$26 \cdot 26 \cdot 26 \cdot 10 \cdot 10 \cdot 10 =$ $26 \cdot 25 \cdot 24 \cdot 10 \cdot 9 \cdot 8 =$
$17,576,000$ license plates $11,232,000$ license plates

45. O_1: Choosing 5 spades out of 13 possible (order is not important)
N_1: $C_{13,5}$
O_2: Choosing 2 hearts out of 13 possible (order is not important)
N_2: $C_{13,2}$
Using the multiplication principle, we have:
Number of hands $= C_{13,5} \cdot C_{13,2} = \dfrac{13!}{5!(13-5)!} \cdot \dfrac{13!}{2!(13-2)!}$

$\qquad = 1,287 \cdot 78$
$\qquad = 100,386$

47. O_1: Selecting the color N_1: 5 ways
O_2: Selecting the transmission N_2: 3 ways
O_3: Selecting the interior N_3: 4 ways
O_4: Selecting the engine N_4: 2 ways
Applying the multiplication principle, there are $5 \cdot 3 \cdot 4 \cdot 2 = 120$ variations of the car.

49. O_1: Choosing 3 appetizers out of 8 possible (order is not important)
N_1: $C_{8,3}$
O_2: Choosing 4 main courses out of 10 possible (order is not important)
N_2: $C_{10,4}$
O_3: Choosing 2 desserts out of 7 possible (order is not important)
N_3: $C_{7,2}$
Using the multiplication principle, we have:
$$\text{Number of banquets} = C_{8,3} \cdot C_{10,4} \cdot C_{7,2} = \frac{8!}{3!(8-3)!} \cdot \frac{10!}{4!(10-4)!} \cdot \frac{7!}{2!(7-2)!}$$
$$= 56 \cdot 210 \cdot 21$$
$$= 246,960$$

51. O_1: Choosing a left glove out of 12 possible
N_1: 12
O_2: Choosing a right glove out of all the right gloves that do *not* match the left glove already chosen
N_2: 11

Using the multiplication principle, we have:
Number of ways to mismatch gloves = $12 \cdot 11 = 132$.

53. The number of ways to deal a hand containing exactly 1 king is computed as follows:
O_1: Choosing 1 king out of 4 possible (order is not important)
N_1: $C_{4,1}$

O_2: Choosing 4 remaining cards out of 48 cards that are not kings (order is not important)

N_2: $C_{48,4}$

Using the multiplication principle, we have

$$N_1 \cdot N_2 = C_{4,1} \cdot C_{48,4} = \frac{4!}{1!(4-1)!} \cdot \frac{48!}{4!(48-4)!}$$

$$= \frac{4!}{1! \, 3!} \cdot \frac{48!}{4! \, 44!}$$

$$= 4 \cdot 194,580$$
$$= 778,320 \text{ ways}$$

The number of ways to deal a hand containing no hearts is

$$C_{39,5} = \frac{39!}{5!(39-5)!} = \frac{39!}{5! \, 34!} = 575,757 \text{ ways}$$

The hand that contains exactly one king is more likely.

55. To seat two people, we can seat one person, then the second person.
O_1: Seat the first person in any chair
N_1: 5 ways
O_2: Seat the second person in any remaining chair
N_2: 4 ways
Applying the multiplication principle, we can seat two persons in
$N_1 \cdot N_2 = 5 \cdot 4 = 20$ ways.

We can continue this reasoning for a third person.

O_3: Seat the third person in any of the three remaining chairs

N_3: 3 ways

We can seat 3 persons in $5 \cdot 4 \cdot 3 = 60$ ways.

For a fourth person:

O_4: Seat the fourth person in any of the two remaining chairs

N_4: 2 ways

We can seat 4 persons in $5 \cdot 4 \cdot 3 \cdot 2 = 120$ ways.

For a fifth person:

O_5: Seat the fifth person. There will be only one chair remaining.

N_5: 1 way

We can seat 5 persons in $5 \cdot 4 \cdot 3 \cdot 2 \cdot 1 = 120$ ways

57. (A) Order is important, so we use permutations, selecting, in order, 5 persons out of 8:

$$P_{8,5} = 8 \cdot 7 \cdot 6 \cdot 5 \cdot 4 = 6{,}720 \text{ teams.}$$

(B) Order is not important, so we use combinations, selecting 5 persons out of 8

$$C_{8,5} = \frac{8!}{5!(8-5)!} = \frac{8!}{5! \, 3!} = \frac{8 \cdot 7 \cdot 6 \cdot 5!}{5! \cdot 3 \cdot 2 \cdot 1} = 56 \text{ teams}$$

(C) O_1: Selecting either Mike or Ken out of {Mike, Ken}

N_1: $C_{2,1}$

O_2: Selecting the 4 remaining players out of the 6 possibilities that do not include either Mike or Ken.

N_2: $C_{6,4}$

Using the multiplication principle, we have

$$
\begin{aligned}
N_1 \cdot N_2 = C_{2,1} \cdot C_{6,4} &= \frac{2!}{1!(2-1)!} \cdot \frac{6!}{4!(6-4)!} \\
&= \frac{2!}{1! \, 1!} \cdot \frac{6!}{4! \, 2!} \\
&= 2 \cdot 15 \\
&= 30 \text{ teams}
\end{aligned}
$$

59. (A) We are choosing 2 points out of the 8 to join by a chord. Order is not important.

$$C_{8,2} = \frac{8!}{2!(8-2)!} = 28 \text{ chords}$$

(B) No three of the points can be collinear, since no line intersects a circle at more than two points. So we can select any three of the 8 to use as vertices of the triangle. Order is not important.

$$C_{8,3} = \frac{8!}{3!(8-3)!} = 56 \text{ triangles}$$

(C) We can select any four of the eight points to use as vertices of a quadrilateral. Order is not important.

$$C_{8,4} = \frac{8!}{4!(8-4)!} = 70 \text{ quadrilaterals}$$

61. (A) $P_{10,0} = 1$ $0! = 1$

$P_{10,1} = 10$ $1! = 1$

$P_{10,2} = 90$ $2! = 2$

$P_{10,3} = 720$ $3! = 6$

$P_{10,4} = 5{,}040$ $4! = 24$

$$P_{10,5} = 30,240 \qquad 5! = 120$$
$$P_{10,6} = 151,200 \qquad 6! = 720$$
$$P_{10,7} = 604,800 \qquad 7! = 5,040$$
$$P_{10,8} = 1,814,400 \qquad 8! = 40,320$$
$$P_{10,9} = 3,628,800 \qquad 9! = 362,880$$
$$P_{10,10} = 3,628,800 \qquad 10! = 3,628,800$$

We can see that $P_{10,r} \geq r!$ for $r = 0, 1, \dots, 10$

(B) If $r = 0$, $P_{10,0} = 0! = 1$. If $r = 10$, $P_{10,10} = 10! = 3,628,800$

(C) $P_{n,r}$ and $r!$ are each the product of r consecutive integers, the largest of which is n for $P_{n,r}$ and r for $r!$. So if $r \leq n$, $P_{n,r} \geq r!$

Section 7-5

1. Answers will vary.   3. Answers will vary.   5. Answers will vary.

7. Event E is certain to occur.

9. (A) Must be rejected. It violates condition 1; no probability can be negative.
    (B) Must be rejected. It violates condition 2, since $P(R) + P(G) + P(Y) + P(B) \neq 1$.
    (C) This is an acceptable probability assignment.

11. This is a compound event made up of the simple events R and Y.
    $P(E) = P(R) + P(Y) = .26 + .30 = .56$

13. $P(E) \approx \dfrac{f(E)}{n} = \dfrac{25}{250} = .1$

15. $P(E) \approx \dfrac{f(E)}{n} = \dfrac{189}{420} = .45$

17. A probability assignment in this case would be based on observed data, so it would be an empirical probability.

19. A probability assignment in this case would be based on the assumption that each of the combinations has a certain likelihood. This is an example of theoretical probability.

21. The sample space S is the set of all possible sequences of three digits with no repeats. $n(S) = P_{10,3} = 10 \cdot 9 \cdot 8 = 720$. The event is the simple event of one particular sequence. $n(E) = 1$. Then $P(E) = \dfrac{n(E)}{n(S)} = \dfrac{1}{720} \approx .0014$.

23. The sample space S is the set of all possible 5-card hands, chosen from a deck of 52 cards, $n(S) = C_{52,5}$. The event E is the set of all possible 5-card hands that are all black, so are chosen from the 26 black cards. $n(E) = C_{26,5}$.
    $$P(E) = \frac{n(E)}{n(S)} = \frac{C_{26,5}}{C_{52,5}} = \frac{26!}{5!(26-5)!} \div \frac{52!}{5!(52-5)!} \approx .025$$

25. The sample space S is the set of all possible 5-card hands, chosen from a deck of 52 cards. $n(S) = C_{52,5}$. The event E is the set of all possible 5-card hands that are all face cards, so are chosen from the 4 face cards in each of 4 suits, or 16 face cards. $n(E) = C_{16,5}$.

$$P(E) = \frac{n(E)}{n(S)} = \frac{C_{16,5}}{C_{52,5}} = \frac{16!}{5!\,(16-5)!} \div \frac{52!}{5!\,(52-5)!} \approx .0017$$

27. It is acceptable. There are four aces in a deck so a 5-card hand could have anywhere from zero to four aces. But it is not an equally likely sample space. Drawing four aces in a 5-card hand is far less likely than drawing none.

29. The sample space S is the set of all possible four digit numbers less than 5,000 formed from the digits 1, 3, 5, 7, 9.
$n(S) = N_1 \cdot N_2 \cdot N_3 \cdot N_4$
where
  $N_1 = 2$ (digits 1 or 3 only for the first digit since the number is
        less than 5,000)
  $N_2 = N_3 = N_4 = 5$ (digits 1, 3, 5, 7, 9)
$n(S) = 2 \cdot 5 \cdot 5 \cdot 5 = 250$
The event E is the set of those numbers in S which are divisible by 5.
These must end in 5.
$n(E) = N_1 \cdot N_2 \cdot N_3 \cdot N_4'$
where
  $N_4' = 1$ (digit 5 only)
$n(E) = 2 \cdot 5 \cdot 5 \cdot 1 = 50$
$$P(E) = \frac{n(E)}{n(S)} = \frac{50}{250} = .2$$

31. The sample space S is the set of all arrangements of the five notes in the five envelopes. $n(S) = P_{5,5} = 5! = 120$. The event E is the one correct arrangement.

$n(E) = 1.$ $P(E) = \dfrac{n(E)}{n(S)} = \dfrac{1}{120} \approx .008$

In problems 33—47, using Figure 1 of the text, we have that S = the set of all possible ordered pairs of dots showing, $n(S) = 36$.

33. The sum of the dots can be 2 in exactly one way. $E = \{(1, 1)\}$
$n(E) = 1$
$P(E) = \dfrac{n(E)}{n(S)} = \dfrac{1}{36}$

35. The sum of the dots can be 6 in 5 ways:
$E = \{(5, 1), (4, 2), (3, 3), (2, 4),$
     $(1, 5)\}$
$n(E) = 5$
$P(E) = \dfrac{n(E)}{n(S)} = \dfrac{5}{36}$

37. The sum of the dots can be less than 5 in 6 ways. (upper left-hand corner of figure) $n(E) = 6$.
$P(E) = \dfrac{n(E)}{n(S)} = \dfrac{6}{36} = \dfrac{1}{6}$

39. The sum of the dots can be 7 in 6 ways and 11 in 2 ways:
$n(E) = 36 - (6 + 2) = 28$
$P(E) = \dfrac{n(E)}{n(S)} = \dfrac{28}{36} = \dfrac{7}{9}$

41. The sum of the dots cannot be 1: $P(E) = 0$.

43. The sum of the dots is divisible by 3 if the sum is 3, 6, 9, or 12. 3 can occur in 2 ways, 6 in 5 ways, 9 in 4 ways, and 12 in 1 way:
$n(E) = 2 + 5 + 4 + 1 = 12$.
$P(E) = \dfrac{n(E)}{n(S)} = \dfrac{12}{36} = \dfrac{1}{3}$

45. The sum of the dots can be 7 in 6 ways and 11 in 2 ways:

$n(E) = 6 + 2 = 8.$

$$P(E) = \frac{n(E)}{n(S)} = \frac{8}{36} = \frac{2}{9}$$

47. The sum of the dots will *not* be divisible by 2 or 3 if the sum is 5, 7, or 11. 5 can occur in 4 ways, 7 in 6 ways, and 11 in 2 ways. Otherwise the sum will be divisible by 2 or 3:

$n(E) = 36 - (4 + 6 + 2) = 36 - 12 = 24.$

$$P(E) = \frac{n(E)}{n(S)} = \frac{24}{36} = \frac{2}{3}$$

49. The sample space is the set of events consisting of choosing a person with birthday on day 1 through day 365 of the year. (February 29 is neglected.)

$S = \{1, 2, 3, \ldots, 365\}$

Since each of these days is essentially equally likely, $P(e_i) = \dfrac{1}{365}$.

51. (A) Use $P(E) \approx \dfrac{f(E)}{n}$ with $n = 500$, where $f(E)$ is the observed frequency of an event.

$P(2) = \dfrac{11}{500} = .022$ $P(3) = \dfrac{35}{500} = .07$ $P(4) = \dfrac{44}{500} = .088$

$P(5) = \dfrac{50}{500} = .1$ $P(6) = \dfrac{71}{500} = .142$ $P(7) = \dfrac{89}{500} = .178$

$P(8) = \dfrac{72}{500} = .144$ $P(9) = \dfrac{52}{500} = .104$ $P(10) = \dfrac{36}{500} = .072$

$P(11) = \dfrac{26}{500} = .052$ $P(12) = \dfrac{14}{500} = .028$

(B) Use Figure 1 of the text, with S = the set of all possible ordered pairs of dots showing, $n(S) = 36$.

$P(2) = \dfrac{1}{36}$ $P(3) = \dfrac{2}{36}$ $P(4) = \dfrac{3}{36}$

$P(5) = \dfrac{4}{36}$ $P(6) = \dfrac{5}{36}$ $P(7) = \dfrac{6}{36}$

$P(8) = \dfrac{5}{36}$ $P(9) = \dfrac{4}{36}$ $P(10) = \dfrac{3}{36}$

$P(11) = \dfrac{2}{36}$ $P(12) = \dfrac{1}{36}$

(C) The expected frequency of the occurrence of E is $n \cdot P(E)$. So the expected frequency of the occurrence of sum 2 is $500 \cdot P(2) = 500 \cdot \dfrac{1}{36}$ or 13.9.

The expected frequency of the occurrence of sum 3 is $500 \cdot P(3) = 500 \cdot \dfrac{2}{36} = 27.8.$

The expected frequency of the occurrence of sum 4 is $500 \cdot P(4) = 500 \cdot \dfrac{3}{36} = 41.7.$

The expected frequency of the occurrence of sum 5 is $500 \cdot P(5) = 500 \cdot \dfrac{4}{36} = 55.6.$

The expected frequency of the occurrence of sum 6 is $500 \cdot P(6) = 500 \cdot \dfrac{5}{36} = 69.4.$

The expected frequency of the occurrence of sum 7 is $500 \cdot P(7) = 500 \cdot \dfrac{6}{36} = 83.3.$

The expected frequency of the occurrence of sum 8 is $500 \cdot P(8) = 500 \cdot \dfrac{5}{36} = 69.4.$

The expected frequency of the occurrence of sum 9 is $500 \cdot P(9) = 500 \cdot \dfrac{4}{36} = 55.6.$

The expected frequency of the occurrence of sum 10 is $500 \cdot P(10) = 500 \cdot \dfrac{3}{36} = 41.7.$

The expected frequency of the occurrence of sum 11 is $500 \cdot P(11) = 500 \cdot \frac{2}{36} = 27.8$.

The expected frequency of the occurrence of sum 12 is $500 \cdot P(12) = 500 \cdot \frac{1}{36} = 13.9$.

(D) Use the command r and Int (i, k, n). Let 1 to 6 represent the possible outcomes in tossing one die; then rand Int (1, 6, 500) will give a list of 500 possible outcomes in tossing one die. rand Int (1, 6, 500) + rand Int (1, 6, 500) will give a list of 500 possible outcomes in tossing 2 dice. A possible result is outlined here, but the student's result will almost certainly be different.

Generating the random numbers.

Setting up the histogram.

Selecting the window variables.

Histogram

Use trace to display any result.

53. (A) It is possible, but very unlikely, to get 29 heads in 30 flips of a coin. If one thinks of the 30 flips as a sequence of words with 30 letters chosen from $\{H, T\}$, then there are 2 ways to choose each of the letters and therefore 2^{30} such words. Only 30 of these words have exactly 29 H's because only 30 have exactly one T. So the probability of getting 29 heads in 30 flips of a fair coin is

$$P(E) = \frac{n(E)}{n(S)} = \frac{30}{2^{30}} \approx 2.8 \times 10^{-8}$$

(B) There is a strong suspicion that the coin is unfair, because this event is unlikely, although not as unlikely as in part A. If the coin is unfair, the empirical probabilities are

$$P(H) \approx \frac{f(H)}{n} = \frac{42}{50} = .84 \qquad P(T) \approx \frac{f(T)}{n} = \frac{8}{50} = .16$$

55. The sample space for this experiment is $\{HTH, HHH, THH, TTH\}$, assuming that the third coin is the one with a head on both sides:
$n(S) = 4$. E is the event $\{TTH\}$: $n(E) = 1$.
$$P(E) = \frac{n(E)}{n(S)} = \frac{1}{4}$$

57. See problem 55. $n(S) = 4$. E is the event $\{HHH\}$: $n(E) = 1$
$$P(E) = \frac{n(E)}{n(S)} = \frac{1}{4}$$

59. See problem 55. $n(S) = 4$. E is the event $\{HTH, HHH, THH\}$: $n(E) = 3$
$$P(E) = \frac{n(E)}{n(S)} = \frac{3}{4}$$

61. Since it is equally likely that each die shows a 1, a 2 or a 3, we can draw the following figure to represent S, the set of all possibilities:

First Die

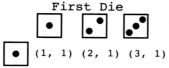

	(1, 1)	(2, 1)	(3, 1)
Second Die	(1, 2)	(2, 2)	(3, 2)
	(1, 3)	(2, 3)	(3, 3)

$n(S) = 9$

The sum of the dots can be 2 in exactly one way. $n(E) = 1$

$$P(E) = \frac{n(E)}{n(S)} = \frac{1}{9}$$

63. See figure, problem 61, $n(S) = 9$. The sum of the dots can be 4 in 3 ways: $E = \{(1, 3), (2, 2), (3, 1)\}$. $n(E) = 3$

$$P(E) = \frac{n(E)}{n(S)} = \frac{3}{9} = \frac{1}{3}$$

65. See figure, problem 61. $n(S) = 9$. The sum of the dots can be 6 in exactly one way: $E = \{(3, 3)\}$. $n(E) = 1$

$$P(E) = \frac{n(E)}{n(S)} = \frac{1}{9}$$

67. See figure, problem 61. $n(S) = 9$. The sum of the dots will be odd if the sum is 3 or 5.

$E = \{(1, 2), (2, 1)(2, 3), (3, 2)\}$. $n(E) = 4$

$$P(E) = \frac{n(E)}{n(S)} = \frac{4}{9}$$

69. The sample space S is the set of all possible 5-card hands, chosen from a deck of 52 cards. $n(S) = C_{52,5} = 2,598,960$. The event E is the set of all possible 5-card hands that are all jacks through aces, so are chosen from the 16 cards (4 jacks, 4 queens, 4 kings, 4 aces) of this type.
$n(E) = C_{16,5} = 4,368.$

$$P(E) = \frac{n(E)}{n(S)} = \frac{C_{16,5}}{C_{52,5}} = \frac{4,368}{2,598,960} \approx .00168$$

71. The sample space S is the set of all possible 5-card hands chosen from a deck of 52 cards. $n(S) = C_{52,5}$. The event E is the set of all possible 5-card hands of the type $AAAAx$ where x is any of the $52 - 4 = 48$ cards that are not aces.
$n(E) = 48$

$$P(E) = \frac{n(E)}{n(S)} = \frac{48}{C_{52,5}} = \frac{48}{2,598,960} \approx .000\ 0185$$

73. The sample space S is the set of all possible 5-card hands, chosen from a deck of 52 cards. $n(S) = C_{52,5}$. The event E is the set consisting of the cards ace, king, queen, jack, ten, in each of the 4 suits, abbreviated:
{club AKQJ10, diamond AKQJ10, heart AKQJ10, spade AKQJ10}. $n(E) = 4$.

$$P(E) = \frac{n(E)}{n(S)} = \frac{4}{C_{52,5}} = \frac{4}{2,598,960} \approx .000\ 0015$$

75. The sample space S is the set of all possible 5-card hands, chosen from a deck of 52 cards. $n(S) = C_{52,5}$. The event E is the set of all possible 5-card hands chosen as follows:

O_1: Choose 2 aces out of the set of 4 aces

N_1: $C_{4,2}$

O_2: Choose 3 queens out of the set of 4 queens

N_2: $C_{4,3}$

Applying the multiplication principle, $n(E) = N_1 N_2 = C_{4,2}C_{4,3}$

$$P(E) = \frac{n(E)}{n(S)} = \frac{C_{4,2} \cdot C_{4,3}}{C_{52,5}} = \frac{6 \cdot 4}{2,598,960} = .000\,009$$

77. (A) P (this event) $= \dfrac{15}{1,000} = .015$

(B) P (this event) $= \dfrac{130 + 80 + 12}{1,000} = .222$

(C) P (this event) $= \dfrac{30 + 32 + +28 + 25 + 20 + 21 + 12 + 1}{1,000} = .169$

(D) P (this event) $= 1 - P$ (owning zero television sets)

$$= 1 - \frac{2 + 10 + 30}{1,000} = .958$$

Section 7-6

1. Answers will vary.   **3.** Answers will vary.

For Problems 5—15, the first eight lines of Pascal's triangle are as follows:

```
0                       1
1                    1    1
2                 1    2    1
3              1    3    3    1
4           1    4    6    4    1
5        1    5   10    10    5    1
6      1    6   15    20    15   6   1
7    1    7   21   35    35   21  7   1
```

5. Using the row of the triangle marked "4", we get

$(a + b)^4 = a^4 + 4a^3b + 6a^2b^2 + 4ab^3 + b^4$

7. Using the row of the triangle marked "5", we get

$(x + 2)^5 = x^5 + 5x^4 \cdot 2^1 + 10x^3 \cdot 2^2 + 10x^2 \cdot 2^3 + 5x \cdot 2^4 + 2^5$

$= x^5 + 10x^4 + 40x^3 + 80x^2 + 80x + 32$

9. $\dbinom{5}{3}$ is the fourth entry in line 5 (start counting from 0) of the triangle, 10.

11. $\dbinom{4}{2}$ is the third entry in line 4 (start counting from 0) of the triangle, 6.

13. $C_{6,3}$ is the fourth entry in line 6 (start counting from 0) of the triangle, 20.

15. $C_{7,4}$ is the fifth entry in line 7 (start counting from 0) of the triangle, 35.

Problems 17 and 19 are
shown on one screen of a
graphing calculator.

```
9 nCr 3
                84
12 nCr 10
                66
```

Problems 21 and 23 are
shown on one screen of a
graphing calculator.

```
17 nCr 13
              2380
50 nCr 4
            230300
```

25. $(m + n)^3 = \sum_{k=0}^{3} \binom{3}{k} m^{3-k} n^k$

$\qquad = \binom{3}{0} m^3 + \binom{3}{1} m^2 n + \binom{3}{2} mn^2 + \binom{3}{3} n^3$

$\qquad = m^3 + 3m^2 n + 3mn^2 + n^3$

27. $(2x - 3y)^3 = [2x + (-3y)]^3$

$\qquad = \sum_{k=0}^{3} \binom{3}{k} (2x)^{3-k} (-3y)^k$

$\qquad = \binom{3}{0} (2x)^3 + \binom{3}{1} (2x)^2 (-3y)^1 + \binom{3}{2} (2x)(-3y)^2 + \binom{3}{3} (-3y)^3$

$\qquad = 8x^3 + 3(4x^2)(-3y) + 3(2x)(9y^2) + (-27y^3)$

$\qquad = 8x^3 - 36x^2 y + 54xy^2 - 27y^3$

29. $(x - 2)^4 = [x + (-2)]^4 = \sum_{k=0}^{4} \binom{4}{k} x^{4-k} (-2)^k$

$\qquad = \binom{4}{0} x^4 + \binom{4}{1} x^3 (-2)^1 + \binom{4}{2} x^2 (-2)^2 + \binom{4}{3} x(-2)^3 + \binom{4}{4} (-2)^4$

$\qquad = x^4 - 8x^3 + 24x^2 - 32x + 16$

31. $(m + 3n)^4 = \sum_{k=0}^{4} \binom{4}{k} m^{4-k} (3n)^k$

$\qquad = \binom{4}{0} m^4 + \binom{4}{1} m^3 (3n)^1 + \binom{4}{2} m^2 (3n)^2 + \binom{4}{3} m(3n)^3 + \binom{4}{4} (3n)^4$

$\qquad = m^4 + 12m^3 n + 54m^2 n^2 + 108mn^3 + 81n^4$

33. $(2x - y)^5 = [2x + (-y)]^5 = \sum_{k=0}^{5} \binom{5}{k} (2x)^{5-k} (-y)^k$

$\qquad = \binom{5}{0} (2x)^5 + \binom{5}{1} (2x)^4 (-y)^1 + \binom{5}{2} (2x)^3 (-y)^2 + \binom{5}{3} (2x)^2 (-y)^3$

$\qquad\qquad + \binom{5}{4} (2x)(-y)^4 + \binom{5}{5} (-y)^5$

$\qquad = 32x^5 - 80x^4 y + 80x^3 y^2 - 40x^2 y^3 + 10xy^4 - y^5$

35. $(m + 2n)^6 = \sum\limits_{k=0}^{6} \binom{6}{k} m^{6-k}(2n)^k$

$= \binom{6}{0} m^6 + \binom{6}{1} m^5(2n)^1 + \binom{6}{2} m^4(2n)^2 + \binom{6}{3} m^3(2n)^3 + \binom{6}{4} m^2(2n)^4$

$+ \binom{6}{5} m(2n)^5 + \binom{6}{6}(2n)^6$

$= m^6 + 12m^5n + 60m^4n^2 + 160m^3n^3 + 240m^2n^4 + 192mn^5 + 64n^6$

37. In the expansion

$(x + 1)^7 = \sum\limits_{k=0}^{7} \binom{7}{k} x^{7-k} 1^k$

the exponent of x is 4 when $k = 3$. So the term containing x^4 is

$\binom{7}{3} x^4 1^3 = 35x^4$

39. In the expansion

$(2x - 1)^{11} = [2x + (-1)]^{11} = \sum\limits_{k=0}^{11} \binom{11}{k}(2x)^{11-k}(-1)^k$

the exponent of x is 6 when $k = 5$. So the term containing x^6 is

$\binom{11}{5}(2x)^6(-1)^5 = (462)(64x^6)(-1) = -29,568x^6$

41. In the expansion

$(2x + 3)^{18} = \sum\limits_{k=0}^{18} \binom{18}{k}(2x)^{18-k}(3)^k$

the exponent of x is 14 when $k = 4$. So the term containing x^{14} is

$\binom{18}{4}(2x)^{14}(3)^4 = 3,060(2^{14}x^{14})(3^4) = 4,060,938,240x^{14}$

43. In the expansion

$(x^2 - 1)^6 = \sum\limits_{k=0}^{6} \binom{6}{k}(x^2)^{6-k}(-1)^k = \sum\limits_{k=0}^{6} \binom{6}{k} x^{12-2k}(-1)^k$

the exponent of x, $12 - 2k$, is 8, when $12 - 2k = 8$ and $2k = 4$, $k = 2$. So the term containing x^8 is

$\binom{6}{2} x^8(-1)^2 = 15x^8$

45. In the expansion

$(x^2 + 1)^9 = \sum\limits_{k=0}^{9} \binom{9}{k}(x^2)^{9-k}(1)^k = \sum\limits_{k=0}^{9} \binom{9}{k} x^{18-2k}(1)^k$

the exponent of x, $18 - 2k$, is even and can never equal 11.

47. In the expansion of $(a + b)^n$, the exponent of b in the rth term is $r - 1$ and the exponent of a is $n - (r - 1)$. Here, $r = 7$, $n = 15$.

$$\text{Seventh term} = \binom{15}{6} u^9 v^6$$

$$= \frac{15!}{9!\,6!} u^9 v^6$$

$$= \frac{15 \cdot 14 \cdot 13 \cdot 12 \cdot 11 \cdot 10}{6 \cdot 5 \cdot 4 \cdot 3 \cdot 2 \cdot 1} u^9 v^6$$

$$= 5,005 u^9 v^6$$

49. In the expansion of $(a + b)^n$, the exponent of b in the rth term is $r - 1$ and the exponent of a is $n - (r - 1)$. Here, $r = 11$, $n = 12$.

$$\text{Eleventh term} = \binom{12}{10} (2m)^2 n^{10}$$

$$= \frac{12!}{10!\,2!} 4m^2 n^{10}$$

$$= \frac{12 \cdot 11}{2 \cdot 1} 4m^2 n^{10}$$

$$= 264 m^2 n^{10}$$

51. In the expansion of $(a + b)^n$, the exponent of b in the rth term is $r - 1$ and the exponent of a is $n - (r - 1)$. Here, $r = 7$, $n = 12$.

$$\text{Seventh term} = \binom{12}{6} \left(\frac{w}{2}\right)^6 (-2)^6$$

$$= \frac{12!}{6!\,6!} \frac{w^6}{2^6} 2^6$$

$$= \frac{12 \cdot 11 \cdot 10 \cdot 9 \cdot 8 \cdot 7 \cdot 6!}{6 \cdot 5 \cdot 4 \cdot 3 \cdot 2 \cdot 1 \cdot 6!} w^6$$

$$= 924 w^6$$

53. In the expansion of $(a + b)^n$, the exponent of b in the rth term is $r - 1$ and the exponent of a is $n - (r - 1)$. Here, $r = 6$, $n = 8$.

$$\text{Sixth term} = \binom{8}{5} (3x)^3 (-2y)^5$$

$$= \frac{8!}{5!\,3!} (3x)^3 (-2y)^5$$

$$= \frac{8 \cdot 7 \cdot 6 \cdot 5!}{5! \cdot 3 \cdot 2 \cdot 1} (27x^3)(-32y^5)$$

$$= -(56)(27)(32) x^3 y^5$$

$$= -48,384 x^3 y^5$$

55. $\dfrac{f(x + h) - f(x)}{h} = \dfrac{(x + h)^3 - x^3}{h} = \dfrac{x^3 + 3x^2h + 3xh^2 + h^3 - x^3}{h}$

$$= \dfrac{3x^2h + 3xh^2 + h^3}{h} = \dfrac{h(3x^2 + 3xh + h^2)}{h} = 3x^2 + 3xh + h^2$$

As h approaches 0, this quantity approaches $3x^2$.

57. $\dfrac{f(x + h) - f(x)}{h} = \dfrac{(x + h)^5 - x^5}{h} = \dfrac{x^5 + 5x^4h + 10x^3h^2 + 10x^2h^3 + 5xh^4 + h^5 - x^5}{h}$

$$= \dfrac{5x^4h + 10x^3h^2 + 10x^2h^3 + 5xh^4 + h^5}{h}$$

$$= \dfrac{h(5x^4 + 10x^3h + 10x^2h^2 + 5xh^3 + h^4)}{h}$$

$$= 5x^4 + 10x^3h + 10x^2h^2 + 5xh^3 + h^4$$

As h approaches 0, this approaches $5x^4$.

59. Graphing the sequence, we obtain

The solid line indicates $\dfrac{1}{2}(184,756) = 92,378$, that is, half the largest term. The graph shows five terms greater than half the largest term. The table display confirms this:

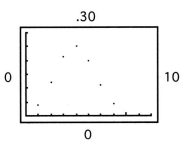

$\dbinom{20}{8} = 125,970$ $\qquad$ $\dbinom{20}{9} = 167,960$ $\qquad$ $\dbinom{20}{10} = 184,756$ $\qquad$ $\dbinom{20}{11} = 167,960$ $\qquad$ $\dbinom{20}{12} = 125,970$

are all greater than $92,378$, but the other terms are less than $92,378$.

61. (A) Graphing the sequence, we obtain the graph at the right.

From the table display we find the largest term to be $\dbinom{10}{4}(.6)^6(.4)^4 = 0.251$ as displayed in the graph.

(B) According to the binomial formula,

$$\sum_{k=0}^{n} \binom{n}{k} a^{n-k}b^k = (a + b)^n$$

Then $a_0 + a_1 + a_2 + \cdots + a_{10} = \displaystyle\sum_{k=0}^{10} \binom{10}{k}(0.6)^{10-k}(0.4)^k = (0.6 + 0.4)^{10} = 1^{10} = 1$

63. $(1.01)^{10} = (1 + 0.01)^{10} = \sum_{k=0}^{10} \binom{10}{k} 1^{10-k}(0.01)^k, \; 1^{10-k} = 1$

$$= \sum_{k=0}^{10} \binom{10}{k}(0.01)^k$$

$$= \binom{10}{0} + \binom{10}{1}(0.01)^1 + \binom{10}{2}(0.01)^2 + \binom{10}{3}(0.01)^3$$

$$+ \binom{10}{4}(0.01)^4 + \binom{10}{5}(0.01)^5 + \binom{10}{6}(0.01)^6 + \binom{10}{7}(0.01)^7$$

$$+ \binom{10}{8}(0.01)^8 + \binom{10}{9}(0.01)^9 + \binom{10}{10}(0.01)^{10}$$

$$= 1 + 0.1 + 0.0045 + 0.00012 + 0.0000021 + \text{(other terms}$$
$$\text{with no effect in fourth decimal place)}$$

$$= 1.1046$$

65. $\binom{n}{r} = \dfrac{n!}{r!(n-r)!} = \dfrac{n!}{(n-r)!\,r!} = \dfrac{n!}{(n-r)![n-(n-r)]!} = \binom{n}{n-r}$

67. $\binom{k}{r-1} + \binom{k}{r} = \dfrac{k!}{(r-1)!(k-r+1)!} + \dfrac{k!}{r!(k-r)!}$

$$= \dfrac{rk! + (k-r+1)k!}{r!(k-r+1)!} = \dfrac{(r+k-r+1)k!}{r!(k-r+1)!}$$

$$= \dfrac{(k+1)k!}{r!(k-r+1)!} = \dfrac{(k+1)!}{r!(k-r+1)!}$$

$$= \binom{k+1}{r}$$

69. $\binom{k}{k} = \dfrac{k!}{k!(k-k)!} = 1 = \dfrac{(k+1)!}{(k+1)![(k+1)-(k+1)]!} = \binom{k+1}{k+1}$

71. $2^n = (1+1)^n = \sum_{k=0}^{n} \binom{n}{k} 1^{n-k}1^k, \; 1^{n-k} = 1^k = 1$

$$= \sum_{k=0}^{n} \binom{n}{k}$$

$$= \binom{n}{0} + \binom{n}{1} + \binom{n}{2} + \cdots + \binom{n}{n}$$

CHAPTER 7 REVIEW

1. (A) Since $\dfrac{-8}{16} = \dfrac{4}{-8} = -\dfrac{1}{2}$, this could start a geometric sequence.

(B) Since $7 - 5 = 9 - 7 = 2$, this could start an arithmetic sequence.

(C) Since $-5 - (-8) = -2 - (-5) = 3$, this could start an arithmetic sequence.

(D) Since $\dfrac{3}{2} \neq \dfrac{5}{3}$ and $3 - 2 \neq 5 - 3$, this could start neither an arithmetic nor a geometric sequence.

(E) Since $\dfrac{2}{-1} = \dfrac{-4}{2} = -2$, this could start a geometric sequence.　　　　(7-1, 7-3)

2. $a_n = 2n + 3$

    (A) $\begin{aligned} a_1 &= 2 \cdot 1 + 3 = 5 \\ a_2 &= 2 \cdot 2 + 3 = 7 \\ a_3 &= 2 \cdot 3 + 3 = 9 \\ a_4 &= 2 \cdot 4 + 3 = 11 \end{aligned}$

    (B) This is an arithmetic sequence with $d = 2$.
$$a_n = a_1 + (n - 1)d$$
$$a_{10} = 5 + (10 - 1)2$$
$$= 23$$

    (C) $S_n = \dfrac{n}{2}(a_1 + a_n)$

        $S_{10} = \dfrac{10}{2}(5 + 23)$

        $= 140$

        (7-1, 7-3)

3. $a_n = 32\left(\dfrac{1}{2}\right)^n$

    (A) $a_1 = 32\left(\dfrac{1}{2}\right)^1 = 16$

        $a_2 = 32\left(\dfrac{1}{2}\right)^2 = 8$

        $a_3 = 32\left(\dfrac{1}{2}\right)^3 = 4$

        $a_4 = 32\left(\dfrac{1}{2}\right)^4 = 2$

    (B) This is a geometric sequence with $r = \dfrac{1}{2}$.
$$a_n = a_1 r^{n-1}$$
$$a_{10} = 16\left(\dfrac{1}{2}\right)^{10-1}$$
$$= \dfrac{1}{32}$$

    (C) $S_n = \dfrac{a_1 - r a_n}{1 - r}$

        $S_{10} = \dfrac{16 - \frac{1}{2}\left(\frac{1}{32}\right)}{\frac{1}{2}} = \dfrac{16 - \frac{1}{64}}{\frac{1}{2}} = 31\frac{31}{32}$

        (7-1, 7-3)

4. $a_1 = -8, \ a_n = a_{n-1} + 3, \ n \geq 2$

    (A) $\begin{aligned} a_1 &= -8 \\ a_2 &= a_1 + 3 = -8 + 3 = -5 \\ a_3 &= a_2 + 3 = -5 + 3 = -2 \\ a_4 &= a_3 + 3 = -2 + 3 = 1 \end{aligned}$

    (B) This is an arithmetic sequence with $d = 3$.
$$a_n = a_1 + (n - 1)d$$
$$a_{10} = -8 + (10 - 1)3$$
$$= 19$$

    (C) $S_n = \dfrac{n}{2}(a_1 + a_n)$

        $S_{10} = \dfrac{10}{2}(-8 + 19)$

        $= 55$

        (7-1, 7-3)

5. $a_1 = -1, \ a_n = (-2)a_{n-1}, \ n \geq 2$

    (A) $\begin{aligned} a_1 &= -1 \\ a_2 &= (-2)a_1 = (-2)(-1) = 2 \\ a_3 &= (-2)a_2 = (-2)2 = -4 \\ a_4 &= (-2)a_3 = (-2)(-4) = 8 \end{aligned}$

    (B) This is a geometric sequence with $r = -2$.
$$a_n = a_1 r^{n-1}$$
$$a_{10} = (-1)(-2)^{10-1}$$
$$= 512$$

(C) $S_n = \dfrac{a_1 - ra_n}{1 - r}$

$S_{10} = \dfrac{-1 - (-2)(512)}{1 - (-2)}$

$= 341$

(7-1, 7-3)

6. This is a geometric sequence with $a_1 = 16$ and $r = \dfrac{1}{2} < 1$, so the sum exists:

$S_\infty = \dfrac{a_1}{1 - r}$

$= \dfrac{16}{1 - \frac{1}{2}}$

$= 32$

(7-3)

7. $6! = 6 \cdot 5 \cdot 4 \cdot 3 \cdot 2 \cdot 1 = 720$ (7-4)

8. $\dfrac{22!}{19!} = \dfrac{22 \cdot 21 \cdot 20 \cdot 19!}{19!} = 9,240$ (7-4)

9. $\dfrac{7!}{2!(7-2)!} = \dfrac{7!}{2! \, 5!} = \dfrac{7 \cdot 6 \cdot 5!}{2 \cdot 1 \cdot 5!} = 21$ (7-4)

10. $C_{6,2} = \dfrac{6!}{2!(6-2)!} = \dfrac{6!}{2! \, 4!} = \dfrac{6 \cdot 5 \cdot 4!}{2 \cdot 1 \cdot 4!} = 15 \quad P_{6,2} = 6 \cdot 5 = 30$ (7-5)

11. (A) The outcomes can be displayed in a tree diagram as follows:

(B) O_1: Rolling the die
N_1: 6 outcomes
O_2: Flipping the coin
N_2: 2 outcomes

Applying the multiplication principle, there are $6 \cdot 2 = 12$ combined outcomes.

(7-5)

12. O_1: Seating the first person N_1: 6 ways
O_2: Seating the second person N_2: 5 ways
O_3: Seating the third person N_3: 4 ways
O_4: Seating the fourth person N_4: 3 ways
O_5: Seating the fifth person N_5: 2 ways
O_6: Seating the sixth person N_6: 1 way

Applying the multiplication principle, there are $6 \cdot 5 \cdot 4 \cdot 3 \cdot 2 \cdot 1 = 720$ arrangements.

(7-5)

13. Order is important here. We use permutations to determine the number of arrangements of 6 objects. $P_{6,6} = 6! = 720$ (7-5)

14. The sample space S is the set of all possible 5-card hands, chosen from a deck of 52 cards. $n(S) = C_{52,5}$
The event E is the set of all possible 5-card hands that are all clubs, so are chosen from the 13 clubs. $n(E) = C_{13,5}$

$$P(E) = \frac{n(E)}{n(S)} = \frac{C_{13,5}}{C_{52,5}} = \frac{13!}{5!(13-5)!} \div \frac{52!}{5!(52-5)!} \approx .0005$$ (7-5)

15. The sample space S is the set of all possible arrangements of two persons, chosen from a set of 15 persons. Order is important here, so we use permutations. $n(S) = P_{15,2}$. The event E is one of these arrangements. $n(E) = 1$.

$$P(E) = \frac{n(E)}{n(S)} = \frac{1}{P_{15,2}} = \frac{1}{15 \cdot 14} \approx .005$$ (7-5)

16. $P(E) \approx \dfrac{f(E)}{n} = \dfrac{50}{1,000} = .05$ (7-5)

17. P_1: $5 = 1^2 + 4 \cdot 1 = 5$
P_2: $5 + 7 = 2^2 + 4 \cdot 2$
$\quad\quad 12 = 12$
P_3: $5 + 7 + 9 = 3^2 + 4 \cdot 3$
$\quad\quad\quad 21 = 21$ (7-2)

18. P_1: $2 = 2^{1+1} - 2 = 4 - 2 = 2$
P_2: $2 + 4 = 2^{2+1} - 2$
$\quad\quad 6 = 6$
P_3: $2 + 4 + 8 = 2^{3+1} - 2$
$\quad\quad\quad 14 = 14$ (7-2)

19. P_1: $49^1 - 1$ is divisible by 6
$48 = 6 \cdot 8$ true
P_2: $49^2 - 1$ is divisible by 6
$2,400 = 6 \cdot 400$ true
P_3: $49^3 - 1$ is divisible by 6
$117,648 = 19,608 \cdot 6$ true (7-2)

20. P_k: $5 + 7 + 9 + \cdots + (2k + 3) = k^2 + 4k$
P_{k+1}: $5 + 7 + 9 + \cdots + (2k + 3) + (2k + 5) = (k + 1)^2 + 4(k + 1)$ (7-2)

21. P_k: $2 + 4 + 8 + \cdots + 2^k = 2^{k+1} - 2$
P_{k+1}: $2 + 4 + 8 + \cdots + 2^k + 2^{k+1} = 2^{k+2} - 2$ (7-2)

22. P_k: $49^k - 1 = 6r$ for some integer r
P_{k+1}: $49^{k+1} - 1 = 6s$ for some integer s (7-2)

23. Although 1 is less than 4, $1 + \dfrac{1}{2}$ is less than 4, $1 + \dfrac{1}{2} + \dfrac{1}{3}$ is less than 4, and so on, the statement is false. In fact,
$$1 + \frac{1}{2} + \frac{1}{3} + \cdots + \frac{1}{31} \approx 4.027245$$
so $n = 31$ is a counterexample. (7-2)

24. $S_{10} = (2\cdot1 - 8) + (2\cdot2 - 8) + (2\cdot3 - 8) + (2\cdot4 - 8) + (2\cdot5 - 8) + (2\cdot6 - 8)$
$$+ (2\cdot7 - 8) + (2\cdot8 - 8) + (2\cdot9 - 8) + (2\cdot10 - 8)$$

$= (-6) + (-4) + (-2) + 0 + 2 + 4 + 6 + 8 + 10 + 12$
$= 30$

(7-3)

25. $S_7 = \dfrac{16}{2^1} + \dfrac{16}{2^2} + \dfrac{16}{2^3} + \dfrac{16}{2^4} + \dfrac{16}{2^5} + \dfrac{16}{2^6} + \dfrac{16}{2^7}$

$= 8 + 4 + 2 + 1 + \dfrac{1}{2} + \dfrac{1}{4} + \dfrac{1}{8}$

$= 15\dfrac{7}{8}$

(7-3)

26. This is an infinite geometric sequence with $a_1 = 27$.

$r = \dfrac{-18}{27} = -\dfrac{2}{3}$ $\left|-\dfrac{2}{3}\right| < 1$, so the sum exists

$S_\infty = \dfrac{a_1}{1 - r}$

$= \dfrac{27}{1 - \left(-\dfrac{2}{3}\right)}$

$= \dfrac{81}{5}$

(7-3)

27. $S_n = \displaystyle\sum_{k=1}^{n} \dfrac{(-1)^{k+1}}{3^k}$

This geometric sequence has $a_1 = \dfrac{1}{3}$.

$r = \left(-\dfrac{1}{9}\right) \div \dfrac{1}{3} = -\dfrac{1}{3}$ $\left|-\dfrac{1}{3}\right| < 1$, so the sum exists.

$S_\infty = \dfrac{a_1}{1 - r}$

$= \dfrac{\dfrac{1}{3}}{1 - \left(-\dfrac{1}{3}\right)}$

$= \dfrac{1}{4}$

(7-3)

28. The probability of an event cannot be negative, but $P(e_2)$ is given as negative. The sum of the probabilities of the simple events must be 1, but it is given as 2.5. The probability of an event cannot be greater than 1, but $P(e_4)$ is given as 2.

(7-5)

29. We can select any three of the six points to use as vertices of the triangle. Order is not important. $C_{6,3} = \dfrac{6!}{3!(6 - 3)!} = 20$ triangles

(7-5)

30. First, find d:
$a_n = a_1 + (n - 1)d$
$31 = 13 + (7 - 1)d$
$31 = 13 + 6d$
$d = 3$
So $a_5 = 13 + (5 - 1)3$
$= 25$

(7-3)

31.

	Case 1	Case 2	Case 3
O_1: select the first letter	N_1: 8 ways	8 ways	8 ways
O_2: select the second letter	N_2: 7 ways	8 ways	7 ways
O_3: select the third letter	N_3: 6 ways (exclude first and second letter)	8 ways	7 ways (exclude second letter.)
	$8\cdot7\cdot6 = 336$ words	$8\cdot8\cdot8 = 512$ words	$8\cdot7\cdot7 = 392$ words

(7-4)

32. (A) $P(2 \text{ heads}) \approx \dfrac{210}{1,000} = .21$

$P(1 \text{ head}) \approx \dfrac{480}{1,000} = .48$

$P(0 \text{ heads}) \approx \dfrac{310}{1,000} = .31$

(B) A sample space of equally likely events is:

$$S = \{HH, \ HT, \ TH, \ TT\}$$

Let $E_1 = 2 \text{ heads} = \{HH\}$

$E_2 = 1 \text{ head} = \{HT, \ TH\}$

$E_3 = 0 \text{ heads} = \{TT\}$

then $P(E_1) = \dfrac{n(E_1)}{n(S)} = \dfrac{1}{4} = .25$

$P(E_2) = \dfrac{n(E_2)}{n(S)} = \dfrac{2}{4} = .5$

$P(E_3) = \dfrac{n(E_3)}{n(S)} = \dfrac{1}{4} = .25$

(C) Expected frequency = $P(E) \cdot$ number of trials
Expected frequency of 2 heads = $P(E_1) \cdot 1,000 = .25(1,000) = 250$
Expected frequency of 1 head = $P(E_2) \cdot 1,000 = .5(1,000) = 500$
Expected frequency of 0 heads = $P(E_3) \cdot 1,000 = .25(1,000) = 250$ (7-5)

33. The sample space S is the set of all possible 5-card hands chosen from a deck of 52 cards. $n(S) = C_{52,5}$.

(A) The event E is the set of all possible 4-card hands that are all diamonds, so are chosen from the 13 diamonds. $n(E) = C_{13,5}$

$$P(E) = \frac{n(E)}{n(S)} = \frac{C_{13,5}}{C_{52,5}}$$

(B) The event E is the set of all possible 5-card hands chosen as follows:
O_1: Choose 3 diamonds out of the 13 in the deck. N_1: $C_{13,3}$
O_2: Choose 2 spades out of the 13 in the deck. N_2: $C_{13,2}$
Applying the multiplication principle, $n(E) = N_1 \cdot N_2 = C_{13,3} \cdot C_{13,2}$

$$P(E) = \frac{n(E)}{n(S)} = \frac{C_{13,3} \cdot C_{13,2}}{C_{52,5}}$$ (7-5)

34. The sample space S is the set of all possible 4-person choices out of the 10 people. $n(S) = C_{10,4}$
The event E is the set of all of those choices that include 2 particular people, so
O_1: Choose the 2 married people N_1: 1
O_2: Choose 2 more people out of the 8 remaining N_2: $C_{8,2}$
Applying the multiplication principle, $n(E) = N_1 \cdot N_2 = 1 \cdot C_{8,2} = C_{8,2}$

$$P(E) = \frac{n(E)}{n(S)} = \frac{C_{8,2}}{C_{10,4}} = \frac{8!}{2!(8-2)!} \div \frac{10!}{4!(10-4)!} = 28 \div 210 = \frac{2}{15}$$ (7-5)

35. The sample space S is the set of all possible results of spinning the device twice:
$\{(1, \ 1), \ (1, \ 2), \ (1, \ 3), \ (2, \ 1), \ (2, \ 2), \ (2, \ 3), \ (3, \ 1), \ (3, \ 2), \ (3, \ 3)\}$.
$n(S) = 9$.

(A) The event E is the set of all possible results in which both spins are the same.
$E = \{(1, \ 1), \ (2, \ 2), \ (3, \ 3)\}$. $n(E) = 3$.
$$P(E) = \frac{n(E)}{n(S)} = \frac{3}{9} = \frac{1}{3}$$

(B) The event E is the set of those spin sequences that add up to 5:
$E = \{(2, 3), (3, 2)\}$. $n(E) = 2$.

$P(E) = \dfrac{n(E)}{n(S)} = \dfrac{2}{9}$

$(7\text{-}5)$

36. $0.\overline{72} = 0.72 + 0.0072 + 0.000072 + \cdots$

This is an infinite geometric sequence with $a_1 = 0.72$ and $r = 0.01$.

$0.\overline{72} = S_\infty = \dfrac{a_1}{1 - r}$

$= \dfrac{0.72}{1 - 0.01}$

$= \dfrac{0.72}{0.99}$ or $\dfrac{72}{99}$

$= \dfrac{8}{11}$

$(7\text{-}3)$

37. (A) Order is important here. We are selecting 3 digits out of 6 possible.
$P_{6,3} = 6 \cdot 5 \cdot 4 = 120$ lock combinations

(B) Order is not important here. We are selecting 2 players out of 5.

$C_{5,2} = \dfrac{5!}{2!(5 - 2)!} = 10$ games

$(7\text{-}4)$

38. $\dfrac{20!}{18!(20 - 18)!} = \dfrac{20!}{18!\,2!}$

$= \dfrac{20 \cdot 19 \cdot 18!}{18!\,2 \cdot 1} = 190$

$(7\text{-}6)$

39. $\dbinom{16}{12} = \dfrac{16!}{12!(16 - 12)!} = \dfrac{16!}{12!\,4!}$

$= \dfrac{16 \cdot 15 \cdot 14 \cdot 13 \cdot 12!}{12! \cdot 4 \cdot 3 \cdot 2 \cdot 1} = 1{,}820$

$(7\text{-}6)$

40. $\dbinom{11}{11} = \dfrac{11!}{11!(11 - 11)!} = \dfrac{11!}{11!\,0!} = 1$

$(7\text{-}6)$

41. $(x - y)^5 = [x + (-y)]^5 = \displaystyle\sum_{k=0}^{5} \binom{5}{k}(x)^{5-k}(-y)^k$

$= \dbinom{5}{0}x^5 + \dbinom{5}{1}x^4(-y)^1 + \dbinom{5}{2}x^3(-y)^2 + \dbinom{5}{3}x^2(-y)^3$

$+ \dbinom{5}{4}x(-y)^4 + \dbinom{5}{5}(-y)^5$

$= x^5 - 5x^4y + 10x^3y^2 - 10x^2y^3 + 5xy^4 - y^5$

$(7\text{-}6)$

42. In the expansion

$(x + 2)^9 = \displaystyle\sum_{k=0}^{9} \binom{9}{k}x^{9-k}\,2^k$

the exponent of x is 6 when $k = 3$, and the term containing x^6 is

$\dbinom{9}{3}x^6 2^3 = 84x^6 \cdot 8 = 672x^6$

$(7\text{-}6)$

43. In the expansion of $(a + b)^n$, the exponent of b in the rth term is $r - 1$ and the exponent of a is $n - (r - 1)$. Here, $r = 10$, $n = 12$.

$$\text{Tenth term} = \binom{12}{9}(2x)^3(-y)^9$$

$$= \frac{12!}{9! \, 3!}(8x^3)(-y^9)$$

$$= \frac{12 \cdot 11 \cdot 10 \cdot 9!}{9! \cdot 3 \cdot 2 \cdot 1}(-8x^3y^9)$$

$$= -1{,}760x^3y^9 \qquad\qquad (7\text{-}6)$$

44. Write: P_n: $5 + 7 + 9 + \cdots + (2n + 3) = n^2 + 4n$

Proof: Part 1: Show that P_1 is true:

P_1: $5 = 1^2 + 4 \cdot 1$

$ = 1 + 4$ Clearly true.

Part 2: Show that if P_k is true, then P_{k+1} is true:

Write out P_k and P_{k+1}.

$\quad P_k$: $5 + 7 + 9 + \cdots + (2k + 3) = k^2 + 4k$

$\quad P_{k+1}$: $5 + 7 + 9 + \cdots + (2k + 3) + (2k + 5) = (k + 1)^2 + 4(k + 1)$

We start with P_k:

$\quad 5 + 7 + 9 + \cdots + (2k + 3) = k^2 + 4k$

Adding $2k + 5$ to both sides:

$\quad 5 + 7 + 9 + \cdots + (2k + 3) + (2k + 5) = k^2 + 4k + 2k + 5$

$$= k^2 + 6k + 5$$

$$= k^2 + 2k + 1 + 4k + 4$$

$$= (k + 1)^2 + 4(k + 1)$$

We have shown that if P_k is true, then P_{k+1} is true.

Conclusion: P_n is true for all positive integers n. $\qquad\qquad (7\text{-}2)$

45. Write: P_n: $2 + 4 + 8 + \cdots + 2^n = 2^{n+1} - 2$

Proof: Part 1: Show that P_1 is true:

P_1: $2 = 2^{1+1} - 2$

$ = 4 - 2$ Clearly true.

Part 2: Show that if P_k is true, then P_{k+1} is true:

Write out P_k and P_{k+1}.

$\quad P_k$: $2 + 4 + 8 + \cdots + 2^k = 2^{k+1} - 2$

$\quad P_{k+1}$: $2 + 4 + 8 + \cdots + 2^k + 2^{k+1} = 2^{k+2} - 2$

We start with P_k:

$\quad 2 + 4 + 8 + \cdots + 2^k = 2^{k+1} - 2$

Adding 2^{k+1} to both sides:

$\quad 2 + 4 + 8 + \cdots + 2^k + 2^{k+1} = 2^{k+1} - 2 + 2^{k+1}$

$$= 2^{k+1} + 2^{k+1} - 2$$

$$= 2 \cdot 2^{k+1} - 2$$

$$= 2^{k+2} - 2$$

We have shown that if P_k is true, then P_{k+1} is true.

Conclusion: P_n is true for all positive integers n. $\qquad\qquad (7\text{-}2)$

46. Write: P_n: $49^n - 1 = 6r$ for some r in N.

Part 1: Show that P_1 is true:

P_1: $49^1 - 1 = 6r$ is true $(r = 8)$.

Part 2: Show that if P_k is true, then P_{k+1} is true:

Write out P_k and P_{k+1}.

P_k: $49^k - 1 = 6r$, for some integer r

P_{k+1}: $49^{k+1} - 1 = 6s$ for some integer s

We start with P_k:

$49^k - 1 = 6r$ for some integer r

Now, $49^{k+1} - 1 = 49^{k+1} - 49^k + 49^k - 1$
$$= 49^k(49 - 1) + 49^k - 1$$
$$= 49^k \cdot 8 \cdot 6 + 6r$$
$$= 6(49^k \cdot 8 + r)$$
$$= 6s \text{ with } s = 49^k \cdot 8 + r$$

We have shown that if P_k is true, then P_{k+1} is true.

Conclusion: P_n is true for all positive integers n.   *(7-2)*

47.

Enter the sequences
in the sequence
editor.

Graph the sequences.

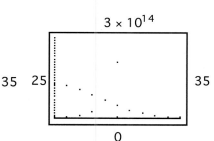

Display the points
(in the appropriate
interval) in a table.

From the second graph, 29 is the least positive integer n such that $a_n < b_n$.
The table confirms this.   *(7-6)*

48.

```
Plot1 Plot2 Plot3
nMin=1
\u(n)■.99u(n-1)+
5
 u(nMin)■{100}
\v(n)■9+7n
 v(nMin)■
\w(n)=
```

Enter the sequences in
the sequence editor.

300

1 ⟍⟍⟍ 35

0

Graph the sequences.

n	u(n)	v(n)
22	176.11	163
23	179.35	170
24	182.55	177
25	185.73	184
26	188.87	191
27	191.98	198
28	195.06	205

$n=26$

Display the points in a
table.

From the graph, the least positive integer n such that $a_n < b_n$ is between 25
and 30. The table confirms that $n = 26$.   *(7-1)*

49. In the first case, the order matters. We have five successive events, each of
which can happen two ways (girl or boy). Applying the multiplication principle,
there are $2 \cdot 2 \cdot 2 \cdot 2 \cdot 2 = 32$ possible families. In the second case, the possible
families can be listed as {0 girls, 1 girl, 2 girls, 3 girls, 4 girls, 5
girls}, so there are 6 possibilities.   *(7-4)*

50. An arithmetic sequence is involved, with $a_1 = \dfrac{g}{2}$, $d = \dfrac{3g}{2} - \dfrac{g}{2} = g$.

Distance fallen during the twenty-fifth second = a_{25}.

$a_n = a_1 + (n - 1)d$

$a_{25} = \dfrac{g}{2} + (25 - 1)g$

$\quad = \dfrac{49g}{2}$ feet

Total distance fallen after twenty-five seconds = $a_1 + a_2 + a_3 + \cdots + a_{25} = S_{25}$

$S_n = \dfrac{n}{2}(a_1 + a_n)$

$S_{25} = \dfrac{25}{2}\left(\dfrac{g}{2} + \dfrac{49g}{2}\right)$

$\quad = \dfrac{625g}{2}$ feet

$(7\text{-}3)$

51. To seat two people, we can seat one person, then the second person.

O_1: Seat the first person in any chair.

N_1: 4 ways

O_2: Seat the second person in any remaining chair.

N_2: 3 ways

Applying the multiplication principle, there are $4 \cdot 3 = 12$ ways to seat two persons.

$(7\text{-}4)$

52. $(x + i)^6 = \displaystyle\sum_{k=0}^{6} \binom{6}{k} x^{6-k} i^k$

$= \binom{6}{0} x^6 + \binom{6}{1} x^5 i^1 + \binom{6}{2} x^4 i^2 + \binom{6}{3} x^3 i^3 + \binom{6}{4} x^2 i^4 + \binom{6}{5} x i^5 + \binom{6}{6} i^6$

$= x^6 + 6ix^5 - 15x^4 - 20ix^3 + 15x^2 + 6ix - 1$

$(7\text{-}6)$

53. Since there are several ways in which at least one woman can be selected, it is simplest to note: probability of selecting at least one woman = 1 - probability of selecting 0 women. To calculate the probability of selecting 0 women, that is, three men, we note:

The sample space S is the set of all 3-person subsets, chosen from the ten people. $n(S) = C_{10,3}$

The event E is the set of all 3-man subsets, chosen from the 7 men.

$n(E) = C_{7,3}$

Then $P(E) = \dfrac{C_{7,3}}{C_{10,3}}$ and

$P(\text{at least one woman}) = 1 - P(0\text{ women}) = 1 - P(E)$

$= 1 - \dfrac{C_{7,3}}{C_{10,3}} = 1 - \dfrac{7}{24} = \dfrac{17}{24}$

$(7\text{-}5)$

54. (A) $P(E) \approx \dfrac{f(E)}{n} = \dfrac{350}{1,000} = .350$

(B) A sample space of equally likely events is:

$S = \{HHH,\ HHT,\ HTH,\ HTT,\ THH,\ THT,\ TTH,\ TTT\}$

Let $E = 2$ heads = $\{HHT,\ HTH,\ THH\}$

then $P(E) = \dfrac{n(E)}{n(S)} = \dfrac{3}{8} = .375$

(C) Expected frequency = $P(E) \cdot$ number of trials = $.375(1,000) = 375$

$(7\text{-}5)$

55. Write: P_n: $\displaystyle\sum_{k=1}^{n} k^3 = \left(\sum_{k=1}^{n} k\right)^2$

Proof: Part 1: Show that P_1 is true.

P_1: $\displaystyle\sum_{k=1}^{1} k^3 = 1^3 = 1^2 = \left(\sum_{k=1}^{1} k\right)^2$

Part 2: Show that if P_j is true, then P_{j+1} is true.
Write out P_j and P_{j+1}.

P_j: $\displaystyle\sum_{k=1}^{j} k^3 = \left(\sum_{k=1}^{j} k\right)^2$

P_{j+1}: $\displaystyle\sum_{k=1}^{j+1} k^3 = \left(\sum_{k=1}^{j+1} k\right)^2$

We start with P_j:

$$\sum_{k=1}^{j} k^3 = \left(\sum_{k=1}^{j} k\right)^2$$

Adding $(j + 1)^3$ to both sides:

$$\sum_{k=1}^{j} k^3 + (j + 1)^3 = \left(\sum_{k=1}^{j} k\right)^2 + (j + 1)^3$$

$$\sum_{k=1}^{j+1} k^3 = (1 + 2 + 3 + \dots + j)^2 + (j + 1)^3$$

$$= \left[\frac{\cancel{j}(j + 1)}{2}\right]^2 + (j + 1)^3 \text{ using Matched Problem 2, Section 7-2}$$

$$= (j + 1)^2 \frac{j^2}{4} + (j + 1)^2(j + 1)$$

$$= (j + 1)^2 \left[\frac{j^2}{4} + j + 1\right]$$

$$= (j + 1)^2 \left[\frac{j^2 + 4j + 4}{4}\right]$$

$$= (j + 1)^2 \frac{(j + 2)^2}{2^2}$$

$$= \left[\frac{(j + 1)(j + 2)}{2}\right]^2$$

$$= [1 + 2 + 3 + \dots + (j + 1)]^2 \text{ using Matched Problem 2, Section 7-2}$$

$$= \left(\sum_{k=1}^{j+1} k\right)^2$$

We have shown that if P_j is true, then P_{j+1} is true.

Conclusion: P_n is true for all positive integers n.   *(7-2)*

56. Write: P_n: $x^{2n} - y^{2n} = (x - y)Q_n(x, y)$, where $Q_n(x, y)$ denotes some polynomial in x and y.

Proof: Part 1: Show that P_1 is true.

P_1: $x^{2 \cdot 1} - y^{2 \cdot 1} = (x - y)(x + y) = (x - y)Q_1(x, y)$ P_1 is true.

Part 2: Show that if P_k is true, then P_{k+1} is true.
Write out P_k and P_{k+1}.

P_k: $x^{2k} - y^{2k} = (x - y)Q_k(x, y)$

P_{k+1}: $x^{2k+2} - y^{2k+2} = (x - y)Q_{k+1}(x, y)$

We start with P_k:

$x^{2k} - y^{2k} = (x - y)Q_k(x, y)$

Now, $x^{2k+2} - y^{2k+2}$ $= x^{2k+2} - x^{2k}y^2 + x^{2k}y^2 - y^{2k+2}$

$= x^{2k}(x^2 - y^2) + y^2(x^{2k} - y^{2k})$

$= (x - y)x^{2k}(x + y) + y^2(x - y)Q_k(x, y)$ by P_k

$= (x - y)[x^{2k}(x + y) + y^2 Q_k(x, y)]$

$= (x - y)Q_{k+1}(x, y)$

We have shown that if P_k is true, then P_{k+1} is true.

Conclusion: P_n is true for all positive integers n.   (7-2)

57. Write: P_n: $\dfrac{a^n}{a^m} = a^{n-m}$, m an arbitrary positive integer, $n > m$.

Proof: Part 1: Show P_{m+1} is true.

P_{m+1}: $\dfrac{a^{m+1}}{a^m} = a^{m+1-m}$

$\dfrac{a^m a}{a^m} = a^1$ by the recursive definition of a^n

$a = a$ P_{m+1} is true.

Part 2: Show that if P_k is true, then P_{k+1} is true.
Write out P_k and P_{k+1}.

P_k: $\dfrac{a^k}{a^m} = a^{k-m}$

P_{k+1}: $\dfrac{a^{k+1}}{a^m} = a^{k-m+1}$

We start with P_k:

$\dfrac{a^k}{a^m} = a^{k-m}$

Multiplying both sides by a:

$\dfrac{a^k}{a^m} a = a^{k-m} a$

$\dfrac{a^k a}{a^m} = a^{k-m+1}$

$\dfrac{a^{k+1}}{a^m} = a^{k-m+1}$

We have shown that if P_k is true, then P_{k+1} is true, for m an arbitrary positive integer. Conclusion: P_n is true for all positive integers m, n.   (7-2)

58. To prove $a_n = b_n$, n a positive integer, write:

P_n: $a_n = b_n$

Proof: Part 1: Show P_1 is true.

$a_1 = -3$ $\qquad$ $b_1 = -5 + 2 \cdot 1 = -3$.

$a_1 = b_1$

Part 2: Show that if P_k is true, then P_{k+1} is true.

Write out P_k and P_{k+1}.

$\qquad$ P_k: $a_k = b_k$

$\qquad$ P_{k+1}: $a_{k+1} = b_{k+1}$

We start with P_k:

$\qquad$ $a_k = b_k$

Now, $a_{k+1} = a_k + 2 = b_k + 2$

$$\begin{aligned} &= -5 + 2k + 2 \\ &= -5 + 2(k + 1) \\ &= b_{k+1} \end{aligned}$$

Therefore, $a_{k+1} = b_{k+1}$.

So if P_k is true, then P_{k+1} is true.

Conclusion: P_n is true for all $n \in N$. That is $\{a_n\} = \{b_n\}$ $\hspace{2cm}$ (7-2)

59. Write: P_n: $(1!)1 + (2!)2 + (3!)3 + \cdots + (n!)n = (n + 1)! - 1$.

Proof: Part 1: Show that P_1 is true.

$\qquad$ P_1: $(1!)1 = 1 = 2 - 1 = 2! - 1$ is true.

Part 2: Show that if P_k is true, then P_{k+1} is true.

Write out P_k and P_{k+1}.

$\qquad$ P_k: $(1!)1 + (2!)2 + (3!)3 + \cdots + (k!)k = (k + 1)! - 1$

$\qquad$ P_{k+1}: $(1!)1 + (2!)2 + (3!)3 + \cdots + (k!)k + (k + 1)!(k + 1) = (k + 2)! - 1$

We start with P_k:

$\qquad$ $(1!)1 + (2!)2 + (3!)3 + \cdots + (k!)k = (k + 1)! - 1$

Adding $(k + 1)!(k + 1)$ to both sides:

$\qquad$ $(1!)1 + (2!)2 + (3!)3 + \cdots + (k!)k + (k + 1)!(k + 1)$

$$\begin{aligned} &= (k + 1)! - 1 + (k + 1)!(k + 1) \\ &= (k + 1)!(1 + k + 1) - 1 \\ &= (k + 2)(k + 1)! - 1 \\ &= (k + 2)! - 1 \end{aligned}$$

So if P_k is true, then P_{k+1} is true.

Conclusion: P_n is true for all positive integers n. $\hspace{2cm}$ (7-2)

60. The unpaid balance starts at $7,200, and decreases each month by $300. So the interest starts at $7,200(0.01)$ and decreases each month by $300(0.01)$. This is an arithmetic sequence with

$a_1 = 7,200(0.01)$, $a_{24} = 300(0.01)$, $n = 24$.

Then the total interest paid is

$$S_n = \frac{n}{2}(a_1 + a_n)$$

$$S_{24} = \frac{24}{2}[7,200(0.01) + 300(0.01)]$$

$$= \$900$$

$\hspace{12cm}$ (7-3)

61. This involves an infinite geometric series. $a_1 = (0.75)2,400.$ $r = 0.75.$

$|r| < 1$, so the series has a sum.

$$S_\infty = \frac{a_1}{1-r} = \frac{(0.75)2400}{1-0.75} = \$7,200 \qquad\qquad (7\text{-}3)$$

62. Since $A(n)$ is a geometric sequence with common ratio 1.06, we can write

$$A(n) = a_n = (1.06)^n 500$$

If $n = 10$, $a_{10} = (1.06)^{10} 500 = \$895.42$

If $n = 20$, $a_{10} = (1.06)^{20} 500 = \$1,603.57 \qquad\qquad (7\text{-}3)$

63. A route plan can be regarded as a series of choices of stores, which is really an arrangement of the 5 stores. Since the order matters, we use permutations:
$P_{5,5} = 5! = 120$ route plans. $\qquad\qquad (7\text{-}4)$

64. (A) $P(\text{this event}) = \dfrac{40}{1,000} = .04$

(B) $P(\text{this event}) = \dfrac{100 + 60}{1,000} = .16$

(C) $P(\text{this event}) = 1 - P\ (\text{not } 12 - 18 \text{ and buys 0 or 1 cassette annually})$

$$= 1 - \frac{60 + 70 + 70 + 110 + 100 + 50}{1,000} = 1 - \frac{460}{1,000} = .54 \qquad (7\text{-}5)$$

65. $P\ (\text{shipment returned}) = 1 - P\ (\text{no substandard part found}).$
The sample space S is the set of all possible 4-part subsets of the 12 part set.
$n(S) = C_{12,4}$
The event, no substandard part found, is the set of all possible 4-part subsets of the 10 acceptable parts. $n(E) = C_{10,4}$.

Then $P(\text{shipment returned}) = 1 - \dfrac{n(E)}{n(S)} = 1 - \dfrac{C_{10,4}}{C_{12,4}} = 1 - \dfrac{14}{33} \approx .576. \qquad (7\text{-}5)$

CHAPTER 8   Additional Topics in Analytic Geometry

Section 8-1

1. Answers will vary.   **3.** Answers will vary.   **5.** Answers will vary.

7. To graph $y^2 = 4x$, assign x values that make the right side a perfect square (x must be non-negative for y to be real) and solve for y. Since the coefficient of x is positive, a must be positive, and the parabola opens right.

x	0	1	4
y	0	±2	±4

To find the focus and directrix, solve
$4a = 4$
$a = 1$
Focus: (1, 0) Directrix: $x = -1$

9. To graph $x^2 = 8y$, assign y values that make the right side a perfect square (y must be non-negative for x to be real) and solve for x. Since the coefficient of y is positive, a must be positive, and the parabola opens up.

x	0	±4	±2
y	0	2	$\frac{1}{2}$

To find the focus and directrix, solve
$4a = 8$
$a = 2$
Focus: (0, 2) Directrix: $y = -2$

11. To graph $y^2 = -12x$, assign x values that make the right side a perfect square (x must be zero or negative for y to be real) and solve for y. Since the coefficient of x is negative, a must be negative, and the parabola opens left.

x	0	-3	$-\frac{1}{3}$
y	0	±6	±2

To find the focus and directrix, solve
$4a = -12$
$a = -3$
Focus: (-3, 0)
Directrix: $x = -(-3) = 3$

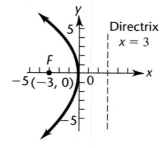

13. To graph $x^2 = -4y$, assign y values that make the right side a perfect square (y must be zero or negative for x to be real) and solve for x. Since the coefficient of y is negative, a must be negative, and the parabola opens down.

x	0	±2	±4
y	0	-1	-4

To find the focus and directrix, solve
$4a = -4$
$a = -1$
Focus: (0, -1)
Directrix: $y = -(-1) = 1$

15. To graph $y^2 = -20x$, we may proceed as in problem 11. Alternatively, after noting that x must be zero or negative for y to be real, we may pick convenient values for x and solve for y using a calculator. Since the coefficient of x is negative, a must be negative, and the parabola opens left.

x	0	-1	-2
y	0	$\pm\sqrt{20} \approx \pm 4.5$	$\pm\sqrt{40} \approx \pm 6.3$

To find the focus and directrix, solve
$4a = -20$
$a = -5$
Focus: $(-5, 0)$
Directrix: $x = -(-5) = 5$

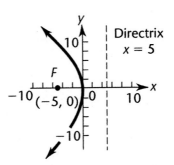

17. To graph $x^2 = 10y$, we may proceed as in Problem 9. Alternatively, after noting that y must be non-negative for x to be real, we may pick convenient values for y and solve for x using a calculator. Since the coefficient of y is positive, a must be positive, and the parabola opens up.

x	0	$\pm\sqrt{10} \approx \pm 3.2$	$\pm\sqrt{20} \approx \pm 4.5$
y	0	1	2

To find the focus and directrix, solve

$4a = 10$
$a = 2.5$
Focus: $(0, 2.5)$
Directrix: $y = -2.5$

19. Comparing $y^2 = 39x$ with $y^2 = 4ax$, the standard equation of a parabola symmetric with respect to the x axis, we have
$4a = 39$ Focus on x axis
$a = 9.75$
Focus: $(9.75, 0)$

21. Comparing $x^2 = -105y$ with $x^2 = 4ay$, the standard equation of a parabola symmetric with respect to the y axis, we have
$4a = -105$ Focus on y axis
$a = -26.25$
Focus: $(0, -26.25)$

23. Comparing $y^2 = -77x$ with $y^2 = 4ax$, the standard equation of a parabola symmetric with respect to the x axis, we have
$4a = -77$ Focus on x axis
$a = -19.25$
Focus: $(-19.25, 0)$

25. Comparing directrix $y = -3$ with the information in the chart of standard equations for a parabola, we see:
$a = 3$. Axis: the y axis. Equation: $x^2 = 4ay$.
The equation of the parabola must be $x^2 = 4 \cdot 3y$, or $x^2 = 12y$.

27. Comparing focus $(0, -7)$ with the information in the chart of standard equations for a parabola, we see:
$a = -7$ Axis: the y-axis. Equation: $x^2 = 4ay$.
The equation of the parabola must be $x^2 = 4(-7)y$, or $x^2 = -28y$.

29. Comparing directrix $x = 6$ with the information in the chart of standard equations for a parabola, we see:

$6 = -a.$ $a = -6.$ Axis: the x-axis.   Equation: $y^2 = 4ax.$

The equation of the parabola must be $y^2 = 4(-6)x$, or $y^2 = -24x.$

31. Comparing focus $(2, 0)$ with the information in the chart of standard equations for a parabola, we see:

$a = 2$ Axis: the x-axis.   Equation: $y^2 = 4ax.$

The equation of the parabola must be $y^2 = 4(2)x$, or $y^2 = 8x.$

33. The parabola is opening up and has an equation of the form $x^2 = 4ay$. Since $(4, 2)$ is on the graph, we have:

$$x^2 = 4ay$$
$$(4)^2 = 4a(2)$$
$$16 = 8a$$
$$2 = a$$

The equation of the parabola is
$$x^2 = 4(2)y$$
$$x^2 = 8y$$

35. The parabola is opening left and has an equation of the form $y^2 = 4ax$. Since $(-3, 6)$ is on the graph, we have:

$$y^2 = 4ax$$
$$(6)^2 = 4a(-3)$$
$$-3 = a$$

The equation of the parabola is
$$y^2 = 4(-3)x$$
$$y^2 = -12x$$

37. The parabola is opening down and has an equation of the form $x^2 = 4ay$. Since $(-6, -9)$ is on the graph, we have:

$$x^2 = 4ay$$
$$(-6)^2 = 4a(-9)$$
$$36 = -36a$$
$$-1 = a$$

The equation of the parabola is $x^2 = 4(-1)y$
$$x^2 = -4y$$

39. $x^2 = 4y$
$y^2 = 4x$

Solve for y in the first equation, then substitute into the second equation.

$$y = \frac{x^2}{4}$$

$$\left(\frac{x^2}{4}\right)^2 = 4x$$

$$\frac{x^4}{16} = 4x$$

$$x^4 = 64x$$
$$x^4 - 64x = 0$$
$$x(x^3 - 64) = 0$$
$$x(x - 4)(x^2 + 4x + 8) = 0$$
$$x = 0 \quad x - 4 = 0 \quad x^2 + 4x + 8 = 0$$
$$x = 4 \quad \text{No real solutions}$$

For $x = 0$ For $x = 4$

$$y = 0 \qquad y = \frac{4^2}{4}$$
$$y = 4$$

Solutions: $(0, 0), (4, 4)$

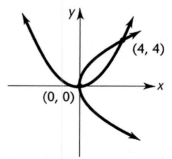

These exact solutions yield more than 3-digit accuracy.

41. $y^2 = 6x$
$x^2 = 5y$

Solve for x in the first equation, then substitute into the second equation.

$$x = \frac{y^2}{6}$$

$$\left(\frac{y^2}{6}\right)^2 = 5y$$

$$\frac{y^4}{36} = 5y$$

$$y^4 = 180y$$

$$y^4 - 180y = 0$$

$$y(y^3 - 180) = 0$$

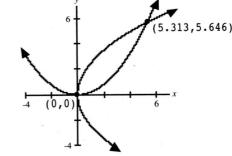

$$y(y - \sqrt[3]{180})(y^2 + y\sqrt[3]{180} + (\sqrt[3]{180})^2) = 0$$

$y = 0 \qquad y - \sqrt[3]{180} = 0 \qquad\qquad y^2 + y\sqrt[3]{180} + (\sqrt[3]{180})^2 = 0$

$\qquad\qquad\qquad y = \sqrt[3]{180}$ \qquad No real solutions

$\qquad\qquad\qquad \approx 5.646$

For $y = 0$ \qquad For $y = \sqrt[3]{180}$

$\qquad x = 0$ \qquad\qquad $x = \dfrac{(\sqrt[3]{180})^2}{6}$

$\qquad\qquad\qquad\qquad \approx 5.313$

Solutions: $(0, 0)$, $(5.313, 5.646)$

43. (A) The line $x = 0$ intersects the parabola $x^2 = 4ay$ only at $(0, 0)$.
The line $y = 0$ intersects the parabola $x^2 = 4ay$ only at $(0, 0)$.
Only these 2 lines intersect the parabola at exactly one point (see part (B)).

(B) A line through $(0, 0)$ with slope $m \neq 0$ has equation $y = mx$.
Solve the system:
$$y = mx$$
$$x^2 = 4ay$$
by substituting y from the first equation into the second equation.
$$x^2 = 4amx$$
$$x^2 - 4amx = 0$$
$$x(x - 4am) = 0$$
$$x = 0 \qquad x = 4am$$

For $x = 0$ \qquad For $x = 4am$

$\qquad y = 0$ \qquad\qquad $y = m(4am)$

$\qquad\qquad\qquad\qquad = 4am^2$

Solutions: $(0, 0)$, $(4am, 4am^2)$ are the required coordinates.

45. Since A and B lie on the curve $x^2 = 4ay$, their coordinates must satisfy the equation of the curve. Clearly, the y coordinate of A, F, and B is a. Substituting a for y, we have

$$x^2 = 4aa$$
$$x^2 = 4a^2$$
$$x = \pm 2a$$

Therefore, A has coordinates $(-2a, a)$ and B has coordinates $(2a, a)$.

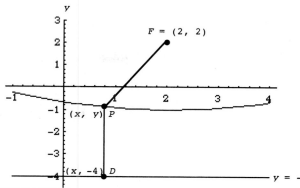

47. Let $P = (x, y)$ be a point on the parabola. Then, by the definition of the parabola, the distance from $P = (x, y)$ to the focus $F = (2, 2)$ must equal the perpendicular distance from P to the directrix at $D = (x, -4)$. Applying the distance formula, we have

$$d(P, F) = d(P, D)$$
$$\sqrt{(x - 2)^2 + (y - 2)^2} = \sqrt{(x - x)^2 + [y - (-4)]^2}$$
$$(x - 2)^2 + (y - 2)^2 = (x - x)^2 + (y + 4)^2$$
$$x^2 - 4x + 4 + y^2 - 4y + 4 = 0 + y^2 + 8y + 16$$
$$x^2 - 4x - 12y - 8 = 0$$

49. Let $P = (x, y)$ be a point on the parabola.

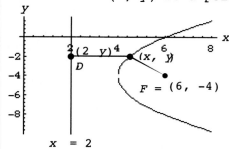

$x = 2$

Then, by the definition of the parabola, the distance from $P = (x, y)$ to the focus $F = (6, -4)$ must equal the perpendicular distance from P to the directrix at $D = (2, y)$.

Applying the distance formula, we have
$$d(P, F) = d(P, D)$$
$$\sqrt{(x - 6)^2 + [y - (-4)]^2} = \sqrt{(x - 2)^2 + (y - y)^2}$$
$$(x - 6)^2 + (y + 4)^2 = (x - 2)^2 + (y - y)^2$$
$$x^2 - 12x + 36 + y^2 + 8y + 16 = x^2 - 4x + 4 + 0$$
$$y^2 + 8y - 8x + 48 = 0$$

51. Using Figure 5 in Section 8-1, we note:
The point $P = (x, y)$ is a point on the parabola if and only if
$$d_1 = d_2$$
$$d(P, N) = d(P, F)$$
$$\sqrt{(x - x)^2 + (y + a)^2} = \sqrt{(x - 0)^2 + (y - a)^2}$$
$$(y + a)^2 = x^2 + (y - a)^2$$
$$y^2 + 2ay + a^2 = x^2 + y^2 - 2ay + a^2$$
$$x^2 = 4ay$$

53. From the figure, we see that the coordinates of P must be $(-100, -50)$. The parabola is opening down with axis the y axis, so it has an equation of the form $x^2 = 4ay$. Since $(-100, -50)$ is on the graph, we have

$$(-100)^2 = 4a(-50)$$
$$10,000 = -200a$$
$$a = -50$$

The equation of the parabola is
$$x^2 = 4(-50)y$$
$$x^2 = -200y$$

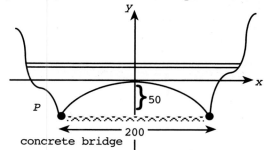

concrete bridge

55. (A) From the figure, we see that the parabola is opening up with axis the y axis, so it has an equation of the form $x^2 = 4ay$. Since the focus is at $(0, a) = (0, 100)$, $a = 100$ and the equation of the parabola is $x^2 = 400y$ or $y = 0.0025x^2$

(B) Since the depth represents the y coordinate y_1 of a point on the parabola with $x = 100$, we have $y_1 = 0.0025(100)^2$
   depth = 25 feet

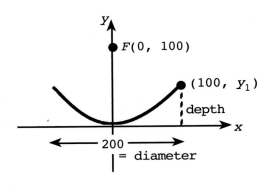

Section 8-2

1. No. Explanations will vary.

3. No. Explanations will vary.

5. Answers will vary.

Except as otherwise noted, the graphing calculator drawings in this and the following sections of this chapter have been plotted in a squared window (zoom menu 5 on the TI-84) to avoid distorting the shapes of the conic sections.

7. When $y = 0$, $\dfrac{x^2}{25} = 1$. x intercepts: ± 5

When $x = 0$, $\dfrac{y^2}{4} = 1$. y intercepts: ± 2

So $a = 5$, $b = 2$, and the major axis is on the x axis.
Foci: $c^2 = a^2 - b^2$
$\qquad c^2 = 25 - 4$
$\qquad c^2 = 21$
$\qquad\ c = \sqrt{21}$
Foci: $F' = (-\sqrt{21}, 0)$, $F = (\sqrt{21}, 0)$
Major axis length = $2(5) = 10$; Minor axis length = $2(2) = 4$

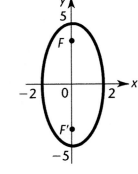

9. When $y = 0$, $\dfrac{x^2}{4} = 1$. x intercepts: ± 2

When $x = 0$, $\dfrac{y^2}{25} = 1$. y intercepts: ± 5

So $a = 5$, $b = 2$, and the major axis is on the y axis.
Foci: $c^2 = a^2 - b^2$
$\qquad c^2 = 25 - 4$
$\qquad c^2 = 21$
$\qquad\ c = \sqrt{21}$ Foci: $F' = (0, -\sqrt{21})$, $F = (0, \sqrt{21})$
Major axis length = $2(5) = 10$ Minor axis length = $2(2) = 4$

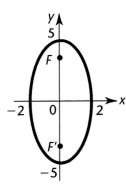

11. First, write the equation in standard form by dividing both sides by 9.
$x^2 + 9y^2 = 9$

$\dfrac{x^2}{9} + \dfrac{y^2}{1} = 1$

Locate the intercepts.

When $y = 0$, $\dfrac{x^2}{9} = 1$. x intercepts: ± 3

When $x = 0$, $\dfrac{y^2}{1} = 1$. y intercepts: ± 1

So $a = 3$, $b = 1$, and the major axis is on the x axis.
Foci: $c^2 = a^2 - b^2$
$\qquad c^2 = 9 - 1$
$\qquad c^2 = 8$
$\qquad\ c = \sqrt{8}$ Foci: $F' = (-\sqrt{8}, 0)$, $F = (\sqrt{8}, 0)$
Major axis length = $2(3) = 6$ Minor axis length = $2(1) = 2$

13. When $y = 0$, $9x^2 = 144$. x intercepts: ± 4
When $x = 0$, $16y^2 = 144$. y intercepts: ± 3
This corresponds to graph (b).

15. When $y = 0$, $4x^2 = 16$. x intercepts: ± 2
When $x = 0$, $y^2 = 16$. y intercepts: ± 4
This corresponds to graph (a).

17. First, write the equation in standard form by dividing both sides by 225.

$25x^2 + 9y^2 = 225$

$$\frac{x^2}{9} + \frac{y^2}{25} = 1$$

Locate the intercepts.

When $y = 0$, $\dfrac{x^2}{9} = 1$. x intercepts: ± 3

When $x = 0$, $\dfrac{y^2}{25} = 1$. y intercepts: ± 5

So $a = 5$, $b = 3$, and the major axis is on the y axis.
Foci: $c^2 = a^2 - b^2$
$c^2 = 25 - 9$
$c^2 = 16$
$c = 4$ Foci: $F' = (0, -4)$, $F = (0, 4)$
Major axis length $= 2(5) = 10$ Minor axis length $= 2(3) = 6$

19. First, write the equation in standard form by dividing both sides by 12.

$2x^2 + y^2 = 12$

$$\frac{x^2}{6} + \frac{y^2}{12} = 1$$

Locate the intercepts.

When $y = 0$, $\dfrac{x^2}{6} = 1$. x intercepts: $\pm \sqrt{6}$

When $x = 0$, $\dfrac{y^2}{12} = 1$. y intercepts: $\pm \sqrt{12}$

So $a = \sqrt{12}$, $b = \sqrt{6}$, and the major axis is on the
y axis.
Foci: $c^2 = a^2 - b^2$
$c^2 = 12 - 6$
$c^2 = 6$
$c = \sqrt{6}$ Foci: $F' = (0, -\sqrt{6})$, $F = (0, \sqrt{6})$
Major axis length $= 2\sqrt{12} \approx 6.93$ Minor axis length $= 2\sqrt{6} \approx 4.90$

21. First, write the equation in standard form by dividing both sides by 28.

$4x^2 + 7y^2 = 28$

$$\frac{x^2}{7} + \frac{y^2}{4} = 1$$

Locate the intercepts.

When $y = 0$, $\dfrac{x^2}{7} = 1$. x intercepts: $\pm \sqrt{7}$

When $x = 0$, $\dfrac{y^2}{4} = 1$. y intercepts: ± 2

So $a = \sqrt{7}$, $b = 2$, and the major axis is on the
x axis.

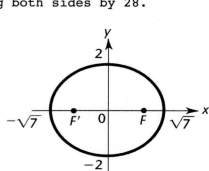

Foci: $c^2 = a^2 - b^2$
$\qquad c^2 = 7 - 4$
$\qquad c^2 = 3$
$\qquad c = \sqrt{3}$ Foci: $F' = (-\sqrt{3}, 0)$, $F = (\sqrt{3}, 0)$
Major axis length $= 2\sqrt{7} \approx 5.29$
Minor axis length $= 2(2) = 4$

23. The x intercepts are ± 5; if $y = 0$, $\dfrac{(\pm 5)^2}{M} = 1$, so $M = 25$.

The y intercepts are ± 4; if $x = 0$, $\dfrac{(\pm 4)^2}{N} = 1$, so $N = 16$.

Equation: $\dfrac{x^2}{25} + \dfrac{y^2}{16} = 1$

25. The x intercepts are ± 3; if $y = 0$, $\dfrac{(\pm 3)^2}{M} = 1$, so $M = 9$.

The y intercepts are ± 6; if $x = 0$, $\dfrac{(\pm 6)^2}{N} = 1$, so $N = 36$.

Equation: $\dfrac{x^2}{9} + \dfrac{y^2}{36} = 1$

27. Make a rough sketch of the ellipse and compute x and y intercepts.
$\dfrac{x^2}{a^2} + \dfrac{y^2}{b^2} = 1$
$a = \dfrac{10}{2} = 5$, $b = \dfrac{6}{2} = 3$
$\dfrac{x^2}{25} + \dfrac{y^2}{9} = 1$

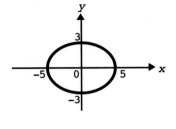

29. Make a rough sketch of the ellipse and compute x and y intercepts.
$\dfrac{x^2}{b^2} + \dfrac{y^2}{a^2} = 1$
$a = \dfrac{22}{2} = 11$, $b = \dfrac{16}{2} = 8$
$\dfrac{x^2}{64} + \dfrac{y^2}{121} = 1$

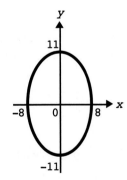

31. Make a rough sketch of the ellipse, locate focus and x intercepts, then determine y intercepts using the Pythagorean Theorem.
$\dfrac{x^2}{a^2} + \dfrac{y^2}{b^2} = 1$
$a = \dfrac{16}{2} = 8$
$b^2 = 8^2 - 6^2 = 64 - 36 = 28$
$b = \sqrt{28}$
$\dfrac{x^2}{64} + \dfrac{y^2}{28} = 1$

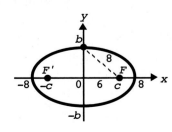

33. Make a rough sketch of the ellipse, locate focus and x intercepts, then determine y intercepts using the Pythagorean Theorem.

$$\frac{x^2}{b^2} + \frac{y^2}{a^2} = 1$$

$$b = \frac{20}{2} = 10$$

$$a^2 = 10^2 + (\sqrt{70})^2 = 100 + 70 = 170$$

$$a = \sqrt{170}$$

$$\frac{x^2}{100} + \frac{y^2}{170} = 1$$

35. The graph does not pass the vertical line test; most vertical lines that intersect an ellipse do so in two places. The equation does not define a function.

37. From the figure, we see that the point $P = (x, y)$ is a point on the curve if and only if

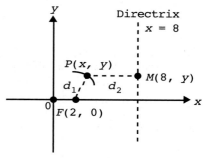

$$d_1 = \frac{1}{2}d_2$$

$$d(P, F) = \frac{1}{2}d(P, M)$$

$$\sqrt{(x - 2)^2 + (y - 0)^2} = \frac{1}{2}\sqrt{(x - 8)^2 + (y - y)^2}$$

$$(x - 2)^2 + y^2 = \frac{1}{4}(x - 8)^2$$

$$x^2 - 4x + 4 + y^2 = \frac{1}{4}(x^2 - 16x + 64)$$

$$4x^2 - 16x + 16 + 4y^2 = x^2 - 16x + 64$$

$$3x^2 + 4y^2 = 48$$

$$\frac{x^2}{16} + \frac{y^2}{12} = 1$$

The curve must be an ellipse since its equation can be written in standard form for an ellipse.

39. The only points that could satisfy this condition are those on the line segment $F'F$, and all of those would qualify.

41. $d_1 + d_2 = 2a$ Use the distance formula:

$$\sqrt{(x + c)^2 + y^2} = 2a - \sqrt{(x - c)^2 + y^2}$$ Square both sides:

$$(x + c)^2 + y^2 = 4a^2 - 4a\sqrt{(x - c)^2 + y^2} + (x - c)^2 + y^2$$ Isolate the square root:

$$\sqrt{(x - c)^2 + y^2} = a - \frac{cx}{a}$$ Square both sides:

$$(x - c)^2 + y^2 = a^2 - 2cx + \frac{c^2x^2}{a^2}$$ Simplify and collect terms:

$$\left(1 - \frac{c^2}{a^2}\right)x^2 + y^2 = a^2 - c^2$$ Use $a^2 - c^2 = b^2$, $1 - \frac{c^2}{a^2} = \frac{b^2}{a^2}$, and divide both sides by b^2:

$$\frac{x^2}{a^2} + \frac{y^2}{b^2} = 1$$

43. From the figure we see that the x and y intercepts of the ellipse must be 20 and 12 respectively. The equation of the ellipse must be

$$\frac{x^2}{(20)^2} + \frac{y^2}{(12)^2} = 1$$

or

$$\frac{x^2}{400} + \frac{y^2}{144} = 1$$

To find the clearance above the water 5 feet from the bank, we need the y coordinate y_1 of the point P whose x coordinate is $a - 5 = 15$. Since P is on the ellipse, we have

$$\frac{15^2}{400} + \frac{y_1^2}{144} = 1$$

$$\frac{225}{400} + \frac{y_1^2}{144} = 1$$

$$0.5625 + \frac{y_1^2}{144} = 1$$

$$\frac{y_1^2}{144} = 0.4375$$

$$y_1^2 = 144(0.4375)$$

$$y_1^2 = 63$$

$$y_1 \approx \pm7.94$$

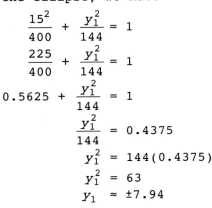

Therefore, the clearance is 7.94 feet, approximately.

45. (A) From the figure we see that the x intercept of the ellipse must be 24.0. The equation of the ellipse must have the form

$$\frac{x^2}{(24.0)^2} + \frac{y^2}{b^2} = 1$$

Since the point (23.0, 1.14) is on the ellipse, its coordinates must satisfy the equation of the ellipse. So

$$\frac{(23.0)^2}{(24.0)^2} + \frac{(1.14)^2}{b^2} = 1$$

$$\frac{(1.14)^2}{b^2} = 1 - \frac{(23.0)^2}{(24.0)^2}$$

$$\frac{(1.14)^2}{b^2} = \frac{47}{576}$$

$$b^2 = \frac{576(1.14)^2}{47}$$

$$b^2 = 15.9$$

The equation of the ellipse must be $\dfrac{x^2}{576} + \dfrac{y^2}{15.9} = 1$

(B) From the figure, we can see that the width of the wing must equal
$1.14 + b = 1.14 + \sqrt{15.9} = 5.13$ feet.

Section 8-3

1. Answers will vary.   **3.** Answers will vary.   **5.** Answers will vary.

7. When $y = 0$, $x^2 = 1$. x intercepts: ±1
When $x = 0$, $-y^2 = 1$. There are no y intercepts.
This corresponds to graph (d).

9. When $y = 0$, $-x^2 = 4$. There are no x intercepts.
When $x = 0$, $y^2 = 4$. y intercepts: ± 2
This corresponds to graph (c).

11. When $y = 0$, $\dfrac{x^2}{9} = 1$. x intercepts: ± 3 $a = 3$

When $x = 0$, $-\dfrac{y^2}{4} = 1$. There are no y intercepts, but $b = 2$.

Sketch the asymptotes using the
asymptote rectangle, then sketch in
the hyperbola.
Foci: $c^2 = 3^2 + 2^2$
$\qquad c^2 = 13$
$\qquad\quad c = \sqrt{13}$
$F' = (-\sqrt{13}, \ 0)$, $F = (\sqrt{13}, \ 0)$

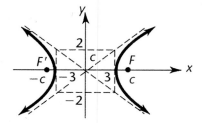

Transverse axis length = 2(3) = 6
Conjugate axis length = 2(2) = 4

13. When $y = 0$, $-\dfrac{x^2}{9} = 1$. There are no x intercepts, but $b = 3$.

When $x = 0$, $\dfrac{y^2}{4} = 1$. y intercepts: ± 2, $a = 2$

Sketch the asymptotes using the
asymptote rectangle, then sketch in
the hyperbola.
Foci: $c^2 = 2^2 + 3^2$
$\qquad c^2 = 13$
$\qquad\quad c = \sqrt{13}$
$F' = (0, \ -\sqrt{13})$, $F = (0, \ \sqrt{13})$

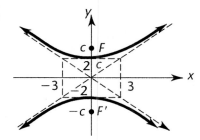

Transverse axis length = 2(2) = 4
Conjugate axis length = 2(3) = 6

15. First, write the equation in standard form by dividing both sides by 16.
$\qquad 4x^2 - y^2 = 16$
$\qquad \dfrac{x^2}{4} - \dfrac{y^2}{16} = 1$
Locate intercepts:
When $y = 0$, $x = \pm 2$. x intercepts: ± 2 $a = 2$

When $x = 0$, $-\dfrac{y^2}{16} = 1$. There are no y intercepts,
$\qquad\qquad\qquad\qquad\qquad$ but $b = 4$.
Sketch the asymptotes using the asymptote
rectangle, then sketch in the hyperbola.
Foci: $c^2 = 2^2 + 4^2$
$\qquad c^2 = 20$
$\qquad\quad c = \sqrt{20}$
$F' = (-\sqrt{20}, \ 0)$, $F = (\sqrt{20}, \ 0)$

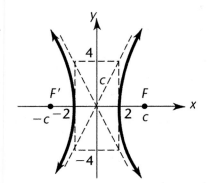

Transverse axis length = 2(2) = 4
Conjugate axis length = 2(4) = 8

17. First, write the equation in standard form by dividing both sides by 144.

$9y^2 - 16x^2 = 144$

$$\frac{y^2}{16} - \frac{x^2}{9} = 1$$

Locate intercepts:

When $y = 0$, $-\dfrac{x^2}{9} = 1$. There are no x intercepts, but $b = 3$.

When $x = 0$, $\dfrac{y^2}{16} = 1$, $y = \pm 4$.

y intercepts: ± 4 $a = 4$

Sketch the asymptotes using the asymptote rectangle, then sketch in the hyperbola.

Foci: $c^2 = 4^2 + 3^2$

$c^2 = 25$

$c = 5$

$F' = (0, -5)$, $F = (0, 5)$

Transverse axis length = $2(4) = 8$. Conjugate axis length = $2(3) = 6$.

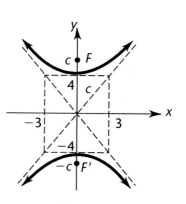

19. First, write the equation in standard form by dividing both sides by 12.

$3x^2 - 2y^2 = 12$

$$\frac{x^2}{4} - \frac{y^2}{6} = 1$$

Locate intercepts: When $y = 0$, $\dfrac{x^2}{4} = 1$. $x = \pm 2$.

x intercepts: ± 2 $a = 2$

When $x = 0$, $-\dfrac{y^2}{6} = 1$. There are no y intercepts, but $b = \sqrt{6}$.

Sketch the asymptotes using the asymptote rectangle, then sketch in the hyperbola.

Foci: $c^2 = 2^2 + (\sqrt{6})^2$

$c^2 = 10$

$c = \sqrt{10}$

$F' = (-\sqrt{10}, 0)$, $F = (\sqrt{10}, 0)$

Transverse axis length = $2(2) = 4$ Conjugate axis length = $2\sqrt{6} \approx 4.90$

21. First, write the equation in standard form by dividing both sides by 28.

$7y^2 - 4x^2 = 28$

$$\frac{y^2}{4} - \frac{x^2}{7} = 1$$

Locate intercepts:

When $y = 0$, $-\dfrac{x^2}{7} = 1$. There are no x intercepts, but $b = \sqrt{7}$.

When $x = 0$, $\dfrac{y^2}{4} = 1$, $y = \pm 2$. y intercepts: ± 2 $a = 2$

Sketch the asymptotes using the asymptote rectangle, then sketch in the hyperbola.

Foci: $c^2 = 2^2 + (\sqrt{7})^2$

$c^2 = 11$

$c = \sqrt{11}$

$F' = (0, -\sqrt{11})$, $F = (0, \sqrt{11})$

Transverse axis length = $2(2) = 4$

Conjugate axis length = $2\sqrt{7} \approx 5.29$

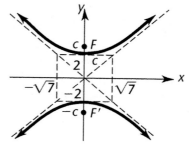

23. Since the graph has x intercepts ($x = \pm3$) but no y intercepts, the equation must be in the form

$$\frac{x^2}{M} - \frac{y^2}{N} = 1$$

and, when $y = 0$, $x = \pm3$, so

$$\frac{9}{M} - \frac{0}{N} = 1$$
$$M = 9$$

Since the point $(5, 4)$ is on the graph, its coordinates must satisfy the equation, which is known to be of form

$$\frac{x^2}{9} - \frac{y^2}{N} = 1$$

So we get

$$\frac{5^2}{9} - \frac{4^2}{N} = 1$$
$$\frac{25}{9} - \frac{16}{N} = 1$$
$$-\frac{16}{N} = -\frac{16}{9}$$
$$N = 9$$

The equation is therefore

$$\frac{x^2}{9} - \frac{y^2}{9} = 1$$

25. Since the graph has y intercepts ($y = \pm4$) but no x intercepts, the equation must be in the form

$$\frac{y^2}{N} - \frac{x^2}{M} = 1$$

and, when $x = 0$, $y = \pm4$, so

$$\frac{16}{N} - \frac{0}{M} = 1$$
$$N = 16$$

Since the point $(3, 5)$ is on the graph, its coordinates must satisfy the equation, which is known to be of form

$$\frac{y^2}{16} - \frac{x^2}{M} = 1$$

So we get

$$\frac{5^2}{16} - \frac{3^2}{M} = 1$$
$$\frac{25}{16} - \frac{9}{M} = 1$$
$$-\frac{9}{M} = -\frac{9}{16}$$
$$M = 16$$

The equation is therefore

$$\frac{y^2}{16} - \frac{x^2}{16} = 1$$

27. Since the transverse axis is on the x axis, start with

$$\frac{x^2}{a^2} - \frac{y^2}{b^2} = 1$$

and find a and b

$$a = \frac{14}{2} = 7 \text{ and } b = \frac{10}{2} = 5$$

The equation is

$$\frac{x^2}{49} - \frac{y^2}{25} = 1$$

29. Since the transverse axis is on the y axis, start with

$$\frac{y^2}{a^2} - \frac{x^2}{b^2} = 1$$

and find a and b

$$a = \frac{24}{2} = 12 \text{ and } b = \frac{18}{2} = 9$$

The equation is

$$\frac{y^2}{144} - \frac{x^2}{81} = 1$$

31. Since the transverse axis is on the x axis, start with

$$\frac{x^2}{a^2} - \frac{y^2}{b^2} = 1$$

and find a and b

$$a = \frac{18}{2} = 9$$

To find b, sketch the asymptote rectangle, label known parts, and use the Pythagorean Theorem.

$$b^2 = 11^2 - 9^2$$
$$b^2 = 40$$
$$b = \sqrt{40}$$

The equation is

$$\frac{x^2}{81} - \frac{y^2}{40} = 1$$

33. Since the conjugate axis is on the x axis, start with

$$\frac{y^2}{a^2} - \frac{x^2}{b^2} = 1$$

and find a and b

$$b = \frac{14}{2} = 7$$

To find a, sketch the asymptote rectangle, label known parts, and use the Pythagorean Theorem.

$$a^2 = (\sqrt{200})^2 - 7^2$$
$$a^2 = 151$$
$$a = \sqrt{151}$$

The equation is

$$\frac{y^2}{151} - \frac{x^2}{49} = 1$$

35. The equation $\dfrac{x^2}{25} - \dfrac{y^2}{4} = 1$ is in standard form with $a^2 = 25$, $b^2 = 4$. Since the intercepts are on the x axis, the asymptotes have equations $y = \pm\dfrac{b}{a}x = \pm\dfrac{2}{5}x$.

37. The equation $\dfrac{y^2}{4} - \dfrac{x^2}{16} = 1$ is in standard form with $a^2 = 4$, $b^2 = 16$. Since the intercepts are on the y axis, the asymptotes have equations $y = \pm\dfrac{a}{b}x = \pm\dfrac{2}{4}x$ or $y = \pm\dfrac{1}{2}x$.

39. $9x^2 - y^2 = 9$. In standard form, this becomes
$$\dfrac{x^2}{1} - \dfrac{y^2}{9} = 1$$
Then $a^2 = 1$, $b^2 = 9$. Since the intercepts are on the x axis, the asymptotes have equations $y = \pm\dfrac{b}{a}x = \pm\dfrac{3}{1}x$ or $y = \pm3x$.

41. $2y^2 - 3x^2 = 1$. In standard form, this becomes
$$\dfrac{y^2}{1/2} - \dfrac{x^2}{1/3} = 1$$
Then $a^2 = \dfrac{1}{2}$, $b^2 = \dfrac{1}{3}$. Since the intercepts are on the y axis, the asymptotes have equations $y = \pm\dfrac{a}{b}x$. Write $a = \sqrt{\dfrac{1}{2}}$, $b = \sqrt{\dfrac{1}{3}}$, $\dfrac{a}{b} = \sqrt{\dfrac{1}{2}} \div \sqrt{\dfrac{1}{3}} = \sqrt{\dfrac{3}{2}}$ or $\dfrac{\sqrt{3}}{\sqrt{2}}$.

Asymptotes: $y = \pm\dfrac{\sqrt{3}}{\sqrt{2}}x$.

43. (A) If a hyperbola has center at $(0, 0)$ and a focus at $(1, 0)$, its equation must be of form
$$\dfrac{x^2}{a^2} - \dfrac{y^2}{b^2} = 1$$
with $c = 1$, so $a^2 + b^2 = 1$ or $b^2 = 1 - a^2$
Therefore there are an infinite number of such hyperbolas.
Each has an equation of form
$$\dfrac{x^2}{a^2} - \dfrac{y^2}{1 - a^2} = 1 \qquad \text{Note that since } 0 < a < c, \text{ we require } 0 < a < 1.$$

(B) If an ellipse has center at $(0, 0)$ and a focus at $(1, 0)$, its equation must be of form
$$\dfrac{x^2}{a^2} + \dfrac{y^2}{b^2} = 1$$
with $c = 1$, so $a^2 - 1 = b^2$
Therefore there are an infinite number of such ellipses. Each has an equation of form
$$\dfrac{x^2}{a^2} + \dfrac{y^2}{a^2 - 1} = 1$$
Note that since $a > c > 0$, we require $a > 1$.

(C) If a parabola has vertex at $(0, 0)$ and focus at $(1, 0)$, its equation must be of form
$$y^2 = 4ax$$
with $a = 1$.

Therefore there is one such parabola; its equation is $y^2 = 4x$.

45. (A) The points of intersection are the solutions of the system
$$x^2 - y^2 = 1$$
$$y = 0.5x = \frac{1}{2}x$$

We solve using substitution:
$$x^2 - \left(\frac{1}{2}x\right)^2 = 1$$
$$x^2 - \frac{1}{4}x^2 = 1$$
$$\frac{3}{4}x^2 = 1$$
$$x^2 = \frac{4}{3}$$
$$x = \pm\frac{2}{\sqrt{3}}$$

$y = \frac{1}{2}x$, so when $x = \frac{2}{\sqrt{3}}$, $y = \frac{1}{2} \cdot \frac{2}{\sqrt{3}} = \frac{1}{\sqrt{3}}$

when $x = \frac{-2}{\sqrt{3}}$, $y = \frac{1}{2} \cdot \frac{-2}{\sqrt{3}} = \frac{-1}{\sqrt{3}}$

The points of intersection are $\left(\frac{2}{\sqrt{3}}, \frac{1}{\sqrt{3}}\right)$ and $\left(\frac{-2}{\sqrt{3}}, \frac{-1}{\sqrt{3}}\right)$.

(B)
$$x^2 - y^2 = 1$$
$$y = 2x$$
$$x^2 - (2x)^2 = 1$$
$$x^2 - 4x^2 = 1$$
$$-3x^2 = 1$$
$$x^2 = -\frac{1}{3}$$

No solution.
The graphs do not intersect.
Repeat the above calculations using $y = mx$ instead of $y = 2x$.
$$x^2 - y^2 = 1$$
$$y = mx$$
$$x^2 - (mx)^2 = 1$$
$$x^2 - m^2x^2 = 1$$
$$x^2(1 - m^2) = 1$$
$$x^2 = \frac{1}{1 - m^2}$$

If the right side is negative the equation will have no solution, so the graphs intersect only if $1 - m^2 > 0$. This occurs only if m^2 is between zero and 1; that is, for $-1 < m < 1$. In this case, $x = \pm\frac{1}{\sqrt{1 - m^2}}$; $y = mx = \pm\frac{m}{\sqrt{1 - m^2}}$.

So if $-1 < m < 1$, the graphs intersect at $\left(\frac{1}{\sqrt{1 - m^2}}, \frac{m}{\sqrt{1 - m^2}}\right)$ and $\left(\frac{-1}{\sqrt{1 - m^2}}, \frac{-m}{\sqrt{1 - m^2}}\right)$.

47. (A) The points of intersection are the solutions of the system
$$y^2 - 4x^2 = 1$$
$$y = x$$
$$x^2 - 4x^2 = 1$$
$$-3x^2 = 1$$
$$x^2 = -\frac{1}{3}$$

No solution. The graphs do not intersect.

(B) $\qquad y^2 - 4x^2 = 1$

$\qquad\qquad y = 3x$

$\qquad (3x)^2 - 4x^2 = 1$

$\qquad\quad 9x^2 - 4x^2 = 1$

$\qquad\qquad\quad 5x^2 = 1$

$\qquad\qquad\quad\ x^2 = \dfrac{1}{5}$

$\qquad\qquad\quad\ x = \pm\dfrac{1}{\sqrt{5}}$

$y = 3x$, so when $x = \dfrac{1}{\sqrt{5}}$, $y = \dfrac{3}{\sqrt{5}}$ and when $x = \dfrac{-1}{\sqrt{5}}$, $y = \dfrac{-3}{\sqrt{5}}$. The points of intersection are $\left(\dfrac{1}{\sqrt{5}}, \dfrac{3}{\sqrt{5}}\right)$ and $\left(\dfrac{-1}{\sqrt{5}}, \dfrac{-3}{\sqrt{5}}\right)$.

Repeat the above calculations with $y = mx$ in place of $y = 3x$.

$\qquad\ y^2 - 4x^2 = 1$

$\qquad\qquad\ y = mx$

$\quad\ (mx)^2 - 4x^2 = 1$

$\quad\ m^2x^2 - 4x^2 = 1$

$\quad\ x^2(m^2 - 4) = 1$

$\qquad\qquad x^2 = \dfrac{1}{m^2 - 4}$

This equation will have solutions only if $m^2 - 4$ is positive; this occurs when $m > 2$ or $m < -2$. In this case,

$$x = \pm\dfrac{1}{\sqrt{m^2 - 4}}; \quad y = mx = \pm\dfrac{m}{\sqrt{m^2 - 4}}.$$

So if $m > 2$ or $m < -2$ the points of intersection are

$$\left(\dfrac{1}{\sqrt{m^2 - 4}}, \dfrac{m}{\sqrt{m^2 - 4}}\right) \text{ and } \left(\dfrac{-1}{\sqrt{m^2 - 4}}, \dfrac{-m}{\sqrt{m^2 - 4}}\right).$$

49. (A) $\dfrac{x^2}{a^2} - \dfrac{y^2}{b^2} = 1 \qquad$ Solve for y.

$$-\dfrac{y^2}{b^2} = 1 - \dfrac{x^2}{a^2}$$

$$\dfrac{y^2}{b^2} = \dfrac{x^2}{a^2} - 1$$

$$y^2 = \dfrac{b^2x^2}{a^2} - b^2$$

$$y^2 = \dfrac{b^2}{a^2}x^2\left(1 - \dfrac{a^2}{x^2}\right)$$

$$y = \pm\dfrac{b}{a}x\sqrt{1 - \dfrac{a^2}{x^2}}$$

(B) As $|x| \to \infty$, $\dfrac{a}{|x|} \to 0$, $\dfrac{a^2}{x^2} \to 0$, $1 - \dfrac{a^2}{x^2} \to 1$, and $\sqrt{1 - \dfrac{a^2}{x^2}} \to 1$.

So as $|x| \to \infty$, $y \to \pm\dfrac{b}{a}x$. Therefore these lines form asymptotes for the hyperbola.

(C) Since $\sqrt{1 - \dfrac{a^2}{x^2}} < 1$ for $|x| > a$, $\dfrac{b}{a}x\sqrt{1 - \dfrac{a^2}{x^2}} < \dfrac{b}{a}x$, for $x > 0$. So in quadrant

I, the hyperbola is below its asymptote $y = \dfrac{b}{a}x$.

However, $-\dfrac{b}{a}x < -\dfrac{b}{a}x\sqrt{1 - \dfrac{a^2}{x^2}}$, for $x > 0$, so, in quadrant IV, the hyperbola is

above its asymptote, $y = -\dfrac{b}{a}x$. Similar arguments applied for $x < 0$ show that in

quadrant II the hyperbola is below its asymptote $y = -\dfrac{b}{a}x$, and in quadrant III,

it is above its asymptote $y = \dfrac{b}{a}x$.

51. By the triangle inequality, no points can satisfy this condition. The set is empty.

53.
$$|d_1 - d_2| = 2a$$
$$d_1 - d_2 = \pm 2a$$
$$d_1 = \pm 2a + d_2 \qquad \text{Use the distance formula}$$
$$\sqrt{(x + c)^2 + y^2} = \pm 2a + \sqrt{(x - c)^2 + y^2} \qquad \text{Square both sides}$$
$$(x + c)^2 + y^2 = 4a^2 \pm 4a\sqrt{(x - c)^2 + y^2} + (x - c)^2 + y^2 \qquad \text{Isolate the square root.}$$
$$\pm\sqrt{(x - c)^2 + y^2} = a - \dfrac{cx}{a} \qquad \text{Square both sides}$$
$$(x - c)^2 + y^2 = a^2 - 2cx + \dfrac{c^2 x^2}{a^2} \qquad \text{Simplify and collect terms}$$
$$\left(1 - \dfrac{c^2}{a^2}\right)x^2 + y^2 = a^2 - c^2 \quad \text{Use } a^2 - c^2 = b^2, \ 1 - \dfrac{c^2}{a^2} = \dfrac{b^2}{a^2} \text{ and divide both sides by } b^2.$$
$$\dfrac{x^2}{a^2} - \dfrac{y^2}{b^2} = 1$$

55. From the figure, we see that the point $P(x, y)$ is a point on the curve if and only if $d_1 = \dfrac{3}{2}d_2$

$$d(P, F) = \dfrac{3}{2}d(P, M)$$

$$\sqrt{(x - 3)^2 + (y - 0)^2} = \dfrac{3}{2}\sqrt{\left(x - \dfrac{4}{3}\right)^2 + (y - y)^2}$$

$$(x - 3)^2 + y^2 = \dfrac{9}{4}\left(x - \dfrac{4}{3}\right)^2$$

$$x^2 - 6x + 9 + y^2 = \dfrac{9}{4}\left(x^2 - \dfrac{8}{3}x + \dfrac{16}{9}\right)$$

$$4x^2 - 24x + 36 + 4y^2 = 9x^2 - 24x + 16$$

$$-5x^2 + 4y^2 = -20$$

$$\dfrac{x^2}{4} - \dfrac{y^2}{5} = 1$$

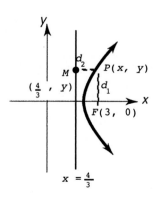

The curve must be a hyperbola since its equation can be written in standard form for a hyperbola.

57. From the figure below, we see that the transverse axis of the hyperbola must be on the y axis and $a = 4$. So the equation of the hyperbola must have the form

$$\frac{y^2}{4^2} - \frac{x^2}{b^2} = 1$$

To find b, we note that the point $(8, 12)$ is on the hyperbola, so its coordinates satisfy the equation. Substituting, we have

$$\frac{12^2}{4^2} - \frac{8^2}{b^2} = 1$$

$$9 - \frac{64}{b^2} = 1$$

$$-\frac{64}{b^2} = -8$$

$$-64 = -8b^2$$

$$b^2 = 8$$

The equation required is

$$\frac{y^2}{16} - \frac{x^2}{8} = 1$$

Using this equation, we can compute y when $x = 6$ to answer the question asked (see figure).

$$\frac{y^2}{16} - \frac{6^2}{8} = 1$$

$$\frac{y^2}{16} - \frac{36}{8} = 1$$

$$y^2 - 72 = 16$$

$$y^2 = 88$$

$$y = 9.38 \text{ to two decimal places}$$

The height above the vertex

$$= y - \text{height of vertex}$$
$$= 9.38 - 4$$
$$= 5.38 \text{ feet}$$

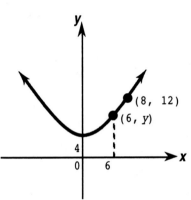

Hyperbola part of dome

59. From the figure below, we can see:

$FF' = 2c = 120 - 20 = 100$

So $c = 50$

$FV = c - a = 120 - 110 = 10$

So $a = c - 10 = 50 - 10 = 40$

Since $c^2 = a^2 + b^2$

$$50^2 = 40^2 + b^2$$

$$b = 30$$

So the equation of the hyperbola, in standard form, is

$$\frac{y^2}{40^2} - \frac{x^2}{30^2} = 1$$

Expressing y in terms of x, we have

$$\frac{y^2}{40^2} = 1 + \frac{x^2}{30^2}$$

$$\frac{y^2}{40^2} = \frac{1}{30^2}(30^2 + x^2)$$

$$y^2 = \frac{40^2}{30^2}(30^2 + x^2)$$

$$y = \frac{4}{3}\sqrt{x^2 + 30^2}$$

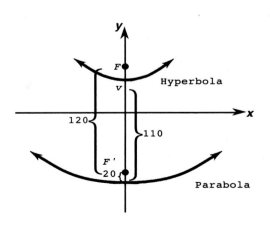

discarding the negative solution, since the reflecting hyperbola is above the x axis.

Section 8-4

1. Answers will vary.

3.
$$x^2 + y^2 = 169$$
$$x = -12$$
$$(-12)^2 + y^2 = 169$$
$$y^2 = 25$$
$$y = \pm 5$$
Solution: $(-12, 5)$, $(-12, -5)$

Check:
$$-12 \overset{\checkmark}{=} -12$$
$$(-12)^2 + (\pm 5)^2 \overset{\checkmark}{=} 169$$

5. $8x^2 - y^2 = 16$
$\qquad y = 2x$

Substitute y from the second equation into the first equation.
$$8x^2 - (2x)^2 = 16$$
$$8x^2 - 4x^2 = 16$$
$$4x^2 = 16$$
$$x^2 = 4$$
$$x = \pm 2$$

For $x = 2$ $\qquad$ For $x = -2$
$\quad y = 2(2)$ $\qquad \quad y = 2(-2)$
$\quad y = 4$ $\qquad \qquad y = -4$
Solutions: $(2, 4)$, $(-2, -4)$

Check:
For $(2, 4)$ $\qquad\qquad$ For $(-2, -4)$
$$4 \overset{\checkmark}{=} 2 \cdot 2 \qquad\qquad -4 \overset{\checkmark}{=} 2(-2)$$
$$8(2)^2 - 4^2 \overset{\checkmark}{=} 16 \qquad 8(-2)^2 - (-4)^2 \overset{\checkmark}{=} 16$$

7. $3x^2 - 2y^2 = 25$
$\quad x + y = 0$

Solve for y in the first degree equation
$y = -x$
Substitute into the second degree equation.
$$3x^2 - 2(-x)^2 = 25$$
$$x^2 = 25$$
$$x = \pm 5$$
For $x = 5$ $\qquad$ For $x = -5$
$\quad y = -5$ $\qquad \quad y = 5$
Solutions: $(5, -5)$, $(-5, 5)$

Check:
For $(5, -5)$ $\qquad\qquad$ For $(-5, 5)$
$$5 + (-5) \overset{\checkmark}{=} 0 \qquad\qquad (-5) + 5 \overset{\checkmark}{=} 0$$
$$3(5)^2 - 2(-5)^2 \overset{\checkmark}{=} 25 \quad 3(-5)^2 - 2(5)^2 \overset{\checkmark}{=} 25$$

From this point on we will not show the checking steps. The student should still perform these checking steps, however.

9.
$$y^2 = x$$
$$x - 2y = 2$$

Solve for x in the first degree equation.
$x = 2y + 2$
Substitute into the second degree equation.
$$y^2 = 2y + 2$$
$$y^2 - 2y - 2 = 0$$

$$y = \frac{-b \pm \sqrt{b^2 - 4ac}}{2a} \qquad a = 1, \; b = -2, \; c = -2$$

$$y = \frac{-(-2) \pm \sqrt{(-2)^2 - 4(1)(-2)}}{2(1)}$$

$$y = \frac{2 \pm \sqrt{12}}{2}$$

$$y = 1 \pm \sqrt{3}$$

For $y = 1 + \sqrt{3}$ For $y = 1 - \sqrt{3}$

$x = 2(1 + \sqrt{3}) + 2$ $x = 2(1 - \sqrt{3}) + 2$

$x = 4 + 2\sqrt{3}$ $x = 4 - 2\sqrt{3}$

Solutions: $(4 + 2\sqrt{3}, 1 + \sqrt{3})$, $(4 - 2\sqrt{3}, 1 - \sqrt{3})$

11. $2x^2 + y^2 = 24$

$x^2 - y^2 = -12$

Solve using elimination by addition. Adding, we obtain:

$3x^2 = 12$

$x^2 = 4$

$x = \pm 2$

For $x = 2$ For $x = -2$

$4 - y^2 = -12$ $4 - y^2 = -12$

$-y^2 = -16$ $-y^2 = -16$

$y^2 = 16$ $y = 16$

$y = \pm 4$ $y = \pm 4$

Solutions: $(2, 4)$, $(2, -4)$, $(-2, 4)$, $(-2, -4)$

13. $x^2 + y^2 = 10$

$16x^2 + y^2 = 25$

Solve using elimination by addition. Multiply the top equation by -1 and add.

$-x^2 - y^2 = -10$

$\underline{16x^2 + y^2 = \ \ 25}$

$15x^2 \qquad\quad = \ \ 15$

$x^2 \qquad = \quad 1$

$x = \pm 1$

For $x = 1$ For $x = -1$

$1 + y^2 = 10$ $1 + y^2 = 10$

$y^2 = 9$ $y = \pm 3$

$y = \pm 3$

Solutions: $(1, 3)$, $(1, -3)$, $(-1, 3)$, $(-1, -3)$

15. $xy - 4 = 0$

$x - y = 2$

Solve for x in the first degree equation.

$x = y + 2$

Substitute into the second degree equation

$(y + 2)y - 4 = 0$

$y^2 + 2y - 4 = 0$

$y = \dfrac{-b \pm \sqrt{b^2 - 4ac}}{2a}$ $a = 1, b = 2,$ $c = -4$

$y = \dfrac{-2 \pm \sqrt{(2)^2 - 4(1)(-4)}}{2(1)}$

$y = \dfrac{-2 \pm \sqrt{20}}{2}$

$y = -1 \pm \sqrt{5}$

For $y = -1 + \sqrt{5}$ For $y = -1 - \sqrt{5}$

$x = -1 + \sqrt{5} + 2$ $x = -1 - \sqrt{5} + 2$

$x = 1 + \sqrt{5}$ $x = 1 - \sqrt{5}$

Solutions: $(1 + \sqrt{5}, -1 + \sqrt{5})$, $(1 - \sqrt{5}, -1 - \sqrt{5})$

17. $x^2 + 2y^2 = 6$

$xy = 2$

Solve for y in the second equation

$y = \dfrac{2}{x}$

Substitute into the first equation

$$x^2 + 2\left(\dfrac{2}{x}\right)^2 = 6$$

$$x^2 + \dfrac{8}{x^2} = 6 \qquad x \neq 0$$

$$x^2 \cdot x^2 + x^2 \cdot \dfrac{8}{x^2} = 6x^2$$

$$x^4 + 8 = 6x^2$$

$$x^4 - 6x^2 + 8 = 0$$
$$(x^2 - 2)(x^2 - 4) = 0$$
$$(x - \sqrt{2})(x + \sqrt{2})(x - 2)(x + 2) = 0$$
$$x = \sqrt{2}, -\sqrt{2}, 2, -2$$

For $x = \sqrt{2}$ For $x = -\sqrt{2}$ For $x = 2$ For $x = -2$

$$y = \frac{2}{\sqrt{2}}$$ $$y = -\frac{2}{\sqrt{2}}$$ $$y = \frac{2}{2}$$ $$y = \frac{2}{-2}$$

$$y = \sqrt{2}$$ $$y = -\sqrt{2}$$ $$y = 1$$ $$y = -1$$

Solutions: $(\sqrt{2}, \sqrt{2})$, $(-\sqrt{2}, -\sqrt{2})$, $(2, 1)$, $(-2, -1)$

19. $2x^2 + 3y^2 = -4$
$\quad\;\; 4x^2 + 2y^2 = 8$

Solve using elimination by addition. Multiply the second equation by $-\frac{1}{2}$ and add.

$$\begin{array}{rcl} 2x^2 + 3y^2 &=& -4 \\ -2x^2 - \;\;y^2 &=& -4 \\ \hline 2y^2 &=& -8 \end{array}$$

$$y^2 = -4$$
$$y = \pm 2i$$

For $y = 2i$ For $y = -2i$

$2x^2 + 3(2i)^2 = -4$ $2x^2 + 3(-2i)^2 = -4$

$\quad 2x^2 - 12 = -4$ $\quad\;\; 2x^2 - 12 = -4$

$\qquad\quad 2x^2 = 8$ $\qquad\qquad 2x^2 = 8$

$\qquad\quad\;\; x^2 = 4$ $\qquad\qquad\;\; x^2 = 4$

$\qquad\qquad x = \pm 2$ $\qquad\qquad\;\; x = \pm 2$

Solutions: $(2, 2i)$, $(-2, 2i)$, $(2, -2i)$, $(-2, -2i)$

21. $x^2 - y^2 = 2$
$\qquad\;\; y^2 = x$

Substitute y^2 from the second equation into the first equation.

$$x^2 - x = 2$$
$$x^2 - x - 2 = 0$$
$$(x - 2)(x + 1) = 0$$
$$x = 2, -1$$

For $x = 2$ For $x = -1$

$\quad y^2 = 2$ $\quad y^2 = -1$

$\quad y = \pm\sqrt{2}$ $\quad y = \pm i$

Solutions: $(2, \sqrt{2})$, $(2, -\sqrt{2})$, $(-1, i)$, $(-1, -i)$

23. $x^2 + y^2 = 9$
$\qquad\;\; x^2 = 9 - 2y$

Substitute x^2 from the second equation into the first equation.

$$9 - 2y + y^2 = 9$$
$$y^2 - 2y = 0$$
$$y(y - 2) = 0$$
$$y = 0, 2$$

For $y = 0$ For $y = 2$

$\quad x^2 = 9 - 2(0)$ $\quad x^2 = 9 - 2(2)$

$\quad x^2 = 9$ $\quad x^2 = 5$

$\quad x = \pm 3$ $\quad x = \pm\sqrt{5}$

Solutions: $(3, 0)$, $(-3, 0)$, $(\sqrt{5}, 2)$, $(-\sqrt{5}, 2)$

25. $x^2 - y^2 = 3$
$\qquad\;\; xy = 2$

Solve for y in the second equation.

$$y = \frac{2}{x}$$

Substitute into the first equation:

$$x^2 - \left(\frac{2}{x}\right)^2 = 3$$

$$x^2 - \frac{4}{x^2} = 3 \qquad x \neq 0$$

$$x^4 - 4 = 3x^2$$
$$x^4 - 3x^2 - 4 = 0$$
$$(x^2 - 4)(x^2 + 1) = 0$$

$$x^2 - 4 = 0 \qquad x^2 + 1 = 0$$
$$x^2 = 4 \qquad\quad x^2 = -1$$
$$x = \pm 2 \qquad\quad x = \pm i$$

For $x = 2$ For $x = -2$ For $x = i$ For $x = -i$

$$y = \frac{2}{2} \qquad\quad y = \frac{2}{-2} \qquad\quad y = \frac{2}{i} \qquad\quad y = \frac{2}{-i}$$
$$y = 1 \qquad\quad y = -1 \qquad\quad y = -2i \qquad\quad y = 2i$$

Solutions: $(2, 1)$, $(-2, -1)$, $(i, -2i)$, $(-i, 2i)$

27. $y = 5 - x^2$
$y = 2 - 2x$

Substitute y from the first equation into the second equation.
$$5 - x^2 = 2 - 2x$$
$$0 = x^2 - 2x - 3$$
$$0 = (x - 3)(x + 1)$$
$$x = 3, -1$$

For $x = 3$ For $x = -1$
$$\quad y = 2 - 2(3) \qquad y = 2 - 2(-1)$$
$$\quad y = -4 \qquad\qquad y = 4$$
Solutions: $(3, -4)$, $(-1, 4)$

29. $y = x^2 - x$
$y = 2x$

Substitute y from the first equation into the second equation.
$$x^2 - x = 2x$$
$$x^2 - 3x = 0$$
$$x(x - 3) = 0$$
$$x = 0, 3$$

For $x = 0$ For $x = 3$
$$\quad y = 2(0) \qquad y = 2(3)$$
$$\quad y = 0 \qquad\quad y = 6$$
Solutions: $(0, 0)$, $(3, 6)$

31. $y = x^2 - 6x + 9$
$y = 5 - x$

Substitute y from the first equation into the second equation.
$$x^2 - 6x + 9 = 5 - x$$
$$x^2 - 5x + 4 = 0$$
$$(x - 1)(x - 4) = 0$$
$$x = 1, 4$$

For $x = 1$ For $x = 4$
$$\quad y = 5 - 1 \qquad y = 5 - 4$$
$$\quad y = 4 \qquad\quad y = 1$$
Solutions: $(1, 4)$, $(4, 1)$

33. $y = 8 + 4x - x^2$
$y = x^2 - 2x$

Substitute y from the first equation into the second equation.
$$8 + 4x - x^2 = x^2 - 2x$$
$$0 = 2x^2 - 6x - 8$$
$$0 = x^2 - 3x - 4$$
$$0 = (x - 4)(x + 1)$$
$$x = 4, -1$$

For $x = 4$ For $x = -1$
$$\quad y = 4^2 - 2(4) \qquad y = (-1)^2 - 2(-1)$$
$$\quad y = 8 \qquad\qquad\quad y = 3$$
Solutions: $(4, 8)$, $(-1, 3)$

35. (A) The lines are tangent to the circle.

(B) To find values of b such that
$$x^2 + y^2 = 5$$
$$2x - y = b$$
has exactly one solution, we solve the system for arbitrary b. Solve for y in the second equation.
$$y = 2x - b$$

Substitute into the first equation:
$$x^2 + (2x - b)^2 = 5$$
$$x^2 + 4x^2 - 4bx + b^2 = 5$$
$$5x^2 - 4bx + b^2 - 5 = 0$$

This quadratic equation will have one solution if the discriminant
$B^2 - 4AC = (-4b)^2 - 4(5)(b^2 - 5)$ is equal to 0.

This will occur when
$$16b^2 - 20b^2 + 100 = 0$$
$$-4b^2 + 100 = 0$$
$$b^2 = 25$$
$$b = \pm 5$$

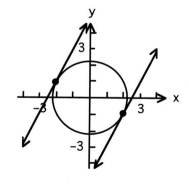

Consider $b = 5$

Then the solution of the system
$$x^2 + y^2 = 5$$
$$2x - y = 5$$
will be given by solving
$$5x^2 - 4bx + b^2 - 5 = 0$$
for $b = 5$.

$$5x^2 - 4 \cdot 5x + 5^2 - 5 = 0$$
$$5x^2 - 20x + 20 = 0$$
$$5(x - 2)^2 = 0$$
$$x - 2 = 0$$
$$x = 2$$

Since $2x - y = 5$
$$2 \cdot 2 - y = 5$$
$$y = -1$$

The intersection point is $(2, -1)$
for $b = 5$.

Consider $b = -5$

Then the solution of the system
$$x^2 + y^2 = 5$$
$$2x - y = -5$$
will be given by solving
$$5x^2 - 4bx + b^2 - 5 = 0$$
for $b = -5$.

$$5x^2 - 4(-5)x + (-5)^2 - 5 = 0$$
$$5x^2 + 20x + 20 = 0$$
$$5(x + 2)^2 = 0$$
$$x + 2 = 0$$
$$x = -2$$

Since $2x - y = -5$
$$2(-2) - y = -5$$
$$y = 1$$

The intersection point is $(-2, 1)$
for $b = -5$.

(C) The line $x + 2y = 0$ is perpendicular to all the lines in the family and
intersects the circle at the intersection points found in part B, since
this line passes through the center of the circle and therefore includes a
diameter of the circle, which is perpendicular to the tangent line at
their mutual point of intersection with the circle. Solving the system
$x^2 + y^2 = 5$, $x + 2y = 0$ would determine the intersection points.

37. $2x + 5y + 7xy = 8$
$$xy - 3 = 0$$
Solve for y in the second equation.
$$xy = 3$$
$$y = \frac{3}{x}$$
Substitute into the first equation.
$$2x + 5\left(\frac{3}{x}\right) + 7x\left(\frac{3}{x}\right) = 8$$
$$2x + \frac{15}{x} + 21 = 8 \quad x \neq 0$$
$$2x^2 + 15 + 21x = 8x$$
$$2x^2 + 13x + 15 = 0$$
$$(2x + 3)(x + 5) = 0$$
$$x = -\frac{3}{2}, -5$$

For $x = -\frac{3}{2}$ For $x = -5$

$$y = 3 \div \left(-\frac{3}{2}\right) \qquad y = \frac{3}{-5}$$

$$y = -2 \qquad\qquad y = -\frac{3}{5}$$

Solutions: $\left(-\frac{3}{2}, -2\right), \left(-5, -\frac{3}{5}\right)$

39. $x^2 - 2xy + y^2 = 1$
$$x - 2y = 2$$
Solve for x in terms of y in the
first-degree equation.
$$x = 2y + 2$$
Substitute into the second-degree
equation.
$$(2y + 2)^2 - 2(2y + 2)y + y^2 = 1$$
$$4y^2 + 8y + 4 - 4y^2 - 4y + y^2 = 1$$
$$y^2 + 4y + 3 = 0$$
$$(y + 1)(y + 3) = 0$$
$$y = -1, -3$$
For $y = -1$ For $y = -3$
$$x = 2(-1) + 2 \qquad x = 2(-3) + 2$$
$$= 0 \qquad\qquad = -4$$
Solutions: $(0, -1), (-4, -3)$

41. $2x^2 - xy + y^2 = 8$
     $x^2 - y^2 = 0$

Factor the left side of the equation that has a zero constant term.
$(x - y)(x + y) = 0$
     $x = y$ or $x = -y$

> **Common Error:**
> It is incorrect to replace
> $x^2 - y^2 = 0$ or $x^2 = y^2$ by $x = y$.
> This neglects the possibility $x = -y$.

The original system is equivalent to the two systems
$2x^2 - xy + y^2 = 8$ $2x^2 - xy + y^2 = 8$
     $x = y$ $x = -y$

These systems are solved by substitution.

First system: Second system:
$2x^2 - xy + y^2 = 8$ $2x^2 - xy + y^2 = 8$
     $x = y$ $x = -y$
$2y^2 - yy + y^2 = 8$ $2(-y)^2 - (-y)y + y^2 = 8$
     $2y^2 = 8$ $2y^2 + y^2 + y^2 = 8$
     $y^2 = 4$ $4y^2 = 8$
     $y = \pm 2$ $y^2 = 2$
                                    $y = \pm\sqrt{2}$

For $y = 2$ For $y = -2$ For $y = \sqrt{2}$ For $y = -\sqrt{2}$
   $x = 2$ $x = -2$ $x = -\sqrt{2}$ $x = \sqrt{2}$

Solutions: $(2, 2)$, $(-2, -2)$, $(-\sqrt{2}, \sqrt{2})$, $(\sqrt{2}, -\sqrt{2})$

43. $x^2 + xy - 3y^2 = 3$
     $x^2 + 4xy + 3y^2 = 0$

Factor the left side of the equation that has a zero constant term.
$(x + y)(x + 3y) = 0$
          $x = -y$ or $x = -3y$

The original system is equivalent to the two systems
$x^2 + xy - 3y^2 = 3$ $x^2 + xy - 3y^2 = 3$
          $x = -y$ $x = -3y$

These systems are solved by substitution.

First system: Second system:
     $x^2 + xy - 3y^2 = 3$ $x^2 + xy - 3y^2 = 3$
          $x = -y$ $x = -3y$
$(-y)^2 + (-y)y - 3y^2 = 3$ $(-3y)^2 + (-3y)y - 3y^2 = 3$
     $y^2 - y^2 - 3y^2 = 3$ $9y^2 - 3y^2 - 3y^2 = 3$
          $-3y^2 = 3$ $3y^2 = 3$
          $y^2 = -1$ $y^2 = 1$
          $y = \pm i$ $y = \pm 1$

For $y = i$ For $y = -i$ For $y = 1$ For $y = -1$
   $x = -i$ $x = i$ $x = -3$ $x = 3$

Solutions: $(-i, i)$, $(i, -i)$, $(-3, 1)$, $(3, -1)$

45. Before we can enter these equations in our graphing calculator, we must solve
for y:
     $-x^2 + 2xy + y^2 = 1$ $3x^2 - 4xy + y^2 = 2$
$y^2 + 2xy - 1 - x^2 = 0$ $y^2 - 4xy + 3x^2 - 2 = 0$

Applying the quadratic formula to each equation, we have

$$y = \frac{-2x \pm \sqrt{4x^2 - 4(-1 - x^2)}}{2} \qquad y = \frac{4x \pm \sqrt{16x^2 - 4(3x^2 - 2)}}{2}$$

$$y = \frac{-2x \pm \sqrt{8x^2 + 4}}{2} \qquad y = \frac{4x \pm \sqrt{4x^2 + 8}}{2}$$

$$y = -x \pm \sqrt{2x^2 + 1} \qquad y = 2x \pm \sqrt{x^2 + 2}$$

Entering each of these four equations into a graphing calculator produces the graph shown at the right.

Using the INTERSECT command (details omitted), yields $(-1.41, -0.82)$, $(-0.13, 1.15)$, $(0.13, -1.15)$, and $(1.41, 0.82)$ to two decimal places.

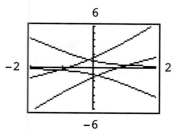

47. Before we can enter these equations in our graphing calculator, we must solve for y:

$$3x^2 - 4xy - y^2 = 2 \qquad\qquad 2x^2 + 2xy + y^2 = 9$$
$$y^2 + 4xy + 2 - 3x^2 = 0 \qquad y^2 + 2xy + 2x^2 - 9 = 0$$

Applying the quadratic formula to each equation, we have

$$y = \frac{-4x \pm \sqrt{16x^2 - 4(2 - 3x^2)}}{2} \qquad\qquad y = \frac{-2x \pm \sqrt{4x^2 - 4(2x^2 - 9)}}{2}$$

$$y = \frac{-4x \pm \sqrt{28x^2 - 8}}{2} \qquad\qquad y = \frac{-2x \pm \sqrt{36 - 4x^2}}{2}$$

$$y = -2x \pm \sqrt{7x^2 - 2} \qquad\qquad y = -x \pm \sqrt{9 - x^2}$$

Entering each of these four equations into a graphing calculator produces the graph shown at the right.

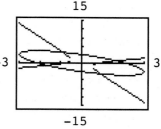

Using the INTERSECT command (details omitted), yields $(-1.66, -0.84)$, $(-0.91, 3.77)$, $(0.91, -3.77)$, and $(1.66, 0.84)$ to two decimal places.

49. Before we can enter these equations in our graphing calculator, we must solve for y:

$$2x^2 - 2xy + y^2 = 9 \qquad\qquad 4x^2 - 4xy + y^2 + x = 3$$
$$y^2 - 2xy + 2x^2 - 9 = 0 \qquad y^2 - 4xy + 4x^2 + x - 3 = 0$$

Applying the quadratic formula to each equation, we have

$$y = \frac{2x \pm \sqrt{4x^2 - 4(2x^2 - 9)}}{2} \qquad\qquad y = \frac{4x \pm \sqrt{16x^2 - 4(4x^2 + x - 3)}}{2}$$

$$y = \frac{2x \pm \sqrt{36 - 4x^2}}{2} \qquad\qquad y = \frac{4x \pm \sqrt{12 - 4x}}{2}$$

$$y = x \pm \sqrt{9 - x^2} \qquad\qquad y = 2x \pm \sqrt{3 - x}$$

Entering each of these four equations into a graphing calculator produces the graph shown at the right.

Using the INTERSECT command (details omitted), yields $(-2.96, -3.47)$, $(-0.89, -3.76)$, $(1.39, 4.05)$, and $(2.46, 4.18)$ to two decimal places.

51. Let x and y equal the two numbers. We have the system

$$x + y = 3$$
$$xy = 1$$

Solve the first equation for y in terms of x, then substitute into the second degree equation.

$$y = 3 - x$$
$$x(3 - x) = 1$$
$$3x - x^2 = 1$$
$$-x^2 + 3x - 1 = 0$$

$$x^2 - 3x + 1 = 0$$

$$x = \frac{-b \pm \sqrt{b^2 - 4ac}}{2a} \qquad a = 1, \ b = -3, \ c = 1$$

$$x = \frac{-(-3) \pm \sqrt{(-3)^2 - 4(1)(1)}}{2(1)}$$

$$x = \frac{3 \pm \sqrt{5}}{2}$$

For $x = \dfrac{3 + \sqrt{5}}{2}$ For $x = \dfrac{3 - \sqrt{5}}{2}$

$$y = 3 - x \qquad\qquad\qquad y = 3 - x$$

$$= 3 - \frac{3 + \sqrt{5}}{2} \qquad\qquad = 3 - \frac{3 - \sqrt{5}}{2}$$

$$= \frac{6 - 3 - \sqrt{5}}{2} \qquad\qquad = \frac{6 - 3 + \sqrt{5}}{2}$$

$$= \frac{3 - \sqrt{5}}{2} \qquad\qquad\quad = \frac{3 + \sqrt{5}}{2}$$

The two numbers are $\dfrac{1}{2}(3 - \sqrt{5})$ and $\dfrac{1}{2}(3 + \sqrt{5})$.

53. Sketch a figure.
Let x and y represent the lengths of the two legs.

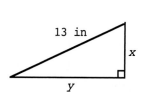

13 in

x

y

From the Pythagorean Theorem we have
$$x^2 + y^2 = 13^2$$
From the formula for the area of a triangle we have
$$\frac{1}{2}xy = 30$$
So the system of equations is
$$x^2 + y^2 = 169$$
$$\frac{1}{2}xy = 30$$

Solve the second equation for y in terms of x, then substitute into the first equation.

$$xy = 60$$
$$y = \frac{60}{x}$$
$$x^2 + \left(\frac{60}{x}\right)^2 = 169$$
$$x^2 + \frac{3600}{x^2} = 169 \qquad x \neq 0$$
$$x^4 + 3600 = 169x^2$$
$$x^4 - 169x^2 + 3600 = 0$$
$$(x^2 - 144)(x^2 - 25) = 0$$
$$(x - 12)(x + 12)(x - 5)(x + 5) = 0$$
$$x = \pm12, \ \pm5$$

Discarding the negative solutions, we have
$x = 12$ or $x = 5$

For $x = 12$ For $x = 5$

$$y = \frac{60}{x} \qquad\qquad y = \frac{60}{x}$$
$$y = 5 \qquad\qquad\quad y = 12$$

The lengths of the legs are 5 inches and 12 inches.

55. Let x = width of screen.
y = height of screen.
From the Pythagorean Theorem, we have
$x^2 + y^2 = (7.5)^2$
From the formula for the area of a rectangle we have
$xy = 27$
So the system of equations is:
$x^2 + y^2 = 56.25$
$xy = 27$

Solve the second equation for y in terms of x, then substitute into the first equation.

$$y = \frac{27}{x}$$

$$x^2 + \left(\frac{27}{x}\right)^2 = 56.25 \quad x \neq 0$$

$$x^2 + \frac{729}{x^2} = 56.25 \quad x \neq 0$$

$$x^4 + 729 = 56.25x^2$$

$$x^4 - 56.25x^2 + 729 = 0 \quad \text{quadratic in } x^2$$

$$x^2 = \frac{-b \pm \sqrt{b^2 - 4ac}}{2a} \quad a = 1, \ b = -56.25, \ c = 729$$

$$x^2 = \frac{-(-56.25) \pm \sqrt{(-56.25)^2 - 4(1)(729)}}{2(1)}$$

$$x^2 = \frac{56.25 \pm 15.75}{2}$$

$$x = 6, \ 4.5 \ \text{(discarding the negative solutions)}$$

For $x = 6$ For $x = 4.5$

$$y = \frac{27}{6} = 4.5 \qquad y = \frac{27}{4.5} = 6$$

The dimensions of the screen must be 6 inches by 4.5 inches.

57.

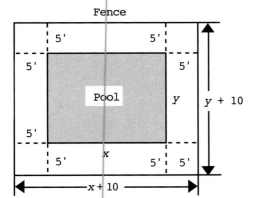

Redrawing and labeling the figure as shown, we have

Area of pool = 572
$$xy = 572$$
Area enclosed by fence = 1,152
$$(x + 10)(y + 10) = 1,152$$

We solve this system by solving for y in terms of x in the first equation, then substituting into the second equation.

$$y = \frac{572}{x}$$

$$(x + 10)\left(\frac{572}{x} + 10\right) = 1,152$$

$$572 + 10x + \frac{5,720}{x} + 100 = 1,152$$

$$10x + \frac{5,720}{x} - 480 = 0 \quad x \neq 0$$

$$10x^2 + 5,720 - 480x = 0$$

$$x^2 - 48x + 572 = 0$$

$$(x - 26)(x - 22) = 0$$

$$x = 26, \ 22$$

For $x = 26$ For $x = 22$

$$y = \frac{572}{26} \qquad\qquad\qquad y = \frac{572}{22}$$

$$y = 22 \qquad\qquad\qquad\quad y = 26$$

The dimensions of the pool are 22 feet by 26 feet.

59.   Let x = average speed of Boat B

Then $x + 5$ = average speed of Boat A

Let y = time of Boat B, then $y - \frac{1}{2}$ = time of Boat A

Using Distance = rate $\times$ time, we have

$$75 = xy$$

$$75 = (x + 5)\left(y - \frac{1}{2}\right)$$

Note: The *faster* boat, A, has the *shorter* time. It is a common error to confuse the signs here. Another common error: if rates are expressed in miles per hour, then $y - 30$ is not the correct time for boat A. Times must be expressed in hours.

Solve the first equation for y in terms of x, then substitute into the second equation.

$$y = \frac{75}{x}$$

$$75 = (x + 5)\left(\frac{75}{x} - \frac{1}{2}\right)$$

$$75 = 75 - \frac{1}{2}x + \frac{375}{x} - \frac{5}{2}$$

$$0 = -\frac{1}{2}x + \frac{375}{x} - \frac{5}{2} \quad x \neq 0$$

$$2x(0) = 2x\left(-\frac{1}{2}x\right) + 2x\left(\frac{375}{x}\right) - 2x\left(\frac{5}{2}\right)$$

$$0 = -x^2 + 750 - 5x$$

$$x^2 + 5x - 750 = 0$$

$$(x - 25)(x + 30) = 0$$

$$x = 25, \ -30$$

Discarding the negative solution, we have

     $x = 25$ mph = average speed of Boat B

$x + 5 = 30$ mph = average speed of Boat A

CHAPTER 8 REVIEW

1. First write the equation in standard form by dividing both sides by 225.

   $9x^2 + 25y^2 = 225$

   $$\frac{x^2}{25} + \frac{y^2}{9} = 1$$

   In this form the equation is identifiable as that of an ellipse. Locate the intercepts.

   When $y = 0$, $\dfrac{x^2}{25} = 1$. x intercepts: ± 5

   When $x = 0$, $\dfrac{x^2}{9} = 1$. y intercepts: ± 3

   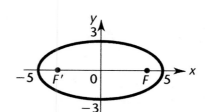

   So $a = 5$, $b = 3$, and the major axis is on the x axis.

   Foci: $c^2 = a^2 - b^2$
   $\quad\quad c^2 = 25 - 9$
   $\quad\quad c^2 = 16$
   $\quad\quad\ \ c = 4$

   Foci: $F'(-4, 0)$, $F(4, 0)$
   Major axis length $= 2(5) = 10$
   Minor axis length $= 2(3) = 6$

   > **Common Error:**
   > The relationship $c^2 = a^2 + b^2$ applies to a, b, c as defined for hyperbolas but not for ellipses.

   $(8-2)$

2. $x^2 = -12y$ is the equation of a parabola. To graph, assign y values that make the right side a perfect square (y must be zero or negative for x to be real) and solve for x. Since the coefficient of y is negative, a must be negative, and the parabola opens down.

x	0	± 6	± 2
y	0	-3	$-\frac{1}{3}$

   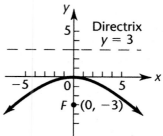

   To find the focus and directrix, solve
   $4a = -12$
   $\ \ a = -3$
   Focus: $(0, -3)$ Directrix: $y = -(-3) = 3$

   $(8-1)$

3. First, write the equation in standard form by dividing both sides by 225.

   $25y^2 - 9x^2 = 225$

   $$\frac{y^2}{9} - \frac{x^2}{25} = 1$$

   In this form the equation is identifiable as that of a hyperbola.

   When $y = 0$, $-\dfrac{x^2}{25} = 1$. There are no x intercepts, but $b = 5$.

   When $x = 0$, $\dfrac{y^2}{9} = 1$. y intercepts: ± 3

   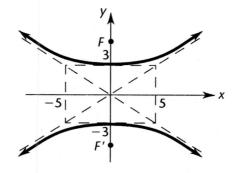

   Sketch the asymptotes using the asymptote rectangle, then sketch in the hyperbola.
   Foci: $c^2 = 3^2 + 5^2$
   $\quad\quad c^2 = 34$
   $\quad\quad\ \ c = \sqrt{34}$
   Foci: $F'(0, -\sqrt{34})$, $F(0, \sqrt{34})$
   Transverse axis length $= 2(3) = 6$
   Conjugate axis length $= 2(5) = 10$

   $(8-3)$

4. First, write the equation in standard form by dividing both sides by 16.

$$x^2 - y^2 = 16$$

$$\frac{x^2}{16} - \frac{y^2}{16} = 1$$

In this form the equation is identifiable as a hyperbola.

When $x = 0$, $-\dfrac{y^2}{16} = 1$. There are no y intercepts, but $b = 4$.

When $y = 0$, $\dfrac{x^2}{16} = 1$. x intercepts: ± 4.

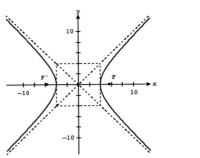

Sketch the asymptotes using the asymptote rectangle, then sketch in the hyperbola.

Foci: $c^2 = 4^2 + 4^2$

$\quad\quad c^2 = 32$

$\quad\quad c = 4\sqrt{2}$

Foci: $F' = (-4\sqrt{2}, 0)$, $F = (4\sqrt{2}, 0)$

Transverse axis length $= 2(4) = 8$

Conjugate axis length $= 2(4) = 8$

$\qquad\qquad\qquad\qquad\qquad\qquad\qquad$ (8-3)

5. $y^2 = 8x$ is the equation of a parabola. To graph, assign x values that make the right side a perfect square (x must be zero or positive for y to be real) and solve for y.
Since the coefficient of x is positive, a must be positive, and the parabola opens right.

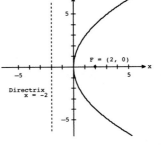

x	0	2	8
y	0	±4	±8

To find the focus and directrix, solve $4a = 8$

$\qquad\qquad\qquad\qquad\qquad\qquad a = 2.$

Focus: $(2, 0)$ Directrix: $x = -2$

$\qquad\qquad\qquad\qquad\qquad\qquad\qquad\qquad$ (8-1)

6. First, write the equation in standard form by dividing both sides by 8.

$$2x^2 + y^2 = 8$$

$$\frac{x^2}{4} + \frac{y^2}{8} = 1$$

In this form the equation is identifiable as that of an ellipse.
Locate the intercepts.

When $y = 0$, $\dfrac{x^2}{4} = 1$ x intercepts: ± 2

When $x = 0$, $\dfrac{y^2}{8} = 1$ y intercepts: $\pm 2\sqrt{2}$

So $a = 2\sqrt{2}$, $b = 2$, and the major axis is on the y axis.

Foci: $c^2 = a^2 - b^2$

$\quad\quad c^2 = 8 - 4$

$\quad\quad c^2 = 4$

$\quad\quad c = 2$

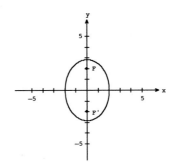

Foci: $F' = (0, -2)$, $F = (0, 2)$

Major axis length $= 2(2\sqrt{2}) = 4\sqrt{2}$

Minor axis length $= 2(2) = 4$

$\qquad\qquad\qquad\qquad\qquad\qquad\qquad\qquad$ (8-2)

7. Substitute the second equation in for y in the first equation.

$$y = x^2 - 5x - 3$$
$$y = -x + 2$$
$$-x + 2 = x^2 - 5x - 3$$
$$0 = x^2 - 4x - 5$$
$$0 = (x - 5)(x + 1)$$
$$x - 5 = 0 \quad \text{or} \quad x + 1 = 0$$
$$x = 5 \qquad\qquad x = -1$$

When $x = 5$, $y = -5 + 2 = -3$. When $x = -1$, $y = -(1) + 2 = 3$. The solutions are (5, −3) and (−1, 3).   (8-4)

8. Solve the second equation for y:

$$2x - y = 3$$
$$2x = 3 + y$$
$$y = 2x - 3$$

Now substitute into the first equation:

$$x^2 + y^2 = 2$$
$$x^2 + (2x - 3)^2 = 2$$
$$x^2 + 4x^2 - 12x + 9 = 2$$
$$5x^2 - 12x + 7 = 0$$
$$(5x - 7)(x - 1) = 0$$
$$5x - 7 = 0 \quad \text{or} \quad x - 1 = 0$$
$$5x = 7 \qquad\qquad x = 1$$
$$x = \frac{7}{5}$$

When $x = \frac{7}{5}$, $y = 2\left(\frac{7}{5}\right) - 3 = \frac{14}{5} - 3 = \frac{14}{5} - \frac{15}{5} = -\frac{1}{5}$.

When $x = 1$, $y = 2(1) - 3 = -1$. The solutions are $\left(\frac{7}{5}, -\frac{1}{5}\right)$ and (1, −1).   (8-4)

9. Multiply the first equation by 3 and add to the second to eliminate y:

$$3x^2 - y^2 = -6$$
$$2x^2 + 3y^2 = 29$$

$$9x^2 - 3y^2 = -18$$
$$\underline{2x^2 + 3y^2 = 29}$$
$$11x^2 = 11$$
$$x^2 = 1$$
$$x = \pm 1$$

When $x = 1$, $3(1)^2 - y^2 = -6$ When $x = -1$, $3(-1)^2 - y^2 = -6$
$$-y^2 = -9 \qquad\qquad\qquad\qquad -y^2 = -9$$
$$y^2 = 9 \qquad\qquad\qquad\qquad\quad y^2 = 9$$
$$y = 3, -3 \qquad\qquad\qquad\qquad y = 3, -3$$

The solutions are (1, 3), (1, −3), (−1, 3), and (−1, −3).   (8-4)

10. The parabola is opening either left or right and has an equation of the form $y^2 = 4ax$. Since $(-4, -2)$ is on the graph, we have:

$$(-2)^2 = 4a(-4)$$
$$4 = -16a$$
$$-\frac{1}{4} = a$$

The equation of the parabola is

$$y^2 = 4\left(-\frac{1}{4}\right)x$$
$$y^2 = -x$$

$(8\text{-}4)$

11. Make a rough sketch of the ellipse and compute x and y intercepts:

$$\frac{x^2}{a^2} + \frac{y^2}{b^2} = 1$$

$$a = \frac{12}{2} = 6 \qquad b = \frac{10}{2} = 5$$

$$\frac{x^2}{36} + \frac{y^2}{25} = 1$$

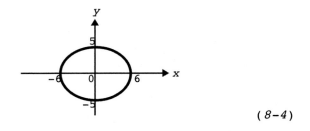

$(8\text{-}4)$

12. Make a rough sketch of the ellipse, locate focus and x intercepts, then determine y intercepts using the Pythagorean theorem.

$$\frac{x^2}{b^2} + \frac{y^2}{a^2} = 1$$

$$b = \frac{12}{2} = 6$$
$$a^2 = 6^2 + 8^2 = 100$$
$$a = 10$$
$$\frac{x^2}{36} + \frac{y^2}{100} = 1$$

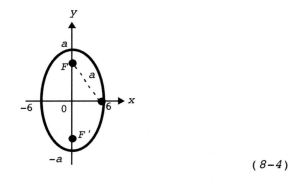

$(8\text{-}4)$

13. Since the transverse axis is on the y axis, start with

$$\frac{y^2}{a^2} - \frac{x^2}{b^2} = 1$$

and find a and b.

$$b = \frac{6}{2} = 3, \quad c = \frac{8}{2} = 4$$

To find a, sketch the asymptote rectangle, label known parts, and use the Pythagorean Theorem.

$$a^2 = c^2 - b^2$$
$$a^2 = 4^2 - 3^2$$
$$a^2 = 7$$
$$a = \sqrt{7}$$

The equation is

$$\frac{y^2}{7} - \frac{x^2}{9} = 1$$

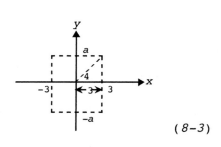

$(8\text{-}3)$

14. Since the transverse axis is on the x axis, start with

$$\frac{x^2}{a^2} - \frac{y^2}{b^2} = 1$$

and find a and b.

$$a = \frac{14}{2} = 7 \qquad b = \frac{16}{2} = 8$$

The equation is

$$\frac{x^2}{49} - \frac{y^2}{64} = 1 \tag{8-3}$$

15. $x^2 + 4y^2 = 32$
$x + 2y = 0$

We solve for x in terms of y in the second equation, then substitute the expression for x into the first equation.

$$x = -2y$$
$$(-2y)^2 + 4y^2 = 32$$
$$4y^2 + 4y^2 = 32$$
$$8y^2 = 32$$
$$y^2 = 4$$
$$y = \pm 2$$

For $y = 2$ For $y = -2$
   $x = -2(2)$ $x = -2(-2)$
     $= -4$ $= 4$

Solutions: $(-4, 2)$, $(4, -2)$ $\tag{8-4}$

16. $16x^2 + 25y^2 = 400$
$16x^2 - 45y = 0$

We eliminate x by multiplying the second equation by -1, then adding the two equations.

$$16x^2 + 25y^2 = 400$$
$$\underline{-16x^2 + 45y = 0}$$
$$25y^2 + 45y = 400 \qquad \text{(Divide both sides by 5)}$$
$$5y^2 + 9y - 80 = 0$$
$$(5y - 16)(y + 5) = 0$$
$$y = \frac{16}{5} \quad \text{or} \quad y = -5$$

For $y = \dfrac{16}{5}$ For $y = -5$
$16x^2 = 45y$ $16x^2 = 45(-5)$
$16x^2 = 45\left(\dfrac{16}{5}\right)$ $x^2 = -\dfrac{225}{16}$
   $x^2 = 9$ No real solution
   $x = \pm 3$

Solutions: $\left(3, \dfrac{16}{5}\right)$, $\left(-3, \dfrac{16}{5}\right)$ $\tag{8-4}$

17. $x^2 + y^2 = 10$
$16x^2 + y^2 = 25$

We eliminate y by multiplying the first equation by -1, then adding the two equations.

$$-x^2 - y^2 = -10$$
$$\underline{16x^2 + y^2 = 25}$$
$$15x^2 = 15$$

$$x^2 = 1$$
$$x = \pm 1$$

For $x = \pm 1$

$$x^2 + y^2 = 10$$
$$1 + y^2 = 10$$
$$y^2 = 9$$
$$y = \pm 3$$

Solutions: $(1, 3)$, $(1, -3)$, $(-1, 3)$, $(-1, -3)$ (8-4)

18. $x^2 - y^2 = 2$
   $y^2 = x$

Substitute y^2 in the second equation into the first equation.

$$x^2 - x = 2$$
$$x^2 - x - 2 = 0$$
$$(x - 2)(x + 1) = 0$$
$$x = 2, -1$$

For $x = 2$ For $x = -1$

$\quad y^2 = 2$ $y^2 = -1$

$\quad y = \pm\sqrt{2}$ $y = \pm i$

Solutions: $(2, \sqrt{2})$, $(2, -\sqrt{2})$, $(-1, i)$, $(-1, -i)$ (8-4)

19. $x^2 + 2xy + y^2 = 1$
   $xy = -2$

Solve for y in the second equation.

$$y = \frac{-2}{x}$$

Substitute into the first equation

$$x^2 + 2x\left(\frac{-2}{x}\right) + \left(\frac{-2}{x}\right)^2 = 1$$

$$x^2 - 4 + \frac{4}{x^2} = 1 \quad\quad x \neq 0$$

$$x^4 - 4x^2 + 4 = x^2$$
$$x^4 - 5x^2 + 4 = 0$$
$$(x^2 - 4)(x^2 - 1) = 0$$
$$(x - 2)(x + 2)(x - 1)(x + 1) = 0$$
$$x = 2, -2, 1, -1$$

For $x = 2$ For $x = -2$ For $x = 1$ For $x = -1$

$y = \dfrac{-2}{2}$ $y = \dfrac{-2}{-2}$ $y = \dfrac{-2}{1}$ $y = \dfrac{-2}{-1}$

$y = -1$ $y = 1$ $y = -2$ $y = 2$

Solutions: $(2, -1)$, $(-2, 1)$, $(1, -2)$, $(-1, 2)$ (8-4)

20. $2x^2 + xy + y^2 = 8$
   $x^2 - y^2 = 0$

We factor the left side of the equation that has a zero constant term.

$(x - y)(x + y) = 0$

$x = y$ or $x = -y$

So the original system is equivalent to the two systems

$2x^2 + xy + y^2 = 8$ $2x^2 + xy + y^2 = 8$

$\quad\quad\quad x = y$ $x = -y$

These systems are solved by substitution.

First System:

$2x^2 + xy + y^2 = 8$

$x = y$

$2y^2 + yy + y^2 = 8$

$4y^2 = 8$

$y^2 = 2$

$y = \pm\sqrt{2}$

For $y = \sqrt{2}$ For $y = -\sqrt{2}$

$x = \sqrt{2}$ $x = -\sqrt{2}$

Second System:

$2x^2 + xy + y^2 = 8$

$x = -y$

$2(-y)^2 + (-y)y + y^2 = 8$

$2y^2 = 8$

$y^2 = 4$

$y = \pm2$

For $y = 2$ For $y = -2$

$x = -2$ $x = 2$

Solutions: $(\sqrt{2}, \sqrt{2})$, $(-\sqrt{2}, -\sqrt{2})$, $(2, -2)$, $(-2, 2)$ (8-4)

21. Comparing focus $(0, -5)$ with the information in the chart of standard equations for a parabola, we see:

$a = -5$ Axis of symmetry: the y axis. Equation: $x^2 = 4ay$.

The equation of the parabola must be $x^2 = 4(-5)y$ or $x^2 = -20y$.   (8-1)

22. Since the major axis is on the x axis, the foci must be at $(\pm c, 0)$ and the equation of the ellipse is of form

$$\frac{x^2}{a^2} + \frac{y^2}{b^2} = 1$$

Since the major axis has twice the length of the minor axis, $a = 2b$. The equation then becomes

$$\frac{x^2}{4b^2} + \frac{y^2}{b^2} = 1$$

Since $(-6, 0)$ is on the graph, its coordinates satisfy the equation, so

$$\frac{(-6)^2}{4b^2} + \frac{0^2}{b^2} = 1$$

$$\frac{36}{4b^2} = 1$$

$$b^2 = 9$$

Then $a^2 = 4b^2 = 36$ and $c^2 = a^2 - b^2 = 36 - 9 = 27$; $c = \sqrt{27}$. The foci are at $(\pm\sqrt{27}, 0)$, that is, $(-3\sqrt{3}, 0)$ and $(3\sqrt{3}, 0)$.   (8-2)

23. Since the conjugate axis has length 4, $2b = 4$, and $b = 2$. Since $(0, -3)$ is a focus, $c = 3$. Then $a^2 = c^2 - b^2 = 9 - 4 = 5$ and $a = \sqrt{5}$. The y intercepts are $(0, \pm a)$, that is $(0, \sqrt{5})$ and $(0, -\sqrt{5})$.   (8-3)

24. Since the focus is at $(-4, 0)$, the axis of symmetry is the x axis, the parabola opens left, and the directrix has equation $x = -(-4)$ or $x = 4$.   (8-1)

25. $x^2 = 8y$

$y^2 = -x$

Solve for y in the first equation, then substitute into the second equation

$$y = \frac{x^2}{8}$$

$$\left(x^2/8\right)^2 = -x$$

$$\frac{x^4}{64} = -x$$

$$x^4 = -64x$$

$x^4 + 64x = 0$

$x(x^3 + 64) = 0$

$x(x + 4)(x^2 - 4x + 16) = 0$

$x = 0$ $x + 4 = 0$ $x^2 - 4x + 16 = 0$

$x = -4$ No real solutions.

For $x = 0$ For $x = -4$

$y = 0$ $y = \frac{(-4)^2}{8}$

$y = 2$

Solutions: $(0, 0)$, $(-4, 2)$

(8-1)

26. Since the major axis has length 14, $2a = 14$, and $a = 7$. Since $(0, -1)$ is a focus, $c = 1$. Then $b^2 = a^2 - c^2 = 49 - 1 = 48$; $b = \sqrt{48}$. The x intercepts are at $(\pm\sqrt{48}, 0)$, that is $(4\sqrt{3}, 0)$ and $(-4\sqrt{3}, 0)$.   *(8-2)*

27. Since the transverse axis is on the y axis, the foci must be at $(0, \pm c)$ and the equation of the hyperbola is of the form

$$\frac{y^2}{a^2} - \frac{x^2}{b^2} = 1$$

Since the conjugate axis has twice the length of the transverse axis, $b = 2a$. The equation then becomes

$$\frac{y^2}{a^2} - \frac{x^2}{4a^2} = 1$$

Since $(0, -4)$ is on the graph,, its coordinates satisfy the equation:

$$\frac{(-4)^2}{a^2} - \frac{0^2}{4a^2} = 1$$

$$\frac{16}{a^2} = 1$$

$$a^2 = 16$$

Then $b^2 = 4a^2 = 64$ and $c^2 = a^2 + b^2 = 16 + 64 = 80$; $c = \sqrt{80}$. The foci are at $(0, \pm\sqrt{80})$, that is $(0, 4\sqrt{5})$ and $(0, -4\sqrt{5})$.   *(8-3)*

28. From the figure, we see that the point $P = (x, y)$ is a point on the curve if and only if

$$d_1 = d_2$$
$$d(F, P) = d(M, P)$$
$$\sqrt{(x-2)^2 + (y-4)^2} = \sqrt{(x-6)^2 + (y-y)^2}$$
$$(x-2)^2 + (y-4)^2 = (x-6)^2$$
$$x^2 - 4x + 4 + (y-4)^2 = x^2 - 12x + 36$$
$$(y-4)^2 = -8x + 32$$
$$(y-4)^2 = -8(x-4) \text{ or}$$
$$y^2 - 8y + 16 = -8x + 32$$
$$y^2 - 8y + 8x - 16 = 0$$

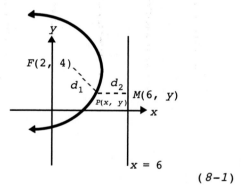

(8-1)

29. From the figure, we see that the point $P = (x, y)$ is a point on the curve if and only if

$$d_1 = 2d_2$$
$$d(F, P) = 2d(N, P)$$
$$\sqrt{(x-4)^2 + (y-0)^2} = 2\sqrt{(x-1)^2 + (y-y)^2}$$
$$(x-4)^2 + y^2 = 4(x-1)^2$$
$$x^2 - 8x + 16 + y^2 = 4(x^2 - 2x + 1)$$
$$x^2 - 8x + 16 + y^2 = 4x^2 - 8x + 4$$
$$-3x^2 + y^2 = -12$$
$$\frac{x^2}{4} - \frac{y^2}{12} = 1$$

This is the equation of a hyperbola.

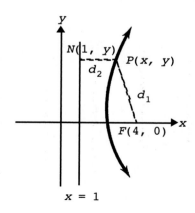

(8-3)

30. From the figure, we see that the point $P = (x, y)$ is a point on the curve if and only if

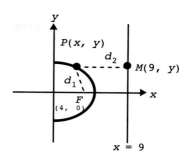

$$d_1 = \frac{2}{3}d_2$$

$$d(F, P) = \frac{2}{3}d(M, P)$$

$$\sqrt{(x-4)^2 + (y-0)^2} = \frac{2}{3}\sqrt{(x-9)^2 + (y-y)^2}$$

$$(x-4)^2 + y^2 = \frac{4}{9}(x-9)^2$$

$$9[(x-4)^2 + y^2] = 4(x-9)^2$$

$$9(x^2 - 8x + 16 + y^2) = 4(x^2 - 18x + 81)$$

$$9x^2 - 72x + 144 + 9y^2 = 4x^2 - 72x + 324$$

$$5x^2 + 9y^2 = 180$$

$$\frac{x^2}{36} + \frac{y^2}{20} = 1$$

This is the equation of an ellipse.

(8-2)

31. The equation $\dfrac{x^2}{49} - \dfrac{y^2}{25} = 1$ is in standard form with $a^2 = 49$, $b^2 = 25$. Since the intercepts are on the x axis, the asymptotes have equations $y = \pm\dfrac{b}{a}x = \pm\dfrac{5}{7}x$.

(8-3)

32. The equation $\dfrac{y^2}{64} - \dfrac{x^2}{4} = 1$ is in standard form with $a^2 = 64$, $b^2 = 4$. Since the intercepts are on the y axis, the asymptotes have equations $y = \pm\dfrac{a}{b}x = \pm\dfrac{8}{2}x$ or $y = \pm 4x$.

(8-3)

33. $4x^2 - y^2 = 1$. In standard form, this becomes

$$\frac{x^2}{1/4} - \frac{y^2}{1} = 1$$

Then $a^2 = \dfrac{1}{4}$, $b^2 = 1$. Since the intercepts are on the x axis, the asymptotes have equations $y = \pm\dfrac{b}{a}x = \pm\dfrac{1}{1/2}x$ or $y = \pm 2x$.

(8-3)

34. From the figure, we see that the parabola opens up, so its equation must be of the form $x^2 = 4ay$. Since $(4, 1)$ is on the graph we have

$4^2 = 4a \cdot 1$

$16 = 4a$

$a = 4$

So a, the distance of the focus from the vertex, is 4 feet.

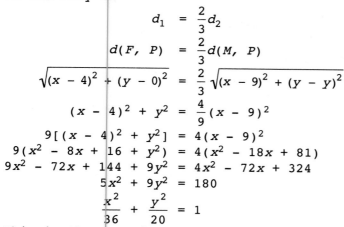

(8-1)

35. From the figure, we see that the x intercepts must be at $(-5, 0)$ and $(5, 0)$, the foci at $(-4, 0)$ and $(4, 0)$. So $a = 5$ and $c = 4$. We can determine the y intercepts using the special triangle relationship

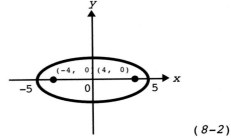

$$5^2 = 4^2 + b^2$$
$$25 = 16 + b^2$$
$$b = 3$$

The equation of the ellipse is

$$\frac{x^2}{5^2} + \frac{y^2}{3^2} = 1$$

(8-2)

36. From the figure, we can see:

$d + a = y_1 =$ the y coordinate of the point on the hyperbola with x coordinate 15. From the equation of the hyperbola $\dfrac{y^2}{40^2} - \dfrac{x^2}{30^2} = 1$, we have $a = 40$, and

$d + 40 = y_1$, or $d = y_1 - 40$. Since the point $(15, y_1)$ is on the hyperbola, its coordinates must satisfy the equation of the hyperbola.

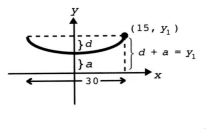

$$\frac{y_1^2}{40^2} - \frac{15^2}{30^2} = 1$$
$$\frac{y_1^2}{40^2} = 1 + \frac{15^2}{30^2}$$
$$\frac{y_1^2}{40^2} = \frac{5}{4}$$
$$y_1^2 = 2000$$
$$y_1 = 44.72$$

$$\text{depth} = y_1 - 40 = 4.72 \text{ feet.}$$

(8-3)

Chapters 7 & 8 Cumulative Review

1. (A) Since $15 - 20 = 10 - 15 = -5$, this could start an arithmetic sequence.

   (B) Since $\dfrac{25}{5} = \dfrac{125}{25} = 5$, this could start a geometric sequence.

   (C) Since $\dfrac{25}{5} \neq \dfrac{50}{25}$ and $25 - 5 \neq 50 - 25$, this could start neither an arithmetic nor a geometric sequence.

   (D) Since $\dfrac{-9}{27} = \dfrac{3}{-9} = -\dfrac{1}{3}$, this could start a geometric sequence.

   (E) Since $(-6) - (-9) = (-3) - (-6) = 3$, this could start an arithmetic sequence. *(7-3)*

2. $a_n = 2 \cdot 5^n$

   (A) $a_1 = 2 \cdot 5^1 = 10$
   $a_2 = 2 \cdot 5^2 = 50$
   $a_3 = 2 \cdot 5^3 = 250$
   $a_4 = 2 \cdot 5^4 = 1{,}250$

   (B) This is a geometric sequence with $r = 5$.
   $a_n = a_1 r^{n-1}$
   $a_8 = 10(5)^{8-1}$
   $\quad = 781{,}250$

   (C) $S_n = \dfrac{a_1 - r a_n}{1 - r}$
   $S_8 = \dfrac{10 - 5(781{,}250)}{1 - 5}$
   $\quad = 976{,}560$ *(7-3)*

3. $a_n = 3n - 1$

   (A) $a_1 = 3 \cdot 1 - 1 = 2$
   $a_2 = 3 \cdot 2 - 1 = 5$
   $a_3 = 3 \cdot 3 - 1 = 8$
   $a_4 = 3 \cdot 4 - 1 = 11$

   (B) This is an arithmetic sequence with $d = 3$.
   $a_n = a_1 + (n - 1)d$
   $a_8 = 2 + (8 - 1)3$
   $\quad = 23$

   (C) $S_n = \dfrac{n}{2}(a_1 + a_n)$
   $S_8 = \dfrac{8}{2}(2 + 23)$
   $\quad = 100$ *(7-3)*

4. $a_1 = 100 \quad a_n = a_{n-1} - 6 \quad n \geq 2$

   (A) $a_1 = 100$
   $a_2 = a_1 - 6 = 94$
   $a_3 = a_2 - 6 = 88$
   $a_4 = a_3 - 6 = 82$

   (B) This is an arithmetic sequence with $d = -6$.
   $a_n = a_1 + (n - 1)d$
   $a_8 = 100 + (8 - 1)(-6)$
   $\quad = 58$

   (C) $S_n = \dfrac{n}{2}(a_1 + a_n)$
   $S_8 = \dfrac{8}{2}(100 + 58)$
   $\quad = 632$ *(7-3)*

5. (A) $8! = 8 \cdot 7 \cdot 6 \cdot 5 \cdot 4 \cdot 3 \cdot 2 \cdot 1 = 40{,}320$.

   (B) $\dfrac{32!}{30!} = \dfrac{32 \cdot 31 \cdot 30!}{30!} = 992$

   (C) $\dfrac{9!}{3!(9-3)!} = \dfrac{9!}{3! \, 6!}$
   $\qquad = \dfrac{9 \cdot 8 \cdot 7 \cdot 6!}{3 \cdot 2 \cdot 1 \cdot 6!} = 84$ *(7-4)*

6. (A) $\dbinom{7}{2} = \dfrac{7!}{2!(7-2)!} = \dfrac{7!}{2! \, 5!} = \dfrac{7 \cdot 6 \cdot 5!}{2 \cdot 1 \cdot 5!} = 21$

   (B) $C_{7,2} = \dfrac{7!}{2!(7-2)!} = 21$

   (C) $P_{7,2} = \dfrac{7!}{(7-2)!} = \dfrac{7!}{5!} = \dfrac{7 \cdot 6 \cdot 5!}{5!} = 42$ *(7-4, 7-5)*

7. First, write the equation in standard form by dividing both sides by 900.
   $$25x^2 - 36y^2 = 900$$
   $$\frac{x^2}{36} - \frac{y^2}{25} = 1$$
   In this form the equation is identifiable as that of a hyperbola.

   When $x = 0$, $-\frac{y^2}{25} = 1$. There are no y intercepts, but $b = 5$.

   When $y = 0$, $\frac{x^2}{36} = 1$. x intercepts: ±6.

   Sketch the asymptotes using the asymptote rectangle, then sketch in the hyperbola.
   Foci: $c^2 = 5^2 + 6^2$
   $$c^2 = 61$$
   $$c = \sqrt{61}$$
   Foci: $F' = (-\sqrt{61}, 0)$, $F = (\sqrt{61}, 0)$
   Transverse axis length = 2(6) = 12
   Conjugate axis length = 2(5) = 10

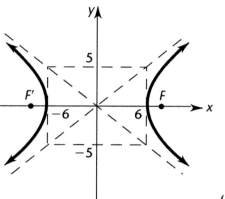

   (8-3)

8. First, write the equation in standard form by dividing both sides by 900.
   $$25x^2 + 36y^2 = 900$$
   $$\frac{x^2}{36} + \frac{y^2}{25} = 1$$
   In this form the equation is identifiable as that of an ellipse. Locate the intercepts:

   When $y = 0$, $\frac{x^2}{36} = 1$. x intercepts: ±6.

   When $x = 0$, $\frac{y^2}{25} = 1$. y intercepts: ±5

   So $a = 6$, $b = 5$, and the major axis is on the x axis.
   Foci: $c^2 = a^2 - b^2$
   $$c^2 = 36 - 25$$
   $$c^2 = 11$$
   $$c = \sqrt{11}$$
   Foci: $F' = (-\sqrt{11}, 0)$, $F = (\sqrt{11}, 0)$
   Major axis length = 2(6) = 12
   Minor axis length = 2(5) = 10

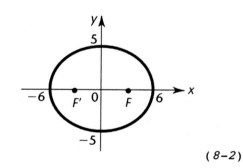

   (8-2)

9. $25x^2 - 36y = 0$ is the equation of a parabola. For convenience, we rewrite this as $25x^2 = 36y$. To graph, assign y values that make $25x^2$ a perfect square (y must be positive or zero for x to be real) and solve for x. Since the coefficient of x is positive, a must be positive, and the parabola opens up.

x	0	$\pm\frac{6}{5}$	$\pm\frac{12}{5}$
y	0	1	4

   To find the focus and directrix, solve
   $$4a = \frac{36}{25}$$
   $$a = \frac{9}{25}$$

   Focus: $\left(0, \frac{9}{25}\right)$. Directrix: $y = -\frac{9}{25}$

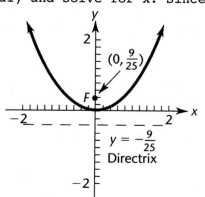

   (8-1)

10. (A) The outcomes can be displayed in a tree diagram as follows:

HHH
HHT
HTH
HTT
THH
THT
TTH
TTT

(B) O_1: First flip of the coin
N_1: 2 outcomes
O_2: Second flip of the coin
N_2: 2 outcomes
O_3: Third flip of the coin
N_3: 2 outcomes

Applying the multiplication principle, there are $2 \cdot 2 \cdot 2 = 8$ possible outcomes.

(7-5)

11. (A) O_1: Place first book N_1: 4 ways
O_2: Place second book N_2: 3 ways
O_3: Place third book N_3: 2 ways
O_4: Place fourth book N_4: 1 way

Applying the multiplication principle, there are $4 \cdot 3 \cdot 2 \cdot 1$, or 24, arrangements.

(B) Order is important here. We use permutations to determine the number of arrangements of 4 objects. $P_{4,4} = 4! = 24$.   *(7-5)*

12. The sample space S is the set of all possible 3-card hands, chosen from a deck of 52 cards. $n(S) = C_{52,3}$.

The event E is the set of all possible 3-card hands that are all diamonds, and are chosen from the 13 diamonds. $n(E) = C_{13,3}$

$$P(E) = \frac{n(E)}{n(S)} = \frac{C_{13,3}}{C_{52,3}} = \frac{13!}{3!(13-3)!} \div \frac{52!}{3!(52-3)!} \approx .0129 \qquad (7\text{-}5)$$

13. In the first case, order is important. The sample space is the set of all possible arrangements of four objects, drawn from a set of 10 objects. $n(S) = P_{10,4}$.

The event E is one of these arrangements. $n(E) = 1$.

$$P(E) = \frac{n(E)}{n(S)} = \frac{1}{P_{10,4}} = \frac{1}{10 \cdot 9 \cdot 8 \cdot 7} \approx .0002$$

In the second case, order is not important. The sample space is the set of all possible ways of drawing four objects from a set of 10 objects $n(S) = C_{10,4}$.

The event E is one of these ways. $n(E) = 1$.

$$P(E) = \frac{n(E)}{n(S)} = \frac{1}{C_{10,4}} = 1 \div \frac{10!}{4!(10-4)!} \approx .0048 \qquad (7\text{-}5)$$

14. $P(E) \approx \dfrac{f(E)}{n} = \dfrac{100 - 38}{100} = .62$ *(7-5)*

15. P_1: $1 = 1(2 \cdot 1 - 1) = 1 \cdot 1 = 1$
P_2: $1 + 5 = 2(2 \cdot 2 - 1)$
$\ \ 6 = 6$
P_3: $1 + 5 + 9 = 3(2 \cdot 3 - 1)$
$15 = 15$ *(7-2)*

16. P_1: $1^2 + 1 + 2$ is divisible by 2
$\ \ 4 = 2 \cdot 2$ true
P_2: $2^2 + 2 + 2$ is divisible by 2
$\ \ 8 = 2 \cdot 4$ true
P_3: $3^2 + 3 + 2$ is divisible by 2
$\ \ 14 = 2 \cdot 7$ true   *(7-2)*

17. P_k: $1 + 5 + 9 + \cdots + (4k - 3) = k(2k - 1)$

    P_{k+1}: $1 + 5 + 9 + \cdots + (4k - 3) + (4k + 1) = (k + 1)(2k + 1)$ *(7-2)*

18. P_k: $k^2 + k + 2 = 2r$ for some integer r

    P_{k+1}: $(k + 1)^2 + (k + 1) + 2 = 2s$ for some integer s *(7-2)*

19. The parabola is opening either up or down and has an equation of the form $x^2 = 4ay$. Since $(2, -8)$ is on the graph, we have:

    $2^2 = 4a(-8)$

    $4 = -32a$

    $a = -\dfrac{1}{8}$

    The equation of the parabola is

    $x^2 = 4\left(-\dfrac{1}{8}\right)y$

    $x^2 = -\dfrac{1}{2}y$

    $y = -2x^2$ *(8-1)*

20. Make a rough sketch of the ellipse, locate focus and x intercepts, then determine y intercepts using the Pythagorean theorem.

    $\dfrac{x^2}{a^2} + \dfrac{y^2}{b^2} = 1$

    $a = \dfrac{10}{2} = 5$

    $b^2 = a^2 - c^2 = 5^2 - 3^2 = 25 - 9 = 16$

    $b = 4$

    $\dfrac{x^2}{25} + \dfrac{y^2}{16} = 1$

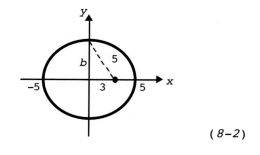

(8-2)

21. Start with $\dfrac{x^2}{a^2} - \dfrac{y^2}{b^2} = 1$ and find a and b.

    $a = \dfrac{16}{2} = 8$

    To find b, sketch the asymptote rectangle, label known parts, and use the Pythagorean Theorem.

    $b^2 = (\sqrt{89})^2 - 8^2$

    $\quad = 89 - 64$

    $b^2 = 25$

    $b = 5$

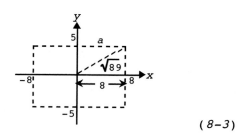

    The equation is $\dfrac{x^2}{64} - \dfrac{y^2}{25} = 1$

(8-3)

22. $x^2 - 3xy + 3y^2 = 1$
$xy = 1$

Solve for y in the second equation.

$$y = \frac{1}{x}$$

$$x^2 - 3x\left(\frac{1}{x}\right) + 3\left(\frac{1}{x}\right)^2 = 1$$

$$x^2 - 3 + \frac{3}{x^2} = 1 \quad x \neq 0$$

$$x^4 - 3x^2 + 3 = x^2$$
$$x^4 - 4x^2 + 3 = 0$$
$$(x^2 - 1)(x^2 - 3) = 0$$

$x^2 - 1 = 0 \qquad\qquad x^2 - 3 = 0$
$\quad x^2 = 1 \qquad\qquad\quad x^2 = 3$
$\quad\; x = \pm 1 \qquad\qquad\quad x = \pm\sqrt{3}$

For $x = 1$ For $x = -1$ For $x = \sqrt{3}$ For $x = -\sqrt{3}$

$\quad y = \dfrac{1}{1}$ $y = \dfrac{1}{-1}$ $y = \dfrac{1}{\sqrt{3}}$ $y = \dfrac{1}{-\sqrt{3}}$

$\quad y = 1$ $y = -1$ $y = \dfrac{\sqrt{3}}{3}$ $y = -\dfrac{\sqrt{3}}{3}$

Solutions: $(1, 1)$, $(-1, -1)$, $\left(\sqrt{3}, \dfrac{\sqrt{3}}{3}\right)$, $\left(-\sqrt{3}, -\dfrac{\sqrt{3}}{3}\right)$ *(7-3)*

23. $x^2 - 3xy + y^2 = -1$
$x^2 - xy = 0$

Factor the left side of the equation that has a zero constant term.
$x(x - y) = 0$
$\quad x = 0 \quad$ or $\quad x = y$
The original system is equivalent to the two systems
$x^2 - 3xy + y^2 = -1 \qquad\qquad x^2 - 3xy + y^2 = -1$
$\qquad\qquad x = 0 \qquad\qquad\qquad\qquad\qquad x = y$

These systems are solved by substitution.
First system: Second system:
$\quad x^2 - 3xy + y^2 = -1 \qquad\quad x^2 - 3xy + y^2 = -1$
$\qquad\qquad\quad x = 0 \qquad\qquad\qquad\qquad\quad x = y$
$0^2 - 3(0)y + y^2 = -1 \qquad\quad y^2 - 3yy + y^2 = -1$
$\qquad\qquad\quad y^2 = -1 \qquad\qquad\qquad\quad -y^2 = -1$
$\qquad\qquad\quad\; y = \pm i \qquad\qquad\qquad\qquad y^2 = 1$
$\qquad\qquad\quad\; x = 0 \qquad\qquad\qquad\qquad\quad y = \pm 1$

                         For $y = 1$ For $y = -1$
                            $x = 1$ $x = -1$

Solutions: $(1, 1)$, $(-1, -1)$, $(0, i)$, $(0, -i)$ *(7-3)*

24. Before we can enter these equations in our graphing calculator, we must solve for y:

$\quad x^2 + 2xy - y^2 = 1 \qquad\qquad\qquad 9x^2 + 4xy + y^2 = 15$
$y^2 - 2xy - x^2 + 1 = 0 \qquad\qquad\quad y^2 + 4xy + 9x^2 - 15 = 0$

Applying the quadratic formula to each equation, we have

$$y = \frac{2x \pm \sqrt{4x^2 - 4(-x^2 + 1)}}{2} \qquad\qquad y = \frac{-4x \pm \sqrt{16x^2 - 4(9x^2 - 15)}}{2}$$

$$y = \frac{2x \pm \sqrt{8x^2 - 4}}{2} \qquad\qquad\qquad\quad y = \frac{-4x \pm \sqrt{60 - 20x^2}}{2}$$

$$y = x \pm \sqrt{2x^2 - 1} \qquad\qquad\qquad\qquad y = -2x \pm \sqrt{15 - 5x^2}$$

Entering each of these four equations into a graphing calculator produces the graph shown at the right.

Using the INTERSECT command (details omitted), yields (-1.35, 0.28), (-0.87, -1.60), (0.87, 1.60), and (1.35, -0.28) to two decimal places.

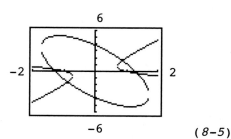

(8-5)

25. $\sum_{k=1}^{5} k^k = 1^1 + 2^2 + 3^3 + 4^4 + 5^5 = 1 + 4 + 27 + 256 + 3,125 = 3,413$ (7-1)

26. $S_6 = \dfrac{2}{2!} - \dfrac{2^2}{3!} + \dfrac{2^3}{4!} - \dfrac{2^4}{5!} + \dfrac{2^5}{6!} - \dfrac{2^6}{7!}$

$k = 1, 2, 3, 4, 5, 6$

Noting that the terms alternate in sign, we can rewrite as follows:

$S_6 = (-1)^2 \dfrac{2}{2!} + (-1)^3 \dfrac{2^2}{3!} + (-1)^4 \dfrac{2^3}{4!} + (-1)^5 \dfrac{2^4}{5!} + (-1)^6 \dfrac{2^5}{6!} + (-1)^7 \dfrac{2^6}{7!}$

In this form, we can see that $a_k = (-1)^{k+1} \dfrac{2^k}{(k+1)!}$

$S_6 = \sum_{k=1}^{6} (-1)^{k+1} \dfrac{2^k}{(k+1)!}$ (7-1)

27. $\dfrac{a_2}{a_1} = \dfrac{-36}{108} = -\dfrac{1}{3} = r.$ $|r| < 1.$

Therefore, this infinite geometric series has a sum.

$S_\infty = \dfrac{a_1}{1-r}$

$= \dfrac{108}{1 - \left(-\frac{1}{3}\right)}$

$= 81$ (7-3)

28.

	Case 1	Case 2	Case 3
O_1: Select the first letter	N_1: 6 ways	6 ways	6 ways
O_2: Select the second letter	N_2: 5 ways	6 ways	5 ways (exclude first letter)
O_3: Select the third letter	N_3: 4 ways	6 ways	5 ways (exclude second letter)
O_4: Select the fourth letter	N_4: 3 ways	6 ways	5 ways (exclude third letter)
	$6 \cdot 5 \cdot 4 \cdot 3 =$	$6 \cdot 6 \cdot 6 \cdot 6 =$	$6 \cdot 5 \cdot 5 \cdot 5 = 750$ words
	360 words	1,296 words	

(7-4)

29. The sample space is the set of all possible ways to choose 5 players out of 12.

$n(S) = C_{12,5}$

The event E is the set of all those choices that include 2 particular people, so

O_1: Choose the two centers N_1: 1

O_2: Choose 3 more people out N_2: $C_{10,3}$
     of the 10 remaining

Applying the multiplication principle, $n(E) = N_1 \cdot N_2 = 1 \cdot C_{10,3} = C_{10,3}$

$P(E) = \dfrac{n(E)}{n(S)} = \dfrac{C_{10,3}}{C_{12,5}} = \dfrac{10!}{3!(10-3)!} \div \dfrac{12!}{5!(12-5)!} = 120 \div 792 = \dfrac{5}{33} = .\overline{15}$ (7-5)

30. (A) P (this event) $= \dfrac{195 + 170}{1000} = .365$

(B) A sample space of equally likely events is:
$$S = \{1, 2, 3, 4, 5, 6\} \quad n(S) = 6$$
The event E is the set of those outcomes that are divisible by 3.
$$E = \{3, 6\}. \quad n(E) = 2$$
Then $P(E) = \dfrac{n(E)}{n(S)} = \dfrac{2}{6} = \dfrac{1}{3}$

$(7\text{-}5)$

31.

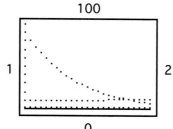

Enter the sequences in the sequence editor.	Graph the sequences.	Display the points in a table.

From the graph, 22 is the least positive integer n such that $a_n < b_n$.
The table confirms this.

$(7\text{-}3)$

32. (A) $P_{25,5} = 25 \cdot 24 \cdot 23 \cdot 22 \cdot 21 = 6{,}375{,}600$

(B) $C_{25,5} = \dfrac{25!}{5!\,(25-5)!} = \dfrac{25!}{5!\,20!} = \dfrac{25 \cdot 24 \cdot 23 \cdot 22 \cdot 21 \cdot 20!}{5 \cdot 4 \cdot 3 \cdot 2 \cdot 1 \cdot 20!} = 53{,}130$

(C) $\dbinom{25}{5} = \dfrac{25!}{5!\,(25-5)!} = 53{,}130$

$(7\text{-}4, \ 7\text{-}6)$

33.
$$\left(a + \tfrac{1}{2}b\right)^6 = \sum_{k=0}^{6} \binom{6}{k} a^{6-k}\left(\tfrac{1}{2}b\right)^k$$
$$= \binom{6}{0}a^6 + \binom{6}{1}a^5\left(\tfrac{1}{2}b\right)^1 + \binom{6}{2}a^4\left(\tfrac{1}{2}b\right)^2 + \binom{6}{3}a^3\left(\tfrac{1}{2}b\right)^3 + \binom{6}{4}a^2\left(\tfrac{1}{2}b\right)^4$$
$$+ \binom{6}{5}a\left(\tfrac{1}{2}b\right)^5 + \binom{6}{6}\left(\tfrac{1}{2}b\right)^6$$
$$= a^6 + 3a^5b + \tfrac{15}{4}a^4b^2 + \tfrac{5}{2}a^3b^3 + \tfrac{15}{16}a^2b^4 + \tfrac{3}{16}ab^5 + \tfrac{1}{64}b^6$$

$(7\text{-}6)$

34. In the expansion of $(a + b)^n$, the exponent of b in the rth term is $r - 1$ and the exponent of a is $n - (r - 1)$. In the first case, $r = 5$, $n = 10$.

Fifth term $= \dbinom{10}{4}(3x)^6(-y)^4$

$= 153{,}090x^6y^4$

In the second case, $r = 8$, n is still 10.

Eighth term $= \dbinom{10}{7}(3x)^3(-y)^7$

$= -3{,}240x^3y^7$

$(7\text{-}6)$

35. P_n: $1 + 5 + 9 + \cdots + (4n - 3) = n(2n - 1)$

Part 1: Show that P_1 is true.

P_1: $1 = 1(2 \cdot 1 - 1) = 1$ True

Part 2: Show that if P_k is true, then P_{k+1} is true.

Write out P_k and P_{k+1}.

P_k: $1 + 5 + 9 + \cdots + (4k - 3) = k(2k - 1)$

P_{k+1}: $1 + 5 + 9 + \cdots + (4k - 3) + (4k + 1) = (k + 1)(2k + 1)$

We start with P_k:

$1 + 5 + 9 + \cdots + (4k - 3) = k(2k - 1)$

Adding $4k + 1$ to both sides:

$$1 + 5 + 9 + \cdots + (4k - 3) + (4k + 1) = k(2k - 1) + 4k + 1$$
$$= 2k^2 - k + 4k + 1$$
$$= 2k^2 + 3k + 1$$
$$= (k + 1)(2k + 1)$$

We have shown that if P_k is true, then P_{k+1} is true.

Conclusion: P_n is true for all positive integers n.   ($7\text{-}2$)

36. P_n: $n^2 + n + 2 = 2p$ for some integer p.

Part 1: Show that P_1 is true.

P_1: $1^2 + 1 + 2 = 4 = 2 \cdot 2$ is true.

Part 2: Show that if P_k is true, then P_{k+1} is true.

Write out P_k and P_{k+1}.

P_k: $k^2 + k + 2 = 2r$ for some integer r

P_{k+1}: $(k + 1)^2 + (k + 1) + 2 = 2s$ for some integer s

We start with P_k:

$k^2 + k + 2 = 2r$ for some integer r

Now, $(k + 1)^2 + (k + 1) + 2 = k^2 + 2k + 1 + k + 1 + 2$

$$= k^2 + k + 2 + 2k + 2$$
$$= 2r + 2k + 2$$
$$= 2(r + k + 2)$$

Therefore,

$(k + 1)^2 + (k + 1) + 2 = 2s$ for some integer s $(= r + k + 2)$

$(k + 1)^2 + (k + 1) + 2$ is divisible by 2.

We have shown that if P_k is true, then P_{k+1} is true.

Conclusion: P_n is true for all positive integers n.   ($7\text{-}2$)

37. We are to find $51 + 53 + \cdots + 499$. This is the sum of an arithmetic sequence, S_n, with $d = 2$.

First, find n:

$a_n = a_1 + (n - 1)d$

$499 = 51 + (n - 1)2$

$n = 225$

Now, find S_{225}

$S_n = \dfrac{n}{2}(a_1 + a_n)$

$S_{225} = \dfrac{225}{2}(51 + 499) = 61{,}875$ ($7\text{-}3$)

38. $2.\overline{45} = 2.454545\ldots$
$= 2 + 0.454545\ldots$

$0.454545\ldots = 0.45 + 0.0045 + 0.000045 + \cdots$
This is an infinite geometric series with $a_1 = 0.45$ and $r = 0.01$.

$S_\infty = \dfrac{a_1}{1 - r} = \dfrac{0.45}{1 - 0.01} = \dfrac{0.45}{0.99} = \dfrac{5}{11}$

Then $2.\overline{45} = 2 + \dfrac{5}{11} = \dfrac{27}{11}$

($7\text{-}3$)

39. Graphing the sequence, we obtain the graph at the right. The solid line indicates 0.01. The largest term of the sequence is a_{27}. From the table display,

$$a_{27} = \binom{30}{27}(0.1)^{30-27}(0.9)^{27} = 0.236088$$

There are 8 terms larger than 0.01, as can be seen from the graph or the table display (details omitted).

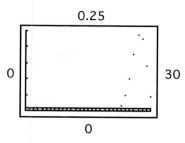

$(7\text{-}6)$

40.

	Allowing digits to repeat	No digit can be repeated
O_1: Selecting first digit	N_1: 10 ways	N_1: 10 ways
O_2: Selecting second digit	N_2: 10 ways	N_2: 9 ways
$\vdots$	$\vdots$	$\vdots$
O_9: Selecting ninth digit	N_9: 10 ways	N_9: 2 ways

$10 \cdot 10 \cdot 10 \cdot 10 \cdot 10 \cdot 10 \cdot 10 \cdot 10 \cdot 10$
$= 10^9$ zip codes

$P_{10,9} = 10 \cdot 9 \cdot 8 \cdot 7 \cdot 6 \cdot 5 \cdot 4 \cdot 3 \cdot 2$
$= 3,628,800$ zip codes

$(7\text{-}1)$

41. Prove P_n: $\dfrac{1}{1 \cdot 3} + \dfrac{1}{3 \cdot 5} + \dfrac{1}{5 \cdot 7} + \cdots + \dfrac{1}{(2n-1)(2n+1)} = \dfrac{n}{2n+1}$

Part 1: Show that P_1 is true.

P_1: $\dfrac{1}{1 \cdot 3} = \dfrac{1}{2 \cdot 1 + 1}$

$\dfrac{1}{3} = \dfrac{1}{3}$ P_1 is true

Part 2: Show that if P_k is true, then P_{k+1} is true.
Write out P_k and P_{k+1}:

P_k: $\dfrac{1}{1 \cdot 3} + \dfrac{1}{3 \cdot 5} + \dfrac{1}{5 \cdot 7} + \cdots + \dfrac{1}{(2k-1)(2k+1)} = \dfrac{k}{2k+1}$

P_{k+1}: $\dfrac{1}{1 \cdot 3} + \dfrac{1}{3 \cdot 5} + \dfrac{1}{5 \cdot 7} + \cdots + \dfrac{1}{(2k-1)(2k+1)} + \dfrac{1}{(2k+1)(2k+3)} = \dfrac{k+1}{2k+3}$

We start with P_k:

$$\frac{1}{1 \cdot 3} + \frac{1}{3 \cdot 5} + \frac{1}{5 \cdot 7} + \cdots + \frac{1}{(2k-1)(2k+1)} = \frac{k}{2k+1}$$

Adding $\dfrac{1}{(2k+1)(2k+3)}$ to both sides, we get

$$\frac{1}{1 \cdot 3} + \frac{1}{3 \cdot 5} + \frac{1}{5 \cdot 7} + \cdots + \frac{1}{(2k-1)(2k+1)} + \frac{1}{(2k+1)(2k+3)}$$

$$= \frac{k}{2k+1} + \frac{1}{(2k+1)(2k+3)}$$

$$= \frac{k(2k+3)}{(2k+1)(2k+3)} + \frac{1}{(2k+1)(2k+3)}$$

$$= \frac{k(2k+3) + 1}{(2k+1)(2k+3)}$$

$$= \frac{2k^2 + 3k + 1}{(2k + 1)(2k + 3)}$$

$$= \frac{(2k + 1)(k + 1)}{(2k + 1)(2k + 3)}$$

$$= \frac{k + 1}{2k + 3}$$

We have shown that if P_k is true, then P_{k+1} is true.

Conclusion: P_n is true for positive integers n.   (7-2)

42. Case 1: No repetition.
The sample space S is the set of all possible sequences of three digits, chosen from {1, 2, 3, 4, 5} with no repeats.
$n(S) = P_{5,3} = 5 \cdot 4 \cdot 3 = 60$
The event E is the set of those sequences that end in 2 or 4.
$n(E) = N_1 \cdot N_2 \cdot N_3$
where
$N_1 = 2$ (digits 2 or 4 only for the last digit)
$N_2 = 4$ (digits remaining for middle digit after last digit is chosen)
$N_3 = 3$ (digits remaining for first digit after others are chosen)
$n(E) = 2 \cdot 4 \cdot 3 = 24$
$P(E) = \dfrac{n(E)}{n(S)} = \dfrac{24}{60} = \dfrac{2}{5}$

Case 2: Allowing repetition
The sample space S is the set of all possible three digit numbers chosen from {1, 2, 3, 4, 5}.
$n(S) = 5 \cdot 5 \cdot 5 = 125$
The event E is the set of those numbers that end in 2 or 4.
$n(E) = 5 \cdot 5 \cdot 2 = 50$
$P(E) = \dfrac{n(E)}{n(S)} = \dfrac{50}{125} = \dfrac{2}{5}$ (7-5)

43. $(x - 2i)^6 = [x + (-2i)]^6 = \displaystyle\sum_{k=0}^{6} \binom{6}{k} x^{6-k}(-2i)^k$

$$= \binom{6}{0} x^6 + \binom{6}{1} x^5(-2i)^1 + \binom{6}{2} x^4(-2i)^2 + \binom{6}{3} x^3(-2i)^3$$

$$+ \binom{6}{4} x^2(-2i)^4 + \binom{6}{5} x(-2i)^5 + \binom{6}{6}(-2i)^6$$

$$= x^6 - 12ix^5 - 60x^4 + 160ix^3 + 240x^2 - 192ix - 64 \quad (7\text{-}6)$$

44. Let $P = (x, y)$ be a point on the parabola. Then by the definition of the parabola, the distance from $P = (x, y)$ to the focus $F = (6, 1)$ must equal the perpendicular distance from P to the directrix at $D = (x, 3)$. Applying the distance formula, we have

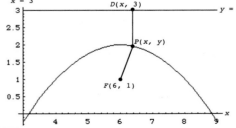

$$d(P, F) = d(P, D)$$

$$\sqrt{(x - 6)^2 + (y - 1)^2} = \sqrt{(x - x)^2 + (y - 3)^2}$$

$$(x - 6)^2 + (y - 1)^2 = (x - x)^2 + (y - 3)^2$$

$$x^2 - 12x + 36 + y^2 - 2y + 1 = 0 + y^2 - 6y + 9$$

$$x^2 - 12x + 4y + 28 = 0$$ (8-1)

45. Make a rough sketch of the ellipse, locate focus and x intercepts, then determine y intercepts using the special triangle relationship.

$a = 4$

$b^2 = a^2 - c^2$

$\quad = 4^2 - 2^2$

$\quad = 12$

$b = \sqrt{12} = 2\sqrt{3}$

y intercepts: $\pm 2\sqrt{3}$

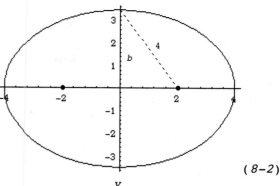

(8-2)

46. Sketch the asymptote rectangle, label known parts, and use the Pythagorean Theorem.

$a = 3 \qquad c = 5$

$b^2 = c^2 - a^2$

$\quad = 5^2 - 3^2$

$\quad = 16$

$b = 4$

So the length of the conjugate axis is $2b = 8$.

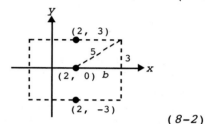

(8-2)

47. We can select any three of the seven points to use as vertices of the triangle. Order is not important.

$C_{7,3} = \dfrac{7!}{3!(7-3)!} = 35$ triangles

(8-4)

48. P_n: $2^n < n!$ $n \geq 4$ (for n an integer, this is identical with the condition $n > 3$)

Part 1: Show that P_4 is true

P_4: $2^4 < 4!$

$\qquad 16 < 24$ True.

Part 2: Show that if P_k is true, then P_{k+1} is true.

Write out P_k and P_{k+1}:

$\quad P_k$: $2^k < k!$

$\quad P_{k+1}$: $2^{k+1} < (k+1)!$

We start with P_k:

$\quad 2^k < k!$

Multiplying both sides by 2: $2 \cdot 2^k < 2 \cdot k!$

Now, $k > 3$, so $1 < k$, $2 < k + 1$, so $2 \cdot k! < (k+1)k!$

Therefore, $2^{k+1} = 2 \cdot 2^k < 2 \cdot k! < (k+1)k! = (k+1)!$

$\qquad\qquad 2^{k+1} < (k+1)!$

So if P_k is true, then P_{k+1} is true.

Conclusion: P_n is true for all $n \geq 4$.

(7-2)

49. To prove $a_n = b_n$ for all positive integers n, write:

P_n: $a_n = b_n$

Proof: Part 1: Show P_1 is true.

P_1: $a_1 = b_1$. $a_1 = 3$. $b_1 = 2^1 + 1 = 3$.

So, $a_1 = b_1$

Part 2: Show that if P_k is true, then P_{k+1} is true.

Write out P_k and P_{k+1}.

$\quad P_k$: $a_k = b_k$

$\quad P_{k+1}$: $a_{k+1} = b_{k+1}$

We start with P_k:

$$a_k = b_k$$

Now, $a_{k+1} = 2a_{k+1-1} - 1 = 2a_k - 1 = 2b_k - 1 = 2(2^k + 1) - 1$

$$= 2 \cdot 2^k + 2 - 1 = 2^1 \cdot 2^k + 1 = 2^{k+1} + 1 = b_{k+1}$$

Therefore, $a_{k+1} = b_{k+1}$

So if P_k is true, then P_{k+1} is true.

Conclusion: P_n is true for all positive integers n. So, $\{a_n\} = \{b_n\}$. $\qquad$ (7-2)

50. Make a rough sketch of the situation.

We are given $d(P, A) = 3d(P, B)$. Applying the distance formula, we have

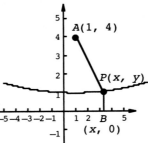

$$\sqrt{(x - 1)^2 + (y - 4)^2} = 3\sqrt{(x - x)^2 + (y - 0)^2}$$
$$(x - 1)^2 + (y - 4)^2 = 9[(x - x)^2 + (y - 0)^2]$$
$$x^2 - 2x + 1 + y^2 - 8y + 16 = 9y^2$$
$$x^2 - 2x - 8y^2 - 8y + 17 = 0$$

is the equation of the curve. Completing the square relative to x and y, we have

$$x^2 - 2x + 1 - 8\left(y^2 + y + \frac{1}{4}\right) - 1 + 2 + 17 = 0$$

$$(x - 1)^2 - 8\left(y + \frac{1}{2}\right)^2 = -18$$

$$\frac{\left(y + \frac{1}{2}\right)^2}{\frac{18}{8}} - \frac{(x - 1)^2}{18} = 1$$

This is the equation of a hyperbola. $\qquad$ (8-3)

51. The sample space S is the set of all possible ways of selecting 3 bulbs out of 12.

$n(S) = C_{12,3}$

P(at least one defective bulb) = 1 - P(no defective bulbs chosen)

The event, no defective bulbs chosen, is the set of all possible 3-bulb subsets of the 8 acceptable bulbs. $n(E) = C_{8,3}$

Then P(at least one defective bulb) $= 1 - \dfrac{n(E)}{n(S)} = 1 - \dfrac{C_{8,3}}{C_{12,3}}$

$$= 1 - \frac{56}{220} = \frac{41}{55} = .7\overline{45} \qquad (7-5)$$

52. We are asked for the sum of an infinite geometric series.

$a = \$2,000,000(0.75)$

$r = 0.75 \quad |r| \leq 1$, so the series has a sum,

$$S_\infty = \frac{a_1}{1 - r}$$

$$= \frac{\$2,000,000(0.75)}{1 - 0.75}$$

$$= \frac{\$2,000,000(0.75)}{0.25}$$

$$= \$6,000,000 \qquad (7-3)$$

53. Let x = the length and y = the width.
The perimeter is $2x + 2y$ so $2x + 2y = 24$.
The area is xy so $xy = 32$.

We solve the second equation for y and substitute into the first:

$$xy = 32 \qquad\qquad 2x + 2\left(\frac{32}{x}\right) = 24$$

$$y = \frac{32}{x} \qquad\qquad 2x + \frac{64}{x} = 24 \qquad \text{(Multiply both sides by } x.)$$

$$2x^2 + 64 = 24x$$
$$2x^2 - 24x + 64 = 0$$
$$2(x^2 - 12x + 32) = 0$$
$$2(x - 8)(x - 4) = 0$$
$$x = 8 \qquad \text{or} \qquad x = 4$$

If $x = 8$, $y = \dfrac{32}{8} = 4$. If $x = 4$, $y = \dfrac{32}{4} = 8$. The dimensions are 4m × 8m.   (8-5)

54.

From the figure, we see that the parabola can be positioned so that it is opening up with axis the y axis, and it has an equation of the form $x^2 = 4ay$. Since the focus is at $(0, a) = (0, 1)$, $a = 1$ and the equation of the parabola is $x^2 = 4y$.

Since the depth represents the y coordinate y_1 of a point on the parabola with $x = 4$, we have
$$4^2 = 4y_1$$
$$y_1 = \text{depth} = 4 \text{ in.}$$
(8-1)

55.

From the figure, we see that the x and y intercepts of the ellipse must be 40 and 24 respectively. The equation of the ellipse must be

$$\frac{x^2}{(40)^2} + \frac{y^2}{(24)^2} = 1$$

or $\dfrac{x^2}{1,600} + \dfrac{y^2}{576} = 1$

To find the distance c of each focus from the center of the arch, we use the special triangle relationship. $a = 40$, $b = 24$
$$c^2 = a^2 - b^2$$
$$= 40^2 - 24^2$$
$$c^2 = 1024$$
$$c = 32 \text{ ft.}$$

To find the height of the arch above each focus, we need the y coordinate y_1 of the point P whose x coordinate is 32. Since P is on the ellipse, we have

$$\frac{32^2}{1,600} + \frac{y_1^2}{576} = 1$$

$$\frac{1,024}{1,600} + \frac{y_1^2}{576} = 1$$

$$0.64 + \frac{y_1^2}{576} = 1$$

$$\frac{y_1^2}{576} = 0.36$$

$$y_1^2 = 576(0.36)$$

$$y_1^2 = 207.36$$

$$y_1 = 14.4 \text{ ft.}$$
(8-2)

56. (A) P (this event) $= \dfrac{130}{1,000} = .13$

(B) P (this event) $= \dfrac{80 + 90}{1,000} = .17$

(C) P (this event) $= P$ (being an independent) $+ P$ (being over 59) $- P$ (being both)

$\qquad\qquad\qquad\;\; = \dfrac{120}{1,000} + \dfrac{230}{1,000} - \dfrac{30}{1,000} = .32$ $\hspace{3cm}$ $(7\text{--}5)$

APPENDIX A   BASIC ALGEBRA REVIEW

Section A-1

1. True   **3.** False   **5.** False   **7.** True

9. False   **11.** True   **13.** 1

15. Division by zero is not defined. Undefined.

17. $0 \div (9 \cdot 8) = 0 \div 72 = 0$

19. $0 \cdot \left(100 + \dfrac{1}{100}\right) = 0$ since $a \cdot 0 = 0 \cdot a = 0$ for all real numbers a.

21. $\dfrac{4}{9} + \dfrac{12}{5} = \dfrac{4 \cdot 5 + 12 \cdot 9}{9 \cdot 5} = \dfrac{20 + 108}{45} = \dfrac{128}{45}$

23. $-\left(\dfrac{1}{100} + \dfrac{4}{25}\right) = -\left(\dfrac{1}{100} + \dfrac{16}{100}\right) = -\dfrac{17}{100}$ or $\dfrac{-17}{100}$

25. $\left(\dfrac{3}{8}\right)^{-1} + 2^{-1} = \dfrac{8}{3} + \dfrac{1}{2} = \dfrac{8 \cdot 2 + 1 \cdot 3}{3 \cdot 2} = \dfrac{16 + 3}{6} = \dfrac{19}{6}$

27. Commutative ($\cdot$)   **29.** Distributive   **31.** Inverse ($\cdot$)

33. Inverse (+)   **35.** Identity (+)   **37.** Negatives (Theorem 1)

39. The even integers between −3 and 5 are −2, 0, 2, and 4. The set is written {−2, 0, 2, 4}.

41. The letters in "status" are: s, t, a, u. The set is written $\{s, t, a, u\}$, or, equivalently, $\{a, s, t, u\}$.

43. Since there are no months starting with B, the set is empty. $\varnothing$

45. (A) The empty set, $\varnothing$, is a subset of every set.
   $\{a\}$ and $\{b\}$ are subsets of S_2.
   S_2 is a subset of itself.
   There are four subsets of S_2.

   (B) The empty set, $\varnothing$, is a subset of every set.
   $\{a\}$, $\{b\}$, and $\{c\}$ are one-member subsets of S_3.
   $\{b, c\}$, $\{a, c\}$, and $\{a, b\}$ are two-member subsets of S_3.
   S_3 is a subset of itself.
   There are $1 + 3 + 3 + 1 = 8$ subsets of S_3.

   (C) The empty set, $\varnothing$, is a subset of every set.
   $\{a\}$, $\{b\}$, $\{c\}$, and $\{d\}$ are one-member subsets of S_4.
   $\{a, b\}$, $\{a, c\}$, $\{a, d\}$, $\{b, c\}$, $\{b, d\}$, and $\{c, d\}$ are two-member subsets of S_4.
   $\{b, c, d\}$, $\{a, c, d\}$, $\{a, b, d\}$, and $\{b, c, d\}$ are three-member subsets of S_4.
   S_4 is a subset of itself.
   There are $1 + 4 + 6 + 4 + 1 = 16$ subsets of S_4.

47. Yes. This restates Zero Property (2). (Theorem 2, Part 2)

49. (A) True.

   (B) False, $\dfrac{2}{3}$ is an example of a real number that is not irrational.

   (C) True.

51. $\frac{3}{5}$ and -1.43 are two examples of infinitely many.

53. (A) $\{1, \sqrt{144}\}$ (B) $\{-3, 0, 1, \sqrt{144}\}$ (C) $\left\{-3, -\frac{2}{3}, 0, 1, \frac{9}{5}, \sqrt{144}\right\}$ (D) $\{\sqrt{3}\}$

55. (A) $0.888\ 888\ldots$; repeating; repeated digit: 8
(B) $0.272\ 727\ldots$; repeating; repeated digits: 27
(C) $2.236\ 067\ 977\ldots$; nonrepeating and nonterminating
(D) 1.375; terminating

57. (A) True; commutative property for addition.
(B) False; for example $3 - 5 \neq 5 - 3$.
(C) True; commutative property for multiplication.
(D) False; for example $9 \div 3 \neq 3 \div 9$.

59. F $3 - 8 = -5$ is one of many counterexamples. **61.** T

63. F $\sqrt{2} \cdot \frac{\sqrt{2}}{2} = 1$ is one of many counterexamples. **65.** T

67. (A) List each element of A. Follow these with each element of B that is not yet listed. $\{1,2,3,4,6\}$
(B) List each element of A that is also an element of B $\{2,4\}$.

69. Let $c = 0.090909\ldots$
Then $100c = 9.0909\ldots$
$100c - c = (9.0909\ldots) - (0.090909\ldots)$
$99c = 9$
$c = \frac{9}{99} = \frac{1}{11}$

71.
$$23$$
$$\underline{12}$$
$$46 \quad 23 \cdot 2$$
$$\underline{230} \quad 23 \cdot 10$$
$$276$$

$23 \cdot 12 = 23(2 + 10)$
$= 23 \cdot 2 + 23 \cdot 10$
$= 46 + 230$
$= 276$

Section A-2

1. 256

3. $\left(\frac{2}{3}\right)^4 = \frac{2^4}{3^4} = \frac{16}{81}$

5. $4^{-4} = \frac{1}{4^4} = \frac{1}{256}$

7. 625

9. $(-3)^{-1} = \frac{1}{-3} = \frac{-1}{3}$

11. $-7^{-2} = -\frac{1}{7^2} = \frac{-1}{49}$

13. 1

15. 10

17. 5

19. $9^{-3/2} = \frac{1}{9^{3/2}} = \frac{1}{\left(9^{1/2}\right)^3} = \frac{1}{3^3} = \frac{1}{27}$

21. Irrational

23. $1^{-3} + 3^{-1} = \frac{1}{1^3} + \frac{1}{3^1} = \frac{1}{1} + \frac{1}{3} = \frac{3+1}{3} = \frac{4}{3}$

25. $\frac{5}{5^4} = \frac{1}{5^3} = \frac{1}{125}$

27. $x^5 x^{-2} = x^{5+(-2)} = x^3$

29. $(2y)(3y^2)(5y^4) = 2 \cdot 3 \cdot 5 yy^2 y^4 = 30y^{1+2+4} = 30y^7$

31. $(a^2 b^3)^5 = (a^2)^5 (b^3)^5 = a^{10} b^{15}$

33. $u^{1/3} u^{5/3} = u^{(1/3)+(5/3)} = u^{6/3} = u^2$

35. $(x^{-3})^{1/6} = x^{-3/6} = x^{-1/2} = \frac{1}{x^{1/2}}$

37. $(49a^4 b^{-2})^{1/2} = 49^{1/2}(a^4)^{1/2}(b^{-2})^{1/2} = 7a^2 b^{-1} = \frac{7a^2}{b}$

39. $45{,}320{,}000 = \boxed{4.5320000.} \times 10^7$

7 places left

$\qquad = 4.532 \times 10^7$

41. $0.066 = 0.06.6 \times 10^{-2} = 6.6 \times 10^{-2}$

2 places right

negative exponent

43. $0.000\,000\,084 = 8.4 \times 10^{-8}$

8 places right

45. $9 \times 10^{-5} = 0.00009. = 0.000\,09$

5 places left

47. $3.48 \times 10^6 = \boxed{3.480\,000.} = 3{,}480{,}000$

6 places right

49. $4.2 \times 10^{-9} = \boxed{0.000\,000\,004.2} = 0.000\,000\,004\,2$

9 places left

51. $\left(\dfrac{x^4 y^{-1}}{x^{-2} y^3}\right)^2 = \dfrac{x^8 y^{-2}}{x^{-4} y^6} = x^{8-(-4)} y^{-2-6} = x^{12} y^{-8} = \dfrac{x^{12}}{y^8}$

53. $\left(\dfrac{2x^{-3} y^2}{4xy^{-1}}\right)^{-2} = \left(\dfrac{x^{-3} y^2}{2xy^{-1}}\right)^{-2} = \dfrac{x^6 y^{-4}}{2^{-2} x^{-2} y^2} = 2^2 x^{6-(-2)} y^{-4-2} = 4x^8 y^{-6} = \dfrac{4x^8}{y^6}$

55. $\left(\dfrac{a^{-3}}{b^4}\right)^{1/12} = \dfrac{a^{-3/12}}{b^{4/12}} = \dfrac{a^{-1/4}}{b^{1/3}} = \dfrac{1}{a^{1/4} b^{1/3}}$

57. $\left(\dfrac{4x^{-2}}{y^4}\right)^{-1/2} = \dfrac{4^{-1/2} x^1}{y^{-2}} = \dfrac{xy^2}{4^{1/2}} = \dfrac{xy^2}{2}$

> **Common Error:** $\dfrac{4x^1}{y^{-2}}$. This is wrong; the exponent $\left(-\tfrac{1}{2}\right)$ applies to constant as well as variable factors.

59. $\left(\dfrac{8a^{-4} b^3}{27 a^2 b^{-3}}\right)^{1/3} = \dfrac{8^{1/3} a^{-4/3} b^1}{27^{1/3} a^{2/3} b^{-1}} = \dfrac{2b^{1-(-1)}}{3a^{2/3-(-4/3)}} = \dfrac{2b^2}{3a^2}$

61. $-3(x^3 + 3)^{-4}(3x^2) = -3 \cdot \dfrac{1}{(x^3 + 3)^4} \cdot 3x^2$

> **Common Error:** $(x^3 + 3)^{-4} \neq x^{-12} + 3^{-4}$

$\qquad = \dfrac{-9x^2}{(x^3 + 3)^4}$

63. $2^{3^2} = 64$ on the assumption that 2^3^2 is entered, without parentheses.

65. The identity property for multiplication states that for any real number x, $x(1) = x$, and 1 is the only real number with that property. Therefore, if $a^m a^0 = a^m$, since $a^m(1) = a^m$, a^0 should equal 1.

67. $\dfrac{(32.7)(0.000\,000\,008\,42)}{(0.0513)(80{,}700{,}000{,}000)} = \dfrac{(32.7)(8.42 \times 10^{-9})}{(0.0513)(8.07 \times 10^{10})} = 6.65 \times 10^{-17}$

69. $\dfrac{(5{,}760{,}000{,}000)}{(527)(0.000\,007\,09)} = \dfrac{5.76 \times 10^9}{(527)(7.09 \times 10^{-6})} = 1.54 \times 10^{12}$

71. $15^{5/4} = 15^{1.25} = 29.52$

73. $103^{-3/4} = 103^{-0.75} = 0.03093$

75. $2.876^{8/5} = 2.876^{1.6} = 5.421$

77. $(0.000\ 000\ 077\ 35)^{-2/7} = (7.735 \times 10^{-8})^{(-2\div 7)} = 107.6$

79. There are many examples; one simple choice is $x = y = 1$. Then $(x + y)^{1/2} = (1 + 1)^{1/2} = 2^{1/2} = \sqrt{2}$, and $x^{1/2} + y^{1/2} = 1^{1/2} + 1^{1/2} = 1 + 1 = 2$; the left side is not equal to the right side.

81. There are many examples; one simple choice is $x = y = 1$. Then $(x + y)^{1/3} = (1 + 1)^{1/3} = 2^{1/3} = \sqrt[3]{2}$ and $\dfrac{1}{(x + y)^3} = \dfrac{1}{(1 + 1)^3} = \dfrac{1}{2^3} = \dfrac{1}{8}$; the left side is not equal to the right side.

83. $(a^{3/n}b^{3/m})^{1/3} = a^{(1/3)(3/n)}b^{(1/3)(3/m)} = a^{1/n}\,b^{1/m}$

85. $(x^{m/4}y^{n/3})^{-12} = x^{-12(m/4)}y^{-12(n/3)} = x^{-3m}\,y^{-4n} = \dfrac{1}{x^{3m}y^{4n}}$

87. (A) Since $(x^2)^{1/2}$ represents the positive real square root of x^2, it will not equal x if x is negative. So any negative value of x, for example $x = -2$, will make the left side $\ne$ right side. $[(-2)^2]^{1/2} = 4^{1/2} = 2 \ne -2$.

(B) Similarly any positive (or zero) value of x, for example $x = 2$, will make the left side = right side. $[2^2]^{1/2} = 4^{1/2} = 2 = 2$.

(C) Any value of x will make the left side = right side.
$[2^3]^{1/3} = 8^{1/3} = 2 \quad [(-2)^3]^{1/3} = [-8]^{1/3} = -2 \quad [0^3]^{1/3} = 0^{1/3} = 0$
There is no real value of x such that $(x^3)^{1/3} \ne x$.

89. No. Any negative value of b, for example -4, will make $(b^m)^{1/n}$ not a real number if n is even and m is odd. So $[(-4)^3]^{1/2} = [-64]^{1/2}$ is not a real number.

91. mass of earth in pounds
= mass of earth in grams × number of pounds/gram
= $6.1 \times 10^{27} \times 2.2 \times 10^{-3}$
= 13.42×10^{24}
= 1.342×10^{25}
= 1.3×10^{25} pounds to two significant digits

93. 1 operation in 10^{-8} seconds means $1 \div 10^{-8}$ operations/second, that is 10^8 operations in 1 second, that is, 100,000,000 or 100 million. Similarly, 1 operation in 10^{-10} seconds means $1 \div 10^{-10}$, that is 10,000,000,000 or 10 billion operations in 1 second.

Since 1 minute is 60 seconds, multiply each number by 60 to get $60 \times 10^8 = 6 \times 10^9$ or 6,000,000,000 or 6 billion operations in 1 minute and $60 \times 10^{10} = 6 \times 10^{11}$ or 600,000,000,000 or 600 billion operations in 1 minute.

95. One person's share of national debt
= amount of national debt ÷ number of persons = 6,760,000,000,000 ÷ 291,000,000
= $6.76 \times 10^{12} \div 2.91 \times 10^8$
= 2.32×10^4 or \$23,200 per person to three significant digits

97. We are asked to calculate
$N = 10x^{3/4}y^{1/4}$ given $x = 256$ units of
labor and $y = 81$ units of capital.

$N = 10(256)^{3/4}(81)^{1/4}$

$\quad = 10(256^{1/4})^3(81)^{1/4}$

$\quad = 10(4^3)(3)$

$\quad = 1{,}920$ units of finished product.

99. We are asked to calculate
$d = 0.0212v^{7/3}$ given $v = 70$
miles/hour.

$d = 0.0212(70)^{7/3}$

$\quad = 0.0212(70)^{(7 \div 3)}$

$\quad = 428$ feet (to the nearest ft)

Section A-3

1. $\sqrt[5]{32}$

3. $8\sqrt[3]{x^2}$

5. $4x^{-1/2} = 4\sqrt{x^{-1}}$ or $\dfrac{4}{\sqrt{x}}$

7. $\sqrt[3]{x} - \sqrt[3]{y}$

9. $361^{1/2}$

11. $4xy^{3/5}$

13. $(x^2 + y^2)^{1/3}$

15. -2

17. $-\sqrt{128} = -\sqrt{64 \cdot 2} = -\sqrt{8^2 \cdot 2} - \sqrt{8^2}\sqrt{2} = -8\sqrt{2}$

19. $\sqrt{27} - 5\sqrt{3} = \sqrt{9 \cdot 3} - 5\sqrt{3} = \sqrt{3^2 \cdot 3} - 5\sqrt{3} = \sqrt{3^2}\sqrt{3} - 5\sqrt{3} = 3\sqrt{3} - 5\sqrt{3} = -2\sqrt{3}$

21. $\sqrt[3]{5} - \sqrt[3]{25} + \sqrt[3]{125 \cdot 5} = \sqrt[3]{5} - \sqrt[3]{25} + \sqrt[3]{5^3 \cdot 5} = \sqrt[3]{5} - \sqrt[3]{25} + \sqrt[3]{5^3}\sqrt[3]{5}$

$\qquad = \sqrt[3]{5} - \sqrt[3]{25} + 5\sqrt[3]{5} = 6\sqrt[3]{5} - \sqrt[3]{25}$

23. $\sqrt[3]{25}\sqrt[3]{10} = \sqrt[3]{250} = \sqrt[3]{125 \cdot 2} = \sqrt[3]{5^3 \cdot 2} = \sqrt[3]{5^3}\sqrt[3]{2} = 5\sqrt[3]{2}$

25. $\sqrt{9x^8y^4} = \sqrt{9}\sqrt{x^8}\sqrt{y^4} = 3x^4y^2$

27. $\sqrt[4]{16m^4n^8} = \sqrt[4]{16}\sqrt[4]{m^4}\sqrt[4]{n^8} = 2mn^2$

29. $\sqrt[4]{m^2} = \sqrt[2\cdot 2]{m^{2\cdot 1}} = \sqrt[2]{m^1} = \sqrt{m}$

31. $\sqrt[5]{\sqrt[3]{xy}} = \sqrt[15]{xy}$

33. $\sqrt[3]{9x^2}\sqrt[3]{9x} = \sqrt[3]{(9x^2)(9x)} = \sqrt[3]{81x^3} = \sqrt[3]{27x^3 \cdot 3} = \sqrt[3]{27x^3}\sqrt[3]{3} = 3x\sqrt[3]{3}$

35. $\dfrac{1}{2\sqrt{5}} = \dfrac{1\sqrt{5}}{2\sqrt{5}\sqrt{5}} = \dfrac{\sqrt{5}}{2 \cdot 5} = \dfrac{\sqrt{5}}{10}$

37. $\dfrac{1}{\sqrt[3]{7}} = \dfrac{1\sqrt[3]{7^2}}{\sqrt[3]{7}\sqrt[3]{7^2}} = \dfrac{\sqrt[3]{49}}{\sqrt[3]{7^3}} = \dfrac{\sqrt[3]{49}}{7}$

39. $\dfrac{2}{\sqrt{44}} = \dfrac{2}{\sqrt{2^2 \cdot 11}} = \dfrac{2\sqrt{11}}{\sqrt{2^2}\sqrt{11}\sqrt{11}} = \dfrac{2\sqrt{11}}{2 \cdot 11} = \dfrac{\sqrt{11}}{11}$

41. $\dfrac{6x}{\sqrt{3x}} = \dfrac{6x}{\sqrt{3x}}\dfrac{\sqrt{3x}}{\sqrt{3x}} = \dfrac{6x\sqrt{3x}}{3x} = 2\sqrt{3x}$

43. $\dfrac{2}{\sqrt{2} - 1} = \dfrac{2}{(\sqrt{2} - 1)}\dfrac{(\sqrt{2} + 1)}{(\sqrt{2} + 1)} = \dfrac{2(\sqrt{2} + 1)}{2 - 1} = \dfrac{2(\sqrt{2} + 1)}{1} = 2(\sqrt{2} + 1) = 2\sqrt{2} + 2$

45. $\dfrac{\sqrt{2}}{\sqrt{6} + 2} = \dfrac{\sqrt{2}}{(\sqrt{6} + 2)}\dfrac{(\sqrt{6} - 2)}{(\sqrt{6} - 2)} = \dfrac{\sqrt{2}(\sqrt{6} - 2)}{6 - 4} = \dfrac{\sqrt{2}\sqrt{6} - 2\sqrt{2}}{2} = \dfrac{\sqrt{12} - 2\sqrt{2}}{2} = \dfrac{2\sqrt{3} - 2\sqrt{2}}{2}$

$\qquad = \dfrac{2(\sqrt{3} - \sqrt{2})}{2} = \sqrt{3} - \sqrt{2}$

47. $x\sqrt[5]{3^6 x^7 y^{11}} = x\sqrt[5]{(3^5 x^5 y^{10})(3x^2 y)} = x\sqrt[5]{3^5 x^5 y^{10}}\sqrt[5]{3x^2 y} = x(3xy^2)\sqrt[5]{3x^2 y} = 3x^2 y^2\sqrt[5]{3x^2 y}$

49. $\dfrac{\sqrt[4]{32m^7 n^9}}{2mn} = \dfrac{\sqrt[4]{16m^4 n^8 \cdot 2m^3 n}}{2mn} = \dfrac{2mn^2\sqrt[4]{2m^3 n}}{2mn} = n\sqrt[4]{2m^3 n}$

51. $\sqrt[3]{\sqrt[4]{a^9 b^3}} = \sqrt[3 \cdot 4]{a^{3 \cdot 3} b^{3 \cdot 1}} = \sqrt[4]{a^3 b}$

53. $\sqrt[3]{2x^2 y^4} \sqrt[3]{3x^5 y} = \sqrt[3]{(2x^2 y^4)(3x^5 y)} = \sqrt[3]{6x^7 y^5} = \sqrt[3]{(x^6 y^3)(6xy^2)} = \sqrt[3]{x^6 y^3} \sqrt[3]{6xy^2} = x^2 y \sqrt[3]{6xy^2}$

55. $\dfrac{\sqrt{2m}\sqrt{5}}{\sqrt{20m}} = \dfrac{\sqrt{10m}}{\sqrt{20m}} = \sqrt{\dfrac{10m}{20m}} = \sqrt{\dfrac{1}{2}} = \sqrt{\dfrac{1}{2} \cdot \dfrac{2}{2}} = \sqrt{\dfrac{2}{4}} = \dfrac{\sqrt{2}}{2}$ or $\dfrac{1}{2}\sqrt{2}$

57. $\dfrac{3\sqrt{y}}{2\sqrt{y} - 3} = \dfrac{3\sqrt{y}}{(2\sqrt{y} - 3)} \dfrac{(2\sqrt{y} + 3)}{(2\sqrt{y} + 3)} = \dfrac{3\sqrt{y}(2\sqrt{y} + 3)}{(2\sqrt{y})^2 - (3)^2} = \dfrac{6y + 9\sqrt{y}}{4y - 9}$

59. $\dfrac{2\sqrt{5} + 3\sqrt{2}}{5\sqrt{5} + 2\sqrt{2}} = \dfrac{(2\sqrt{5} + 3\sqrt{2})}{(5\sqrt{5} + 2\sqrt{2})} \dfrac{(5\sqrt{5} - 2\sqrt{2})}{(5\sqrt{5} - 2\sqrt{2})}$

$= \dfrac{2\sqrt{5} \cdot 5\sqrt{5} - 2\sqrt{5} \cdot 2\sqrt{2} + 3\sqrt{2} \cdot 5\sqrt{5} - 3\sqrt{2} \cdot 2\sqrt{2}}{(5\sqrt{5})^2 + (2\sqrt{2})^2}$

$= \dfrac{(10)(5) - 4\sqrt{10} + 15\sqrt{10} - (6)(2)}{(25)(5) - (4)(2)} = \dfrac{50 + 11\sqrt{10} - 12}{125 - 8} = \dfrac{38 + 11\sqrt{10}}{117}$

61. Here we want to remove the radical terms from the *numerator*.

$\dfrac{\sqrt{t} - \sqrt{x}}{t - x} = \dfrac{(\sqrt{t} - \sqrt{x})}{t - x} \cdot \dfrac{(\sqrt{t} + \sqrt{x})}{\sqrt{t} + \sqrt{x}} = \dfrac{(\sqrt{t})^2 - (\sqrt{x})^2}{(t - x)(\sqrt{t} + \sqrt{x})}$

$= \dfrac{\overset{1}{\cancel{t - x}}}{\underset{1}{\cancel{(t - x)}}(\sqrt{t} + \sqrt{x})} = \dfrac{1}{\sqrt{t} + \sqrt{x}}$

63. Here we want to remove the radical terms from the *numerator*.

$\dfrac{\sqrt{x + h} - \sqrt{x}}{h} = \dfrac{(\sqrt{x + h} - \sqrt{x})}{h} \cdot \dfrac{(\sqrt{x + h} + \sqrt{x})}{(\sqrt{x + h} + \sqrt{x})} = \dfrac{(\sqrt{x + h})^2 - (\sqrt{x})^2}{h(\sqrt{x + h} + \sqrt{x})} = \dfrac{x + h - x}{h(\sqrt{x + h} + \sqrt{x})}$

$= \dfrac{\overset{1}{\cancel{h}}}{\underset{1}{\cancel{h}}(\sqrt{x + h} + \sqrt{x})} = \dfrac{1}{\sqrt{x + h} + \sqrt{x}}$

65. $\sqrt{0.032\,965} = 0.1816$ **67.** $\sqrt[3]{45.0218} = (45.0218)^{1/3} = 45.0218^{(1 \div 3)} = 3.557$

69. $\sqrt[8]{5.477 \times 10^{-9}} = (5.477 \times 10^{-9})^{(1 \div 8)} = 0.09275$

71. $\sqrt[5]{9 + \sqrt[5]{9}} = (9 + 9^{(1 \div 5)})^{(1 \div 5)} = 1.602$

73. $\sqrt{x^2}$ is the principal square root of x^2, that is, the positive (or zero) square root of x^2. The two square roots of x^2 are x and $-x$. $\sqrt{x^2} = -x$ if and only if $-x$ is positive or zero. So $\sqrt{x^2} = -x$ if and only if x is negative or zero. $x \leq 0$.

75. $\sqrt[3]{x^3}$ is the real third root of x^3. This is always equal to x if x is real. All real numbers.

77. (A) $\sqrt{3} + \sqrt{5} \approx 3.968\,118\,785$ (B) $\sqrt{2 + \sqrt{3}} + \sqrt{2 - \sqrt{3}} \approx 2.449\,489\,743$

(C) $1 + \sqrt{3} \approx 2.732\,050\,808$ (D) $\sqrt[3]{10 + 6\sqrt{3}} \approx 2.732\,050\,808$

(E) $\sqrt{8 + \sqrt{60}} \approx 3.968\ 118\ 785$ (F) $\sqrt{6} \approx 2.449\ 489\ 743$

So (A) and (E), (B) and (F), and (C) and (D) have the same value to 9 decimal places. To show that they are actually equal:

(A) and (E): $(\sqrt{3} + \sqrt{5})^2 = (\sqrt{3})^2 + 2\sqrt{3}\sqrt{5} + (\sqrt{5})^2 = 3 + \sqrt{4}\sqrt{3}\sqrt{5} + 5$ $= 8 + \sqrt{60}$ Therefore, since $\sqrt{3} + \sqrt{5}$ is positive, $\sqrt{3} + \sqrt{5}$ is the positive square root of $8 + \sqrt{60}$, and $\sqrt{8 + \sqrt{60}} = \sqrt{3} + \sqrt{5}$.

(B) and (F):

$$\left(\sqrt{2 + \sqrt{3}} + \sqrt{2 - \sqrt{3}}\right)^2 = \left(\sqrt{2 + \sqrt{3}}\right)^2 + 2\sqrt{2 + \sqrt{3}}\sqrt{2 - \sqrt{3}} + \left(\sqrt{2 - \sqrt{3}}\right)^2$$

$$= 2 + \sqrt{3} + 2\sqrt{(2 + \sqrt{3})(2 - \sqrt{3})} + 2 - \sqrt{3}$$

$$= 4 + 2\sqrt{2^2 - (\sqrt{3})^2}$$

$$= 4 + 2\sqrt{4 - 3}$$

$$= 4 + 2 \cdot 1$$

$$= 6$$

Therefore, since $\sqrt{2 + \sqrt{3}} + \sqrt{2 - \sqrt{3}}$ is positive, $\sqrt{2 + \sqrt{3}} + \sqrt{2 - \sqrt{3}}$ is the positive square root of 6, and $\sqrt{2 + \sqrt{3}} + \sqrt{2 - \sqrt{3}} = \sqrt{6}$.

(C) and (D): $(1 + \sqrt{3})^3 = (1 + \sqrt{3})(1 + \sqrt{3})(1 + \sqrt{3})$

$$= (1^2 + 2\sqrt{3} + \sqrt{3}^2)(1 + \sqrt{3})$$

$$= (4 + 2\sqrt{3})(1 + \sqrt{3})$$

$$= 4 + 4\sqrt{3} + 2\sqrt{3} + 2\sqrt{3}\sqrt{3}$$

$$= 10 + 6\sqrt{3}$$

Therefore, $1 + \sqrt{3}$ is the real cube root of $10 + 6\sqrt{3}$, and $1 + \sqrt{3} = \sqrt[3]{10 + 6\sqrt{3}}$.

79. $\dfrac{1}{\sqrt[3]{a} - \sqrt[3]{b}} = \dfrac{1}{\left(\sqrt[3]{a} - \sqrt[3]{b}\right)} \dfrac{\left[(\sqrt[3]{a})^2 + \sqrt[3]{a}\sqrt[3]{b} + (\sqrt[3]{b})^2\right]}{\left[(\sqrt[3]{a})^2 + \sqrt[3]{a}\sqrt[3]{b} + (\sqrt[3]{b})^2\right]}$

using $(x - y)(x^2 + xy + y^2) = x^3 - y^3$

$$= \dfrac{1\left[\sqrt[3]{a^2} + \sqrt[3]{ab} + \sqrt[3]{b^2}\right]}{(\sqrt[3]{a})^3 - (\sqrt[3]{b})^3} = \dfrac{\sqrt[3]{a^2} + \sqrt[3]{ab} + \sqrt[3]{b^2}}{a - b}$$

81. $\dfrac{1}{\sqrt{x} - \sqrt{y} + \sqrt{z}} = \dfrac{1}{\left[(\sqrt{x} - \sqrt{y}) + \sqrt{z}\right]} \dfrac{\left[(\sqrt{x} - \sqrt{y}) - \sqrt{z}\right]}{\left[(\sqrt{x} - \sqrt{y}) - \sqrt{z}\right]} = \dfrac{\sqrt{x} - \sqrt{y} - \sqrt{z}}{(\sqrt{x} - \sqrt{y})^2 + (\sqrt{z})^2}$

$$= \dfrac{\sqrt{x} - \sqrt{y} - \sqrt{z}}{(\sqrt{x})^2 - 2\sqrt{x}\sqrt{y} + (\sqrt{y})^2 - (\sqrt{z})^2} = \dfrac{\sqrt{x} - \sqrt{y} - \sqrt{z}}{x - 2\sqrt{xy} + y - z}$$

$$= \dfrac{\sqrt{x} - \sqrt{y} - \sqrt{z}}{x - 2\sqrt{xy} + y - z} = \dfrac{\sqrt{x} - \sqrt{y} - \sqrt{z}}{x + y - z - 2\sqrt{xy}}$$

$$= \dfrac{(\sqrt{x} - \sqrt{y} - \sqrt{z})}{[(x + y - z) - 2\sqrt{xy}]} \dfrac{[(x + y - z) + 2\sqrt{xy}]}{[(x + y - z) + 2\sqrt{xy}]}$$

$$= \dfrac{(\sqrt{x} - \sqrt{y} - \sqrt{z})(x + y - z) + 2\sqrt{xy}}{(x + y - z)^2 - (2\sqrt{xy})^2}$$

$$= \dfrac{(\sqrt{x} - \sqrt{y} - \sqrt{z})(x + y - z) + 2\sqrt{xy}}{(x + y - z)^2 - 4xy}$$

83.
$$\frac{\sqrt[3]{x+h} - \sqrt[3]{x}}{h} = \frac{\left(\sqrt[3]{x+h} - \sqrt[3]{x}\right)}{h} \frac{\left[(\sqrt[3]{x+h})^2 + \sqrt[3]{x+h}\,\sqrt[3]{x} + (\sqrt[3]{x})^2\right]}{\left[(\sqrt[3]{x+h})^2 + \sqrt[3]{x+h}\,\sqrt[3]{x} + (\sqrt[3]{x})^2\right]}$$

$$= \frac{(\sqrt[3]{x+h})^3 - (\sqrt[3]{x})^3}{h\left[\sqrt[3]{(x+h)^2} + \sqrt[3]{x(x+h)} + \sqrt[3]{x^2}\right]}$$

Using $(a-b)(a^2 + ab + b^2) = a^3 - b^3$

$$= \frac{x+h-x}{h\left[\sqrt[3]{(x+h)^2} + \sqrt[3]{x(x+h)} + \sqrt[3]{x^2}\right]}$$

$$= \frac{h}{h\left[\sqrt[3]{(x+h)^2} + \sqrt[3]{x(x+h)} + \sqrt[3]{x^2}\right]}$$

$$= \frac{1}{\sqrt[3]{(x+h)^2} + \sqrt[3]{x(x+h)} + \sqrt[3]{x^2}}$$

85. $\sqrt[kn]{x^{km}} = (x^{km})^{1/kn} = x^{km/kn} = x^{m/n} = \sqrt[n]{x^m}$

87.
$$\frac{M_0}{\sqrt{1 - \frac{v^2}{c^2}}} = M_0 \div \sqrt{1 - \frac{v^2}{c^2}}$$

$$= M_0 \div \sqrt{\frac{c^2 - v^2}{c^2}} = M_0 \div \frac{\sqrt{c^2 - v^2}}{c}$$

$$= M_0 \cdot \frac{c}{\sqrt{c^2 - v^2}}$$

Now rationalize the denominator

$$= M_0 \cdot \frac{c}{\sqrt{c^2 - v^2}} \frac{\sqrt{c^2 - v^2}}{\sqrt{c^2 - v^2}}$$

$$= \frac{M_0 c \sqrt{c^2 - v^2}}{c^2 - v^2}$$

Section A-4

1. 4

3. $(x^4 - 2x^2 + 3) + (x^3 - 1) = x^4 - 2x^2 + 3 + x^3 - 1 = x^4 + x^3 - 2x^2 + 2$
This polynomial has degree 4.

5. $(x^4 - 2x^2 + 3)(x^3 - 1) = x^7 - x^4 - 2x^5 + 2x^2 + 3x^3 - 3$
$\qquad\qquad\qquad\qquad\qquad = x^7 - 2x^5 - x^4 + 3x^3 + 2x^2 - 3$

7. $(x^4 - 2x^2 + 3) - (x^3 - 1) = x^4 - 2x^2 + 3 - x^3 + 1 = x^4 - x^3 - 2x^2 + 4$

9. Yes. 2 　　　　　　　**11.** No 　　　　　　　**13.** Yes. 3

15. $2(x - 1) + 3(2x - 3) - (4x - 5) = 2x - 2 + 6x - 9 - 4x + 5 = 4x - 6$

17. $2y - 3y[4 - 2(y - 1)] = 2y - 3y[4 - 2y + 2]$
$\qquad\qquad\qquad\qquad = 2y - 3y[6 - 2y]$
$\qquad\qquad\qquad\qquad = 2y - 18y + 6y^2$
$\qquad\qquad\qquad\qquad = 6y^2 - 16y$

19. $m^2 - n^2$

21. $(4t - 3)(t - 2) = 4t^2 - 8t - 3t + 6 = 4t^2 - 11t + 6$

23. $(5y - 1)(3 - 2y) = 15y - 10y^2 - 3 + 2y = -10y^2 + 17y - 3$

25. $(3x + 2y)(x - 3y) = 3x^2 - 9xy + 2xy - 6y^2 = 3x^2 - 7xy - 6y^2$

27. $4m^2 - 49$

29. $(6x - 4y)(5x + 3y) = 30x^2 + 18xy - 20xy - 12y^2 = 30x^2 - 2xy - 12y^2$

31. $9x^2 - 4y^2$

33. $(4x - y)^2 = (4x)^2 - 2(4x)y + y^2 = 16x^2 - 8xy + y^2$

35. $(a + b)(a^2 - ab + b^2) = a^3 - a^2b + ab^2 + a^2b - ab^2 + b^3 = a^3 + b^3$

37. $2x - 3\{x + 2[x - (x + 5)] + 1\} = 2x - 3\{x + 2[x - x - 5] + 1\}$
$$= 2x - 3\{x + 2(-5) + 1\}$$
$$= 2x - 3\{x - 10 + 1\}$$
$$= 2x - 3\{x - 9\}$$
$$= 2x - 3x + 27$$
$$= -x + 27$$

39. $2\{3[a - 4(1 - a)] - (5 - a)\} = 2\{3[a - 4 + 4a] - 5 + a\}$
$$= 2\{3[5a - 4] - 5 + a\}$$
$$= 2\{15a - 12 - 5 + a\}$$
$$= 2\{16a - 17\}$$
$$= 32a - 34$$

41. $(2x^2 - 3x + 1)(x^2 + x - 2) = 2x^4 + 2x^3 - 4x^2 - 3x^3 - 3x^2 + 6x + x^2 + x - 2$
$$= 2x^4 - x^3 - 6x^2 + 7x - 2$$

43. $(x - 2y)^2(x + 2y)^2 = [(x - 2y)(x + 2y)]^2$
$$= [x^2 - 4y^2]^2$$
$$= x^4 - 8x^2y^2 + 16y^4$$

45. $(3u - 2v)^2 - (2u - 3v)(2u + 3v) = (9u^2 - 12uv + 4v^2) - (4u^2 - 9v^2)$
$$= 9u^2 - 12uv + 4v^2 - 4u^2 + 9v^2$$
$$= 5u^2 - 12uv + 13v^2$$

47. $(z + 2)(z^2 - 2z + 3) + z - 7 = z^3 - 2z^2 + 3z + 2z^2 - 4z + 6 + z - 7$
$$= z^3 - 1$$

49. $(2m - n^3) = (2m - n)^2(2m - n)$
$$= (4m^2 - 4mn + n^2)(2m - n)$$
$$= 8m^3 - 4m^2n - 8m^2n + 4mn^2 + 2mn^2 - n^3$$
$$= 8m^3 - 12m^2n + 6mn^2 - n^3$$

51. $3(x + h) - 7 - (3x - 7) = 3x + 3h - 7 - 3x + 7 = 3h$

53. $2(x + h)^2 - 3(x + h) - (2x^2 - 3x) = 2(x^2 + 2xh + h^2) - 3x - 3h - 2x^2 + 3x$
$$= 2x^2 + 4xh + 2h^2 - 3x - 3h - 2x^2 + 3x$$
$$= 4xh + 2h^2 - 3h$$

55. $2(x + h)^2 - 4(x + h) - 9 - (2x^2 - 4x - 9)$
$$= 2(x^2 + 2xh + h^2) - 4x - 4h - 9 - 2x^2 + 4x + 9$$
$$= 2x^2 + 4xh + 2h^2 - 4x - 4h - 9 - 2x^2 + 4x + 9$$
$$= 4xh + 2h^2 - 4h$$

57. $(x + h)^3 - 2(x + h)^2 - (x^3 - 2x^2)$
$$= (x + h)^2(x + h) - 2(x + h)^2 - (x^3 - 2x^2)$$
$$= (x^2 + 2xh + h^2)(x + h) - 2(x^2 + 2xh + h^2) - x^3 + 2x^2$$
$$= x^3 + hx^2 + 2xh^2 + 2h^2x + h^2x + h^3 - 2x^2 - 4xh - 2h^2 - x^3 + 2x^2$$
$$= 3hx^2 + 3h^2x + h^3 - 4xh - 2h^2$$

59. The sum of the first two polynomials:
$(3m^2 - 2m + 5) + (4m^2 - m) = 3m^2 - 2m + 5 + 4m^2 - m = 7m^2 - 3m + 5$

The sum of the last two polynomials:
$(3m^2 - 3m - 2) + (m^3 + m^2 + 2) = 3m^2 - 3m - 2 + m^3 + m^2 + 2 = m^3 + 4m^2 - 3m$

Subtract from the sum of the last two polynomials the sum of the first two.
$$(m^3 + 4m^2 - 3m) - (7m^2 - 3m + 5) = m^3 + 4m^2 - 3m - 7m^2 + 3m - 5$$
$$= m^3 - 3m^2 - 5$$

61. The area of the entire region is given by $(a + b)(c + d)$. The sum of the areas of the four small rectangles is given by $ac + ad + bc + bd$. Since the area of the entire region must equal the sum of the areas of the four small rectangles, $(a + b)(c + d) = ac + ad + bc + bd$.

63. The area of the large square is given by c^2. The area of the small square is given by $(a - b)^2$. The area of each of the four triangles is given by $\frac{1}{2}ab$.
Therefore,
$$(a - b)^2 + 4\left(\frac{1}{2}ab\right) = c^2$$
Simplifying, we obtain
$$a^2 - 2ab + b^2 + 2ab = c^2$$
$$a^2 + b^2 = c^2$$
This is the Pythagorean Theorem.

65. $2(x - 2)^3 - (x - 2)^2 - 3(x - 2) - 4$
$$= 2(x - 2)^2(x - 2) - (x - 2)^2 - 3(x - 2) - 4$$
$$= 2(x^2 - 4x + 4)(x - 2) - (x^2 - 4x + 4) - 3(x - 2) - 4$$
$$= 2(x^3 - 2x^2 - 4x^2 + 8x + 4x - 8) - (x^2 - 4x + 4) - 3(x - 2) - 4$$
$$= 2x^3 - 4x^2 - 8x^2 + 16x + 8x - 16 - x^2 + 4x - 4 - 3x + 6 - 4$$
$$= 2x^3 - 13x^2 + 25x - 18$$

67. $-3x\{x[x - x(2 - x)] - (x + 2)(x^2 - 3)\}$
$$= -3x\{x[x - 2x + x^2] - (x^3 - 3x + 2x^2 - 6)\}$$
$$= -3x\{x[-x + x^2] - (x^3 - 3x + 2x^2 - 6)\}$$
$$= -3x\{-x^2 + x^3 - x^3 + 3x - 2x^2 + 6\}$$
$$= -3x\{-3x^2 + 3x + 6\}$$
$$= 9x^3 - 9x^2 - 18x$$

69. One example is given by choosing $a = 1$ and $b = 1$. Then $(a + b)^2 = (1 + 1)^2 = 2^2 = 4$, but $a^2 + b^2 = 1^2 + 1^2 = 1 + 1 = 2$. So in general, $(a + b)^2 \neq a^2 + b^2$. In fact, since $(a + b)^2 = a^2 + 2ab + b^2$, this quantity can only equal $a^2 + b^2$ if $2ab = 0$. By the properties of 0, either $a = 0$ or $b = 0$. In these cases $(a + b)^2$ would equal $a^2 + b^2$, but only in these.

71. The non-zero term with the highest degree in the polynomial of degree m has degree m. This term is not changed during the addition, since there is no term in the polynomial of degree n with the degree $m(m > n)$. It will still be the highest degree term in the sum, so the sum will be a polynomial of degree m.

73. Now the non-zero term with the highest degree in each polynomial has degree *m*. It is possible for the coefficients of these terms to be equal in absolute value but opposite in sign, in which case they would add to 0, leaving a term of degree less than *m* as the term with the highest degree in the sum. Otherwise the terms of degree *m* will combine to give another term of degree *m*, which will be the highest degree term in the sum. So $(2x^4 + x^2 + 1) + (-2x^4 + x^2 + 1) = 2x^2 + 2$, but $(2x^4 + x^2 + 1) + (2x^4 + 2x^2 + 1) = 4x^4 + 3x^2 + 2$. Summarizing, the degree of the sum may be less than or equal to *m*.

75. There are three quantities in this problem, perimeter, length, and width. They are related by the perimeter formula $P = 2\ell + 2w$. Since x = length of the rectangle, and the width is 5 meters less than the length, $x - 5$ = width of the rectangle. So $P = 2x + 2(x - 5)$ represents the perimeter of the rectangle. Simplifying: $P = 2x + 2x - 10 = 4x - 10$ (meters)

77. There are several quantities involved in this problem. It is important to keep them distinct by using enough words. We write:

$$x = \text{number of nickels}$$
$$x - 5 = \text{number of dimes}$$
$$(x - 5) + 2 = \text{number of quarters}$$

This follows because there are five fewer dimes than nickels ($x - 5$) and 2 more quarters than dimes (2 more than $x - 5$). Each nickel is worth 5 cents, each dime worth 10 cents, and each quarter worth 25 cents. So the value of the nickels is 5 times the number of nickels, the value of the dimes is 10 times the number of dimes, and the value of the quarters is 25 times the number of quarters.

$$\text{value of nickels} = 5x$$
$$\text{value of dimes} = 10(x - 5)$$
$$\text{value of quarters} = 25[(x - 5) + 2]$$

The value of the pile = (value of nickels) + (value of dimes)
$$+ \text{ (value of quarters)}$$
$$= 5x + 10(x - 5) + 25[(x - 5) + 2]$$

Simplifying this expression, we get:
$$\text{The value of the pile} = 5x + 10x - 50 + 25[x - 5 + 2]$$
$$= 5x + 10x - 50 + 25[x - 3]$$
$$= 5x + 10x - 50 + 25x - 75$$
$$= 40x - 125 \text{ (cents)}$$

79. The volume of the plastic shell is equal to the volume of the larger sphere ($V = \frac{4}{3}\pi r^3$) minus the volume of the hole. Since the radius of the hole is x cm and the plastic is 0.3 cm thick, the radius of the larger sphere is $x + 0.3$ cm.

$$\begin{pmatrix} \text{Volume of} \\ \text{shell} \end{pmatrix} = \begin{pmatrix} \text{Volume of} \\ \text{larger sphere} \end{pmatrix} - \begin{pmatrix} \text{Volume of} \\ \text{hole} \end{pmatrix}$$

$$\begin{aligned} \text{Volume} &= \frac{4}{3}\pi(x + 0.3)^3 - \frac{4}{3}\pi x^3 \\ &= \frac{4}{3}\pi(x + 0.3)(x + 0.3)^2 - \frac{4}{3}\pi x^3 \\ &= \frac{4}{3}\pi(x + 0.3)(x^2 + 0.6x + 0.09) - \frac{4}{3}\pi x^3 \\ &= \frac{4}{3}\pi(x^3 + 0.6x^2 + 0.09x + 0.3x^2 + 0.18x + 0.027) - \frac{4}{3}\pi x^3 \\ &= \frac{4}{3}\pi(x^3 + 0.9x^2 + 0.27x + 0.027) - \frac{4}{3}\pi x^3 \\ &= \frac{4}{3}\pi x^3 + 1.2\pi x^2 + 0.36\pi x + 0.036\pi - \frac{4}{3}\pi x^3 \\ &= 1.2\pi x^2 + 0.36\pi x + 0.036\pi \text{ (cm}^3) \end{aligned}$$

Section A-5

1. $2x^2(3x^2 - 4x - 1)$ 3. $5xy(2x^2 + 4xy - 3y^2)$ 5. $(x + 1)(5x - 3)$ 7. $(y - 2z)(2w - x)$

9. $x^2 - 2x + 3x - 6 = (x^2 - 2x) + (3x - 6)$
$$= x(x - 2) + 3(x - 2)$$
$$= (x - 2)(x + 3)$$

11. $6m^2 + 10m - 3m - 5 = (6m^2 + 10m) - (3m + 5) = 2m(3m + 5) - 1(3m + 5)$
$$= (3m + 5)(2m - 1)$$

13. $2x^2 - 4xy - 3xy + 6y^2 = (2x^2 - 4xy) - (3xy - 6y^2) = 2x(x - 2y) - 3y(x - 2y)$
$$= (x - 2y)(2x - 3y)$$

15. $8ac + 3bd - 6bc - 4ad = 8ac - 4ad - 6bc + 3bd$
$$= (8ac - 4ad) - (6bc - 3bd)$$
$$= 4a(2c - d) - 3b(2c - d)$$
$$= (2c - d)(4a - 3b)$$

17. $(2x + 3)(x - 1)$ 19. Prime

21. $(x + 7y)(x - 2y)$

23. $(2a + 3b)(2a - 3b)$

25. $4x^2 - 20x + 25 = (2x)^2 - 2(2x)(5) + 5^2 = (2x - 5)^2$ 27. Prime

29. Prime

31. $6x^2 + 48x + 72 = 6(x^2 + 8x + 12) = 6(x + 2)(x + 6)$

33. $2y^3 - 22y^2 + 48y = 2y(y^2 - 11y + 24) = 2y(y - 3)(y - 8)$

35. $16x^2y - 8xy + y = y(16x^2 - 8x + 1) = y[(4x)^2 - 2(4x) + 1] = y(4x - 1)^2$

37. $(3m + 4n)(2m - 3n)$

39. $x^3y - 9xy^3 = xy(x^2 - 9y^2) = xy(x - 3y)(x + 3y)$

41. $3m(m^2 - 2m + 5)$ 43. $(m + n)(m^2 - mn + n^2)$

45. $8x^3 - 125 = (2x)^3 - 5^3 = (2x - 5)[(2x)^2 + (2x)5 + 5^2]$
$$= (2x - 5)(4x^2 + 10x + 25)$$

47. $2x(x + 1)^4 + 4x^2(x + 1)^3 = 2x(x + 1)^3[(x + 1) + 2x] = 2x(x + 1)^3(3x + 1)$

49. $6(3x - 5)(2x - 3)^2 + 4(3x - 5)^2(2x - 3) = 2(3x - 5)(2x - 3)[3(2x - 3) + 2(3x - 5)]$
$$= 2(3x - 5)(2x - 3)[6x - 9 + 6x - 10]$$
$$= 2(3x - 5)(2x - 3)(12x - 19)$$

51. $5x^4(9 - x)^4 - 4x^5(9 - x)^3 = x^4(9 - x)^3[5(9 - x) - 4x]$
$$= x^4(9 - x)^3[45 - 5x - 4x]$$
$$= x^4(9 - x)^3(45 - 9x) \quad \text{or} \quad 9x^4(9 - x)^3(5 - x)$$

53. $2(x + 1)(x^2 - 5)^2 + 4x(x + 1)^2(x^2 - 5) = 2(x + 1)(x^2 - 5)[x^2 - 5 + 2x(x + 1)]$
$$= 2(x + 1)(x^2 - 5)[x^2 - 5 + 2x^2 + 2x]$$
$$= 2(x + 1)(x^2 - 5)(3x^2 + 2x - 5)$$
$$= 2(x + 1)(x^2 - 5)(3x + 5)(x - 1)$$

55. $(a - b)^2 - 4(c - d)^2 = (a - b)^2 - [2(c - d)]^2$
$$= [(a - b) - 2(c - d)][(a - b) + 2(c - d)]$$

57. $2am - 3an + 2bm - 3bn = (2am - 3an) + (2bm - 3bn) = a(2m - 3n) + b(2m - 3n)$
$$= (2m - 3n)(a + b)$$

59. Prime

61. $x^3 - 3x^2 - 9x + 27 = (x^3 - 3x^2) - (9x - 27) = x^2(x - 3) - 9(x - 3) = (x - 3)(x^2 - 9)$
$$= (x - 3)(x - 3)(x + 3) = (x - 3)^2(x + 3)$$

63. $a^3 - 2a^2 - a + 2 = (a^3 - 2a^2) - (a - 2) = a^2(a - 2) - 1(a - 2) = (a - 2)(a^2 - 1)$
$$= (a - 2)(a + 1)(a - 1)$$

65. Prime

67. $(x^2 + 2)(x^2 + 4)$

69. $m^4 - n^4 = (m^2)^2 - (n^2)^2 = (m^2 - n^2)(m^2 + n^2) = (m - n)(m + n)(m^2 + n^2)$

71. $y^4 - 11y^2 + 18 = (y^2 - 2)(y^2 - 9) = (y^2 - 2)(y - 3)(y + 3)$

73. $m^2 + 2mn + n^2 - m - n = (m^2 + 2mn + n^2) - (m + n) = (m + n)^2 - 1(m + n)$
$$= (m + n)(m + n - 1)$$

75. $18a^3 - 8a(x^2 + 8x + 16) = 2a[9a^2 - 4(x^2 + 8x + 16)] = 2a\{(3a)^2 - [2(x + 4)]^2\}$
$$= 2a[3a - 2(x + 4)][3a + 2(x + 4)]$$

77. $x^4 + 2x^2 + 1 - x^2 = (x^4 + 2x^2 + 1) - x^2 = (x^2 + 1)^2 - x^2$
$$= (x^2 + 1 - x)(x^2 + 1 + x) = (x^2 - x + 1)(x^2 + x + 1)$$

79. $u^7 - v^7 = (u - v)(u^6 + u^5v + u^4v^2 + u^3v^3 + u^2v^4 + uv^5 + v^6)$
Verifying: The right side =
$(u - v)u^6 + (u - v)u^5v + (u - v)u^4v^2 + (u - v)u^3v^3 + (u - v)u^2v^4$
$$+ (u - v)uv^5 + (u - v)v^6$$
$= u^7 - u^6v + u^6v - u^5v^2 + u^5v^2 - u^4v^3 + u^4v^3 - u^3v^4 + u^3v^4 - u^2v^5 + u^2v^5 - uv^6$
$$+ uv^6 - v^7$$
$= u^7 - v^7$

81. 2, 3, 5, 7, 11, 13, 17, 19, 23, 29, 31, 37, 41, 43, 47, 53, 59, 61, 67, 71, 73, 79, 83, 89, 97, 101, 103, 107, 109, 113, 127, 131, 137, 139, 149, 151, 157, 163, 167, 173, 179, 181, 191, 193, 197, 199.

83. (A) If $\sqrt{2} = \dfrac{a}{b}$, than $2 = \dfrac{a^2}{b^2}$ and $a^2 = 2b^2$.

 (B) Any factor of a must appear in the factorization of a^2 twice for each time it appears in the factorization of a. This means that it appears an even number of times (possibly 0 times).

 (C) Any factor of b^2 appears an even number of times by the reasoning of (B). Then the number of times 2 appears must be 1 more than this even number, that is, an odd number of times.

 (D) To have two different numbers of appearances of a factor contradicts the uniqueness of a prime factorization.

85. (A) The area of the cardboard can be written as (Original area) - (Removed area), where the original area = 20^2 = 400 and the removed area consists of 4 squares of area x^2 each;

(Original area) - (Removed area) = $400 - 4x^2$ in expanded form.

In factored form, $400 - 4x^2 = 4(100 - x^2) = 4(10 - x)(10 + x)$.

(B) See figure.

The volume of the box = length × width × height = $x(20 - 2x)(20 - 2x)$
$$= 4x(10 - x)(10 - x) \text{ in factored form}$$

In expanded form, $4x(10 - x)(10 - x) = 4x[10^2 - 2(10)x + x^2]$
$$= 4x[100 - 20x + x^2]$$
$$= 400x - 80x^2 + 4x^3$$

Section A-6

1. $\dfrac{42}{105} = \dfrac{21 \cdot 2}{21 \cdot 5} = \dfrac{2}{5}$ **3.** $\dfrac{x + 1}{x^2 + 3x + 2} = \dfrac{\overset{1}{\cancel{x + 1}}}{\underset{1}{\cancel{(x + 1)}} (x + 2)} = \dfrac{1}{x + 2}$

5. $\dfrac{x^2 - 9}{x^2 + 3x - 18} = \dfrac{\overset{1}{\cancel{(x - 3)}} (x + 3)}{\underset{1}{\cancel{(x - 3)}} (x + 6)} = \dfrac{x + 3}{x + 6}$

7. $\dfrac{3x^2 y^3}{x^4 y} = \dfrac{x^2 y \cdot 3y^2}{x^2 y \cdot x^2} = \dfrac{3y^2}{x^2}$

9. $\dfrac{5}{6} - \dfrac{11}{15} = \dfrac{25}{30} - \dfrac{22}{30} = \dfrac{3}{30} = \dfrac{1}{10}$

11. $\left(\dfrac{b^2}{2a} \div \dfrac{b}{a^2}\right) \cdot \dfrac{a}{3b} = \left(\dfrac{b^2}{2a} \cdot \dfrac{a^2}{b}\right) \cdot \dfrac{a}{3b} = \left(\dfrac{\overset{b}{\cancel{b^2}}}{2a} \cdot \dfrac{\overset{a}{\cancel{a^2}}}{\underset{1}{\cancel{b}}}\right) \cdot \dfrac{a}{3b}$

$$= \dfrac{ba}{2} \cdot \dfrac{a}{3b}$$

$$= \dfrac{a^2}{6}$$

13. $\dfrac{1}{n} - \dfrac{1}{m} = \dfrac{m}{mn} - \dfrac{n}{mn} = \dfrac{m - n}{mn}$

15. $\dfrac{x^2 - 1}{x + 2} \div \dfrac{x + 1}{x^2 - 4} = \dfrac{x^2 - 1}{x + 2} \cdot \dfrac{x^2 - 4}{x + 1}$

$$= \dfrac{(x - 1)(x + 1)}{x + 2} \cdot \dfrac{(x - 2)(x + 2)}{x + 1}$$

$$= \dfrac{(x - 1)\overset{1}{\cancel{(x + 1)}}}{\underset{1}{\cancel{x + 2}}} \cdot \dfrac{(x - 2)\overset{1}{\cancel{(x + 2)}}}{\underset{1}{\cancel{x + 1}}}$$

$$= (x - 1)(x - 2)$$

17. $\dfrac{1}{c} + \dfrac{1}{b} + \dfrac{1}{a} = \dfrac{ab}{abc} + \dfrac{ac}{abc} + \dfrac{bc}{abc} = \dfrac{ab + ac + bc}{abc}$

19. $\dfrac{2a - b}{a^2 - b^2} - \dfrac{2a + 3b}{a^2 + 2ab + b^2} = \dfrac{2a - b}{(a + b)(a - b)} - \dfrac{2a + 3b}{(a + b)(a + b)}$

$$= \dfrac{(2a - b)(a + b)}{(a + b)(a + b)(a - b)} - \dfrac{(2a + 3b)(a - b)}{(a + b)(a + b)(a - b)}$$

$$= \dfrac{(2a - b)(a + b) - (2a + 3b)(a - b)}{(a + b)^2 (a - b)}$$

$$= \dfrac{2a^2 + ab - b^2 - (2a^2 + ab - 3b^2)}{(a + b)^2 (a - b)}$$

$$= \dfrac{2a^2 + ab - b^2 - 2a^2 - ab + 3b^2}{(a + b)^2 (a - b)}$$

$$= \dfrac{2b^2}{(a + b)^2 (a - b)}$$

21. $m + 2 - \dfrac{m - 2}{m - 1} = \dfrac{m + 2}{1} - \dfrac{m - 2}{m - 1} = \dfrac{(m + 2)(m - 1)}{m - 1} - \dfrac{(m - 2)}{m - 1}$

$$= \dfrac{(m^2 + m - 2) - (m - 2)}{m - 1} = \dfrac{m^2 + m - 2 - m + 2}{m - 1}$$

$$= \dfrac{m^2}{m - 1}$$

23. $\dfrac{3}{x - 2} - \dfrac{2}{2 - x} = \dfrac{3}{x - 2} - \dfrac{-2}{x - 2} = \dfrac{3 + 2}{x - 2} = \dfrac{5}{x - 2}$

25. $\dfrac{3}{y + 2} + \dfrac{2}{y - 2} - \dfrac{4y}{y^2 - 4} = \dfrac{3(y - 2)}{(y + 2)(y - 2)} + \dfrac{2(y + 2)}{(y + 2)(y - 2)} - \dfrac{4y}{(y + 2)(y - 2)}$

$$= \dfrac{3(y - 2) + 2(y + 2) - 4y}{(y + 2)(y - 2)} = \dfrac{3y - 6 + 2y + 4 - 4y}{(y + 2)(y - 2)}$$

$$= \dfrac{y - 2}{(y + 2)(y - 2)} = \dfrac{\overset{1}{\cancel{(y - 2)}}}{(y + 2)\underset{1}{\cancel{(y - 2)}}} = \dfrac{1}{y + 2}$$

27. $\dfrac{\dfrac{x^2}{y^2} - 1}{\dfrac{x}{y} + 1} = \dfrac{y^2 \left(\dfrac{x^2}{y^2} - 1 \right)}{y^2 \left(\dfrac{x}{y} + 1 \right)} = \dfrac{x^2 - y^2}{xy + y^2} = \dfrac{(x - y)\overset{1}{\cancel{(x + y)}}}{y\underset{1}{\cancel{(x + y)}}} = \dfrac{x - y}{y}$

29.
$$\frac{6x^3(x^2+2)^2 - 2x(x^2+2)^3}{x^4} = \frac{2x(x^2+2)^2[3x^2 - (x^2+2)]}{x^4}$$

$$= \frac{2x(x^2+2)^2[3x^2 - x^2 - 2]}{x^4}$$

$$= \frac{2\overset{1}{\cancel{x}}(x^2+2)^2(2x^2-2)}{\underset{x^3}{\cancel{x^4}}}$$

$$= \frac{2(x^2+2)^2(2x^2-2)}{x^3}$$

$$= \frac{2(x^2+2)^2\,2(x^2-1)}{x^3}$$

$$= \frac{4(x^2+2)^2(x+1)(x-1)}{x^3}$$

31.
$$\frac{2x(1-3x)^3 + 9x^2(1-3x)^2}{(1-3x)^6} = \frac{x(1-3x)^2[2(1-3x) + 9x]}{(1-3x)^6}$$

$$= \frac{x(1-3x)^2[2 - 6x + 9x]}{(1-3x)^6}$$

$$= \frac{x\cancel{(1-3x)}^2(2+3x)}{\underset{(1-3x)^4}{\cancel{(1-3x)^6}}}$$

$$= \frac{x(2+3x)}{(1-3x)^4}$$

33.
$$\frac{-2x(x+4)^3 - 3(3-x^2)(x+4)^2}{(x+4)^6} = \frac{(x+4)^2[-2x(x+4) - 3(3-x^2)]}{(x+4)^6}$$

$$= \frac{(x+4)^2[-2x^2 - 8x - 9 + 3x^2]}{(x+4)^6}$$

$$= \frac{\cancel{(x+4)}^2(x^2 - 8x - 9)}{\underset{(x+4)^4}{\cancel{(x+4)^6}}}$$

$$= \frac{x^2 - 8x - 9}{(x+4)^4}$$

$$= \frac{(x+1)(x-9)}{(x+4)^4}$$

35.
$$\frac{y}{y^2 - 2y - 8} - \frac{2}{y^2 - 5y + 4} + \frac{1}{y^2 + y - 2}$$

$$= \frac{y}{(y-4)(y+2)} - \frac{2}{(y-4)(y-1)} + \frac{1}{(y+2)(y-1)}$$

$$= \frac{y(y-1)}{(y-1)(y-4)(y+2)} - \frac{2(y+2)}{(y-4)(y-1)(y+2)} + \frac{1(y-4)}{(y-4)(y-1)(y+2)}$$

$$= \frac{y^2 - y - 2y - 4 + y - 4}{(y - 4)(y - 1)(y + 2)} \quad = \frac{y^2 - 2y - 8}{(y - 4)(y - 1)(y + 2)} \quad = \frac{\overset{1}{\cancel{(y - 4)}}\,\overset{1}{\cancel{(y + 2)}}}{\underset{1}{\cancel{(y - 4)}}(y - 1)\underset{1}{\cancel{(y + 2)}}}$$

$$= \frac{1}{y - 1}$$

37. $\dfrac{16 - m^2}{m^2 + 3m - 4} \cdot \dfrac{m - 1}{m - 4} = \dfrac{\overset{-1}{\cancel{(4 - m)}}\,\overset{1}{\cancel{(4 + m)}}}{\underset{1}{\cancel{(m + 4)}}\,\underset{1}{\cancel{(m - 1)}}} \cdot \dfrac{\overset{1}{\cancel{m - 1}}}{\underset{1}{\cancel{m - 4}}} = \dfrac{-1}{1} = -1$

39. $\dfrac{x + 7}{ax - bx} + \dfrac{y + 9}{by - ay} = \dfrac{x + 7}{x(a - b)} + \dfrac{y + 9}{y(b - a)} = \dfrac{y(x + 7)}{xy(a - b)} + \dfrac{-x(y + 9)}{xy(a - b)}$

$$= \frac{xy + 7y - xy - 9x}{xy(a - b)} = \frac{7y - 9x}{xy(a - b)}$$

41. $\dfrac{x^2 - 16}{2x^2 + 10x + 8} \div \dfrac{x^2 - 13x + 36}{x^3 + 1} = \dfrac{x^2 - 16}{2x^2 + 10x + 8} \cdot \dfrac{x^3 + 1}{x^2 - 13x + 36}$

$$= \frac{\overset{1}{\cancel{(x - 4)}}\,\overset{1}{\cancel{(x + 4)}}}{2\underset{1}{\cancel{(x + 1)}}\,\underset{1}{\cancel{(x + 4)}}} \cdot \frac{\overset{1}{\cancel{(x + 1)}}(x^2 - x + 1)}{\underset{1}{\cancel{(x - 4)}}(x - 9)} = \frac{x^2 - x + 1}{2(x - 9)}$$

43. $\dfrac{x^2 - xy}{xy + y^2} \div \left(\dfrac{x^2 - y^2}{x^2 + 2xy + y^2} \div \dfrac{x^2 - 2xy + y^2}{x^2y + xy^2} \right)$

$$= \frac{x^2 - xy}{xy + y^2} \div \left(\frac{x^2 - y^2}{x^2 + 2xy + y^2} \cdot \frac{x^2y + xy^2}{x^2 - 2xy + y^2} \right)$$

$$= \frac{x^2 - xy}{xy + y^2} \div \left(\frac{\overset{1}{\cancel{(x - y)}}\,\overset{1}{\cancel{(x + y)}}}{\underset{1}{\cancel{(x + y)}}\,(x + y)} \cdot \frac{xy\,\overset{1}{\cancel{(x + y)}}}{\underset{1}{\cancel{(x - y)}}\,(x - y)} \right)$$

$$= \frac{x^2 - xy}{xy + y^2} \div \frac{xy}{x - y} = \frac{x^2 - xy}{xy + y^2} \cdot \frac{x - y}{xy}$$

$$= \frac{\overset{1}{\cancel{x}}(x - y)}{y(x + y)} \cdot \frac{x - y}{\underset{1}{\cancel{xy}}} = \frac{(x - y)^2}{y^2(x + y)}$$

45. $\left(\dfrac{x}{x^2 - 16} - \dfrac{1}{x + 4} \right) \div \dfrac{4}{x + 4} = \left(\dfrac{x}{(x - 4)(x + 4)} - \dfrac{1}{x + 4} \right) \div \dfrac{4}{x + 4}$

$$= \left(\frac{x}{(x - 4)(x + 4)} - \frac{(x - 4)}{(x - 4)(x + 4)} \right) \div \frac{4}{x + 4}$$

$$= \frac{x - x + 4}{(x - 4)(x + 4)} \div \frac{4}{x + 4} = \frac{4}{(x - 4)(x + 4)} \div \frac{4}{x + 4}$$

$$= \frac{\overset{1}{\cancel{4}}}{(x - 4)\underset{1}{\cancel{(x + 4)}}} \cdot \frac{\overset{1}{\cancel{x + 4}}}{\underset{1}{\cancel{4}}} = \frac{1}{x - 4}$$

47. $\dfrac{1 + \frac{2}{x} - \frac{15}{x^2}}{1 + \frac{4}{x} - \frac{5}{x^2}} = \dfrac{x^2\left(1 + \frac{2}{x} - \frac{15}{x^2}\right)}{x^2\left(1 + \frac{4}{x} - \frac{5}{x^2}\right)} = \dfrac{x^2 + 2x - 15}{x^2 + 4x - 5} = \dfrac{\overset{1}{\cancel{(x+5)}}(x-3)}{\underset{1}{\cancel{(x+5)}}(x-1)} = \dfrac{x-3}{x-1}$

49. $\dfrac{\frac{1}{x+h} - \frac{1}{x}}{h} = \dfrac{\frac{x}{x(x+h)} - \frac{x+h}{x(x+h)}}{h} = \dfrac{\frac{x-x-h}{x(x+h)}}{h} = \dfrac{\frac{-h}{x(x+h)}}{h} = \dfrac{-h}{x(x+h)} \div h$

$ = \dfrac{-h}{x(x+h)} \div \dfrac{h}{1} = \dfrac{-\overset{1}{\cancel{h}}}{x(x+h)} \cdot \dfrac{1}{\underset{1}{\cancel{h}}} = \dfrac{-1}{x(x+h)}$

51. $\dfrac{\frac{(x+h)^2}{x+h+2} - \frac{x^2}{x+2}}{h} = \dfrac{\frac{(x+h)^2(x+2)}{(x+h+2)(x+2)} - \frac{x^2(x+h+2)}{(x+h+2)(x+2)}}{h} = \dfrac{\frac{(x+h)^2(x+2) - x^2(x+h+2)}{(x+h+2)(x+2)}}{h}$

$ = \dfrac{\frac{(x^2 + 2xh + h^2)(x+2) - x^3 - x^2h - 2x^2}{(x+h+2)(x+2)}}{h}$

$ = \dfrac{\frac{x^3 + 2x^2h + h^2x + 2x^2 + 4xh + 2h^2 - x^3 - x^2h - 2x^2}{(x+h+2)(x+2)}}{h}$

$ = \dfrac{\frac{x^2h + h^2x + 4xh + 2h^2}{(x+h+2)(x+2)}}{h} = \dfrac{x^2h + h^2x + 4xh + 2h^2}{(x+h+2)(x+2)} \div h$

$ = \dfrac{x^2h + h^2x + 4xh + 2h^2}{(x+h+2)(x+2)} \div \dfrac{h}{1} = \dfrac{\overset{1}{\cancel{h}}(x^2 + hx + 4x + 2h)}{(x+h+2)(x+2)} \cdot \dfrac{1}{\underset{1}{\cancel{h}}}$

$ = \dfrac{x^2 + hx + 4x + 2h}{(x+h+2)(x+2)}$

53. **(A)** The solution is incorrect, because in the first step the quantity 4 has been removed from numerator and denominator. But 4 is a term, and only factors can be cancelled. A correct solution would factor numerator and denominator, then cancel common factors, if any.

(B) $\dfrac{x^2 + 5x + 4}{x + 4} = \dfrac{(x+1)\overset{1}{\cancel{(x+4)}}}{\underset{1}{\cancel{x+4}}} = x + 1$

55. **(A)** The solution is incorrect, because in the first step the quantity h has been removed from numerator and denominator. Although h is a factor of the denominator it is not a factor of the numerator as shown and only common factors of numerator and denominator can be cancelled. A correct solution would factor numerator and denominator, revealing that h is actually a factor of the numerator in factored form and allowing a correct cancellation of h.

(B) $\dfrac{(x+h)^2 - x^2}{h} = \dfrac{[(x+h) - x][(x+h) + x]}{h}$ using the difference of two squares

$ = \dfrac{[x + h - x][x + h + x]}{h}$

$ = \dfrac{\overset{1}{\cancel{h}}[2x + h]}{\underset{1}{\cancel{h}}} = 2x + h$

57. (A) The solution is incorrect. In adding a quantity to a fractional expression, the quantity cannot simply be placed in the numerator. A correct solution would rewrite $x - 2$ as a fractional expression, first with denominator 1, and then with denominator equal to the LCD of all denominators in the addition.

(B)
$$\frac{x^2 - 2x}{x^2 - x - 2} + x - 2 = \frac{\overset{1}{x(x - 2)}}{(x + 1)\,\overset{1}{(x - 2)}} + \frac{x - 2}{1}$$

$$= \frac{x}{x + 1} + \frac{x - 2}{1}$$

$$= \frac{x}{x + 1} + \frac{(x - 2)(x + 1)}{x + 1}$$

$$= \frac{x + x^2 - x - 2}{x + 1}$$

$$= \frac{x^2 - 2}{x - 1}$$

59. (A)(B) The solution is correct.

61.
$$\frac{y - \dfrac{y^2}{y - x}}{1 + \dfrac{x^2}{y^2 - x^2}} = \frac{(y^2 - x^2)\left[y - \dfrac{y^2}{y - x}\right]}{(y^2 - x^2)\left[1 + \dfrac{x^2}{y^2 - x^2}\right]} = \frac{(y^2 - x^2)y - \overset{(y+x)}{(y^2 - x^2)}\dfrac{y^2}{\overset{y - x}{1}}}{y^2 - x^2 + \overset{1}{(y^2 - x^2)}\dfrac{x^2}{\overset{y^2 - x^2}{1}}}$$

$$= \frac{y^3 - x^2 y - y^3 - xy^2}{y^2 - x^2 + x^2} = \frac{-x^2 y - xy^2}{y^2} = \frac{-x\overset{1}{y}(x + y)}{\underset{y}{y^2}} = \frac{-x(x + y)}{y}$$

63.
$$2 - \frac{1}{1 - \dfrac{2}{a + 2}} = 2 - \frac{(a + 2)\cdot 1}{(a + 2)\left[1 - \dfrac{2}{a + 2}\right]} = 2 - \frac{a + 2}{a + 2 - \overset{1}{(a + 2)}\dfrac{2}{\underset{1}{a+2}}} = 2 - \frac{a + 2}{a + 2 - 2}$$

$$= 2 - \frac{a + 2}{a} = \frac{2}{1} - \frac{a + 2}{a} = \frac{2a}{a} - \frac{(a + 2)}{a} = \frac{2a - a - 2}{a} = \frac{a - 2}{a}$$

65. (A) The multiplicative inverse of x is the unique real number $\frac{1}{x}$ such that $x\left(\frac{1}{x}\right) = 1$. Since $\frac{c}{d} \cdot \frac{d}{c} = \frac{cd}{dc} = \frac{cd}{cd} = 1$, $\frac{d}{c}$ is the multiplicative inverse of $\frac{c}{d}$.

(B) By Definition 1 of Section A-1 of the appendix
$a \div b = a\left(\frac{1}{b}\right) = a \cdot$ (multiplicative inverse of b).

$\quad \frac{a}{b} \div \frac{c}{d} = \frac{a}{b} \cdot$ (multiplicative inverse of $\frac{c}{d}$)

$\qquad\quad = \frac{a}{b} \cdot \frac{d}{c}$ by part A.

APPENDIX A REVIEW

1. (A) Since 3 is an element of {1, 2, 3, 4, 5}, the statement is true. T.

(B) Since 5 is not an element of {4, 1, 2}, the statement is true. T.

(C) Since B is not an element of {1, 2, 3, 4, 5}, the statement is false. F.

(D) Since each element of the set {1, 2, 4} is also an element of the set {1, 2, 3, 4, 5}, the statement is true. T.

(D) Since sets B and C have exactly the same elements, the sets are equal. The statement is false. F.

(E) The statement is false, since not every element of the set {1, 2, 3, 4, 5} is an element of the set {1, 2, 4}. For example, 3 is an element of the first, but not the second. F.

(A-1)

2. (A) The commutative property ($\cdot$) states that, in general, $xy = yx$.
Comparing this with
$x(y + z) = ?$
we see that $x(y + z) = (y + z)x$
$(y + z)x$

(B) The associative property (+) states that, in general,
$(x + y) + z = x + (y + z)$
Comparing this with
$? = 2 + (x + y)$
we see that
$(2 + x) + y = 2 + (x + y)$
$(2 + x) + y$

(C) The distributive property states that, in general,
$(x + y)z = xz + yz$
Comparing this with
$(2 + 3)x = ?$
we see that
$(2 + 3)x = 2x + 3x$
$2x + 3x$

(A-1)

3. $\left(\dfrac{2}{3} \cdot \dfrac{15}{14}\right)^{-1} = \left(\dfrac{5}{7}\right)^{-1} = \dfrac{7}{5}$ *(A-1)*

4. $\dfrac{9}{10} + \dfrac{5}{12} = \dfrac{54}{60} + \dfrac{25}{60} = \dfrac{79}{60}$ *(A-1)*

5. $-\left(\dfrac{8}{7} + 2^{-1}\right) = -\left(\dfrac{8}{7} + \dfrac{1}{2}\right) = -\left(\dfrac{16}{14} + \dfrac{7}{14}\right) = -\left(\dfrac{23}{14}\right) = -\dfrac{23}{14}$ *(A-1)*

6. $(4^{-1}9^{-1})^{-1} = \left(\dfrac{1}{4} \cdot \dfrac{1}{9}\right)^{-1} = \left(\dfrac{1}{36}\right)^{-1} = 36$ *(A-1)*

7. 3   *(A-4)*

8. 5   *(A-4)*

9. $(x^3 + 2) + (x^5 - x^3 + 1) = x^3 + 2 + x^5 - x^3 + 1 = x^5 + 3$. This polynomial has degree 5.   *(A-4)*

10. $(x^3 + 2)(x^5 - x^3 + 1) = x^8 + 2x^5 - x^6 - 2x^3 + x^3 + 2 = x^8 - x^6 + 2x^5 - x^3 + 2$. This polynomial has degree 8.   *(A-4)*

11. See problem 10. $x^8 - x^6 + 2x^5 - x^3 + 2$.   *(A-4)*

12. See problem 9. $x^5 + 3$ *(A-4)*

13. 17   *(A-2)*   14. 6   *(A-2)*   15. $8^{-2/3} = \dfrac{1}{8^{2/3}} = \dfrac{1}{(8^{1/3})^2} = \dfrac{1}{2^2} = \dfrac{1}{4}$ *(A-2)*

16. $(-64)^{5/3} = [(-64)^{1/3}]^5 = (-4)^5 = -1,024$ *(A-2)*

17. $\left(\dfrac{9}{16}\right)^{-1/2} = 1 \div \left(\dfrac{9}{16}\right)^{1/2} = 1 \div \dfrac{3}{4} = 1 \cdot \dfrac{4}{3} = \dfrac{4}{3}$ *(A-2)*

18. $(121^{1/2} + 25^{1/2})^{-3/4} = (11 + 5)^{-3/4} = 16^{-3/4} = \dfrac{1}{16^{3/4}} = \dfrac{1}{(16^{1/4})^3} = \dfrac{1}{2^3} = \dfrac{1}{8}$ *(A-2)*

19.
$$\begin{aligned}
5x^2 - 3x[4 - 3(x - 2)] &= 5x^2 - 3x[4 - 3x + 6] \\
&= 5x^2 - 3x[10 - 3x] \\
&= 5x^2 - 30x + 9x^2 \\
&= 14x^2 - 30x
\end{aligned}$$ *(A-4)*

20. $9m^2 - 25n^2$ *(A-4)*

21. $(2x + y)(3x - 4y) = 6x^2 - 8xy + 3xy - 4y^2 = 6x^2 - 5xy - 4y^2$ *(A-4)*

22. $(2a - 3b)^2 = (2a)^2 - 2(2a)(3b) + (3b)^2 = 4a^2 - 12ab + 9b^2$ *(A-4)*

23. $(3x - 2)^2$ *(A-5)*

24. Prime   *(A-5)*

25. $6n^3 - 9n^2 - 15n = 3n(2n^2 - 3n - 5) = 3n(2n - 5)(n + 1)$ *(A-5)*

26. $\dfrac{2}{5b} - \dfrac{4}{3a^3} - \dfrac{1}{6a^2b^2} = \dfrac{12a^3b}{30a^3b^2} - \dfrac{40b^2}{30a^3b^2} - \dfrac{5a}{30a^3b^2} = \dfrac{12a^3b - 40b^2 - 5a}{30a^3b^2}$ *(A-6)*

27. $\dfrac{3x}{3x^2 - 12x} + \dfrac{1}{6x} = \dfrac{3x}{3x(x - 4)} + \dfrac{1}{6x} = \dfrac{2(3x)}{6x(x - 4)} + \dfrac{1(x - 4)}{6x(x - 4)} = \dfrac{6x + x - 4}{6x(x - 4)} = \dfrac{7x - 4}{6x(x - 4)}$ *(A-6)*

28. $\dfrac{y - 2}{y^2 - 4y + 4} \div \dfrac{y^2 + 27}{y^2 + 4y + 4} = \dfrac{y - 2}{y^2 - 4y + 4} \cdot \dfrac{y^2 + 4y + 4}{y^2 + 2y}$

$$= \dfrac{\overset{1}{\cancel{(y - 2)}}}{\underset{1}{\cancel{(y - 2)}}(y - 2)} \cdot \dfrac{\overset{1}{\cancel{(y + 2)}}(y + 2)}{y\underset{1}{\cancel{(y + 2)}}}$$

$$= \dfrac{y + 2}{y(y - 2)}$$ *(A-6)*

29. $\dfrac{u - \frac{1}{u}}{1 - \frac{1}{u^2}} = \dfrac{u^2\left(u - \frac{1}{u}\right)}{u^2\left(1 - \frac{1}{u^2}\right)} = \dfrac{u^3 - u}{u^2 - 1} = \dfrac{u\overset{1}{\cancel{(u^2 - 1)}}}{\underset{1}{\cancel{u^2 - 1}}} = u$ *(A-6)*

30. $6(xy^3)^5 = 6x^5(y^3)^5 = 6x^5y^{15}$ *(A-2)*

31. $\dfrac{9u^8v^6}{3u^4v^8} = \dfrac{3u^4}{v^2}$ *(A-2)*

32. $(2 \times 10^5)(3 \times 10^{-3}) = (2 \cdot 3) \times 10^{5+(-3)} = 6 \times 10^2$ *(A-2)*

33. $(x^{-3}y^2)^{-2} = (x^{-3})^{-2}(y^2)^{-2} = x^6y^{-4} = \dfrac{x^6}{y^4}$ *(A-2)*

34. $u^{5/3}u^{2/3} = u^{5/3+2/3} = u^{7/3}$ *(A-2)*

35. $(9a^4b^{-2})^{1/2} = 9^{1/2}(a^4)^{1/2}(b^{-2})^{1/2} = 3a^2b^{-1} = \dfrac{3a^2}{b}$ *(A-3)*

36. $3\sqrt[5]{x^2}$ *(A-3)*

37. $-3(xy)^{2/3}$ *(A-3)*

38. $3x\sqrt[3]{x^5y^4} = 3x\sqrt[3]{x^3y^3 \cdot x^2y} = 3x\sqrt[3]{x^3y^3}\sqrt[3]{x^2y} = 3x(xy)\sqrt[3]{x^2y} = 3x^2y\sqrt[3]{x^2y}$ *(A-3)*

39. $\sqrt{2x^2y^5}\ \sqrt{18x^3y^2} = \sqrt{(2x^2y^5)(18x^3y^2)} = \sqrt{36x^2y^7} = \sqrt{36x^4y^6 \cdot xy} = \sqrt{36x^4y^6}\ \sqrt{xy}$

$\qquad = 6x^2y^3\sqrt{xy}$ $\qquad\qquad$ *(A-3)*

40. $\dfrac{6ab}{\sqrt{3a}} = \dfrac{6ab}{\sqrt{3a}}\ \dfrac{\sqrt{3a}}{\sqrt{3a}} = \dfrac{\overset{2}{\cancel{6ab}}\sqrt{3a}}{\underset{1}{\cancel{3a}}} = 2b\sqrt{3a}$ $\qquad\qquad$ *(A-3)*

41. $\dfrac{\sqrt{5}}{3-\sqrt{5}} = \dfrac{\sqrt{5}}{(3-\sqrt{5})}\dfrac{(3+\sqrt{5})}{(3+\sqrt{5})} = \dfrac{\sqrt{5}(3+\sqrt{5})}{3^2-(\sqrt{5})^2} = \dfrac{\sqrt{5}(3+\sqrt{5})}{9-5} = \dfrac{\sqrt{5}(3+\sqrt{5})}{4} = \dfrac{3\sqrt{5}+5}{4}$ $\quad$ *(A-3)*

42. $\sqrt[8]{y^6} = \sqrt[2\cdot4]{y^{2\cdot3}} = \sqrt[4]{y^3}$ $\qquad\qquad$ *(A-3)*

43. The odd integers between -4 and 2 are -3, -1, and 1. The set is written
{-3, -1, 1} $\qquad\qquad$ *(A-1)*

44. Subtraction $\quad$ *(A-1)* $\qquad$ **45.** Commutative (+) $\quad$ *(A-1)* $\qquad$ **46.** Distributive $\quad$ *(A-1)*

47. Associative ($\cdot$) $\quad$ *(A-1)* $\qquad$ **48.** Negatives $\quad$ *(A-1)* $\qquad$ **49.** Identity (+) $\quad$ *(A-1)*

50. (A) T $\quad$ (B) F $\quad$ *(A-1)* $\qquad$ **51.** 0 and -3 are two examples of infinitely many. *(A-1)*

52. (A) a and d $\quad$ (B) None $\quad$ *(A-4)*

53. $(2x - y)(2x + y) - (2x - y)^2 = 4x^2 - y^2 - (4x^2 - 4xy + y^2)$
$\qquad\qquad\qquad\qquad\qquad\ = 4x^2 - y^2 - 4x^2 + 4xy - y^2$
$\qquad\qquad\qquad\qquad\qquad\ = 4xy - 2y^2$ $\qquad\qquad$ *(A-4)*

54. This can be simplified directly using the distributive property or:
$(m^2 + 2mn - n^2)(m^2 - 2mn - n^2) = [(m^2 - n^2) + 2mn][(m^2 - n^2) - 2mn]$
$\qquad\qquad\qquad\qquad\qquad\qquad\quad = (m^2 - n^2)^2 - (2mn)^2$
$\qquad\qquad\qquad\qquad\qquad\qquad\quad = m^4 - 2m^2n^2 + n^4 - 4m^2n^2$
$\qquad\qquad\qquad\qquad\qquad\qquad\quad = m^4 - 6m^2n^2 + n^4$ $\qquad\qquad$ *(A-4)*

55. $5(x + h)^2 - 7(x + h) - (5x^2 - 7x) = 5(x^2 + 2hx + h^2) - 7(x + h) - (5x^2 - 7x)$
$\qquad\qquad\qquad\qquad\qquad\qquad\quad = 5x^2 + 10hx + 5h^2 - 7x - 7h - 5x^2 + 7x$
$\qquad\qquad\qquad\qquad\qquad\qquad\quad = 10hx + 5h^2 - 7h$ $\qquad\qquad$ *(A-4)*

56. $-2x\{(x^2 + 2)(x - 3) - x[x - x(3 - x)]\}$
$\qquad\quad = -2x\{(x^2 + 2)(x - 3) - x[x - 3x + x^2]\}$
$\qquad\quad = -2x\{(x^2 + 2)(x - 3) - x[-2x + x^2]\}$
$\qquad\quad = -2x\{x^3 - 3x^2 + 2x - 6 + 2x^2 - x^3\}$
$\qquad\quad = -2x\{-x^2 + 2x - 6\}$
$\qquad\quad = 2x^3 - 4x^2 + 12x$ $\qquad\qquad$ *(A-4)*

57. $(x - 2y)^3 = (x - 2y)^2(x - 2y)$
$\qquad\qquad\ = (x^2 - 4xy + 4y^2)(x - 2y)$
$\qquad\qquad\ = x^3 - 2x^2y - 4x^2y + 8xy^2 + 4xy^2 - 8y^3$
$\qquad\qquad\ = x^3 - 6x^2y + 12xy^2 - 8y^3$ $\qquad\qquad$ *(A-4)*

58. $(4x - y)^2 - 9x^2 = (4x - y)^2 - (3x)^2$
$= [(4x - y) - 3x][(4x - y) + 3x]$
$= (x - y)(7x - y)$ *(A-5)*

59. Prime   *(A-5)*   **60.** $3xy(2x^2 + 4xy - 5y^2)$ *(A-5)*

61. $(y - b)^2 - y + b = (y - b)(y - b) - 1(y - b) = (y - b)(y - b - 1)$ *(A-5)*

62. $3x^3 + 24y^3 = 3(x^3 + 8y^3) = 3[x^3 + (2y)^3] = 3(x + 2y)[x^2 - x(2y) + (2y)^2]$
$= 3(x + 2y)(x^2 - 2xy + 4y^2)$ *(A-5)*

63. $y^3 + 2y^2 - 4y - 8 = y^2(y + 2) - 4(y + 2)$
$= (y + 2)(y^2 - 4) = (y + 2)(y + 2)(y - 2)$
$= (y - 2)(y + 2)^2$ *(A-5)*

64. $2x(x - 4)^3 + 3x^2(x - 4)^2 = x(x - 4)^2[2(x - 4) + 3x]$
$= x(x - 4)^2[2x - 8 + 3x]$
$= x(x - 4)^2(5x - 8)$ *(A-5)*

65. $\dfrac{3x^2(x + 2)^2 - 2x(x + 2)^3}{x^4} = \dfrac{x(x + 2)^2[3x - 2(x + 2)]}{x^4}$

$= \dfrac{\overset{1}{\cancel{x}}(x + 2)^2[3x - 2x - 4]}{\underset{x^3}{\cancel{x^4}}}$

$= \dfrac{(x + 2)^2(x - 4)}{x^3}$ *(A-6)*

66. $\dfrac{m - 1}{m^2 - 4m + 4} + \dfrac{m + 3}{m^2 - 4} + \dfrac{2}{2 - m}$

$= \dfrac{(m - 1)}{(m - 2)(m - 2)} + \dfrac{m + 3}{(m - 2)(m + 2)} + \dfrac{-2}{m - 2}$

$= \dfrac{(m - 1)(m + 2)}{(m - 2)(m - 2)(m + 2)} + \dfrac{(m + 3)(m - 2)}{(m - 2)(m - 2)(m + 2)}$
$+ \dfrac{-2(m - 2)(m + 2)}{(m - 2)(m - 2)(m + 2)}$

$= \dfrac{(m - 1)(m + 2) + (m + 3)(m - 2) - 2(m - 2)(m + 2)}{(m - 2)^2(m + 2)}$

$= \dfrac{m^2 + m - 2 + m^2 + m - 6 - 2(m^2 - 4)}{(m - 2)^2(m + 2)}$

$= \dfrac{2m^2 + 2m - 8 - 2m^2 + 8}{(m - 2)^2(m + 2)} = \dfrac{2m}{(m - 2)^2(m + 2)}$ *(A-6)*

67. $\dfrac{y}{x^2} \div \left(\dfrac{x^2 + 3x}{2x^2 + 5x - 3} \div \dfrac{x^3 y - x^2 y}{2x^2 - 3x + 1} \right) = \dfrac{y}{x^2} \div \left(\dfrac{x^2 + 3x}{2x^2 + 5x - 3} \cdot \dfrac{2x^2 - 3x + 1}{x^3 y - x^2 y} \right)$

$$= \dfrac{y}{x^2} \div \left(\dfrac{\overset{1}{\cancel{x}}(\overset{1}{\cancel{x + 3}})}{(2x - 1)(x + 3)} \cdot \dfrac{(\overset{1}{\cancel{2x - 1}})(\overset{1}{\cancel{x - 1}})}{\underset{x}{\cancel{x}^2} y(\underset{1}{\cancel{x - 1}})} \right)$$

$$= \dfrac{y}{x^2} \div \dfrac{1}{xy}$$

$$= \dfrac{y}{\underset{x}{\cancel{x}^2}} \cdot \dfrac{\overset{1}{\cancel{x}} y}{1}$$

$$= \dfrac{y^2}{x} \qquad\qquad (A\text{-}6)$$

68. $\dfrac{1 - \dfrac{1}{1+\frac{x}{y}}}{1 - \dfrac{1}{1-\frac{x}{y}}} = \dfrac{1 - \dfrac{y(1)}{y\left(1+\frac{x}{y}\right)}}{1 - \dfrac{y(1)}{y\left(1-\frac{x}{y}\right)}} = \dfrac{1 - \dfrac{y}{y+x}}{1 - \dfrac{y}{y-x}} = \dfrac{\dfrac{y+x}{y+x} - \dfrac{y}{y+x}}{\dfrac{y-x}{y-x} - \dfrac{y}{y-x}}$

$$= \dfrac{\dfrac{x}{y+x}}{\dfrac{-x}{y-x}} = \dfrac{x}{y+x} \div \dfrac{-x}{y-x} = \dfrac{\overset{-1}{\cancel{x}}}{y+x} \cdot \dfrac{y-x}{\underset{1}{-\cancel{x}}} = \dfrac{-1(y-x)}{y+x} = \dfrac{x-y}{x+y} \qquad (A\text{-}6)$$

69. $\dfrac{a^{-1} - b^{-1}}{ab^{-2} - ba^{-2}} = \dfrac{\dfrac{1}{a} - \dfrac{1}{b}}{\dfrac{a}{b^2} - \dfrac{b}{a^2}} = \dfrac{a^2 b^2 \left(\dfrac{1}{a} - \dfrac{1}{b} \right)}{a^2 b^2 \left(\dfrac{a}{b^2} - \dfrac{b}{a^2} \right)}$

$$= \dfrac{ab^2 - a^2 b}{a^3 - b^3} = \dfrac{ab(\overset{-1}{\cancel{b - a}})}{\underset{1}{\cancel{(a - b)}}(a^2 + ab + b^2)} = \dfrac{-ab}{a^2 + ab + b^2} \qquad (A\text{-}6, \ A\text{-}2)$$

70. The solution is incorrect. In adding a quantity to a fractional expression, the quantity cannot simply be placed in the numerator. A correct solution would rewrite $x + 2$ as a fractional expression, first with denominator 1, and then with denominator equal to the LCD of all denominators in the expression.

$$\dfrac{x^2 + 2x}{x^2 + x - 2} + x + 2 = \dfrac{x(x + 2)}{(x - 1)(x + 2)} + \dfrac{x + 2}{1}$$

$$= \dfrac{x}{x - 1} + \dfrac{x + 2}{1}$$

$$= \dfrac{x}{x - 1} + \dfrac{(x + 2)(x - 1)}{x - 1}$$

$$= \dfrac{x + x^2 + x - 2}{x - 1}$$

$$= \dfrac{x^2 + 2x - 2}{x - 1} \qquad\qquad (A\text{-}6)$$

71. $\left(\dfrac{8u^{-1}}{2^2 u^2 v^0} \right)^{-2} \left(\dfrac{u^{-5}}{u^{-3}} \right)^3 = \left(\dfrac{2^3 u^{-1}}{2^2 u^2} \right)^{-2} \left(\dfrac{u^{-5}}{u^{-3}} \right)^3 = (2u^{-3})^{-2} \left(\dfrac{u^{-5}}{u^{-3}} \right)^3 = 2^{-2} u^6 \dfrac{u^{-15}}{u^{-9}}$

$$= \dfrac{1}{2^2} u^6 u^{-15-(-9)} = \dfrac{1}{2^2} u^6 u^{-6} = \dfrac{1}{4} \qquad\qquad (A\text{-}2)$$

72. $\dfrac{5^0}{3^2} + \dfrac{3^{-2}}{2^{-2}} = \dfrac{1}{3^2} + \dfrac{2^2}{3^2} = \dfrac{1}{9} + \dfrac{4}{9} = \dfrac{5}{9}$ (A-2)

73. $\left(\dfrac{27x^2y^{-3}}{8x^{-4}y^3}\right)^{1/3} = \left(\dfrac{3^3 x^{2-(-4)} y^{-3-3}}{2^3}\right)^{1/3} = \left(\dfrac{3^3 x^6 y^{-6}}{2^3}\right)^{1/3} = \dfrac{3x^2 y^{-2}}{2} = \dfrac{3x^2}{2y^2}$ (A-2)

74. $(a^{-1/3}b^{1/4})(9a^{1/3}b^{-1/2})^{3/2} = a^{-1/3}b^{1/4}9^{3/2}a^{1/2}b^{-3/4}$

$$= a^{-1/3+1/2}b^{1/4-3/4}9^{3/2} = 27a^{1/6}b^{-1/2}$$

$$= \dfrac{27a^{1/6}}{b^{1/2}}$$

                                                        (A-2)

75. $(x^{1/2} + y^{1/2})^2 = (x^{1/2})^2 + 2(x^{1/2})(y^{1/2}) + (y^{1/2})^2$

$$= x + 2x^{1/2}y^{1/2} + y$$

> **Common Error:**
> $(x^{1/2} + y^{1/2})^2 \neq x + y$.
> (Exponents do not distribute over addition.)

                           (A-2)

76. $(3x^{1/2} - y^{1/2})(2x^{1/2} + 3y^{1/2}) = (3x^{1/2})(2x^{1/2}) + (3x^{1/2})(3y^{1/2})$

$$- (2x^{1/2})(y^{1/2}) - (y^{1/2})(3y^{1/2})$$

$$= 6x + 9x^{1/2}y^{1/2} - 2x^{1/2}y^{1/2} - 3y$$

$$= 6x + 7x^{1/2}y^{1/2} - 3y$$

                       (A-2)

77. $\dfrac{0.000\,000\,000\,52}{(1,\,300)(0.000\,002)} = \dfrac{5.2 \times 10^{-10}}{(1.3 \times 10^3)(2 \times 10^{-6})}$

$$= \dfrac{\overset{2}{\cancel{5.2}} \times 10^{-10}}{\underset{1}{\cancel{2.6}} \times 10^{-3}} = 2 \times 10^{-10-(3)}$$

$$= 2 \times 10^{-7}$$

                       (A-2)

78. $\dfrac{(20,\,410)(0.000\,003\,477)}{0.000\,000\,022\,09} = \dfrac{(2.041 \times 10^4)(3.477 \times 10^{-6})}{(2.209 \times 10^{-8})} = 3.213 \times 10^6$

Note that the input is given to 4 significant digits, so we round the calculator answer to 4 significant digits.   (A-2)

79. 4.434×10^{-5} (A-2)

80. -4.541×10^{-6} (A-2)

81. $82.45^{8/3} = (82.45)^{(8 \div 3)} = 128,800$ (A-2)

82. $(0.000\,000\,419\,9)^{2/7} = (4.199 \times 10^{-7})^{(2 \div 7)} = 0.01507$ (A-2)

83. $\sqrt[5]{0.006\,604} = (6.604 \times 10^{-3})^{1/5} = (6.604 \times 10^{-3})^{0.2} = 0.3664$ (A-3)

84. $\sqrt[3]{3 + \sqrt{2}} = (3 + \sqrt{2})^{(1 \div 3)} = 1.640$ (A-3)

85. $\dfrac{2^{-1/2} - 3^{-1/2}}{2^{-1/3} + 3^{-1/3}} = [2^{(-0.5)} - 3^{(-0.5)}] \div [2^{(-1 \div 3)} + 3^{(-1 \div 3)}] = 0.08726$ (A-2)

86. $-2x\sqrt[5]{3^6 x^7 y^{11}} = -2x\sqrt[5]{3^5 x^5 y^{10} \cdot 3x^2 y} = -2x\sqrt[5]{3^5 x^5 y^{10}}\,\sqrt[5]{3x^2 y}$

$$= -2x \cdot 3xy^2\,\sqrt[5]{3x^2 y} = -6x^2 y^2\,\sqrt[5]{3x^2 y}$$

                       (A-3)

87. $\dfrac{2x^2}{\sqrt[3]{4x}} = \dfrac{2x^2}{\sqrt[3]{4x}}\,\dfrac{\sqrt[3]{2x^2}}{\sqrt[3]{2x^2}} = \dfrac{2x^2\,\sqrt[3]{2x^2}}{\sqrt[3]{8x^3}} = \dfrac{\overset{x}{\cancel{2x}}\!^2\,\sqrt[3]{2x^2}}{\underset{1}{\cancel{2x}}} = x^3\sqrt[3]{2x^2}$ (A-3)

88. $\sqrt[5]{\dfrac{3y^2}{8x^2}} = \sqrt[5]{\dfrac{3y^2}{8x^2}\dfrac{4x^3}{4x^3}} = \sqrt[5]{\dfrac{12x^3y^2}{32x^5}} = \dfrac{\sqrt[5]{12x^3y^2}}{\sqrt[5]{32x^5}} = \dfrac{\sqrt[5]{12x^3y^2}}{2x}$ *(A-3)*

89. $\sqrt[9]{8x^6y^{12}} = \sqrt[9]{2^3x^6y^{12}} = \sqrt[3\cdot3]{2^{3\cdot1}x^{3\cdot2}y^{3\cdot4}} = \sqrt[3]{2x^2y^4} = \sqrt[3]{y^3\cdot2x^2y} = \sqrt[3]{y^3}\sqrt[3]{2x^2y} = y\sqrt[3]{2x^2y}$ *(A-3)*

90. $\sqrt{\sqrt[3]{4x^4}} = \sqrt[2\cdot3]{2^{2\cdot1}x^{2\cdot2}} = \sqrt[3]{2x^2}$ *(A-3)*

91. $(2\sqrt{x} - 5\sqrt{y})(\sqrt{x} + \sqrt{y}) = 2\sqrt{x}\sqrt{x} + 2\sqrt{x}\sqrt{y} - 5\sqrt{x}\sqrt{y} - 5\sqrt{y}\sqrt{y}$
$= 2x - 3\sqrt{x}\sqrt{y} - 5y = 2x - 3\sqrt{xy} - 5y$ *(A-3)*

92. $\dfrac{3\sqrt{x}}{2\sqrt{x} - \sqrt{y}} = \dfrac{3\sqrt{x}}{(2\sqrt{x} - \sqrt{y})}\dfrac{(2\sqrt{x} + \sqrt{y})}{(2\sqrt{x} + \sqrt{y})} = \dfrac{3\sqrt{x}(2\sqrt{x} + \sqrt{y})}{(2\sqrt{x})^2 - (\sqrt{y})^2}$

$= \dfrac{3\sqrt{x}(2\sqrt{x} + \sqrt{y})}{4x - y} = \dfrac{6x + 3\sqrt{xy}}{4x - y}$ *(A-3)*

93. $\dfrac{2\sqrt{u} - 3\sqrt{v}}{2\sqrt{u} + 3\sqrt{v}} = \dfrac{(2\sqrt{u} - 3\sqrt{v})}{(2\sqrt{u} + 3\sqrt{v})}\dfrac{(2\sqrt{u} - 3\sqrt{v})}{(2\sqrt{u} - 3\sqrt{v})} = \dfrac{(2\sqrt{u})^2 - 2(2\sqrt{u})(3\sqrt{v}) + (3\sqrt{v})^2}{(2\sqrt{u})^2 - (3\sqrt{v})^2} = \dfrac{4u - 12\sqrt{uv} + 9v}{4u - 9v}$ *(A-3)*

94. $\dfrac{y^2}{\sqrt{y^2 + 4} - 2} = \dfrac{y^2}{(\sqrt{y^2 + 4} - 2)}\dfrac{(\sqrt{y^2 + 4} + 2)}{(\sqrt{y^2 + 4} + 2)} = \dfrac{y^2(\sqrt{y^2 + 4} + 2)}{(\sqrt{y^2 + 4})^2 - (2)^2} = \dfrac{y^2(\sqrt{y^2 + 4} + 2)}{y^2 + 4 - 4}$

$= \dfrac{\cancel{y^2}(\sqrt{y^2 + 4} + 2)}{\cancel{y^2}} = \sqrt{y^2 + 4} + 2$ *(A-3)*

95. $\dfrac{\sqrt{t} - \sqrt{5}}{t - 5} = \dfrac{(\sqrt{t} - \sqrt{5})}{(t - 5)}\dfrac{(\sqrt{t} + \sqrt{5})}{\sqrt{t} + \sqrt{5}} = \dfrac{(\sqrt{t})^2 - (\sqrt{5})^2}{(t - 5)(\sqrt{t} + \sqrt{5})} = \dfrac{\overset{1}{\cancel{t - 5}}}{\underset{1}{\cancel{(t - 5)}}(\sqrt{t} + \sqrt{5})} = \dfrac{1}{\sqrt{t} + \sqrt{5}}$ *(A-3)*

96. $\dfrac{4\sqrt{x} - 3}{2\sqrt{x}} = \dfrac{4\sqrt{x}}{2\sqrt{x}} - \dfrac{3}{2\sqrt{x}} = 2 - \dfrac{3}{2x^{1/2}} = 2 - \dfrac{3}{2}x^{-1/2} = 2x^0 - \dfrac{3}{2}x^{-1/2}$ *(A-3)*

97. Let $\quad\quad c = 0.54545454\ldots$ Then
$\quad\quad\quad 100c = 54.545454\ldots$
So $(100c - c) = (54.545454\ldots) - (0.54545454\ldots)$
$\quad\quad\quad\quad\quad = 54$
$\quad\quad\quad 99c = 54$
$\quad\quad\quad\quad c = \dfrac{54}{99} = \dfrac{6}{11}$

The number can be written as the quotient of two integers, so it is rational.
(A-1)

98. (A) List each element of M. Follow these with each element of N that is not yet listed. $\{-4,-3,2,0\}$, or, in increasing order, $\{-4,-3,0,2\}$.
(B) List each element of M that is also an element of N. $\{-3,2\}$. *(A-1)*

99. $x^2 - 4x + 1 = (2 - \sqrt{3})^2 - 4(2 - \sqrt{3}) + 1 = (2)^2 - 2(2)\sqrt{3} + (\sqrt{3})^2 - 8 + 4\sqrt{3} + 1$
$= 4 - 4\sqrt{3} + 3 - 8 + 4\sqrt{3} + 1 = 0$ *(A-3)*

100.
$$
\begin{aligned}
x(2x - 1)(x + 3) &= x(2x^2 + 6x - x - 3) \\
&= x(2x^2 + 5x - 3) \\
&= 2x^3 + 5x^2 - 3x
\end{aligned}
$$

$$
\begin{aligned}
(x - 1)^3 &= (x - 1)^2(x - 1) \\
&= (x^2 - 2x + 1)(x - 1) \\
&= x^3 - x^2 - 2x^2 + 2x + x - 1 \\
&= x^3 - 3x^2 + 3x - 1
\end{aligned}
$$

Therefore,

$$
\begin{aligned}
x(2x - 1)(x + 3) - (x - 1)^3 &= (2x^3 + 5x^2 - 3x) - (x^3 - 3x^2 + 3x - 1) \\
&= 2x^3 + 5x^2 - 3x - x^3 + 3x^2 - 3x + 1 \\
&= x^3 + 8x^2 - 6x + 1
\end{aligned}
$$
(A-4)

101.
$$
\begin{aligned}
4x(a^2 - 4a + 4) - 9x^3 &= x[4(a^2 - 4a + 4) - 9x^2] \\
&= x[2^2(a - 2)^2 - (3x)^2] \\
&= x[2(a - 2) + 3x][2(a - 2) - 3x] \\
&= x(2a - 4 + 3x)(2a - 4 - 3x) \\
&= x(2a + 3x - 4)(2a - 3x - 4)
\end{aligned}
$$
(A-5)

102. (A) $\sqrt{3 + \sqrt{5}} + \sqrt{3 - \sqrt{5}} \approx 3.162\ 277\ 660$

(B) $\sqrt{4 + \sqrt{15}} + \sqrt{4 - \sqrt{15}} \approx 3.162\ 277\ 660$

(C) $\sqrt{10} \approx 3.162\ 277\ 660$

(A), (B), and (C) all have the same value to 9 decimal places. To show that they are actually equal:

$$
\begin{aligned}
(\sqrt{3 + \sqrt{5}} + \sqrt{3 - \sqrt{5}})^2 &= (\sqrt{3 + \sqrt{5}})^2 + 2\sqrt{3 + \sqrt{5}}\sqrt{3 - \sqrt{5}} + (\sqrt{3 - \sqrt{5}})^2 \\
&= 3 + \sqrt{5} + 2\sqrt{(3 + \sqrt{5})(3 - \sqrt{5})} + 3 - \sqrt{5} \\
&= 6 + 2\sqrt{3^2 - (\sqrt{5})^2} \\
&= 6 + 2\sqrt{9 - 5} \\
&= 6 + 4 \\
&= 10
\end{aligned}
$$

Similarly,

$$
\begin{aligned}
(\sqrt{4 + \sqrt{15}}\sqrt{4 - \sqrt{15}})^2 &= (\sqrt{4 + \sqrt{15}})^2 + 2\sqrt{4 + \sqrt{15}}\sqrt{4 - \sqrt{15}} + (\sqrt{4 - \sqrt{15}})^2 \\
&= 4 + \sqrt{15} + 2\sqrt{(4 + \sqrt{15})(4 - \sqrt{15})} + 4 - \sqrt{15} \\
&= 8 + 2\sqrt{4^2 - (\sqrt{15})^2} \\
&= 8 + 2\sqrt{16 - 15} \\
&= 8 + 2 \\
&= 10
\end{aligned}
$$

Therefore, since both $\sqrt{3 + \sqrt{5}} + \sqrt{3 - \sqrt{5}}$ and $\sqrt{4 + \sqrt{15}} + \sqrt{4 - \sqrt{15}}$ are positive, they are both equal to the positive square root of 10, and

$$
\sqrt{3 + \sqrt{5}} + \sqrt{3 - \sqrt{5}} = \sqrt{4 + \sqrt{15}} + \sqrt{4 - \sqrt{15}} = \sqrt{10}
$$
(A-3)

103.
$$
\frac{8(x - 2)^{-3}(x + 3)^2}{12(x - 2)^{-4}(x + 3)^{-2}} = \frac{2(x - 2)^{-3-(-4)}(x + 3)^{2-(-2)}}{3} = \frac{2(x - 2)^1(x + 3)^4}{3} = \frac{2(x - 2)(x + 3)^4}{3}
$$

$$
= \frac{2}{3}(x - 2)(x + 3)^4
$$
(A-2)

104. $\left(\dfrac{a^{-2}}{b^{-1}} + \dfrac{b^{-2}}{a^{-1}}\right)^{-1} = \left(\dfrac{b}{a^2} + \dfrac{a}{b^2}\right)^{-1} = \left(\dfrac{b^3}{a^2b^2} + \dfrac{a^3}{a^2b^2}\right)^{-1} = \left(\dfrac{b^3 + a^3}{a^2b^2}\right)^{-1} = \dfrac{a^2b^2}{b^3 + a^3}$ or $\dfrac{a^2b^2}{a^3 + b^3}$

(A-2)

105. $(x^{1/3} - y^{1/3})(x^{2/3} + x^{1/3}y^{1/3} + y^{2/3}) = (x^{1/3})^3 - (y^{1/3})^3 = x - y$ (A-2)

106. $\left(\dfrac{x^{m^2}}{x^{2m-1}}\right)^{1/(m-1)} = (x^{m^2-(2m-1)})^{1/(m-1)} = (x^{m^2-2m+1})^{1/(m-1)} = x^{(m^2-2m+1)/(m-1)} = x^{(m-1)^2/(m-1)}$

$= x^{(m-1)^2/\cancel{m-1}} = x^{m-1}$ (A-2)

107. $\dfrac{1}{1 - \sqrt[3]{x}} = \dfrac{1}{(1 - \sqrt[3]{x})} \dfrac{[1 + \sqrt[3]{x} + (\sqrt[3]{x})^2]}{[1 + \sqrt[3]{x} + (\sqrt[3]{x})^2]} = \dfrac{1 + \sqrt[3]{x} + \sqrt[3]{x^2}}{(1)^3 - (\sqrt[3]{x})^3}$

Common Error:
$(1 - \sqrt[3]{x})(1 + \sqrt[3]{x}) \neq 1 - x$
This would not be a correct
application of
$(a - b)(a + b) = a^2 - b^2$

$= \dfrac{1 + \sqrt[3]{x} + \sqrt[3]{x^2}}{1 - x}$ (A-3)

108. $\dfrac{\sqrt[3]{t} - \sqrt[3]{5}}{t - 5} = \dfrac{(\sqrt[3]{t} - \sqrt[3]{5})}{(t - 5)} \dfrac{[(\sqrt[3]{t})^2 + \sqrt[3]{t}\sqrt[3]{5} + (\sqrt[3]{5})^2]}{[(\sqrt[3]{t})^2 + \sqrt[3]{t}\sqrt[3]{5} + (\sqrt[3]{5})^2]} = \dfrac{(\sqrt[3]{t})^3 - (\sqrt[3]{5})^3}{(t - 5)[\sqrt[3]{t^2} + \sqrt[3]{5t} + \sqrt[3]{5^2}]}$

$= \dfrac{\cancel{t - 5}}{\cancel{(t - 5)}[\sqrt[3]{t^2} + \sqrt[3]{5t} + \sqrt[3]{25}]} = \dfrac{1}{\sqrt[3]{t^2} + \sqrt[3]{5t} + \sqrt[3]{25}}$ (A-3)

109. $\sqrt[(n+1)]{x^{n^2}x^{2n+1}} = \sqrt[(n+1)]{x^{n^2}x^{2n+1}}$

$= (x^{n^2+2n+1})^{1/(n+1)} = x^{(n^2+n+1)/(n+1)} = x^{(n+1)^2/(n+1)} = x^{\cancel{(n+1)}^2/\cancel{(n+1)}} = x^{n+1}$ (A-3)

110. The volume of the concrete wall is equal to the volume of the outer cylinder ($V = \pi r^2 h$) minus the volume of the basin. Since the radius of the basin is x ft and the concrete is 2 ft thick, the radius of the outer cylinder is $x + 2$ ft.

$$\begin{pmatrix}\text{Volume of} \\ \text{concrete} \\ \text{wall}\end{pmatrix} = \begin{pmatrix}\text{Volume of} \\ \text{outer} \\ \text{cylinder}\end{pmatrix} - \begin{pmatrix}\text{Volume} \\ \text{of} \\ \text{basin}\end{pmatrix}$$

$$\text{Volume} = \pi(x + 2)^2\,3 - \pi x^2 3$$
$$= \pi(x^2 + 4x + 4)3 - \pi x^2 3$$
$$= 3\pi x^2 + 12\pi x + 12\pi - 3\pi x^2$$
$$= 12\pi x + 12\pi \ (\text{ft}^3)$$

(A-4)

111. average personal income = total personal income ÷ number of persons
$$= 9{,}208{,}000{,}000{,}000 \div 291{,}000{,}000$$
$$= 9.208 \times 10^{12} \div 2.91 \times 10^8$$
$$= 3.16 \times 10^4 \quad \text{or} \quad \$31{,}600 \text{ to three significant digits}$$

(A-2)

112. (A) We are asked to estimate $N = 20x^{1/2}y^{1/2}$ given $x = 1,600$ units of capital and $y = 900$ units of labor.
$N = 20(1,600)^{1/2}(900)^{1/2} = 20(40)(30) = 24,000$ units produced.

(B) Given $x = 3,200$ units of capital and $y = 1,800$ units of labor, then

$$
\begin{aligned}
N = 20(3,200)^{1/2}(1,800)^{1/2} &= 20(2 \cdot 1,600)^{1/2}(2 \cdot 900)^{1/2} \\
&= 20 \cdot 2^{1/2}(1,600)^{1/2}\, 2^{1/2}(900)^{1/2} \\
&= 20 \cdot 2^{1/2}(40) \cdot 2^{1/2}(30) \\
&= 20(40)(30) \cdot 2^{1/2} \cdot 2^{1/2} \\
&= 24,000 \cdot 2^1 \\
&= 48,000 \text{ units produced}
\end{aligned}
$$

(C) The effect of changing x to $2x$ and y to $2y$ is to replace N by

$$
\begin{aligned}
20(2x)^{1/2}(2y)^{1/2} &= 20 \cdot 2^{1/2}x^{1/2}2^{1/2}y^{1/2} \\
&= 20 \cdot 2^{1/2} \cdot 2^{1/2}x^{1/2}y^{1/2} \\
&= 2^1 \cdot 20x^{1/2}y^{1/2} \\
&= 2N
\end{aligned}
$$

The production is doubled at any production level.   (A-6)

113. $\dfrac{1}{\dfrac{1}{R_1} + \dfrac{1}{R_2} + \dfrac{1}{R_3}} = \dfrac{R_1 R_2 R_3 \cdot 1}{R_1 R_2 R_3 \left(\dfrac{1}{R_1} + \dfrac{1}{R_2} + \dfrac{1}{R_3}\right)} = \dfrac{R_1 R_2 R_3}{R_2 R_3 + R_1 R_3 + R_1 R_2}$ (A-4)

114. (A) The area of the cardboard can be written as (Original area) − (Removed area), where the original area = 16 × 30 = 480 and the removed area consists of 6 squares of area x^2 each;

(Original area) − (Removed area) = $480 - 6x^2$ in expanded form.

In factored form, $480 - 6x^2 = 6(80 - x^2)$

(B) See figure.

The volume of the box = length × width × height = $x(16 - 2x)\frac{1}{2}(30 - 3x)$

$$= x(16 - 2x)(15 - 1.5x)$$

or

$$3x(8 - x)(10 - x) \quad \text{in factored form.}$$

In expanded form, $3x(8 - x)(10 - x) = 3x(80 - 18x + x^2)$

$$= 3x^3 - 54x^2 + 240x$$ (A-5)

APPENDIX B Review of Equations and Graphing

Section B-1

1. $x + 5 = 12$
    $x + 5 - 5 = 12 - 5$
    $x = 7$

3. $2s - 7 = -2$
    $2s - 7 + 7 = -2 + 7$
    $2s = 5$
    $\dfrac{2s}{2} = \dfrac{5}{2}$
    $s = 2.5$

5. $2m + 8 = 5m - 7$
    $2m + 8 - 5m = 5m - 7 - 5m$
    $8 - 3m = -7$
    $8 - 3m - 8 = -7 - 8$
    $-3m = -15$
    $\dfrac{-3m}{-3} = \dfrac{-15}{-3}$
    $m = 5$

7. $-8 \le x \le 7$

8.

9. $-6 \le x < 6$

11. $x \ge -6$

13. $(-2, 6]$

15. $(-7, 8)$

17. $(-\infty, -2]$

19. $[-7, 2)$; $-7 \le x < 2$

21. $(-\infty, 0]$; $x \le 0$

23. $12 > 6$ and $12 + 5 > 6 + 5$

25. $-6 > -8$ and $-6 - 3 > -8 - 3$

27. $2 > -1$ and $-2(2) < -2(-1)$

29. $2 < 6$ and $\dfrac{2}{2} < \dfrac{6}{2}$

31. $7x - 8 < 4x + 7$
    $3x < 15$
    $x < 5$ or $(-\infty, 5)$

33. $3 - x \ge 5(3 - x)$
    $3 - x \ge 15 - 5x$
    $4x \ge 12$
    $x \ge 3$ or $[3, \infty)$

35. $\dfrac{N}{-2} > 4$
    $N < -8$ or $(-\infty, -8)$

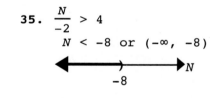

37. $3 - m < 4(m - 3)$
    $3 - m < 4m - 12$
    $-5m < -15$
    $m > 3$ or $(3, \infty)$

Common Error :
Neglecting to reverse the
order after division by -5.

39. $-2 - \dfrac{B}{4} \le \dfrac{1 + B}{3}$
    $12\left(-2 - \dfrac{B}{4}\right) \le 12\dfrac{(1 + B)}{3}$
    $-24 - 3B \le 4(1 + B)$
    $-24 - 3B \le 4 + 4B$
    $-7B \le 28$
    $B \ge -4$ or $[-4, \infty)$

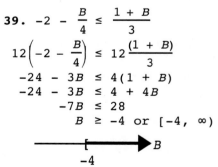

41. $3 - \dfrac{2x - 3}{3} = \dfrac{5 - x}{2}$ LCD $= 6$
    $6 \cdot 3 - 6\dfrac{(2x - 3)}{3} = 6\dfrac{(5 - x)}{2}$
    $18 - 2(2x - 3) = 3(5 - x)$
    $18 - 4x + 6 = 15 - 3x$
    $-4x + 24 = 15 - 3x$
    $-x = -9$
    $x = 9$

Common Error :
After line 2, students often write
$18 \quad - \quad 2\dfrac{2x \quad - \quad 3}{3} \quad = \quad ...$
$18 \quad - \quad 4x \quad - \quad 3 \quad = \quad ...$
forgetting to distribute the -2.
Put compound numerators in parentheses to avoid this.

43.

$$0.1(x - 7) + 0.05x = 0.8$$
$$0.1x - 0.7 + 0.05x = 0.8$$
$$0.15x - 0.7 = 0.8$$
$$0.15x = 1.5$$
$$x = \frac{1.5}{0.15}$$
$$x = 10$$

Problems of this type can also be solved by elimination of decimals: we could multiply every term on both sides by 100 to get $10(x - 7) + 5x = 80$ and proceed from here.

45.

$$0.3x - 0.04(x + 1) = 2.04$$
$$0.3x - 0.04x - 0.04 = 2.04$$
$$0.26x - 0.04 = 2.04$$
$$0.26x = 2.08$$
$$x = \frac{2.08}{0.26}$$
$$x = 8$$

47.

$[4, 7]$

$(-5, 5)$

$(-5, 5) \cup [4, 7] = (-5, 7]$ $-5 < x \le 7$

49.

$[-1, 4)$

$(2, 6]$

$[-1, 4) \cap (2, 6] = (2, 4)$ $2 < x < 4$

51.

$(-\infty, 1)$

$(-2, \infty)$

$(-\infty, 1) \cup (-2, \infty) = (-\infty, \infty)$ $-\infty < x < \infty$

53.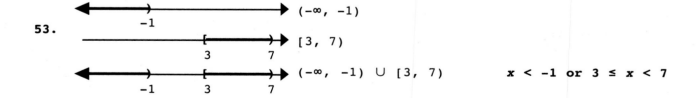

$(-\infty, -1)$

$[3, 7)$

$(-\infty, -1) \cup [3, 7)$ $x < -1$ or $3 \le x < 7$

55.

$[2, 3]$

$(1, 5)$

$[2, 3] \cup (1, 5) = (1, 5)$ $1 < x < 5$

57.

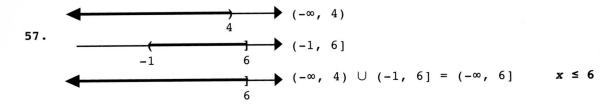

$(-\infty, 4)$

$(-1, 6]$

$(-\infty, 4) \cup (-1, 6] = (-\infty, 6]$ $\qquad x \le 6$

59. $\dfrac{q}{7} - 3 > \dfrac{q - 4}{3} + 1$

$21\left(\dfrac{q}{7} - 3\right) > 21\left(\dfrac{(q - 4)}{3}\right)$

$3q - 63 > 7(q - 4) + 21$
$3q - 63 > 7q - 28 + 21$
$3q - 63 > 7q - 7$
$\quad -4q > 56$
$\qquad q < -14$ or $(-\infty, -14)$

61. $\dfrac{2x}{5} - \dfrac{1}{2}(x - 3) \le \dfrac{2x}{3} - \dfrac{3}{10}(x + 2)$

LCD = 30
$12x - 15(x - 3) \le 20x - 9(x + 2)$
$12x - 15x + 45 \le 20x - 9x - 18$
$\quad -3x + 45 \le 11x - 18$
$\qquad -14x \le -63$
$\qquad\quad x \ge 4.5$ or $[4.5, \infty)$

63. $-4 \qquad \le \dfrac{9}{5}x + 32 \le 68$

$-36 \le \dfrac{9}{5}x \le 36$

$\dfrac{5}{9}(-36) \le x \le \dfrac{5}{9}(36)$

$\quad -20 \le x \le 20$ or $[-20, 20]$

65. $16 < 7 - 3x \le 31$
$\quad 9 < -3x \le 24$
$-3 > x \ge -8$
$-8 \le x < -3$ or $[-8, -3)$

67. $-6 < -\dfrac{2}{5}(1 - x) \le 4$

$-30 < -2(1 - x) \le 20$
$-30 < -2 + 2x \le 20$
$-28 < 2x \le 22$
$-14 < x \le 11$ or $(-14, 11]$

69. (A) F (B) T (C) T

71. If $a < b$, then by definition of $<$, there exists a positive number p such that $a + p = b$. Then, adding c to both sides, we obtain $(a + c) + p = b + c$, where p is positive. So by definition of $<$, $a + c < b + c$

73. (A) If $a < b$, then by definition of $<$, there exists a positive number p such that $a + p = b$. If we multiply both sides of this by the positive number c, we obtain $(a + p)c = bc$, or $ac + pc = bc$, where pc is positive. So by definition of $<$, $ac < bc$.

(B) If $a < b$, then by definition of $<$, there exists a positive number p such that $a + p = b$. If we multiply both sides of this by the negative number c, we obtain $(a + p)c = bc$, or $ac + pc = bc$, where pc is negative. So by definition of $<$, we have $ac > bc$.

Section B-2

1.

3.

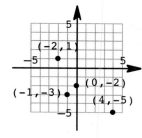

5. $A = (2, 4)$, $B = (3, -1)$, $C = (-4, 0)$, $D = (-5, 2)$

7. $A = (-3, -3)$, $B = (0, 4)$, $C = (-3, 2)$, $D = (5, -1)$

9.

x	$y = x + 1$
-3	-2
-2	-1
-1	0
0	1
1	2
2	3
3	4

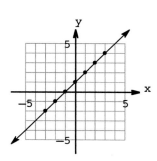

11.

x	$y = x^2 - 5$
-3	4
-2	-1
-1	-4
0	-5
1	-4
2	-1
3	4

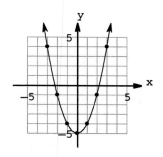

13.

x	$y = 3 + x - 0.5x^2$
-3	-4.5
-2	-1
-1	1.5
0	3
1	3.5
2	3
3	1.5

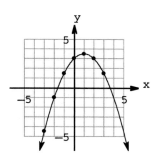

15. (A) When $x = 8$, the corresponding y value on the graph is 6, to the nearest integer.

(B) When $x = -5$, the corresponding y value on the graph is -5, to the nearest integer.

(C) When $x = 0$, the corresponding y value on the graph is -1, to the nearest integer.

(D) When $y = 6$, the corresponding x value on the graph is 8, to the nearest integer.

(E) When $y = -5$, the corresponding x value on the graph is -5, to the nearest integer.

(F) When $y = 0$, the corresponding x value on the graph is 5, to the nearest integer.

17. (A) When $x = 1$, the corresponding y value on the graph is 6, to the nearest integer.

(B) When $x = -8$, the corresponding y value on the graph is 4, to the nearest integer.

(C) When $x = 0$, the corresponding y value on the graph is 4, to the nearest integer.

(D) When $y = -6$, the corresponding x value on the graph is 8, to the nearest integer.

(E) Three values of x correspond to $y = 4$ on the graph. To the nearest integer they are -8, 0, and 6.

(F) Three values of x correspond to $y = 0$ on the graph. To the nearest integer they are -7, -2, and 7.

19. (A) Plot the points $(-2, -2)$, $(0, 0)$, and $(2, 2)$. There are many possible graphs connecting these points. The simplest is:

(B) Plot the points $(-1, 2)$, $(0, 0)$, and $(1, -2)$. Again, the simplest graph connecting the points is a line:

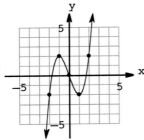

(C)

x	-2	-1	0	1	2
y	-2	2	0	-2	2

Plot the points $(-2, -2)$, $(-1, 2)$, $(0, 0)$, $(1, -2)$, and $(2, 2)$. The graph of $y = x^3 - 3x$ is shown below.

(D) All of the points on the two graphs in (A) and (B) are on the graph of $y = x^3 - 3x$. If you only plot certain points, it's possible, even likely, to draw an incorrect conclusion about the shape of the graph.

21.

x	$y = x^{1/3}$
-5	$\sqrt[3]{5} \approx -1.7$
-3	$\sqrt[3]{-3} \approx -1.4$
-1	-1
0	0
1	1
3	$\sqrt[3]{3} \approx 1.4$
5	$\sqrt[3]{5} \approx 1.7$

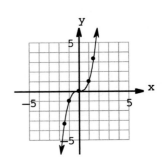

23.

x	$y = x^3$
-2	-8
-1.5	-3.375
-1	-1
0	0
1	1
1.5	3.375
2	8

25.

x	$y = \sqrt{x - 1}$
1	0
2	1
3	$\sqrt{2} \approx 1.4$
4	$\sqrt{3} \approx 1.7$
5	2

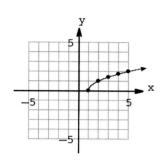

27.

x	$y = \sqrt{1 + x^2}$
−5	$\sqrt{26} \approx 5.1$
−3	$\sqrt{10} \approx 3.2$
−1	$\sqrt{2} \approx 1.4$
0	1
1	$\sqrt{2} \approx 1.4$
3	$\sqrt{30} \approx 3.2$
5	$\sqrt{26} \approx 5.1$

29. (A) and (B)

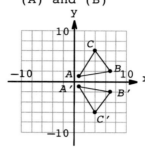

(C) Triangle $A'B'C'$ is the reflection of triangle ABC in the x axis, and triangle ABC is the reflection of triangle $A'B'C'$ in the x axis. In general, changing the sign of the y coordinate of all the points on a graph has the effect of reflecting the graph in the x axis.

31. (A) and (B)

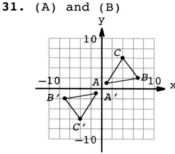

(C) Triangle $A'B'C'$ is the reflection of triangle ABC through the origin, and triangle ABC is the reflection of triangle $A'B'C'$ through the origin. In general, changing the signs of the x and y coordinates of all the points on a graph has the effect of reflecting the graph through the origin.

33.

P	$R = (10 - p)p$
5	25
6	24
7	21
8	16
9	9
10	0

35. (A)

x	$v = 0.5\sqrt{2 - x}$
0	0.7
0.5	0.6
1	0.5
1.5	0.35
2	0

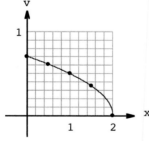

(B) When the ball is at the top of its swing (maximum displacement) $x = 2$, the velocity $v = 0$. As the ball moves to the bottom of its swing $x = 0$, the velocity increases to approximately 0.7 meters per second. As the ball swings past the bottom up to its maximum displacement, the velocity decreases to 0 again.

37. (A) 3,000 cases

(B) If the price is increased from \$6 to \$6.30 a case, the demand decreases from 3,000 cases to 2,600 cases, that is, by 400 cases.

(C) If the price is decreased from \$6 to \$5.70 a case, the demand increases from 3,000 cases to 3,600 cases, that is, by 600 cases.

(D) As the price varies from \$5.60 to \$6.90 a case, the demand varies from 4,000 to 2,000 cases, increasing with decreasing price and decreasing with increasing price.

39. (A) The temperature at 9 AM is approximately 53°.

(B) The highest temperature occurs at approximately 3 PM. This temperature is approximately 68°.

(C) The temperature is 49° at approximately 1 AM, 7 AM and 11 PM.

Section B-3

1. $d = \sqrt{(4 - 1)^2 + (4 - 0)^2} = \sqrt{9 + 16} = \sqrt{25} = 5$

$M = \left(\dfrac{1 + 4}{2}, \dfrac{0 + 4}{2}\right) = \left(\dfrac{5}{2}, 2\right)$

3. $d = \sqrt{(5 - 0)^2 + (10 - (-2))^2} = \sqrt{5^2 + 12^2} = \sqrt{25 + 144} = \sqrt{169} = 13$

$M = \left(\dfrac{0 + 5}{2}, \dfrac{-2 + 10}{2}\right) = \left(\dfrac{5}{2}, 4\right)$

5. $d = \sqrt{(3 - (-6))^2 + (4 - (-4))^2} = \sqrt{9^2 + 8^2} = \sqrt{81 + 64} = \sqrt{145}$

$M = \left(\dfrac{-6 + 3}{2}, \dfrac{-4 + 4}{2}\right) = \left(\dfrac{-3}{2}, 0\right)$

7. $d = \sqrt{(4 - 6)^2 + (-2 - 6)^2} = \sqrt{(-2)^2 + (-8)^2} = \sqrt{4 + 64} = \sqrt{68}$

$M = \left(\dfrac{6 + 4}{2}, \dfrac{6 + (-2)}{2}\right) = (5, 2)$

9. $(x - 0)^2 + (y - 0)^2 = 7^2$
$x^2 + y^2 = 49$

11. $(x - 2)^2 + (y - 3)^2 = 6^2$
$(x - 2)^2 + (y - 3)^2 = 36$

13. $[x - (-4)]^2 + (y - 1)^2 = (\sqrt{7})^2$
$(x + 4)^2 + (y - 1)^2 = 7$

Common Error: *not* $(x - 4)^2$

15. $[x - (-3)]^2 + [y - (-4)]^2 = (\sqrt{2})^2$
$(x + 3)^2 + (y + 4)^2 = 2$

17. $\dfrac{b}{2} = \dfrac{8}{2} = 4; \ (x + 4)^2 = x^2 + 8x + 16$

19. $\dfrac{b}{2} = \dfrac{-3}{2}; \ \left(x - \dfrac{3}{2}\right)^2 = x^2 - 3x + \dfrac{9}{4}$

21. This is a circle with center $(0, 0)$ and radius 2.
$$x^2 + y^2 = 4$$

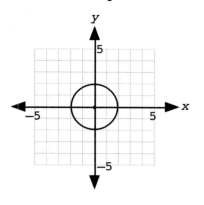

23. This is a circle with center $(1, 0)$ and radius 1.
$$(x - 1)^2 + y^2 = 1$$

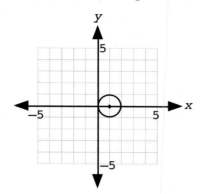

25. This is a circle with center $(-2, 1)$ and radius 3.
$$(x - (-2))^2 + (y - 1)^2 = 9$$
$$\text{or } (x + 2)^2 + (y - 1)^2 = 9$$

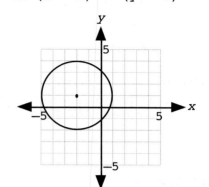

27. (A) $-2 = \dfrac{a_1 + 1}{2}$

$-4 = a_1 + 1$

$-5 = a_1$

(B) $6 = \dfrac{a_2 + 3}{2}$

$12 = a_2 + 3$

$9 = a_2$

(C) From parts (A) and (B), $A = (-5, 9)$

$d(A, M) = \sqrt{(-2(-5)^2 + (6 - 9)^2} = \sqrt{3^2 + (-3)^2} = \sqrt{9 + 9} = \sqrt{18}$

$d(M, B) = \sqrt{(1 - (-2)^2 + (3 - 6)^2} = \sqrt{3^2 + (-3)^2} = \sqrt{9 + 9} = \sqrt{18}$

As expected, the distances are the same.

29. This is a circle with center $(0, 2)$ and radius 2. That is, the set of all points that are 2 units away from $(0, 2)$.
$$x^2 + (y - 2)^2 = 4$$

31. This is a circle with center $(1, 1)$ and radius 4. That is, the set of all points that are 4 units away from $(1, 1)$.
$$(x - 1)^2 + (y - 1)^2 = 16$$

33. $M = \left(\dfrac{-4.3 + 9.6}{2}, \dfrac{5.2 + (-1.7)}{2} \right) = \left(\dfrac{5.3}{2}, \dfrac{3.5}{2} \right) = (2.65, 1.75)$

$d(A, M) = \sqrt{(2.65 - (-4.3))^2 + (1.75 - 5.2)^2} = \sqrt{6.95^2 + (-3.45)^2}$
$$= \sqrt{48.3025 + 11.9025}$$
$$= \sqrt{60.205} = 7.76$$

$d(M, B) = \sqrt{(4.6 - 2.65)^2 + (-1.7 - 1.75)^2} = \sqrt{6.95^2 + (-3.45)^2}$
$$= \sqrt{48.3025 + 11.9025}$$
$$= \sqrt{60.205} = 7.76$$

$d(A, B) = \sqrt{(9.6 - (-4.3))^2 + (-1.7 - 5.2)^2} = \sqrt{13.9^2 + (-6.9)^2}$
$$= \sqrt{193.21 + 47.61}$$
$$= \sqrt{240.82} = 15.52$$

$\dfrac{1}{2} d(A, B) = \dfrac{1}{2}(15.52) = 7.76$

35. Write $B = (b_1, b_2)$. -5 is the average of 25 and b_1, so $-5 = \dfrac{25 + b_1}{2}$

$$-10 = 25 + b_1$$
$$-35 = b_1$$

-2 is the average of 10 and b_2, so $-2 = \dfrac{10 + b_2}{2}$

$$-4 = 10 + b_2$$
$$-14 = b_2$$

So $B = (-35, -14)$.

$d(A, M) = \sqrt{(-5 - 25)^2 + (-2 - 10)^2} = \sqrt{(-30)^2 - (12)^2} = \sqrt{900 + 144} = \sqrt{1,044} = 32.3$

$d(M, B) = \sqrt{(-35 - (-5))^2 + (-14 - (-2))^2} = \sqrt{(-30)^2 - (12)^2} = \sqrt{900 + 144} = \sqrt{1,044} = 32.3$

$d(A, B) = \sqrt{(-35 - 25)^2 + (-14 - 10)^2} = \sqrt{(-60)^2 + (-24)^2} = \sqrt{3,600 + 576} = \sqrt{4,176} = 64.6$

$\dfrac{1}{2} d(A, B) = \dfrac{1}{2}(64.6) = 32.3$

37. Write $A = (a_1, a_2)$. -8 is the average of a_1 and 2, so $-8 = \dfrac{a_1 + 2}{2}$

$$-16 = a_1 + 2$$
$$-18 = a_1$$

-6 is the average of a_2 and 4, so $-6 = \dfrac{a_2 + 4}{2}$

$$-12 = a_2 + 4$$
$$-16 = a_2$$

So $A = (-18, -16)$.

$d(A, M) = \sqrt{(-8 - (-18))^2 + (-6 - (-16))^2} = \sqrt{10^2 + 10^2} = \sqrt{100 + 100} = \sqrt{200} = 14.14$

$d(M, B) = \sqrt{(2 - (-8))^2 + (4 - (-6))^2} = \sqrt{10^2 + 10^2} = \sqrt{100 + 100} = \sqrt{200} = 14.14$

$d(A, B) = \sqrt{(2 - (-18))^2 + (4 - (-16))^2} = \sqrt{20^2 + 20^2} = \sqrt{400 + 400} = \sqrt{800} = 28.28$

$\dfrac{1}{2} d(A, B) = \dfrac{1}{2}(28.28) = 14.14$

39. We need to complete the square with respect to y.

$x^2 + y^2 - 2y + ? = 0 \qquad \left(\dfrac{-2}{2}\right)^2 = (-1)^2 = 1$

$x^2 + y^2 - 2y + 1 = 1$
$x^2 + (y - 1)^2 = 1$
center $= (0, 1)$, $r = 1$

41. We need to complete the square with respect to x and y.

$x^2 - 2x + ? + y^2 + 6y + ? = 6 \qquad \left(\dfrac{-2}{2}\right)^2 = 1; \ \left(\dfrac{6}{2}\right)^2 = 9$

$x^2 - 2x + 1 + y^2 + 6y + 9 = 6 + 1 + 9$
$(x - 1)^2 + (y + 3)^2 = 16$
center $= (1, -3)$, $r = 4$

43. We need to complete the square with respect to x and y.

$$x^2 + x + ? + y^2 + 3y + ? = -2 \qquad \left(\frac{1}{2}\right)^2 = \frac{1}{4}; \ \left(\frac{3}{2}\right)^2 = \frac{9}{4}$$

$$x^2 + x + \frac{1}{4} + y^2 + 3y + \frac{9}{4} = -2 + \frac{1}{4} + \frac{9}{4}$$

$$\left(x + \frac{1}{2}\right)^2 + \left(y + \frac{3}{2}\right)^2 = -\frac{8}{4} + \frac{1}{4} + \frac{9}{4} = \frac{2}{4} = \frac{1}{2}$$

$$\left(x + \frac{1}{2}\right)^2 + \left(y + \frac{3}{2}\right)^2 = \frac{1}{2}$$

center $= \left(-\frac{1}{2}, -\frac{3}{2}\right), \ r = \sqrt{\frac{1}{2}}$

45. Let $A = (-3, 2)$, $B = (1, -2)$, $C = (8, 5)$

$$d(A, B) = \sqrt{(1 - (-3))^2 + (-2 - 2)^2} = \sqrt{4^2 + (-4)^2} = \sqrt{16 + 16} = \sqrt{32}$$

$$d(B, C) = \sqrt{(8 - 1)^2 + (5 - (-2))^2} = \sqrt{7^2 + 7^2} = \sqrt{49 + 49} = \sqrt{98}$$

$$d(A, C) = \sqrt{(8 - (-3))^2 + (5 - 2)^2} = \sqrt{11^2 + 3^2} = \sqrt{121 + 9} = \sqrt{130}$$

Notice that $(d(A, B))^2 + (d(B, C))^2 = 32 + 98 = 130 = (d(A, C))^2$

Since these distances satisfy the Pythagorean Theorem, the three points are vertices of a right triangle. The segment connecting A and C is the hypotenuse (it's the longest side) so we need to find its midpoint.

$$M = \left(\frac{-3 + 8}{2}, \frac{2 + 5}{2}\right) = \left(\frac{5}{2}, \frac{7}{2}\right)$$

The vertex opposite the hypotenuse is B.

$$d(M, B) = \sqrt{\left(1 - \frac{5}{2}\right)^2 + \left(-2 - \frac{7}{2}\right)^2} = \sqrt{\left(\frac{-3}{2}\right)^2 + \left(-\frac{11}{2}\right)^2} = \sqrt{\frac{9}{4} + \frac{121}{4}} = \sqrt{\frac{130}{4}} = \sqrt{32.5}$$

47. Perimeter = sum of lengths of all three sides

$$= \sqrt{[1 - (-3)]^2 + [(-2) - 1]^2} + \sqrt{(4 - 1)^2 + [3 - (-2)]^2} + \sqrt{[4 - (-3)]^2 + (3 - 1)^2}$$

$$= \sqrt{16 + 9} + \sqrt{9 + 25} + \sqrt{49 + 4}$$

$$= \sqrt{25} + \sqrt{34} + \sqrt{53}$$

$$= 18.11 \text{ to two decimal places}$$

49. $d(P_1, M) = \sqrt{\left(\frac{x_1 + x_2}{2} - x_1\right)^2 + \left(\frac{y_1 + y_2}{2} - y_1\right)^2}$

$$= \sqrt{\left(\frac{x_1 + x_2 - 2x_1}{2}\right)^2 + \left(\frac{y_1 + y_2 - 2y_1}{2}\right)^2}$$

$$= \sqrt{\left(\frac{x_2 - x_1}{2}\right)^2 + \left(\frac{y_2 - y_1}{2}\right)^2}$$

$d(M, P_2) = \sqrt{\left(x_2 - \frac{x_1 + x_2}{2}\right)^2 + \left(y_1 - \frac{y_1 + y_2}{2}\right)^2}$

$$= \sqrt{\left(\frac{2x_2 - x_1 - x_2}{2}\right)^2 + \left(\frac{2y_2 - y_1 - y_2}{2}\right)^2}$$

$$= \sqrt{\left(\frac{x_2 - x_1}{2}\right)^2 + \left(\frac{y_2 - y_1}{2}\right)^2}$$

$$\frac{1}{2} d(P_1, P_2) = \frac{1}{2} \sqrt{(x_2 - x_1)^2 + (y_2 - y_1)^2}$$

$$= \sqrt{\frac{1}{4} \left[(x_2 - x_1)^2 + (y_2 - y_1)^2 \right]} \quad \text{Note: The } \tfrac{1}{2} \text{ becomes } \tfrac{1}{4} \text{ when moved inside the root.}$$

$$= \sqrt{\frac{(x_2 - x_1)^2}{4} + \frac{(y_2 - y_1)^2}{4}}$$

$$= \sqrt{\left(\frac{x_2 - x_1}{2}\right)^2 + \left(\frac{y_2 - y_1}{2}\right)^2}$$

All three of these distances are equal.

51. The center of a circle is the midpoint of its diameter. The radius of a circle is half the length of its diameter. Using the midpoint formula, the center of this circle has coordinates given by

$$\left(\frac{x_1 + x_2}{2}, \frac{y_1 + y_2}{2}\right) = \left(\frac{7 + 1}{2}, \frac{-3 + 7}{2}\right) = (4, 2)$$

The radius of the circle is given by

$$r = \frac{1}{2} \sqrt{(x_2 - x_1)^2 + (y_2 - y_1)^2}$$

$$= \frac{1}{2} \sqrt{(1 - 7)^2 + [7 - (-3)]^2}$$

$$= \frac{1}{2} \sqrt{36 + 100} = \frac{1}{2} \sqrt{136} = \frac{1}{2} \sqrt{4 \cdot 34} = \sqrt{34}$$

So the equation of the circle is given by

$$(x - 4)^2 + (y - 2)^2 = (\sqrt{34})^2$$
$$(x - 4)^2 + (y - 2)^2 = 34$$

53. The radius of a circle is the distance from the center to any point on the circle. Since the center of this circle is (2, 2) and (3, -5) is a point on the circle, the radius is given by

$$r = \sqrt{(3 - 2)^2 + [(-5) - 2]^2} = \sqrt{1 + 49} = \sqrt{50}$$

So the equation of the circle is given by

$$(x - 2)^2 + (y - 2)^2 = (\sqrt{50})^2$$
$$(x - 2)^2 + (y - 2)^2 = 50$$

55. (A)

(B) $d(B, D) = \sqrt{(60 - 0)^2 + (27 - 13.5)^2} = \sqrt{60^2 + 13.5^2} = \sqrt{3600 + 182.25}$

$$= \sqrt{3,782.25} = 66 \text{ ft.}$$

$d(F, C) = \sqrt{(78 - 0)^2 + (13.5 - 27)^2} = \sqrt{78^2 + (-13.5)^2} = \sqrt{6,084 + 182.25}$

$$= \sqrt{6,266.25} = 79 \text{ ft.}$$

57. Using the hint, we note that $(2, r - 1)$ must satisfy $x^2 + y^2 = r^2$, that is

$$2^2 + (r - 1)^2 = r^2$$
$$4 + r^2 - 2r + 1 = r^2$$
$$-2r + 5 = 0$$
$$r = \frac{5}{2} \text{ or } 2.5 \text{ ft.}$$

59. (A) From the drawing, we can write:

$$\begin{pmatrix} \text{Distance from tower} \\ \text{to town } B \end{pmatrix} = 2 \times \begin{pmatrix} \text{Distance from tower} \\ \text{to town } A \end{pmatrix}$$

$$\begin{pmatrix} \text{Distance from } (x, y) \\ \text{to } (36, 15) \end{pmatrix} = 2 \times \begin{pmatrix} \text{Distance from } (x, y) \\ \text{to } (0, 0) \end{pmatrix}$$

$$\sqrt{(36 - x)^2 + (15 - y)^2} = 2\sqrt{(0 - x)^2 + (0 - y)^2}$$
$$\sqrt{(36 - x)^2 + (15 - y)^2} = 2\sqrt{x^2 + y^2}$$
$$(36 - x)^2 + (15 - y)^2 = 4(x^2 + y^2)$$
$$1{,}296 - 72x + x^2 + 225 - 30y + y^2 = 4x^2 + 4y^2$$
$$1{,}521 = 3x^2 + 3y^2 + 72x + 30y$$
$$507 = x^2 + y^2 + 24x + 10y$$
$$144 + 25 + 507 = x^2 + 24x + 144 + y^2 + 10y + 25$$
$$676 = (x + 12)^2 + (y + 5)^2$$

The circle has center $(-12, -5)$ and radius 26.

(B) All points due east of Town A have y coordinate 0 in this coordinate system. The points on the circle for which $y = 0$ are found by substituting $y = 0$ into the equation of the circle and solving for x.

$$(x + 12)^2 + (y + 5)^2 = 676$$
$$(x + 12)^2 + 25 = 676$$
$$(x + 12)^2 = 651$$
$$x + 12 = \pm\sqrt{651}$$
$$x = -12 \pm \sqrt{651}$$

Choosing the positive square root so that x is greater than -12 (east rather than west) we have $x = -12 + \sqrt{651} \approx 13.5$ miles.

APPENDIX C   Special Topics

Section C-1

1. $\underline{123,005}$ **3.** $\underline{20,040}$ **5.** $\underline{6.0}$ **7.** $\underline{80.000}$ **9.** $0.0\underline{12}$ **11.** $0.000\ \underline{960}$

13. 3.08   **15.** 924,000   **17.** 23.7   **19.** 2.82×10^3 **21.** 6.78×10^{-4}

23.

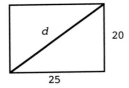

Using the Pythagorean Theorem,
$$25^2 + 20^2 = d^2$$
$$625 + 400 = d^2$$
$$1,025 = d^2$$
$$32.0156 = d$$
Rounded to 1 significant digit, which is the smallest number of significant digits in the original measurements, the diagonal is 30 ft.

Section C-2

1. $\dfrac{7x - 14}{(x - 4)(x + 3)} = \dfrac{A}{x - 4} + \dfrac{B}{x + 3} = \dfrac{A(x + 3) + B(x - 4)}{(x - 4)(x + 3)}$

The numerators must be equal, so
$$7x - 14 = A(x + 3) + B(x - 4)$$

Graphical Solution:

$7x - 14 = A(x + 3) + B(x - 4) \Rightarrow 7x - 14 = (A + B)x + 3A - 4B$

For this equation to be true, the coefficients of x on each side must be equal; that is, $7 = A + B$. Also, the constant terms must be equal; that is, $-14 = 3A - 4B$. Now solve each of these equations for B and graph, treating A as independent variable and B as dependent variable.

$$\begin{array}{l} A + B = 7 \\ 3A - 4B = -14 \end{array} \Rightarrow \begin{array}{l} B = 7 - A \\ B = \dfrac{14 + 3A}{4} \end{array}$$

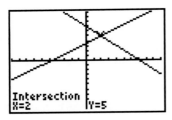

$A = 2$, $B = 5$

Algebraic Solution:

If $x = -3$, then
$$-35 = -7B$$
$$B = 5$$
If $x = 4$, then
$$14 = 7A$$
$$A = 2$$
$$A = 2, \ B = 5$$

3. $\dfrac{17x - 1}{(2x - 3)(3x - 1)} = \dfrac{A}{2x - 3} + \dfrac{B}{3x - 1} = \dfrac{A(3x - 1) + B(2x - 3)}{(2x - 3)(3x - 1)}$

The numerators must be equal, so

$17x - 1 = A(3x - 1) + B(2x - 3)$

<u>Graphical Solution:</u>

(See Problem 1 for details.)

$17x - 1 = A(3x - 1) + B(2x - 3) \Rightarrow 17x - 1 = (3A + 2B)x - A - 3B$

$\begin{aligned} 3A + 2B &= 17 \\ -A - 3B &= -1 \end{aligned} \Rightarrow \begin{aligned} B &= \dfrac{17 - 3A}{2} \\ B &= \dfrac{1 - A}{3} \end{aligned}$

$A = 7, \ B = -2$

<u>Algebraic Solution:</u> If $x = \dfrac{1}{3}$, then

$17\left(\dfrac{1}{3}\right) - 1 = B\left[2\left(\dfrac{1}{3}\right) - 3\right]$

$\qquad \dfrac{14}{3} = -\dfrac{7}{3}B$

$\qquad B = -2$

If $x = \dfrac{3}{2}$, then

$17\left(\dfrac{3}{2}\right) - 1 = A\left[3\left(\dfrac{3}{2}\right) - 1\right]$

$\qquad \dfrac{49}{2} = \dfrac{7}{2}A$

$\qquad A = 7$

$A = 7, \ B = -2$

5. $\dfrac{3x^2 + 7x + 1}{x(x + 1)^2} = \dfrac{A}{x} + \dfrac{B}{x + 1} + \dfrac{C}{(x + 1)^2} = \dfrac{A(x + 1)^2 + Bx(x + 1) + Cx}{x(x + 1)^2}$

$3x^2 + 7x + 1 = A(x + 1)^2 + Bx(x + 1) + Cx$

$3x^2 + 7x + 1 = A(x^2 + 2x + 1) + Bx^2 + Bx + Cx$

$3x^2 + 7x + 1 = Ax^2 + 2AX + A + Bx^2 + Bx + Cx$

$3x^2 + 7x + 1 = (A + B)x^2 + (2A + B + C)x + A$

Equating coefficients, we know $A = 1$. Also, $A + B = 3$, so $1 + B = 3$ and $B = 2$. Finally, $2A + B + C = 7$, so $2 + 2 + C = 7$ and $C = 3$.

Check by addition:

$\dfrac{1}{x} + \dfrac{2}{x + 1} + \dfrac{3}{(x + 1)^2} = \dfrac{(x + 1)^2}{x(x + 1)^2} + \dfrac{2x(x + 1)}{x(x + 1)^2} + \dfrac{3x}{x(x + 1)^2}$

$$= \frac{x^2 + x + 1 + 2x^2 + 2x + 3x}{x(x + 1)^2}$$

$$= \frac{3x^2 + 7x + 1}{x(x + 1)^2}$$

Check with graphing calculator:

The graphs appear to be identical and the tables are the same.

7. $\dfrac{3x^2 + x}{(x - 2)(x^2 + 3)} = \dfrac{A}{x - 2} + \dfrac{Bx + C}{x^2 + 3} = \dfrac{A(x^2 + 3) + (Bx + C)(x - 2)}{(x - 2)(x^2 + 3)}$

$3x^2 + x = A(x^2 + 3) + (Bx + C)(x - 2)$

First, plug in $x = 2$:

$$3(2)^2 + 2 = A(4 + 3) + 0$$
$$14 = 7A$$
$$A = 2$$
$$3x^2 + x = 2(x^2 + 3) + (Bx + C)(x - 2)$$
$$3x^2 + x = 2x^2 + 6 + Bx^2 - 2Bx + Cx - 2C$$
$$3x^2 + x = (2 + B)x^2 + (C - 2B)x + 6 - 2C$$

Equating coefficients,
$2 + B = 3$, so $B = 1$
$6 - 2C = 0$, so $6 = 2C$ and $C = 3$.

Check by addition:

$$\frac{x}{x - 2} + \frac{x + 3}{x^2 + 3} = \frac{2(x^2 + 3) + (x + 3)(x - 2)}{(x - 2)(x^2 + 3)}$$

$$= \frac{2x^2 + 6 + x^2 + x - 6}{(x - 2)(x^2 + 3)}$$

$$= \frac{3x^2 + x}{(x - 2)(x^2 + 3)}$$

Check with graphing calculator:

The graphs appear to be identical and the tables are the same.

9. $\dfrac{2x^2 + 4x - 1}{(x^2 + x + 1)^2} = \dfrac{Ax + B}{x^2 + x + 1} + \dfrac{Cx + D}{(x^2 + x + 1)^2} = \dfrac{(Ax + B)(x^2 + x + 1) + (Cx + D)}{(x^2 + x + 1)^2}$

$$2x^2 + 4x - 1 = (Ax + B)(x^2 + x + 1) + Cx + D$$
$$2x^2 + 4x - 1 = Ax^3 + Ax^2 + Ax + Bx^2 + Bx + B + Cx + D$$
$$2x^2 + 4x - 1 = Ax^3 + (A + B)x^2 + (A + B + C)x + B + D$$

Equating coefficients,
$A = 0$
$A + B = 2$, so $B = 2$
$A + B + C = 4$, so $0 + 2 + C = 4$, and $C = 2$
$B + D = -1$, so $2 + D = -1$, and $D = -3$

Check by addition:

$$\frac{2}{x^2 + x + 1} + \frac{2x - 3}{(x^2 + x + 1)^2} = \frac{2(x^2 + x + 1)}{(x^2 + x + 1)^2} + \frac{2x - 3}{(x^2 + x + 1)^2}$$

$$= \frac{2x^2 + 2x + 2 + 2x - 3}{(x^2 + x + 1)^2}$$

$$= \frac{2x^2 + 4x - 1}{(x^2 + x + 1)^2}$$

Check with graphing calculator:

The graphs appear to be identical and the tables are the same.

11. Since $x^2 - 2x - 8 = (x + 2)(x - 4)$, we write

$$\frac{-x + 22}{x^2 - 2x - 8} = \frac{A}{x + 2} + \frac{B}{x - 4} = \frac{A(x - 4) + B(x + 2)}{(x + 2)(x - 4)}$$

The numerators must be equal, so

$-x + 22 = A(x - 4) + B(x + 2)$
If $x = 4$
$18 = 6B$
$B = 3$

If $x = -2$
$24 = -6A$
$A = -4$

So $\dfrac{-x + 22}{x^2 - 2x - 8} = \dfrac{-4}{x + 2} + \dfrac{3}{x - 4}$.

13. Since $6x^2 - x - 12 = (3x + 4)(2x - 3)$, we write

$$\frac{3x - 13}{6x^2 - x - 12} = \frac{A}{3x + 4} + \frac{B}{2x - 3} = \frac{A(2x - 3) + B(3x + 4)}{(3x + 4)(2x - 3)}$$

The numerators must be equal, so
$3x - 13 = A(2x - 3) + B(3x + 4)$
If $x = \dfrac{3}{2}$

$$3\left(\frac{3}{2}\right) - 13 = B\left[3\left(\frac{3}{2}\right) + 4\right]$$

$$-\frac{17}{2} = \frac{17}{2}B$$

$$B = -1$$

If $x = -\dfrac{4}{3}$

$$3\left(-\dfrac{4}{3}\right) - 13 = A\left[2\left(-\dfrac{4}{3}\right) - 3\right]$$

$$-17 = -\dfrac{17}{3}A$$

$$A = 3$$

So $\dfrac{3x - 13}{6x^2 - x - 12} = \dfrac{3}{3x + 4} - \dfrac{1}{2x - 3}$

15. Since $x^3 - 6x^2 + 9x = x(x^2 - 6x + 9) = x(x - 3)^2$, we write

$$\dfrac{x^2 - 12x + 18}{x^3 - 6x^2 + 9x} = \dfrac{A}{x} + \dfrac{B}{x - 3} + \dfrac{C}{(x - 3)^2} = \dfrac{A(x - 3)^2 + Bx(x - 3) + Cx}{x(x - 3)^2}$$

The numerators must be equal, so

$$x^2 - 12x + 18 = A(x - 3)^2 + Bx(x - 3) + Cx$$

If $x = 3$
$-9 = 3C$
$C = -3$

If $x = 0$
$18 = 9A$
$A = 2$

If $x = 2$
$-2 = A - 2B + 2C$
$-2 = 2 - 2B - 6$ using $A = 2$ and $C = -3$
$2 = -2B$
$B = -1$

So $\dfrac{x^2 - 12x + 18}{x^3 - 6x^2 + 9x} = \dfrac{2}{x} - \dfrac{1}{x - 3} - \dfrac{3}{(x - 3)^2}$

17. Since $x^3 + 2x^2 + 3x = x(x^2 + 2x + 3)$, we write

$$\dfrac{5x^2 + 3x + 6}{x^3 + 2x^2 + 3x} = \dfrac{A}{x} + \dfrac{Bx + C}{x^2 + 2x + 3} = \dfrac{A(x^2 + 2x + 3) + (Bx + C)x}{x(x^2 + 2x + 3)}$$

> **Common Error:**
> Writing $\dfrac{B}{x^2 + 2x + 3}$. Since $x^2 + 2x + 3$ is quadratic, the numerator must be first degree.

The numerators must be equal, so
$$5x^2 + 3x + 6 = A(x^2 + 2x + 3) + (Bx + C)x = (A + B)x^2 + (2A + C)x + 3A$$
Equating coefficients of like terms, we have
$5 = A + B$
$3 = 2A + C$
$3A = 6$
We can now see that $A = 2$; use the 2nd and 3rd equations to find B and C: $B = 3$ and $C = -1$.

So $\dfrac{5x^2 + 3x + 6}{x^3 + 2x^2 + 3x} = \dfrac{2}{x} + \dfrac{3x - 1}{x^2 + 2x + 3}$

19. Since $x^4 + 4x^2 + 4 = (x^2 + 2)^2$, we write

$$\frac{2x^3 + 7x + 5}{x^4 + 4x^2 + 4} = \frac{Ax + B}{x^2 + 2} + \frac{Cx + D}{(x^2 + 2)^2} = \frac{(Ax + B)(x^2 + 2) + Cx + D}{(x^2 + 2)^2}$$

The numerators must be equal, so

$$2x^3 + 7x + 5 = (Ax + B)(x^2 + 2) + Cx + D = Ax^3 + Bx^2 + (2A + C)x + 2B + D$$

Equating coefficients of like terms, we have

$2 = A$
$0 = B$
$7 = 2A + C$
$5 = 2B + D$

Now we know that $A = 2$ and $B = 0$; use the 3rd and 4th equations to find C and D: $C = 3$ and $D = 5$.

$$\frac{2x^3 + 7x + 5}{x^4 + 4x^2 + 4} = \frac{2x}{x^2 + 2} + \frac{3x + 5}{(x^2 + 2)^2}$$

21. First we divide to obtain a polynomial plus a proper fraction.

$$
\begin{array}{r}
x - 2 \\
x^2 - 5x + 6 \overline{\smash{\big)}\ x^3 - 7x^2 + 17x - 17} \\
\underline{x^3 - 5x^2 + 6x} \\
-2x^2 + 11x - 17 \\
\underline{-2x^2 + 10x - 12} \\
x - 5
\end{array}
$$

So, $\dfrac{x^3 - 7x^2 + 17x - 17}{x^2 - 5x + 6} = x - 2 + \dfrac{x - 5}{x^2 - 5x + 6}$

To decompose the proper fraction, we note $x^2 - 5x + 6 = (x - 2)(x - 3)$ and we write:

$$\frac{x - 5}{x^2 - 5x + 6} = \frac{A}{x - 2} + \frac{B}{x - 3} = \frac{A(x - 3) + B(x - 2)}{(x - 2)(x - 3)}$$

The numerators must be equal, so

$x - 5 = A(x - 3) + B(x - 2)$
If $x = 3$
$-2 = B$
If $x = 2$
$-3 = -A$
$A = 3$

So $\dfrac{x^3 - 7x^2 + 17x - 17}{x^2 - 5x + 6} = x - 2 + \dfrac{3}{x - 2} - \dfrac{2}{x - 3}$

23. First, we must factor $x^3 - 6x - 9$. Possible rational zeros of this polynomial are ± 1, ± 3, ± 9. Forming a synthetic division table, we see

	1	0	-6	-9
1	1	1	-5	-14
3	1	3	3	0

So $x^3 - 6x - 9 = (x - 3)(x^2 + 3x + 3)$
$x^2 + 3x + 3$ cannot be factored further in the real numbers.

So $\dfrac{4x^2 + 5x - 9}{x^3 - 6x - 9} = \dfrac{A}{x - 3} + \dfrac{Bx + C}{x^2 + 3x + 3} = \dfrac{A(x^2 + 3x + 3) + (Bx + C)(x - 3)}{(x - 3)(x^2 + 3x + 3)}$

The numerators must be equal, so

$$4x^2 + 5x - 9 = A(x^2 + 3x + 3) + (Bx + C)(x - 3)$$
$$= Ax^2 + 3Ax + 3A + Bx^2 - 3Bx + Cx - 3C$$
$$= (A + B)x^2 + (3A - 3B + C)x + 3A - 3C$$

Before equating coefficients of like terms, we note that if $x = 3$

$$4(3)^2 + 5(3) - 9 = A(3^2 + 3 \cdot 3 + 3)$$
$$\text{So} \quad 42 = 21A$$
$$A = 2$$
$$\text{Since} \quad 4 = A + B, \; B = 2$$
$$5 = 3A - 3B + C = 6 - 6 + C, \text{ so } C = 5$$

We have $A = 2$, $B = 2$, $C = 5$

So $\dfrac{4x^2 + 5x - 9}{x^3 - 6x - 9} = \dfrac{2}{x - 3} + \dfrac{2x + 5}{x^2 + 3x + 3}$

25. First, we must factor $x^3 + 2x^2 - 15x - 36$. The possible rational zeros are ± 1, ± 2, ± 3, ± 4, ± 6, ± 9, ± 12, ± 18, ± 36. Forming a synthetic division table, we see

	1	2	-15	-36
1	1	3	-12	-48
2	1	4	-7	-50
3	1	5	0	-36
4	1	6	9	0

So $x^3 + 2x^2 - 15x - 36 = (x - 4)(x^2 + 6x + 9) = (x - 4)(x + 3)^2$.

So $\dfrac{x^2 + 16x + 18}{x^3 + 2x^2 - 15x - 36} = \dfrac{A}{x - 4} + \dfrac{B}{x + 3} + \dfrac{C}{(x + 3)^2}$

$$= \dfrac{A(x + 3)^2 + B(x - 4)(x + 3) + C(x - 4)}{(x - 4)(x + 3)^2}$$

The numerators must be equal, so

$$x^2 + 16x + 18 = A(x + 3)^2 + B(x - 4)(x + 3) + C(x - 4)$$

If $x = -3$
$$-21 = -7C$$
$$C = 3$$
If $x = 4$
$$98 = 49A$$
$$A = 2$$
If $x = 5$
$$123 = 64A + 8B + C$$
$$= 128 + 8B + 3 \text{ using } A = 2 \text{ and } C = 3$$
$$-8 = 8B$$
$$B = -1$$

So $\dfrac{x^2 + 16x + 18}{x^3 + 2x^2 - 15x - 36} = \dfrac{2}{x - 4} - \dfrac{1}{x + 3} + \dfrac{3}{(x + 3)^2}$

27. First, we must factor $x^4 - 5x^3 + 9x^2 - 8x + 4$. The possible rational zeros are ± 1, ± 2, ± 4. We form a synthetic division table:

	1	-5	9	-8	4
1	1	-4	5	-3	1
2	1	-3	3	-2	0

We examine $x^3 - 3x^2 + 3x - 2$ to see if 2 is a double zero.

	1	-3	3	-2
2	1	-1	1	0

So $x^4 - 5x^3 + 9x^2 - 8x + 4 = (x - 2)^2(x^2 - x + 1)$. $x^2 - x + 1$ cannot be factored further in the real numbers, so

$$\frac{-x^2 + x - 7}{x^4 - 5x^3 + 9x^2 - 8x + 4} = \frac{A}{x - 2} + \frac{B}{(x - 2)^2} + \frac{Cx + D}{x^2 - x + 1}$$

$$= \frac{A(x - 2)(x^2 - x + 1) + B(x^2 - x + 1) + (Cx + D)(x - 2)^2}{(x - 2)^2(x^2 - x + 1)}$$

The numerators must be equal, so

$-x^2 + x - 7 = A(x - 2)(x^2 - x + 1) + B(x^2 - x + 1) + (Cx + D)(x - 2)^2$

If $x = 2$

$-9 = 3B$

$B = -3$

$-x^2 + x - 7 = A(x^3 - 3x^2 + 3x - 2) + B(x^2 - x + 1) + (Cx + D)(x^2 - 4x + 4)$

$\qquad = (A + C)x^3 + (-3A + B - 4C + D)x^2 + (3A - B + 4C - 4D)x - 2A + B + 4D$

We have already $B = -3$, so equating coefficients of like terms,

$0 = A + C$

$-1 = -3A - 3 - 4C + D$

$1 = 3A + 3 + 4C - 4D$

$-7 = -2A - 3 + 4D$

Since $C = -A$, we can write

$-1 = -3A - 3 + 4A + D$

$1 = 3A + 3 - 4A - 4D$

$-7 = -2A - 3 + 4D$

So $D = 0$ (adding the first equations), $A = 2$, $C = -2$.

$$\frac{-x^2 + x - 7}{x^4 - 5x^3 + 9x^2 - 8x + 4} = \frac{2}{x - 2} - \frac{3}{(x - 2)^2} - \frac{2x}{x^2 - x + 1}$$

29. First we divide to obtain a polynomial plus a proper fraction.

$$
\begin{array}{r}
x + 2 \\
4x^4 + 4x^3 - 5x^2 + 5x - 2\overline{\smash{\big)}\,4x^5 + 12x^4 - x^3 + 7x^2 - 4x + 2} \\
\underline{4x^5 + 4x^4 - 5x^3 + 5x^2 - 2x} \\
8x^4 + 4x^3 + 2x^2 - 2x + 2 \\
\underline{8x^4 + 8x^3 - 10x^2 + 10x - 4} \\
- 4x^3 + 12x^2 - 12x + 6
\end{array}
$$

Now we must decompose $\dfrac{-4x^3 + 12x^2 - 12x + 6}{4x^4 + 4x^3 - 5x^2 + 5x - 2}$, starting by factoring $4x^4 + 4x^3 - 5x^2 + 5x - 2$.

Possible rational zeros are ± 1, ± 2, $\pm\frac{1}{2}$, $\pm\frac{1}{4}$. We form a synthetic division table:

	4	4	-5	5	-2	
1	4	8	3	8	6	1 is an upper bound, eliminating 2
-1	4	0	-5	10	-12	
-2	4	-4	3	-1	0	-2 is a zero

We investigate $4x^3 - 4x^2 + 3x - 1$; the only remaining rational zeros are $\pm\frac{1}{2}$, $\pm\frac{1}{4}$.

We form a synthetic division table:

	4	-4	3	-1
$\frac{1}{2}$	4	-2	2	0

So $4x^4 + 4x^3 - 5x^2 + 5x - 2 = (x + 2)\left(x - \frac{1}{2}\right)(4x^2 - 2x + 2)$

$$= (x + 2)\left(x - \frac{1}{2}\right)2(2x^2 - x + 1)$$

$$= (x + 2)(2x - 1)(2x^2 - x + 1)$$

$2x^2 - x + 1$ cannot be factored further in the real numbers, so we write:

$$\frac{-4x^3 + 12x^2 - 12x + 6}{4x^4 + 4x^3 - 5x^2 + 5x - 2}$$

$$= \frac{A}{x + 2} + \frac{B}{2x - 1} + \frac{Cx + D}{2x^2 - x + 1}$$

$$= \frac{A(2x - 1)(2x^2 - x + 1) + B(x + 2)(2x^2 - x + 1) + (Cx + D)(x + 2)(2x - 1)}{(x + 2)(2x - 1)(2x^2 - x + 1)}$$

The numerators must be equal, so

$-4x^3 + 12x^2 - 12x + 6$
$= A(2x - 1)(2x^2 - x + 1) + B(x + 2)(2x^2 - x + 1) + (Cx + D)(x + 2)(2x - 1)$

If $x = -2$

$-4(-2)^3 + 12(-2)^2 - 12(-2) + 6 = A(-5)[2(-2)^2 - (-2) + 1]$

$$32 + 48 + 24 + 6 = -55A$$

$$110 = -55A$$

$$A = -2$$

If $x = \frac{1}{2}$

$$-4\left(\frac{1}{2}\right)^3 + 12\left(\frac{1}{2}\right)^2 - 12\left(\frac{1}{2}\right) + 6 = B\left(\frac{5}{2}\right)\left[2\left(\frac{1}{2}\right)^2 - \frac{1}{2} + 1\right]$$

$$-\frac{1}{2} + 3 - 6 + 6 = B\left(\frac{5}{2}\right)(1)$$

$$B = 1$$

If $x = 0$
$6 = A(-1)(1) + B(2)(1) + D(2)(-1)$
$6 = -A + 2B - 2D$
$6 = 2 + 2 - 2D$
$D = -1$
If $x = 1$
$-4 + 12 - 12 + 6 = A(1)(2) + B(3)(2) + (C + D)(3)(1)$
$$2 = 2A + 6B + 3C + 3D$$
$$= -4 + 6 + 3C - 3$$
$$C = 1$$

So $\dfrac{4x^5 + 12x^4 - x^3 + 7x^2 - 4x + 2}{4x^4 + 4x^3 - 5x^2 + 5x - 2} = x + 2 - \dfrac{2}{x + 2} + \dfrac{1}{2x - 1} + \dfrac{x - 1}{2x^2 - x + 1}$.

Section C-3

1. $P(x) = x^3 + 2x + 7$; signs of coefficients: $+ + +$
   variations in sign: none

   $P(-x) = (-x)^3 + 2(-x) + 7 = -x^3 - 2x + 7$; signs of coefficients: $- - +$
   variations in sign: one

3. $P(x) = x^3 - 3x^2 - 9$; signs of coefficients: + - -
variations in sign: one

$P(-x) = (-x)^3 - 3(-x)^2 - 9 = -x^3 - 3x^2 - 9$; signs of coefficients: - - -
variations in sign: none

5. $P(x) = x^4 + 2x^3 + 3x - 5$; signs of coefficients: + + + -
variations in sign: one

$P(-x) = (-x)^4 + 2(-x)^3 + 3(-x) - 5 = x^4 - 2x^3 - 3x - 5$;
signs of coefficients: + - - -
variations in sign: one

7. $P(x) = x^3 + 2x - 4$ One variation in sign
$P(-x) = (-x)^3 + 2(-x) - 4 = -x^3 - 2x - 4$ No variations in sign

$P(x)$ has either 1 positive zero or the number of positive zeros is less than 1
by an even number. Since there obviously can't be less than zero positive
zeros, we know for sure there is one. Since $P(-x)$ has no variations in sign,
$P(x)$ has no negative zeros. As a third degree polynomial, $P(x)$ has a total of
three zeros: one is real and two are imaginary.

```
WINDOW
Xmin=-10
Xmax=10
Xscl=1
Ymin=-40
Ymax=40
Yscl=10
Xres=1
```

9. $f(x) = x^3 + 2x^2 + 1$ No variations in sign.
$f(-x) = (-x)^3 + 2(-x)^2 + 1 = -x^3 + 2x^2 + 1$ One variation in sign.

Since $f(x)$ has no variations in sign, $f(x)$ has no positive zeros. Since $f(-x)$
has one variation in sign, $f(x)$ has exactly one negative zero. (The number of
negative zeros can't be less than 1 by an even number because then it would be
negative.) As a third degree polynomial, $f(x)$ has a total of three zeros: one
is real and two are imaginary.

```
WINDOW
Xmin=-10
Xmax=10
Xscl=1
Ymin=-10
Ymax=10
Yscl=1
Xres=1
```

11. $S(x) = x^4 - 2x^3 - 7x - 8$ One variation in sign
$S(-x) = (-x)^4 - 2(-x)^3 - 7(-x) - 8 = x^4 + 2x^3 + 7x - 8$ One variation in sign

Since both $S(x)$ and $S(-x)$ have one variation in sign, $S(x)$ has exactly one
positive and one negative zero. (The number of either can't be less than 1 by
an even number because then it would be negative.) As a fourth degree
polynomial, $S(x)$ has a total of four zeros: two are real and two are imaginary.

```
WINDOW
Xmin=-10
Xmax=10
Xscl=1
Ymin=-30
Ymax=30
Yscl=10
Xres=1
```

13. $P(x) = x^3 - 3x^2 - 2x + 4$ Two variations in sign; two or zero positive zeros

$P(-x) = (-x)^3 - 3(-x)^2 - 2(-x) + 4$

$= -x^3 - 3x^2 + 2x + 4$ One variation in sign; one negative zero

Zeros	Pos.	Neg.	Imag.
	2	1	0
	0	1	2

15. $t(x) = x^4 - 2x^3 - 4x^2 - 2x + 3$ Two variations in sign; two or zero positive zeros

$t(-x) = (-x)^4 - 2(-x)^3 - 4(-x)^2 - 2(-x) + 3$

$= x^4 + 2x^3 - 4x^2 + 2x + 3$ Two variations in sign; two or zero negative zeros

Zeros	Pos.	Neg.	Imag.
	2	2	0
	2	0	2
	0	2	2
	0	0	4

17. $f(x) = x^5 - x^4 + 3x^3 + 9x^2 - x + 5$

Four variations in sign; four, two, or zero positive zeros

$f(-x) = (-x)^5 - (-x)^4 + 3(-x)^3 + 9(-x)^2 - (-x) + 5$

$= -x^5 - x^4 - 3x^3 + 9x^2 + x + 5$

One variation in sign; one negative zero

Zeros	Pos.	Neg.	Imag.
	4	1	0
	2	1	2
	0	1	4

19. $P(x) = x^6 - 12$ One variation in sign; one positive zero

$P(-x) = (-x)^6 - 12 = x^6 - 12$ One variation in sign; one negative zero

Zeros	Pos.	Neg.	Imag.
	1	1	4

21. $r(x) = x^7 + 32$ No variations in sign; no positive zeros

$r(-x) = (-x)^7 + 32 = -x^7 + 32$ One variation in sign, one negative zero

Zeros	Pos.	Neg.	Imag.
	0	1	6

23. $P(x) = x^2 + ax + b$

If $a > 0$ and $b > 0$, then $P(x)$ has no variations in sign, so there are no positive zeros.

$P(-x) = (-x)^2 + a(-x) + b = x^2 - ax + b$

In this case, the coefficients look like + − +, so there are two variations in sign and either two or zero negative zeros. If there are no negative zeros, there are two imaginary zeros.

25. $P(x) = x^2 + ax + b$

If $a > 0$ and $b < 0$, the signs of the coefficients are + + −, so there is one variation in sign and one positive zero.

$P(-x) = (-x)^2 + a(-x) + b = x^2 - ax + b$

In this case, the signs of the coefficients are + − −, and there is one variation in sign, so one negative zero. In summary, there is one positive zero and one negative zero.

27. $f(x) = x^6 - 3x^5 + 4x^4 + 3x^3 - 2x + 5$ Four variations in sign;
four, two, or zero positive zeros.

$f(-x) = (-x)^6 - 3(-x)^5 + 4(-x)^4 + 3(-x)^3 - 2(-x) + 5$
$= x^6 + 3x^5 + 4x^4 - 3x^3 + 2x + 5$ Two variations in sign;
two or zero negative zeros.

Zeros	Pos.	Neg.	Imag.
	4	2	0
	4	0	2
	2	2	2
	2	0	4
	0	2	4
	0	0	6

29. $S(x) = x^7 + 3x^5 + 4x^2 - 3x + 5$ Two variations in sign;
two or zero positive zeros

$S(-x) = (-x)^7 + 3(-x)^5 + 4(-x)^2 - 3(-x) + 5$
$= -x^7 - 3x^5 + 4x^2 + 3x + 5$ One variation in sign; one negative zero

Zeros	Pos.	Neg.	Imag.
	2	1	4
	0	1	6

31. $P(x) = x^8 - x - 1$ One variation in sign; one positive zero
$P(-x) = (-x)^8 - (-x) - 1$
$= x^8 + x - 1$ One variation in sign; one negative zero

Zeros	Pos.	Neg.	Imag.
	1	1	6

33. $P(x) = x^3 + ax + b$
If a and b are both positive, there are no variations in sign so there are no positive zeros.
$P(-x) = (-x)^3 + a(-x) + b = -x^3 - ax + b$
In this case, there is one variation in sign so there is one negative zero. The remaining two zeros are imaginary.

35. $P(x) = x^3 + ax + b$
If $a < 0$ and $b > 0$, the signs of the coefficients are + - +, so there are two variations in sign and either two or zero positive zeros.
$P(-x) = (-x)^3 + a(-x) + b = -x^3 - ax + b$ with a negative, the signs of the coefficients are - + +, so there is one variation in sign and one negative zero. The remaining two zeros are either positive or imaginary.

Section C-4

1. Note that if you eliminate the parameter from the parametric equations by substituting x for t^2 in $y = t^2 - 2$, the result is the equation $y = x - 2$. Since $x = t^2$, x must be non-negative, and the graph of the parametric equations will consist only of a ray, the portion of the line $y = x - 2$ for $x \geq 0$.

3. Using trace and the forward and backward arrows, we can move the cursor along the graph to locate each of the points identified in the table.

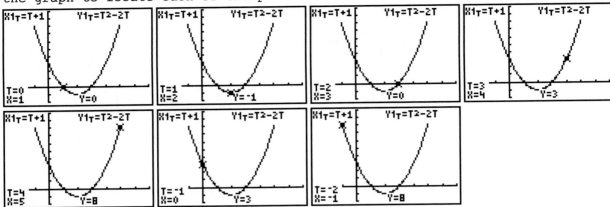

5. After applying ZDecimal, the "window" screen looks like this:

```
WINDOW
 Tmin=-2
 Tmax=4
 Tstep=.05
 Xmin=-4.7
 Xmax=4.7
 Xscl=1
↓Ymin=-3.1
```
```
WINDOW
↑Tstep=.05
 Xmin=-4.7
 Xmax=4.7
 Xscl=1
 Ymin=-3.1
 Ymax=3.1
 Yscl=1
```

The values of x and y were affected but the values of t are unchanged.

7. After re-entering the original window variables, choosing Zoom In, and zooming on the portion of the graph near the y-intercept, the graph and "window" screens look like this:

```
WINDOW
 Tmin=-2
 Tmax=4
 Tstep=.05
 Xmin=-1.271276…
 Xmax=1.2287234…
 Xscl=1
↓Ymin=1.9193548…
```
```
WINDOW
↑Tstep=.05
 Xmin=-1.271276…
 Xmax=1.2287234…
 Xscl=1
 Ymin=1.9193548…
 Ymax=4.9193548…
 Yscl=1
```

The values of x and y were affected but the values of t are unchanged.

9. Construct a table and graph.

t	0	1	2	3	−1	−2	−3
x	0	−1	−2	−3	1	2	3
y	−2	0	2	4	−4	−6	−8

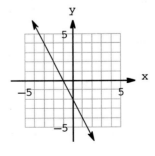

To eliminate the parameter t we solve
$x = -t$ for t to obtain
$$t = -x$$
Then we substitute the expression for t into
$y = 2t - 2$ to obtain
$$y = 2(-x) - 2$$
$$y = -2x - 2$$
This is the equation of a straight line.

11. Construct a table and graph.

t	0	1	2	3	−1	−2	−3	
x	0	−1	−4	−9	−1	−4	−9	Note: $x \leq 0$
y	−2	0	6	16	0	6	16	

To eliminate the parameter t we solve
$x = -t^2$ for t^2 (we do not need t) to obtain
$$t^2 = -x$$
Then we substitute the expression for t^2 into
$y = 2t^2 - 2$ to obtain
$$y = 2(-x) - 2$$
$$y = -2x - 2$$

This is the equation of a straight line. However,
since $x = -t^2$, x is restricted so that $x \leq 0$.
Therefore the equation of the curve is actually
$$y = -2x - 2, \quad x \leq 0$$
This is the equation of a ray (part of a straight line).

13. Construct a table and graph.

t	0	1	2	3	−1	−2	−3
x	0	3	6	9	−3	−6	−9
y	0	−2	−4	−6	2	4	6

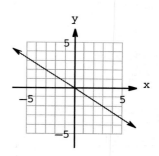

To eliminate the parameter t we solve
$x = 3t$ for t to obtain
$$t = \frac{x}{3}$$
Then we substitute the expression for t into
$y = -2t$ to obtain
$$y = -2\left(\frac{x}{3}\right)$$
$$y = -\frac{2}{3}x$$
This is the equation of a straight line.

15. Construct a table of values and graph.

t	0	1	2	3	−1	−2	−3
x	0	$\frac{1}{4}$	1	$\frac{9}{4}$	$\frac{1}{4}$	1	$\frac{9}{4}$
y	0	1	2	3	−1	−2	−3

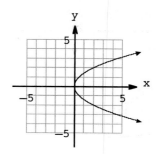

To eliminate the parameter t we substitute $t = y$
(from $y = t$) into the expression for x to obtain
$$x = \frac{1}{4}y^2$$
$$y^2 = 4x$$
This is the equation of a parabola.

17. Construct a table of values and graph.

t	0	1	2	3	-1	-2	-3
x	0	$\frac{1}{4}$	4	$\frac{81}{4}$	$\frac{1}{4}$	4	$\frac{81}{4}$
y	0	1	4	9	1	4	9

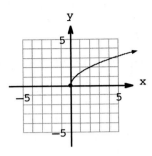

To eliminate the parameter we substitute $t^2 = y$
(from $y = t^2$) into the expression for x to obtain

$$x = \frac{1}{4}t^4$$
$$x = \frac{1}{4}(t^2)^2$$
$$x = \frac{1}{4}y^2$$
$$y^2 = 4x$$

This is the equation of a parabola. However, since $y = t^2$, y is restricted to nonnegative values only, and the equation of the curve is actually
$$y^2 = 4x, \ y \geq 0$$
This is the equation of the upper half of a parabola.

19. We eliminate the parameter t as follows:
$$x = t - 2$$
$$x + 2 = t$$
Substituting this expression for t into the expression for y, we obtain
$$y = 4 - 2t$$
$$y = 4 - 2(x + 2)$$
$$y = 4 - 2x - 4$$
$$y = -2x$$
This is the equation of a line.

21. We eliminate the parameter t as follows:
Solve $x = t - 1$ for t to obtain $t = x + 1$. Substitute the expression for t into $y = \sqrt{t}$ to obtain
$$y = \sqrt{x + 1} \ \text{or}$$
$$y^2 = x + 1, \ y \geq 0.$$
Since t is restricted so that $t \geq 0$, we also have
$x + 1 \geq 0$ or $x \geq -1$. This graph is a parabola with vertex at $(-1, 0)$, opening right. Since $y \geq 0$ only the upper half of the parabola is traced out.

23. We eliminate the parameter t as follows:
Solve $x = \sqrt{t}$ for t to obtain $t = x^2$.
Substitute the expression for t into $y = 2\sqrt{16 - t}$ to obtain
$$y = 2\sqrt{16 - x^2} \ \text{or}$$
$$y^2 = 4(16 - x^2)$$
$$y^2 = 64 - 4x^2$$
$$4x^2 + y^2 = 64$$
$$\frac{x^2}{16} + \frac{y^2}{64} = 1 \quad \text{(ellipse in standard form)}$$

Since $0 \leq t \leq 16$ and $x = \sqrt{t}$, it follows that $0 \leq x \leq 4$.
Since $0 \leq t \leq 16$ and $t = x^2$, it follows that $0 \leq x^2 \leq 16$, $0 \leq 16 - x^2 \leq 16$,
$0 \leq \sqrt{16 - x^2} \leq 4$, and $0 \leq 2\sqrt{16 - x^2} \leq 8$. So $0 \leq y \leq 8$.

The graph of $\frac{x^2}{16} + \frac{y^2}{64} = 1$, $0 \leq x \leq 4$ and $0 \leq y \leq 8$, is the portion of an ellipse lying in the first quadrant.

25. We eliminate the parameter t as follows:
Solve $x = -\sqrt{t + 1}$ for t to obtain
$$x^2 = t + 1$$
$$t = x^2 - 1$$

Substitute the expression for t into $y = -\sqrt{t - 1}$ to obtain
$$y = -\sqrt{x^2 - 1 - 1}$$
$$y = -\sqrt{x^2 - 2}$$
$$y^2 = x^2 - 2$$
$$x^2 - y^2 = 2 \quad \text{(hyperbola)}$$

Since $t \geq 1$ it follows that $t + 1 \geq 2$, $\sqrt{t + 1} \geq \sqrt{2}$ and $x = -\sqrt{t + 1} \leq -\sqrt{2}$.
Also it follows that $t - 1 \geq 0$, $\sqrt{t - 1} \geq 0$, and $y = -\sqrt{t - 1} \leq 0$.
The graph of $x^2 - y^2 = 2$, $x \leq -\sqrt{2}$ and $y \leq 0$, is the portion of a hyperbola
lying in the third quadrant.

27. If $A \neq 0$, $C = 0$, and $E \neq 0$, write
$$Ax^2 + Dx + Ey + F = 0$$
There are many possible parametric equations for this curve. A simple approach
is to set $x = t$ and solve for y in terms of t.
$$At^2 + Dt + Ey + F = 0$$
$$At^2 + Dt + F = -Ey$$
$$\frac{At^2 + Dt + F}{-E} = y$$

Then $x = t$, $y = \dfrac{At^2 + Dt + F}{-E}$, $-\infty < t < \infty$, are parametric equations for this

curve. The curve is a parabola.

29. We eliminate the parameter t as follows:
Square the expressions for x and y.
$$x^2 = t^2 + 1 \qquad y^2 = t^2 + 9$$

Solve the first equation for t^2 and substitute the result into the second
equation.
$$t^2 = x^2 - 1$$
$$y^2 = x^2 - 1 + 9$$
$$y^2 = x^2 + 8$$
$$y^2 - x^2 = 8$$

This is the equation of a hyperbola.

Since $t^2 \geq 0$, $t^2 + 1 \geq 1$, $\sqrt{t^2 + 1} \geq 1$. It follows that $x \geq 1$.

Since $t^2 \geq 0$, $t^2 + 9 \geq 9$, $\sqrt{t^2 + 9} \geq 3$. It follows that $y \geq 3$.

The curve is the part of the hyperbola $y^2 - x^2 = 8$ for $x \geq 1$ and $y \geq 3$.

31. We eliminate the parameter t as follows:
Square the expressions for x and y.

$$x^2 = \frac{4}{t^2 + 1} \qquad y^2 = \frac{4t^2}{t^2 + 1}$$

$$x^2 + y^2 = \frac{4}{t^2 + 1} + \frac{4t^2}{t^2 + 1}$$

$$x^2 + y^2 = \frac{4 + 4t^2}{t^2 + 1}$$

$$x^2 + y^2 = \frac{4(1 + t^2)}{t^2 + 1}$$

$$x^2 + y^2 = 4$$

This is the equation of a circle.
Since $t^2 \geq 0$, $t^2 + 1 \geq 1$, $\sqrt{t^2 + 1} \geq 1$, $0 < \dfrac{1}{\sqrt{t^2 + 1}} \leq 1$, $0 < \dfrac{2}{\sqrt{t^2 + 1}} \leq 2$.

It follows that $0 < x \leq 2$. Therefore $0 < x^2 \leq 4$, $0 > -x^2 \geq -4$, $4 > 4 - x^2 \geq 0$, $2 > \sqrt{4 - x^2} \geq 0$, $-2 < -\sqrt{4 - x^2} \leq 0$. Since $y = \pm\sqrt{4 - x^2}$ on the circle $x^2 + y^2 = 4$, it follows that $-2 < y < 2$ (since $x \neq 0$). So the curve is the portion of the circle $x^2 + y^2 = 4$ for which $0 < x \leq 2$, $-2 < y < 2$. This is the right-hand semicircle, excluding the points $(0, \pm 2)$.

33. We eliminate the parameter t as follows: First note that since $x = \dfrac{8}{t^2 + 4}$, $(t^2 + 4)x = 8$ or $t^2 + 4 = \dfrac{8}{x}$. Therefore x cannot be 0. Now square the expressions for x and y.

$$x^2 = \frac{64}{(t^2 + 4)^2}$$

$$y^2 = \frac{16t^2}{(t^2 + 4)^2}$$

$$x^2 + y^2 = \frac{64}{(t^2 + 4)^2} + \frac{16t^2}{(t^2 + 4)^2}$$

$$x^2 + y^2 = \frac{64 + 16t^2}{(t^2 + 4)^2}$$

$$x^2 + y^2 = \frac{16(4 + t^2)}{(t^2 + 4)^2}$$

$$x^2 + y^2 = \frac{16}{t^2 + 4}$$

Since $x = \dfrac{8}{t^2 + 4}$, $2x = \dfrac{16}{t^2 + 4}$, so

$$x^2 + y^2 = 2x \quad x \neq 0$$

This is the equation of a circle. Completing the square we see
$$x^2 - 2x + y^2 = 0$$
$$x^2 - 2x + 1 + y^2 = 1$$
$$(x - 1)^2 + y^2 = 1, \quad x \neq 0$$

So the circle has center $(1, 0)$ and radius 1. The origin is not part of the curve since x cannot equal zero, so there is a hole in the graph there.

35. Here are the graphs in a squared
viewing window (shown together with
the line $y = x$).

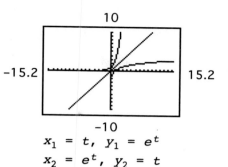

(A) The graphs are reflections of
each other in the line $y = x$.

(B) Since $t = x_1$, $y_1 = e^{x_1}$. Since
$t = y_2$, $x_2 = e^{y_2}$, $y_2 = \ln x_2$.
The functions are inverses of each
other.

$x_1 = t, \; y_1 = e^t$
$x_2 = e^t, \; y_2 = t$

37. At the instant the supplies are dropped, the vertical speed is 0, the
horizontal speed is 125 meters per second, and the altitude is 1,000 meters.
Substituting these values in
$$x = h_0 t$$
$$y = a_0 + v_0 t - 4.9 t^2 \qquad 0 \le t \le b$$

the parametric equations for the path of the supplies are
$$x = 125t$$
$$y = 1000 - 4.9 t^2$$

The supplies strike the ground when $y = 0$. Using the parametric equations for
y, we obtain
$$y = 1,000 - 4.9 t^2 = 0$$
$$-4.9 t^2 = -1,000$$
$$t = \sqrt{\frac{-1,000}{-4.9}} \approx 14.2857 \text{ sec.}$$

The distance inland required is the value of x for this value of t.
Substituting $t = 14.2857$ in $x = 125t$, the distance is
$x = 125(14.2857) \approx 1,786$ meters

Notes

Notes

Notes